Handbook of Ceramic Composites

Handbook of Ceramic Composites

Edited by

Narottam P. Bansal
NASA Glenn Research Center
USA

KLUWER ACADEMIC PUBLISHERS
Boston / Dordrecht / London

Library of Congress Cataloging-in-Publication Data

A C.I.P. Catalogue record for this book is available
from the Library of Congress.

ISBN 978-1-4419-5482-4
e-ISBN 978-0-387-23986-6

Printed on acid-free paper.

©2010 Kluwer Academic Publishers
All rights reserved. This work may not be translated or copied in whole or in part without the written permission of the publisher (Springer Science+Business Media, Inc., 233 Spring Street, New York, NY 10013, USA), except for brief excerpts in connection with reviews or scholarly analysis. Use in connection with any form of information storage and retrieval, electronic adaptation, computer software, or by similar or dissimilar methodology now know or hereafter developed is forbidden.
The use in this publication of trade names, trademarks, service marks and similar terms, even if the are not identified as such, is not to be taken as an expression of opinion as to whether or not they are subject to proprietary rights.

Printed in the United States of America.

9 8 7 6 5 4 3 2 1

springeronline.com

Contents

PART I. Ceramic Fibers   1

1. Oxide Fibers   3
   A. R. Bunsell

2. Non-oxide (Silicon Carbide) Fibers   33
   J. A. DiCarlo and H.-M. Yun

PART II. Non-oxide/Non-oxide Composites   53

3. Chemical Vapor Infiltrated SiC/SiC Composites (CVI SiC/SiC)   55
   J. Lamon

4. SiC/SiC Composites for 1200°C and Above   77
   J. A. DiCarlo, H.-M. Yun, G. N. Morscher, and R. T. Bhatt

5. Silicon Melt Infiltrated Ceramic Composites (HiPerCompTH)   99
   G. S. Corman and K. L. Luthra

6. Carbon Fibre Reinforced Silicon Carbide Composites (C/SiC, C/C-SiC)   117
   W. Krenkel

7. Silicon Carbide Fiber-Reinforced Silicon Nitride Composites   149
   R. T. Bhatt

8. MoSi$_2$-Base Composites   173
   M. G. Hebsur

9. Ultra High Temperature Ceramic Composites   197
   M. J. Gasch, D. T. Ellerby and S. M. Johnson

PART III. Non-oxide/Oxide Composites 225

10. SiC Fiber-Reinforced Celsian Composites 227
N. P. Bansal

11. In Situ Reinforced Silicon Nitride – Barium Aluminosilicate Composite 251
K. W. White, F. Yu and Y. Fang

12. Silicon Carbide and Oxide Fiber Reinforced Alumina Matrix Composites Fabricated Via Directed Metal Oxidation 277
A. S. Fareed

13. SiC Whisker Reinforced Alumina 307
T. Tiegs

14. Mullite-SiC Whisker and Mullite-ZrO_2-SiC Whisker Composites 325
R. Ruh

15. Nextel™ 312/Silicon Oxycarbide Ceramic Composites 347
S. T. Gonczy and J. G. Sikonia

PART IV. Oxide/Oxide Composites 375

16. Oxide-Oxide Composites 377
K. A. Keller, G. Jefferson and R. J. Kerans

17. WHIPOX All Oxide Ceramic Matrix Composites 423
M. Schmücker and H. Schneider

18. Alumina-Reinforced Zirconia Composites 437
S. R. Choi and N. P. Bansal

PART V. Glass and Glass-Ceramic Composites 459

19. Continuous Fibre Reinforced Glass and Glass-Ceramic Matrix Composites 461
A. R. Boccaccini

20. Dispersion-Reinforced Glass and Glass-Ceramic Matrix Composites 485
J. A. Roether and A. R. Boccaccini

21. Glass Containing Composite Materials: Alternative Reinforcement Concepts 511
A. R. Boccaccini

Index 533

List of contributors

Narottam P. Bansal
NASA Glenn Research Center
Cleveland, Ohio
USA

Ramakrishna T. Bhatt
NASA Glenn Research Center
Cleveland, Ohio
USA

Aldo R. Boccaccini
Department of Materials
Imperial College London
London
United Kingdom

Anthony R. Bunsell
Ecole des Mines de Paris
Centre des Materiaux
Evry Cedex, France

Sung R. Choi
NASA Glenn Research Center
Cleveland, Ohio
USA

G. S. Corman
GE Global Research Center
Niskayuna, New York
USA

James A. DiCarlo
NASA Glenn Research Center
Cleveland, Ohio
USA

Donald T. Ellerby
NASA Ames Research Center
Moffett Field, California
USA

Yi Fang
Department of echanical Engineering
University of Houston
Houston, Texas
USA

Ali S. Fareed
Power Systems Composites, LLC
Newark, Delaware
USA

Matthew J. Gasch
ELORET – NASA Ames Research Center
Moffett Field, California
USA

Stephen T. Gonczy
Gateway Materials Technology, Inc.
Mt. Prospect, Illinois
USA

Mohan G. Hebsur
NASA Glenn Research Center
Cleveland, Ohio
USA

George Jefferson
National Research Council
Washington, DC
USA

Sylvia M. Johnson
NASA Ames Research Center
Moffett Field, California
USA

Kristin A. Keller
Air Force Research Laboratory
Materials and Manufacturing Directorate
AFRI/MLLN
Wright-Patterson AFB, Ohio
USA

Ronald J. Kerans
Air Force Research Laboratory
Materials and Manufacturing Directorate
AFRI/MLLN
Wright-Patterson AFB, Ohio
USA

Dr.-Ing. Walter Krenkel
University of Bayreuth
Ceramic Materials Engineering
Bayreuth
Germany

Jacques Lamon
Laboratoire des Composites
Thermostructuraux
Pessac
France

K. L. Luthra
GE Global Research Center
Niskayuna, New York
USA

G. N. Morscher
NASA Glenn Research Center
Cleveland, Ohio
USA

Judith A. Roether
Department of Dental Biomaterials
Science
GKT Dental Institute
London
United Kingdom

Robert Ruh
Universal Technology Corporation
Beavercreek, Ohio
USA

Martin Schmücker
German Aerospace Center (DLR)
Institute of Materials Research
Koln
Germany

Hartmut Schneider
German Aerospace Center (DLR)
Institute of Materials Research
Koln
Germany

John G. Sikonia
Sikonia Consulting
Bend, Oregon
USA

Terry Tiegs
Oak Ridge National Laboratory
Oak Ridge, Tennessee
USA

Kenneth W. White
Department of Mechanical Engineering
University of Houston
Houston, Texas
USA

Feng Yu
Department of echanical Engineering
University of Houston
Houston, Texas
USA

Hee-Mann Yun
NASA Glenn Research Center
Cleveland, Ohio
USA

Preface

Metallic materials, including superalloys, have reached the upper limit in their use temperatures. Alternative materials, such as ceramics, are needed for significant increase in service temperatures. Advanced ceramics generally possess, low density, high strength, high elastic modulus, high hardness, high temperature capability, and excellent chemical and environmental stability. However, monolithic ceramics are brittle and show catastrophic failure limiting their applications as structural engineering materials. This problem is alleviated in ceramic-ceramic composites where the ceramic matrix is reinforced with ceramic particles, platelets, whiskers, chopped or continuous fibers. Ceramic matrix composites (CMCs) are at the forefront of advanced materials technology because of their light weight, high strength and toughness, high temperature capabilities, and graceful failure under loading. This key behavior is achieved by proper design of the fiber-matrix interface which helps in arresting and deflecting the cracks formed in the brittle matrix under load and preventing the early failure of the fiber reinforcement.

Ceramic composites are considered as enabling technology for advanced aeropropulsion, space propulsion, space power, aerospace vehicles, space structures, ground transportation, as well as nuclear and chemical industries. During the last 25 years, tremendous progress has been made in the development and advancement of CMCs under various research programs funded by the U.S. Government agencies: National Aeronautics and Space Administration (NASA), Department of Defense (DoD), and Department of Energy (DOE). Some examples are NASA's High Temperature Engine Materials Technology Program (HiTEMP), National Aerospace Plane (NASP), High Speed Civil Transport (HSCT), Ultra Efficient Engine Technology (UEET), and Next Generation Launch Technology (NGLT) programs; DoD's Integrated High Performance Turbine Engine Technology (IHPTET), Versatile Affordable Advanced Turbine Engines (VAATE), and Integrated High Performance Rocket Propulsion Technology (IHPRPT) programs; and DOE's Continuous Fiber Ceramic Composites (CFCC) program. CMCs would find applications in advanced aerojet engines, stationary gas turbines for electrical power generation, heat exchangers, hot gas filters, radiant burners, heat treatment and materials growth furnaces, nuclear fusion reactors, automobiles, biological implants, etc. Other applications of CMCs are as machinery wear parts, cutting and forming tools, valve seals, high precision ball bearing for corrosive environments, and plungers for chemical pumps. Potential applications of various ceramic composites are described in individual chapters of the present handbook.

This handbook is markedly different than the other books available on Ceramic Matrix Composites. Here, a ceramic composite system or a class of composites has been covered in a separate chapter, presenting a detailed description of processing, properties, and applications. Each chapter is written by internationally renowned researchers in the field. The handbook is organized into five sections. The first section "**Ceramic Fibers**" gives details of commercially available oxide fibers and non-oxide (silicon carbide) fibers which are used as reinforcements for ceramic matrices in two separate chapters. The next section "**Non-oxide/Non-oxide Composites**" consists of seven chapters describing various composite systems where both the matrix and the reinforcement are non-oxide ceramics. Special attention has been given to silicon carbide fiber-reinforced silicon carbide matrix (SiC_f/SiC) composite system because of its great commercial importance. This CMC system has been covered in three separate chapters as it has been investigated extensively during the last thirty years and is the most advanced composite material system which is commercially available. The section "**Non-oxide/Oxide Composites**" comprises of six chapters presenting the details of various composites which consist of oxide matrix and non-oxide reinforcement or vice versa. The composites where both the matrix and the reinforcements are oxides are covered in three chapters in the section "**Oxide/Oxide Composites**". The final section "**Glass and Glass-Ceramic Composites**" contains three chapters describing composites where the matrix is either glass or glass-ceramic.

This handbook is intended for use by scientists, engineers, technologists, and researchers interested in the field of ceramic matrix composites and also for designers to design parts and components for advanced engines and various other industrial applications. Students and educators will also find the information presented in this book useful. The reader would be able to learn state-of-the-art about ceramic matrix composites from this handbook. Like any other compilation where individual chapters are contributed by different authors, the present handbook may have some duplication of material and non-uniformity of symbols and nomenclature in different chapters.

I am grateful to all the authors for their valuable and timely contributions as well as for their cooperation during the publication process. Thanks are due to Mr. Gregory T. Franklin, Senior Editor, Kluwer Academic Publishers, for his help and guidance during the production of this handbook. I would also like to express my gratitude to Professor Robert H. Doremus for helpful suggestions and valuable advice.

Narottam P. Bansal
Cleveland, OH

Part I

Ceramic Fibers

1
Oxide Fibers

Anthony R. Bunsell
Ecole des Mines de Paris
Centre des Matériaux, BP 87, 91003 Evry Cedex, France
Tel. +33 (0) 160763015; E-mail :anthony.bunsell@ensmp.fr

ABSTRACT

Oxide fibers find uses both as insulation and as reinforcements. Glass fibers, based on silica, possess a variety of compositions in accordance with the characteristics desired. They represent the biggest market for oxide fibers. Unlike other oxide fibers, glass fibers are continuously spun from the melt and are not used at temperatures above 250°C. Short oxide fibers can be melt blown whilst other aluminasilicate and alumina based continuous fibers are made by sol-gel processes. Initial uses for these fibers were as refractory insulation, up to 1600°C, but they are now also produced as reinforcements for metal matrix composites. Continuous oxide fibers are candidates as reinforcements for use up to and above 1000°C.

1.0. INTRODUCTION

Synthetic fibers, both organic and inorganic, were developed in the twentieth century and represent an enormous market. Their development has had a marked effect on the textile industry, initially in long established industrial nations and increasingly in developing countries. The processing and handling techniques of synthetic fibers are often related to traditional textile processes but a considerable fraction of even organic fibers are used for industrial end products. This fraction is considerably greater for inorganic fibers. More than 99% of the reinforcements of resin matrix composites are glass fibers and most of these are of one type of glass. The diameters of glass fibers are of the order of 10 μm or about one eighth the diameter of a human hair. The fineness of the filaments makes them very flexible despite the inherent brittleness and stiffness of the material. It is the development of glass fibers which has laid the foundations for the present composite materials market. The fibers are produced as tows of continuous filaments which are then converted into many different products. The fibers can be woven by the same techniques as other continuous synthetic fibers. The fibers can also be wound around a mandrel and, impregnated with a resin, made into filament wound tubes, for example. Alternatively they can be formed into a non woven mat which can then be draped around a form and impregnated with a resin or put into a mould and impregnated. The resin then can be cured to form a structural composite material. Glass fibers can be chopped into short lengths and mixed with an uncured resin which can then be placed into a mould and formed into a structure or mixed with a resin to be injected into a mould so as to form a structure. Glass fibers are also chopped and projected with the resin against a mould to make cheap large scale structures. Glass fiber reinforced resin composites are ubiquitous materials which find uses in applications such as pipelines, parts of car bodies, boats, pressure vessels and a thousand and one other applications. It is particularly useful as it resists many corrosive environments and so is used for chemical storage tanks and for other applications for which chemical inertness is required. Glass is however limited in its use as it has a low Young's modulus, about the same as that of aluminum and it has limited high temperature capabilities. It is also sensitive to extreme variations in pH.

Glass fibers are predominantly formed with silica but also contain alumina. Fibers which are rich in alumina have been produced since the late 1940s. This type of fiber was initially produced in a low cost discontinuous form and used for refractory insulation, typically in furnace linings, and has found a very large market. Alumina is about five times stiffer than silica so that, in the form of fine filaments, it is attractive as a potential reinforcement for light alloys and even vitreous ceramics. The development of ceramic matrix composites, in the 1980s, originally based on silicon carbide based fibers, opened up other horizons to oxide fibers. Unlike SiC based fibers they were insensitive to oxidation and held out the promise of enhanced properties far above the best metal alloys and even silicon carbide ceramics. Such fibers are used as reinforcements for light alloys such as aluminum but also with matrices such as mullite.

2.0. DEVELOPMENT OF OXIDE FIBERS

The primary component of glass filaments is SiO_2, followed by CaO, Al_2O_3 and other oxides. A number of types of glass fiber exist with different compositions according to the desired characteristics. Glass filaments have probably been formed since or before Roman times and more recently the production of fine filaments was demonstrated in Great Britain in the nineteenth century and used as a substitute for asbestos in Germany during the first World War. In the latter application molten glass was poured onto a spinning disc to produce discontinuous fibers. In 1931 two American firms, Owen Illinois Glass Co. and Corning Glass Works developed a method of spinning glass filaments from the melt through spinnerets. The two firms combined in 1938 to form Owens Corning Fiberglas Corporation. Since that time extensive use of glass fibers has been made and there are major producers in several countries. Initially the glass fibers were destined for filters and textile uses however the development of heat setting resins opened up the possibility of fiber reinforced composites and in the years following the Second World War the fiber took a dominant role in this type of material. Today, by far the greatest volume of composite materials is reinforced with glass fibers.

The development of more refractory fibers dates from 1942 and in 1949 a patent was awarded to Babcock and Wilcox in the USA for the melt blown production of aluminosilicate filaments (1). Refractory insulation is most usually produced in the form of a felt consisting of discontinuous fibers and other non fibrous forms, depending on the manufacturing process used. The usual starting material for production is kaolin, also known as china clay. It is a natural form of hydrated aluminum silicate ($Al_2Si_2O_5(OH)_4$). An alternative route is to use mixtures of alumina and silica. The fibers are known collectively at aluminosilicate Refractory Ceramic Fibers or simply RCFs. The progressive replacement, in the earlier fibers, of silica by alumina improved their refractory characteristics but made manufacture more difficult. The fibers made from kaolin contain around 47% by weight of alumina. Shot, or non fibrillar particles, levels are high and can be of the order of 50% of product mass. These products continue to find important markets and are continuing to develop. A concern for these classes of fibers is the possibility of risks to health. This concern comes from the proven carcinogenic effects of asbestos fibers and which cause all fiber producers to take the possibility of health hazards seriously. An important consideration is the diameter of the fibers being made which if they are similar to the alveolar cellular structure of the lungs can mean that they can become blocked in the lungs. Even if no long term morbidity occurs the efficiency of the lungs would be reduced. The critical size seems to be one micron however even if no effects are proven the industry is developing low biopersistent fibers, to be used as thermal insulation. These are vitreous fibers containing calcium oxide, CaO, magnesia, MgO, and silica, SiO_2, in variable proportions. Other oxides may be added to optimize temperature resistance or other properties. The fibers are more soluble than the traditional RCFs and would reside for less time in the lungs if inhaled.

The aluminosilicate RCF fibers are most widely used in the form of a non-woven blanket or board for furnace linings in the metallurgical, ceramic and chemical industries. An alternative refractory brick would be up to ten times heavier. The use of aluminosilicate felts allows fast heating and cooling cycles of furnaces, because of the reduced mass which has to be heated or cooled and this allows considerable cost savings to be made compared to other types of insulation.

Producing oxide fibers by sol-gel processes is more expensive than the melt blown process but greater control of the final product is possible and the fibers can be made with a much higher alumina content. Another advantage is that the precursor is spun at low temperatures before being pyrolysed. A British patent was awarded to Babcock and Wilcox in 1968 for the production of oxide fibers by this process and since then a considerable number of other companies, mostly in the USA, UK and Japan have made fibers using the sol-gel route (2). ICI developed a short fiber with a diameter of 3 μm called Saffil in 1974 (3). This fiber is 97% alumina and 3% silica and was originally developed for high temperature insulation up to 1600°C. The increased interest during the late 1970s for metal matrix composites saw Saffil used to reinforce aluminum and it remains the most widely used fibrous reinforcement for light alloys. The successful use of Saffil fiber reinforced aluminum by Toyota to replace nickel based alloy inserts to maintain oil rings in diesel engines has encouraged other firms to produce similar products.

The first alumina based continuous fiber was produced in 1974 by 3M and is sold under the name Nextel 312. It contains only 62% alumina together with boria and silica.

It has an essentially amorphous structure and is limited to use below 1000°C because of the volatility of boria but it remains the foundation of the 3M Nextel range of oxide fibers. Later in that same decade DuPont produced the first continuous polycrystalline 99.9% alpha-alumina fiber called Fiber FP (4). The fiber was made by spinning in air, a slurry, composed of an aqueous suspension of α-alumina particles and aluminum salts. The as-obtained fiber was then dried and fired in two steps. The incentive for producing this fiber was the possibility of reinforcing aluminum connecting rods in, initially, Toyota engines. The fiber had the high modulus of bulk alumina and this, coupled with its relatively large grain size of around 0.5 μm and a diameter of 20 μm, meant that it could not be easily handled. The fiber had a failure strain of approximately 0.3%. Fiber FP was not developed commercially but is seen as a model fiber against which other polycrystalline oxide fibers can be compared. In an attempt to improve handleability DuPont produced a fiber, called PRD-166, containing 80% by weight of α-alumina and 20% zirconia (5). The presence of the second phase, in the form of grains of 0.1 μm, reduced the grain size of the alumina to 0.3 μm. The presence of tetragonal zirconia in bulk alumina increased room temperature strength by phase transformation toughening and also limited grain boundary mobility, grain sliding and growth at high temperatures. The zirconia phase also reduced the overall Young's modulus of the fiber. However the improvement of the tensile properties was not sufficient to allow commercial development of the PRD-166 fiber. During the 1980s and 1990s a number of companies in Japan and the USA developed oxide fibers which overcame the difficulties encountered by the fibers produced by DuPont. Sumitomo Chemicals produced the continuous Altex fiber in which the 15% of amorphous silica stabilized the alumina grains in the γ-phase which meant that the grain size was 25 nm (6). The Altex fiber had only half the Young's modulus of a pure, dense α-alumina fiber and so could be more easily handled and woven. Mitsui Mining produced the Almax fiber, which in composition and grain size, was very similar to the Fiber FP, however it had only half its diameter (7). The reduction in diameter meant an eight times increase in flexibility and so the fiber could be woven. Later 3M produced the Nextel 610 fiber with the same diameter as that of the Almax fiber but with grain sizes of 0.1 μm which doubled the fiber strength (8).

During this period 3M produced a range of oxide fibers with increasingly high performance properties. The sol-gel process used to produce the Nextel 312 was modified to produce the Nextel 440 fiber. The composition of 3 mol of alumina for 2 mol of silica was maintained but the boria content was reduced to increase its high temperature stability. The Nextel 440 fiber is formed of nano-sized γ-alumina grains in an amorphous silica phase. The fiber has been successfully used to reinforce mullite. The Nextel 720 fiber from 3M is made up of aggregates of mullite grains in which are embedded α-alumina grains (9). Although the grains of each phase are small the aggregates of similarly aligned mullite grains act like single grains of 0.5 μm and this gives the Nextel 720 fiber the lowest creep rate of any oxide fiber at temperatures above 1000°C (10). The fiber is however sensitive to alkaline contamination (11). 3M also produces the Nextel 650 fiber which is reminiscent of the PRD-166 fiber as it contains zirconia as a second phase (12).

The initial interest in small diameter oxide fibers as rivals to small diameter SiC fibers for use in ceramic matrix composites has been largely unfulfilled. Although the oxide fibers do not suffer from oxidation, as do the SiC fibers, they are inherently less mechanically stable above 1000°C. Whereas the co-valent bonds in SiC resist creep the ionic bonds in oxides allow easier movement of the structure. The complexity of the crystal structures of some oxides, such as mullite, does impart good inherent creep properties but ultimately grain boundary sliding and also the metastable state of some of the more complex systems means that oxide fibers are primarily limited to uses below 1200°C if they have to carry loads.

Removing grain boundaries by growing single crystal oxide filaments from the melt either by heating the ceramic in a crucible or by laser has been explored since the 1960s (13). This technique involves a single seeding grain touching the surface of the molten ceramic and slowly being drawn away from it. Such fibers were investigated by Tyco Laboratories (14) and developed commercially by Saphikon in the USA (15). It has been shown that such α-alumina fibers with their C-axis aligned parallel to the fiber axis can resist creep up to 1600°C (16). Saphikon produced fibers composed of single crystal α-alumina and also YAG-alumina, however the large diameters of 100 μm, and above, coupled with their prohibitive cost means that there seems to be no prospect of these fibers leaving the laboratory. A much cheaper process developed in Russia at the turn of this century consists of infiltrating the molten oxide along channels formed by sandwiching molybdenum wires between sheets of molybdenum (17). When the filaments are formed the molybdenum is etched away. The fibers so formed are inevitably of large diameter and are not circular in cross-section but may show the way for this type of fiber being developed in a commercially viable way.

Diameters over 20 μm have been seen to be too great for easy transformation and processing into structures but in the future very fine fibers may also be produced with nanometric sized diameters and these will also require some innovative processing procedures. It has been known since the 1950s that single crystal filaments, of oxides and other materials, with micron size diameters can be grown (18). These filaments, which are known as whiskers, possess very high strengths because of the lack of defects which otherwise weaken larger diameter fibers. Whiskers have diameters in the range of 0.5 to 1.5 μm and lengths which can range from tens of microns to centimeters. The large aspect ratio of length to diameter makes them theoretically interesting as reinforcements for composite materials but difficulties due to their toxicity and simply handling them have meant that they have been little exploited. A technology which is still in the laboratory electrospins sol-gel precursors which can then be pyrolysed to form even finer, nano-oxide fibers. Little is known about the properties which can be expected of such fibers but their development shows that the evolution of oxide fibers is far from over.

3.0. PROCESSING

1.1. Glass Fibers

The basic material for making glass is sand, or silica, which has a melting point around 1750°C, too high to be extruded through a spinneret. However combining silica with other elements can reduce the melting point of the glass which is produced. Fibers of glass are produced by extruding molten glass, at a temperature around 1300°C, through holes in a spinneret, made of a platinum-rhodium alloy, with diameters of one or two millimetres and then drawing the filaments to produce fibers having diameters usually between 5 and

TABLE 1. Compositions (% wt) of various glasses used in fiber production. Soda lime glass is known as A-glass. The type E is the most widely used glass fiber, types S and R are glasses with enhanced mechanical properties, type C resists corrosion in an acid environment, type Z in an alkaline environment and type D is used for its dielectric properties.

Glass type	A	E	S	R	C	Z	D
SiO_2	72	54	65	60	65	70	74
Al_2O_3	1	15	25	25	4	1	
CaO	10	18		9	14		0.2
MgO	3	4	10	6	3		0.2
B_2O_3		8			5.5		23
Li_2O						1	
F_2		0.3					
Fe_2O_3		0.3					
TiO_2						2	0.1
Na_2O	14				8	11	1.2
K_2O		0.4			0.5		1.3
ZrO_2						15	

15 μm. The spinnerets usually contain several hundred holes so that a strand of glass fibers is produced.

Several types of glass exist but all are based on silica (SiO_2) which is combined with other elements to create specialty glasses. The compositions of the most common types of glass fibers are shown in Table 1. A-glass is alkali or soda lime glass and is most usually used for bottles and not in fiber form. The most widely used glass for fiber reinforced composites is called E-glass, glass fibers with superior mechanical properties are known as S- and R-glasses which contain a higher amount of alumina. However the higher the content of refractory solids such as alumina and silica the more difficult it is to obtain a homogenous melt and this is reflected in the cost of the final product. C-glass is resistant to acid environments and Z-glass to alkaline environments. Type D-glass is produced so as to have a low dielectric constant. The temperature of the molten glass is chosen so that a viscosity of around 500 P (slightly less viscous than molasses) is achieved. The best production temperature is that which gives the desired viscosity and is at least 100°C higher than the liquidus temperature, which is the temperature above which devitrification cannot occur and is around 1100°C for type E glass. This ensures that any slight variation in the temperature of the spinneret bushings does not lead to them being blocked. A lower temperature risks causing breaks in the fiber however a lower viscosity could induce instabilities into the glass stream. The cost of glass fiber production is sensitive to the purity of the raw materials, for which only very small amounts of iron are desired, for example, and to the use of expensive batch materials, such as materials containing boron oxide and sodium oxide (19). Typical values of forming parameters for glass fiber spinning are given in Table 2.

Drawing takes place at high speed and as the glass leaves the spinneret it is cooled by a water spray so that by the time it is wound onto a spool its temperature has dropped to around 200°C in between 0.1 and 0.3 seconds. An open atomic network results from the rapid cooling and the structure of the glass fibers is vitreous with no definite compounds

TABLE 2. Typical drawing conditions for forming glass fibers.

Typical drawing speeds (Upper limit)	450 to 4500 m min^{-1} < 5000 m min^{-1}
Typical nozzle bore diameters in the spinneret	1 to 2 mm
Nozzle lengths	2 to 6 mm
Typical draw-down diameter ratios	96 to 321
Extension ratios	9230 to 103,224

being formed and no crystallization taking place. Despite this rapid rate of cooling there appear to be no appreciable residual stresses within the fiber and the structure is isotropic. The glass fibers which are produced have slightly lower densities than the equivalent bulk glass. The difference is approximately 0.04 g/cc. The higher the draw speed used the lower the density of the glass fiber which is produced. Heating glass fibers above around 250°C will produce an increase in density.

The strength of glass fibers depends on the size of flaws, most usually at the surface, and as the fibers would be easily damaged by abrasion, either with other fibers or by coming into contact with machinery in the manufacturing process, they are coated with a size. The purpose of this coating is both to protect the fiber and to hold the strand together. The size may be temporary, usually a starch-oil emulsion, to aid handling of the fiber, which is then removed and replaced with a finish to help fiber matrix adhesion in the composite. Alternatively the size may be of a type which has several additional functions which are to act as a coupling agent, lubricant and to eliminate electrostatic charges.

Continuous glass fibers may be woven, as are textile fibers, made into a non-woven mat in which the fibers are arranged in a random fashion, used in filament winding or chopped into short fibers. In this latter case the fibers are chopped into lengths of up to 5 cm and lightly bonded together to form a mat, or chopped into shorter lengths of a few millimeters for inclusion in molding resins.

1.2. Discontinuous Oxide Fibers

1.2.1. Melt-Spun Aluminosilicate Fibers

The Chemical Abstract Service has defined these materials under the CAS number 142844-00-6 as: Refractories, fibers, aluminosilicates. Amorphous man-made fibers produced from melting, blowing or spinning of calcinated kaolin clay or a combination of alumina (Al_2O_3) and silica (SiO_2). Oxides such as zirconia, ferric oxide, magnesium oxide, calcium oxide and alkalines may also be added.

These aluminosilicate fibers are produced by a melt-spun process in which the starting material is melted, at around 2000°C, by passing an electric current through it. The molten ceramic is poured into a stream of compressed air which carries the ceramic with it, producing drawing. The molten ceramic should be viscous but have a low surface tension in order to be drawn into fiber form, even so a considerable fraction of the ceramic is not drawn and is known as 'shot'. Turbulence breaks the filaments which are formed into discontinuous lengths with irregular cross sections but a mean diameter would be of the range of 2.5 to 3.5 μm. The need for a low surface tension restricts the alumina/silica ratio to an upper limit of 60/40 and pure alumina is not drawn into filament form if produced by this technique (20).

Alternatively the molten ceramic can be fed to a rapidly rotating disk, or series of disks, from which short fibers are thrown by centrifugal force. This latter process is similar to that used in Germany during WWI to produce short glass fibers to replace asbestos. It produces longer fibers with a slightly larger diameter (3–5 μm) than the first process, which however is more common. Both techniques produce fibers of great variability in diameter which however are generally within the range of (1–8 μm) and lengths (up to several centimeters) and a considerable fraction of non-fibrous shot. The specific surface area of these fibers is in the range of 0.4–0.8 m^2/g.

Shot is undesirable as it does not contribute to the strength and insulation properties of the product. It is of irregular shape and size and is considerably larger than the fibers which are formed, ranging from tens of microns to several hundred microns. Shot content can be reduced to less than 25% by sifting using a standard 212 μm mesh.

The range of compositions of melt-spun aluminosilicate fibers is 45–60 wt% Al_2O_3 with $Al_2O_3SiO_2$ as the other major component together with minor amounts of Fe_2O_3, TiO_2, CaO and other oxides (21). The limit to the composition is the resistance of the material to devitrification of the glass with, for example, the nucleation and growth of mullite ($3Al_2O_3 \cdot 2SiO_2$) which reduces strength dramatically. Strength at temperature increases with alumina content so that some compositions have 52 wt% Al_2O_3, for use as an insulation up to 1250°C. The highest levels of alumina allow insulation blankets to be produced for use up to 1400°C. Small additions of Cr_2O_3 improve temperature resistance.

These melt blown aluminosilicate fibers are produced in several forms by companies such as Morgan Thermal Ceramics and the Unifrax Corporation : they can be a loose collection of fibers which is known as 'bulk fiber' and are used as fillers; 'blankets which can be needle punched felts; the fibers can be made as a laminated felt or paper; stronger 'boards' or 'modules' are formed by a wet vacuum process to produce a felt in which the fibers are held together by an organic binder and these products are typically used in electrical furnaces; 'blocks' are made by stacking squares of blanket material, typically twelve layers 300 × 300 × 25mm are stacked to form blocks of 300 × 300 × 100 mm with the fibers aligned normal to the larger surfaces to give higher strength in the thickness direction. The fibers can also be mixed with binders to form product which can be cast or molded or used as a reinforced refractory cement.

1.2.2. The Saffil Fiber

The Saffil fiber which contains 4% of silica is produced by the blow extrusion of partially hydrolyzed solutions of some aluminum salts with a small amount of silica, in which the liquid is extruded through apertures into a high velocity gas stream. The fiber contains mainly small δ-alumina grains of around 50 nm but also some α-alumina grains of 100 nm. The widest use of the Saffil type fiber in composites is in the form of a mat which can be shaped to the form desired and then infiltrated with molten metal, usually aluminium alloy. It is the most successful fiber reinforcement for metal matrix composite.

For refractory insulation applications heat treatments of the fiber above 1000°C induce the delta alumina to progressively change into alpha alumina. After 100 hours at 1200°C, or one hour at 1400°C, acicular alpha alumina grains can be seen on the surface of the fiber and mullite is detected. After 2 hours at 1400°C the transformation is complete and the equilibrium mullite concentration of 13% is established. Shrinkage of the fiber and hence dimension of bricks are controlled up to at least 1500°C (21). Saffil was originally produced as a refractory insulation but, in addition, has become the most widely used reinforcement for light alloys.

1.3. Fine Continuous Oxide Fibers

1.3.1. Manufacture

Continuous fine oxide fibers are based on alumina in one of its forms, often combined with silica or other phases such as zirconia or mullite (22). Precursors of alumina (Al_2O_3) can be obtained from viscous aqueous solutions of aluminum salts Al X_n $(OH)_{3-n}$, where X can be an inorganic ligand (Cl^-, NO^{3-}...) or an organic ligand ($HCOOH^-$...). The precursor gel fibers which are spun are then dried and heat treated. Heating these precursors causes the precipitation of aluminum hydroxides, such as boehmite (AlO(OH) and the outgassing of large volumes of residual compounds. The associated volume change and porosity at this step has to be carefully controlled if useful fibers are to be produced. It is also possible to spin aqueous sols based on aluminum hydroxide directly. Heating the precursor fibers induces the sequential development of transition phases of alumina which if heated to a high enough temperature all convert to the most stable form which is alpha alumina. Above 400°C and up to around 1000°C transitional phases of alumina are produced with grain sizes in the range of 10 to 100 nm. Above 1100°C α-alumina is formed. However this transformation is followed by a rapid growth of porous α-alumina grains, of micron sizes and above, giving rise to weak fibers. It is essential that this rapid grain growth is controlled or retarded if fibers with useful properties are to be obtained. Applications of alumina fibers above 1100°C requires that the nucleation and growth of the α-alumina grains be controlled and porosity limited. This is achieved by either adding silica precursors or seeds for α-alumina formation to the fiber precursors. This has led to the development of two families of alumina based fibers, one consisting of primarily of α-alumina grains and the other of transitional alumina phases together with another phase.

If alumina is combined with silica (SiO_2) the transformation to alpha alumina can be retarded and controlled. The microstructures of such fibers depend on the highest temperature the fibers have seen during the ceramisation. Very small grains of η, γ or δ alumina in an amorphous silica continuum are obtained with temperatures below 1000–1100°C. The combination of alumina and silica phases changes the inherent rigidity of the fibers as the Young's modulus of alumina is around 400 GPa and that of silica approximately 70 GPa, as can be seen in Figure 1.

Strength is not effected by the silica content, as can be seen by Figure 2. The differences in the strengths of the α-alumina fibers are principally due to differences in grain size. The Nextel 610 fiber is composed of α-alumina of around 0.1 μm whereas the other two fibers shown have grains or 0.5 μm. The lower strength of the Almax fiber compared to the Fiber FP is due to porosity although the former's smaller diameter makes it easier to handle (22). Silica softens at around 1000°C so that alumina fibers which contain amorphous silica are not suitable for applications at higher temperatures. However the fibers are inherently resistant to oxidation and are stable in molten metals. They have been used successfully in reinforcing light metal alloys. It should be noted however that alumina is not easily wetted by many molten metals so that attention has to be taken to improve fiber-matrix interface.

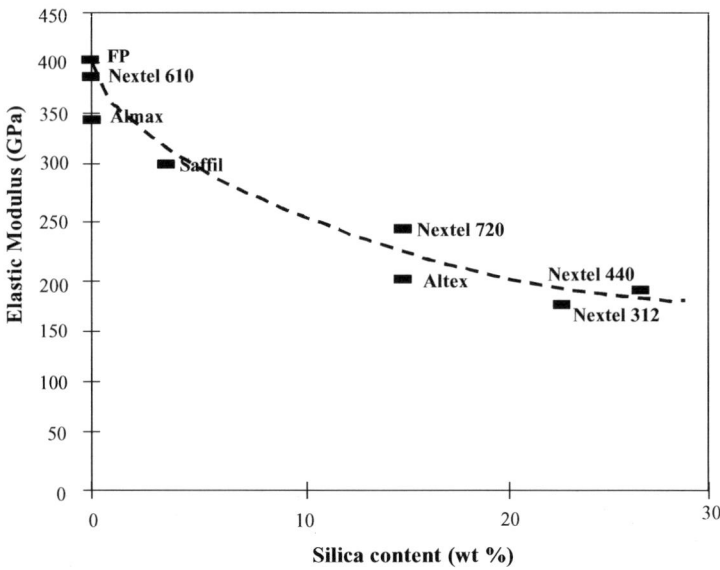

FIGURE 1. The variation of Young's modulus as a function of silica content for a number of alumina based fibers.

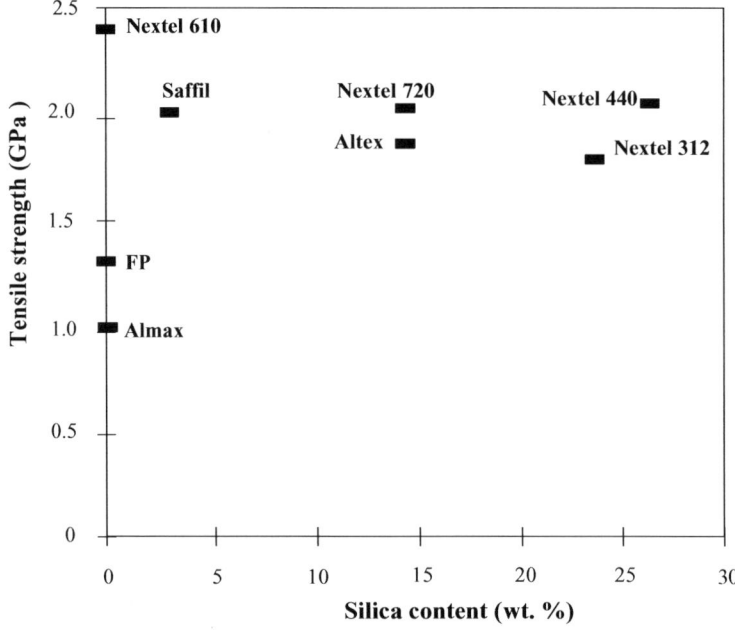

FIGURE 2. There is no direct link between the silica content and the strength of alumina fibers.

Many metal matrix composites are made by squeeze casting in which the molten metal is infiltrated under pressure into a fiber perform. The applied pressure is sufficient to achieve good interfacial bonding with fibers composed of γ- or δ-alumina because of the small sizes of the grains leading to large active contact surfaces.

When heated to around 1200°C alumina combined with silica is partially converted to mullite which can have a range of compositions from $2Al_2O_3 \cdot SiO_2$ to $3Al_2O_3 \cdot 2SiO_2$. The interatomic bonds governing creep in alumina which are ionic and covalent lead to creep at temperatures above 1000°C. The development of fibers combining both alumina and another phase, such as mullite or zirconia, which can hinder creep processes has encouraged interest in the possibility of oxide fibers being used to reinforce ceramics.

1.3.2. Continous Alumino-Silicate Fibers

1.3.2.1. The Altex Fiber   The Altex fiber is a fiber produced by Sumitomo Chemicals. The fiber is circular in cross section and has a smooth surface. The fiber is obtained by the chemical conversion of a polymeric precursor fiber, made from a polyaluminoxane dissolved in an organic solvent to give a viscous product with an alkyl silicate added to provide silica (18). The precursor is then heated in air to 760°C, a treatment which carbonises the organic groups to give a ceramic fiber composed of 85% alumina and 15% amorphous silica. The fiber is then heated to 970°C and its microstructure consists of small γ-alumina grains of a few tens nanometres intimately dispersed in an amorphous silica phase. Subsequent heat treatment produces mullite above 1100°C. At 1400°C the conversion to mullite is completed and the fiber is composed of 55% mullite and 45% alpha alumina by weight.

1.3.2.2. The Nextel Fibers   The 3M corporation produces a range of ceramic fibers under the general name of Nextel. The Nextel 312 and 440 series of fibers are produced by a sol-gel process. They are composed of 3 moles of alumina for 2 moles of silica with various amounts of boria to restrict crystal growth. Solvent loss and shrinkage during the drying of the filament produces oval cross sections with the major diameter up to twice the minor diameter. They are available with average calculated equivalent diameters of 8–9 μm and 10–12 μm A more crystalline version of the Nextel 440 fiber was produced under the name of Nextel 480 but appears to no longer to be available.

The Nextel 312 fiber, which first appeared in 1974, is composed of 62% wt Al_2O_3, 24% SiO_2 and 14% B_2O_3 and appears mainly amorphous from transmission electron microscopy observations although small crystals of aluminium borate have been reported. It has the lowest production cost of the three fibers and is widely used but has a mediocre thermal stability as boria compounds volatilise from 1000°C inducing some severe shrinkage above 1200°C. To improve the high temperature stability in the Nextel 440 and 480 fiber, the amount of boria has been reduced. These latter fibers have the same compositions: 70% Al_2O_3, 28% SiO_2 and 2% B_2O_3 in weight but their microstructures are different. Nextel 440 is formed in the main of small γ-alumina in amorphous silica whereas Nextel 480 was composed of mullite. These differences may be due to different heat treatments of similar initial fibers, the Nextel 440 fiber being heated below the temperature of mulitisation.

The Nextel 720 contains the same alumina to silica ratio as in the Altex fiber, that is around 85% wt Al_2O_3 and 15% wt SiO_2. The fiber has a circular cross section and a diameter of 12 μm. The sol-gel route and higher processing temperatures have induced the growth of alumina rich mullite and alpha alumina. Unlike other alumina-silica fibers the Nextel

720 fiber is composed of a mosaic of mullite grains of around 0.5 μm consisting of several slightly mutually misoriented grains in which elongated α-alumina grains are found. This structure results in the Nextel 720 being the oxide fiber with the lowest creep rate. Post heat treatment leads to an enrichment of α-alumina in the fiber as mullite rejects alumina to evolve towards a 3:2 equilibrium composition. Grain growth occurs from 1300°C (10).

1.3.3. Alpha Alumina Fibers

Alpha alumina is the most stable and crystalline form of alumina to which all other phases are converted upon heating above around 1000°C. As we have seen above, fibers based on alumina can contain silica as its presence allows the rapid growth of alpha-alumina grains to be controlled. However the presence of silica reduces the Young's moduli of the fibers and reduces their creep strength. High creep resistance implies the production of almost pure alpha alumina fibers however to obtain a fine and dense microstructure is difficult. The control of grain growth and porosity in the production of alpha-alumina fibers is obtained by using a slurry consisting of alpha alumina particles, of strictly controlled granulometry, in an aqueous solution of aluminium salts. The rheology of the slurry is controlled through its water content. The precursor filament which is then produced by dry spinning is pyrolysed to give an alpha-alumina fiber.

1.3.3.1. Fiber FP

The FP-fiber, manufactured by Du Pont in 1979, was the first wholly α-alumina fiber to be produced (4). Its production involved the spinning in air of a slurry composed of an aqueous suspension of Al_2O_3 particles and aluminium salts. The as obtained fiber was then dried and fired in two steps, the first to control shrinkage and followed, at a higher temperature, by flame firing to obtain a dense microstructure of α-alumina. A final step, involving a brief exposure to a high temperature flame to produce a fine surface layer of silica, had the effect of improving fiber strength and aiding wettability with metal matrices. It was a continuous fiber with a diameter of 18 μm This fiber was composed of 99.9% alpha alumina and had a density of 3.92 g/cm^3 and a polycrystalline microstructure with a grain size of 0.5 μm, a high Young's modulus of 410 GPa, a tensile strength of 1.55 GPa at 25 mm but a strain to failure of only 0.4%.This brittleness made it unsuitable for weaving and although showing initial success as a reinforcement for light alloys, production did not progress beyond the pilot plant stage and commercial production ceased. Never-the-less Fiber FP represents an example of an almost pure alumina in filament form and as such allows the fundamental mechanisms in this class of fiber to be investigated (23). This fiber was seen to be chemically stable at high temperature in air, however its isotropic fine grained microstructure led to easy grain sliding and creep, excluding any application as a reinforcement for ceramic structures.

Other manufacturers have modified the production technique to reduce the diameter of the alpha-alumina fibers that they have produced. This reduction of diameter has an immediate advantage of increasing the flexibility and hence the weaveability of the fibers. Mitsui Mining and 3M Corporation have introduced polycrystalline fibers, the Almax and the Nextel 610 fibers with diameters of 10 μm, that is half the diameter of Fiber FP.

1.3.3.2. Almax Fiber

An alpha-alumina fiber which is still commercially available was produced first in the early 1990s by Mitsui Mining (7). It is composed of almost pure alpha alumina and has a diameter of 10 μm. The fiber has a lower density of 3.60 g/cm^3 compared to Fiber FP. Like Fiber FP the Almax fiber consists of one population of grains of around 0.5 μm however the fiber exhibits a large amount of intragranular porosity, and associated with numerous intragranular dislocations without any periodic arrangement. This indicates rapid grain growth of alpha alumina grains during the fiber fabrication process without elimination of porosity and internal stresses. As a consequence, grain growth at 1300°C is activated without an applied load and reaches 40% after 24 hours, unlike that with the other pure alpha alumina fibers, for which grain growth is related to the accommodation of the slip by diffusion.

1.3.3.3. Nextel 610   A continuous alpha-alumina fiber, with a diameter of 10 μm, was introduced by 3M in the early 1990s with the trade-name of Nextel 610 fiber (24). It is composed of around 99% alpha alumina although a more detailed chemical analysis gives 1.15% total impurities including 0.67% Fe_2O_3 used as a nucleating agent and 0.35% SiO_2 as grain growth inhibitor. It is believed that the silica which is introduced does not form a second phase at grain boundaries although the suggestion of a very thin second phase separating most of the grains has been observed by transmission electron microscopy. The fiber is polycrystalline with a grain size of 0.1 μm, five time smaller than in Fiber FP.

1.3.4. Alumina Zirconia Fibers

1.3.4.1. PRD-166 Fiber   Du Pont synthesised the PRD-166 fiber in which 20% wt of partially stabilised tetragonal zirconia was added to increase the elongation to failure of the fiber (5). The intention was to produce a fiber which, compared to Fiber FP, was easier to weave. The dispersion of zirconia intergranular particles of 0.15 μm limited grain growth of the alumina grains which had a mean diameter of 0.3 μm instead of 0.5 μm for Fiber FP for a similar initial alumina powder granulometry. These particles underwent a martinsitic reaction in the vicinity of the crack tips, which in a similar bulk ceramic results in the partial closure of cracks and in an increase of the fiber strength. It is not clear if this process was significant in the fiber form or if the reduction in grain size was more important but the PRD-166 fiber was stronger than the Fiber FP with a failure strength of 1.8 GPa at a gauge length of 25 mm. The resulting stiffness of the reinforced alumina was lower than that of Fiber FP, E = 344 GPa, due to the lower Young's modulus of zirconia ($\approx$200 GPa) compared to that of alumina ($\approx$400 GPa). However the increase in strain to failure was not sufficient to allow weaving with the PRD-166 fiber and production of the PRD-166 fiber did not progress beyond the pilot stage, however the study of this fiber permits a greater understanding of the mechanisms of toughening and the enhancement of creep behaviour of alumina fibers.

1.3.4.2. The Nextel 650 Fiber   The Nextel 650 fiber is produced by 3M, in order to combine the properties of a fiber which was above all resistant to alkaline contamination, as this was important for a reinforcement for high temperature composites, and second have improved creep resistance compared to the Nextel 610 pure α-alumina fiber (25). The Nextel 650 fiber has been produced, like others in the Nextel series, by sol-gel processing with the use of iron compounds as nucleating agents and SiO_2 to restrict grain growth compared to the PRD-166 fiber. The addition of α-Fe_2O_3, which has a similar structure to that of α-alumina, to the sol is to lower the temperature at which transitional phases of alumina are converted to α-alumina and this helps in the production of a low porous structure.

The Nextel 650 fiber is continuous and circular in cross-section with a diameter of 11.2 μm and is composed of α-alumina and 10%wt of cubic zirconia stabilized by 1%wt of

Y_2O_3. The zirconia has been added to increase the elongation to failure and to limit alumina grain growth. The microstructure obtained is very fine with alumina grains of 0.1 μm and a bimodal zirconia grain size distribution: 5 to 10 nm in size for intra-granular grains and 20–30 nm for inter-granular grains. No other phase can be detected in the as-received fiber and inter-granular porosity was quasi-inexistent.

1.3.5. Continuous Monocrystalline Filaments

The techniques for producing single ceramic crystals as filaments has been known since the 1960s and they offer the possibility of producing filaments containing no grain boundaries with associated high strength (13, 14). Continuous α-alumina monocrystalline filaments have been commercially produced by the Saphikon company in the USA. (15). This type of filament is grown from molten alumina by a modified Czochralsky-Stepanov edge-defined film-fed growth method in which an oriented seed crystal is slowly drawn from the molten ceramic. The ceramic is either heated by radio frequency induction furnace in a molybdenum crucible or by a technique known as laser-heated float zone in which the surface only of a ceramic feed rod is melted by a laser beam. The crystal orientation which is induced in the filament is that of the seed crystal, which is attached to a molybdenum rod, although regular patterns of bubbles can be seen in filaments produced by the induction furnace heating process and which are due to convection in the molten ceramic. The production rate is extremely slow and the high cost of producing the filaments together with their large diameters, usually in excess of 100 μm, means that they are not being considered for industrial use. The stoichiometric composition of these fibers with the absence of grain boundaries ensures that they should be able to better withstand high temperatures above 1600°C. Careful orientation of the seed crystal enables the C-axis of the α-alumina fiber can be aligned parallel to the fiber axis so that creep resistance can be optimized.

The same manufacturing processes have been employed to produce an eutectic fiber consisting of interpenetrating phases of -alumina and $Y_3Al_5O_{12}$ (YAG) (25). The structure depends on the conditions of manufacture, in particular the drawing speed but can be lamellar and oriented parallel to the fiber axis. This fiber does not show the same fall in strength seen with the single phase alumina fiber. However such fibers are seen to relax from 1100°C but their strengths do not have as strong a dependence on temperature as with the polycrystalline oxide fibers.

The growth process is extremely slow, typically 100 mm/hr but can easily be adapted to a wide range of ceramic systems for growing single crystal and directionally solidified eutectic filaments (27, 28). Although the crystal structure is continuous and Saphikon produced lengths of up to 3000 m of fiber, usually the lengths of filaments produced in laboratories are short, being typically tens of centimeters.

An alternative approach to making single crystal fibers and one which is potentially much cheaper than the technique described above is to infiltrate the molten ceramic in the channels formed by sandwiching molybdenum wires between molybdenum sheets (17). These dies are prepared and the wires and sheets diffusion bonded together. The processing zone is heated using a 8 kHz induction heated graphite susceptor and molybdenum crucible. Seeds, which are used to control crystal growth and orientation are placed on the top surface of the molybdenum die. The crucible is filled with the raw material which is melted. As the raw material becomes molten the molybdenum die is lowered into it and the molten ceramic is drawn up the channels by capillary forces. The die is then withdrawn and the ceramic solidifies from the top down. The seed crystal determines the crystal orientation of the filaments formed. The technique has been used to produce eutectic ceramic filaments for which it is not necessary to use a seed (26). Systems which have been made by this technique include Al_2O_3, Al_2O_3-$Al_5Y_3O_{12}$, Al_2O_3-$ZrO_2(Y_2O_3)$ and Al_2O_3-$AlGdO_3$. Mullite filaments have also been produced by this process. Many fibers can be produced simultaneously by this technique. Drawing rates are around 10mm/min.

The length of the zone which can be heated limits the length of filament which can be produced but lengths of up to 250 mm have been announced. Processing is carried out under inert gas or vacuum. Several batches of fiber can be processed simultaneously to produce up to 150 g of filaments at a time.

The molybdenum die material is finally removed by etching. Two routes are possible to remove the molybdenum using the following chemical reactions:

$$Mo + 3H_2O_2 = H_2MoO_4 + 2H_2O \qquad (1)$$

Reaction (1), in which hydrogen peroxide is used is more ecologically friendly however the reaction takes up to three times longer than the second. The cost of hydrogen peroxide is

$$3Mo + 6HNO_3 \xrightarrow{HCl} \xrightarrow{H_2SO_4} 3H_2MoO_4 + 6NO \qquad (2)$$

higher than for the acids in the second reaction. Both reactions produce a rise in temperature which however is easier to control in the second reaction. Reaction (2) requires approximately thirty hours to remove the molybdenum so as to release the fibers.

1.3.6. Whiskers

Whiskers are fine high purity monocrystals in the form of filaments. The potential of whiskers as reinforcements has been discussed for many years as their small diameters, usually between 0.5 and 1.5 microns means that they contain very few defects and must posses extremely high strengths, perhaps up to the theoretical strength for matter, which is approximately one tenth of its Young's modulus. In addition their aspect ratios of length to diameter can be considerable as they can be produced with lengths between 20 μm and it is claimed, several centimeters. A high aspect ratio is just what is required to achieve reinforcement in composites.

Amongst the oxides which have been produced as whiskers are Al_2O_3, MgO, MgO-Al_2O_3, Fe_2O_3, BeO, MoO_3, NiO, Cr_2O_3 and ZnO. Typically the whiskers are produced by heating the metal in a suitable atmosphere such as wet hydrogen, a moist inert gas or air. The most commonly produced whiskers are of alumina and silicon carbide.

Alumina whiskers are produced by high temperature chemical vapor deposition at the tip of a substrate particle (21). The temperature has to be high enough for the vapor pressure of the whisker or whisker forming material to become significant, in which case the atoms become attached to the tip of the whisker and contribute to its growth. Alumina whiskers can be produced by passing a stream of moist hydrogen over aluminum powder heated to around 1400°C. A mass of acicular α-alumina crystals are deposited in the cooler part of the furnace. This commonly used reaction is as follows:

$$2Al(g) + 3H_2O(g) \longrightarrow Al_2O_3(s) + 3H_2(g)$$

Alternatively the following reaction can be used :

$$2AlC_3(g) + 3CO_2(g) + 3H_2(g) \Rightarrow Al_2O_3(s) + 6HCl(g) + 3CO(g)$$

Considerable difficulties have to be overcome if whiskers are to be used as reinforcements however. They are extremely small so that plastic bag containing whiskers seems to contain dust. This means that alignment of the whiskers in a matrix is very difficult. There are potential uses for whiskers combined with more conventional fibers so as to provide some reinforcement of the matrix in the transverse direction. Their fineness is also another handicap in their exploitation as one micron is just the size to block up the alveolar structure of the lungs. For this reason, above all, whiskers remain an intriguing possibility as reinforcements but one which is little exploited.

1.3.7. Nano-oxide fibers

An emerging technology is the production of fibers of very small diameter, of the order of 50nm. These fibers are produced by the spinning of a precursor organic fiber from a pipette to a collecting plate. A high voltage (tens of kilovolts) is passed between the pipette and the plate and the polymer is drawn from the pipette to the plate. The fibers are generally collected on the plate to form a random array although work is proceeding to align the fibers. The fibers are too fine to be tested by conventional techniques but can be tested as bundles. The fibers can also be subjected to the same cycles of pyrolysis that have been used to produce larger diameter ceramic fibers. At present this technology is still at the laboratory stage so that few data are available however the nanometric diameter could be expected to confer on the filaments perhaps exceptional properties which are not obtained with larger diameter fibers. This is primarily because dislocation movement should be restricted so that high strengths and low creep rates could be expected. At the laboratory scale oxide fibers such as Al_2O_3, ZrO_2 and TiO_2 as well as carbon fibers have been made but they are far from having being fully evaluated.

4.0. PROPERTIES

1.1. Glass Fibers

The mechanical properties of a range of glass fibers are shown in Table 3. As strength and failure strains are not intrinsic properties of a material it is possible to find a range of values in the literature (29). Physical properties of glass fibers are shown in Table 4.

Glass fibers are known to fail when subjected to steady unvarying loads. This is not creep failure but is sometimes known as static fatigue. In this process microscopic defects grow under the influence of the applied stress. When the defect attains the critical flaw size for the applied stress the fiber breaks. The growth of flaws in glass was first treated by Griffith who discussed the energy necessary to propagate a crack in an elastic medium (30). The energy necessary to separate the two fracture surfaces can be modified by the environment in which the glass is held. Water, for example can reduce the threshold stresses for crack propagation. Table 5 gives details about the chemical resistance of the glass fibers in different environments.

TABLE 3. Mechanical properties of a number of different types of glass fibers.

Glass type	A	E	S	R	C	D
Strength (20°C) GPa	3.3	3.5	4.65	4.65	2.8	2.45
Elastic Modulus (20°C) GPa	69	73.5	86.5	86.5	70	52.5
Failure Strain (20°C) %	4.8	4.5	5.3	5.3	4.0	4.5
Specific Gravity	2.44	2.54	2.48	2.48	2.49	2.16
Poisson's ratio	–	0.2	–	–	–	–
Coefficient of Thermal Expansion ($\times 10^7 \cdot C^{-1}$)	73	54	16	33	63	25

TABLE 4. Physical properties of a number of different types of glass fibers.

Properties	A	E	S	R	C	D
Refractive Index	1.538	1.558	1.521	1.546	1.533	1.465
Dielectric constant 1 MHz	6.2	6.6	5.3	6.4	6.9	3.8
Specific heat cal g^{-1}°C^{-1}	0.796	0.810	0.737	–	0.787	0.733
Softening temperature °C	705	846	1056	952	750	771
Annealing temperature °C	–	657	816	–	588	521
Volume resistivity (ohms-cm)	1.0E + 10	4.20E + 14	9.03E + 12	2.03E + 14	–	–

TABLE 5. Chemical resistance of glass fibers immersed in different environments.

Durability (% weight loss)	A-Glass	E-Glass	E-Glass	R-Glass	C-Glass	D-Glass
H_2O : 24 hr	1.8	0.7	0.5	0.4	1.1	0.7
168 hr	4.7	0.9	0.7	0.6	2.9	5.7
10% HCl : 24 hr	1.4	42	3.8	9.5	4.1	21.6
168 hr		43	5.1	10.2	7.5	21.8
10% H_2SO_4 : 24 hr	0.4	39	4.1	9.9	2.2	18.6
168 hr	2.3	42	5.7	10.9	4.9	19.5
10% Na_2CO_2 : 24 hr		2.1	2.0	3.0	24	13.6
168 hr		2.1	2.1		31	36.3

TABLE 6. Examples of typical melt spun short fibers and their characteristics.

Property	Kaowool	Kaowool 1400	Kaowool 1500	Cera-chrome
Maximum service temperature	1260°C	1400°C	1500°C	1427°C
Melting point	1760°C	1815°C	–	>1760°C
Average fiber diameter	2.8 μm	2.5 μm	2.6 μm	3.5
Specific gravity	2.6	2.8	2.65	2.65
Chemical analysis (%)				
Al_2O_3	47.3	56.3	41.2	42.5
SiO_2	52.3	43.3	56.6	55.0
Cr_2O_3			2.1	2.5

1.2. Aluminosilicate Fibers

The Young's moduli of these fibers are lower compared to that of pure alumina fibres, and such fibers are produced at a lower cost. This, added to easier handling due to their lower stiffness, makes them attractive for thermal insulation applications, in the absence of significant load, in the form of consolidated felts or bricks up to at least 1500°C. Such fibers are also used to reinforce aluminum alloys in the temperature range of 300–350°C. Continuous fibers of this type can be woven due to their lower Young's moduli. However microstructural changes occur if the fibers are heated to sufficiently high temperatures. All transitional phases are changed in to α-alumina around 1100°C. If any amorphous silica is present in the fiber it will begin to soften at these temperatures and facilitate grain boundary sliding and creep of the fiber. In addition mullite may begin to be formed around 1000°C and cristobalite around 1200°C. The ionic bonds which occur in oxides allow faster creep rates than are found in ceramics which possess only co-valent bonds such as silicon carbide.

1.2.1. Melt Spun Aluminosilicate Fibers

The irregular shapes and fine diameters of these fibers make them difficult to characterize and most available data concerns the properties of finished products. However, Table 6 gives some data on typical fibers made by the melt spun process (20).

The standard grade products account for most of the ceramic fibers produced and are made into a variety of products. Their composition of 47–51% alumina and 49–53 silica allows fibers to be produced with small diameters. Within this composition range there is little variation in thermal resistance however when natural kaolin is the starting material there is a possibility of contamination by alkaline oxides (Na_2O and K_2O) which has a detrimental effect on the thermal insulation of the fibers produced. Thes alkaline contaminants can combine with the silica in the fiber to form a low melting point phase and this can cause failure of the fibers. The presence of vanadia (V_2O_3) can further exacerbate the fiber damage by producing a low melting point phase and reacting with the alkaline contaminants. Increased thermal resistance is achieved by increasing the alumina content although the increased difficulty in producing this grade means that a higher

TABLE 7. Property and microstructural changes during the processing of Al_2O_3 fibers containing approximately 4% SiO_2.

Major phases	η	η/γ	γ	γ/δ	δ	δ/θ	θ-mullite	α-mullite
Grain size	6 nm				50 nm		100 nm	>200
Crystallinity (%)	50	62	68	77	79	86	97	100
α-alumina (%)					7	16	20–50	100
Pore volume (mm^3/g)	200		187	121	73	46	0	0
Shrinkage at 1400°C after 1h (%)	18	17	14	8	6.5	3.5	0.5	0
Tensile strength for 3.5 μm diameter fiber (GPa)	1.8				1.5			0.5

shot content should be expected as well as a higher specific gravity. Both of the former materials can suffer from significant shrinkage upon heating above 1200°C due to temperature induced phase changes. The addition of chromia (Cr_2O_3) retards these phase changes and reduces shrinkage. The low biopersistent fibers which contain calcia (CaO), magnesia (MgO) and silica (SiO_2) show lower thermal resistance and are probably limited to around 1000°C.

All of these fibers are subject to changes to their microstructures at high temperature and these ultimately limit their use. Alumina and silica will combine to form mullite from around 970°C and crystoballite, which is a crystalline form of silica, is formed at around 1260°C. These changes can occur progressively over a period of time when the fibers are subjected to temperatures in these ranges. The standard RCF starts to precipitate crystobalite at 1100°C after around 3000 hours, but at 1200°C this is reduced to 300 hours and 50 hours at 1300°C. The 1400°C grade takes two to three times longer at the same temperatures. The effects of these phase changes are to cause the mullite grains to grow from, initially, around 30 nm at 1100°C and 100 nm at 1300°C. The development of these mullite grains is to reduce the flexibility of the fibers and eventually to lead to the fusing together of the fibers and the product becomes brittle. Shrinkage of the fiber structure also occurs when crystallization is initiated and the higher the alumina content the greater the shrinkage. The standard grade heated for 100 hours at 1100°C will shrink 2% and 2.8% at 1200°C whereas the 1400 grade will shrink 2.1% and 3.6% under the same respective conditions.

1.2.2. Alumina based fibers produced from precursors

The alumina based fibers discussed in section 3.2 possess a range of compositions. They can be short, as with the Saffil fibers or continuous, as with the others described. Their properties at room temperature depend on the α-alumina content and at high temperature, the presence of any second phase (31). The Saffil fiber contains a few percent of silica with the remainder of the composition being alumina in one of its transition phases or as a mixture of transition phases and α-alumina. Table 7 shows the changes in processing of fibers of this type (32).The properties of alumina based fibers areshown in Table 8. Figure 3 shows the tensile curves of a pure α-alumina fiber, the Fiber FP, which had a grain size of 0.5 μm (23).

TABLE 8. Properties and compositions of alumina based fibers

Fiber Type	Manufacturer	Trade Mark	Composition (wt%)	Diameter (μm)	Density (g/cm^3)	Strength (GPa)	Strain to failure (%)	Young's Modulus (GPa)	CTE $10^{-6}/°C$
α-Al$_2$O$_3$ based fibers	Du Pont	FP	99.9% Al$_2$O$_3$	20	3.92	1.2	0.29	414	5.9 (25–900°C) 9.6 (900–1500°C)
	Mitsui Mining	Almax	99.9% Al$_2$O$_3$	10	3.6	1.02	0.3	344	7
	3M	Nextel 610	99% Al$_2$O$_3$ 0.2%–0.3SiO$_2$ 0.4–0.7 Fe$_2$O$_3$	10–12	3.75	1.9	0.5	370	8 (100–1100°C)
	Du Pont	PRD 166	80% Al$_2$O$_3$ 20% ZrO$_2$	20	4.2	1.46	0.4	366	9.0
	3M	Nextel 650	90.4 Al$_2$O$_3$ 7.9% ZnO$_2$ 1.1% Y$_2$O$_2$ 0.6% Fe$_2$O$_3$	11	4.1	2.3	0.6	370	8.0
Alumina silica based fibers	Saffil	Saffil	95% Al$_2$O$_3$ 5% SiO$_2$	1–5	3.2	2	0.67	300	6
	Sumitomo Chemicals	Altex	85% Al$_2$O$_3$ 15% SiO$_2$	15	3.2	1.8	0.8	210	6
	3M	Nextel 312	62% Al$_2$O$_3$ 24% SiO$_2$ 14% B$_2$O$_3$	10–12 and 8–9	2.7	1.7	1.12	152	3 (25–500°C)
	3M	Nextel 440	70% Al$_2$O$_3$ 28% SiO$_2$ 2% B$_2$O$_3$	10–12	3.05	2.1	1.11	190	5.3
	3M	Nextel 720	85% Al$_2$O$_3$ 15% SiO$_2$	12	3.4	2.1	0.81	260	6 (100–1100°C)

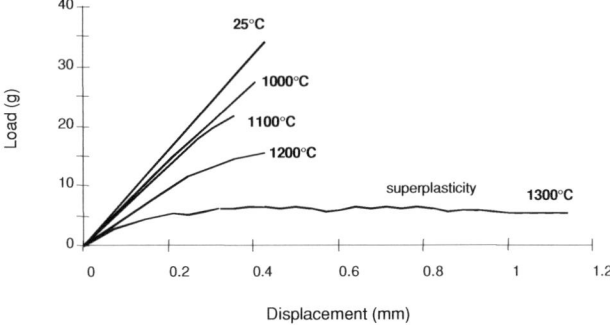

FIGURE 3. Tensile behavior of a pure α-alumina fiber, Fiber FP, as a function of temperature.

It can be seen that not only does the strength fall but the behavior changes from being linearly elastic below 1000°C to showing plastic deformation at 1100°C and above. At 1300°C the fiber deforms greatly to the point where, on occasions, the tensile testing machine would reach its end stops without the fiber breaking. The fall in strength of this fiber as a function of temperature is shown in Figure 4.

Fiber FP creeps from around 1000°C and the rate of creep increases with increasing temperature, as is shown in Figure 5. The addition of zirconia, in the PRD-166 fiber, to the coarse grain structure of the α-alumina, Fiber FP, did improve its room temperature strength (23, 33) but the improvement achieved by the Nextel 610 fiber due to a smaller grain size was much greater. However the advantage in strength achieved by both mechanisms was

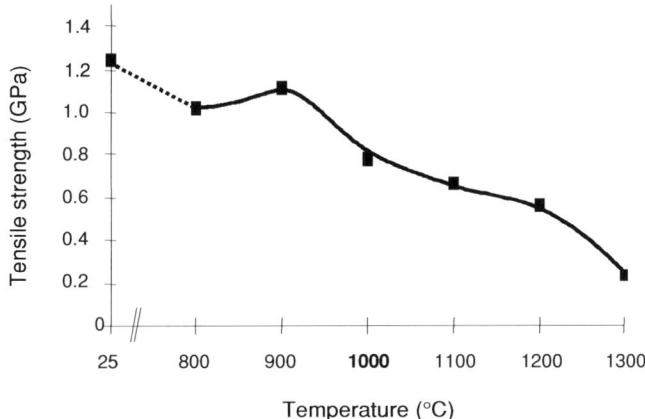

FIGURE 4. Strength of a pure α-alumina fiber, Fiber FP, as a function of temperature. *(Reprinted, from reference 23, with kind permission of Kluwer Academic/Plenum Publishers).*

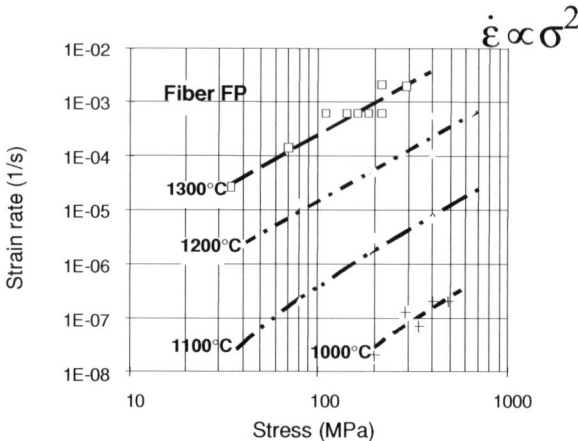

FIGURE 5. The pure α-alumina fiber, Fiber FP, crept from 1000°C and the creep rate increased with temperature. *(Reprinted, from reference 23, with kind permission of Kluwer Academic/Plenum Publishers).*

lost at temperatures of around 1200°C, at which temperature even the Nextel 610 fiber retained only 30% of its room temperature strength. The α-alumina Almax fiber, which has a similar grain size to that of Fiber FP, possesses a lower Young's modulus than the other two α-alumina fibers because of about 9% porosity and this accounts for its lower tensile strength, as can be seen in Figure 6.

Figure 7 compares the creep behavior of the three nearly pure α-alumina fibers. It can be seen that Fiber FP crept at the lowest rate at each temperature. The smaller grains in the Nextel 610 fibers allowed them to creep faster whilst the porosity and residual stresses in the Almax fibers meant that they crept the fastest (23).

Figure 8 shows the creep behavior from 1100 to 1300°C of several fibers: FP and Nextel 610 represent pure alumina fibers, PRD-166 and Nextel 650 zirconia-alumina fibers and

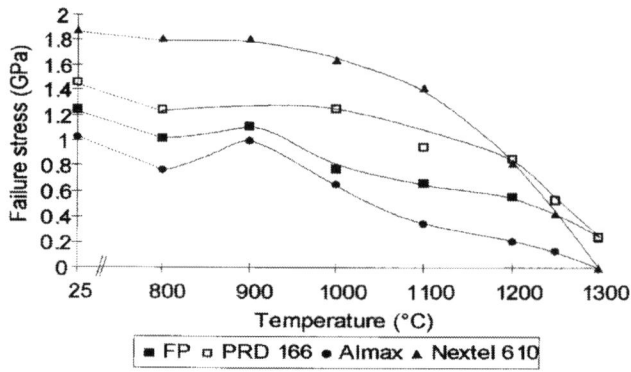

FIGURE 6. Comparison of fiber strengths as a function of temperature. *(Reprinted, from reference 23, with kind permission of Kluwer Academic/Plenum Publishers).*

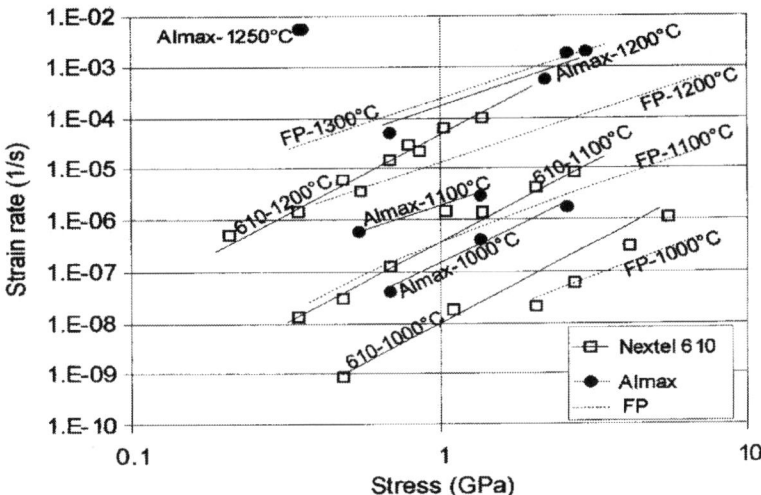

FIGURE 7. Comparison of creep behavior of the three α-alumina alumina fibers. *(Reprinted, from reference 22, with kind permission of Marcel Dekker Inc.).*

Nextel 720, alumina-mullite fibers (12). There are three different parameters distinguishing these fibers: grain sizes, second phase and inter-granular segregation at Al_2O_3/Al_2O_3 grain boundaries. At 1100°C, Figure 8.a, the advantages of yttrium stabilized zirconia additions are obvious. The pure alumina fiber Nextel 610 and the Nextel 650 fibers which contains zirconia have identical α-alumina grain sizes but the latter fiber creeps at a significantly lower rate. The effect is sufficient to overcome the effects of the smaller α-alumina grain sizes of the Nextel fibers (0.1μm), compared to Fiber FP (0.5 μm). Up to 1200°C, Figure 8.b shows that the improvement of creep rates by the addition of zirconia is clearly maintained by a comparison of the Nextel fibers and the FP and PRD-166 fibers. However, at 1200°C, the difference of creep rates seen with the Nextel 610 and 650 is reduced compared to that seen at 1100°C and this is also the case between the FP and PRD-166 fibers. At 1300°C, Figure 8.c reveals that the presence of zirconia has no effect on the creep phenomena, and the difference between the grain sizes can explain why the PRD-166 fiber shows creep rates of around one order of magnitude lower than those of Nextel 650 fiber. Although the Nextel 720 (alumina-mullite) fiber shows high sensitivity to its environment above 1100°C, which renders it chemically unstable, this fiber shows, at 1300°C, creep rates 5000 times lower than those of the Nextel 650 fiber. These excellent creep properties are due to the presence of a mullite phase but also to the evolution of the microstructure towards a finally composed structure of elongated α-alumina grains.

The combination of silica with alumina can retain transitional forms of alumina, as in the Altex fiber but the combination in the Nextel 480 fiber gives a mullite structure whereas the combination in the Nextel 720 fiber gives a mullite structure in which α-alumina grains are embedded. All three fibers, however lose strength above 1100°C, as shown in Figure 9. The fibers show very different creep behavior, as can be seen from Figure 10, with the Nextel 720 fiber showing the lowest creep rate of all oxide fibers.

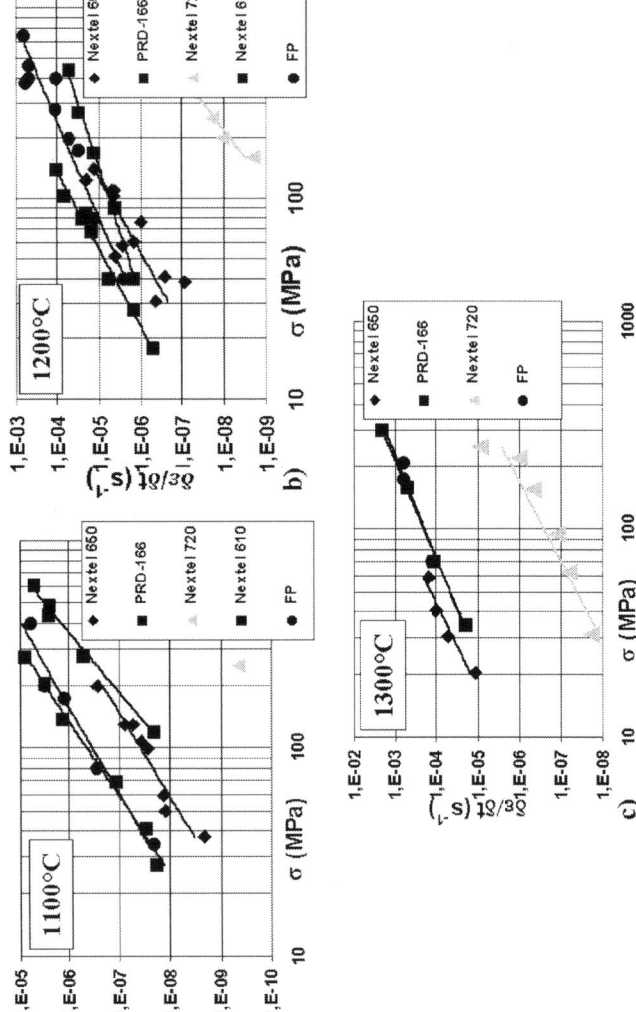

FIGURE 8. A comparison of the creep rate of the Nextel 650 and PRD-166 fibers, both of which are composed of α-alumina and zirconia, the Nextel 720 which is composed of α-alumina and mullite and the pure α-alumina Fiber FP. *(Reprinted, from reference 12, "Mechanical and Microstructural Characterization of Nextel 650 Alumina-Zirconia Fibres", with kind permission of Elsevier).*

OXIDE FIBERS 27

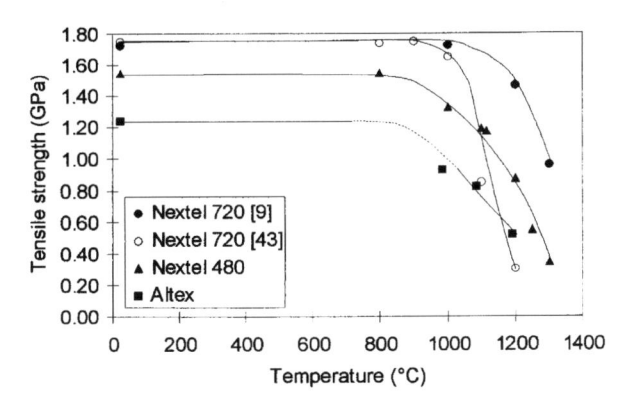

FIGURE 9. Temperature dependence of tensile strength of the Nextel 720, Nextel 480 and Altex fibers; all contain alumina and silica in various forms. *(Reprinted, from reference 22, with kind permission of Marcel Dekker Inc.).*

1.2.3. Continuous single crystal oxide fibers

Figure 11 shows tensile failure stress results obtained in the author's laboratory on single crystal α-alumina fibers, with the c-axis aligned with the fiber axis, supplied by Saphikon. The fibers had a diameter of 125 μm. The limited number of points confirm the observations of Shaninian (34), and others, on filaments having a diameter of 250 μm, that strength variation is not a single function of temperature. The observed fall in strength around 300°C, which is then followed by an increase in strength around 500°C, could be due to stress corrosion followed by crack blunting. These fibers are not without defects and characteristic patterns of bubbles of around 1 μm in diameter can be seen in the fibers most probably due to convection during fiber growth at the meniscus point between the solid and the

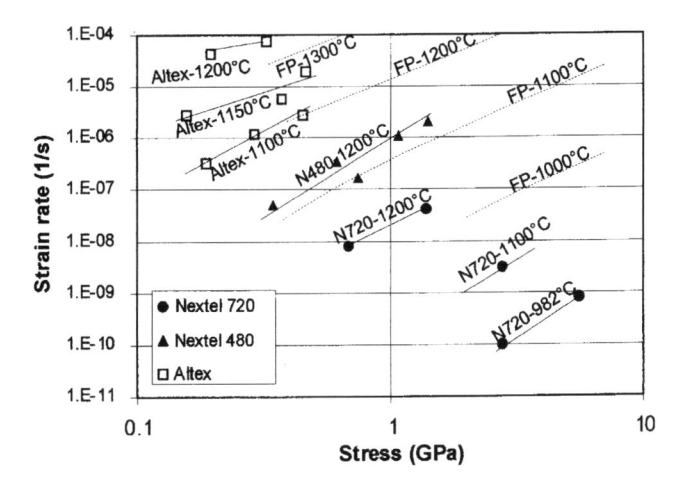

FIGURE 10. A comparison of creep strain rate vs. applied stress at different temperatures for three fibers which are formed by a combination of alumina and silica. *(Reprinted, from reference 22, with kind permission of Marcel Dekker Inc.).*

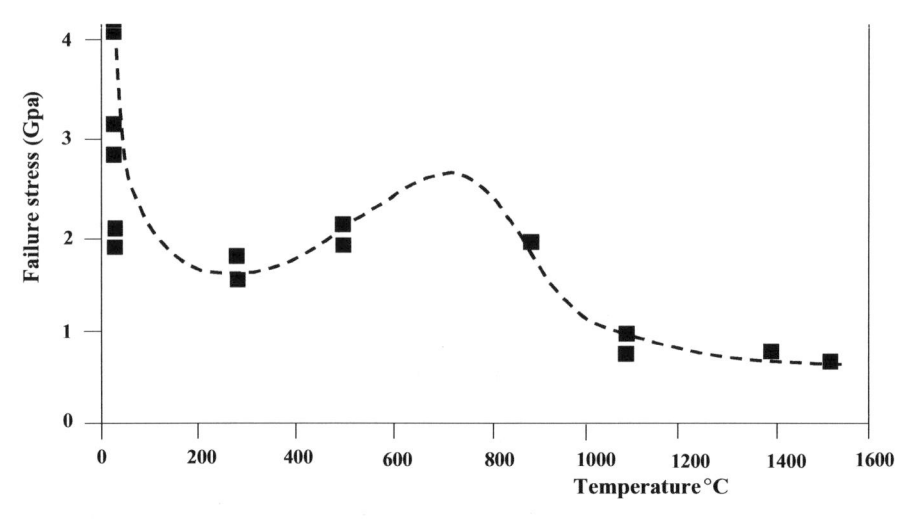

FIGURE 11. Variation of tensile strength of Saphikon single crystal α-alumina fibers as a function of temperature.

melt. The fracture morphology of the α-alumina Saphikon fiber shows clearly the cleavage planes of the crystalline structure. Alpha-alumina has a rhomboidal crystallographic structure which is highly anisotropic so that the tensile and high temperature creep characteristics are strongly dependent on the crystallographic orientation of the fiber with respect to the applied load.

The anisotropy and fall in strength at low temperatures of single crystal α-alumina fibers prompted Saphikon to produce directionally solidified YAG-alumina ($Y_3Al_5O_{12}$–Al_2O_3) fibers by the edge defined film method. Similar fibers have been produced using a laser heated float zone technique (35). YAG belongs to a group of minerals which have a complex cubic crystallographic structure. The structure of YAG is particularly resistant to creep and is much less sensitive to orientation. The combination of α-alumina with YAG in a bulk eutectic structure shows excellent stability. Directionally solidified YAG-alumina fibers have a mosaic microstructure, which can be aligned parallel to the axis. It is composed of the two interpenetrating phases with average lateral dimensions being around 320 nm. Each phase does not necessarily show a single crystal orientation but several orientations may coexist. Filaments of YAG-alumina do not show the fall in strength at low temperatures observed with single crystal α-alumina fibers (26). The room temperature value of Young's modulus is 344 GPa and values of strength in the literature range from 1.35 to 2.4 GPa with good short term strength retention up to 1500°C with less than a ten percent fall at this temperature compared to room temperature strength. Table 9 gives values of tensile strengths of directionally solidified YAG-alumina fibers tested in the author's laboratory (36). The strengths of these fibers were found to be rate dependent at 1400°C due to a phenomenon of slow crack growth so that whereas the strength was measured as being 0.48 GPa with a loading speed of 0.08 MPa/s at a speed of 660 MPa/s the strength was seen to be 0.90 GPa.

A comparison of creep rates of single crystal α-alumina fibers and directionally solidified YAG-alumina fibers, at 1400°C, reveals that the former shows no creep at a rate

TABLE 9. Tensile properties of directionally solidified YAG-alumina fibers as a function of temperature.

Temperature	21°C	900°C	1000°C	1300°C	1400°C
Young's modulus (GPa)	344	312	290	241	240
Failure stress (GPa)	1.35	1.28	1.05	0.99	0.77

whilst the latter creeps at a rate of 10^{-9} s^{-1}. The structure is stable to 1400°C. Prolonged heat treatment of the directionally solidified YAG-alumina fibers, at 1500°C, lasting several days, produces a loss of alignment in the microstructure and growth, by diffusion, of the individual phases.

Other directionally grown eutectics have been produced and evaluated however all of these fibers remain unexploited due to their very high costs.

1.2.4. Whiskers

Table 10 gives data on oxide whiskers (37).

5.0. CONCLUSIONS

Oxides, like all ceramics, are brittle materials however they are made into a wide range of often flexible fibers which find use in many industrial applications. The flexibility is a function of the fineness of the filaments which are produced so that most have diameters less than 20 μm. Glass fibers are the most important reinforcement for composite materials however their high silica content results in a relatively low stiffness. Other fibers with a greater alumina content find uses as refractory insulation and as reinforcements. Oxide fibers are made in a variety of ways, some from the melt, as with glass fibers and short aluminosilicate fibers. These latter fibers are used above all for high temperature insulation up to 1600°C. Prolonged exposure to temperatures above 1200°C produce microstructural changes, however which degrade the fibers. Oxide fibers made by a sol-gel route are produced for high temperature use, often as reinforcements. These fibers retain their properties up to 1000°C but above this temperature creep and lose strength by a variety of processes. Attempts to improve the characteristics of these fibers above 1000°C have produced better control of creep but strength loss remains a major issue.

Large diameter oxide fibers have been produced but their very high cost and inherent bending rigidity, which is directly related to their diameter, have severely limited their

TABLE 10. Properties of oxide whiskers.

Material	Specific gravity	Melting point (°C)	Tensile strength (GPA)	Young's modulus (GPa)
Al_2O_3	3.9	2082	14–28	550
BeO	1.8	2549	14–21	700
B_2O_3	2.5	2449	7	450
MgO	3.6	2799	7–14	310

interest. New manufacturing processes which are at present being explored could substantially reduce their cost and could offer a future for these filaments.

Other forms of very fine oxide fibers exist or are being developed. Single crystal whiskers which pose health risks and also present difficulties for handling, because of their very small dimensions, have not found wide use. Nano-oxide fibers are being developed which may be fine enough for liquid crystal technology to be used so as to handle them but as yet remain a possibility for the future.

REFERENCES

1. I. Harter, C.L. Norton Jr. and L.D. Christie (US patent awarded to the Babcock and Wilcox Co.) U.S. 2467889, 19th July 1949.
2. Babcock and Wilcox Company, U.K. Patent Specification 1098595, 10th January 1968.
3. M.J. Morton, J.D. Birchall and J.E. Cassidy, U.K Patent Specification 1360197, 17th July 1974.
4. A.K. Dhingra, Phil. Trans. R. Soc. Ldn, **294** 411–417 (1980).
5. J.C. Romine, Ceram. Eng. Sci. Proc. **8** 755–776 (1987).
6. Y. Abe, S. Horikiri, F. Fujimara and E. Ichiki *Proc. 4th Int.Con. Composite Mater.*, Tokyo, Japan, 142, (1982).
7. Y. Saitow, K. Iwanaga, S. Itou, T. Fukumoto, and T. Utsunomiya, *Proc.37th Int. SAMPE Symp. and Exhib.*, **35** 808–819 (1992).
8. M-H. Berger and A.R. Bunsell, Oxide Fibers in *Comprehensive Composite Materials*, Ed. A. Kelly and C. Zweben, **Vol. I**, Ed. Tsu-Wei Chou 147–173 (2000).
9. D.M. Wilson, S.L. Lieder and D.C. Lueneburg Ceram. Eng. Sci. Proc. **16** 1005–1014 (1995).
10. F. Deleglise, M-H. Berger and A.R. Bunsell, pp. 84–89 in *High Temperature Ceramic Matrix Composites* Ed. W. Krenkel, R. Naslain and H. Schneider, Wiley VCH (2001).
11. F. Deleglise, M-H. Berger, D. Jeulin and A.R. Bunsell, J. European Ceramic Society **21** 569–580 (2001).
12. A. Poulon-Quintin, M-H. Berger and A.R. Bunsell J. European Ceramics Society **24,** 2769–2783 (2004).
13. H.E. Labelle Jr. Method of growing crystalline materials. US patent No 3,591,348 (1971).
14. J.T.A. Pollock, J.Mat. Sci. **7** 631–348 (1972).
15. Data sheets of Saphikon, 33 Milford, New Hampshire 03055.
16. D.J. Gooch and G.W. Groves, J. Mat. Sci. **8** 1238–1246 (1973).
17. S.T. Mileiko, V.M. Kiiko, N.S. Sarkissyan, M. Yu. Starostin, S.I. Gvozdeva, A.A. Kolchin, and G.K. Strukova. Composites Sci. & Tech. **59** 1763–1772 (1999).
18. A.R. Bunsell, G. Simon, Y. ABe, M. Akiyama in "Ceramic Fibres", ch. 9 pp. 427–478 in *Fiber Reinforcements for Composite Materials*, Edited by A.R. Bunsell, Elsevier, Amsterdam (1988).
19. K. Gupta, "Glass Fibers for Composite Materials", ch. 2 pp. 19–71 in *Fibre Reinforcements for Composite Materials*, Edited by A.R. Bunsell, Elsevier, Amsterdam (1988).
20. Eiji Horie, Ed., *Ceramic Fiber Insulation Theory and Practice*, The Energy Conservaton Center, Japan (1986).
21. R.J. Brook, Ed., *Concise Encyclopedia of Advanced Ceramic Materials*, Pergamon Press (1991).
22. A.R. Bunsell and M-H. Berger, *Fine Ceramic Fibers*, Marcel Dekker, N.Y. (1999).
23. V. Lavaste, M-H. Berger, A.R. Bunsell and J. Besson, J. Mat. Sci. **30** 4215–4225 (1995).
24. D.M. Wilson, D.C. Lueneburg and S.L. Lieder, Ceram.Eng.Sci.Proc. **14**, 7–8, 609–621 (1993).
25. D.M. Wilson and L.R. Visser, Ceram. Eng. Sci. Proc. **21**, 4, 363–373 (2000).
26. T. Mah, T.A. Parthasarathy and M.D. Petry, Ceram. Eng. Sci. Proc. **14**, 7–8, 622–638 (1993).
27. J. Sigalovsky, K.C. Wills, J.S. Haggerty and J.E. Sheenhan, Ceramic Eng. & Sci. Proc. pp. 183–189, Jul–Aug. (1992).
28. H.E. Bates Ceramic Eng. & Sci. Proc. pp. 190–197, Jul–Aug. (1992).
29. D.R. Hartman, M.E. Greenwood and D.M. Miller, High Strength Glass Fibers, Owens Corning Technical Paper (March 1994).
30. A.A. Griffith, Phil. Trans. Roy. Soc. **A221** 163–198 (1921).
31. M-H. Berger, Fracture Processes in Oxide Ceramic Fibres, pp. in *Fiber Fracture* Ed. M. Elices and J. Llorca, Elsevier (2002).

32. J.D. Birchall, J.A.A. Bradbury and J. Dinwoodie, 'Alumina Fibres' pp. 115–155 in *Handbood of Composites*, Ed. W. Watt and B.V. Perov, North Holland, Amsterdam, Netherlands (1985).
33. D.J. Pysher and R.E. Tressler, J. Mat. Sci. **27** 423–428 (1992).
34. P. Shahinian, J. American Cer. Soc-Discussions and Notes. **54** 67–68 (1971).
35. S.C. Farmer, A. Sayir, P.O. Dickerson and S.L. Draper, Ceramic Eng. & Sci. Proc. pp. 969–976, Sept–Oct. (1995).
36. F. Deleglise, Ph.D. thesis Ecole des Mines de Paris, *Fibres Ceramiques Oxydes Biphasées*, July 2000.
37. J.V. Milewski, Whiskers pp. 281–284 in *Concise Encyclopedia of Composite Materials*, Ed. A. Kelly, Pergamon Press, Oxford, UK 1989.

2

Non-oxide (Silicon Carbide) Fibers

James A. DiCarlo and Hee-Mann Yun
NASA Glenn Research Center
21000 Brookpark Road
Cleveland, Ohio 44135
Phone 216-433-5514
James.A.DiCarlo@NASA.gov

ABSTRACT

Non-oxide ceramic fibers are being considered for many applications, but are currently being developed and produced primarily as continuous-length structural reinforcement for ceramic matrix composites (CMC). Since only those fiber types with compositions based on silicon carbide (SiC) have demonstrated their general applicability for this application, this chapter focuses on commercially available SiC-based ceramic fiber types of current interest for CMC and on our current state of experimental and mechanistic knowledge concerning their production methods, microstructures, physical properties, and mechanical properties at room and high temperatures. Particular emphasis is placed on those properties required for successful implementation of the SiC fibers in high-temperature CMC components. It is shown that significant advances have been made in recent years concerning SiC fiber production methods, thereby resulting in pure near-stoichiometric small-diameter fibers that provide most of the CMC fiber property requirements, except for low cost.

1. INTRODUCTION

Continuous-length polycrystalline ceramic fibers with non-oxide compositions are currently being developed and used for a variety of low and high-temperature structural applications. Recent literature reviews detail the process methods, properties, and applications for the many non-oxide fiber types that have been developed over the last 30 years [1–5].

In practically all cases, the non-oxide fiber types have compositions based on silicon compounds such as silicon carbide (SiC) and silicon nitride (Si_3N_4). The application currently envisioned as the most important for these fiber types is that of structural reinforcement of ceramic matrix composites (CMC) with upper use temperatures well above the current capability of structural metallic alloys (>1100°C). However, only the SiC-based types have reached the stage in which they have been used to reinforce different high-temperature CMC systems, and as such have composite property databases available in the open literature in order to verify their reinforcement capabilities. Thus the focus of this chapter is to present key performance-related properties of these commercial SiC-based fiber types since they have demonstrated their general applicability as CMC reinforcement. To this end, the first section presents information on the key production details and as-produced properties of the SiC types as individual fibers and on our current understanding concerning the underlying process-microstructure-property relationships. The second section presents available data concerning how the deformation and fracture properties of the individual fibers are affected by temperature, time, stress, and oxidizing environments, again with a discussion of underlying mechanisms. Finally, the last section discusses how these and other SiC fiber properties are affecting the technical implementation of advanced high-temperature ceramic matrix composites.

2. APPLICATIONS

Research and development efforts over recent years have focused on achieving SiC-based ceramic fibers with the optimum intrinsic and extrinsic thermostructural properties required as reinforcement of high-temperature CMC components, such as those used in the hot-sections of engines for power and propulsion. As such, general first-level fiber property requirements are the ability to display high tensile strength in its as-produced condition, and the ability to retain as much of this strength as possible at high temperatures for long times under high stresses and oxidizing environmental conditions. For strength retention, key goals are achieving thermally stable microstructures that display high creep resistance and high oxidation resistance. However, there are also a variety of second-level fiber property requirements that are related to CMC fabrication and service. These include the need for small diameter, carbon-free surfaces, low roughness surfaces, high intrinsic thermal conductivity, and low acquisition cost.

3. PROCESSING

Today the most common approach for producing polycrystalline ceramic fibers is by spinning and heat treating chemically derived precursors. In the case of SiC-based fibers, "green" fibers are spun from organo-metallic "preceramic" polymer precursors, such as polycarbosilane, followed by cross-linking (curing) and heat treatment steps, such as pyrolysis and sintering, to convert the fibers to ceramic materials typically based on β-phase SiC [4]. Particular process and compositional details for a variety of SiC fiber types are discussed in the next section.

A key characteristic of the polymer-derived SiC fibers is their ultra-fine microstructure with grain sizes in the nanometer range. Fine grains are typically required for good tensile strength, but can be detrimental to creep resistance [6]. The use of polymer technology also allows the commercial preparation of SiC fibers with small diameters (<15 μm) and in the form of continuous-length multi-fiber tows, which are then typically coated with thin polymer-based sizings before being supplied to customers on ~100 gram spools. The sized tows with the small diameter fibers are flexible and generally easy to handle so that they can be textile-formed into fabrics, tapes, braids, and other complex shapes. Many small-diameter SiC fiber manufacturers also supply fibers in fabric form since most composite fabrication processes used today generally require an initial step of fabric stacking or lay-up into a near net-shaped architectural preform of the final CMC product. Although there are no current methods for fabricating continuous-length single-crystal SiC fibers, it is expected that the small-diameter polycrystalline SiC fibers will always display better handle-ability and significantly lower production costs, but reduced creep resistance and temperature capability. Thermostructural capability and thermal conductivity may also be degraded in the polycrystalline SiC fibers by the presence of second phases in the fiber grain boundaries.

Polycrystalline SiC fibers in the form of continuous-length monofilaments are also being produced by the chemical vapor deposition (CVD) route, which typically uses methyltrichlorosilane to vapor deposit fine columnar-grained (~100 nm long) SiC onto a resistively-heated small-diameter (~30 μm) carbon monofilament that continuously passes through a long glass reactor [1]. CVD SiC fibers on small-diameter (~13 μm) tungsten monofilaments are also commercially available, but due to high-temperature reactions between the SiC and tungsten, these fibers are more suitable as reinforcement of low and intermediate temperature composite systems. Although the CVD SiC fibers have displayed very high strengths (~6 GPa for the Ultra SCS fiber), the final fibers currently display diameters greater than 50 μm, which are not conducive to fiber shaping into complex architectural preforms. Laboratory attempts have been made to reduce their diameters and to produce multi-fiber tows, but issues still exist concerning finding substrate filaments with the proper composition and diameter, methods for spreading these filaments during deposition to avoid fiber-to-fiber welding, and low-cost gas precursors for the CVD SiC.

4. PROPERTIES

4.1. As-Produced Physical and Mechanical Properties

Table 1 lists some key production and compositional details for a variety of SiC-based fiber types of current interest and availability as CMC reinforcement. The polymer-derived types range from first-generation fibers with very high percentages of oxygen and excess carbon, such as Nicalon and Tyranno Lox M, to the more recent near-stoichiometric (atomic C/Si ≈ 1) fibers, such as Tyranno SA and Sylramic. For the CVD-derived types, such as the SCS family with carbon cores, the only compositional variables in the SiC sheaths are slight excesses of free silicon or free carbon. Table 2 lists some of the key physical and mechanical properties of the SiC fiber types in their as-produced condition, as well as estimated commercial cost per kilogram, all properties important to fiber application as CMC reinforcement. These SiC fiber properties in Tables 1 and 2 are in most part those published by the indicated commercial vendors. It should be noted that the Sylramic fiber

TABLE 1. Production and Compositional Details for SiC-Based Fiber Types

Trade-Name	Manufacturer	Production Method	Elemental Composition, weight %	Approx. Maximum Production Temp.	~Grain Size, nm (~surface roughness)	Surface composition	Avg. Diam. μm	Fibers Per Tow
Nicalon, NL200	Nippon Carbon	Polymer	56 Si + 32 C + 12 O	1200°C	2	Thin carbon	14	500
Hi-Nicalon	Nippon Carbon	Polymer + Electron irr.	62 Si + 37 C + 0.5 O	1300°C	5	Thin carbon	14	500
Hi-Nicalon Type S	Nippon Carbon	Polymer + Electron irr.	69 Si + 31 C + 0.2 O	1600°C	100	Thin carbon	12	500
Tyranno Lox M	Ube Industries	Polymer	55 Si + 32 C + 10 O + 2.0 Ti	1200°C	1	Thin carbon	11	400/800
Tyranno ZMI	Ube Industries	Polymer	57 Si + 35 C + 7.6 O + 1.0 Zr	1300°C	2	Thin carbon	11	400/800
Tyranno SA 1-3	Ube Industries	Polymer + Sintering	68 Si + 32 C + 0.6 Al	>1700°C	200	Thin carbon	10 – 7.5	800/1600
Sylramic	(Dow Corning) COI Ceramics	Polymer +t Sintering	67 Si + 29 C + 0.8 O +2.3 B + 0.4 N +2.1 Ti	>1700°C	100	Thin carbon + B + Ti	10	800
Sylramic-iBN	COI Ceramics + NASA	Polymer + Sintering + Treatment	Sylramic	>1700°C	>100	Thin in-situ BN (~100 nm)	10	800
SCS-6-9/Ultra SCS	Specialty Materials	CVD on ~30 μm C core	70 Si + 30 C + trace Si + C	1300°C	~100 by ~10	SiC/ Thin carbon	140 – 70/140	mono-filament

TABLE 2. Property and Cost Details for SiC-Based Fiber Types

Trade-Name	Manufacturer	Density, g/cm³	Avg. R.T. Tensile Strength GPa	R.T. Tensile Modulus GPa	R.T. Axial Thermal Conduct. W/m K	Thermal Expansion, ppm/°C (to 1000°C)	Current Cost, (<5Kg), $/Kg
Nicalon, NL200	Nippon Carbon	2.55	3.0	220	3	3.2	~2000
Hi-Nicalon	Nippon Carbon	2.74	2.8	270	8	3.5	8000
Hi-Nicalon Type S	Nippon Carbon	3.05	~2.5	400–420	18		13000
Tyranno Lox M	Ube Industries	2.48	3.3	187	1.5	3.1	1500/1000
Tyranno ZMI	Ube Industries	2.48	3.3	200	2.5		1600/1000
Tyranno SA 1-3	Ube Industries	3.02	2.8	375	65		~5000
Sylramic	(Dow Corning) COI Ceramics	3.05	3.2	~400	46	5.4	10000
Sylramic-iBN	COI Ceramics + NASA	3.05	3.2	~400	>46	5.4	>10000
SCS-6-9/Ultra SCS	Specialty Materials	~3	~3.5/~6	390–350/390	~70	4.6	~9000

type is no longer being produced by Dow Corning, but by ATK COI Ceramics, who is also producing its derivative, the Sylramic-iBN fiber.

Examining the SiC fiber properties in more detail, it can be seen that there exists a strong correlation for each fiber type between its production conditions, its final as-produced microstructure, and its as-produced physical and mechanical properties. For example, the approximate maximum production temperature of 1200°C for the first generation polymer-derived Nicalon and Tyranno Lox M fibers is typically dictated by the fact that during the green fiber curing stage, a significant amount of oxygen is introduced into the fibers. This gives rise to oxy-carbide impurity phases in the fiber microstructures, which tend to decompose into gases that volatize from the fiber near 1200°C, thereby creating porosity and less than optimum fiber tensile strength [4]. Thus for the first generation fiber types, the high oxygen content of ~10 weight % tends to limit fiber thermal capability for strength retention and for CMC fabrication to temperatures less than 1300°C, and for CMC long-term service to temperatures less than 1200°C. The low production temperature also results in very fine grains (<5 nm), thereby contributing to high creep rates for these fiber types, which are even further enhanced by the oxygen content in the grain boundaries. In addition, in comparison to the purer near-stoichiometric SiC fibers, the oxygen and carbon content of these first generation fibers contribute to lower fiber density, elastic modulus, thermal expansion, and thermal conductivity, which is further degraded by the small grain size. On the positive side, the simplicity of the production methods for these fiber types in combination with their low production temperatures, allow these fibers to be available at the lowest cost in comparison to the other higher performing SiC fiber types.

Since it is often desirable to fabricate and to achieve long-term use of CMC above 1200°C, advanced production methods that limit the oxide phases in polymer-derived SiC fibers have recently been developed. In one method, oxygen pickup in the microstructure is reduced significantly by green fiber curing under electron irradiation. The Hi-Nicalon and Hi-Nicalon Type S fiber types from Nippon Carbon are produced by this method, and thus show much reduced oxygen content in comparison to the original Nicalon fiber [4]. The lower oxygen content allows the Hi-Nicalon Type S fiber to be produced near 1600°C, but the Hi-Nicalon fiber production temperature is limited to ~1300°C because of higher oxygen content and significant carbon content. Reducing oxygen increases fiber density, modulus, expansion, and conductivity over the original Nicalon fiber, but the high carbon content of the Hi-Nicalon fiber limits these increases. Also the higher production temperatures for both fiber types are revealed in larger grain sizes, which also impacts fiber thermal conductivity. Finally, removal of oxygen by electron irradiation appears to introduce added production costs since the Hi-Nicalon types are 4 to 6 times more expensive than the Nicalon fiber.

In another low-oxygen method, residual oxide-based impurities in the green fibers are allowed to decompose during subsequent fiber heat-treatment above 1200°C. But by introducing boron or aluminum sintering aids into the low-strength fiber microstructure and heating to temperatures above 1600°C, the remaining small SiC grains are allowed to sinter and grow, forming a fairly dense fiber with reasonable tensile strength [4]. The resulting fibers such as the Tyranno SA fiber (aluminum aids) and the Sylramic fiber (boron aids) are nearly oxygen-free and stoichiometric, with the latter type displaying the higher as-produced tensile strength. In comparison to the other fiber types with lower production temperatures, the final sintered fibers contain larger grains that are beneficial for improved fiber creep resistance and thermal conductivity, but can cause fiber strength to degrade below 3 GPa if allowed to grow beyond ~200 nm in size. Also, the larger grain sizes typically result in rougher fiber surfaces, which can cause fiber-fiber abrasion in multi-fiber tows. In addition, the sintering aids can be somewhat detrimental to fiber creep and oxidation resistance. Nevertheless, the sintering approach has the major advantage of producing strong fibers at very high production temperatures, which can significantly open the temperature window for CMC fabrication and service without experiencing microstructural instabilities in the fiber reinforcement. In addition, for the Sylramic fiber type, the boron sintering aids can be removed from the fiber surface and bulk by a post-process thermal treatment in nitrogen-containing gas, giving rise to the "Sylramic-iBN" fiber that is not only creep and oxidation resistant, but contains a thin protective in-situ grown BN coating on the fiber surface [7]. But for all the sintered fiber types, fiber density and modulus are near the maximum values for SiC, which can have a negative effect, respectively, on CMC weight and on the formability of CMC fiber architectures. In addition, there is the general increase in cost for the sintered fibers that typically accompanies their higher production temperatures.

In contrast to the polymer-derived SiC fibers, the CVD SiC monofilament fibers have the production advantage of rapid high-temperature layer-by-layer formation of the SiC microstructure without the introduction of deleterious oxide phases. However, the mechanisms for SiC formation from silicon and carbon-containing chemical precursors are very complicated and quite sensitive to a number of parameters including reactant gas species, carrier and/or reducing gases, deposition pressure, temperature, gas phase concentrations, flow rate, etc.. A stoichiometric SiC microstructure may be produced locally, but is difficult to achieve throughout the entire fiber. For example, current-production CVD SiC monofilament fibers on carbon generally experience process temperatures that decrease from ~1500 to ~1200°C as they pass through the reactor. These conditions result in a slightly silicon-rich microstructure in the outer layers of the SiC sheath of the SCS-6 fiber, which in turn leads to fiber creep and strength instabilities at temperatures above 800°C [1]. Recently, the newer Ultra SCS fiber was developed at Specialty Materials, which displays slightly carbon-rich sheath compositions, leading to improved as-produced strength, creep resistance, and strength retention at high temperatures. Despite these small variations in stoichiometry, the SiC sheaths of the CVD fibers display physical and mechanical properties that are closest to those expected from pure β-SiC for any of the SiC fiber types. However, due to the carbon monofilament substrate, the CVD SiC fibers are ceramic composites in themselves; so that some of the fiber properties, such as density and axial modulus, are dependent on substrate size and composition. In addition, costs for current-production CVD SiC monofilaments are quite high at about $10,000 per kilogram.

Finally, it is interesting to note in Tables 1 and 2 that, whereas fiber physical and deformation properties can be correlated with as-produced fiber composition and grain size, there is much less correlation between fiber microstructural conditions and fiber tensile strength, which is typically measured at a gauge length of 25 mm.. This non-correlation is generally to be expected since the as-produced strengths of ceramic materials are controlled by any pores or other flaws introduced into the ceramic bulk or surface during consolidation. However, in many cases for large-grained ceramics, these flaws relate in a crude manner with the inverse square root of the grain size (Griffith's law), so that ceramic strength should increase with smaller grain size. For the SiC fibers, this relationship appears to hold for average grain sizes above ~100 nm, based on the limited grain size variation between the Tyranno SA fibers at ~200 nm and the Sylramic and SCS fibers at ~100 nm. However, below 100 nm, all polymer-derived fibers, despite nanometer grain sizes, show a wide variation in strength with none ever exceeding ∼3 GPa. This observation has generally been explained by the fact that inherent in the polymer preparation and filtering processes, potential flaw-related particles with sizes of the order of ∼100 nm are allowed to pass into the green fibers, so that these particles rather than the fiber grains are strength limiting. Another related point is that these fiber types typically have carbon-rich surface layers (see Table 1) that are capable of blunting strength-controlling fiber surface flaws; so that if the carbon is removed during CMC fabrication or service, fiber strength losses up to 20% have been observed. Thus although it would be highly desirable to develop stronger small-diameter SiC fibers, upper limit strengths of ∼3 GPA may be all that can be currently achieved by current polymer-derived fiber production methods.

4.2. Temperature-Dependent Mechanical Properties

With increasing temperature, SiC fiber bulk properties such as elastic modulus and thermal conductivity decrease slowly in a relative manner closely equivalent to the same properties of bulk monolithic SiC. Thus it can be assumed that fiber elastic modulus **E** should decrease with temperature **T** (Celsius) by the following relationship for monolithic SiC:

$$\mathbf{E(T) = E_o - [A \cdot T]} \qquad \mathbf{A = 0.020\,GPa/°C} \tag{1}$$

where E_o is the fiber modulus at 0°C. This modulus behavior should extend to high temperatures, but perceived deviations could begin as low as 800°C where fiber creep strain begins to appear and to add to the fiber elastic strain. The tensile strength of SiC fibers also shows a slow monotonic decrease with temperature up to the beginning of the creep regime. However, within this regime, factors such as grain size, impurity content, and prior thermostructural history have a significant impact on the rate of fiber strength degradation with time and temperature. Thus, to better understand SiC fiber capability as structural reinforcement of high-temperature CMC, one needs to examine available data concerning the effects of temperature, time, stress, and environment on the creep and rupture behavior of the various SiC fiber types.

A convenient approach for evaluating both fiber creep behavior and rupture behavior is by the stress rupture test in which individual small-diameter fibers are dead-weight loaded under constant stress, temperature, environment, and gauge-length test conditions, and fiber creep strain is measured versus time until rupture occurs [8]. Under these conditions at high temperatures, most SiC fibers in their as-produced condition display a large transient creep stage. This can be seen in Fig. 1 for high-temperature SiC fiber types under an applied stress of 275 MPa at 1400°C in air [7]. These curves show that the creep rates for many of the fibers continuously decrease with time until rupture (letter F), giving the appearance of only primary stage creep. Although detailed mechanistic studies still need to be performed, it would appear that the transient behavior is caused by a variety of process-related factors. For example, due to production temperatures that do not exceed ∼1300°C, the Hi-Nicalon fibers contain very fine grains and thermally unstable oxy-carbide second phases. During long-term creep testing at 1200°C and above, it is likely that the grains grow and the creep-prone oxy-carbide phases decompose, resulting in a concurrent reduction in creep rate. These mechanisms are supported by studies in which a reduced transient stage and overall

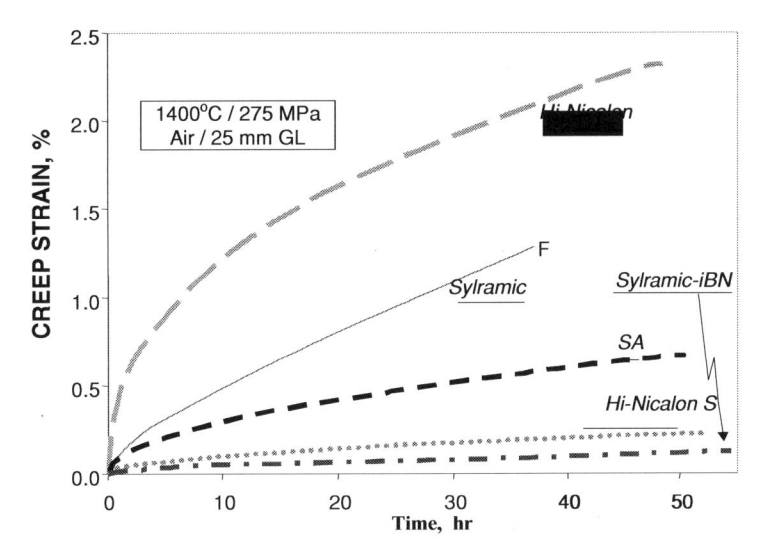

FIGURE 1. Typical creep strain versus time curves for high-temperature SiC fibers tested in air under stress-rupture conditions (F = rupture).

improvement in creep resistance were observed after thermal treatment of the Hi-Nicalon fiber type at its production temperature and above [9, 10].

For the SiC fibers with the largest grains (~100 to 200 nm), a variety of creep recovery and annealing studies have been performed to better understand their transient behavior. For example, recovery tests on the SCS-6 fiber show that the total creep strain typically consists of an anelastic or recoverable component plus a viscous or non-recoverable component [11]. Generally the viscous component continues to increase with time, while the anelastic component saturates with time to a value no greater than ~2 times the initial elastic component [12]. Based on this behavior, the anelastic component can be attributed to Zener grain-boundary sliding [13]. Thus at least a part of the transient behavior observed in Fig. 1 is due to a recoverable creep component. When the large-grained SCS-6 and Sylramic fibers were annealed for extended times near their production temperatures, grain size did not change. However, total creep strain decreased for the same test conditions, with the primary stage diminishing more rapidly than the secondary stage. Since these fiber types contain creep-prone phases, such as free silicon and boron, respectively (see Table 1), the improved creep resistance can be attributed to the elimination or minimization of these second phases from grain boundaries. Indeed, the improved creep behavior of the Ultra SCS fiber relative to the first-generation SCS-6 fiber [6] can be explained by elimination of free silicon. Likewise, the improved creep resistance of the Sylramic-iBN fiber can be attributed to boron removal from the Sylramic fiber [7]. Thus for the Fig. 1 test conditions, the transient creep behavior of as-produced SiC-based fibers can in part be explained by anelastic grain-boundary sliding and in part by unstable grain sizes and mobile second phases.

To gain a mechanistic understanding of SiC fiber creep, it is of interest to examine the property of fiber creep strength. This property is defined as the applied stress required to allow a certain creep strain limit to be reached in a given time or service life at a given service temperature. *Thus the higher the fiber creep strength, the greater the fiber's creep resistance and ability to retain dimensional stability for a particular service application.* Under conditions where only one grain-boundary diffusion mechanism dominates, fiber creep strength σ_c can generally be described by the following relationship [14]:

$$\sigma_c^n = A_o^{-1} d^m \dot{\varepsilon}_c \exp[Q_c/RT]. \qquad (2)$$

Here A_o is a composition-dependent diffusion constant; **n** is the creep stress exponent, **d** is average grain size, **m** is the grain size exponent, Q_c is the controlling creep activation energy, **R** is the universal gas constant (8.314 J/mol-K), and **T** is absolute temperature. If the concentration of vacancies supporting grain-boundary diffusion is not limited, creep occurs by Coble creep in which **n** = 1 and **m** = 3. However, if the concentration of vacancies is limited in the grain boundaries, for example, by the presence of second phases, interface-controlled diffusional (ICD) creep would occur with **n** values ranging from 2 to as high as 3. Also ICD is characterized by a much smaller dependence on grain size with **m** ≈ 1 and energy Q_c greater than that for Coble creep.

Not withstanding the existence of transient behavior for the as-produced SiC-based fibers, Fig. 2 can be constructed to examine the creep strength versus grain size dependence of these fibers both in their as-produced condition (open points) and after creep strength improvement by second phase removal (closed points) [6]. Here the selected conditions are 0.4 % total creep strain in 10 hours at 1400°C in air. In the previous discussion, it was seen that **n** values greater than unity are an indication of ICD creep, which in turn implies **m** = 1. Assuming this to be the case for the SiC-based fibers, one can draw the solid line in Fig. 2 near the points of the most creep-resistant fibers to suggest that ICD behavior exists in these fibers with **m/n** = 1/2, so that **n** = 2, a stress exponent typically observed for creep of polymer-derived SiC fibers [6]. Thus at the present time, the solid line in Fig. 2 represents the best creep strength behavior to be expected for SiC-based fiber types under the selected test conditions. This line shows that the annealed Sylramic fiber at ∼100 nm grain size (i.e., the Sylramic-iBN fiber), probably represents the best SiC fiber type in terms of displaying the highest values for both creep strength and as-produced tensile strength.

For some applications, large fiber creep strains may be tolerable so that the prime concern is to assure that the fiber does not fracture or rupture during service. Typically during stress-rupture testing, the rupture time results show considerable scatter [15], so that data trends become difficult to analyze mechanistically. Nevertheless, it is possible to average the applied stresses that cause a given fiber type to rupture at a given time to develop best fit curves of applied stress or rupture strength versus time at a given temperature. For example Fig. 3 presents such stress-rupture curves for many of the same fiber types in Fig. 1 [7]. Here the test conditions are again 1400°C, air environment, and single fiber gauge lengths of ∼25 mm. Comparing Figs. 1 and 3, one can see a general correlation between good creep-resistance and high rupture strength or long rupture time. But as explained below, this correlation may not always exist.

To simplify SiC fiber rupture behavior for technical application and mechanistic understanding, two simple empirical approaches often used for metals and ceramics have been successfully developed. One approach is by use of Larson-Miller (LM) master curves or, equivalently, thermal-activation q-maps [16]. For this approach, measurements have been made on single fibers across a time range from ∼0.01 to over 100 hours using three types

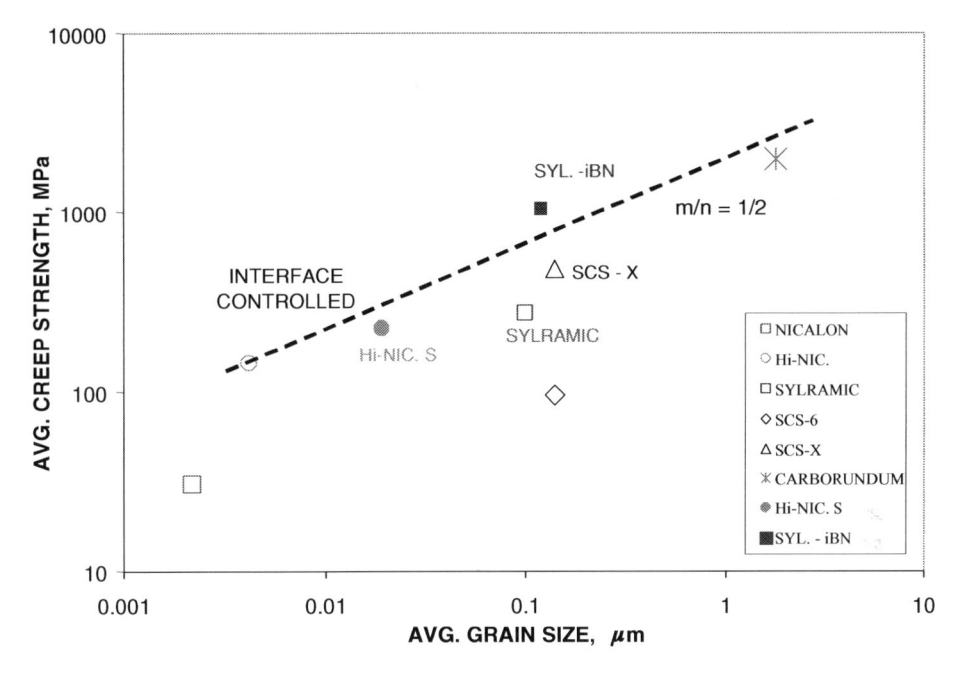

FIGURE 2. Creep strength of SiC fibers for strain limit of 0.4% at 1400°C for 10 hrs in air: open points = as-produced condition; closed points = after second phase removal.

of tests: stress rupture (constant stress and constant temperature), slow warm-up (constant stress, constant rate of temperature change), and fast-fracture (constant temperature and constant rate of stress change). It was found that by use simple thermal-activation theory, the rupture results of the three tests for each fiber type could be combined into a single LM master curve or q-map which describes the applied stress at rupture (fiber rupture strength) versus the time-temperature dependent parameter **q** given by

$$q \equiv Q_r/2.3R = T(\log t_r + 22). \tag{3}$$

Here Q_r is the effective stress-dependent activation energy for fiber rupture; **T** (kelvin) is the absolute temperature for the rupture test; and t_r (hours) is the fiber rupture time. Complete **q**-maps covering a wide range of temperatures and stresses are shown in Fig. 4a for two types of oxide fibers: Nextel 610 and Nextel 720, and for three types of SiC fibers: Hi-Nicalon, Sylramic, Sylramic-iBN. Here the curves represent best-fit averages of the fiber rupture times as measured for a ~25 mm gauge length.

The rupture results of Fig. 3a have many important basic and practical implications. First, on the basic level, all curves display approximately the same shape with increasing **q**; that is, an initial section with a small negative slope (Region I), and a remaining section with a much larger negative slope (Region II). This behavior is typical of the rupture of monolithic ceramics, where as-produced flaws grow slowly in size (slow crack growth) in Region I; whereas in Region II, creep mechanisms aid in the more rapid growth of the same flaws or in the nucleation and growth of new micro-cracks and cavities. On the practical side,

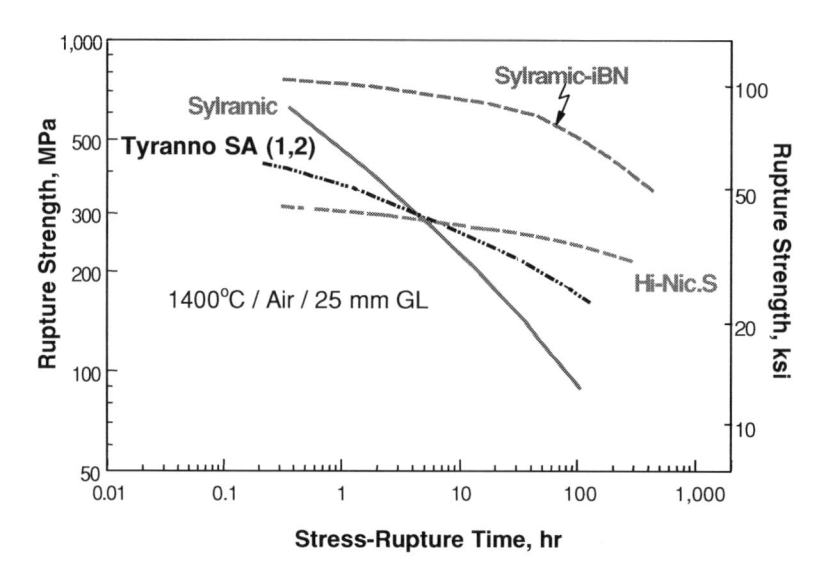

FIGURE 3. Best-fit rupture strength versus time curves for high-temperature SiC fibers tested at 1400°C in air under stress-rupture conditions at 25-mm gauge length.

the Fig. 3a curves indicate that fiber strength values throughout Region I depend directly on the fiber's as-fabricated strength at room temperature ($q \approx 7000$). That is, the entire Region I section moves up in strength when as-produced flaws are reduced in size or frequency. Alternatively, the section would move up or down if the test gauge length was smaller or greater, respectively, than the ~25 mm length used to generate the curves. In addition, Fig. 3a clearly indicates the greater thermostructural capability of the SiC fibers over the oxide-based fibers both in Regions I and II. The Region I advantage is related to the higher fracture toughness of SiC; while the Region II advantage is primarily due to slower diffusion processes in the SiC-based fibers. Finally, the Fig. 3 curves allow prediction of fiber rupture behavior if any four of the following five application variables are known: stress, stress rate, temperature, temperature rate, and time [16].

Besides single fibers, Larson-Miller master curves have also been generated for single multi-fiber tows of various ceramic fiber types [17]. Some of these are shown in Fig. 4b where again the environment is air and the gauge length is ~25 mm. Because tow strengths are typically controlled by the weakest fibers in the tow bundle, the tow LM curves are lower than the single fiber LM curves in Region I. However, in Region II the tows are generally stronger. The source of this strength reversal in Region II currently remains unknown.

One drawback of the LM approach for single fibers is that, as the rupture curves in Region II become steeper, the **q**-maps or LM curves begin to lose sensitivity in predicting fiber rupture strength at high temperatures. As discussed elsewhere [16], this is a consequence of fiber rupture being controlled by creep-induced flaws so that effective rupture energy $\mathbf{Q_r}$ is now controlled by the stress-independent fiber creep energy $\mathbf{Q_c}$. Also the LM

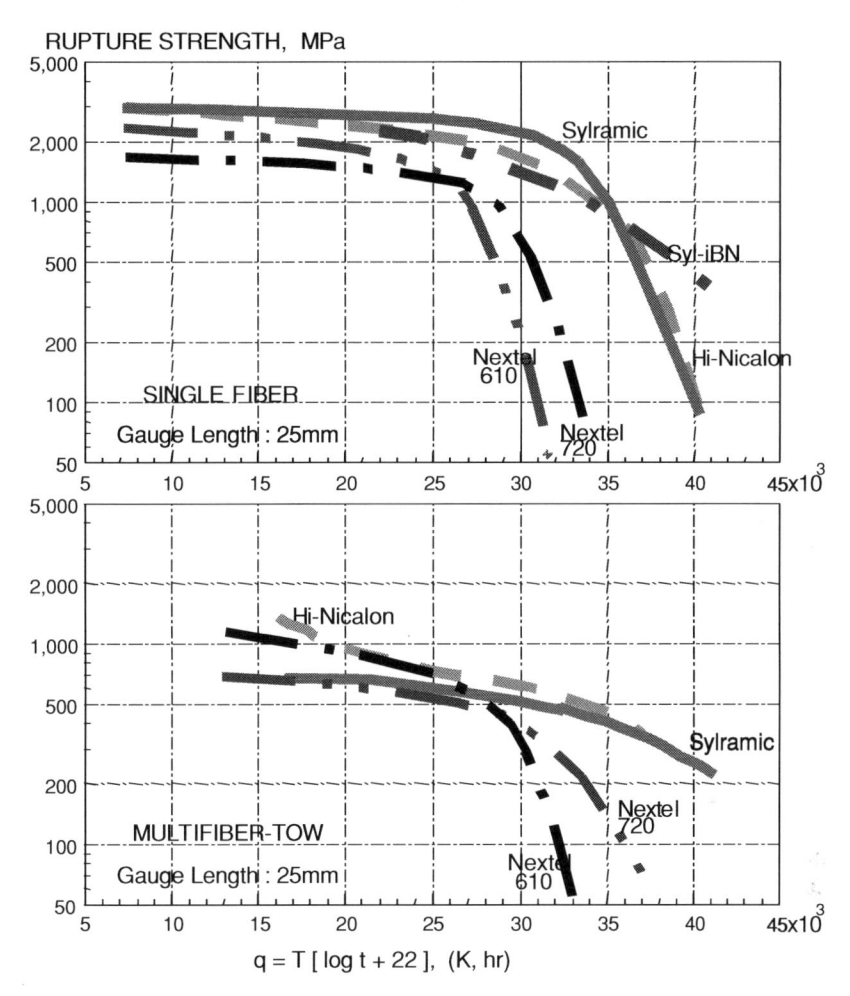

FIGURE 4. Best-fit Larson Miller or **q**-plot master curves measured in air for SiC and oxide fibers as (a) single fibers and (b) multi-fiber tows (T = kelvin, t = hours).

curves do not allow a direct quantitative assessment of fiber creep resistance. For example, the Hi-Nicalon and Sylramic fibers behave similarly in rupture at the high **q** values, but the Sylramic fiber creeps much less. To address these limitations, another convenient approach has been developed for describing SiC fiber rupture in the creep regime. This approach uses Monkman-Grant (MG) diagrams that plot at a given temperature the log of material rupture time versus the log of material creep rate at rupture [18]. For ceramic materials, the log-log results typically fall on one straight-line master curve. At higher temperatures, the MG lines usually retain the same slope, but rupture times increase with temperature for a given creep rate. Although SiC fibers appear to rupture in the primary creep stage (see

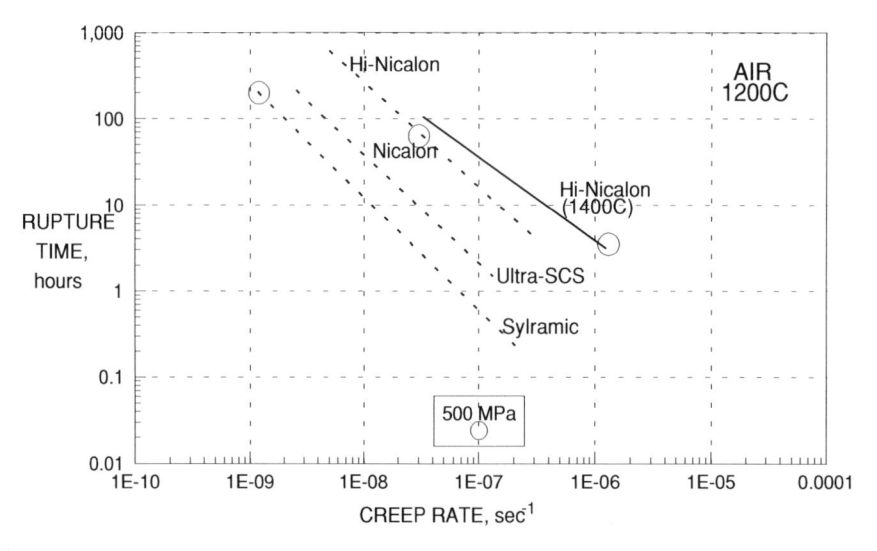

FIGURE 5. Best-fit Monkman-Grant lines measured in air for SiC single fibers at 1200°C and the Hi-Nicalon fiber at 1400°C.

Fig. 1), fairly consistent MG lines can still be constructed for the fibers based on their minimum creep rate at rupture. For example, Fig. 5 shows best-fit MG lines for average rupture time versus minimum creep rate (or instantaneous rate at rupture) for four SiC fibers at 1200°C in air [19]. Also included are the MG results for the Hi-Nicalon fiber tested at 1400°C in air.

As indicated in Fig. 5, the slopes for the MG lines for the various SiC-based fibers are similar to each other; and as temperature increases, rupture times (and strains) at a given creep rate also increase. However, *at a given creep rate and temperature*, rupture time data for the different SiC fiber types do not fall on the same MG line as might be expected from their somewhat similar compositions. This effect may be due to the anelastic component contributing to the measured creep rate, or to the observation that the smaller grained fibers, such as Hi-Nicalon, typically display larger rupture strains than the larger-grained fibers, such as Sylramic. On the practical side, this effect suggests that the rupture strains of the SiC fibers can possibly be improved by microstructural manipulation. Nevertheless, Fig. 5 shows that for a particular application temperature, one cannot select fiber rupture time independently of fiber creep rate. For example, up to 1200°C the only approach for obtaining a 1000-hour fiber lifetime is to assure that the application conditions do not create fiber creep rates more than $\sim 10^{-8}$ sec^{-1} for the creep–prone fibers (Hi-Nicalon), or more than $\sim 10^{-9}$ sec^{-1} for the more creep–resistant fibers (Sylramic).

In general, the maximum temperature/time/stress capability of the more creep-prone fibers is limited by the fiber tendency to display excessive creep strains (for example, >1%) before fracture. On the other hand, the temperature/time/stress capability of the more creep-resistant fibers is limited by fiber fracture at low creep strains (<1%), the values of which are often dependent on the environment. These limitations are illustrated in Table 3, which shows the approximate upper use-temperature for some SiC fibers, as determined from the

TABLE 3. 1000-hr upper use-temperatures for SiC ceramic fibers as estimated from single fiber creep-rupture results in air (and argon)

Fiber Stress 1000-Hr Limit Condition	100 MPa		500 MPa	
	1% Creep	Fiber Fracture*	1% Creep	Fiber Fracture*
Non-Stoichiometric Types				
Tyranno Lox M	1100°C	1250°C	<1000°C	1100°C
Tyranno ZMI, Nicalon	1150°C	1300°C	1000°C	1100°C
	(1150°C)	(1250°C)	(1000°C)	(1100°C)
Hi-Nicalon	1300°C	1350°C	1150°C	1200°C
	(1300°C)	(1300°C)	(1150°C)	(1150°C)
Near-Stoichiometric Types				
Tyranno SA	1350°C	>1400°C	1150°C	1150°C
	(1300°C)	(1400°C)	(NA)	(1150°C)
Hi-Nicalon Type S	(NA)	>1400°C	NA	1150°C
		(1400°C)	(NA)	(1150°C)
Sylramic	(NA)	1350°C	(NA)	1150°C
	(NA)	(1250°C)	(NA)	(1150°C)
Sylramic-iBN	(NA)	>1400°C	NA	1300°C
		(1300°C)	(NA)	(1150°C)
Ultra SCS			1350°C	>1400°C

* For ~25 mm gauge length.

single-fiber creep and rupture data discussed above. These upper use-temperatures were determined based on the assumption that the maximum temperature limiting condition occurred either when the fiber creep strain exceeded 1% in 1000 hours or when the fiber fractured in 1000 hours. Fiber stresses of 100 and 500 MPa were assumed, which are typical of the range of stresses experienced by fibers within structural CMC. A "not applicable" (NA) notation in the table indicates that the more creep-resistant fibers fractured before reaching a creep strain of 1%.

To put the rupture behavior of SiC fibers in further perspective, Fig. 6 compares the estimated 1000-hr upper use-temperatures for both commercially available oxide and SiC ceramic fibers based on single fiber rupture data at 500 MPa in air. These upper use-temperatures clearly indicate the greater thermostructural capability of the SiC fibers over the oxide fibers. Another important observation is that the fracture-limited upper use-temperatures of the more creep-resistant fibers are not measurably better than those of their more creep-prone counter-part fibers; e.g., compare Hi-Nicalon Type S versus Hi-Nicalon, and Tyranno SA versus Tyranno Lox M. Also as expected, the stoichiometric and purer SiC fiber types display the best temperature capability, with the Sylramic-iBN type currently the best polymer-derived fiber type in this regard. However, some of the more creep-resistant fiber types display better behavior in air than argon, with the Sylramic fiber types showing the largest difference [20]. This environmental effect can be attributed in part to a measurable reduction in intrinsic creep rate for some fiber types in air [20], and in part to reaction with oxygen to form a thin silica layer on the fiber surface. This layer minimizes vaporization of thermally unstable phases and also is capable of by blunting surface flaws,

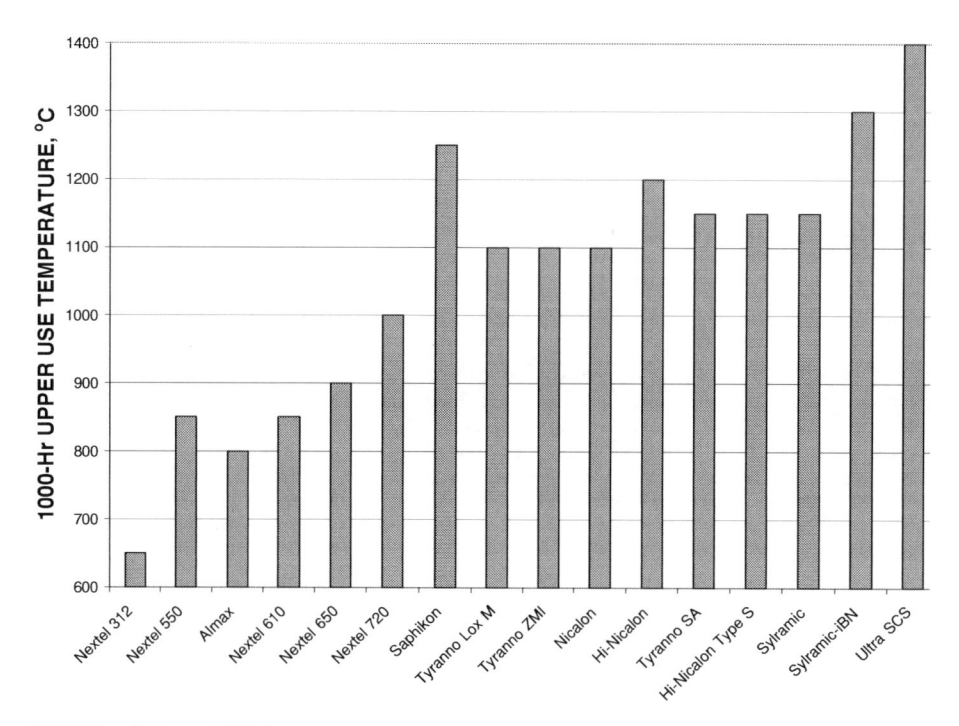

FIGURE 6. Estimated 1000-hour upper use-temperatures for rupture of small and large-diameter single ceramic fibers at 500 MPa in air at 25-mm gauge length.

thereby increasing the creep-rupture strain of all fiber types by as much as 100%. For CMC service under oxidizing conditions, this environmental effect can be important, since air may be the effective fiber environment if the CMC matrix is cracked and inert gas the effective environment if the matrix is uncracked.

Besides potential environmental effects on SiC fiber creep and rupture, one should also realize that the fiber conditions within a CMC during high-temperature application could be considerably different than those employed during simple stress rupture testing. For example, the condition of tensile stress being held constant both in time and along the fiber length will probably never exist during CMC application due to fiber curvature induced by the fiber architecture and to the typical occurrence of CMC matrix cracking, where fiber stresses are highest within the matrix cracks and then drop off within the intact segments of the matrix. It should also be clear that under these CMC conditions, fiber gauge length is considerably smaller than the 25 mm length typically used for the stress rupture testing. Thus, using the data presented here to model SiC fiber high-temperature mechanical behavior in an accurate quantitative manner is probably not warranted at the present time. Nevertheless, as shown by available SiC fiber-reinforced CMC data discussed in other chapters of this book [21, 22], the *relative differences* between various ceramic fiber types as described in this section are generally observed in the high-temperature fiber-controlled properties of the CMC.

4.3. Properties needed for CMC Applications

Composite applications for SiC fibers typically center on CMC for high-temperature structural applications (>1100°C), where lower creep and grain-growth rates in comparison to metallic alloys and oxide fiber composites allow better dimensional stability and strength retention under the combined conditions of temperature and stress. The SiC fibers can also provide CMC with greater thermal and electrical conductivity, higher as-produced strength, and lower density. However, under oxygen and moisture-containing environmental conditions, the exposed surfaces of Si-based fibers will degrade slowly due to silica growth and surface recession. Nevertheless, silica is among the most protective of scales, so that in a general sense, SiC fibers display good oxidation resistance in the short term. It follows then that SiC-based fibers are generally preferred for CMC applications that require (1) structural service under environmental conditions that minimally expose the fibers to oxygen and (2) upper use temperatures higher than possible with oxide/oxide CMC or state-of-the-art metallic alloys. Minimal oxygen exposure is typically achieved by incorporating the fibers in dense protective matrices of similar composition and thermal expansion that can remain un-cracked after CMC fabrication, such as in SiC/SiC composites.

Today extensive developmental efforts are underway to utilize SiC/SiC CMC in land- and aero-based gas turbine engines for hot-section components that require service for many hundreds of hours under combustion gas environments [21, 22]. For the purpose of achieving high performance high-temperature SiC/SiC components, CMC experience has shown that a variety of issues exist which relate to retaining the as-produced properties of the reinforcing SiC fibers during component fabrication and service. Many of these issues arise in the fabrication stage during the various steps of (1) shaping the continuous length fibers into architectural arrays or preforms that yield near net-shape component structures, (2) coating the fibers within the architectural preforms with thin fiber coatings or interphase materials that are required for matrix crack deflection, and (3) infiltrating the coated-fiber architectural preforms with SiC-based matrix material, which is often performed at temperatures of 1400°C and above. Issues also arise during CMC service when the SiC matrix may crack due to unforeseen stresses. Since these issues dictate additional second-level property requirements for the SiC fibers, the remainder of this section discusses these property needs in more detail and the ability of current SiC fibers to achieve them.

For the CMC architecture formation step, some issues that can arise during fiber shaping include fiber-fiber abrasion within the multi-fiber tows and excess fiber bending stresses, which may even cause fiber fracture during this step or provide new fiber surface flaws and residual bending stresses in the architecture that eventually cause premature fiber fracture during component structural service. Fiber-fiber abrasion can be minimized by fiber types with surface layers that are smooth and have abrasion resistant compositions such as carbon or boron nitride. Sizing can also be helpful in this regard, but since the sizing typically covers the outer fibers of the tow, it is probably not as effective as abrasion-resistant coating layers on each fiber surface. Likewise, fiber bending stresses can be minimized by fiber types that have small diameters and low elastic moduli. Thus for architecture formation, as well as for low acquisition cost, the first-generation polymer-derived fibers such as Nicalon and Tyranno Lox M are generally the first choice for component fabrication due to their carbon-rich surfaces, small grain size or surface roughness, and lower elastic moduli. But as described above, these fiber types are not desirable for the higher temperature components.

Since achieving these components necessitates use of high-performance SiC fibers with their concomitant high modulus and large surface roughness, current approaches for high-performance SiC/SiC component are focusing on (1) near-stoichiometric SiC fiber types with either carbon-rich surfaces (Hi-Nicalon Type S) or BN-rich surfaces (Sylramic-iBN), and (2) textile formation processes that provide abrasion-resisting liquids on the fiber surfaces during the architecture formation step.

During the CMC fabrication step in which chemical vapor infiltration (CVI) is typically used to deposit thin BN or carbon-containing crack-deflecting interfacial coatings on the fiber surfaces, potential fiber strength-degrading issues include the risk that chemically aggressive gases such as halogens, hydrogen, or oxygen may reach the SiC fiber surface before the protective BN and carbon interfacial materials are formed. The halogens and hydrogen have been demonstrated to cause fiber weakening by surface flaw etching [23]; whereas oxygen allows the growth of silica on the fiber surfaces, which in turn causes strong mechanical bonds to be formed between contacting fibers in the fiber architectures [24]. The detrimental consequence of fiber-fiber bonding is that if one fiber should fracture prematurely, all others to which it is bonded will prematurely fracture, causing composite fracture or rupture at stresses much lower than those that would be needed if the fibers were able to act independently. This oxidation issue is also serious during SiC/SiC service where the possibility exists that cracks may form in the SiC matrix, thereby allowing oxygen from the service environment to reach the reinforcing fibers. Because of the high reactivity of carbon with oxygen above ~500°C and subsequent volatility of the bi-products, cracking of the matrix can be especially serious for those SiC fiber types with carbon-rich surfaces or for fibers and interphase materials based on carbon. For this reason, many SiC/SiC component development programs in the U.S. are utilizing BN-based interfacial coatings, as well as the Sylramic-iBN SiC fiber type with its in-situ grown BN surface layer [22].

Finally, during the CMC matrix formation step, current SiC/SiC fabrication trends are progressing toward SiC-based matrices that are formed at 1400°C and above in order to improve matrix and composite creep-rupture resistance and thermal conductivity [22]. In these cases, the matrix formation times and temperatures are high enough to cause microstructural changes and strength degradation in the non-stoichiometric SiC fibers that are produced at temperatures below 1400°C. These effects can also occur in a near-stoichiometric type if its maximum production temperature is below that for matrix processing. Thus the high performance SiC fibers with the highest production temperatures (see Table 1) are generally preferred for these new matrix formation approaches at higher temperatures.

5. SUMMARY AND CONCLUSIONS

This paper has examined the current state of experimental and mechanistic knowledge concerning the production methods, microstructures, physical properties, and mechanical properties at room and high temperature for a variety of fine-grained SiC-based ceramic fibers of current interest for CMC reinforcement. It has been shown (1) that good correlations exist between the fiber production methods, microstructures, and properties, and (2) that fiber production methods over recent years have significantly improved the key fiber properties needed for implementing SiC fibers in advanced high-temperature CMC components. In particular, these methods have eliminated such performance degrading impurities as excess oxy-carbides, carbon, boron, and silicon to yield dense, high purity, near-stoichiometric SiC fibers with grain sizes optimized for fiber tensile strength, creep-rupture resistance, thermal conductivity, and intrinsic temperature capability. However, along with these advances, fiber acquisition costs have risen to the point that SiC fiber use in the near term may be limited only to those applications where their usage is enabling, rather than enhancing.

Because of their lower atomic diffusion, higher fracture toughness, lower density, and higher thermal conductivity in comparison to oxide fibers, pure near-stoichiometric SiC fibers are currently the preferred reinforcement for CMC products that are required to operate for long times at temperatures greater than state-of-the-art metal alloys (>1100°C). While reduction in production costs and further improvement in high-temperature creep-rupture resistance are on-going developmental issues for future SiC-based fibers, another important issue is improvement of the fiber surfaces against environmental attack. In this area, possibilities exist for the development of oxidation-resistant fiber coatings that are deposited on tows after fiber processing, or better yet in terms of cost reduction, are formed in-situ during fiber production. These new coating approaches should also be beneficial for reducing fiber abrasion and strength degradation during the complex weaving and braiding processes typically needed for some CMC products. As described here, the new Sylramic-iBN fiber type with an in-situ grown BN layer goes a long way in this direction, as well as providing most of the other key properties needed for the fiber reinforcement of high-temperature SiC/SiC composites.

6. REFERENCES

1. J.A. DiCarlo and S. Dutta, Continuous Ceramic Fibers for Ceramic Composites, *Handbook On Continuous Fiber Reinforced Ceramic Matrix Composites*," Eds R. Lehman, S. El-Rahaiby, and J. Wachtman, Jr., CIAC, Purdue University, West Lafayette, Indiana, 1995, p. 137–183.
2. *Ceramic Fibers and Coatings*, National Materials Advisory Board, Publication NMAB-494, National Academy Press, Washington, D.C., 1998.
3. A.R. Bunsell and M-H. Berger, *Fine Ceramic Fibers*, Marcel Dekker, New York, 1999.
4. H. Ichikawa and T. Ishikawa, Silicon Carbide Fibers (Organometallic Pyrolysis), *Comprehensive Composite Materials*, Vol. 1, Eds. A. Kelly, C. Zweben, and T. Chou, Elsevier Science Ltd., Oxford, England, 2000, p. 107–145.
5. D.M. Wilson, J.A. DiCarlo, and H-M. Yun, Ceramic Fibers, *ASM Handbook, Volume 21 Composites*, ASM International, Materials Park, Ohio, 2001, pp. 46–50.
6. J.A. DiCarlo and H-M. Yun, Microstructural Factors Affecting Creep-Rupture Failure of Ceramic Fibers and Composites, *Ceramic Transaction.* **99,** 1998, p. 119–134.
7. H-M. Yun and J.A. DiCarlo, Comparison of the Tensile, Creep, and Rupture Strength Properties of Stoichiometric SiC Fibers, *Cer. Eng. Sci. Proc.*, **20** [3], 1999, p. 259–272.
8. J.A. DiCarlo, Property Goals and Test Methods for High Temperature Ceramic Fibre Reinforcement, *Ceramics International* **23** (1997) 283.
9. R. Bodet, X. Bourant, J. Lamon, and R. Naslain, Tensile Creep Behavior of A Silicon-Carbide-Based Fibre with Low Oxygen Content, *J. Mater. Sci.,* **30** (1995), 661–677.
10. H-M. Yun, J.C. Goldsby, and J.A. DiCarlo, Effects of Thermal Treatment on Tensile Creep-Rupture Behavior of Hi-Nicalon SiC Fibers, *Cer. Eng. Sci. Proc.*, **16** [5] (1995), 987–996.
11. J.A. DiCarlo, Creep of Chemically Vapour Deposited SiC Fibers, *J. Mater. Sci.*, **21** (1986), 217–224.
12. G.N. Morscher, H-M. Yun, and J.C. Goldsby, Bend Stress relaxation and Tensile Primary Creep of a Polycrystalline α-SiC Fiber, *Plastic Deformation of Ceramics*, eds. R. Bradt, C. Brooks, and J. Routbort (New York: Plenum Publishing, 1995), pp. 467–478.
13. A.S. Nowick and B.S. Berry, *Anelastic Relaxation in Crystalline Solids* (New York: Academic Press, 1972).

14. J.A. DiCarlo and H-M. Yun, Creep of Ceramic Fibers: Mechanisms, Models, and Composite Implications, *Creep Deformation: Fundamentals and Applications*, eds. R.S. Mishra, J.C. Earthman, and S.V. Raj, The Minerals, Metals, and Materials Society, Warrendale, PA, 2002, pp. 195–208.
15. H-M. Yun, J.C. Goldsby, and J.A. DiCarlo, Tensile creep and Stress-Rupture Behavior of Polymer Derived SiC Fibers", *Ceramic Transactions*, **46** (1994), 17–28.
16. H-M. Yun and J.A. DiCarlo, Time/Temperature Dependent Tensile Strength of SiC and Al_2O_3-Based Fibers", *Ceramic Transactions*, **74**, 1996, p. 17–26.
17. H-M. Yun and J.A. DiCarlo, Thermo-mechanical Properties of Ceramic Fibers for Structural Ceramic Matrix Composites, *Proceedings of CIMTEC '02*, Florence, Italy, 2002.
18. F.C. Monkman, and N.J Grant, An Empirical Relationship between Rupture Life and Minimum Creep Rate", *Proc. ASTM*, **56** (1956), 593–620.
19. J.A. DiCarlo and H-M. Yun, "Creep and Stress Rupture Behavior of Advanced SiC Fibers," *Proceedings of ICCM-10, vol. VI* (Cambridge, England: Woodhead Publishing Limited, 1995), 315–322.
20. H-M. Yun, J.C. Goldsby, and J.A. DiCarlo, Environmental Effects on Tensile Creep and Rupture Behavior of Advanced SiC Fibers, *Ceramic Transactions*, **57** (1995), 331–336.
21. G.S. Corman and K.L. Luthra, Silicon Melt Infiltrated Ceramic Composites (HiPerComp^{TM}), in *Handbook of Ceramic Composites,* N.P. Bansal, Ed., Kluwer Academic Publishers, Bostan, MA, 2004, pp. 99–115.
22. J.A. DiCarlo, H.M. Yun, G.N. Morscher, and R.T. Bhatt, SiC Fiber-Reinforced SiC Matrix Composites for Thermostructural Applications to 1200°C and Above, in *Handbook of Ceramic Composites*, N.P. Bansal, Ed., Kluwer Academic Publishers, Boston, MA, 2004, pp. 77–98.
23. F. Rebillat, A. Guette, L. Espitalier, and R. Naslain, Chemical and Mechanical Degradation of Hi-Nicalon and Hi-Nicalon S Fibers under CVD/CVI BN Processing Conditions, *High Temperature Ceramic Matrix Composites III*, The Ceramic Society of Japan, 1998, pp. 31–34.
24. Morscher, G.N., Tensile Stress Rupture of SiC/SiC Minicomposites with Carbon and Boron Nitride Interphases at Elevated Temperatures in Air, *J. Am. Ceram. Soc.* **80** [8], 1997, pp. 2029–2042.

Part II

Non-oxide/Non-oxide Composites

3

Chemical Vapor Infiltrated SiC/SiC Composites (CVI SiC/SiC)

Jacques Lamon
Laboratoire des Composites Thermostructuraux
UMR 5801 (CNRS-Snecma-UB1-CEA)
3, Allée de la Boétie
33600 PESSAC (France)
lamon@lcts.u-bordeaux.fr
Tel: 33556844703

ABSTRACT

CVI SiC/SiC composites are manufactured via Chemical Vapor Infiltration Process. The SiC-based matrices are deposited from gaseous reactants on to a heated substrate of SiC fiber preforms. An interphase coated on the fibers allows control of damage and mechanical behavior.

The advantageous properties of CVI SiC/SiC composites such as their excellent high temperature strength, creep and corrosion resistances, low density, high toughness, resistance to shocks, fatigue and damage, and reliability make them ideal candidates for the replacement of metals and ceramics in many engineering applications involving loads, high temperatures and aggressive environments. Mechanical properties exhibit features that differentiate CVI SiC/SiC composites from monolithic ceramics and glasses, from other ceramic matrix composites and from other composites.

1. INTRODUCTION

The concept of composite material is very powerful. It covers a wide variety of materials of which one can tailor the properties with respect to end use applications, through the combination of constituents. A large variety of combinations is possible in terms of properties and arrangement.

The ceramic matrix composites (CMCs) contain brittle fibers and a brittle matrix. This combination ends up in a damage tolerant material. CMCs are of interest to thermostructural applications. They consist of ceramics or carbon reinforced with continuous ceramic or carbon fibers. Their mechanical behavior displays several typical features that differentiate them from the other composites (such as polymer matrix composites, metal matrix composites, etc . . .) and from the homogeneous (monolithic) materials.

CMCs exhibit high mechanical properties at high or very high temperatures (400–3000°C), and in severe environments. They were developed initially for military and aerospace applications. Now they are being introduced into new fields and their range of applications will grow when their cost is lowered drastically.

CMCs can be fabricated by different processing techniques, using either liquid or gaseous precursors. The CVI SiC/SiC composites consist of a SiC-based matrix reinforced by SiC fibers. They are produced by Chemical Vapour Infiltration (CVI). The CVI technique has been studied since the 1960s, and it has become quite important commercially since commercialization by SNECMA (former SEP, Société Européenne de Propulsion). This technique derives directly from Chemical Vapour Deposition. In very simple terms, the SiC-based matrix is deposited from gaseous reactants on to a heated substrate of fibrous preforms (SiC). CVI is a slow process, and the obtained composite materials possess some residual porosity and density gradients. Despite these drawbacks, the CVI process presents a few advantages: (i) the strength of reinforcing fibers is not affected during composite manufacture, (ii) the nature of the deposited material can be changed easily, simply by introducing the appropriate gaseous precursors into the infiltration chamber, (iii) a large number of components, and (iv) large complex shapes can be produced in a near net shape.

The CVI SiC/SiC composites exhibit good mechanical properties at room and high temperatures that depend on the fiber/matrix interface. Pyrocarbon (PyC) has proven to be an efficient interphase to control fiber/matrix interactions and composite mechanical behavior. But pyrocarbon is sensitive to oxidation at temperatures above 450°C. A few versions of high temperature resistant CVI SiC/SiC composites are produced. In order to protect the PyC interphase against oxidation, multilayered interphases and matrices have been developed. Multilayered matrices contain phases which produce sealants at high temperatures preventing oxygen from reaching the interphase. This composite is referred to as CVI SiC/Si-B-C. Oxidation resistant interphases such as BN or multilayered materials can also be coated on the fibers. An "oxygen getter" can be added to the matrix to scavenge oxygen that might ingress into the matrix (enhanced CVI SiC/SiC).

The mechanical behavior of CVI SiC/SiC composites exhibits features which differentiate these composites from monolithic ceramics, from other ceramic matrix composites and from other composites. These features depend on composite microstructure, interphases, fiber and matrix properties. The main characteristics of CVI SiC/SiC composites are examined: i.e. the applications, the Chemical Vapor Infiltration process and properties. Main properties are discussed with respect to features of the mechanical behavior.

TABLE 1. Potential industrial nonaerospace applications for continuous fiber-reinforced ceramic composites (after Sheppard [1])

Product area	Examples	Likely industrial market(s)
Advanced heat engines	Combustors, liners, wear parts, etc.	Primarily high-temperature gas turbines; possibly adiabatic diesels, S.I. engines (promising market is gas turbine combustor retrofits)
Heat recovery equipment internals	Air preheaters, recuperators	Any indirect heating uses; energy-intensive industrial processes (e.g. aluminum remelters, steel reheaters, glass melters)
Burners and combustors	Radiant tube burners	Potentially any indirect-fired, high-temperature and/or controlled-atmosphere heating/melting/heat-treating industrial application
Burners and combustors	Catathermal combustors	Low-NO_x clean fuel heating applications – including gas turbine combustors, industrial process heat
Burners and combustors	Low-temperature radiant combustors	Low-NO_x clean fuel heating applications – including small scale (space heating) and large scale (industrial process) applications
Process equipment	Reformers, reactors, HIP equipment	Chemical process industry, petroleum refining
Waste incineration systems	Handling equipment, internals, cleanup	Conventional MSW/RDF facilities, with or without energy recovery
Separation/filtration systems	Filters, substrates, centrifuges	Gas turbine, combined cycle, and IGCC configurations; particulate traps for diesel exhausts; molten metal filters; sewage treatment
Refractories and related	Furnace linings, crucibles, flacks, etc.	High-temperature industrial heating/melting/heat treating processes
Structural components	Beams, panels, decking, containers	Possible niche applications for EMI shielding, corrosive/abrasive environments, fire-protection, missile protection (e.g. gas turbine shrouds); and major infrastructure applications

2. APPLICATIONS

CMCs are candidate materials for many high-temperature structural applications, where their attractive properties can be exploited to increase performances at reasonable cost. CMCs are lightweight, damage tolerant, and they exhibit a much greater resistance to high temperatures and aggressive environments than metals or other conventional engineering materials. Potential applications include heat exchangers, heat engines, gas turbines, structural components in the aerospace industry, in nuclear reactors . . . Table 1 details non aerospace potential applications [1].

Development of CVI SiC/SiC composites began in the 1980s when SEP, Amercorm, Refractory Composites and others began to develop equipment and processes for producing

CVI components for aerospace, defense and other applications. CVI SiC/SiC components were produced and tested. SNECMA (former SEP) is at the forefront of this technology and demonstrated satisfactory component performance in engine and flight tests. A number of CVI SiC/SiC components has performed successfully in engine or full scale tests [2]:

- ramjet chamber for solid propulsion rocket, at 1400°K in kerosene-air,
- combustion chambers and nozzles for liquid propellant rocket engines. A CVI SiC/SiC chamber cumulated 24 000 sec firing duration in 400 thermal cycles,
- stators and disc blades in LOX/ LH$_2$ engines: inlet temperatures over 1600°K, peripherical speed larger than 500 m/sec,
- thermal protection systems for space vehicles,
- nozzle flaps, nozzle cone, afterburner flame holders for jet engines. A Mirage 2000 equipped with CVI SiC/SiC turbine engine flaps flew several times at the 1989 Paris Air show.
- spin discs, representative of turbine wheels: tip speeds larger than 500 m/sec and temperatures higher than 1400°C in an air-kerosene environment.
- recently, nozzle seals made of SiC/Si-B-C composites with a self sealing matrix have been successfully tested in a F100-PW-229 gas turbine engine [3]. These composites also showed a low sensitivity to steam environment during low cycle fatigue at 1200°C [3].

Endurance testing in Solar's Centaur 505 engine has demonstrated that CVI SiC/SiC components can be introduced in Industrial Gas Turbines (Solar Turbines). To date, more than 47,000 hours of engine field test experience has been accumulated with protective environmental barrier coatings (EBCs). The longest single test duration was 15144 hours in Solar's Centaur 505 engines at industrial sites [4].

The CVI SiC/SiC composites are also promising for nuclear applications because of the radiation resistance of the β phase of SiC, their excellent high-temperature fracture, creep, corrosion and thermal shock resistances. Studies on the β phase properties suggest that CVI SiC/SiC composites have the potential for excellent radiation stability [5]. Furthermore, because of excellent thermal fatigue resistance, start-up and shut-down cycles and coolant loss scenarii should not induce significant structural damage [5]. The CVI SiC/SiC are also considered for applications as structural materials in fusion power reactors, because of their low neutron-induced activation characteristics coupled with excellent mechanical properties at high temperature [6–8].

3. HISTORY

Chemical Vapor Deposition (CVD) is a technique of deposition of a solid on a heated substrate, from gaseous precursors. It has been used for many years to produce wear resistant coatings, coatings for nuclear fuels, thin films for electronic circuits ceramic fibers, etc . . . When the CVD technique is used to impregnate rather large amounts of matrix materials in fibrous preforms, it is called chemical vapor impregnation or infiltration

(CVI) [9]. The CVI technique has been studied since the 1960s, as an extension of CVD technology [10,11].

CVI was used first for the synthesis of carbon-carbon composites via pyrolysis of CH_4 at 1000–2000°C. Carbon-carbon composites display several advantageous properties (such as low density, good mechanical properties at high temperatures, etc . . .) but it appeared from 1973 that applications of carbon/carbon composites would be limited because of a very poor oxidation resistance at temperatures higher than 450°C [2]. CVI SiC matrix composites were considered to be a solution to overcome the above shortcoming of carbon/carbon composites for long service life at high temperature in oxidative environments. The feasibility of CVI SiC matrix composites was established in 1977 by Christin et al. [12–14] and confirmed independently by Fitzer in 1978 [15]. CVI of porous carbon substrates was investigated first from 1975 to 1979. CVI SiC/SiC composites reinforced by SiC fibers were manufactured in 1980.

The CVI SiC/SiC composites exhibit good mechanical properties at high temperatures. However, their fatigue behavior at high temperatures may be limited by the oxidation of the pyrocarbon interphase coating on the fibers. In order to protect the PyC interphase against oxidation, CVI SiC/SiC composites with multilayered matrices have been developed (CVI SiC/Si-B-C composites) [16]. Such multilayered matrices contain phases which produce sealants at high temperatures causing healing of the cracks and preventing oxygen from reaching the interphase [17].

4. PROCESSING

Manufacture of CVI SiC/SiC composites requires three main steps:

(i) preparation of a fibrous preform
(ii) infiltration of an interface material
(iii) infiltration of the SiC matrix.

4.1. Fibrous preform

The preforms of CVI SiC/SiC composites are made of refractory SiC based fibers. The fibers exhibit high strength, high stiffness, low density and high thermal and chemical stability to withstand long exposure at high temperatures. Finally fiber diameter must be small (<20 μm) so that the fibers can be woven easily.

The fiber preforms may consist of:

(i) a simple stack of unidirectional fiber layers or of fabrics (1D or 2D preforms)
(ii) a multidirectional fiber architecture (3D preforms). Weaving in four or five directions can also be used.
(iii) a quasi-isotropic felt.

The 2D layers are stacked and kept together using a tool or using fibers in the orthogonal direction (3D preforms).

4.2. Coating of fibers

An interface material is deposited on the fibers. This acts as a debond layer between the fiber and the matrix. This interlayer consists essentially of Pyrocarbon, Boron Nitride or a multilayer ((PyC/SiC)n or (BN/SiC) n sequences). The gas precursor is CH_4 for carbon, BCl_3 and NH_3 for boron nitride. Multilayered interphases may be deposited via pulsed CVI.

4.3. Infiltration of the SiC matrix

The coated preform is densified with a silicon carbide matrix via CVI. The gaseous precursor is methylchlorosilane (MTS). When the CVI process is carried out isothermally (I-CVI), surface pores tend to close, restricting the gas flow to the interior of the preform. This phenomenon requires an intermediate operation of surface machining to obtain an adequate density. The CVI parameters (i.e. total pressure, temperature and gas flow rate) have to be selected according to the preform geometry defined by the pore spectrum and the thickness, the number of preforms present in the chamber and the size of the chamber. Finally several coating systems can be applied to these composites via CVD to provide environmental and oxidation protection.

5. CVI PROCESSES

The basic chemistry of making a coating and a matrix by CVI is the same as that of depositing a ceramic on a substrate by CVD. The reactions consist in cracking of a hydrocarbon for deposition of carbon, cracking of methylchlorosilane for deposition of SiC. In the I-CVI process (Isobaric Isothermal CVI) the preform is kept in a uniformly heated chamber (figure 1). The gaseous precursors flow through the fibrous preforms. The deposition chamber (figure 1) is open at both ends. The preform is heated by a furnace. Temperature and pressure are relatively low ($<1200°C$, <0.5 atm).

The gas species are conveyed through the porous preform mainly by diffusion. The driving force is the concentration gradient between the interior and the surface of the preform, which reduces the densification rate. When CVI conditions that shorten the densification time are selected non uniform deposition of the matrix is enhanced. Intermediate cycles of surface machining are thus required to open the pores that have been sealed. A few alternative CVI techniques have been proposed to increase the infiltration rate [11]. These techniques require more complicated CVI chambers, and are not appropriate to the production of large or complex shapes, or large quantities of pieces.

The forced CVI (F-CVI) technique was proposed in the mid 1980s [18]. The precursor gas is forced through the bottom surface of the preform under a pressure P_1, and the exhaust gases are pumped from the opposite face under a pressure $P_2 < P_1$. The fibrous preform is heated from the top surface and sides, and cooled from the bottom (cold) surface. The F-CVI technique has been used essentially to manufacture 2D preforms made of carbon or silicon carbide based fibers. The densification times are significantly lower than those of I-CVI (10–24 h for a SiC matrix, a few hours for carbon), and the conversion efficiency of the precursor is relatively high. However the technique is not appropriate for complex shapes. Only one preform per run can be processed and complex graphite fixtures are required to generate the temperature and pressure gradients.

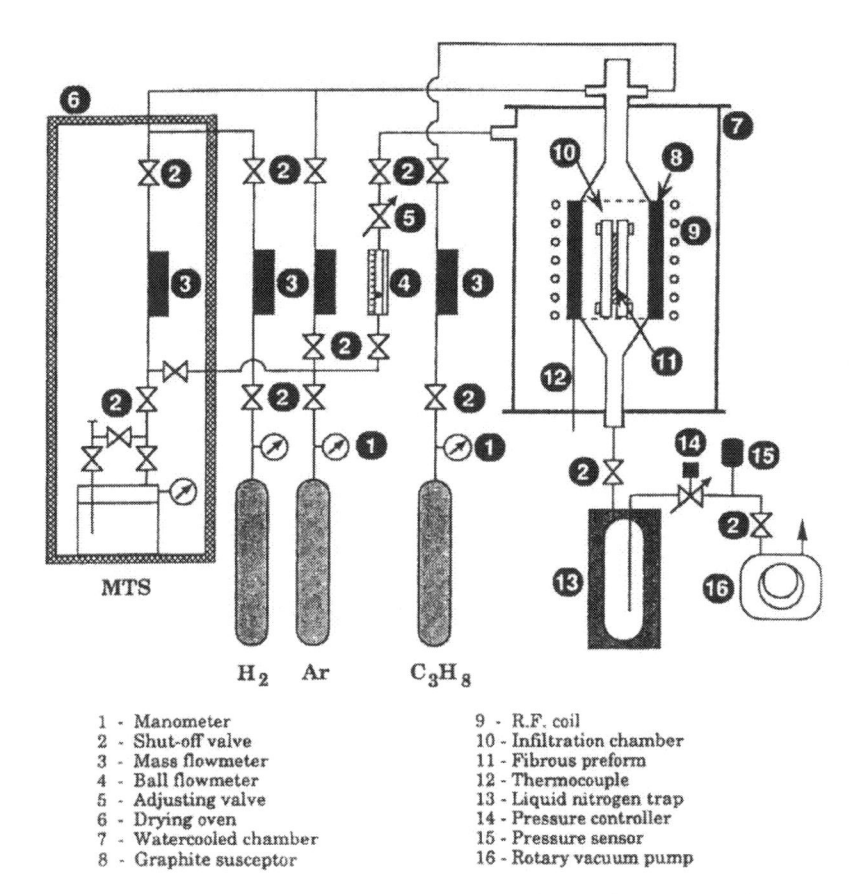

FIGURE 1. An I-CVI reactor for the processing of SiC and/or C matrix composites

1 - Manometer
2 - Shut-off valve
3 - Mass flowmeter
4 - Ball flowmeter
5 - Adjusting valve
6 - Drying oven
7 - Watercooled chamber
8 - Graphite susceptor
9 - R.F. coil
10 - Infiltration chamber
11 - Fibrous preform
12 - Thermocouple
13 - Liquid nitrogen trap
14 - Pressure controller
15 - Pressure sensor
16 - Rotary vacuum pump

In order to overcome the above limitations of the F-CVI technique, alternative techniques using thermal gradients or pressure gradients have been examined for several years [11]. In the thermal gradient process, the core of the fibrous preform is heated in a cold wall reactor. The heat loss by radiation is favorable to get a colder temperature in the external surface. The densification front advances progressively from the internal hot zone toward the cold side of the preform. In the P-CVI process, the source gases are introduced during short pulses [11]. The P-CVI process is appropriate to the deposition of thin films.

6. PROPERTIES

The development of CVI SiC/SiC composites was inspired by the poor oxidation resistance of CVI C/C composites. The CVI SiC/SiC composites are less sensitive to oxidation than their ancestor C/C composites.

TABLE 2. Mechanical and thermophysical properties of 2D SiC/SiC composites reinforced with 0/90 balanced Nicalon™ fabrics (after [2, 26]).

Property	Temperature		
	23°C	1000°C	1400°C
Fiber content (%)	40	40	40
Specific gravity	2.5	2.5	2.5
Porosity (%)	10	10	10
Tensile strength (MPa)	200	200	150
Strain-to-failure (%)	0.3	0.4	0.5
Young's modulus (GPa)	230	200	170
Poisson's ratio			
ν_{12}	0.5		
ν_{13}	0.18		
Flexural strength (MPa)	300	400	280
In-plane compressive strength (MPa)	580	480	300
Thru-the-thickness compressive strength (MPa)	420	380	250
Interlaminar shear strength (MPa)	40	35	25
In-plane thermal diffusivity (10^{-5} m^2/s)	12	5	5
Thru-the-thickness thermal diffusivity (10^{-5} m^2/s)	6	2	2
In-plane coefficient of thermal expansion (10^{-6}/°K)	3	3	
Thru-the-thickness coefficient of thermal expansion (10^{-6}/°K)	1.7	3.4	
Fracture toughness (MPa$\sqrt{m}$)	30	30	30
Specific heat (J/kg°K)	620	1200	
Total emissivity	0.8	0.8	0.8
In-plane thermal conductivity (Wm^{-1} °K^{-1})	19.0	15.2	
Thru-the-thickness thermal conductivity (Wm^{-1} °K^{-1})	9.5	5.7	

The CVI SiC matrix also possesses the superior properties of ceramics over metals:

- high strength at high temperature
- low density
- good resistance to oxidation.

Finally, the CVI SiC/SiC composites provide a solution to overcome the inherent limitations of monolithic SiC for thermostructural service conditions, in terms of:

- toughness
- resistance to thermal shock
- impact resistance
- reliability.

Table 2 gives mechanical and thermophysical properties of 2D CVI SiC/SiC composites. Certain properties may be influenced by various factors including the reinforcing fiber architecture, the SiC fibers used, matrix properties, the fiber/matrix bond strength, etc . . . For instance, high tensile strengths (up to 400 MPa) can be obtained with Hi-Nicalon™ SiC fibers [19], or with rather strong interfaces [17,20]. Larger strains-to-failure (up to 1%) can also be observed [20]. Further details are discussed in subsequent sections. Data on CVI SiC/Si-B-C composites and enhanced CVI SiC/SiC composites are

TABLE 3. Mechanical properties of a CVI SiC/Si-B-C composite with a self healing matrix and a multilayer reinforcement of Hi-Nicalon™ fibers (after [3]).

Property	Temperature	
	Room Temperature	1200°C
Density	2.3	
Porosity (%)	13	
Tensile strength (MPa)	315	
Strain-to-failure (%)	0.5	
Young's modulus (GPa)	220	
Interlaminar shear strength (MPa)	31	23
Flexural strength (MPa)	699	620

reported in tables 3 and 4. The mechanical behavior at room and at high temperature deserves a special attention. CVI SiC/SiC composites exhibit some features that are addressed in the following section.

7. MECHANICAL BEHAVIOR

7.1. Tensile stress-strain behavior

The typical stress-strain curves for 2D CVI SiC/SiC composites shown in figure 2, summarize trends in the mechanical behavior. This composite behaves linearly to a strain of about 0.03%, and then it exhibits a curved domain of non-linear stress-strain relations. The curved domain of deformation results essentially from transverse cracking in the matrix (the cracks are perpendicular to fibers in the loading direction). Saturation of matrix damage is indicated by the end of the curved domain. Then the linear portion of the curve reflects the

TABLE 4. Mechanical properties of 2D CVI enhanced SiC/SiC composite reinforced with 0/90 five harness satin fabrics of Hi-Nicalon™ fibers (source: Power Systems Composites data sheet).

Property	Temperature	
	23°C	1200°C
Fiber content (%)	35	
Density	2.2	
Porosity (%)	10	
Tensile strength (MPa)	324	259
Strain-to-failure (%)	0.74	0.50
Young's modulus (GPa)	207	212

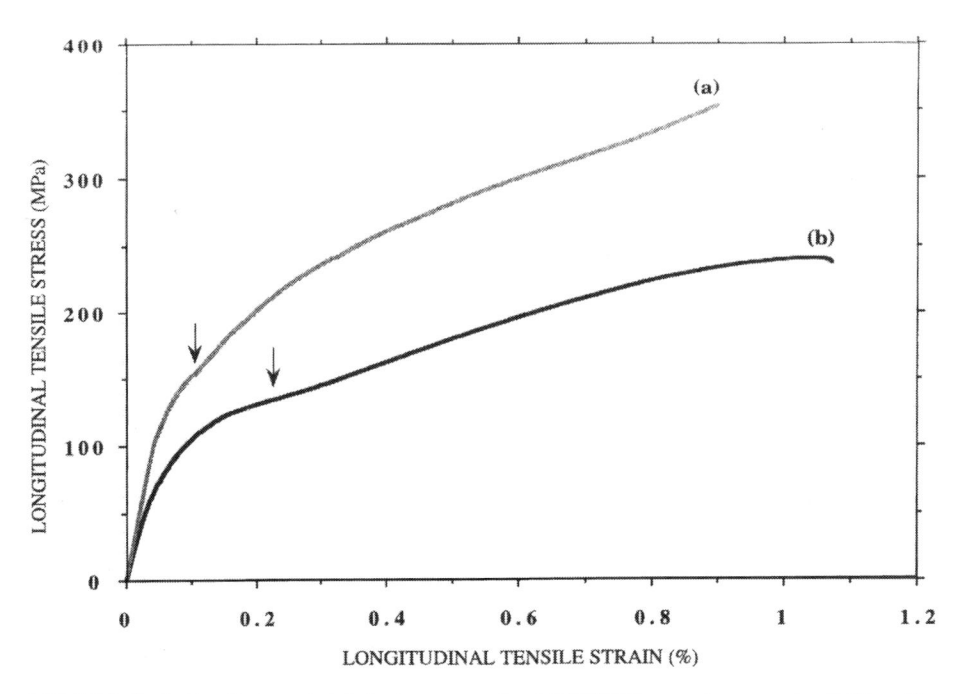

FIGURE 2. Typical tensile stress-strain behaviors measured on 2-D SiC/SiC composites possessing PyC based interphases and fabricated from untreated or treated Nicalon (ceramic grade) fibers: (a) strong fiber/coating interfaces and (b) weak fiber/coating interfaces.

deformation of the fibers. Fiber failures may occur prior to ultimate failure. This mechanical behavior is essentially damage sensitive.

A *damage sensitive* stress-strain behavior is obtained when the load carrying contribution of the matrix is significant. The elastic modulus of the matrix (E_m) is not negligible when compared to that of fiber (E_f) and it contributes significantly to the composite one (E_c), as illustrated by the following mixtures law:

$$E_c = E_m V_m + E_f V_f \tag{1}$$

Where V_m is the volume fraction of matrix, and V_f is the volume fraction of fibers oriented in the loading direction in a 2D woven composite. By contrast, a *damage insensitive* stress-strain behavior is observed when the elastic modulus of the matrix is negligible with respect to that of fibers. As a result, the load carrying contribution of the matrix is negligible.

In 2D CVI SiC/SiC composites E_m ($\approx$410 GPa) $>$ E_f ($\approx$200 GPa). The 2D SiC/SiC composites exhibit an elastic damageable behavior (figure 3). This means that the response of the damaged material is elastic as indicated by the linear portion of the curves on reloading. Figure 4 shows the dependence of elastic modulus on damage.

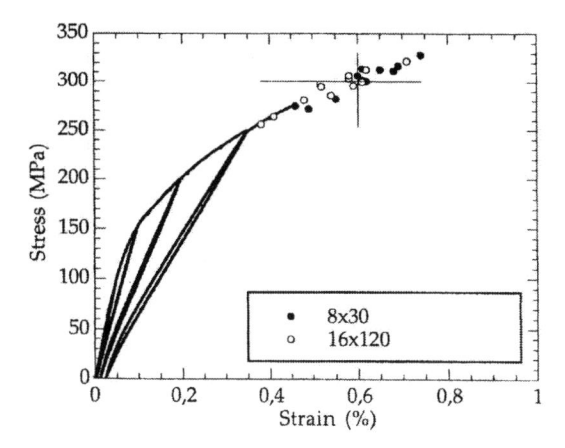

FIGURE 3. Stress-strain curves in tension of 2D SiC/SiC composites test specimens. The open and filled symbols represent ultimate failure data point obtained with the specimens of volumes V_1 and V_2.

7.2. Damage mechanisms

The basic damage phenomena in unidirectional composites involve multiple microcracks or cracks that form in the matrix, perpendicular to the loading direction, and that are arrested by the fibers and deflected at the interface between the fiber and the matrix. In the composites reinforced with fabrics of fiber bundles, matrix damage is influenced by the microstructure [21]. The microstructure of 2D CVI SiC/SiC is heterogeneous, as a result of the presence of woven infiltrated tows, large pores (referred to as macropores) located between the plies or at yarn intersections within the plies and a uniform layer of matrix over

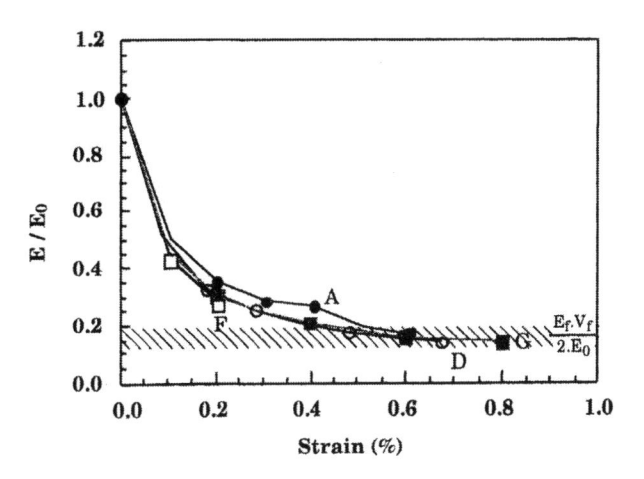

FIGURE 4. Relative elastic modulus versus applied strain during tensile tests on various 2D woven SiC/SiC composites reinforced with treated fibers: (A) Nicalon/(PyC$_{20}$/SiC$_{50}$)$_{10}$/SiC, (D) Nicalon/PyC$_{100}$/SiC, (F) Hi-Nicalon/PyC$_{100}$/SiC, (G) Hi-Nicalon/(PyC$_{20}$/SiC$_{50}$)$_{10}$/SiC.

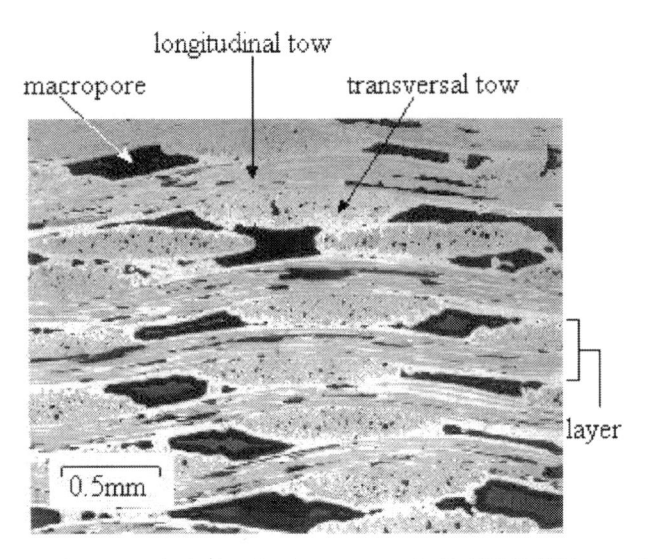

FIGURE 5. Micrograph showing the microstructure of a 2D CVI SiC/SiC composite.

the fiber preform (referred to as the intertow matrix) (figure 5). Much smaller pores are also present within the tows.

Damage in 2D CVI SiC/SiC consists essentially in the formation of transverse cracks in the matrix and their deviation either by the tows (first and second steps) or by the fibers within the tows (third step). The steps in matrix cracking (figure 6) depend on applied deformations and microstructure:

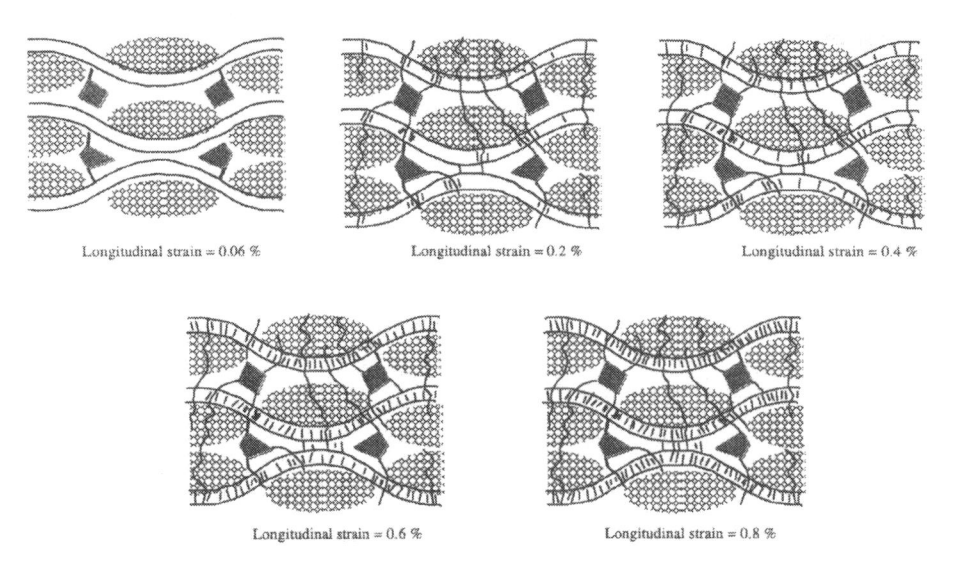

FIGURE 6. Schematic diagram showing matrix cracking in a 2D SiC/SiC composite during a tensile test.

Step 1: cracks initiate at macropores where stress concentrations exist (deformations between 0.025% and 0.12%).

Step 2: cracks form in the transverse yarns and in the interply matrix (deformations between 0.12% and 0.2%).

Step 3: transverse microcracks initiate in the longitudinal tows (deformations larger than 0.2%). These microcracks are confined within the longitudinal tows. They do not propagate in the rest of the composite. The matrix in the longitudinal tows experiences a fragmentation process and the crack spacing decreases as the load increases.

The resulting decreases in Young's modulus illustrate the importance of damage in the mechanical behavior (figure 4). The major modulus loss (70%) is caused by both the first families of cracks located on the outside of the longitudinal tows (deformations <0.2%). By contrast, the microcracks within the longitudinal tows are responsible for only a 10% loss. The big modulus loss reflects important changes in the load sharing: the load becomes carried essentially by the matrix infiltrated longitudinal tows (tow overloading). During microcracking in the longitudinal tows, the load sharing is affected further, and the load becomes carried essentially by the fibers within the tows (fiber overloading). The elastic modulus reaches a minimum described by the following equation (figure 4):

$$E_{min} = \frac{1}{2} E_f V_f \qquad (2)$$

where V_f is the volume fraction of fibers. Equation (2) implies that the matrix contribution is negligible. At this stage matrix damage and debonding are complete (saturation). The fibers only are loaded. The mechanical behavior is controlled by the fiber tows oriented in the direction of loading.

7.3. Ultimate failure

Ultimate failure generally occurs after completion of matrix cracking (saturation). The fibers break under loads close to the maximum. Matrix damage and ultimate failure thus appear to be successive phenomena.

The *ultimate failure of a tow* of parallel fibers involves two steps:

– a first step of stable failure, and
– a second step of unstable failure.

During the first step, fibers fail individually as the load increases. In the absence of fiber interactions, the load is carried by the surviving fibers only (equal load sharing). The ultimate failure of the tow (second step) occurs when the surviving fibers cannot tolerate the load increment resulting from a fiber failure. At this stage, a critical number of fibers have been broken.

The *ultimate failure of matrix infiltrated tows* also involves a two step mechanism and a global load sharing when a fiber fails. In the presence of multiple cracks across the matrix and associated interfacial debond, the load carrying capacity of the matrix is tremendously reduced or annihilated. The matrix infiltrated tows can be assimilated to bundles of fibers

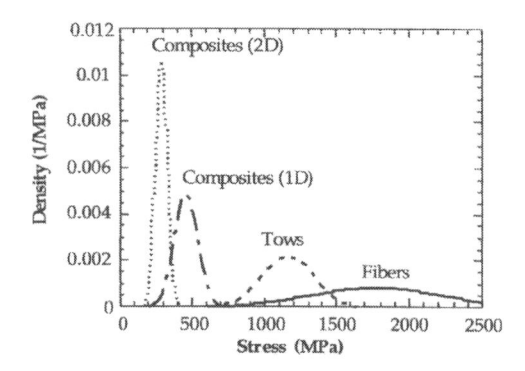

FIGURE 7. Strength density functions for SiC fibers (NLM 202), SiC fiber tows, SiC/SiC (1D) minicomposites and 2D SiC/SiC composites.

subject to the specific stress field induced by the presence of multiple cracks across the matrix. The ultimate failure of a matrix infiltrated tow occurs when a critical number of fibers have failed. This mechanism also operates in infiltrated tows within textile CVI SiC/SiC composites. The ultimate failure of the composite is caused by the failure of a critical number of broken tows (≥ 1).

It is worth pointing out that the failure mechanism of CVI SiC/SiC composites is at variance with that observed in polymer matrix impregnated tows, where a local load sharing occurs when a fiber fails. In these composites, the fibers fail first. Therefore, the uncracked matrix is able to transfer the loads.

7.4. Reliability

The ultimate failure of CVI SiC/SiC composites is highly influenced by stochastic features. Since fibers are brittle, they are highly sensitive to the presence of flaws (stress concentrators), that are distributed randomly. As a consequence, the strength data exhibit a significant scatter, as illustrated by figure 7 [21,22]. Figure 7 shows that the strength magnitude and the strength interval decrease when considering successively single fibers, tows, matrix infiltrated tows and textile composites. As a result of the previously mentioned two step failure mechanism, the ultimate failure of an entity becomes dictated by the lowest extreme of the constituent strength distribution, i.e. the fiber strength distribution for the failure of tows and infiltrated tows, and the infiltrated tow strength distribution for the failure of 2D composites. These extremes correspond respectively to the critical number of individual fiber breaks ($\approx 17\%$ for the SiC Nicalon^{TM} fibers and for the SiC Hi-Nicalon^{TM} fibers) and to the critical number of tow failures (≥ 1). The gap between tows and SiC infiltrated tows results from strength determination: the critical number of individual fiber breaks was taken into account for tow strength determination, whereas the total cross sectional area of specimens was used for determination of strengths of infiltrated tows and composites.

The successive steps involved in damage truncate the flaw populations which leads to a final homogeneous population of flaws [22]: the contribution of the pre-existing flaws in ultimate fracture is reduced as multiple matrix cracking and individual fiber breaks

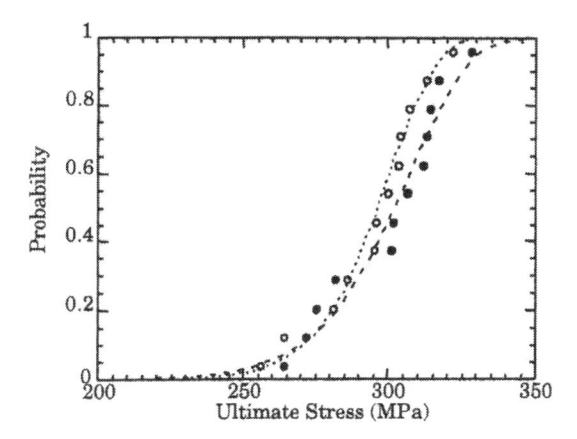

FIGURE 8. Scale effects in 2D woven SiC/SiC composites. Influence of specimen dimensions on ultimate failure in tension: (●) 8×30 mm^2, (o) 160×120 mm^2 [28].

occur. This influences the trends in the mechanical behavior. The tensile stress-strain curves obtained on a batch of several CVI SiC/SiC test specimens coincide quite well (figure 5), whereas the strength data exhibit a certain scatter (figure 5). This scatter is limited (figure 8). Dependence of composite strength on the stressed volume is not significant (figure 8). Furthermore, dependence on the loading conditions is not so large (figure 9): for instance the flexural strength is 1.15 times as large as the tensile strength [22,23] when measured on specimens having comparable sizes (figure 9).

The Weibull model cannot describe the volume dependence of strength data [22], although a Weibull modulus (m) can be extracted from the statistical distribution of strengths: m is in the range 20–29. This value provides an evaluation of the scatter in strength data.

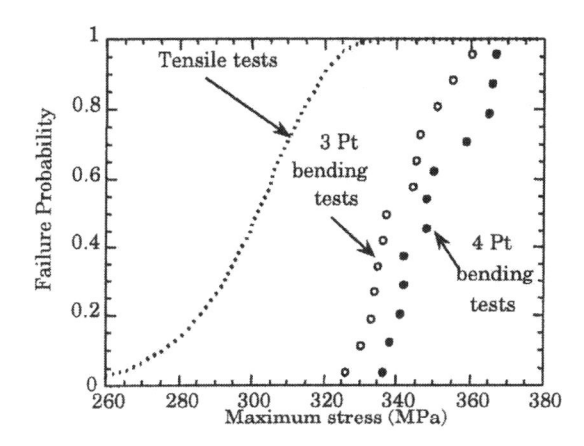

FIGURE 9. Strength distributions for 2D woven SiC/SiC composites tested under various loading conditions: tension, 3-point bending and 4-point bending [28].

TABLE 5. Interfacial shear stresses (MPa) measured using various methods on 2D SiC/SiC composites with PyC based fiber coatings and reinforced with either as-received or treated fibers [25,36–39]

SiC/C/SiC SiC/(C/SiC)n/SiC	Interphase	Crack spacing	Crack spacing	Tensile tests (hysteresis loops)	Tensile tests (curved domain)	Push-out (curved domain)	Push-out tests (plateau)
untreated fibers							
2D woven	PyC (0.1)	12	8	0.7			
microcomposites	PyC (0.1)			3	4–20		
minicomposites	PyC (0.1)			21–115	40–80		
2D woven	PyC (0.5)			4		14–16	12–10
	(PyC/SiC)$_2$			2		31	19.3
	(PyC/SiC)$_4$			9		28	12.5
treated fibers							
2D woven	PyC (0.1)	203	140	190			165–273
	PyC (0.5)			370			100–105
	(PyC/SiC)$_2$			150			133
	(PyC/SiC)$_4$			90			90

7.5. Interface properties – influence on the mechanical behavior

The fiber-matrix interfacial domain is a critical part of composites because load transfers from the matrix to the fiber and vice-versa occur through the interface. Most authors promote the concept of weak interfaces to increase fracture toughness. The major contribution to toughness is attributed to crack bridging and fiber pull-out. Weak interfaces are detrimental to composite strength. A high strength requires efficient load transfers from fibers to the matrix. This is obtained with strong interfaces. This implies short debond cracks and/or significant sliding friction. These latter requirements, to be met for strong composites, are therefore incompatible with the former ones for tough composites, if toughening is based solely upon the above mentioned weak interface-based mechanisms.

Fiber/matrix interfaces exert a profound influence on the mechanical behavior and the lifetime of composites. Efforts have been directed towards optimization of interface properties. Fiber matrix interfaces in CVI SiC/SiC composites consist of a thin coating layer (less than 1 μm thick) of one or several materials deposited on the fiber (interphase). CVI SiC/SiC composites with rather strong interfaces have been obtained using fibers that have been treated in order to increase the fiber/coating bond [20,24]. The concept of strong interfaces has been established on CVI SiC/SiC composites with PyC and multilayered (PyC/SiC)$_n$ fiber coatings. Less interesting results have been achieved with BN interphases [25]. Table 5 gives various values of the interfacial shear stresses measured using various methods on CVI SiC/SiC composites with PyC-based fiber coatings. It can be noticed that the interfacial shear stresses range between 10–20 MPa for the weak interfaces whereas they are larger than 100–300 MPa for the strong interfaces.

In the presence of weak fiber/coating bonds, the matrix cracks generate a single long debond at the surface of fibers (adhesive failure type, figure 10). The associated interface shear stresses are low, and load transfers through the debonded interfaces are poor. The matrix

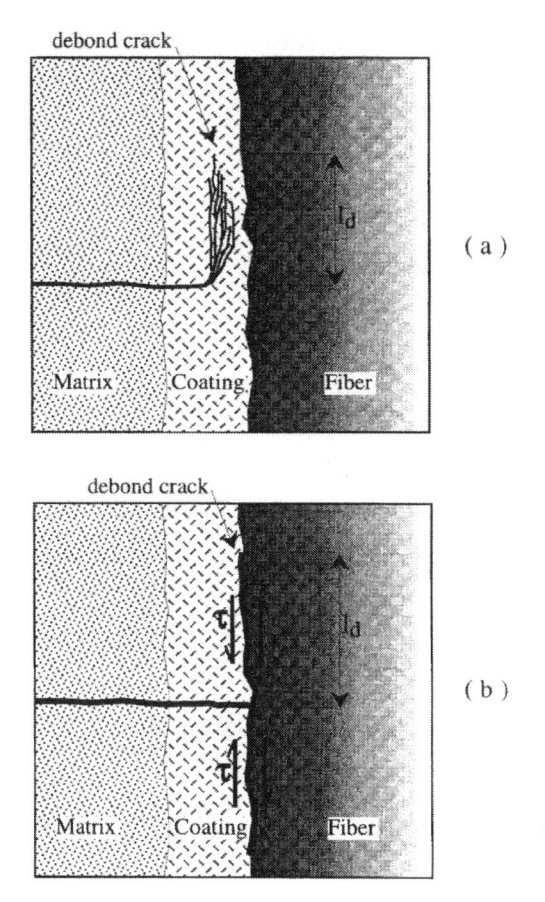

FIGURE 10. Schematic diagram showing crack deflection when the fiber coating/interface is strong (a) or weak (b).

is subjected to low stresses and the volume of matrix that may experience further cracking is reduced by the presence of long debonds. Matrix cracking is not favored. The crack spacing distance at saturation as well as the pull out length tend to be rather long (>100 μm). Toughening results essentially from sliding friction along the debonds. However, as a result of matrix unloading, the fibers carry most of the load, which reduces the composite strength. The corresponding tensile stress-strain curve exhibits a narrow curved domain limited by a stress at matrix saturation which is distinct of ultimate strength (figure 2).

In the presence of stronger fiber/coating bonds, the matrix cracks are deflected within the coating (cohesive failure type, figure 10), into short and branched multiple cracks. Short debonds as well as improved load transfers allow further cracking of the matrix via a scale effect leading to a higher density of matrix cracks (which are slightly opened). Sliding friction within the coating as well as multiple cracking of the matrix increase energy absorption, leading to toughening. Limited debonding and improved load transfers reduce the load carried by the fibers, leading to strengthening. The associated tensile stress strain curve exhibits a wide curved domain and the stress at matrix cracking saturation is close to ultimate failure (figure 2).

7.6. Fracture toughness

The CVI SiC/SiC composites develop a network of matrix cracks under load. Density of matrix cracks is enhanced by rather strong interfaces: the crack spacing distance may be as small as 10–20 μm whereas the crack spacing distance is at least 10 times larger in the presence of rather weak interfaces. Matrix cracking is an alternative mechanism of energy dissipation.

A process zone of diffused microcracking within the matrix is generated at notch tip or at the tip of a preexisting main macroscopic crack. Extension of the main macroscopic crack results from the random failures of fiber bundles located within the process zone. Fracture toughness thus measures the ability of the microcracked composite to resist to ultimate failure from the macroscopic crack [20]. Due to the presence of a more or less large process zone of microcracks, at the tip of a jagged macroscopic crack, toughness cannot refer to an equivalent crack length, and conventional concepts of fracture mechanics are not appropriate (stress intensity factor) or cannot be easily determined (strain energy release rate, J integral). Although the validity of stress intensity factor concept to measure fracture toughness is questionable, this is an interesting characteristic for comparing CVI SiC/SiC composites to other materials. Fracture toughness values on the order of 30 MPa $\sqrt{m}$ have been measured on SENB test specimens [2,26]. Strain energy release rate values ranging from 3 to 8 kJ/m^2 have been determined on CVI SiC/SiC composites respectively with weak or strong interfaces [20]. The corresponding values of J-integral ranged from 11 kJ/m^2 (weak interfaces) to 29 kJ/m^2 (strong interfaces) [20]. These values are quite high. The above mentioned stress intensity factors are maintained up to at least 1400°C [2].

7.7. Fatigue behavior

During cyclic fatigue at room temperature, matrix damage appears during the first cycles. Fatigue resistance is governed by damage of fibers and fiber/matrix bonds. Two different fatigue behavior have been observed: after 1000 cycles, either elastic modulus remains constant and the specimen is running out, or modulus decreases until specimen's ultimate failure [27]. Modulus decreases reflect either wear at the fiber/matrix debonded interface [27] or progressive debonding. No failures are generally observed after 10^6 cycles under stresses smaller than 100 MPa, under tension-tension fatigue.

At high temperatures, additional phenomena contribute to the extension of matrix damage and debond. Oxidation and creep are involved.

7.8. Thermal shock

CVI SiC/SiC composites have been tested in thermal shock with excellent result [2,28]. CVI SiC/SiC generally had good strength retention after thermal shock cycles involving heating up to the desired temperature and then cooling down in water at 20°C.

7.9. High temperature behavior

Non-oxide CMCs are susceptible to degradation by oxidation embrittlement that operates at intermediate temperatures, between 500 and 900°C. This is referred to as "pest phenomenon" by a few authors. The matrix cracks created upon loading become pathways for the ingress of oxygen into the material. The pyrocarbon interphase is consumed (oxygen reacts to form gaseous products) and the SiC fibers are degraded by oxidation.

The fatigue resistance of CVI SiC/SiC is governed by damage of fibers and fiber/matrix bonds. Oxidation-induced degradation of fiber/matrix bonds enhances extension of the matrix cracks generated upon the first loading cycle. This contributes to reducing the fatigue lifetime. In order to protect the PyC interphase against oxidation, multiple coating concepts have been explored and multilayered interphases and matrices have been developed [16]. Such multilayered matrices contain phases which produce sealants at high temperatures, causing healing of the cracks and preventing oxygen from reaching the cracks and the interphases [2,17,29]. Lifetime is also improved with oxidation resistant interphases such as BN or multilayers [30].

7.10. Creep behavior

CVI SiC/SiC and CVI SiC/Si-B-C composites exhibit primary creep only, even during long tests (figures 11 and 12) [31]. Creep of ceramic matrix composites involves local stress transfers depending on the respective creep rates of the fiber and the matrix. Such stress transfers may lead to fiber failures or matrix cracking and debonding and sliding at the interfaces. When the matrix is elastic and creep resistant, fiber creep induces stress transfers from the fibers onto the matrix that may cause matrix cracking. This creep induced matrix damage has been observed on CVI SiC/SiC composites [32–34]. In CVI SiC/SiC

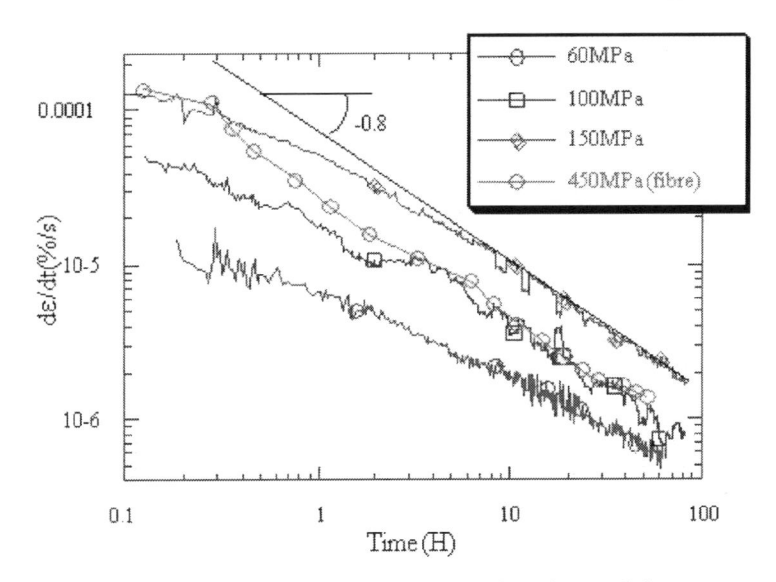

FIGURE 11. Creep rate curves for a damage strain $\varepsilon_0 = 0.8\%$ and for various applied constant stresses for the SiC/Si-B-C composite, and under 450 MPa at 1200°C in argon for a Nicalon NL 202 fiber.

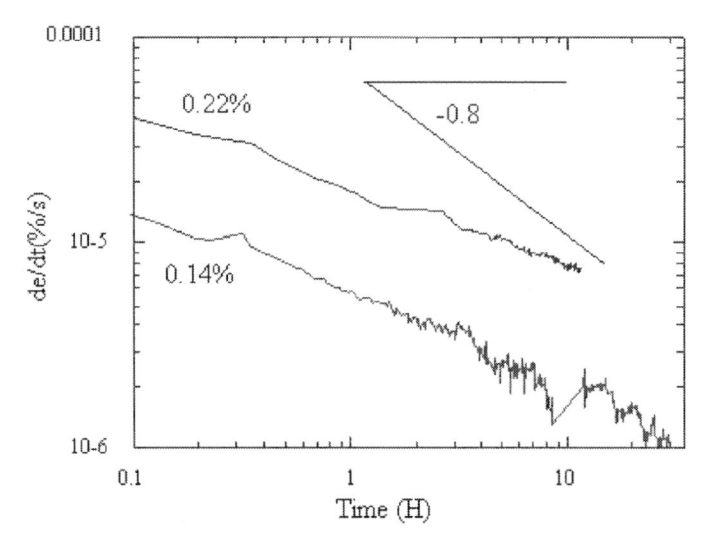

FIGURE 12. Creep rate curves for the SiC/SiC composite under a constant stress of 150 MPa ($\varepsilon_0 = 0.14$ and $\varepsilon_0 = 0.22\%$) at 1200°C.

composites, the SiC matrix is far more creep resistant than the SiC fibers, which creep at 1100°C [32,35]. The creep behavior of CVI SiC/SiC composites with a multilayered matrix (SiC/Si-B-C) is caused by creep of the Nicalon SiC fibers, whatever the extent of initial damage created upon loading (figure 11) [35]. The Si-B-C matrix is less creep resistant and stiff than the SiC matrix.

SUMMARY AND CONCLUDING REMARKS

The CVI process is very flexible and can produce the widest range of chemistries, shapes and dimensions. Properties of CVI SiC/SiC composites have been related to microscopic and macroscopic behaviors. These composites form an interesting class of materials with typical properties that can be improved via microstructure modification or interphase and matrix engineering. In CVI SiC/SiC the matrix is stiffer than the fibers. This characteristic has important consequences on the mechanical behavior and mechanical properties.

The CVI SiC/SiC composites consist of a combination of ceramic materials. However, they are damage tolerant, tough, strong and quite reliable high temperature materials. They can be employed under service conditions involving loads, high temperatures and aggressive environments, in the aerospace as well as in the non aerospace fields.

REFERENCES

1. L. M. Sheppard, Progress in composites processing, *Ceramic Bulletin* **69** 666–673 (1990).
2. J-J. Choury, Thermostructural composite materials in aeronautics and space applications, Proceedings of GIFAS Aeronautical and Space Conference, Bangalore, Delhi, India, 1–18, February 1989.

3. E. Bouillon, G. Habarou, P. Spriet, J. Lecordix, D. Feindel, D. Stetson, G. Ojard, G. Linsey, Characterization and nozzle test experience of a self sealing ceramic matrix composite for gas turbine applications, Proceedings of IGTI/ASME TURBO EXPO Land, Sea and Air 2002, Amsterdam, The Netherlands, June 3–6, 2002.
4. M. Van Roode, Ceramic matrix composite development for combustors for industrial gas turbines, The 27th Annual Cocoa Beach Conference and Exposition on Advanced Ceramics and Composites, January 26–31, 2003, Cocoa Beach, Florida, paper ECD-S1-16-2003.
5. R. H. Jones, SiC/SiC composite for advanced nuclear applications, The 27th Annual Cocoa Beach Conference and Exposition on Advanced Ceramics and Composites, January 26–31, 2003, Cocoa Beach, Florida, paper ECD-S1-18-2003.
6. G. Aiello, CVI SiC/SiC composites as structural components in fusion power reactors ("Utilisation des composites à matrice céramique SiC/SiC comme matériau de structure de composants internes du tore d'un réacteur à fusion"), Ph.D. thesis, University of Evry (France), 2000.
7. P. Fenici, H. W. Scholtz, Advanced low activation materials fiber reinforced ceramic composites, *Journal of Nuclear Materials* 212–215 (1994).
8. R. H. Jones, C. H. Henager, Jr., G. G. Youngblood, H. L. Heinisch, SiC/SiC composites for structural applications in fusion energy systems, *Fusion Technology* **30** (1996).
9. K. K. Chawla, *Ceramic matrix composites*, Chapman & Hall, London (1993).
10. R. Naslain, F. Langlais, CVD-processing of ceramic-ceramic composite materials, in *Tailoring Multiphase and Composite Ceramics*, R. E. Tressler, G. Messing, C. G. Pantano, R. E. Newnham eds., Plenum Publishing Corporation (1986), p.145–164.
11. F. Langlais, Chemical vapor infiltration processing of ceramic matrix composites, in *Comprehensive Composite Materials*, A. Kelly and C. Zweben eds., Elsevier (2000) chap. 4.20, pp. 611–644.
12. F. Christin, R. Naslain, C. Bernard, A thermodynamic and experimental approach to silicon carbide CVD. Application to the CVD-infiltration of porous carbon composites, in Proc. 7th Int. Conf. CVD, T. O. Sedwick and H. Lydin, eds., The Electrochem. Soc., Princeton, (1979) p. 499.
13. F. Christin, R. Naslain, P. Hagenmuller, J-J. Choury, Pièce poreuse carbonée densifiée in-situ par dépôt chimique en phase vapeur de matériaux réfractaires autres que le carbone et procédé de fabrication – French patent, 77/26979, Sept. 1977.
14. L. Heraud, F. Christin, R. Naslain and P. Hagenmuller, Properties and applications of oxidation resistant composite materials obtained by SiC-infiltration, Proc. 8th Int. Conf. CVD, J. M. Blocher et al. eds., The Electrochem. Soc., Pennington (1981), p. 782.
15. E. Fitzer, Chemical vapor deposition of SiC and Si_3N_4, Proc. Int. Symp. on Factors in Densification and Sintering of Oxide and Non Oxide Ceramics, Hakone, Japan (1978), p. 40.
16. F. Lamouroux, R. Pailler, R. Naslain, M. Cataldi, French Patent n° 95 14843 (1995).
17. P. Forio, J. Lamon, Fatigue behavior at high temperatures in air of a 2D SiC/Si-B-C composite with a self-healing multilayered matrix, Ceramic Transactions Vol. 128, American Ceramic Society, Westerville (OH), (2001), pp. 127–141.
18. A. J. Caputo, W. J. Lackey, D. P. Stinton, Development of a new faster process for the fabrication of ceramic fiber-reinforced ceramic composites by chemical vapor infiltration, *Ceramic Engineering and Science Proceedings*, vol. 6, July–August 1984, pp. 694–705.
19. S. Bertrand, Lifetime of SiC/SiC minicomposites with nanometer scale multilayered interphases, Ph.D. Thesis, n° 1927, University of Bordeaux, 28 september 1998.
20. C. Droillard, J. Lamon, Fracture toughness of 2D woven SiC/SiC CVI composites with multilayered interphases, *Journal of the American Ceramic Society* **79** [4] 849–858 (1996).
21. J. Lamon, A micromechanics-based approach to the mechanical behavior of brittle-matrix composites, *Composites Science and Technology* **61** 2259–2272 (2001).
22. V. Calard, J. Lamon, A probabilistic-statistical approach to the ultimate failure of ceramic-matrix composites – Part I: experimental investigation of 2D woven SiC/SiC composites, *Composites Science and Technology* **62** 385–393 (2002).
23. J. C. McNulty, F. W. Zok, Application of weakest-link fracture statistics to fiber-reinforced ceramic-matrix composites, *J. Am. Ceram. Soc.* **80** 1535–1543 (1997).
24. R. Naslain, Fiber-matrix interphases and interfaces in ceramic matrix composites processed by CVI, *Composite Interfaces* **1** 253–258 (1993).
25. F. Rebillat, J. Lamon, A. Guette, The concept of a strong interface applied to SiC/SiC composites with a BN interphase, *Acta Mater.* **48** 4609–4618 (2000).

26. A. Lacombe, J-M. Rougès in AIAA '90, Space program and Technologies Conference '90, Huntsville, AL, September, 1990, Am. Inst. Of Aero. and Astro., Washington, DC, AIAA-90-3837.
27. D. Rouby, P. Reynaud, Fatigue behavior related to interface modification during load cycling in ceramic-matrix fibre composites, *Composites Science and Technology* **48** 109–118 (1993).
28. P. Lamicq, G. A. Bernhart, M. Dauchier, J. Mace, SiC/SiC composite ceramics, *American Ceramic Society Bulletin* **64** 336–338 (1986).
29. P. Carrère, J. Lamon, Fatigue behavior at high temperature in air of a 2D woven SiC/Si-B-C composite with a self healing matrix, *Key Engineering Materials*, Trans. Tech. Publications, Switzerland, **164–165** 357–360 (1999).
30. S. Bertrand, R. Pailler, J. Lamon, Influence of strong fiber–coating interfaces on the mechanical behavior and lifetime of Hi-Nicalon/(PyC-SiC)$_n$/SiC minicomposites, *J. Am. Ceram. Soc.* **84** 787–794 (2001).
31. P. Carrère, J. Lamon, Creep behavior of a SiC/Si-B-C composite with a self-healing multilayered matrix, *J. Eur. Ceram. Soc.* **23** 1105–1114 (2003).
32. F. Abbé, Flexural creep behavior of a 2D SiC/SiC composite. Ph. D. thesis, University of Caen, 1990.
33. J. W. Holmes, J-L. Chermant, Creep behaviour of fiber reinforced ceramic matrix composites, in *High Temperature Ceramic Matrix Composites*, R. Naslain et al. eds., Woodhead, UK (1993), pp. 633–647.
34. A. G. Evans, C. Weber, Creep damage in SiC/SiC composites, *Mater. Sci. Eng.* **A 208** 1–6 (1996).
35. R. Bodet, J. Lamon, N. Jia, R. Tressler, Microstructural stability and creep behavior of Si-C-O (Nicalon) fibers in carbon monoxide and argon environment, *J. Am. Ceram. Soc.* **79** 2673–2686 (1996).
36. J. Lamon, F. Rebillat, A. G. Evans, Microcomposite test procedure for evaluating the interface properties of ceramic matrix composites, *J. Am. Ceram. Soc.*, **78** 401–405 (1995).
37. N. Lissart, J. Lamon, Damage and failure in ceramic matrix minicomposites: experimental study and model, *Acta Metall.* **45** 1025 (1997).
38. F. Rebillat, J. Lamon, R. Naslain, E. Lara-Curzio, M. K. Ferber, T. Besmann, Interfacial bond strength in SiC/C/SiC composite materials as studied by single-fiber push-out tests, *J. Am. Ceram. Soc.*, **81** 965 (1998).
39. F. Rebillat, J. Lamon, R. Naslain, E. Lara-Curzio, M. K. Ferber, T. Besmann, Properties of multilayered interphases in SiC/SiC chemical-vapor-infiltrated composites with "weak" and "strong" interfaces, *J. Am. Ceram. Soc.*, **81** 2315–2326 (1998).

4

SiC/SiC Composites for 1200°C and Above

J.A. DiCarlo, H-M. Yun, G.N. Morscher, and R.T. Bhatt

NASA Glenn Research Center
21000 Brookpark Road
Cleveland, Ohio 44135
Phone 216-433-5514
James.A.DiCarlo@NASA.gov

ABSTRACT

The successful replacement of metal alloys by ceramic matrix composites (CMC) in high-temperature engine components will require the development of constituent materials and processes that can provide CMC systems with enhanced thermal capability along with the key thermostructural properties required for long-term component service. This chapter presents information concerning processes and properties for five silicon carbide (SiC) fiber-reinforced SiC matrix composite systems recently developed by NASA that can operate under mechanical loading and oxidizing conditions for hundreds of hours at 1204, 1315, and 1427°C, temperatures well above current metal capability. This advanced capability stems in large part from specific NASA-developed processes that significantly improve the creep-rupture and environmental resistance of the SiC fiber as well as the thermal conductivity, creep resistance, and intrinsic thermal stability of the SiC matrices.

1. INTRODUCTION

As structural materials for high-temperature components in advanced engines for power and propulsion, fiber-reinforced ceramic matrix composites (CMC) offer a variety of performance advantages over the best metallic alloys with current structural capability to ~1100°C.

These advantages are primarily based on the CMC being capable of displaying higher temperature capability, lower density (∼30–50% metal density), and sufficient toughness for damage tolerance and prevention of catastrophic failure. These properties should in turn result in many important benefits for advanced engines, such as reduced cooling air requirements, simpler component design, reduced weight of support structure, improved fuel efficiency, reduced emissions, longer service life, and higher thrust. However, the successful application of CMC will depend strongly on designing and processing the CMC microstructural constituents so that they can synergistically provide the total CMC system with the key thermostructural properties required by the components. The objectives of this chapter are first to discuss in a general manner these property requirements for typical hot-section engine components, and then to show how in recent years advanced CMC constituent materials and processes have been developed by NASA for fabricating various silicon carbide (SiC) fiber-reinforced SiC-matrix (SiC/SiC) composite systems with increasing temperature capability from ∼1200°C to over 1400°C.

2. APPLICATIONS

Much initial progress in identifying the proper CMC constituent materials and processes to achieve the performance requirements of hot-section components in advanced gas turbine engines was made under the former NASA Enabling Propulsion Materials (EPM) Program, which had as one of its primary goals the development of an advanced CMC combustor liner for a future high speed civil transport (HSCT) [1]. This progress centered on the development of a SiC/SiC CMC system that addresses many of the general performance needs of combustor liners that are required to operate for many hundreds of hours at an upper use temperature of ∼1200°C. In 1999, the NASA EPM Program was terminated due to cancellation of HSCT research. Subsequently the new NASA Ultra Efficient Engine Technologies (UEET) Program was initiated to explore advanced technologies for a variety of low-emission civilian engine systems, including building on NASA EPM success to develop 1315°C SiC/SiC composite systems for potentially hotter components, such as inlet turbine vanes [2]. For hot-section components in space-propulsion engines, the NASA Next Generation Launch Technology (NGLT) program is currently developing SiC/SiC systems with even higher temperature capability since here the primary thermal source is the oxidative combustion of hydrogen fuel rather than jet fuel [3].

Because quantitative property requirements for the various components are engine-specific and often engine company sensitive, the general objective at NASA for all these component development programs has been to develop CMC systems that achieve the upper use temperature goals for hundreds of hours while still displaying, to as high a degree as possible, the key thermo-structural properties needed by a typical hot-section component. To accomplish this objective, a variety of factors had to be optimized within the CMC microstructure, including fiber type, fiber architecture, fiber coating (interphases), and matrix constituents. In order to facilitate this process, NASA selected a short list of first-order property goals that a high-temperature CMC system must display for engine applications. These are listed in the first column of Table 1, which also indicates the technical

TABLE 1. Key CMC Property Goals, Controlling Factors, and Demonstration Tests for CMC Capability

Key CMC Property Goals (*Importance for CMC engine component*)	Key Controlling Constituent Factors	CMC Demonstration Test (*Typically conducted on specimens from thin CMC panels*)
High tensile Proportional Limit Stress (PLS) after CMC processing (*allows high CMC design stress and high environmental resistance*)	Matrix Porosity, Fiber Content	Tensile stress-strain behavior in fiber direction of as-fabricated CMC at 20°C and upper use temperature in air
High Ultimate Tensile Strength (UTS) and strain after CMC processing (*allows good CMC toughness and long life after matrix cracking in aggressive environments*)	Fiber Strength, Fiber Content	Tensile stress-strain behavior in fiber direction of as-fabricated CMC at 20°C and upper use temperature in air
High UTS retention after interphase exposure at intermediate temperatures in wet oxygen (*allows CMC toughness retention when exposed, uncracked or cracked, to combustion gases*)	Fiber Coating Composition	Tensile stress-strain behavior after burner rig exposure near 800°C; Rupture behavior of cracked CMC near 800°C in air
High creep resistance at upper use temperature under high tensile stress (*allows long life, dimensional control, low residual CMC stress*)	Matrix Creep, Fiber Creep	Creep behavior in air at upper use temperature under a constant tensile stress ~60% of matrix cracking stress
Long Rupture life (>500 hours) at upper use temperature under high tensile stress (*allows long-term CMC component service*)	Matrix Rupture, Fiber-Rupture	Rupture life in air at upper use temperature under constant tensile stress ~60% of matrix cracking stress
High thermal conductivity at all service temperatures (*reduces thermal stresses due to thermal gradients and thermal shock*)	Fiber-Coating-Matrix Conductivity, Matrix Porosity	Thermal conductivity from 20°C to upper use temperature

importance of each property goal for a general hot-section CMC component. These goals were specifically selected to address key performance issues for structural CMC in general and for SiC/SiC composites in particular.

Thus for example, it is important that the CMC system display as high a proportional limit stress as possible at all potential service temperatures. This is important for design based on elastic mechanical behavior and for component life since the PLS is closely related to the matrix cracking strength. Therefore, high PLS values will allow the component to carry high combinations of mechanical, thermal, and aerodynamic tensile stresses without cracking. However, unexpectedly higher stress combinations may arise during component service that can locally crack the matrix, thereby causing immediate CMC failure if the fibers are not strong enough or sufficient in volume content to sustain the total stress on the CMC.

In addition, after matrix cracking, CMC failure could occur in undesirably short periods of time if the interphase coating and the fibers were allowed to be degraded by the component service environment as it enters into the CMC through the matrix cracks. For the SiC/SiC components, this attack can be especially serious at intermediate temperatures (~800°C) where oxygen in the engine combustion gases can reach the fibers before being sealed off by slow-growing silica on the matrix crack surfaces. Oxygen primarily attacks the SiC fibers by forming a performance-degrading silica layer on the fiber surfaces, causing fiber-fiber and fiber-matrix bonding. Even a small amount of bonding can eliminate the ability of each fiber to act independently. The detrimental consequence is that if one fiber should fracture, it will cause immediate fracture of other fibers to which it is bonded, thereby causing CMC fracture or rupture at undesirably low stresses and short times.

Also shown in Table 1 are (1) those key constituent factors that CMC theory and practice indicate are the primary elements controlling the various property goals, and (2) the laboratory tests typically employed at NASA to demonstrate CMC system capability for meeting each property goal. For convenience and generality, these tests were usually conducted on specimens machined from thin flat panels fabricated at commercial vendors with the selected CMC constituent materials and processes. NASA's primary objective was not to perform exhaustive testing, but only to use the test results to show directions for advanced CMC systems. As a result, the property databases presented here for the various CMC systems are necessarily limited. It is assumed that by examining the first-level property data, engine designers will be able to select the CMC systems that best meet their component performance requirements, and then initiate with a commercial vendor more extensive efforts for CMC system and component evaluation.

Although not discussed here, NASA has also shown that oxide-based environmental barrier coatings (EBC) need to be applied to the hot surfaces of Si-based (SiC, Si_3N_4, SiC/SiC) components in order to realize long-term service in high temperature combustion environments [4]. Under these wet oxidizing conditions, growing silica on the CMC surface reacts with water to form volatile species, giving rise to paralinear oxidation kinetics and a gas velocity-dependent recession of the Si-based materials [5]. For example, for a lean-burn situation with combustion gases at 10 atm and 90 m/sec velocity, SiC materials are predicted to recess ~250 and 500 μm after 1000 hrs at material temperatures of 2200°F (1204°C) and 2400°F (1315°C), respectively.

3. PROCESSING

Table 2 lists some key constituent material and process data for five SiC fiber-reinforced CMC systems recently developed at NASA. For convenience, these systems have been labeled by the prefix N for NASA, followed by their approximate upper temperature capability in degrees Fahrenheit divided by 100; that is, N22, N24, and N26, with suffix letters A, B, and C to indicate their generation. Also shown in Table 2 are the primary organizations where the different process steps were performed to fabricate each CMC system into a test panel. However, it should be noted that these steps have also been performed at other organizations, resulting in test panels with equivalent properties.

The baseline processing route selected for fabricating the five CMC systems and demonstrating their performance against the Table 1 property goals is shown schematically in

TABLE 2. Key Constituent Material and Process Data for NASA-developed CMC Systems

CMC System	N22	N24-A	N24-B	N24-C	N26-A
Upper Use Temperature	2200°F (1204°C)	2400°F (1315°C)	2400°F (1315°C))	2400°F (1315°C)	2600°F (1427°C)
Fiber Type	Sylramic *(Dow Corning)*	Sylramic-iBN *(Dow Corning + N)*	→	→	→
Interphase Coating	CVI Si-doped BN *(GEPSC)*	→	CVI Si-doped BN outside debond *(GEPSC + N)*	→	→
Matrix	CVI SiC – low content *(GEPSC)*	→	→	CVI SiC – medium content *(GEPSC + N)*	CVI SiC – medium content *(GEPSC)*
	SiC slurry infiltration *(GEPSC)*	→	→		PIP SiC (Polymer Infiltrate and Pyrolysis *(Starfire + N)*
	Silicon melt infiltration *(GEPSC)*	→	→	Silicon melt infiltration *(N)*	

* N = NASA processing.

Fig. 1. As indicated, it involves (1) selecting a high-strength small-diameter SiC fiber type that is commercially available as multi-fiber tows on spools, (2) textile-forming the tows into architectural preforms required by the CMC component or CMC test panel, (3) using conventional chemical vapor infiltration (CVI) methods to deposit thin crack-deflecting interfacial coatings on the fiber surfaces, and (4) over-coating the interfacial coatings with a CVI SiC matrix to a controlled thickness, weight gain, or volume content.

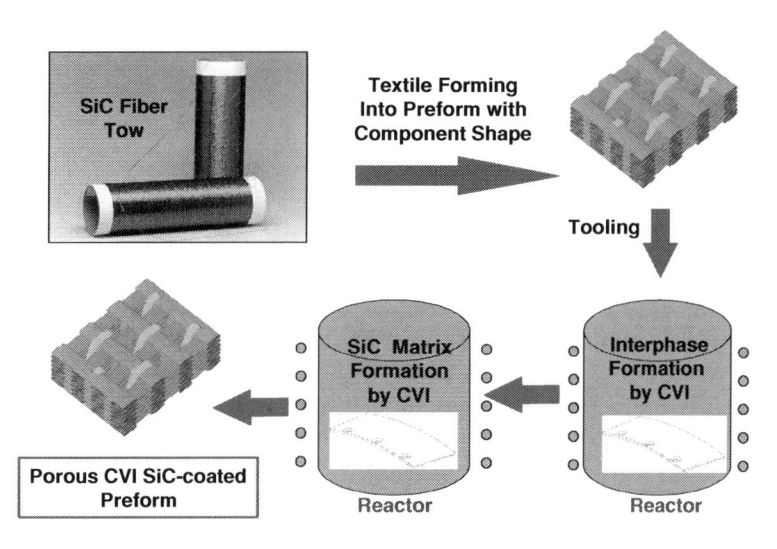

FIGURE 1. Baseline processing route for the NASA CMC systems.

Besides providing environment protection to the interfacial coating, the CVI SiC matrix functions as a strong, creep-resistant, and thermally conductive CMC constituent. However its deposition is not taken to completion because this would trap pores between tows in the fiber architecture, thereby not allowing maximum matrix contribution to the thermal conductivity of the composite system. Depending on the intended CMC upper use temperature, the remaining open porosity in the CVI SiC matrix is then filled with ceramic-based and/or metallic-based materials. Although the filler material in the "hybrid" SiC matrix could serve a variety of functions, its composition and content are typically selected in order to achieve as high a CMC thermal conductivity and as low a CMC porosity as possible.

In general, the baseline processing route of Fig. 1 provides a significant amount of flexibility, particularly in regard to the four key steps involving selection of (1) SiC fiber type, (2) interfacial coating composition, (3) remaining open porosity in the CVI SiC matrix, and (4) infiltration approach(es) to fill this porosity and form the hybrid matrix. As will be discussed in the following, this flexibility was indeed needed in order to optimize the microstructure of the CMC systems in terms of temperature capability and thermostructural properties. Another advantage of this processing route is that it could be used with any textile-formed 2D or 3D architectural preform, which is especially advantageous for fabricating complex-shaped CMC components. In addition, this route can be practiced by any of the many current CMC fabricators who have the capability for interphase and SiC matrix formation by CVI.

3.1. N22 CMC System

During development of the N22 CMC system ($\sim$1997), the only commercially available small-diameter ceramic fiber types with sufficient high-temperature capability were the Sylramic SiC fiber from Dow Corning and the carbon-rich Hi-Nicalon SiC fiber from Nippon Carbon. However, in comparison to the Sylramic fiber, the non-stoichiometry, low process temperature, and carbon-rich surface of the Hi-Nicalon fiber resulted in reduced thermal conductivity, thermal stability, creep resistance, and environmental durability, both for individual fibers [6] and their composites. Thus the selected fiber type for the N22 system was the Sylramic SiC fiber, which is no longer produced by Dow Corning, but by ATK COI Ceramics. This fiber type is fabricated by the polymer route in which precursor fibers based on polycarbosilane are spun into multi-fiber tows and then cured, pyrolyzed, and sintered at high temperature (>1700°C) using boron-containing sintering aids. The sintering process results in very strong fibers (>3 GPa) that are dense, oxygen-free, near stoichiometric, and contain $\sim$1 and $\sim$3 weight % of boron and TiB_2, respectively. To provide enhanced handling and weaving capability, the Sylramic tows were coated by Dow Corning with a polymer-derived Sizing A, which tended to separate contacting fibers in textile-formed preforms. This fiber spreading process typically resulted in better CMC thermostructural properties, such as elastic modulus, ultimate tensile strength (UTS), and rupture strength at intermediate and high temperatures.

For the interfacial coating composition, CVI-produced silicon-doped BN as deposited by GE Power Systems Composites (GEPSC) (formally Honeywell Advanced Composites) was selected because BN not only displays sufficient compliance for matrix crack deflection around the fibers, but also because it is more oxidatively resistant than traditional carbon-based coatings. When doped with silicon, the BN showed little loss in compliance, but an improvement in its resistance to moisture, which is an advantage during removal of the preforms from the CVI BN reactor into ambient air and their subsequent transportation to the CVI SiC matrix reactor.

For the N22 CMC system, remaining open porosity in the CVI SiC matrix was filled by room-temperature infiltration of SiC particulate or slurry casting, followed by the melt-infiltration (MI) of silicon metal near 1400°C. This yielded a final composite with ~2% closed porosity within the fiber tows and ~0% porosity between the tows. The final composite system (often referred to as a slurry-cast MI composite) typically displayed a thermal conductivity about double that of a full CVI SiC composite system in which the CVI matrix process was carried to completion. Also the composite did not require an oxidation-protective over-coating to seal open porosity. Decreasing the porosity of the hybrid matrix also increased the N22 CMC elastic modulus, which in turn contributed to a high proportional limit stress. However, since the filler contained some low-modulus silicon, the modulus increase was not as great as if the filler were completely dense SiC.

3.2. N24-A CMC System

When the property data for the N22 CMC system were analyzed using composite theory and microstructural analysis, certain issues were identified with the fiber, interphase coating, and matrix that indicated that more modifications of these constituents were needed in order to achieve CMC systems for 2400°F (1315°C) components. For example, despite displaying enhanced properties in comparison to other ceramic fiber types, issues related to certain factors existing in the bulk and on the surface of the as-produced Sylramic fiber were found to limit its thermostructural performance, both as individual fibers [6] and as textile-formed architectural preforms for SiC/SiC composites. Most importantly, excess boron in the fiber bulk was typically located on the fiber grain boundaries, thereby inhibiting the fiber from displaying the optimum in creep resistance, rupture resistance, and thermal conductivity associated with its grain size. Also in the presence of oxygen-containing environments during composite fabrication or service, boron on the fiber surface had the potential of promoting detrimental silica-based (SiO_2) glass formation that would bond neighboring fibers together and yield as-produced composites with degraded ultimate tensile strength. This mechanical interaction issue was further compounded by a high surface roughness of the Sylramic fiber, which was related to its grain size and ultimately to its high production temperature [6]. In addition, although Sizing A helped in reducing the roughness issue, during preform warm-up to the temperatures for deposition of the BN interfacial coating, it decomposed and left a continuous carbon-rich char on the fiber surface, which was then trapped under the BN coating. It was found that this continuous carbon layer extended to the composite surface along the 90° tows; so that upon exposure to flowing combustion gas, oxygen was able to enter the CMC microstructure, volatilize the carbon layer throughout the system, and silica-bond the fibers together [7].

For the N24-A system, these issues surrounding the as-produced Sylramic fiber and its sizing were first overcome by using Sylramic fibers with an alternate Sizing B that yielded much less carbon char than Sizing A. In addition, NASA developed a thermal treatment in

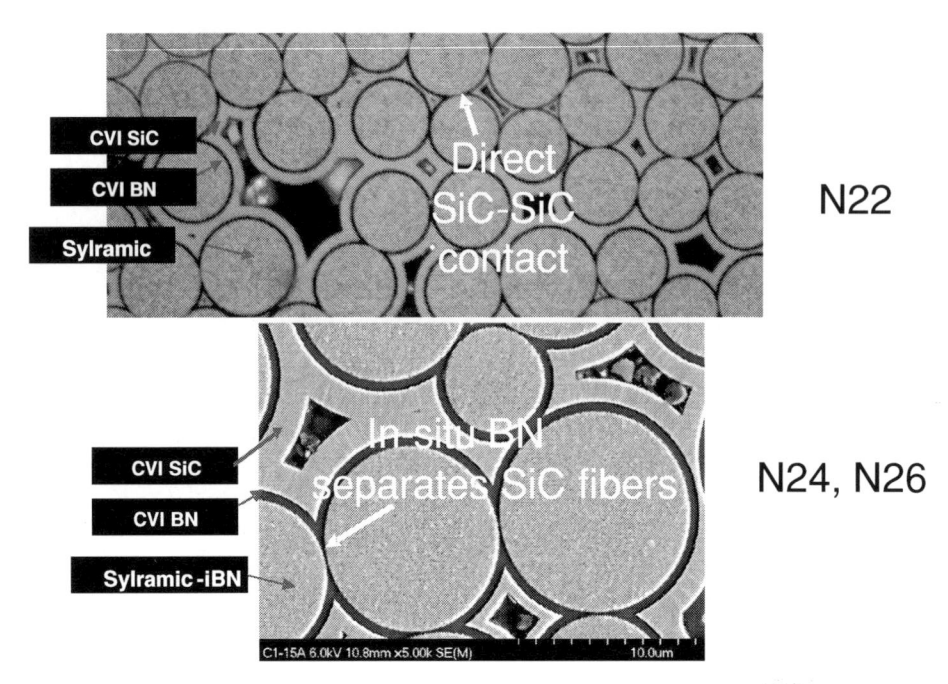

FIGURE 2. SEM micrographs showing that in contrast to the Sylramic N22 CMC system, the in-situ grown BN layer on the Sylramic-iBN fiber is advantageous for physically separating oxidation-prone SiC fiber surfaces within multi-fiber tows in the N24 and N26 systems.

a controlled nitrogen environment that allowed mobile boron sintering aids in the Sylramic fiber to diffuse out of the fiber and to form a thin in-situ grown BN layer on each fiber surface [8]. Removing boron from the fiber bulk significantly improved fiber creep, rupture, and oxidation resistance, while the in-situ BN provided a buffer layer that inhibited detrimental chemical attack from inadvertent oxygen and also reduced detrimental mechanical interactions between contacting fibers. The Scanning Electron Microscopy (SEM) photos in Fig. 2 show that this latter mechanism is indeed a key concern for the as-produced Sylramic fibers in the N22 CMC system, since textile forming of tows typically forces direct contact between neigboring fibers (dark rings are CVI BN interphase coatings). However, as also shown in Fig. 2, this issue is less likely with the Sylramic-iBN fibers in the N24-A system, where direct contact between SiC fiber surfaces cannot be observed due to the thin ($\sim$150 nm) in-situ BN layer that completely surrounds each fiber (dark rings contain both CVI BN and in-situ BN coatings). As will be shown in the Properties section, this in-situ BN layer allows the N24-A CMC system to display enhanced behavior, not only for upper use temperature capability, but also for all key fiber-controlled CMC properties. Thus besides providing SiC fibers with improved performance, another advantage of the NASA fiber thermal treatment was the formation of an in-situ grown BN-based fiber coating, which in effect allows the improved fiber properties to be better retained in textile-formed fiber architectures and CMC structures.

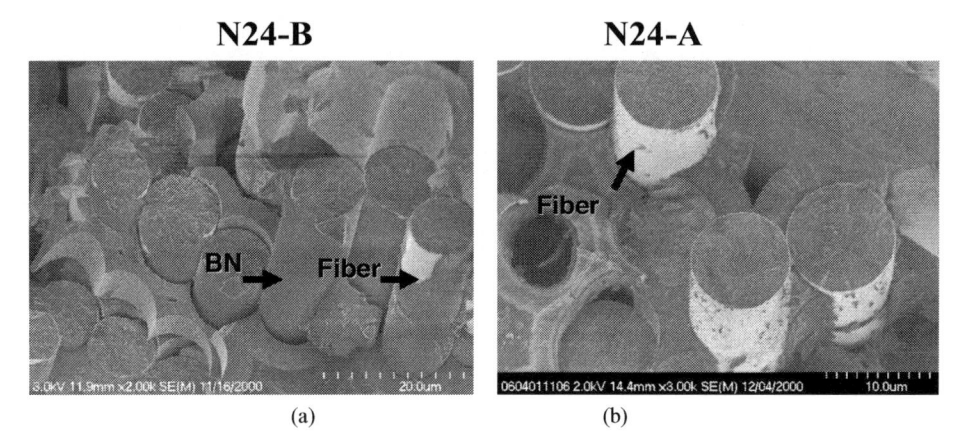

FIGURE 3. SEM micrographs showing (a) outside debonding for the N24-B CMC system and (b) inside debonding for the N24-A system. Note that the BN adheres to the Sylramic-iBN fibers in the outside debonding composites.

3.3. N24-B CMC System

With development of the high performance Sylramic-iBN SiC fiber, the N24-A system showed improvements in practically all the Table 1 properties. Of particular importance in terms of enhanced CMC reliability was an improvement in environmental durability at intermediate temperatures by elimination of the carbon char from Sizing A and by the insertion of an in-situ grown BN surface layer between contacting SiC fibers. As suggested by Fig. 2b, the in-situ grown BN layer delayed SiC-SiC fiber bonding by simply providing an oxidation resistant physical barrier between fibers whenever the fiber tows were exposed to oxygen either during CMC fabrication or during matrix cracking.

Another NASA-developed approach that further improves CMC durability is the basis for the next generation 2400°F CMC system, N24-B. This approach, which is often referred to as "outside debonding", allows the Si-doped BN interphase coating to remain on the fibers during matrix cracking, thereby providing additional environmental protection to the fibers [9]. It is accomplished in a proprietary manner by creating simple constituent and process conditions during composite fabrication that assure that the CVI BN interphase coating is already "outside debonded" from the CVI SiC matrix in the as-fabricated CMC. Even though the interphase coating is attached to the fiber and debonded from the matrix, load transfer between the fibers and matrix is still maintained due to the complex-shaped fiber architectures that allowed the interphases to mechanically slide against the matrix during the application of stress. Figs. 3a and 3b compare, respectively, typical fracture surfaces of the N24-A CMC with an "inside debonding" BN interphase coating and the N24-B with an "outside debonding" BN interphase. In comparison to multi-layer concepts for interphase coatings, this outside debonding approach avoids the fabrication of complex interphase compositions and structures, does not rely on uncertain microstructural conditions for matrix crack deflection outside of the interphase, and provides a more reliable approach for retention of the total interphase on the fiber surface. In addition, this approach also reduces CMC elastic modulus and increases CMC ultimate fracture strain, which can be beneficial, respectively, for reducing thermal stresses within the CMC and increasing its damage tolerance. Thus the N24-B CMC system is more environmentally durable, more damage tolerant, and potentially more resistant to thermal gradients than the N24-A system. However, as a result of "outside debonding, the N24-B system displays a slightly lower thermal conductivity than N24-A.

3.4. N24-C CMC System

Besides improving the fiber and interphase coating for the N24 system, NASA also sought to minimize property limitations associated with the as-produced CVI SiC matrix. These matrix limitations relate to the fact that for best infiltration into the textile-formed fiber tows, the CVI SiC matrix deposition process is typically conducted at temperatures below 1100°C, which is below the application temperatures where the CMC systems will have their greatest practical benefits. Under these processing conditions, although the SiC matrix is fairly dense, its microstructure contains meta-stable atomic defects and is non-stoichiometric due to a small amount of excess silicon. These defects can exist at the matrix grain boundaries where they act as scatterers for thermal phonons and enhance matrix creep by grain-boundary sliding, thereby allowing the matrix and CMC to display less than optimal thermal conductivity and creep-resistance. NASA determined that by using thermal treatments above 1600°C on the CVI SiC-coated preforms prior to the N24 process steps of slurry casting and melt infiltration, excess silicon and process-related defects in the CVI SiC matrix could be removed, yielding the N24-C CMC system with significantly improved creep resistance and thermal conductivity [10]. To maximize these benefits as well as the CMC life, the CVI SiC content of the N24-C preform is increased over that typically used in the N24-A and N24-B systems, but only to the point of avoiding significant trapped porosity. The N24-C preform is then thermally treated in argon, and remaining porosity is filled by the melt infiltration of silicon. As indicated in Table 2, the increase in CVI SiC content for the N24-C CMC system also allows elimination of the slurry infiltration step and its associated production costs.

As shown in Fig. 4, the Sylramic-iBN fiber is the only high-strength SiC fiber type that allows preforms with low CVI SiC content (~20 vol.%) to survive thermal exposure in argon above 1600°C with no loss in ultimate tensile strength of the preform. For the other fiber types in Fig. 4, part of their strength loss could be intrinsic caused by non-optimized microstructures or lower production temperatures [6], and part could be extrinsic caused by fiber attack from inadvertent excess oxygen in the CVI BN fiber coating or from the excess silicon that diffuses out of the CVI SiC matrix during thermal treatment. Since it is well known that BN produced at high temperatures is resistant to oxygen and molten silicon, the better performance of the Sylramic-iBN fiber in Fig. 4 might be expected given its higher production temperature and its in-situ grown BN layer. However, at the higher CVI SiC content used for the N24-C CMC system, thermal exposure resulted in a CMC strength loss of up to 30%. This effect was presumably due to an increase in excess silicon with increasing CVI SiC content, and thus more likelihood of silicon attack of the Sylramic-iBN fiber through the in-situ grown BN layer. In addition, during the high-temperature preform treatment, the BN interphase coating, which was deposited below 1000°C (1830°F), densified and contracted between the fiber and matrix. This in turn caused an automatic "outside debonding" of the BN interphase coating from the matrix, as evidenced by a

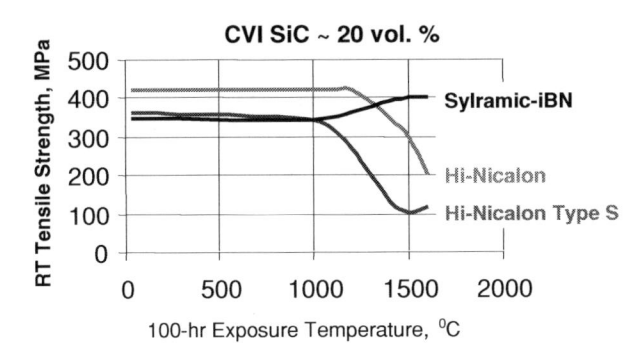

FIGURE 4. Average ultimate tensile strength (UTS) retained at room temperature for various 2D preform panels fabricated by the baseline processing route of Fig. 1 and subjected to 100-hr exposures at high temperatures in argon.

reduced CMC modulus of the final CMC. Thus the N24-C CMC system is more creep resistant, more intrinsically stable, and more thermally conductive than the N24-B system, but at the expense of a lower ultimate strength.

3.5. N26-A CMC System

As described above, small quantities of excess silicon (<1 vol.%) in the as-produced CVI SiC matrix are able to cause a limited attack of the SiC fibers during the high-temperature processing of the N24-C system. The degrading effect of excess silicon from the melt infiltration step was also observed for the N24 CMC systems when they were evaluated for long-term use at potential service temperatures of 2600°F (1427°C) and higher. However in this case, the attack was much more serious and caused primarily by the much greater silicon content (~15 vol.%) introduced by the MI step. The SEM micrograph of Fig. 5 shows an example of this attack for a N24-A CMC system that was thermally exposed at 2552°F (1400°C) in argon for 100 hours under zero stress [11]. It can be seen that silicon from the melt-infiltration step was able to diffuse through the grain boundaries of the CVI SiC matrix, attack the BN coatings and SiC fibers, and severely degrade composite strength. The higher CVI SiC content of the N24-C CMC system helps to slow down this attack [11], but not enough to use this system for over 1000 hours at 2400°F or over 100 hours at 2600°F [12].

To address the temperature issues related to excess silicon, all the same constituents in the N24-C system are used for potential N26 CMC generations, but remaining open pores in the CVI SiC matrix are filled by silicon-free ceramics, rather than by melt infiltration of silicon. In particular, for the N26-A CMC system, a SiC-yielding polymer from Starfire Inc. [13] is infiltrated into the matrix porosity at room temperature and then pyrolyzed at temperatures up to 2912°F (1600°C). This polymer infiltration and pyrolysis (PIP) process was repeated a few times until composite porosity was reduced to ~14 vol.%. At this point, the total CMC system is then thermally treated at NASA to improve its thermal conductivity and creep-resistance. Thus although more porous than the other CMC systems, the N26-A system has no free silicon in the matrix, thereby allowing long-time structural use at 2600°F

FIGURE 5. SEM micrograph showing degradation of the SiC fiber and BN interphase coating after 100-hr thermal exposure of the N24-A CMC system in argon under zero stress at 2552°F (1400°C). Degradation is due to diffusion through the CVI SiC matrix of the silicon from the melt-infiltration step of the N22 and N24 CMC systems.

(1427°C) and higher. Research is on-going at NASA to develop further generations of the N26-A system where porosity is significantly reduced, so that CMC thermal conductivity can be enhanced beyond the matrix annealing step.

4. PROPERTIES

Based on the property goals of Table 1 and the constituent-process data of Table 2, this section presents key physical and mechanical property data for the five high-temperature CMC systems described above. Composite materials for obtaining these data were fabricated as follows. Sylramic SiC fiber tows were woven into 2D orthogonal fabric with equal tow ends per cm in the 0° (warp) and 90° (fill) directions. The fabric was cut into eight 150 × 230 mm pieces or plies, which were then stacked in a balanced manner to form a thin rectangular-shaped architectural preform. For the N22 system, the stacked plies were then provided directly to GEPSC for BN interphase and partial CVI SiC matrix processing as described in Fig. 1. For the N24 and N26 systems, the stacked plies were converted to Sylramic-iBN fibers at NASA prior to sending to GEPSC. After final matrix processing as described above for the five systems, the stacked fabric preforms were converted into flat CMC test panels with approximate dimensions of 2 × 150 × 230 mm and with total fiber content between 32 and 40 vol.%. It should be noted that a variety of potential CMC engine components, such as combustor liners and shrouds [1, 14], are also being made by the laminate or fabric stacking approach, so that the panel property data presented here can be directly used to understand the performance of these components.

For standard measurements of stress-strain and creep-rupture behavior (ASTM C 1337-96), 150 mm long dog-boned shaped tensile specimens with gauge sections of ~10 mm

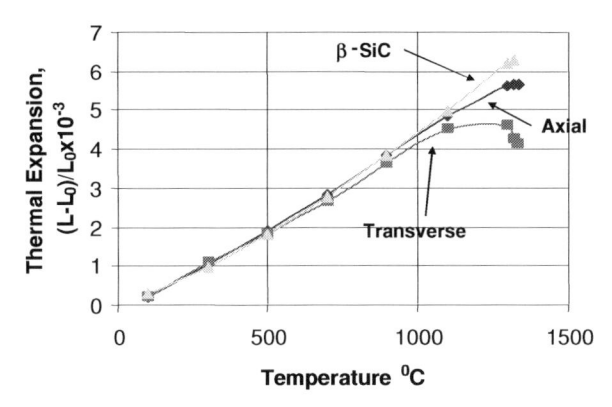

FIGURE 6. Typical linear thermal expansion curves in the axial and transverse directions for the N22 and N24 CMC systems panels fabricated with the silicon melt-infiltration step. Also shown for comparison is the best fit curve for monolithic SiC with the β-phase [16].

width × 25 mm length were machined from each CMC system panel. Each tensile specimen had half of the total fiber fraction aligned along the 150 mm test direction. Composite thermal conductivity was calculated from specimen density data and temperature-dependent data for specimen thermal diffusivity and specific heat. Thermal diffusivity was measured by the thermal flash method in the transverse or through-thickness direction of the test panels. On an absolute basis, the transverse conductivity for a given CMC system was always less than its axial or in-plane conductivity due to the low-conductivity BN interphase.

Table 3 lists some of the important physical properties of the five NASA-developed CMC systems as determined from the as-fabricated test panels. The system densities could be modeled by a simple rule-of-mixtures based on constituent volume fractions and densities [**15**]. All the panels contained an average of ~36 vol.% fiber and ~8 vol.% Si-doped BN, which is thick enough for tough composite behavior, but thin enough for CVI SiC matrix penetration into the fiber tows. As the temperature capability requirement increased, the CVI SiC matrix content increased from ~23 vol.% to ~35 vol.% to take advantage of the silicon-protective nature, creep-resistance, and thermal conductivity of the CVI SiC composition. For the first four systems, the MI silicon content remained at ~15 vol.% to derive the benefits of its thermal conductivity and pore-filling capability, but was removed completely from the N26-A system to allow the long lives needed at the higher upper use temperature. Since all the systems contained up to ~70 vol.% of high-density β-phase SiC grains in the fiber and matrix, their linear thermal expansion behavior was found to be essentially equivalent to that of dense monolithic β-phase SiC. Fig. 6 compares the axial and transverse expansion for the N22 and N24 panels against that of β-phase SiC, which can be described fairly accurately by the following relationship [**16**]:

$$\lambda\,(\%) = T\,[2.62 \times 10^{-4}] + T^2\,[2.314 \times 10^{-7}] - T^3\,[0.518 \times 10^{-10}] \quad (1)$$

where λ is linear thermal expansion strain and T is in degrees Celsius. The CMC showed a slight deviation from β-phase SiC above 1300°C, which is a characteristic of SiC materials that contain a small amount of free silicon [**17**]. It should be noted that Eq. 1 is not only

TABLE 3. Typical Physical Properties for NASA-developed CMC Systems as 2D Test Panels

Property \ CMC System	N22	N24-A	N24-B	N24-C	N26-A
Upper Use Temperature	2200°F (1204°C)	2400°F (1315°C)	2400°F (1315°C)	2400°F (1315°C)	2600°F (1427°C)
Density, gm/cc	2.85	2.85	2.85	2.76	2.52
Constituent Content, ~Vol.%					
SiC Fiber (3.05 gm/cc)	36	36	36	36	36
Si-BN Interphase (1.5 gm/cc)	8	8	8	8	8
CVI SiC (3.2 gm/cc)	23	23	23	35	35
SiC particulate (3.2 gm/cc)	18	18	18	0	6
Silicon (2.35 gm/cc)	13	13	13	18	0
Porosity	2	2	2	2	14
Thermal Linear Expansion, %	$T [2.62 \times 10^{-4}] + T^2 [2.314 \times 10^{-7}] - T^3 [0.518 \times 10^{-10}]$ ($T = °C$)				
Transverse Thermal Conductivity, W/m.C					
204°C (400°F)	24	30	27	41	26
1204°C (2200°F)	15	14	10	17	10

useful for determining CMC linear expansion strain, but also for deriving CMC coefficients of thermal expansion (CTE) at various temperatures.

Typical transverse thermal conductivity data for the CMC panels are listed in Table 3 at 204 and 1204°C, and displayed as a function of temperature in Fig. 7. As with monolithic SiC materials [**18**], the conductivity values increase up to ~200°C and then decrease monotonically with temperature with an approximate inverse temperature dependence. Although both are technically important, transverse rather than in-plane conductivity is generally the first-level property of concern for design of CMC components. This is the case because the components will typically be cooled through their thin wall sections and because the transverse conductivity is typically lower in value, and thus a conservative estimate of material capability. It is affected strongly by matrix composition and porosity, which not only includes the small open and closed pores formed during the matrix infiltration processing steps, but also the long linear pores which are effectively created by poorly conductive interphase coatings on the SiC fibers. The interphase conductivity is dependent both on its composition and its effective contact with the fiber and matrix. Thus wide variation in CMC conductivity is to be expected and has been observed, but qualitative trends are observable. For example, Fig. 7 shows that by changing from the Hi-Nicalon Type S fiber to the Sylramic fiber (N22 system) to the Sylramic-iBN fiber (N24-A system), the transverse conductivities of the Si-containing CMC systems were measurably increased. This reflects on the intrinsic conductivity values for the various fiber types [**6**], and on the ability of the Si-doped BN interphase coating to provide some degree of thermal transport into the fibers. However, as indicated in Table 3, for the N24-B CMC system with an "outside debonding" interphase, a penalty is paid in thermal conductivity due to reduced interphase-matrix contact. This loss was more than recovered when the CVI SiC matrix of the Si-containing N24-C system was annealed. However, another conductivity penalty was taken for the N26-A system when silicon was eliminated and replaced by a hybrid SiC matrix with high porosity.

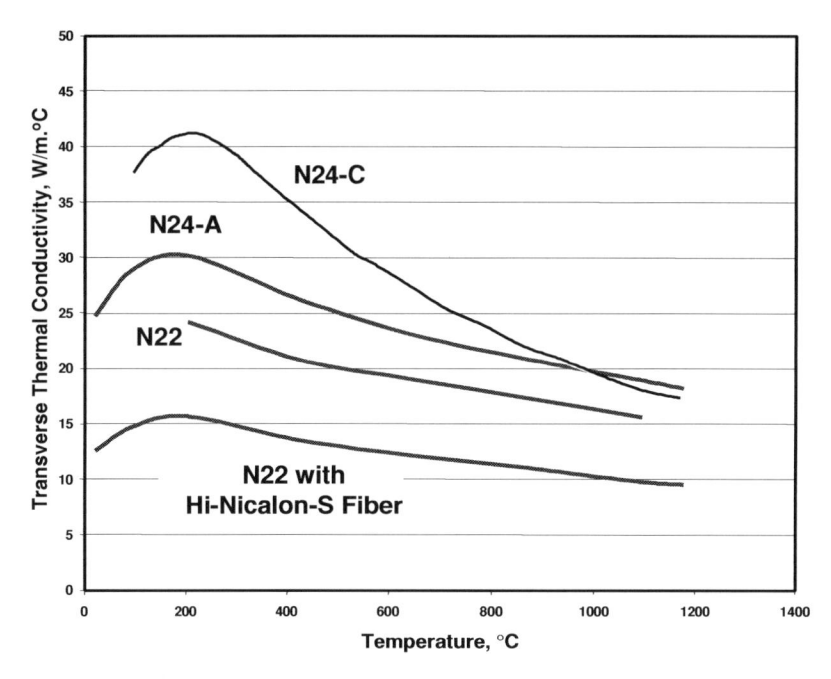

FIGURE 7. Typical transverse thermal conductivity curves for thin panels with CMC systems N22, N24-A, and N24-C. Effect of fiber conductivity is shown by curve for the N22 system with the lower conductivity Hi-Nicalon Type-S fiber type.

Regarding CMC mechanical properties, typical in-plane tensile stress-strain curves at room temperature are shown in Figs. 8 and 9 for the various CMC panels in their as-fabricated condition. Fig. 8 compares the first three systems with inside debonding (N22, N24-A) and with outside debonding (N24-B); whereas Fig. 9 compares the last two systems with annealed matrices and a higher degree of outside debonding (N24-C, N26-A). Using similar curves for multiple test specimens both at room temperature and at their upper use temperature (UUT), Table 4 lists average data for such key mechanical properties as initial elastic modulus, proportional limit stress, ultimate tensile strength, and ultimate tensile strain. It should be noted that since the panels had a range of total fiber content from 32 to 40 vol.%, the data in Table 4 and in the figures, unless otherwise noted, represent average results measured or estimated for a fiber content of ~36 vol.% total fiber content (~18 vol.% in the tensile test direction). It should also be noted that elastic moduli of the various systems are reduced at their UUT, but are not presented in Table 4 because due to CMC creep they are dependent on testing stress-rate.

As predicted by CMC theory, some of the Table 4 properties will change significantly with fiber content and test direction. Nevertheless, for the selected fiber content and panel test conditions, qualitative trends can be observed in the as-fabricated CMC mechanical properties at room temperature. For example, regarding elastic modulus, variations between CMC systems can be correlated to the content of the high modulus CVI SiC matrix and to the existence of matrix porosity and/or an outside debonding interphase, both of which

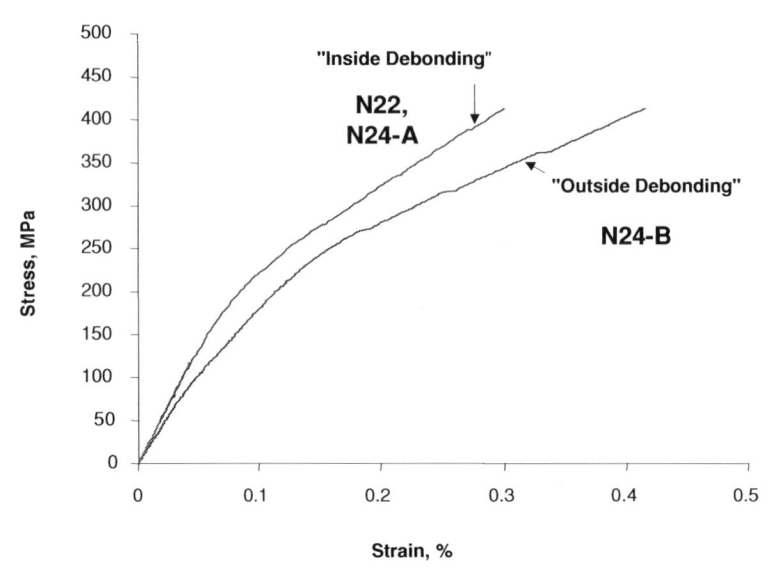

FIGURE 8. Typical room-temperature tensile stress-strain curves for the inside-debonding N22 and N24-A CMC systems, and the outside-debonding N24-B CMC system (total fiber content ~40 vol.%).

will decrease matrix modulus. The role of outside debonding in reducing matrix modulus is related to the fact that the 90° tows are load-bearing elements within the matrix for CMC loads applied in the 0° test direction. Therefore, when contact between the interphase and matrix is reduced by debonding, loads on 90° tows are reduced, effectively introducing new porosity into the matrix. Regarding CMC proportional limit stress, although the underlying mechanisms are quite complex [19], a very crude rule-of-thumb for 2D 0/90 panels is that the first deviation from stress-strain linearity occurs at approximately ~0.07%, so that CMC PLS values are approximately proportional to CMC moduli. Regarding CMC ultimate strength, which can be considered to be directly controlled by the in-situ fiber strength, all the Sylramic-iBN fiber systems that were not annealed displayed the highest value of ~450 MPa.

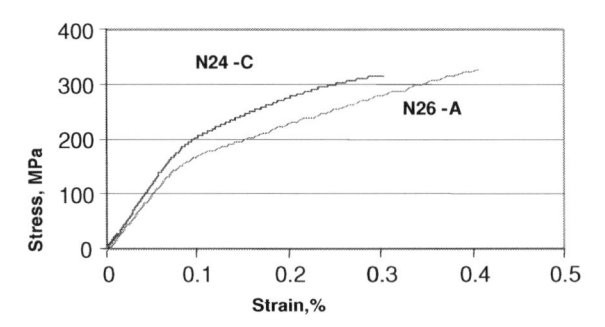

FIGURE 9. Typical room-temperature tensile stress-strain curves for the annealed Sylramic-iBN CMC systems N24-C and N26-A (total fiber content ~34 vol.%).

TABLE 4. Average Mechanical Properties for NASA-developed CMC Systems (2D 0/90 Test Panels with ~36 vol.% Total Fiber Content

Property* \ CMC System	N22	N24-A	N24-B	N24-C	N26-A
Upper Use Temperature	2200°F (1204°C)		2400°F (1315°C)		2600°F (1427°C)
AS-FABRICATED AT 20°C					
Initial Elastic Modulus, GPa	250	250	210	220	200
Proportional Limit Stress, MPa	180	180	170	160	130
Ultimate Tensile Strength, MPa	400	450	450	310	330
Ultimate Tensile Strain	~0.35%	~0.50%	~0.55%	~0.30%	~0.40%
Interfacial Shear Strength, MPa	~70	~70	~7	~7	~7
AT 800°C					
Ultimate Tensile Strength Retention after 100-hr burner rig	60%	100%	100%		
Rupture Strength, 100 hr, air, MPa	200	200	240		
AT OR NEAR UPPER USE TEMPERATURE (Test Temperature)	2200°F (1204°C)		2400°F (1315°C)		2642°F (1450°C)
Proportional Limit Stress, MPa	170	170	160	150	120
Ultimate Tensile Strength, MPa	320	380	380	260	280
Creep Strain, 103 MPa, 500 hr, air	~0.4%	~0.4%	~0.4%	0.2%	
Creep Strain, 69 MPa, 500 hr, air		0.15%	0.15%	0.12%	~0.3%
Rupture Life, 103 MPa, air	~500 hrs	~500 hrs	~500 hrs	>1000 hrs	
Rupture Life, 69 MPa, air					>300 hrs

* Mechanical properties measured in-plane in the 0° direction with a directional fiber content of 18 vol.%.

As discussed earlier, this effect can in large part be attributed to the protective nature of the in-situ grown BN layer on this fiber type since the as-produced strength of the Sylramic-iBN fiber is lower than its precursor Sylramic fiber [6], which was used in the N22 panel. However, at the higher CVI SiC matrix content of the N24-C and N26-A systems, there was enough excess silicon in the CVI product to cause fiber attack and UTS degradation even for the Sylramic-iBN fiber. Finally, regarding CMC ultimate strain, the systems with the highest UTS and greatest degree of outside debonding displayed the highest values. The debonding effect can be related to a lower interfacial shear strength (also listed in Table 4), which allowed greater crack openings and greater strain to develop in the CMC. For a given UTS, outside debonding can increase failure strain by as much as 0.2% over inside debonding.

As indicated in Table 1, a key property need for any SiC fiber-reinforced CMC system is the ability to retain to as high a degree as possible its as-fabricated properties under intermediate and high temperature service conditions. At intermediate temperatures, the key issues are oxygen and moisture attack of the interphase coating that can occur on surface-exposed 90° tows even under zero stress, or within the CMC by environmental ingress along random matrix cracks that are being held open by applied CMC loads. To evaluate the CMC systems against the first interphase issue, tensile test specimens from the various panels were exposed to combustion gases in a low-pressure burner rig with the gauge length held at a constant temperature near 800°C (1472°F) for ~100 hours [7]. Stress-strain curves at room temperature for three CMC systems after the burner rig exposure are shown in Fig. 10. Significant degradation in UTS was seen for the N22 systems with the Sylramic and Hi-Nicalon Type-S fibers; while excellent UTS retention was seen for the Sylramic-iBN N24-B system. As described earlier, this degradation can be attributed to the removal

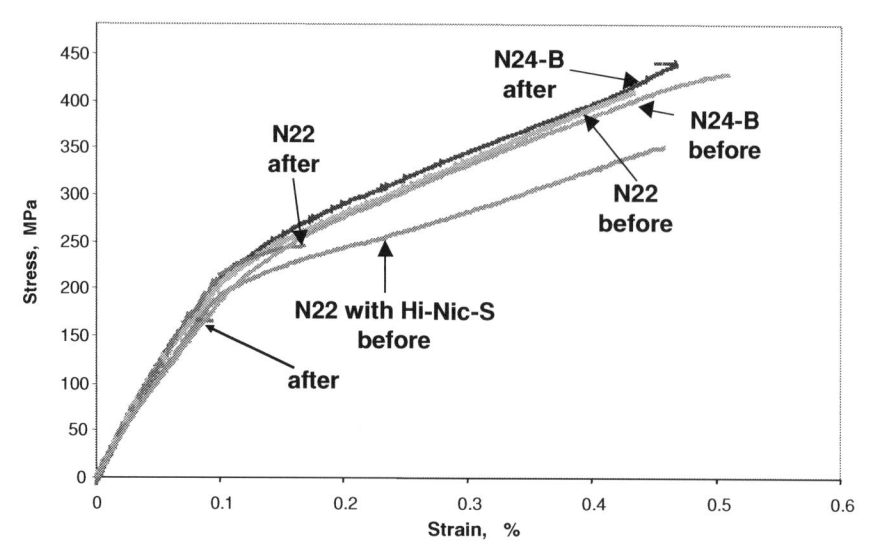

FIGURE 10. Typical room-temperature stress-strain curves for the N22 CMC system with Sylramic and Hi-Nicalon Type-S fibers, and for the N24-B CMC system with Sylramic-iBN fibers before and after combustion gas exposure of the systems in a low-pressure burner rig at ~800°C for ~100 hours. The fibers in the N22 systems each had carbon on their surfaces before BN interphase deposition.

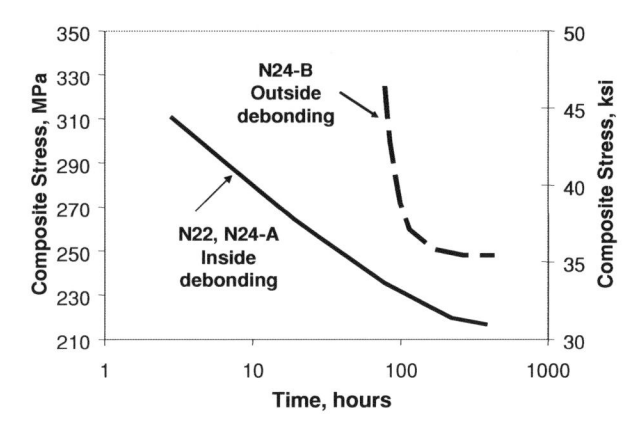

FIGURE 11. Best-fit stress-rupture curves in air at 1500°F (815°C) comparing the inside-debonding N22 and N24-A CMC systems and the outside-debonding N24-B CMC system (total fiber content ~40 vol.%).

of free carbon on the Sylramic fiber surface (char from Sizing A) and on the as-produced Hi-Nicalon Type-S surface, and to the subsequent silica bonding of contacting fibers. The fact that Fig. 10 shows no loss in UTS for the Sylramic-iBN CMC can be attributed in part to the in-situ BN layer, which minimizes direct contact between SiC fibers, but primarily to the non-detection of detrimental carbon at the interface between the fiber and BN. Thus inadvertent free carbon on fiber surfaces must be avoided for CMC components within combustion environments.

To evaluate the CMC systems against the second interphase issue, tensile test specimens from the various CMC panels were subjected to stress-rupture testing in ambient air at 1500°F (815°C) at stresses above matrix cracking. Final CMC rupture typically occurred when sufficient silica bonding existed between crack-bridging fibers so that when one fiber ruptured due to time-dependent slow crack growth, it caused fracture of its neighbors and the CMC [20]. Fig. 11 compares the rupture behavior for the inside debonding systems N22 and N24-A against the first outside debonding system N24-B. Clearly the outside debonding system shows enhanced behavior in that the CMC life is longer for a given applied stress or that a higher stress can be applied to the CMC for a given CMC rupture life. As suggested by Fig. 3 and discussed in the N24-B processing section, this enhanced behavior is due to the Si-doped BN interphase remaining on the Sylramic-iBN fiber surface after matrix cracking, thereby providing additional environmental protection. As shown by Fig. 11, this extra protection enhances CMC life at short times by one or two orders of magnitude, but then eventually loses its ability at longer times. Thus if unpredictable matrix cracking should occur at intermediate and high temperatures, the NASA-developed outside debonding systems will increase CMC life and reliability.

The most important goal for each NASA-developed CMC system was to be able to operate under potential component stress levels for long time at its selected upper use temperature (UUT). To evaluate this capability, tensile test specimens from the various CMC panels were subject to creep-rupture testing in ambient air at their goal UUT and at stresses of ~60% of their room-temperature cracking stress. The primary performance objective was to demonstrate greater than 500-hour life without specimen rupture. Since high-temperature rupture of an initially uncracked CMC is typically controlled by CMC

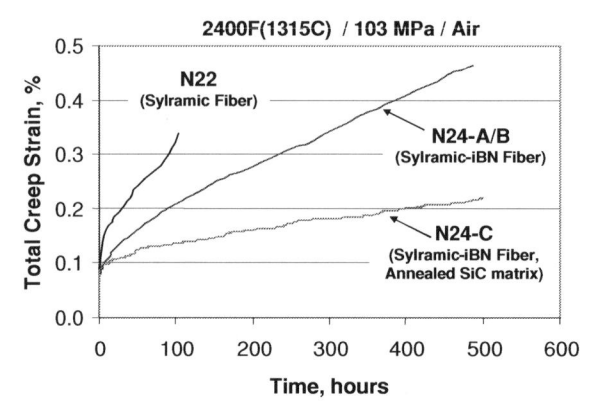

FIGURE 12. Typical total creep strain versus time behavior at 2400°F (1315°C) in air at an applied stress of 103 MPa for the N22 and the N24 CMC systems.

intrinsic creep behavior, another good performance indicator is the ability of a CMC system to display high creep resistance at its UUT and to maintain its creep strain well below a characteristic rupture strain within the 500-hour life goal [**21**]. For example, for test times near 500 hour, the NASA CMC systems typically display a creep-rupture strain of ~0.4% between 2200 and 2600°F, so that staying at or below this strain level after 500-hour testing should demonstrate the desired performance. Thus, as indicated in Table 4, the N22 system with the Sylramic fiber was able to reach this strain level within 500 hours for an applied stress of 103 MPa at its UUT of 2200°F. However, Fig. 12 shows that under the same stress at 2400°F, the N22 system had a life of only 100 hours, whereas the N24-A system with the Sylramic-iBN fiber had a projected life of greater than 500 hours based on its steady state creep rate. Fig. 12 also shows how annealing of the CVI SiC matrix provided the N24-C system with a projected rupture life of over 1000 hours at 2400°F. However, due to the silicon melt infiltration step for all the N24 CMC systems, these time and temperature condition appear to be the upper limit use conditions based on displaying intrinsic thermal stability under zero stress [**12**]. Finally, Fig. 13 shows the difference in creep-rupture

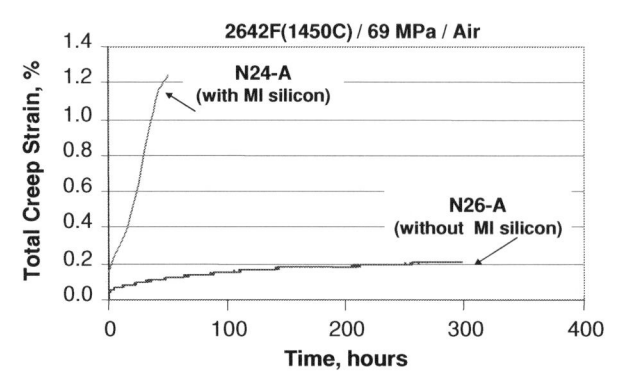

FIGURE 13. Typical total creep strain versus time behavior at 2642°F (1450°C) in air at an applied stress of 69 MPa for the N24-A and N26-A CMC systems.

behavior for a 69 MPa stress at 2642°F between the silicon-containing N24-A system and the silicon-free N26-A system which displayed over 300 hour life capability under these conditions.

5. SUMMARY AND CONCLUSIONS

By working closely with CMC vendors, NASA has been able to identify advanced constituent materials and processes that have yielded various SiC fiber-reinforced SiC matrix composite systems for high temperature structural applications in general and for hot-section engine components in particular. Based on a performance goal of a 500-hour service in air at a stress level of ~60% that for matrix cracking, these systems have demonstrated upper use temperatures of 2200°F (1204°C), 2400°F (1315°F), and 2600°F (1427°C), which are well above the current capability of the best metallic alloys. This progression in temperature capability is related first to the development of the Sylramic-iBN SiC fiber, then to the development of an annealing step for an improved CVI SiC matrix in the baseline NASA processing route, and finally to the elimination of the silicon melt infiltration step for filling pores in the SiC matrix. Advances in CMC environmental durability were also made at NASA by use of in-situ grown BN fiber surface layers, CVI-deposited Si-doped BN interphase coatings, carbon-free interfaces between the fiber and BN, and processing approaches that cause the CVI BN interphase to remain on the fiber and "outside" debond from the matrix during CMC cracking, thereby allowing the interphase coating to provide extra environmental protection to the crack-bridging fibers.

Although the processing information and property results presented here are limited, they should be sufficient for component designers to select the CMC systems and processes that best meet their component performance requirements, and then to initiate more extensive efforts for system scale-up and component evaluation with a variety of commercial CMC vendors. Although not yet demonstrated, it might be expected that as long as environmental effects do not control the performance of these systems, a time-temperature relationship should exist between projected component service life and CMC system upper use temperature capability. For example, based on typical atomic diffusion in SiC materials, rupture life for each CMC system should change by one order of magnitude for every ~180°F (100°C) difference between the component's upper service temperature and the system's upper use temperature capability [12]. Thus the 2600°F system should be able to perform for ~5000 hours at 2420°F, but only ~50 hours at 2780°F. Stress effects will also have an influence CMC component life, but these should be evaluated for sub-elements of the components on a case-by-case basis.

6. ACKNOWLEDGEMENTS

The authors gratefully acknowledge (1) the funding support of the NASA Enabling Propulsion Materials (EPM) program, the NASA Ultra Efficient Engine Technology (UEET) program, and the NASA Glenn Director's Discretionary Fund; (2) the professional support of L. Thomas-Ogbuji and J. Hurst; and (3) the technical support of R. Phillips, R. Babuder, and R. Angus.

7. REFERENCES

1. D. Brewer, HSR/EPM Combustor Materials Development Program, *Materials Science and Engineering*, **A261**, 284–291 (1999).
2. NASA Ultra Efficient Engine Technology (UEET) Program, http://www.grc.nasa.gov/WWW/RT2000/2000/2100shaw.html
3. NASA Next Generation Launch Technology (NGLT) Program, http://www1.msfc.nasa.gov/NEWSROOM/background/facts/ngltfacts.pdf
4. K.N. Lee, D.S. Fox, R.C. Robinson, and N.P. Bansal, Environmental Barrier Coatings for Silicon-Based Ceramics, in *High Temperature Ceramic Matrix Composites*, W. Krenkel, R. Naslain, and H. Schneider, Eds, Wiley-VCH, Weinheim, Germany, (2001), pp. 224–229.
5. J.L. Smialek, R.C. Robinson, E.J. Opila, D.S. Fox, and N.S. Jacobson, SiC and Si_3N_4 Recession Due to SiO_2 Scale Volatility under Combustor Conditions, *Adv. Composite Mater*, **8** [1], 33–45 (1999).
6. J.A. DiCarlo and H-M. Yun, Non-Oxide (Silicon Carbide) Ceramic Fibers, in *Handbook of Ceramic Composites,* N.P. Bansal, Ed., Kluwer Academic Publishers, Boston, MA, 2004, pp. 33–52.
7. L. Thomas-Ogbuji, A Pervasive Mode of Oxidation Degradation in a SiC/SiC Composite, *J. Am. Ceram. Soc.*, **81** [11], 2777–2784 (1998).
8. H-M. Yun, and J.A. DiCarlo, Comparison of the Tensile, Creep, and Rupture Strength Properties of Stoichiometric SiC Fibers, *Cer. Eng. Sci. Proc.*, **20** [3], 259–272 (1999).
9. G.N. Morscher, H-M. Yun, J.A. DiCarlo, and L. Thomas-Ogbuji, Effect of a BN Interphase that Debonds Between the Interphase and the Matrix in SiC/SiC Composites, *J. Am. Ceram. Soc.*, **87**, 104–112 (2004).
10. R.T. Bhatt, NASA Glenn *Research and Technology* 2003, NASA/TM—2004–212729, 20–21 (2004).
11. R.T. Bhatt, T.R. McCue, and J.A. DiCarlo, Thermal Stability of Melt Infiltrated SiC/SiC Composites, *Cer. Eng. Sci. Proc.*, **24** [4B], (2003), 295–300.
12. J.A. DiCarlo, R.T. Bhatt, and T.R. McCue, Modeling the Thermostructural Stability of Melt Infiltrated SiC/SiC Composites, *Cer. Eng. Sci. Proc.,* **24** [4B], (2003), 465–470.
13. Starfire Systems, http://www.starfiresystems.com/
14. G.S. Corman and K.L. Luthra, Silicon Melt Infiltrated Ceramic Composites (HiPerComp^{TM}), in *Handbook of Ceramic Composites*, N.P. Bansal, Ed., Kluwer Academic Publishers, Boston, MA, 2004, pp. 99–115.
15. S.K. Mital, P.L.N. Murthy, and J.A. DiCarlo, Characterizing the Properties of a Woven SiC/SiC Composite, *Journal of Advanced Materials*, **35** [1], 52–60 (2003).
16. Z. Li and R.C. Bradt, Thermal Expansion of the Cubic (3C) Polytype of SiC, *J. Mater. Sci.* **21** (1986), 4366–68.
17. J.A. DiCarlo, Creep of Chemically Vapour Deposited SiC Fibers, *J. Mater. Sci.* **21** (1986), 217–224.
18. *Thermophysical Properties of Matter, Thermal Conductivity, Nonmetallic Solids*, Vol. 2, Y.S. Touloukia el al., Eds., Plenum, New York, (1970), p. 6a.
19. G.N. Morscher, Stress-Dependent Matrix Cracking in 2D Woven SiC-fiber Reinforced Melt-Infiltrated SiC Matrix Composites, *Comp. Sci. Tech.*, in print.
20. G.N. Morscher and J.D. Cawley, Intermediate Temperature Strength Degradation in SiC/SiC Composites, *J. European Ceram. Soc.*, **22**, 2777–2787 (2002).
21. J.A. DiCarlo, H.M. Yun, and J.B. Hurst, Fracture Mechanisms for SiC Fibers and SiC/SiC Composites Under Stress-Rupture Conditions at High Temperatures, *Applied Mathematics and Computation*, **152**, 473–481(2004).

5

Silicon Melt Infiltrated Ceramic Composites (HiPerComp™)

G.S. Corman and K.L. Luthra
GE Global Research Center
Niskayuna, NY 12309
cormang@research.ge.com

ABSTRACT

Silicon melt infiltrated, SiC-based ceramic matrix composites (MI-CMCs) have been developed for use in gas turbine engines. These materials are particularly suited to use in gas turbines due to their low porosity, high thermal conductivity, low thermal expansion, high toughness and high matrix cracking stress. Several variations of the overall fabrication process for these materials are possible, but this paper will focus on "prepreg" and "slurry cast" MI-CMCs with particular reference to applications in power generation gas turbines. These composites have recently been commercialized under the name of HiPerComp™.

1. INTRODUCTION

GE, at its Global Research Center, GE Energy (through Power Systems Composites), and GE Transportation (through Aircraft Engines) divisions, has been actively involved for over 15 years in the development of silicon melt infiltrated ceramic matrix composites (MI-CMCs). These composites offer a unique combination of properties such as high temperature strength, creep resistance, low porosity, low density, high thermal conductivity and low thermal expansion that make them particularly suited for use in gas turbine engines. The history of the development of MI-CMCs has been recently described [1]. An early version of MI-CMC, which was initially developed using monofilament SiC fibers, went by the name of "Toughened Silcomp" [2]. Extensive development of successive generations of

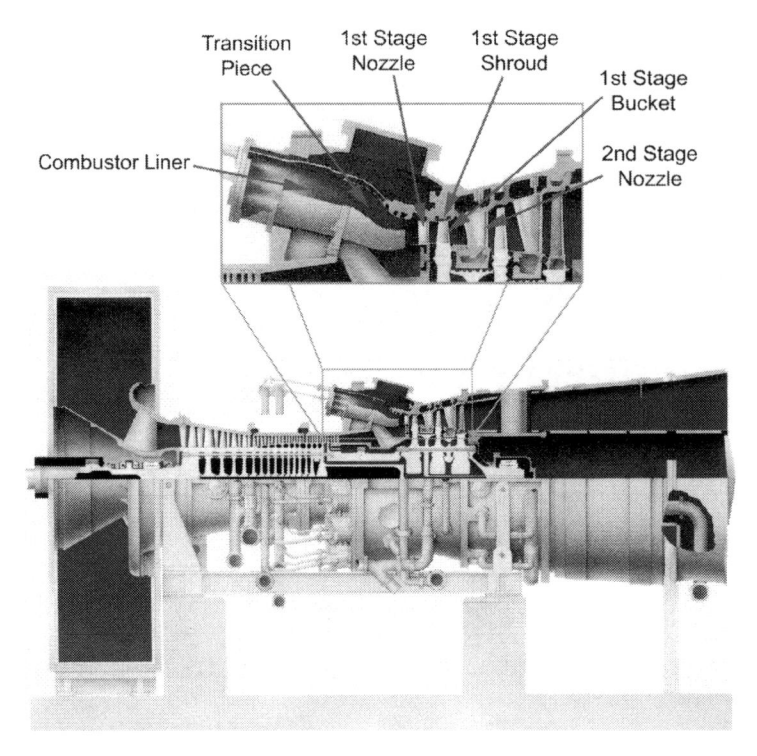

FIGURE 1. Cut-away diagram of a GE 7FA-class (170 MW) industrial gas turbine engine showing the location of several hot gas path components that are candidates for use of HyPerComp™ composites.

this material through the 1990's, including the introduction of fine, tow-based SiC fibers, has lead to the successful rig and engine testing of MI-CMC shroud and combustor liner components [3–7]. Moreover, these materials are now commercially available, under the name HiPerComp™, from GE Power Systems Composites.

2. APPLICATIONS

HiPerComp™ CMCs have been successfully used in several turbine hardware demonstration programs at GE and at Solar Turbines, Inc. At GE 1st and 2nd stage turbine shrouds and combustor liners have been designed, built and tested using high-pressure combustion rig apparatus or in actual turbine engines [4–7]. Most of these tests have been focused on land-based engine applications, primarily for power generation. Figure 1 shows cross section of a 7F class GE gas turbine (simple cycle power of 160–170 MW and combined cycle power output of ~280 MW) and indicates the hot gas path parts for which HiPerComp™ CMCs are being considered. Current programs are developing these applications further, and also investigating the use of HiPerComp™ CMCs as low-pressure turbine blades and vanes in jet engines. Turbine demonstration programs at Solar have focused on application of these materials for combustor liners of small industrial gas turbines [8].

As with any new material, the current costs of HiPerComp™ CMCs are relatively high because of the expensive raw materials, primarily the fiber, and poor economies of scale. At least initially, gas turbine engines, both for power generation and aerospace applications, are one of the few applications where the high cost can be justified in terms of performance benefits [9]. However, as production volumes gradually increase and the prices of the composites come down, various other industrial applications, which can take advantage of the high thermal conductivity, damage tolerance and wear resistance of these CMCs, are expected to develop [10]. Heat exchangers potentially fall into this category of applications.

3. PROCESSING

For this article, the term "melt infiltrated ceramic matrix composite" (MI-CMC) will refer only to continuous fiber composites whose matrices are formed by molten silicon (or silicon alloy) infiltration into a porous SiC- and/or C-containing preform. GE holds numerous patents on the composition and fabrication of these materials, only a few of which are listed in reference 10. Such composites can be made from a variety of constituents and processes. A detailed description of the material variations and processes is also given in reference 1, so only an abbreviated description will be given here.

3.1. MI-CMC Constituents

The three major constituents of any continuous fiber ceramic matrix composite are the reinforcing fibers, the matrix and a fiber-matrix interphase, usually included as a coating on the fibers. HiPerComp™ composites can be processed with various monofilament and multifilament fibers, such as the SCS family of monofilament SiC from Specialty Materials, Inc.; CG-Nicalon™ and Hi-Nicalon Type S™ from Nippon Carbon Company; Tyranno ZE™, Tyranno ZMI™ and Tyranno SA™ from Ube Industries; and Sylramic™ fiber from COI Ceramics. However, the composites described in this paper all utilize Hi-Nicalon™ SiC fiber from Nippon Carbon Company. A companion paper, in this book, by Jim DiCarlo [11] from NASA gives the properties of slurry cast composites reinforced with Sylramic and Sylramic-iBN fibers.

The matrix of HiPerComp™ composite consists primarily of SiC and Si. The coefficient of thermal expansion (CTE) of Si and SiC differ only slightly, such that relatively wide ranges of matrix composition are possible while still providing an adequate match to the CTE of the SiC-based fibers. Typically the HiPerComp™ composites have residual silicon levels of 5 to 15 vol%.

As with most other ceramic composite systems, a coating is applied to the fibers to serve as the fiber-matrix interphase. Such a fiber coating is necessary to prevent chemical attack of the fibers during processing and to provide for a weak mechanical interface between the fiber and matrix for enhanced toughness and graceful failure. The coating most widely used is boron nitride (BN) applied by chemical vapor deposition (CVD). Unfortunately, BN coatings can be degraded by contact with molten silicon during the melt infiltration process, and consequently an over-coating of SiC or Si_3N_4 is commonly used to protect the BN layer.

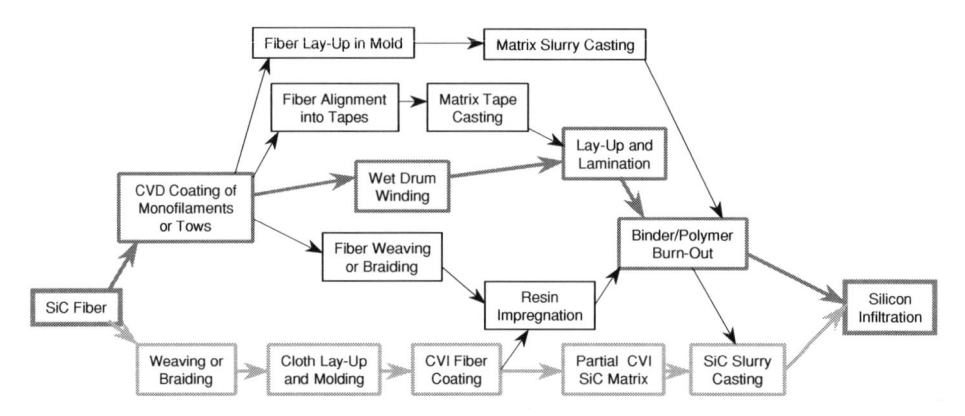

FIGURE 2. Schematic representation of the various methods for producing HiPerComp™ MI CMCs. The bold black highlighted path represents the prepreg process and the highlighted bold gray path represents the slurry cast process. (Reprinted by permission of ASME Press.)

For improved oxidation resistance of the interface, Si-doped BN is often substituted for all or most of the BN interface layer [12].

3.2. The Fabrication Process

The overall preparation of MI composites involves three main steps: application of the fiber coatings, forming a shaped, porous preform containing the fibers, and final densification via silicon infiltration. Preform fabrication itself typically involves many process steps, many of which are similar to those used for fabrication of polymeric, carbon-carbon or other types of ceramic composites. There are a variety of specific process steps that can be used in the production of the porous preform, and a variety of ways in which they can be arranged in the overall process. GE practices two main overall processes for fabricating HiPerComp™ composites, called the "prepreg" and "slurry cast" processes. The prepreg process was developed by GE with funding support primarily from the United States Department of Energy. The slurry cast process was developed by a team of companies, including GE, United Technologies, Carborundum, Goodrich Aerospace and NASA, under the NASA-sponsored High Speed Civil Transport program. Figure 2 shows aschematic of the various processing routes that can be used to fabricate MI CMCs, with the prepreg and slurry cast approaches highlighted.

In the prepreg process the SiC fiber tows are first coated with the BN-based fiber-matrix interphase and protective Si_3N_4 overcoating using a proprietary CVD process, and then formed into unidirectional prepreg sheets via wet drum winding. These prepreg sheets are then laid-up and laminated to form the composite preform, similar to the lay-up of carbon fiber/epoxy prepreg. The matrix of the composite at this stage consists of powders (commercially available SiC and C) in a mixed polymer binder. During the binder burn-out/pyrolysis step part of the polymer is converted to carbon, which maintains the preform shape. Final densification is done by infiltrating this porous preform with a molten silicon alloy, during which the silicon reacts with free C present in the preform to form a continuous SiC phase in the matrix. The resulting composite consists of nominally 20–25 vol% fiber,

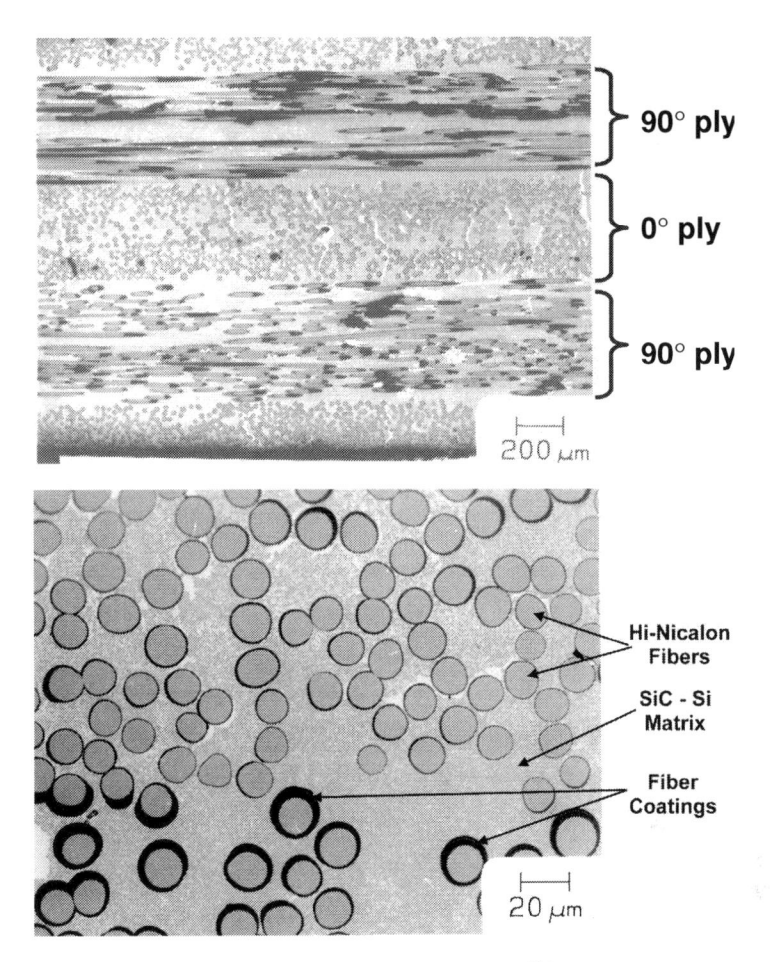

FIGURE 3. Typical microstructure of prepreg HiPerComp™ MI-CMC material.

8–10 vol% fiber coatings, 70–63 vol% matrix and <2 vol% porosity. Typical microstructures of prepreg HiPerComp™ are shown in Figure 3.

In the slurry cast process the fibers are first woven into 2D cloth and laid-up to form the preform, or they can be 3D woven or braided directly into the preform shape. Carbon tooling is then typically used during the first chemical vapor infiltration (CVI) step, where the BN-based interphase coating is applied. The parts can then be removed from the tooling, and are put through a second CVI step to deposit a SiC overcoat. Deposition of the SiC layer is halted while the preform is still porous and a SiC slurry, which may or may not contain an additional source of carbon, is slip-cast into the preform. Again, final densification is accomplished via melt infiltration of a silicon alloy that fills in the remaining porosity between the SiC particles introduced from the slurry. The finished slurry cast composite consists of nominally 35 vol% fiber, 6 vol% fiber coating, 25 vol% CVD SiC, 16 vol% SiC particulate from the slurry, 12 vol% Si alloy, and 6 vol% porosity. Most of the porosity of the

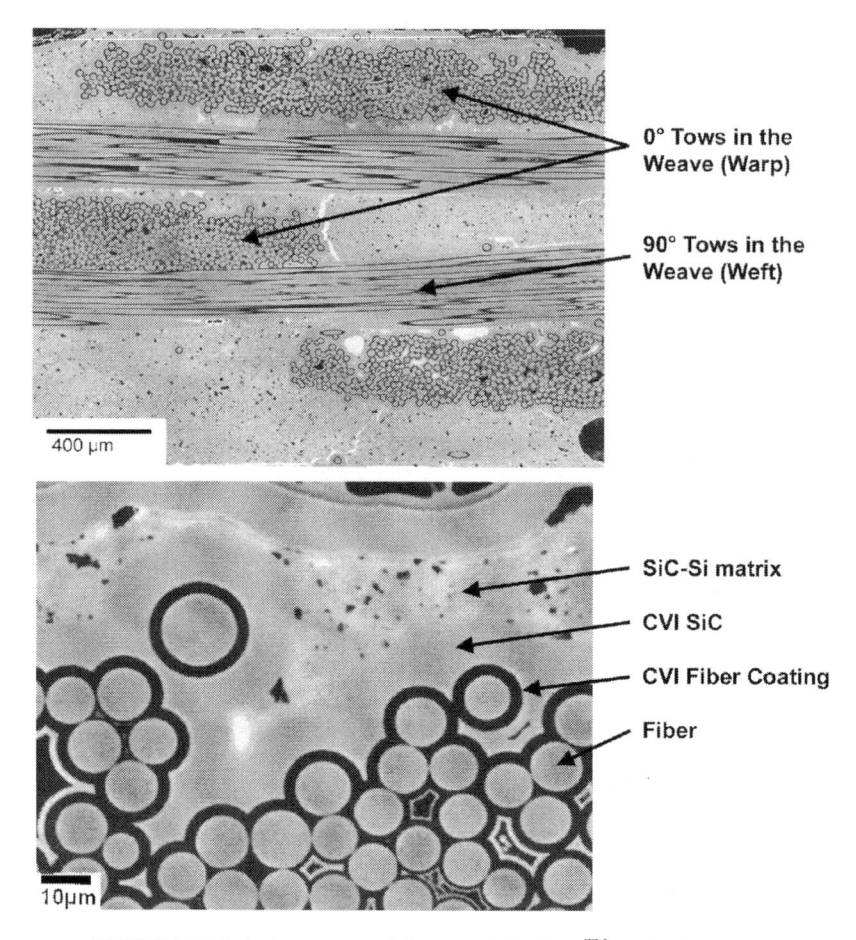

FIGURE 4. Typical microstructure of slurry cast HiPerComp™ MI-CMC material.

slurry cast material is a result of "canning" of the fiber tows during the CVI steps, thereby making it inaccessible to slurry and Si in the later process steps. Typical microstructures of slurry cast HiPerComp™ are shown in Figure 4.

As will be discussed below, the prepreg process results in composites having comparable tensile strength values as the slurry cast process, but with a much lower fiber loading. This represents a potential cost advantage for the prepreg composites, as the fibers are, by far, the highest cost raw material. There are two reasons why the prepreg system is able to better utilize the strength of the fibers. First, in the prepreg system the fibers are all individually coated during the tow-coating step, and thus all of the fibers act independently. In the slurry cast system the fibers are held tight by the fiber weave during CVD coating, and thus the majority of the fibers are in contact with other fibers. Consequently the fibers tend to fail as groups during the fracture process rather than as individual fibers. The second reason is that the fiber weave used in the slurry cast material necessarily introduces some misalignment of the

SILICON MELT INFILTRATED CERAMIC COMPOSITES

fibers with the plane of the composite, thereby loading the crack-bridging fibers in both tension and shear during the fracture process. The prepreg process utilizes unidirectional tapes wherein the fibers can be more precisely aligned in the plane of the composite, and thereby minimize shear loading of the fibers during fracture.

4. MATERIAL PROPERTIES

Thermal and mechanical properties of HiPerComp™ composites have been measured at various stages during their development. Consequently the data presented here differs somewhat from previously published data summaries [1] since the material system and processing techniques have continued to evolve with time. The data presented here for Prepreg HiPerComp™ was mostly measured as part of the DOE-sponsored Continuous Fiber Ceramic Composites (CFCC) program on material fabricated in 2000 and 2001. Data for the Slurry Cast HiPerComp™ material was measured in 2003. Please note that this data is presented for informative purposes only, and should not be construed as, or used for, engineering specifications.

Unless otherwise noted, all data presented for Prepreg HiPerComp™ is for 8-ply laminates, with a balanced [0-90–90-0]s stacking of uniaxial plies, and nominally 22–25% by volume of Hi-Nicalon™ fibers. Data for Slurry Cast HiPerComp™ is for 8-ply laminates made with 0-90, 8 harness satin weave cloth and a nominal volume fraction of Hi-Nicalon™ fiber of 33–38%. Measurement of in-plane properties were generally done in one of the primary fiber directions.

4.1. Thermal and Physical Properties

Some important thermal properties of Prepreg and Slurry Cast HiPerComp™, measured at room temperature and at 1200°C, are listed in Table 1. Overall, the thermal properties

TABLE 1. Thermal Properties of HiPerComp™

Property	Units	Prepreg		Slurry Cast	
		25°C	1200°C	25°C	1200°C
Density	g/cm^3	2.80	(2.76)	2.70	(2.66)
Heat Capacity	J/g-K	0.71	1.14	0.70	(1.2)
Thermal Diffusivity:					
in-plane*	mm^2/s	16.1	4.7	–	–
thru-thickness		11.7	3.8	13.7	4.2
Thermal Conductivity:					
in-plane*	W/m-K	33.8	14.7	(30.8)	(14.8)
thru-thickness		24.7	11.7	22.5	11.8
Average CTE					
in plane* (25–600°C)	$\times 10^{-6}$/°C	3.57		3.74	
in plane* (25–1200°C)		3.73		4.34	
thru thickness (25–600°C)		4.07		3.21	
thru thickness (25–1200°C)		4.15		3.12	

* Measured in one of the primary fiber directions. Property values in parentheses are engineering estimates.

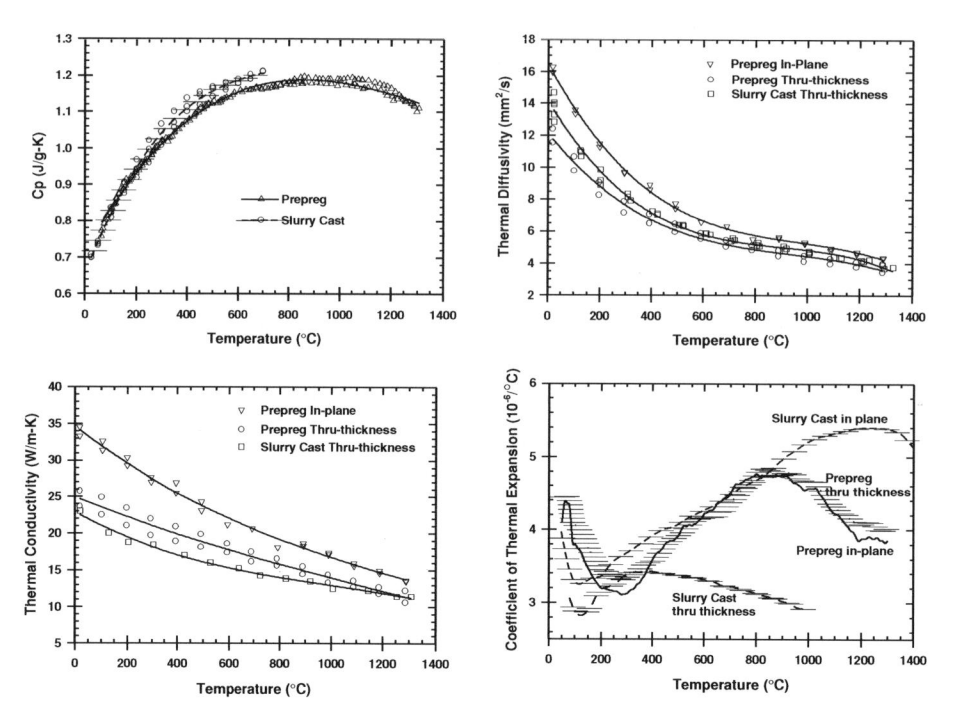

FIGURE 5. Thermal properties of prepreg and slurry cast HiPerComp™ composites with Hi-Nicalon fibers.

for the two types of HiPerComp™ are very similar, with the Prepreg variety having a slightly higher thermal conductivity than the Slurry Cast variety. This difference is due to the lower porosity level and lower content of fibers and BN fiber coatings, which have lower thermal conductivity than the matrix, in the Prepreg material. Figure 5 shows the temperature dependence of these properties in more detail.

4.2. Elastic Properties

The elastic properties of HiPerComp™ are summarized in Table 2. The elastic properties tend to be nearly transversely isotropic for balanced 0-90 lay-ups, and are identical in tension and compression as long as one stays below the matrix cracking (~15 MPa below the proportional limit) stress. These composites are somewhat unusual compared to more traditional polymer matrix or carbon matrix composites in that the matrix modulus (350 to 380 GPa) is actually greater than the modulus of the reinforcing fibers (typically 260–290 GPa). The fact that the composite modulus is lower than that of either the fiber or the matrix reflects the influence of the highly compliant fiber coatings, particularly in the thru-thickness (E_{33}) direction. The low modulus of the slurry cast composite compared to the prepreg composite is also caused by the higher volume fractions of both fiber and porosity in the slurry cast material.

TABLE 2. Summary of the Elastic Properties of HiPerComp™ Composites.

Property	Units	Prepreg		Slurry Cast	
		25°C	1200°C	25°C	1200°C
Tensile Moduli					
$E_{11} = E_{22}$	GPa	285	243	196	144
E_{33}		201	(171)		
Poisson's Ratio:					
ν_{12}	–	0.12	(0.12)		
$\nu_{13} = \nu_{23}$		0.24	(0.24)		
Shear Moduli:					
G_{12}	GPa	14.5	(12.3)		
$G_{13} = G_{23}$		12.8	(10.9)		

Property values in parentheses are engineering estimates.

4.3. Fracture Strength

The in-plane tensile fracture response of HiPerComp™ materials are typically characterized by a stress-strain curve as shown in Figure 6 when measured in a simple displacement-controlled method. In general, the curve can be divided into four sections (shown by the dotted lines in Figure 6), with the first section representing the simple linear elastic loading of the composite. As the stress increases multiple matrix cracks are generated in the composite and the fibers bridging the cracks are shear debonded from the matrix

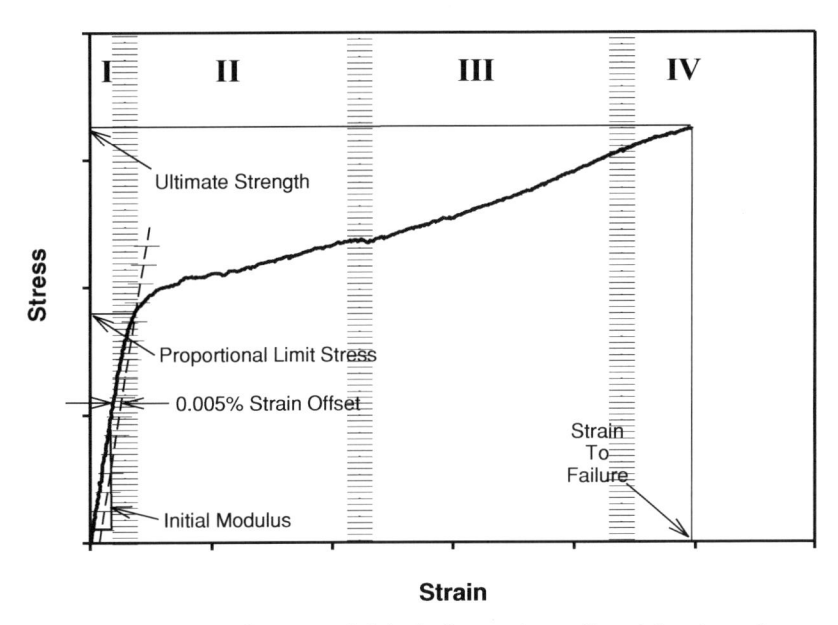

FIGURE 6. Typical in-plane tensile stress-strain behavior for a continuous fiber reinforced ceramic composite.

TABLE 3. Summary of the Fast Fracture Strength Properties of HiPerComp™ Composites.

Property	Units	Prepreg[†]		Slurry Cast[†]	
		25°C	1200°C	25°C	1200°C
In-plane Proportional Limit Stress*	Mpa	167	165	120	130
In-plane Ultimate Tensile Strength*	Mpa	321	224[#]	358	271
In-plane Strain to Failure*	%	0.89	0.31[#]	0.74	0.52
In-plane Compressive Strength*	GPa	1.19	>0.70		
Interlaminar Shear Strength	MPa	135	124		
Interlaminar Tensile Strength	MPa	39.5	–		

[†] Prepreg containing 22–24 vol% Hi-Nicalon™ fiber; Slurry Cast with 35–38 vol% Hi-Nicalon™ fiber
* Measured in one of the primary fiber directions.
[#] Higher values are obtained at higher strain rates.

(Section II). It is this matrix cracking and fiber-matrix debonding that are primarily responsible for the "pseudoplastic" behavior and high toughness of these types of CMCs. Eventually the crack density saturates and the bridging fibers become completely debonded in the regions between the matrix cracks, such that continued loading (Section III) represents the elastic response of the bridging fibers. At still higher loads fiber fracture begins to occur (section IV), which often leads to a slight leveling of the stress-strain curve just before ultimate failure.

Some of the important fracture parameters that are determined from the stress-strain curves are also illustrated in Figure 6, and include the initial modulus, proportional limit stress, ultimate strength and strain to failure. It is often very difficult to determine unambiguously the stress at which the first matrix crack occurs, so the proportional limit stress, i.e. the stress at which the strain deviates by 0.005% from linear loading, is more commonly used to characterize this important stress level. All of the in-plane fracture data reported here was measured at initial strain rates (prior to matrix cracking) between 3×10^{-5} and 10^{-4} s^{-1} unless otherwise noted.

Table 3 summarizes the in-plane tensile fracture behavior of prepreg and slurry cast HiPerComp™ composites at 25° and 1200°C. The temperature dependence of the fracture parameters is shown in more detail in Figure 7. It should be noted that there is a substantial effect of strain rate on the measured strengths and strain to failure of the prepreg composite material at high temperature. When measured at a higher strain rate of 0.002 s^{-1} strain to failure values increase by as much as 2X over the values at the normal strain rate. Strain rate effects are not as pronounced for the slurry cast material. The cause of these varying strain rate effects is under investigation.

4.4. Thermal Stability

Long-term thermal stability is a key requirement for application of CMCs to turbine engines, particularly for power generation turbines where expected component lifetimes range from 24,000 to 48,000 hours (3 to 6 years) of operation. Thermal stability of Prepreg HiPerComp™ composites has been assessed by performing thermal exposure tests for times up to 4000 hours in air, steam and high pressure combustion gas environments. The residual room temperature tensile properties of Prepreg HiPerComp™ following exposure in air at

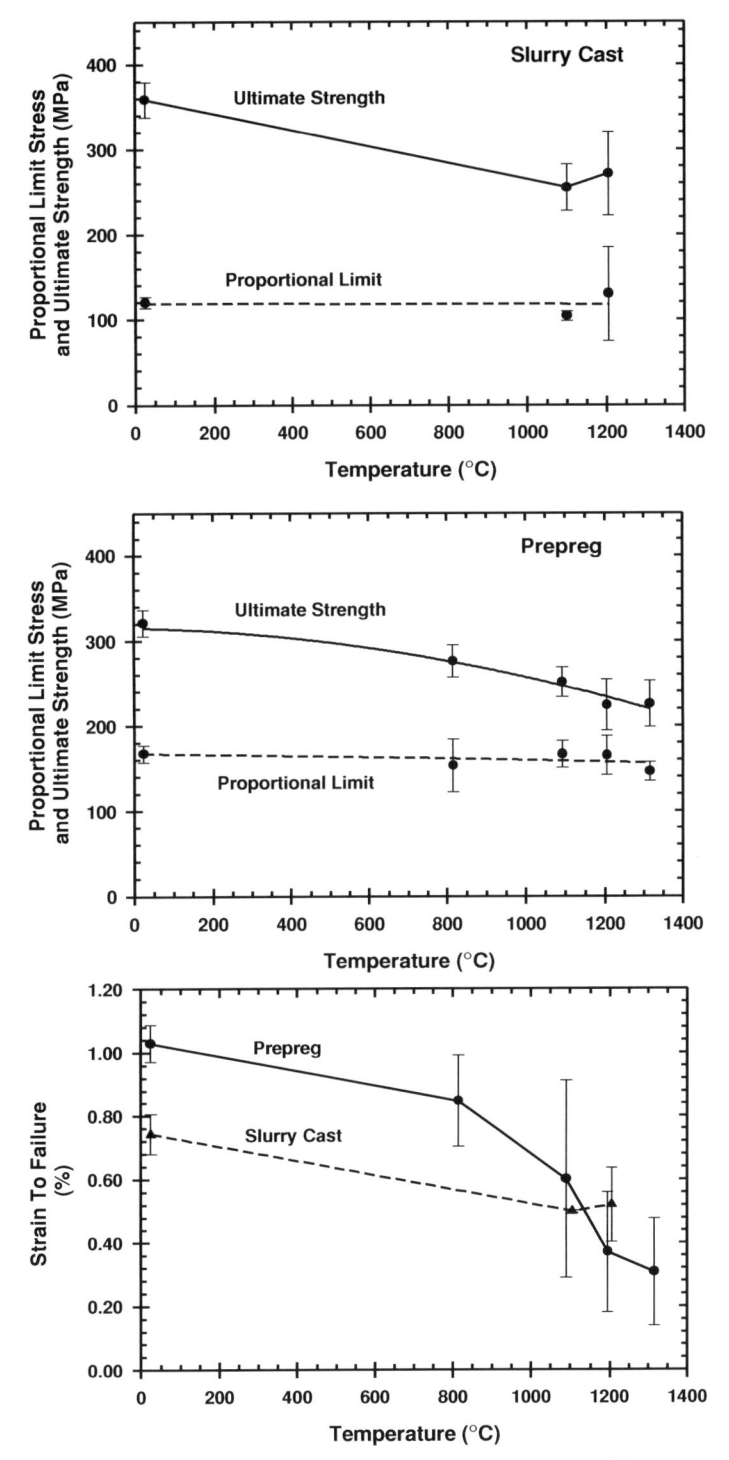

FIGURE 7. Temperature dependence of the tensile fracture properties of HiPerComp™ Composites. Error bars represent 95% confidence intervals on the means.

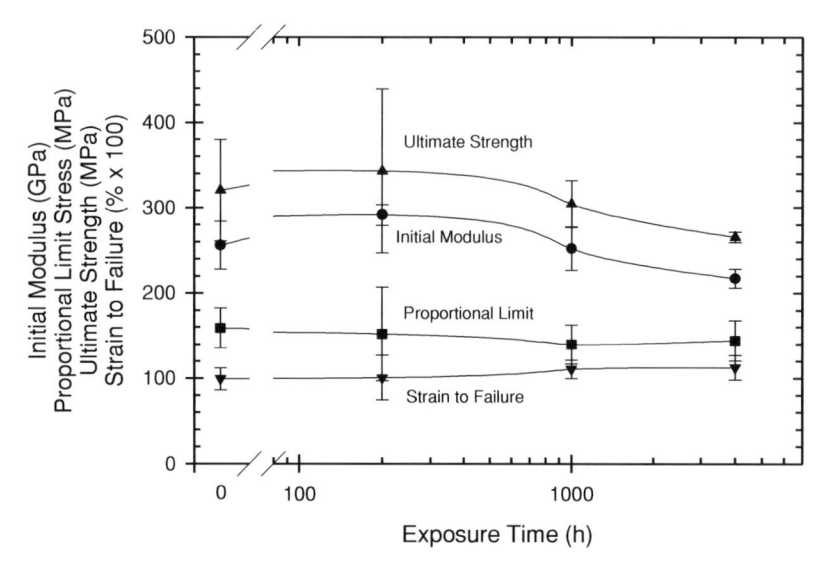

FIGURE 8. Effect of thermal exposure in air at 1200°C on the RT tensile properties of Prepreg HiPerComp™ Composites. Error bars represent +/− 1 standard deviation.

1200°C are shown in Figure 8. Fracture properties were found to be constant with exposure to between 200 and 1000 hours, at which time gradual reductions in the modulus and ultimate strength were observed. Both proportional limit stress and strain to failure values remained relatively constant out to the 4000 hour limit of the test.

4.5. Fatigue and Creep Behavior

Turbine engines are very dynamic environments, with large vibrational loads from combustion dynamics and pressure pulses caused by the rotating components. Consequently the effective life of many turbine components is limited by the fatigue resistance of the current metallic materials used. An understanding of the cyclic fatigue response of HiPerComp™ composites is therefore a prerequisite to their utilization in a gas turbine. Cyclic fatigue behavior of the Prepreg HiPerComp™ composite system has been measured using sinusoidal tension-tension fatigue tests over a range of frequencies (0.33 Hz to 150 Hz) using a load ratio (R-value) of 0.01 to 0.05. The results of these tests are shown in Figure 9.

When combined on a cycles-to-failure basis, as is done in Figure 9, the low and high frequency tests give reasonably consistent trends, whereas if the data are combined on a time-to-failure basis the agreement between low and high frequency tests is rather poor. This suggests that the failure mechanism in these tests is indeed cyclic dependent rather than simply time dependent, as might be expected of a simple thermal degradation mechanism. Additional testing under more traditional high-cycle fatigue conditions, having a constant applied stress with a superimposed cyclic stress at a level of 10–20% of the constant stress, is needed to more fully understand the fatigue behavior.

The tensile creep and creep rupture properties of Prepreg HiPerComp™ composites have also been evaluated in air for times up to 1000 hours. These data are summarized in

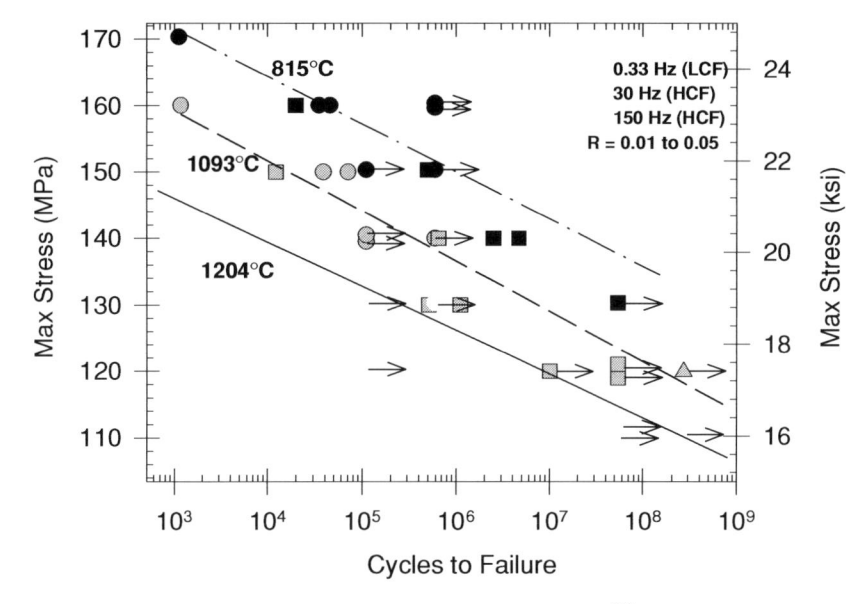

FIGURE 9. Cyclic tension-tension fatigue behavior of Prepreg HiPerComp™ composites tested in air using sinusoidal loading. Black points are for 815°C, gray points for 1093°C and white points for 1204°C. (Arrows indicate test run-outs.)

Table 4. No observable creep strain was obtained at 815°C, though one sample did rupture after 460 hours at 140MPa. At 1093°C and 1204°C measurable strain rates were obtained, though few samples attained a steady state strain rate before the end of the 1000 hour test or at rupture. Over the range of temperature and stress evaluated (1093 and 1204°C and 125 to 160MPa) the measured strain rates at 1000 hours for the test run-out samples ranged from 2×10^{-10} s^{-1} to 2×10^{-11} s^{-1}. This range of strain rates is very low compared to creep rates currently accepted for metallic hardware in turbine engines, and thus creep deformation is not expected to be a limiting factor for the application of HiPerComp™ composites.

The measured creep rupture behavior for Prepreg HiPerComp™ composites is shown in Figure 10. Rupture of samples, when it was observed, generally occurred at strains between 0.1 and 0.3%, although run-out samples also displayed comparable levels of strain without failure. Overall the rupture curves are relatively flat owing to the change in rupture mechanism above and below the matrix cracking stress. The first matrix cracks are generally observed at ~15 MPa below the proportional limit stress, which at these temperatures is ~165 MPa (see Figure 7). Thus at stresses at or above ~150 MPa one would expect the presence of at least one matrix crack and therefore the composite rupture behavior would be controlled by fiber and fiber/matrix interface oxidation and rupture of the bridging fibers themselves. At stresses below ~150 MPa the matrix would remain intact and composite rupture behavior would be governed by the much slower process of subcritical crack growth. The measured rupture data are consistent with this interpretation in that only one of the eight samples tested below 150 MPa actually failed during the test.

TABLE 4. Summary of Creep and Creep Rupture Data for Prepreg HiPerComp™

Temperature (°C)	Stress (MPa)	Time to Failure (h)	Total strain at failure or at 1000 h (%)	Creep Rate at 1000 h (s^{-1})
815	125	>1000	0.07	—
	140	>1000	0.08	—
	140	>1000	0.08	—
	140	460	0.05	—
1093	125	>1000	0.93	2.6×10^{-11}
	140	>1000	0.16	1.9×10^{-10}
	140	>1000	0.14	1.1×10^{-10}
	140	>1000	0.13	8.0×10^{-11}
	150	394	0.20	—
	150	190	0.14	—
	150	18.1	0.12	—
	160	15.4	0.12	—
	160	1.45	0.10	—
1204	125	>1000	0.25	3.0×10^{-10}
	125	>1000	0.20	8.0×10^{-10}
	140	>1000	0.31	2.3×10^{-10}
	140	454	0.28	—
	150	82.1	0.29	—
	150	0.32	0.18	—

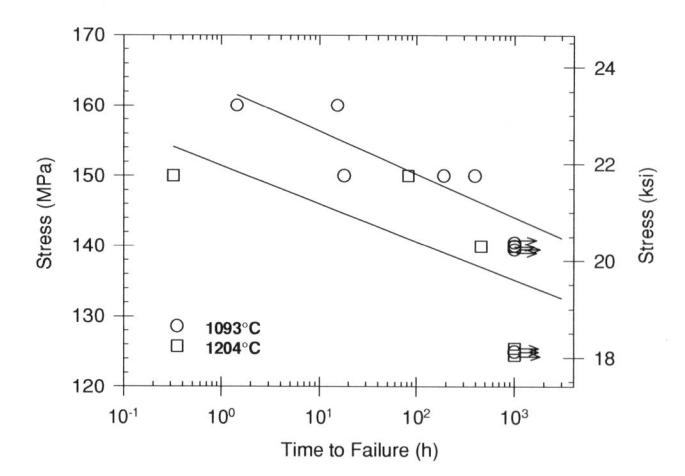

FIGURE 10. Tensile creep rupture behavior of Prepreg HiPerComp™ composites tested in air. (Arrows indicate test run-outs.)

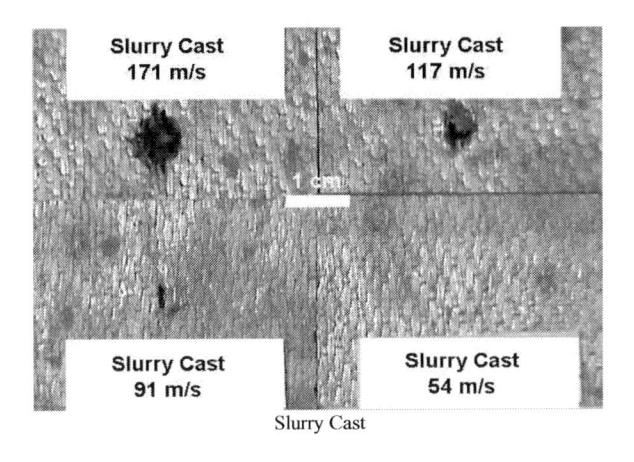

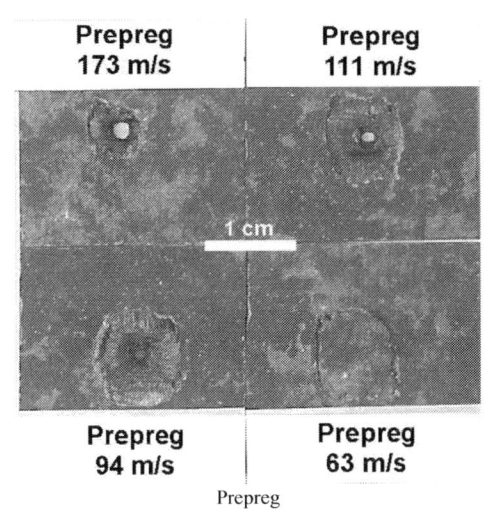

SILICON MELT INFILTRATED CERAMIC COMPOSITES 113

FIGURE 11. Photographs of impact damage in Slurry Cast (top) and Prepreg (bottom) HiPerComp™ composites caused by 4mm chrome-steel projectiles fired at various velocities. The photographs show the back sides of the samples, which generally showed more damage than the impact sides.

4.6. Damage Tolerance

An inevitable occurrence during the operation of any gas turbine engine is foreign object damage, or FOD. FOD denotes damage to the turbine components caused by the passing of solid material through the engine. The source of this material can be from objects, such as sand, pulled into the compressor from outside the engine, or from broken or oxidized pieces of hardware that come from upstream engine components themselves. Since gas velocities in the turbine section approach sonic velocities, such "foreign objects" can be traveling at fairly high velocity, and , depending on their mass, impart quite significant impact energies.

The foreign object damage resistance of HiPerComp™ composites have been evaluated using ballistic impact testing, performed at GE Aircraft Engines and the University of Dayton Research Institute. Nominally 2.5 mm thick panels of prepreg and slurry cast HiPerComp™ have been subjected to direct impacts with 4mm chrome-steel shot at velocities from 54 to 430 m/s (corresponding to impact energies from 0.4 to 24 J, respectively). At 430 m/s the projectile punched a hole entirely through the composites with the entrance hole being roughly equivalent in size to the projectile and the exit hole being approximately twice the area. At 115 m/s the projectile did not fully penetrate the composite panels, but rather caused a hemispherical indent in the front face and ejection of a conical-shaped zone of material from the back face. Non-destructive evaluation (NDE) of the samples following impact was done using both fluorescent dye penetrant and infrared thermography imaging. Both techniques showed the radius of the damage zones to be roughly equal to the size of the exit hole (or ~2X the size of the projectile) on samples impacted at 430 m/s, but the radius of the damage zone was as much as 4X the size of the projectile radius at lower velocities. In general, the size of the damage zone was larger in the prepreg type composites than in the slurry cast type composites, probably reflecting the difference in fiber content between the two types. Photographs of some of the damage zones from impacted samples are shown in Figure 11. It should be noted that when similar sized plates (1–2.5 mm thick) of sintered Si_3N_4 were subjected to a comparable impact event it was found to shatter at all impact energies down to and below 0.5 J.

5. SUMMARY

The HiPerComp™ family of ceramic matrix composites, based on silicon melt infiltration composite technology invented by GE, offer a unique combination of high temperature thermal and mechanical properties that make them highly suited for gas turbine engine applications.

ACKNOWLEDGEMENTS

The authors would like to thank M. Brun, P. Meschter, S. Shirzad, and D. Landini for contributing to the technical work described in this paper. The fabrication of test samples and properties measurements described were partially funded by the U.S. Department of Energy under contracts DE-FC02-92CE4100 and DE-FC02-CH00CH11047.

REFERENCES

1. G.S. Corman, K.L. Luthra, and M.K. Brun, "Silicon Melt Infiltrated Ceramic Composites – Processes and Properties" Chapter 16 in *Progress In Ceramic Gas Turbine Development, Vol. II, Ceramic Gas Turbine Component Development and Evolution, Fabrication, NDE, Testing and Life Prediction*, M. van Roode, M. Ferber, D. Richerson, eds., ASME Press, New York, USA, 2003.
2. K.L. Luthra, R.N. Singh, and M.K. Brun, "Toughened Silcomp Composites – Process and Preliminary Properties," Am. Ceram. Soc. Bull., 72 (7) 79–85, 1993.

3. G.S. Corman, J.T. Heinen, and R.H. Goetz, "Ceramic Composites for Industrial Gas Turbine Engine Applications: DOE CFCC Phase 1 Evaluations," ASME proceedings paper 95-GT-387, presented at the 40th ASME International Gas Turbine and Aeroengine Congress and Exposition, Houston, Texas, June 5–8, 1995.
4. A.J. Dean, G.S. Corman, B. Bagepalli, K.L. Luthra, P.S. DiMascio, and R.M. Orenstein, "Design and Testing of CFCC Shroud and Combustor Components," ASME proceedings paper 99-GT-235, presented at the 44th ASME Gas Turbine and Aeroengine Technical Congress, Exposition and Users Symposium, Indianapolis, Indiana, June 7–10, 1999.
5. G.S. Corman, A.J. Dean, S. Brabetz, M.K. Brun, K.L. Luthra, L. Tognarelli, and M. Pecchioli, "Rig and Engine Testing of Melt Infiltrated Ceramic Composites for Combustor and Shroud Applications," **Transactions of the ASME – Journ. Eng. for Gas Turbines and Power**, vol. 124, Issue 3, July 2002.
6. G.S. Corman, A.J. Dean, S. Brabetz, K. McManus, M.K. Brun, P.J. Meschter, K.L. Luthra, H. Wang, R. Orenstein, M. Schroder, D. Martin, R. De Stefano and L. Tognarelli, "Rig and Gas Turbine Engine Testing of MI-CMC Combustor and Shroud Components," proceedings paper 2001-GT-593, presented at the 46st ASME Gas Turbine and Aeroengine Technical Congress (ASME Turbo Expo, Land, Sea & Air) New Orleans, LA, June 4–7, 2001.
7. P.S. DiMascio, R.M. Orenstein, M.S. Schroder, G.S. Corman, and A.J. Dean, "Ceramic Gas Turbine Programs at GE Power Systems," in 2002 *Ceramic Gas Turbine Design and Test Experience:Progress in Ceramic Gas Turbine Development: Volume 1*, M. van Roode, M. Ferber and D. Richerson, eds., ASME Press, New York, USA, 2002.
8. Miriyala, N., Fahme, A., and van Roode, M., 2001,"Ceramic Stationary Gas Turbine Development – Combustor Liner Development," ASME paper 2001-GT-0512, presented at the 46th ASME TURBO EXPO 2001, LAND, SEA, & AIR, New Orleans, LA, June 4–7, 2001.
9. C.M. Grondahl, "Performance Benefit Assessment of Ceramic Components in an MS9001FA Gas Turbine," ASME Proceedings Paper 98-GT-186, presented at the International Gas Turbine & Aeroengine Congress & Exhibition, Stockholm, Sweden, June 2–5, 1998.
10. D.W. Freitag and D.W. Richerson, "Opportunities for Advanced Ceramics to Meet the Needs of the Industries of the Future," US Department of Energy Report DOE/ORO 2076.
11. J.A. DiCarlo, H.-M. Yun, G.N. Morscher, and R.T. Bhatt, SiC/SiC Composites for 1200°C and above, in *Handbook of Ceramic Composites*, N.P. Bansal, Ed., Kluwer Academic Publishers, Boston, MA, 2004, PP. 77–98.
12. G.S. Corman and K.L. Luthra, "Silicon-Doped Boron Nitride Fiber Coatings for Melt Infiltrated Composites," presented at the 24st annual Conference on Composites, Materials and Structures, Cocoa Beach, FL, January 24–28, 2000.

6

Carbon Fibre Reinforced Silicon Carbide Composites (C/SiC, C/C-SiC)

Prof. Dr.-Ing. Walter Krenkel
University of Bayreuth
Ceramic Materials Engineering
D-95440 Bayreuth, Germany
Phone: +49-921-55-5500
Email: walter.krenkel@uni-bayreuth.de

ABSTRACT

Ceramic matrix composites (CMC), based on reinforcements of carbon fibres and matrices of silicon carbide (called C/SiC or C/C-SiC composites) represent a relatively new class of structural materials. In the last few years new manufacturing processes and materials have been developed. Short fibre reinforcements, cheap polymer precursors and liquid phase processes reduced the costs by almost one order of magnitude in comparison to first generation C/SiC composites which were originally developed for space and military applications. Besides high mass specific properties and high thermal stability, functional properties like low thermal expansion and good tribological behaviour play an increasing importance for new commercial applications like brake disks and pads, clutches, calibration plates or furnace charging devices.

I. INTRODUCTION

First non-oxide CMCs, based on carbon/carbon composites, were developed in the 1970s as lightweight structures for aerospace applications. They had to be designed as limited life structures as the environmental conditions were highly aggressive and the long term behaviour of these composites was still unknown. Typical representatives for such components were rocket nozzles, engine flaps, leading edges of spacecraft and brake disks of aircraft. Their lifetime comprises several minutes to some few hours under highest thermomechanical requirements which can not be fulfilled by any other structural material [1–3].

In order to improve the oxidation resistance and thus the application lifetime of these composites, research has been exerted on using ceramics instead of carbon as the matrix material. Silicon carbide is particularly suitable as a matrix material due to its high oxidation resistance, its superior temperature and thermal shock stability and its high creep resistance. Practically, similar manufacturing techniques can be used for the silicon carbide matrix formation of C/SiC composites as for the manufacture of carbon/carbon composites. Generally, ceramic matrix composites have been developed to combine the advantageous properties of monolithic ceramics with a high damage tolerance, which is known for example from the reinforcing of fibre reinforced polymers. However, the mechanisms which cause high damage tolerance are completely different for both classes of material. Polymers are reinforced with strong and stiff fibres, whereas the matrix is weak and of low strength, stiffness as well as thermal stability. A strong bonding between matrix and fibres is desired as a result of high fibre surface reactions. Based on the differences of stiffness between fibres and polymer, the matrix itself is stressed only slightly and the energy release rate of a matrix crack is low because of the modest matrix strength. Therefore, the highly loaded fibres are able to stop cracks without being damaged.

Ceramic matrix composites are characterized by the fact that the stiffness of both, fibres and matrix, are in the same order of magnitude. High fibre/matrix bonding forces result in stresses which are similar for the matrix as well as for the fibres and the damage tolerance is comparable low to monolithic ceramics. The opposite case with extremely low fibre/matrix bondings leads to a nearly stress-free matrix and a high fracture toughness. However, as the debonding and shear properties mainly depend on frictional effects, such kind of composites are usually not suitable as a structural material. Damage tolerant CMCs therefore require moderate fibre/matrix bondings with adapted interphases. The interphase microstructure can vary from sharp, non-reactive interfaces to in situ reacted interfaces, porous or multilayer interfaces and is responsible for the stopping and deflecting of matrix cracks.

Similar to polymer and metal matrix composites, the CMC's fracture behaviour and properties are dominated by the reinforcing fibres. But to an even higher degree the fibres must show a high stiffness and an extreme thermal stability. Carbon fibres fulfill these requirements in an outstanding way. They are commercially available in various modifications, weavable to preforms and carbon fibres show a very high thermal stability well above 2,000 °C. However, their main disadvantage is the degradation in an oxidising atmosphere beyond 450 °C, resulting in the need for an external oxidation protection. It is known from oxidation kinetics that increasing the final heat treatment temperature (for example by a graphitization step) results in an improved oxidation resistance of the C-fibres. Therefore, the oxidation resistance is also improved by reinforcements with high modulus (HM) or ultra high modulus (UHM) carbon fibres in comparison to high tenacity (HT) fibres [4].

Long-term oxidation protection requires multilayer protection coatings, where the carbon/carbon or carbon/silicon carbide composite is protected for example with SiC layers and additional self-healing glass forming layers, based on oxides like mullite, alumina or silicon. The silicon carbide layer can be performed via pack cementation, but superior oxidation resistance can be achieved with pure β-SiC layers, deposited via the CVD process [5].

Due to the anisotropic coefficient of thermal expansion (CTE) of C/C, C/SiC and C/C-SiC, the oxidation protection of these composites is more difficult than it is for non-reinforced carbon or graphite bulk materials. The mismatch of CTE between the CVD-SiC coating and the carbon fibre reinforcement creates cracks in the SiC coating during the cooling-down period after deposition. Crack formation starts at approximately 100 °C below the CVD coating temperature. Therefore, CVD-SiC coated composites show the highest oxidation rate at about 800 °C, the maximum between crack opening and oxidation kinetics. As a result, sophisticated oxidation and corrosion coatings can only reduce the material degradation in a certain temperature interval under static conditions, but all available protection coatings are not able to prevent oxidation completely under dynamic conditions. The designers and users of C/SiC and C/C-SiC composites have to keep these inherent restrictions under consideration in order to apply these materials adequately, resulting for example in short inspection periods and higher efforts in-service monitoring by NDE methods.

II. APPLICATIONS

C/SiC and C/C-SiC applications lie in fields where conventional materials, due to their insufficient mechanical properties at high temperatures or limited damage tolerance behaviour can no longer be considered and includes in principal all areas of lightweight construction. Some examples are given for high temperature (T > 1,000 °C), medium temperature and low temperature (T < 450 °C) regimes.

(2.1) Space vehicle's TPS and Hot structures

Temperatures of up to 1,800 °C occur during the re-entry phase of orbiters into earth's atmosphere. Thermal protection systems (TPS) are the domain of carbon fibre reinforced SiC-ceramics in spacecraft structures and numerous technology-driven projects have been performed over more than two decades in Europe, the US and Japan. The Hot structures of NASA's experimental space vehicle X-38, which was planed to serve as a technology carrier for a new Crew Return Vehicle (CRV) of the International Space Station (ISS), are regarded as an example for the current stage of C/SiC and C/C-SiC development for thermal protection systems. A nose cap, the adjacent thermal protection panels (nose skirt), two leading edge segments and two body flaps for the steering of the vehicle were manufactured and qualified by ground-tests by a German consortium [6–8].

The nose cap made of LSI-C/C-SiC by DLR is particularly exposed to extreme temperature stresses upon re-entry due to its location directly in the stagnation region of the vehicle. The connection of the nose shell to the fuselage consists of eight individual mounting braces, which are also made of C/C-SiC, or respectively in the cooler areas, of a temperature resistant metal alloy. This lever-type fastening system guarantees high durability against mechanical stress, and it also allows an unhindered thermal expansion of the shell, which may amount to as much as three millimetres at a mean diameter of 700 mm at the expected temperature level of 1,750 °C (Figure 1).

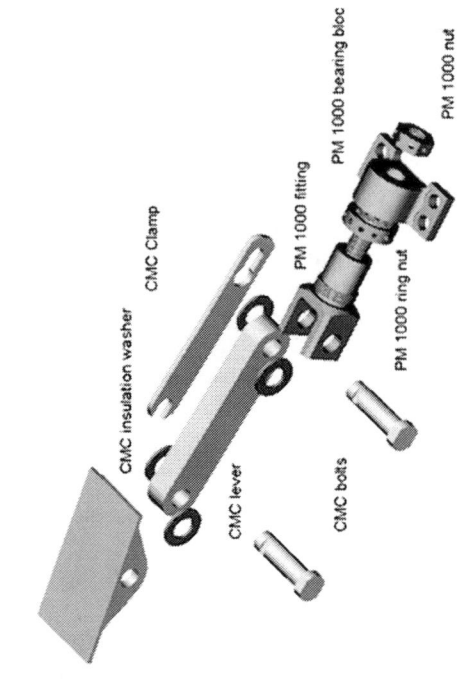

FIGURE 1. Attachment design of the X-38 nosecap (made of C/C-SiC composites)

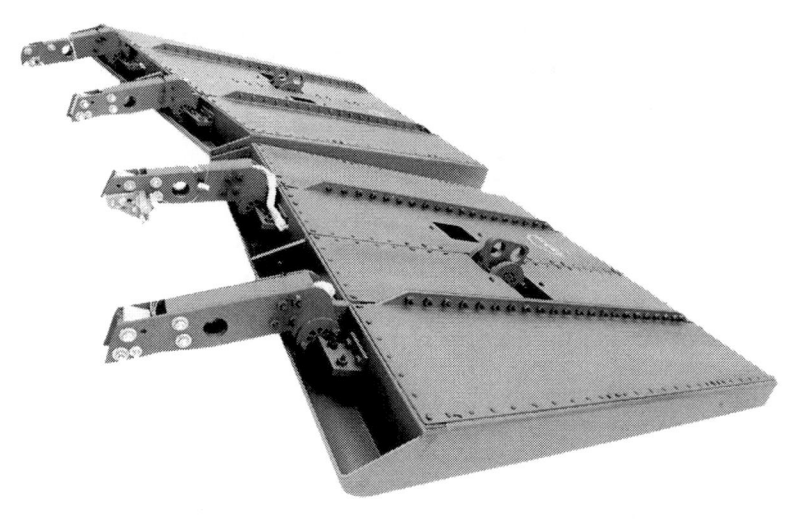

FIGURE 2. CVI-C/SiC body flaps for the X-38, joined with C/SiC screws. The Body Flap was developed by MAN-T in the frame of the German TETRA-Programme which was carried out by order of DLR and sponsored by the BMBF and Bavarian STMWVT

The CVI-C/SiC body flaps, manufactured by MAN Technologie, are build up from four boxes with integral transverse stiffeners and flanges to bear covers. The flaps of 1,600 mm in length and 1,500 mm in breadth are joined with more than 400 screws in total, also made of CVI-C/SiC (Figure 2). The use of C/SiC composites is providing about 50% weight reduction with higher safety margins compared to insulated metallic structures. The C/SiC nose skirt was manufactured by ASTRIUM GmbH via the liquid polymer infiltration process.

(2.2) Vanes, nozzles and flaps of rocket motors and jet engines

Even shorter operational times than those occurring during re-entry are demanded of jet vanes which are used to divert the direction of thrust in solid fuel rockets, but they are loaded by considerably higher stresses, Figure 3. The controllable vanes provide an increased manoeuvrability of the rockets, primarily during the low-speed phase immediately after take-off. Only a few seconds of endurance are required, but these few seconds impose upon the material the utmost demands regarding thermomechanical stability and resistance to abrasion. The C/SiC vane surfaces should be additionally coated with a protective ceramic coating (e.g. CVD-SiC) in order to be able to withstand the immense blast of particles (e.g. Al_2O_3) occurring as the solid fuel burns away. At the same time, the ceramic content of the structure material must be at such a high level, that the unavoidable burn-up consumption takes place only gradually, so that a sufficient residual vane surface is available during the complete burning period. The formulation of the microstructure of the C/SiC composite consequently requires an optimization of the conflicting demands for high fracture toughness (high C contents) and high resistance to abrasion (high SiC contents).

FIGURE 3. C/C-SiC jet vanes for solid fuel rocket propulsion systems

C/SiC composites have also been investigated successfully for expansion nozzles of rocket propulsion systems. Exemplarily, a nozzle demonstrator of the upper stage engine of Ariane 5 was designed and manufactured by filament winding, using LPI technique [9]. The nozzle with a length of 1,360 mm and an exit diameter of 1,330 mm showed a mass of 16 kg. Although considerable thicker wall structures were necessary, weight reductions of 60% and an increase of the allowable temperature of about 500 °C in comparison to superalloy Haynes 25 could be achieved.

Several C/SiC components like flame-holders, exhaust cones and engine flaps have proven their feasibility for military jet engines [10]. Outer flaps in the SNECMA M 88-2 engine provide 50% weight savings over the corresponding superalloy flap (Inconel 718) and almost 300 flaps have been fabricated for ground tests (Figure 4). C/SiC engine flaps are of prime interest for future military engines where they allow the reduction of internal cooling flow, thus yielding benefits in the engine performance.

The general low resistance to oxidation of carbon only permits restricted operational periods of the C/SiC structures at temperatures above about 450 °C. Its utilization in civil aircraft turbines or stationary gas turbines (e.g. as tiles in combustion chambers, for diffusers or for turbine vanes) with operational times of several 10,000 hours is therefore not possible

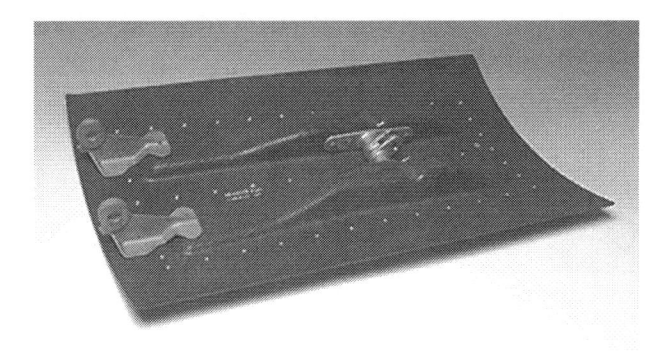

FIGURE 4. C/SiC outer flap of the M 88-2 engine (Snecma)

from today's point of view even with very sophisticated multi-layer protective coatings. The current terrestrial C/SiC developments are therefore concentrating on applications that have a performance range demanding a resistance to high temperature only for a short time, or on products, which make use of other advantageous characteristics of these multiphase composites.

(2.3) Advanced friction systems

C/C-SiC composites, made by liquid silicon infiltration (LSI-process), offer superior tribological properties in terms of high coefficients of friction (CoF) and wear resistance. The carbon fibres lead to an improved damage tolerance in comparison to monolithic SiC, whereas the silicon carbide matrix improves the wear resistance compared to carbon/carbon. C/C-SiC composites are therefore new, outstanding materials for brakes and clutches of high speed cars, trains and emergency brakes in the field of mechanical engineering and conveying.

First attempts to investigate C/C-SiC composites for their use as frictional materials for brake pads and disks started in the early nineties [11]. C/C-SiC materials show, in comparison to carbon/carbon, a considerably lower open porosity (less than 5%), a moderately higher density (about 2 g/cm^3) and a ceramic share of at least 20% in mass. Several activities in institutes as well as in industries now exist to investigate CMC materials for their use as frictional materials for brake pads and disks [12–16]. The resulting materials differ in their constituents (fibres, fillers), microstructure (ceramic content, gradients), properties (density, strength, thermal conductivity) and also in their processing conditions (fibre coating, temperature, etc.). Nevertheless, they are all based on carbon fibres and silicon carbide matrices as the main constituents of the composite material. The carbon fibres generally decrease the brittleness of SiC considerably so that the damage tolerance of C/C-SiC components lies in the same order of magnitude as for grey cast iron.

For automotive use, especially for high performance disks the costs of continuous fibres and the common processing techniques for components used in aerospace are too high for a serial production with high numbers of items. The most promising way to reduce the costs and to simplify the manufacture is to employ short fibre reinforcements and pressing techniques. The use of short fibres reduces the costs of the raw materials primarily by the

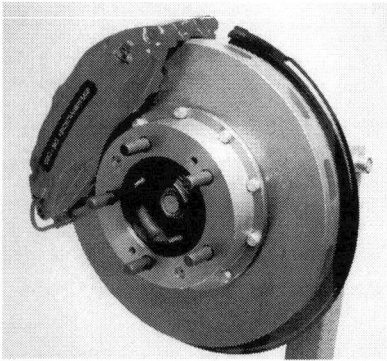

FIGURE 5. Emergency brake system (left) and internally ventilated brake disk for passenger cars (right), made of C/C-SiC

reduction of waste in comparison to bi-directionally woven fabrics. Due to the more isotropic fibre orientation of short fibre reinforced C/C-SiC the thermal conductivity perpendicular to the friction surface of brake disks is generally higher compared to the orthotropic material based on laminated woven fabrics. This leads to lower surface temperatures on the brake disks resulting in an higher and more constant coefficient of friction and lower wear rates. In different tribological test campaigns the performance and the excellent wear resistance were proven and new constructions suitable for these new braking materials were developed (Figure 5). Due to their high thermal stability and their low weight a great leap in brake technology is achievable, combining non-fading characteristics with better driving dynamics. Different passenger cars are already equipped with ceramic brakes and clutches in series and several industrial companies are currently producing or developing C/C-SiC frictional parts in an advanced stage.

Emergency stop brakes of different constructions are used in many fields of engineering, for example in lifts, cranes, electric drives for machine tools and winds. Electromagnetic spring applied brakes are commonly used for braking or holding loads, which are closed in the de-energized condition. The brake system often consists of a rotating and two stationary brake disks similar to heat pack of aircraft brakes. Increasing drive and higher circumferential velocities in modern power transmission necessitate new concepts.

New C/C-SiC composites have been developed and investigated [17, 18] in order to increase the efficiency of emergency brakes. Tribological tests have shown low wear rates and high coefficients of friction even at high energy input, whereas conventional friction materials (metallic disks and organic friction linings) are completely overloaded. The benefits for the customer comprise a higher transmitted braking power and smaller dimensions of the brake systems, equipped with C/C-SiC pads. Presently, different carbon fibre reinforced ceramics are developed and commercially available to be used in high performance brake systems.

(2.4) Low-expansion structures

The low thermal expansion of C/SiC fibre ceramics in combination with their high rigidity and stability is utilized in low-expansion structures. Not the high-temperature properties

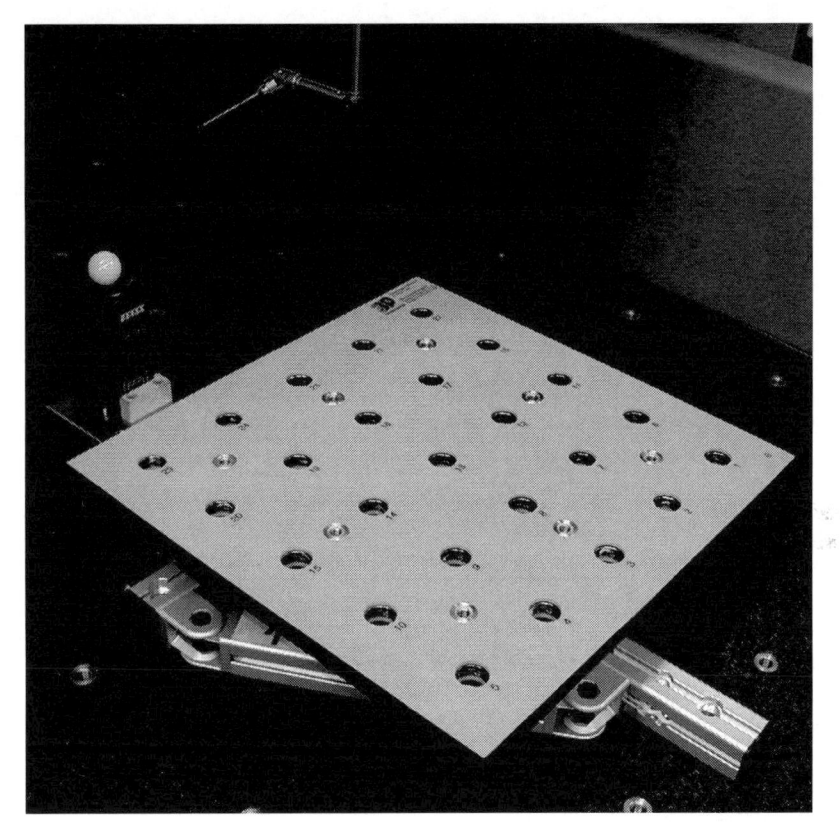

FIGURE 6. C/C-SiC calibrating plate with 25 measurement holes for the calibration of coordinate measuring machines

are relevant for this application, but rather the options to produce large-sized components with high precision as well as the material properties that are not influenced by ambient conditions (e.g. humidity) [19, 20].

Calibrating bodies used in industrial measurement technology are a very interesting application for C/C-SiC composites. Among others, plate-shaped calibrating bodies are used to check coordinate measuring machines, in order to be able to recognize inaccuracies in measuring lengths and angular displacements of the automatic measuring systems in the automotive industry, for example. These high-precision components must have a very low and a constant thermal expansion coefficient within the normal temperature range of -30 to $+50$ °C. In addition, the calibrating plates should be simple to handle, i.e. light-weighted and robust. This requirement profile is impressively fulfilled by C/C-SiC composites. Figure 6 shows such a calibrating plate with an edge length of 420 mm and a thickness of 8 mm with 25 measurement bore holes.

The advantages of C/C-SiC composites compared to conventional low-expansion materials are the absence of any kind of thermal hysteresis, a low heat capacity, a considerably lower weight in combination with a sufficient damage tolerance of the component.

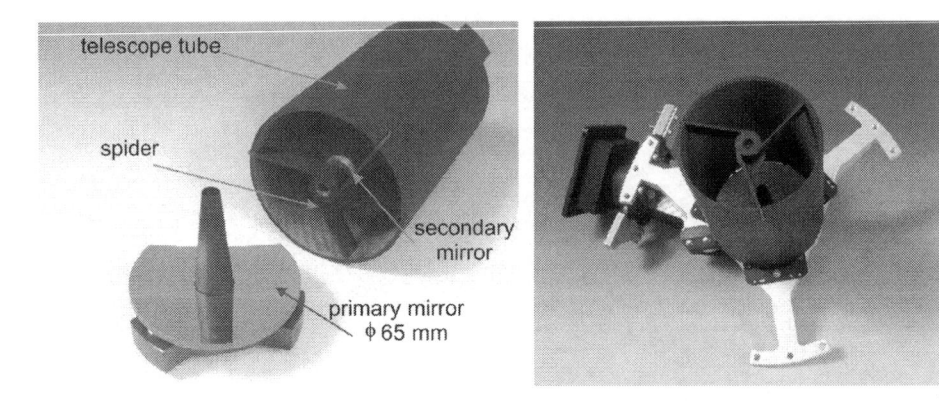

FIGURE 7. Optical unit for laser communication satellite before and after final assembly consisting of low expansion C/C-SiC tube and spider

Advantages for the customer arise here from primarily in mobile use, where the handling of the plates has been made fundamentally easier, and the waiting times for temperature equalization between the calibrating plate and the measuring room has been reduced.

Satellite communication systems based on optical concepts enable high data rates with a relatively low energy consumption compared to systems based on radio waves. However, an optical system has high demands on precision and stability. In this application the demands on thermal and structural stability are extreme. Novel C/C-SiC composites have been designed under the aspect of series manufacture especially for this purpose. Figure 7 shows the optical unit consisting of a primary and secondary mirror, connected with a telescope tube and spider element of low expansion C/C-SiC. In the final assembly the tube is aligned with the mirror optical axis in vertical direction. The secondary mirror faces the primary and the protruding rectangular blades of the tube are connected to the mirror cell by adhesive bonding.

The given telescope tube and spider element design with C/C-SiC composites enables integral manufacture in comparison to conventional non-CMC solutions where parts must be glued together. In addition, the design exhibits no hygroscopicity, demonstrates fracture toughness and is lightweight.

(2.5) Further fields of application

The economically most attractive field of application for C/SiC composites are primary structures in the energy industry. Despite their inherent limitations of oxidation resistance, international research continues to improve manufacture processes and surface coatings for the long term use of C/SiC composites under corrosive conditions. One first step towards ceramic constructions in energy and power station engineering is the development of a double-pipe heat exchanger in bayonet-pipe technique, which is intended for use in combined processes (Externally Fired Combined Cycles, EFCC) with indirectly fired gas turbines using coal, Figure 8. Here the HT heat exchanger serves for the indirect firing of a gas turbine. The air to be heated flows through the coaxially arranged fibre ceramic pipes. In

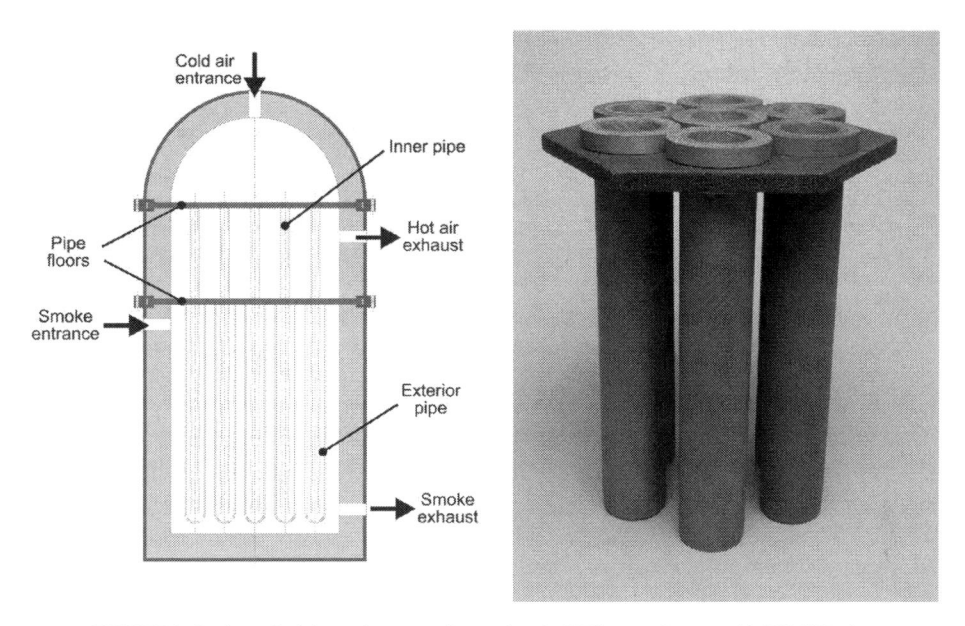

FIGURE 8. Design principles and construction study of a HT-heat exchanger with C/C-SiC tubes

cross-countercurrent direction, the flue gas flows around the pipe rows, which are laid out in a staggered arrangement. The coated C/C-SiC pipes are mounted separately in corresponding pipe floors, thus reducing the tension stresses induced by longitudinal expansions. Because of the micro-porosity of the C/C-SiC materials, the multi-layered coatings for corrosion protection have great significance for this application, since these protective coatings have two functions: to act as a sealing agent for the pressurized pipes, and also to protect the matrix and the carbon fibres from oxidation and corrosion [21, 22].

Components in the furnace chamber of thermal incineration installations require an extreme mechanical, thermal and corrosive resistance, particularly if problematic waste is burned. Moving firing grates made of C/SiC with a high ceramic content, which simultaneously have a forward feed of the material to be burned are already in operation in incineration installations [23]. These C/SiC components are designed as hollow bodies, so that additional air can be introduced via holes, thus achieving an optimization of the incineration.

In the field of furnace engineering, C/C-SiC ceramics are used as charging devices and work piece supports for metal hardening. These ceramic components are considerably lighter in comparison with high-temperature resistant metals, and they have a very low tendency to warping and distortion at high temperatures. Charging devices can be produced from textile preforms that are contoured similar to the final product, so that only very little machining effort is necessary. Further industrial components manufactured in series are C/C-SiC fan blades used to circulate the atmosphere in heat treatment furnaces. Oxygen probe tubes, thermocouple protection tubes, pouring gutters used in metallurgy, or HT-nozzles made of C/C-SiC fibre ceramic are also in practical operation and are here mostly replacing monolithic ceramic materials.

In the field of ballistic protection of vehicles and aircraft, the use of monolithic ceramics in compound systems permits a weight reduction of more than 50% as compared to armor plating steels. Further-reaching mass reductions are possible for future lightweight armor by using light C/SiC or C/C-SiC plates instead of the alumina or silicon carbide materials commonly used. Ballistic tests have shown, that fibre ceramics additionally offer improved protection against multiple hits because of their higher fracture toughness [24].

A typical lightweight armor principally consists of a multi-layered sandwich compound, of which the front side is made of a ceramic material and the rear side is primarily made of energy absorbing materials, such as synthetic fabrics (e.g. aramide) or ductile metals. Aside of the potentially lower weight per unit area of such lightweight armor (goal to be achieved: <25 kg/m^2), the variability in the design of the fibre ceramics is another essential advantage of fibre ceramics in comparison to conventional ceramics. Their manufacture is based on the classic methods used in composites technology, which allows almost any kind of curved, thin-walled structure. The largest sales volume of future fibre ceramic production, besides brake discs, is consequently envisaged in ballistic protection systems based on C/SiC and C/C-SiC composites.

III. PROCESSING

In principal, there exist numerous processing routes to infiltrate the matrix system into the fibre preform. Conventional powder processing techniques used for making monolithic ceramics are mostly not suitable and rather unconventional techniques to avoid damage of the fibre preforms are applied. They can be distinguished between the impregnation either by gas phase or by liquid phase infiltration (Figure 9). Three different techniques are currently used in an industrial scale for the production of C/SiC and C/C-SiC composites, each of them leading to specific microstructures and properties (Figure 10):

– Chemical Vapour Infiltration (CVI)
– Liquid Polymer Infiltration (LPI) or Polymer Infiltration and Pyrolysis (PIP)
– Liquid Silicon Infiltration (LSI)

Historically, the processing routes moved from the isothermal CVI process to more cost-effective techniques such as gradient-CVI and liquid polymer or liquid silicon infiltration. These routes are faster and lead to shorter manufacture cycles than isothermal CVI and, especially the two liquid phase processes LPI and LSI, use technologies already developed for polymer matrix composites (PMC).

One important aspect is to realize that processing should be considered as an integral part of the whole process of designing a CMC component. Fibre orientation, dimensionality of the preform and thermal treatment conditions are important parameters of influence for the performance of the final CMC product.

(3.1) Chemical Vapour Infiltration

Chemical vapour infiltration in general allows the deposit of quite a variety of matrices and the processing of any complex shape. The multi-directional fibre preform is fixed in the furnace with a tooling or build-up with a temporary binding agent, which is removed in a

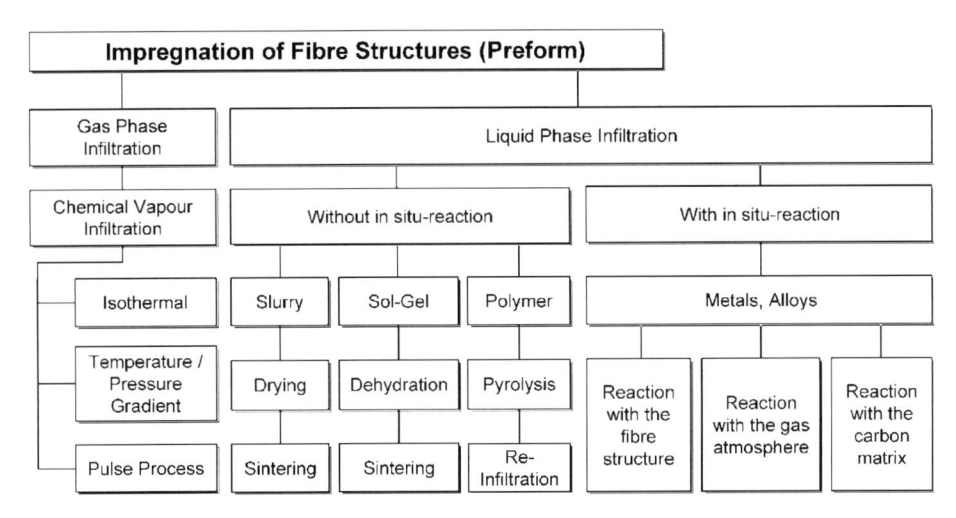

FIGURE 9. General overview of the manufacturing processes for CMC materials

first pyrolysis step. The ceramic matrix is obtained by the decomposition of gaseous species within the open porosity of the preform. Usually, methyltrichlorosilane (MTS, CH_3SiCl_3) as the process gas and hydrogen as the catalyst are used to form the SiC matrix. Two different phenomena control the matrix growth rate within the preform's pores:

- the kinetics of the chemical reaction
- the mass transport of the reaction products into the porosity.

The quality and purity of the deposed silicon carbide is determined by the ratio of mixture between hydrogen and MTS. High portions of H_2 lead to Si-rich matrices, whilst increasing amounts of MTS result in SiC matrices with residual carbon. In order to get a good in-depth deposition, low pressures (50–100 hPa) and low temperatures (800–900 °C) are necessary and must be kept constant over the whole process [25]. This isothermal/isobaric CVI-process

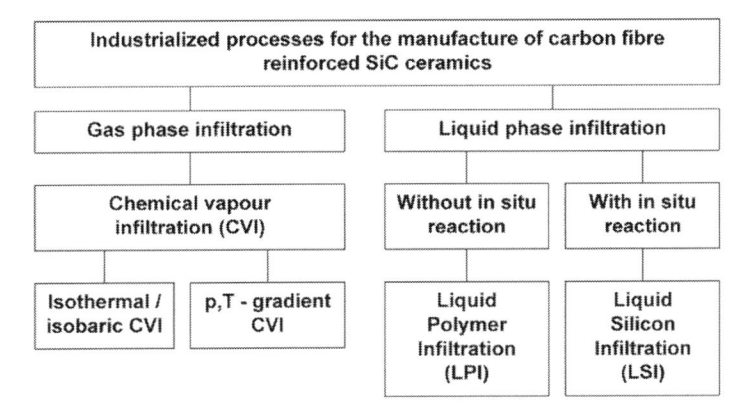

FIGURE 10. Processes for the industrial manufacture of C/SiC and C/C-SiC composites

leads to very good thermomechanical properties and high fracture toughness. The drawbacks lie in long processing times lasting several weeks or months and in restricted preform depths which can be infiltrated in one step.

To overcome these geometrical restrictions and to increase the deposit rate of SiC, thermal and pressure gradient-CVI processes have been developed and industrialized [26, 27]. In contrast to the isothermal CVI process a forced mass flow of MTS, driven by gradients of temperature and pressure within the preform, is applied and allows considerable higher process temperatures and pressures. The higher gas density and higher reaction speed result in manufacture times which are more than one order of magnitude faster than the isothermal/isobaric CVI. Typically, 40 to 60 hours are necessary to infiltrate carbon preforms of 5 mm thickness to a residual open porosity of 12%. As the forced mass flow of gases necessitate the sealing and cooling of the fibre preform, this variant on CVI is focussed on simple and standardized geometries like profiles, tubes and plates.

Besides the good quality of the C/SiC composites processed by chemical vapour infiltration, one of the major advantages of this manufacturing route is that is allows the control of the fibre/matrix interphase. Therefore, C/SiC composites are shortly deposited in a first step with carbon (e.g. by the deposition of methane gas, CH_4) to govern the fibre/matrix bonding forces. In summary, the manufacture of C/SiC components via the gradient-CVI process comprises the following steps:

1. Fibre preform build-up with a polymer binding agent
2. Pyrolysis of the polymer (leads to a self-supporting fibre preform)
3. Deposition of carbon for fibre coating (interphase formation)
4. Deposition of SiC for matrix infiltration
5. Final machining of the near-net shaped C/SiC component

(3.2) Liquid Polymer Infiltration

The manufacture of C/SiC components by using polymeric precursors is called liquid polymer infiltration (LPI) process or polymer infiltration and pyrolysis (PIP) process and represents one of the most advanced manufacturing methods for large and complex shaped CMC parts for the aerospace industry [28–30]. The starting materials to form the silicon carbide matrix are usually polycarbosilane or polysilane polymers which convert from polymer to an amorphous or crystalline ceramic during pyrolysis. The big difference in the densities of the polymeric precursor and the SiC matrix results in a reduction of volume and induces a high amount of pores in the matrix. Inert submicronic SiC particles or reactive fillers can be used to reduce the shrinkage, but usually a number of successive infiltration cycles followed by a respective pyrolysis step are required to obtain a SiC matrix with a sufficient high density. Typically, five to seven impregnations are necessary to obtain a residual porosity of less than 10%.

Figure 11 gives an overview of the LPI-process which is used for the manufacture of integral C/SiC components. Prior to their slurry infiltration the carbon fibres are coated by CVD to reduce the fibre/matrix bonding. In the following steps, the processing route follows the classical methods of manufacturing carbon fibre reinforced plastics (CFRP). The coated fibres are infiltrated with the precursor, for example by filament winding, fabric prepreging or resin transfer moulding (RTM). The resulting laminate is subsequently cured in an autoclave at 200–300 °C.

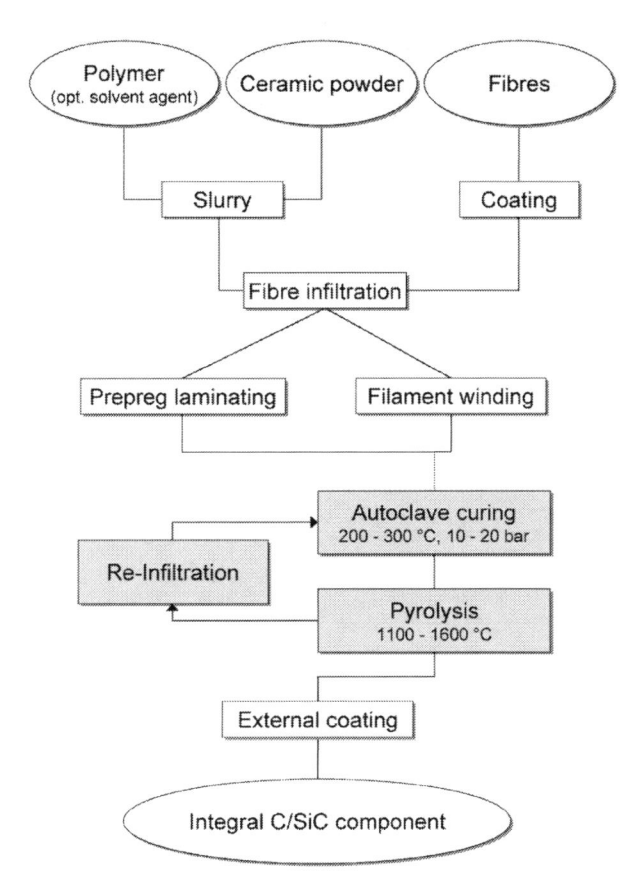

FIGURE 11. The technique for making C/SiC components by liquid polymer infiltration (LPI)

During the following high temperature treatment under inert atmosphere or vacuum (pyrolysis) the polymer is decomposed and the real ceramic matrix is formed within several intermediate steps. The chosen temperature-time-cycle influences heavily the morphology of the resulting SiC matrix. Whilst high heating rates and low final temperatures lead to a high porosity and an amorphous matrix, a crystalline SiC matrix with low and fine distributed pores can be obtained by low heating rates and ultimate temperatures up to 1,600 °C.

To summarize, this polymer route of making C/SiC composites has the advantage of a good matrix composition control, relatively low densification temperatures and the possibility to join together different parts with the original matrix slurry to highly complex components. The disadvantages are the multiple infiltration/densification/pyrolysis cycles and the large shrinkage during pyrolysis, resulting in a cracking of the matrix.

(3.3) Liquid Silicon Infiltration

This technique is based on the impregnation of porous carbon/carbon composites by molten silicon and the reaction of the metallic melt with the solid matrix carbon to silicon carbide. The melt infiltration process can use almost any reinforced geometry and yields a high-density, low-porous matrix. A fibre preform having a network of pores and cracks is infiltrated by liquid silicon, mostly using capillary pressure. Application of pressure or processing in vacuum can support the infiltration process. The temperatures involved are at least beyond the melting point of silicon (1,415 °C) and can lead to deleterious reactions between the fibres and silicon. The melt viscosity, the chemical reactivity, the wetting of the reinforcement and the anomaly of silicon during phase transition (change of density of approximately 8%) are critical processing parameters to be considered. The thermal expansion mismatch between the fibres and the matrix and the rather large temperature interval between the processing temperature and room temperature are problems which have to be taken into account.

First attempts to infiltrate carbon/carbon composites by liquid silicon have been conducted for more than twenty years [31–33]. After these basic investigations, the carbon fibres have to be coated prior to the infiltration of silicon in order to reduce the degree of fibre degradation. Also, highly graphitized carbon fibres like high modulus fibres (HM) are recommended as reinforcement for the fibre preform which are more stable in contact with silicon than only carbonized fibres. Both requirements are in contrast with a cost-efficient processing of CMCs and only poor fracture toughnesses have been achieved by the infiltration of unadapted carbon/carbon preforms. In the last few years, new processes and novel C/C-composites with adapted microstructures which also allow the use of uncoated and chopped high tenacity (HT) carbon fibres were developed [34–37].

Figure 12 gives an overview of the LSI-process which can be split into three major steps. The fibre preform fabrication starts with the manufacture of carbon fibre reinforced plastic composites with polymeric matrices of high carbon yield. Normally, commercially available resins like phenolics or other aromatic polymers are used to fabricate laminates by common CFRP techniques like resin transfer moulding, autoclave, warm pressing or filament winding. After curing, the composites are postcured for the complete polymerization of the matrix. Subsequently, the CFRP composites are pyrolysed under inert atmosphere (e.g. nitrogen or vacuum) at temperature beyond 900 °C to convert the polymer matrix to amorphous carbon. The pyrolysis of the CFRP composite leads to a volumetric contraction of about 50% of the neat polymer. As this polymer contraction is hindered by the embedded fibres the macroscopical shrinkage of the composites is essentially lower. In direction of the fibre alignment the shrinkage is close to zero with the result that the matrix contraction leads to a microscopical network of cracks within the C/C composites.

The fibre/matrix bonding (FMB) strength in the polymer composite plays an important role during the pyrolysis step and can be modified by fibre coatings (e.g. PyC), thermal pre-treatments of the fibres or by variation of the fibre type. Generally, the higher the FMB strength the higher the tendency to form dense segments of fibres, interconnected with a translaminar crack system [38]. These microcracks represent the open porosity of the C/C composite.

During the third and final processing step, the capillary effect of the open pores and the low viscosity of molten silicon enable a quick filling of the microcracks. The simultaneous exothermic reaction between the carbon matrix and the liquid silicon results in silicon carbide encapsulated carbon fibres. The resulting composite comprises of three phases: carbon fibres and residual carbon matrix, silicon carbide and a certain amount of unreacted silicon. As

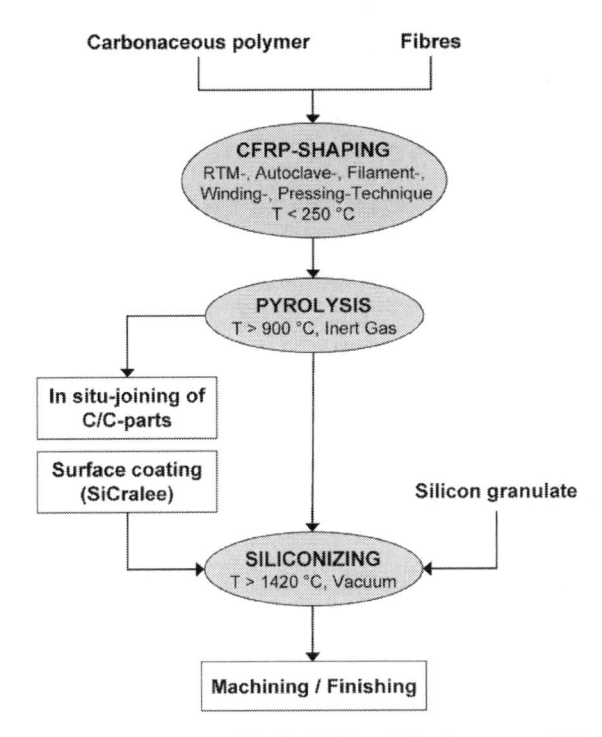

FIGURE 12. Schematic of the liquid silicon infiltration process (LSI)

the load bearing capability derives from the encapsulated C/C-regions, this material is also called C/C-SiC composites.

Simple as well as complex building units can be joined in situ within the siliconizing step using a carbonaceous paste with the optional addition of either carbon felt or carbon fabric as the joining material to form integral components. Porous C/C components are prepared and fixed together and molten silicon is caused to flow between the surfaces and react with the carbon material to convert it into SiC and bond the surfaces together. In situ-joining is desirable as it eliminates expensive and complicated machining as well as the need for additional metallic bolts or ceramic adhesives.

Additionally, the LSI process allows the implementation of reaction bonded SiSiC-coatings on the component's surface. Adding porous carbonaceous layers and additional silicon granulate on the C/C surfaces to be protected, extremely wear resistant coatings can be formed simultaneously to the SiC formation inside the C/C composite in an easy and economic way [39].

(3.4) Combined processing methods

Hybrid processes involving a combination of two or more (standard) processing methods has been described by several authors [5, 40, 41]. Such combined processes pursue the aim to overcome the drawbacks of the individual processes in order to improve the performance of the CMC material or to make the manufacture more cost-efficient. For example, a combined CVI/LPI impregnation technique is used for the industrial manufacture of filament wound furnace parts and crucibles. A pre-densification via CVI leads to the formation of an interphase between fibres and SiC-matrix whilst the subsequent precursor infiltration fills up the matrix. In comparison to a pure LPI-process, this hybrid process reduces the number of reimpregnation steps and improves the mechanical properties of the composite. If the last LPI-impregnation steps are substituted by the infiltration of liquid silicon, the processing time can be reduced further and matrices of low porosity can be obtained.

One other approach is the combination of a first CVI step, followed by a matrix densification via liquid silicon infiltration (LSI). By chemical vapour deposition the carbon fibres are coated with pyro-carbon or boron nitride in order to prevent them from reacting with the molten silicon and to improve the fracture toughness of the composite. The following infiltration of silicon densifies the matrix, resulting in a residual amount of free silicon in the matrix. The combination of LPI and LSI processes pursues a similar goal: The coating of C-fibres with a polymer in an initial step reduces the reaction between the fibres and the subsequently infiltrated silicon.

Despite their heterogeneous microstructure and their complex processing, hybrid processes show a promising route to combine the high material purity and process controllability of the chemical vapour infiltration technique with the less time consuming and in most cases cheaper liquid phase infiltration techniques of LSI and LPI. However, additional research and development have to be done in order to take benefit of the whole potential of combined processes.

IV. PROPERTIES

(4.1) General remarks

Ceramic composite materials are different from all other composite materials due to their microporous and microcracked matrix. In comparison with metals, the fracture behaviour of the CMC-materials is still relatively brittle. However, their strain to failure is up to one order of magnitude greater than monolithic ceramics, and their non-linear stress-strain behaviour make them an engineering material with quasi-ductile breaking behaviour. Even if a comparison with grey cast iron is not permissible because of the completely different structure, the damage tolerance of the fibre ceramics can be compared with that of grey cast iron in first approximation: These materials are still very brittle from the designer's viewpoint familiar with metals, the ceramist, however, will regard CMC materials as a new class of materials, opening completely new fields of application due to their fracture toughness. Low material densities result in mass-specific properties, which are unsurpassed by other structural materials at temperatures above 1,000 °C. Generally, the properties of fibre ceramics depend strongly on their micro-structural composition, and therefore, also on the respective manufacturing method.

Typical properties of two-dimensionally reinforced C/SiC and C/C-SiC composites, fabricated by the isothermal CVI-, gradient CVI-, LPI- and LSI-process, respectively, are shown in Table 1, published by representative manufacturers [9, 35, 42, 43]. The variance of properties depends on the fibre type and on the fibre volume content which is governed by the design of the component and the method of preform manufacture. The given values

TABLE 1. Overview of material data for SiC ceramics with bi-dimensional carbon fibre reinforcements

Property	Unit	Gasphase Infiltration (CVI) Processes		Liquid Infiltration Processes		
		CVI (isothermic) C/SiC	CVI (p,T-Gradient) C/SiC	LPI C/SiC	LPI C/SiC	LSI C/C-SiC
Tensile strength	MPa	350	300–320	250	240–270	80–190
Strain to failure	%	0.9	0.6–0.9	0.5	0.8–1.1	0.15–0.35
Young's modulus	GPa	90–100	90–100	65	60–80	50–70
Compression strength	MPa	580–700	450–550	590	430–450	210–320
Flexural strength	MPa	500–700	450–500	500	330–370	160–300
Interlam. shear strength	MPa	35	45–48	10	35	28–33
Porosity	%	10	10–15	10	15–20	2–5
Fibre content	Vol.%	45	42–47	46	42–47	55–65
Density	g/cm^3	2.1	2.1–2.2	1.8	1.7–1.8	1.9–2.0
Coefficient of thermal expansion ∥	10^{-6} K^{-1}	3[1]	3	1.16[4]	3	−1 bis 2.5[2]
Coefficient of thermal expansion ⊥		5[1]	5	4.06[4]	4	2.5–7[2]
Thermal conductivity ∥	W/mK	14.3–20.6[1]	14	11.3–12.6[2]	–	17.0–22.6[3]
Thermal conductivity ⊥		6.5–5.9[1]	7	5.3–5.5[2]	–	7.5–10.3[3]
Specific heat	J/kgK	620–1400	–	900–1600[2]	–	690–1550
Manufacturer		SNECMA	MAN	Dornier	MAN	DLR

∥ and ⊥ = Fibre orientation (1) = RT − 1000 °C (2) = RT − 1500 °C (3) = 200 − 1650 °C (4) = RT − 700 °C

are derived from a two-dimensional reinforcement with a 0°/90° fibre orientation. C/SiC composites, processed via CVI have a somewhat higher density than the liquid phase derived composites due to their crystalline matrix. The precursor route of the LPI-process leads to an amorphous microporous SiC matrix, derived from the Si-polymer which can be optionally combined with a crystalline phase of primary SiC powder. The matrix of C/C-SiC composites, manufactured by LSI, is much more complex: crystals of silicon carbide in different sizes (<100 μm), unreacted silicon and amorphous carbon form the heterogeneous matrix, in which comparable high amounts of fibres (typical fibre volume contents are 50–65%) are embedded. Figure 13 shows some characteristic SEM-micrographs of CVI-, LPI- and LSI-composites.

All composites in Table 1 are derived from preforms with a fibre coating based on carbon, except for the LSI-C/C-SiC composites. As a consequence, these composites show comparable lower values of strain to failure (0.15 to 0.35%) and lower strength levels. The open porosity of CVI- as well as LPI-C/SiC composites lies between 10 and 20%, whereas melt-infiltrated C/C-SiC composites typically vary between 2 and 5%.

C/SiC composites show a quasi-ductile fracture behaviour, derived from mechanisms like crack deflection and fibre pullout. Figure 14 shows exemplarily these effects within a C/C-SiC composite. The linear-elastic behaviour of C/SiC is less pronounced than for example SiC/SiC composites due to the inherent microcracks in the matrix which occur during cooling-down from processing to room temperature because of the high thermal mismatch between C-fibres and SiC-matrix.

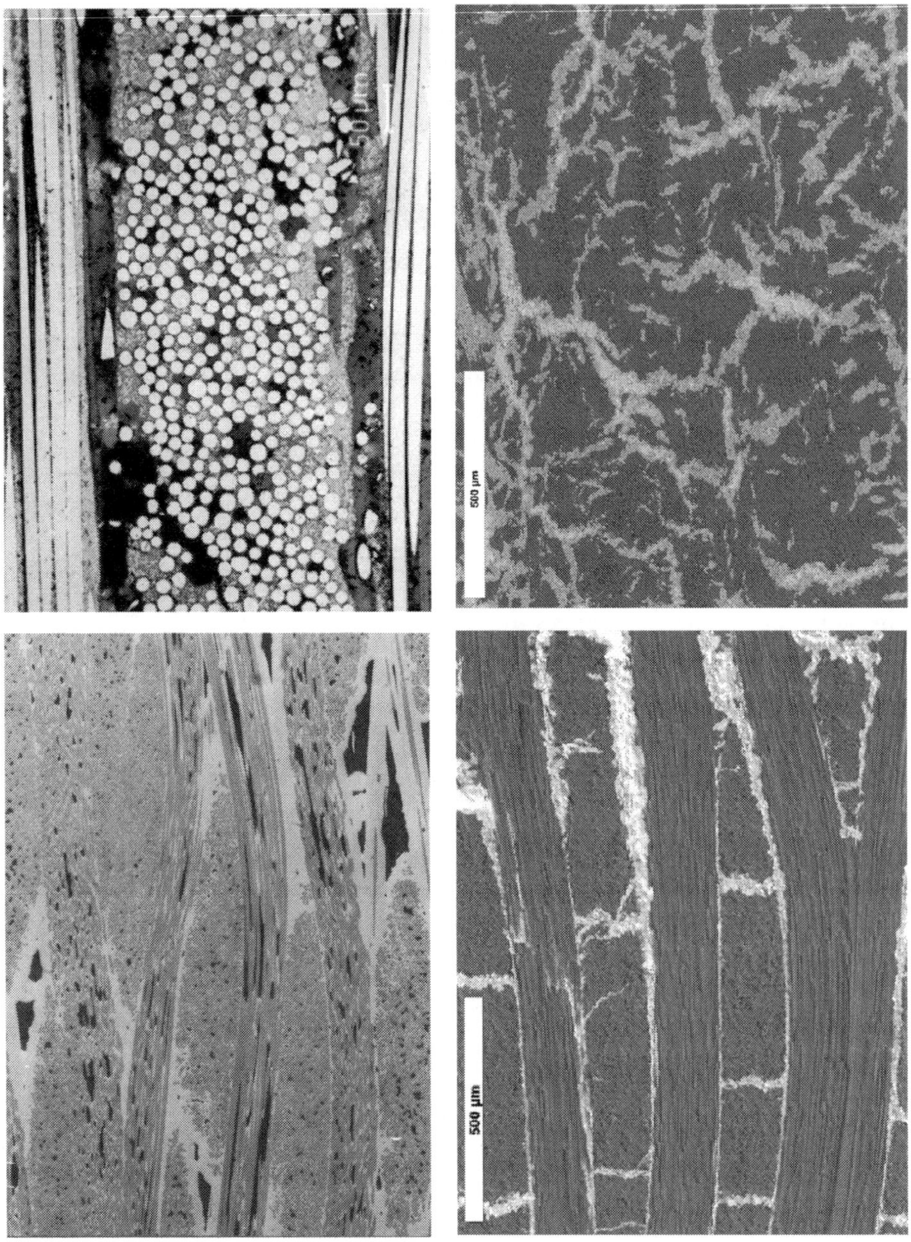
500 µm
500 µm

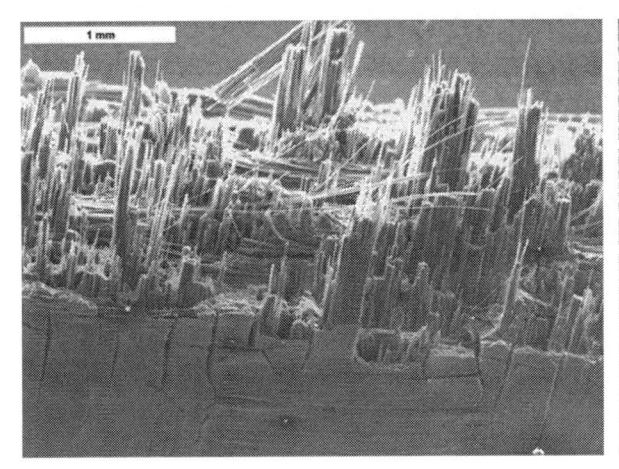

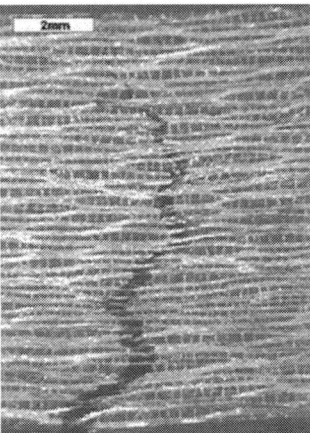

FIGURE 14. Crack deflection and fibre pull-out in a 2D C/C-SiC composite, loaded under bending stresses

One common characteristic of all C/SiC composites is their distinct anisotropy in the mechanical as well as thermophysical properties. Considerable lower values of the tensile strength and the strain to failure have to be considered for an appropriate design if the load direction and the fibre alignment are not congruent. As the carbon fibres show a different physical behaviour in longitudinal and radial direction, the composite's properties like thermal conductivity and coefficient of thermal expansion differ widely with respect to the in-plane or transverse direction.

Designing with C/SiC composites, manufactured via the liquid phase routes LPI and LSI, principally follows the same rules as exist for anisotropic CFRP composite materials. However, due to the high processing temperatures and the irreversible shrinkage of the matrix the influence of the fibre anisotropy on the macroscopic dimensional stability must be taken into account. The matrix shrinkage is impeded by the fibres and, dependent on the fibre/matrix bonding forces, dimensional changes in direction of the fibres are prevented. In contrast, transverse to the fibre alignment the composite's shrinkage is unhindered. As a result, angled 2 D-plates show spring forward effects, which occur irreversibly during pyrolysis or reversibly during their use in intervals of high temperature differences. Moreover, the transverse shrinkage of the matrix in C/SiC components with a closed contour such as wound tubes can result in additional matrix stresses in radial direction. By the use of more isotropic preforms like short fibre or three-dimensional reinforcements these anisotropy effects on the dimensional stability of C/SiC components can be minimized.

←

FIGURE 13. SEM micrographs of different C/SiC composites

a CVI-C/SiC (magnification 100×)
   Ref.: MAN-T

b LPI-C/SiC (0°/90° multilayer)
   Ref.: DaimlerChrysler Research and Technology / Dornier

c LSI-C/C-SiC (woven fibres)
   Ref.: DLR

d LSI-C/C-SiC (chopped fibres)
   Ref.: DLR

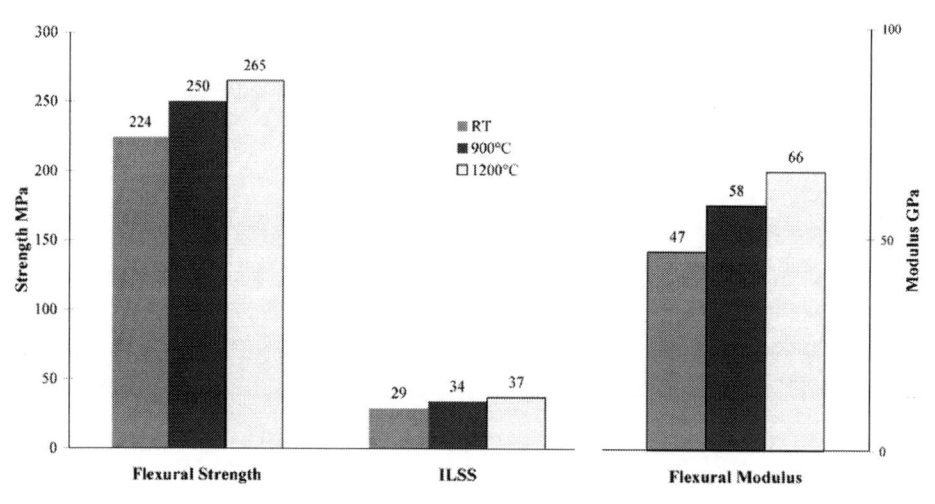

FIGURE 15. High temperature properties of C/C-SiC composites

In contrast to most other structural materials, C/SiC and C/C-SiC composites retain their strength level at elevated temperatures, similar to carbon/carbon materials. Besides, their high temperature strength is superior to the level at room temperature, i.e. the higher the temperature, the higher the strength (Figure 15). The data given in Table 1 are determined at room temperature, but by reason of this aspect they can be used also for a first design approach at high temperatures.

(4.2) Mechanical and thermal properties

There exist several independent manufacturers of C/SiC materials utilizing different processing routes. In the following, representative characteristics for the three main processes CVI, LPI and LSI are summarized in the Tables 2 to 8 to give an overview on the

TABLE 2. Material data of CVI- and PIP-C/SiC composites with different fibre reinforcements [6]

Property	Unit	CVI-C/SiC[1]	PIP-C/SiC[2]	
Fibre content	Vol%	42–47	42–44	45–50
Density	g/cm^3	2.1–2.2	1.7–1.8	1.6–1.8
Porosity	%	10–15	15–20	10–25
Tensile strength	MPa	300–380	240	200–250
Strain to failure	%	0.6–0.9	0.8–1.1	0.3–0.5
Young's modulus	GPa	90–100	60–80	70–80
Flexural strength	MPa	450–500	330	250
ILSS	MPa	44–48	30–35	10–12
Coefficient of thermal expansion ∥	10^{-6} K^{-1}	3	3	2–3
Coefficient of thermal expansion ⊥	10^{-6} K^{-1}	5	4	4–7

[1] Woven fabric reinforcement 0°/90°
[2] UD cross ply

TABLE 3. Mechanical and thermophysical properties of LPI-C/SiC [9]

Fibre	T 800/6K				Density	1.8 g/cm^3
Fibre coating	pyC				Porosity	10%
Manufacture temp.	1,600 °C				Fibre volume content	46%

		RT			RT after 1600 °C/10 h Argon	1600 °C vacuum
Lay-up		UD	0°/90°	45°/45°	0°/90°	0°/90°
Tensile strength	MPa	470	250	55	260	266
Strain to failure	%	0.5	0.5	0.56	0.54	0.37
Young's modulus	GPa	145	65	31	65	75
Flexural strength	MPa	–	500	–	–	–
Compression strength	MPa	1080	590	–	–	–
ILSS	MPa	–	10.4	–	11	–

			100 °C	1000 °C	1500 °C
Thermal conductivity	W/mK	∥	11.33	12.62	12.18
		⊥	5.24	5.48	5.29
Coefficient of thermal expansion	10^{-6} 1/K	∥	1.16	2.19	2.68
		⊥	4.06	5.72	6.23
Heat capacity	J/kg K		900	1400	1600

TABLE 4. Mechanical properties of C/SiC composites reinforced with plain-weave fabrics [44]

			RT	1200 °C
Fibre content	Vol%		45	
Density	g/cm^3		2.1	
Porosity	%		10	
Tensile strength	MPa		518	544
Strain elongation	%		1.06	1.10
Young's modulus	GPa		75.2	114.5
Compression strength	MPa		544	–
Contraction under compression	%		0.38	–
Compressive modulus	GPa		102	–
4 pt-flexural strength	MPa		450	–
ILSS	MPa		34.5	–
Fracture toughness	MPa √m		27.2	–

TABLE 5. Mechanical properties of heat-treated C/SiC composites based on plain-weave fabric from T-300 fibre [44]

			RT	1200 °C
Fibre content	Vol %		40	
Density	g/cm^3		2.1	
Porosity	%		10	
Tensile strength	MPa		411	321
Strain elongation	%		0.79	0.45
Young's modulus	GPa		84.9	142.8
Compressive strength	MPa		497	–
ILSS	MPa		27.6	–

TABLE 6. Design values for five different LSI-C/C-SiC composites [45]

Property	Unit		XB	XT	XD	XG	SF
Density	$10^3 kg/m^3$		1.9	1.9	2.3	2.2	2.0–2.1
Open porosity	%		3.5	3.5	1.0	<5	<3
Interlaminar shear strength	MPa		28	33	–	–	–
Flexural strength	MPa		160	300	80	65–80	90–140
Tensile strength	MPa		80	190	30	–	–
Strain to failure	%		0.15	0.35	0.04	–	0.15–0.25*
Young's modulus	GPa		60	60	100	–	50–70*
Coefficient of thermal expansion	$10^{-6} 1/K$	100 °C ∥	–1	–1	1.5	–	0.5
(Reference temperature 25°C)		100 °C ⊥	2.5	2.5	4.5	–	1.0
Thermal conductivity	W/mK	200 °C ∥	18.5	22.6	33.7	–	–
		200 °C ⊥	9.0	10.3	18.2	–	25–30

* measured from flexural tests

material's potential. In most cases, only RT-values of the usual mechanical and thermal properties are published. If available, also HT-properties and correlations with the fibre orientation of the preform are included. However, only few information on the type of fibre, the phase composition of the matrix, the testing method and sample size are given by the manufacturers. To compare the individual composite materials, this lack of transparency has to be taken into account and additional information must be obtained from the specialist literature or directly from the producer.

The dynamic behaviour of C/SiC composites is generally very high with an outstanding fatigue behaviour under cyclic loading. The inherent microcrack pattern of the matrix prevents the propagation of matrix cracks and results in a high number of load cycles without showing an essential decrease of the residual strength level. The good low cycle fatigue

TABLE 7. Properties of LSI-C/SiC for brake disk applications [46]

Mechanical properties			
Density		g/cm³	2.4
Open porosity		%	<1
Tensile strength (RT)		MPa	40
Failure strain		%	0.4
3 pt-bending strength (RT)		MPa	80
3 pt-bending modulus (RT)		GPa	30
Phase composition	SiC	%	60
	C	%	30
	Si	%	10
Thermophysical properties			
Maximum temperature		°C	1350
Coeff. of therm. expansion	0 °C–300 °C	10^{-6} 1/K	1.8
	300 °C–1200 °C	10^{-6} 1/K	3.0
Thermal conductivity	20 °C	W/mK	40
	1200°C	W/mK	20
Specific heat capacity	20 °C	J/kg K	800
	1200°C	J/kgK	1200

TABLE 8. Properties of LSI-C/C-SiC composites for brake disk applications [5]

Material grade	Unit	CF226 P75	CF226 P76	CF 226/1 P77	CF 226/2 P77	FU 2952 P77	FU 2905/4 P77
Reinforcement	–	fabric	fabric	fabric	fabric	short fibre	felt
Density	g/cm3	>1.8	>1.8	>1.8	>1.95	>2.0	>1.55
Silicon uptake	%	>20	>20	>25	>45	>50	>20
Flexural strength	MPa	180–200	220–240	140–160	130–140	60–80	55–70
Flexural modulus	GPa	55–60	65–70	50–55	55–60	23–27	12–17
Strain to Failure	%	0.30–0.35	0.25–0.30	0.25–0.30	0.23–0.27	0.20–0.26	0.6–0.8
ILLS	MPa	14–18	17–20	15–17	14–17	7–11	–
Thermal conductivity (x/y-direction)	W/mK	12–15	12–15	12–15	18–22	18–23	15–22
Thermal conductivity (z-direction)	W/mK	30–35	30–35	28–33	30–35	28–33	–
CTE 25–800 °C (x/y-direction)	$10^{-6} K^{-1}$	1.0–1.5	1.0–1.5	1.0–1.5	0.8–1.3	1.2–1.6	2.4–3.2
CTE 25–800 °C (z-direction)	$10^{-6} K^{-1}$	5.5–6.0	5.5–6.0	6.0–6.5	5.5–6.0	5.2–5.6	–

(LCF) resistance in combination with a high notch insensitivity underlines the applicability of C/SiC composites as structural ceramic materials.

The thermal properties of C/SiC as well as C/C-SiC composites given in the tables refer to inert ambient conditions to eliminate oxidative influences on the fibres. For a long term use in air the composites have to be protected with an appropriate external coating. Such coating systems are mostly based on brittle materials like CVD-SiC with much lower allowable fracture elongations than the substrate with the result, that the allowable design values of strength and strain must be reduced considerably. Similar to the mechanical properties the thermal properties depend on the fibre architecture of the preform and show a high degree of anisotropy.

Figure 16 shows the relationship between the material's density and the transverse thermal conductivity of a C/C-SiC composite. Generally, the lower values correspond with continuous fibre reinforcements (i.e. fabrics), whereas the highest densities refer to short fibre architectures. As for most other materials, the thermal conductivity of C/C-SiC decreases with higher temperatures. Figure 17 demonstrates this relationship for different material modifications which have been manufactured by varying the fibre preform and the processing conditions. The average decrease of transverse conductivity between 300 °C and 900 °C amounts to approximately 30%.

In the temperature range of up to about 200 °C, the coefficient of thermal expansion (CTE) of the carbon fibres in fibre direction are generally negative, and they offer the option to create composites which are tailored to render a specific expansion behaviour by combining the positive expansion coefficient of the SiC matrix with a varied orientation of the fibres.

The thermal expansion coefficient of two-dimensional C/C-SiC is already generally low over a very large temperature range. By specific modification of the C/C and SiC proportions

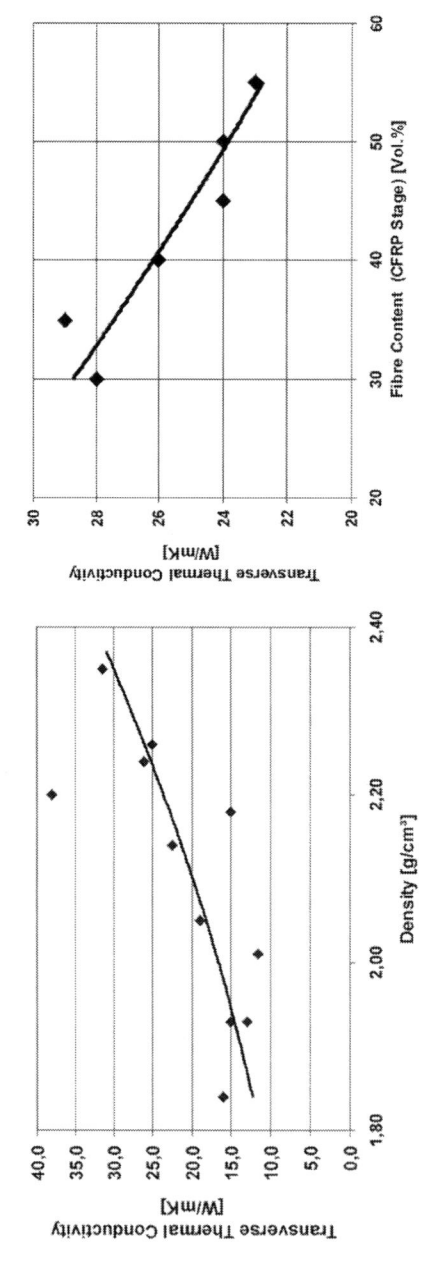

FIGURE 16. Thermal conductivity at 50 °C versus fibre volume content and versus density of short fibre reinforced C/C-SiC composites

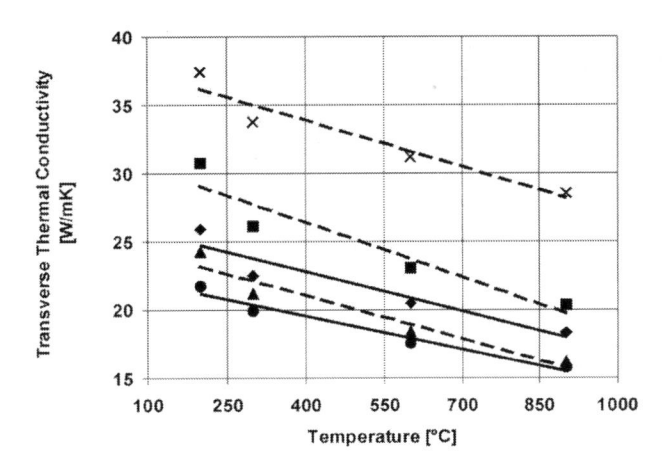

FIGURE 17. Transverse thermal conductivities of different C/C-SiC composites versus temperature

in the C/C-SiC composite, the negative expansion behaviour of the embedded C-fibres may be, for example, compensated by an increase of the ceramic content, thereby rendering an altogether expansion-neutral behaviour.

Figure 18 depicts these influences of the ceramic content on the CTE in fabric-reinforced C/C-SiC composites. Pure carbon/carbon and monolithic SiC establish the respective limits of a potential material modification for very low or high expansion coefficients.

(4.3) Frictional properties

Among all other functional properties of C/SiC and C/C-SiC composites the frictional properties are of major interest because of their high economical potential when used as high performance braking materials. High heat fluxes from the outer friction surface to the centre of the composite are necessary to avoid overheating of the friction surface, i.e. high transverse conductivities must be adjusted in order to achieve high coefficients of friction (CoF) and

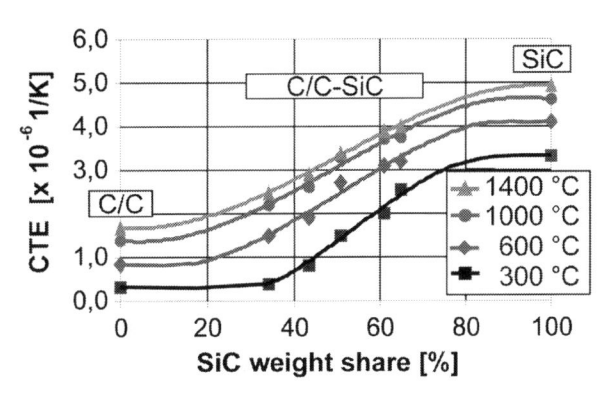

FIGURE 18. Coefficient of thermal expansion of fabric-reinforced C/C-SiC in relation to the SiC content

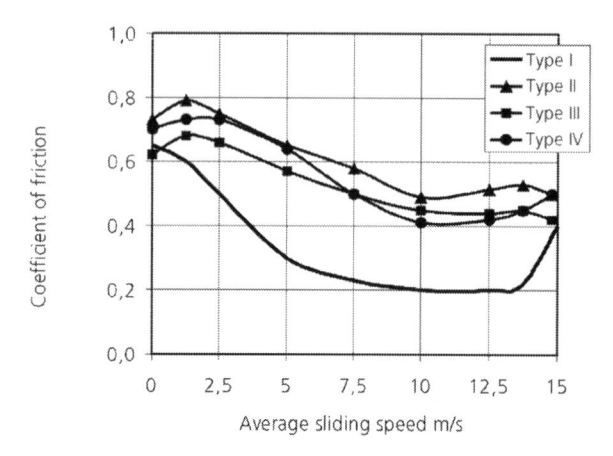

FIGURE 19. CoF of different types of C/C-SiC composites [12] (Type I: 2D reinforcement, Type II: HM-fibres, Type III: axial fibre orientation, Type IV: high SiC-content)

low wear rates. Standard C/C-SiC composites with bidirectional reinforcements have been modified in their composition and microstructure and lead to an essential improvement of heat conductivity and CoF stability (Figure 19). Particularly, by increasing the silicon carbide content of the composite the transverse thermal conductivity was doubled in comparison to the original space materials.

However, high SiC levels and low carbon contents also influence the mechanical properties and decrease the damage tolerance of the brake disk. As high ceramic contents are mainly necessary in the outer region of the friction surfaces, two approaches for a further material improvement were pursued [47]:

- Gradient C/C-SiC composites with a gradual increase of SiC from the centre to the surfaces
- Homogeneous C/C-SiC composites with SiC-rich coatings on the surfaces

Both approaches successfully fulfill the conflicting requirements on high ductility, high hardness and good wear resistance. The coefficient of thermal expansion of the ceramic-rich surface is considerably higher than the CTE of the C/C-SiC substrate. Depending on this CTE mismatch, a more or less microcracked surface occurs as a result of the tensile stresses within the outer region during cooling after processing. The formation of the microcrack pattern depends on the fibre architecture of the core material. The most pronounced crack pattern with plenty of randomly orientated cracks can be observed for bi-directionally reinforced C/C-SiC composites whose thermal shrinkage during cooling is close to zero. Short fibre reinforcements with their higher CTE can lead to nearly crack-free surfaces. The width of the microcracks depends on the thickness of the SiC-layer. In general, the thicker the layer, the wider the surface cracks. During braking, when the coating is heated up, the cracks get narrower as the outer region expands more than the substrate. All cracks normally run through the total thickness of the layer, but stop in the ductile core region and no breakage of the fibres occurs.

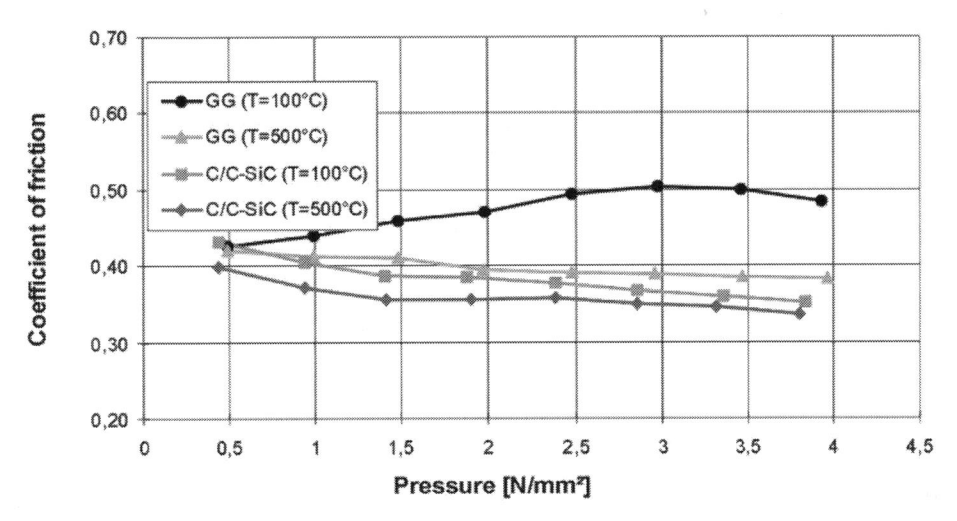

FIGURE 20. Coefficient of friction for grey cast iron (GG) and SiSiC coated C/C-SiC brake disks in combination with organic pads. Sliding velocity between 3.2 and 8.5 m/s

In the last few years, many tribological tests for different brake applications have been conducted in order to optimize this tribosystem. In particular for disk brakes of high performance cars combinations with organic and sintermetallic pads were developed which allow operating times comparable to the lifetime of a car (300,000 km). Most of the test results as well as the material's compositions are confidential and not published. At least, some values are depicted in the Figures 20 and 21 showing some representative frictional properties of C/C-SiC composites and demonstrating their superior stability against fading.

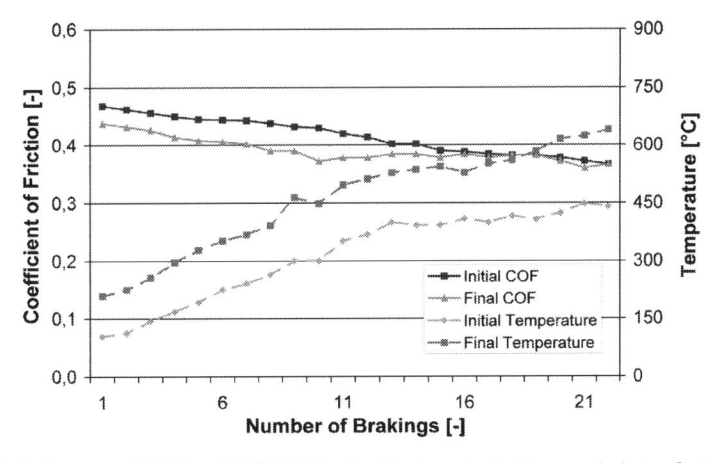

FIGURE 21. Fading tests of SiSiC coated C/C-SiC brake disks (repeated brakings on the hot surfaces by vehicle's deceleration from 134 to 44 km/h)

SUMMARY

Originally, the demands in space technology played the decisive role in the development of carbon fibre reinforced silicon carbide composites (C/SiC, C/C-SiC). High mass-specific characteristics and extreme temperature resistance are important selection criteria for materials used in jet engines and thermal protection systems of new spacecraft and rockets. Within the last few years, the properties and the manufacturing methods were consistently improved, so that now also the industry in general can share in the profits of this new class of materials. The liquid silicon infiltration (LSI) process is regarded as the most promising process for industrial products, especially if the aspect of costs is considered.

C-fibre reinforced SiC composites differ widely in their microstructures and properties, depending on the applied processing method. Typically, their matrix is microporous and microcracked with a considerably heterogeneous structure. The microstructures can be tailored by different methods, e.g. fibre coatings, fibre surface preparations or variations of the fibre alignment covering the whole range from nearly dense to fairly graded composites. In particular, C/C-SiC composites show superior tribological properties in comparison to grey cast iron or carbon/carbon. In combination with their low density, high thermal shock resistance and good abrasive resistance, these CMCs are promising candidates for advanced brake and clutch systems. High improvements in wear resistance were achieved by SiSiC coatings. Almost wear-free brake disks in combination with acceptable low wear rates for pads show a high potential for lifetime brake disks in passenger cars.

The utilization of C/SiC and C/C-SiC materials has been expanded to also cover quite different fields of lightweight constructions, such as casings and construction elements for optical systems, charging devices for heat treatment, calibrating plates or lightweight armor for ballistic protection.

Due to the beginning industrialization of the manufacturing processes, which has just started out, the designers and also the users are lacking practical experience with this new structural ceramic material. Design rules, as have been developed for other composite materials, metals or ceramics without reinforcement, are only available in limited extent until now. This will, however, change in the near future, if one succeed in making full use of the production capacities built up so far, and in gaining a broad basic knowledge as well as application-specific experience.

Until now, low empirical information has been gained regarding the production of C/C-SiC components in larger quantities. The manufactured quantities are still modest (approximately 10,000 parts per year), and they are burdened with high specific component-costs of about 250 Dollar/kg. The growth of production facilities among several manufacturers and cost-reducing technologies are measures already being applied, which allow considerably lower material costs to be anticipated for the future.

REFERENCES

1. G. Savage, Carbon/Carbon Composites, Chapman & Hall (1993).
2. V. I. Trefilov, Ceramic- and Carbon-Matrix Composites, Chapman & Hall (1995).
3. K. K. Chawla, Ceramic Matrix Composites, Chapman & Hall (1993).

4. D. B. Fischbach and D. R. Uptegrove, in *Proceedings of 13th Biennial Confernce on Carbon* (1977).
5. R. Weiss, Carbon Fibre Reinforced CMCs: Manufacture, Properties, Oxidation Protection, *High Temperature Ceramic Matrix Composites* (Eds.: W. Krenkel, R. Naslain, H. Schneider), WILEY-VCH, Weinheim, Germany, (2001), p. 440–456.
6. A. Mühlratzer and M. Leuchs, Applications of Non-Oxide CMCs, *High Temperature Ceramic Matrix Composites* (Eds.: W. Krenkel, R. Naslain, H. Schneider), WILEY-VCH, Weinheim, Germany (2001), p. 288–298.
7. H. Hald, H. Weihs, B. Benitsch, I. Fischer, T. Reimer, P. Winkelmann and A. Gülhan, Development of a Nose Cap System for X-38, in *Proceedings of International Symposium Atmospheric Reentry Vehicles and Systems*, Arcachon, France (1999).
8. U. Trabandt and W. Fischer, in *Proceedins of the Int. Congress on Environmental Systems*, Orlando, FL, USA, paper 01-ICES-184 (2001).
9. W. Schäfer and W. D. Vogel, Faserverstärkte Keramiken hergestellt durch Polymerinfiltration, *Keramische Verbundwerkstoffe* (Ed.: W. Krenkel), WILEY-VCH, Weinheim, Germany (2003), p. 76–94.
10. F. Christin, Design, Fabrication and Application of Thermostructural Composites (TSC) like C/C, C/SiC and SiC/SiC Composites, *Advanced Engineering Materials*, **4**, No. 12 (2002), p. 903–912.
11. W. Krenkel, CMC Materials for High Performance Brakes, in *Proceedings ISATA Conference on Supercars*, Aachen (1994).
12. W. Krenkel, B. Heidenreich and R. Renz, C/C-SiC Composites for Advanced Friction Systems, *Advanced Engineering Materials* **4,** No. 7 (2002) p. 427–436.
13. S. Vaidyaraman, M. Purdy, T. Walker and S. Horst, C/SiC Material Evaluation for Aircraft Brake Applications, *High Temperature Ceramic Matrix Composites* (Eds.: W. Krenkel, R. Naslain, H. Schneider), WILEY-VCH, Weinheim, Germany (2001), p. 802–808.
14. M. Krupka and A. Kienzle, Fiber Reinforced Ceramic Composites for Brake Discs, *SAE Technical Paper Series 2000-01-2761* (2000).
15. R. Gadow and M. Speicher, Manufacturing of Ceramic Matrix Composites for Automotive Applications, Ceramic Transactions **Vol. 128** (2001), p. 25–41.
16. Z. Rak, C_F/SiC/C Composites for Tribological Application, *High Temperature Ceramic Matrix Composites* (Eds.: W. Krenkel, R. Naslain, H. Schneider), WILEY-VCH, Weinheim, Germany (2001), p. 820–825.
17. R. Renz, W. Krenkel, C/C-SiC Composites for High Performance Emergency Brake Systems, in *Proceedings of 9th European Conference on Composite Materials (ECCM-9)*, Brighton, UK (2000).
18. B. Heidenreich, R. Renz and W. Krenkel, Short Fibre Reinforced CMC Materials for High Performance Brakes, *High Temperature Ceramic Matrix Composites* (Eds.: W. Krenkel, R. Naslain, H. Schneider), WILEY-VCH, Weinheim, Germany (2001), p. 809–815.
19. R. Kochendörfer and N. Lützenburger, Applications of CMCs made via the Liquid Silicon Infiltration (LSI) Technique, *High Temperature Ceramic Matrix Composites* (Eds.: W. Krenkel, R. Naslain, H. Schneider), WILEY-VCH, Weinheim, Germany (2001), p. 277–287.
20. R. Renz, B. Heidenreich, W. Krenkel, A. Schöppach and F. Richter, CMC Materials for Lightweight and low CTE Applications, *High Temperature Ceramic Matrix Composites* (Eds.: W. Krenkel, R. Naslain, H. Schneider), WILEY-VCH, Weinheim, Germany (2001), p. 839–845.
21. J. Schmidt, M. Scheiffele and W. Krenkel, Engineering of CMC Tubular Components, *High Temperature Ceramic Matrix Composites* (Eds.: W. Krenkel, R. Naslain, H. Schneider), WILEY-VCH, Weinheim, Germany (2001), p. 826–831.
22. M. Labanti, G. Martignani, C. Mingazzini, G. L. Minoccari, L. Pilotti, A. Ricci and R. Weiss, Evaluation of Damage by Oxidation Corrosion at High Temperatures of Coated C/C-SiC Ceramic Composite, *High Temperature Ceramic Matrix Composites* (Eds.: W. Krenkel, R. Naslain, H. Schneider), WILEY-VCH, Weinheim, Germany (2001), p. 218–223.
23. ECM, Cesic® *Kohlefaserverstrktes Siliciumcarbid, Produktinformation*.
24. B. Heidenreich, W. Krenkel and B. Lexow, Development of CMC-Materials for Lightweight Armor, in *Ceram. Eng. Sci. Proc.*, **24** [3] (2003), p. 375–381.
25. R. Naslain and F. Langlais, CVD Processing of Ceramic-Ceramic Composites Materials, *Mat. Science Research*, **Vol. 20**, Plenum Press New York (1985), p. 145–164.
26. D. P. Stinton, A. J. Caputo and R. A. Lowden, Synthesis of Fibre-Reinforced SiC Composites by Chemical Vapour Infiltration, *J. Amer. Ceram. Society Bulletin*, **Vol. 65**, No 2 (1986), p. 347.

27. M. Leuchs and A. Mühlratzer, CVI-Verfahren zur Herstellung faserverstrkter Keramik – Herstellung, Eigenschaften, Anwendungen, *Keramische Verbundwerkstoffe* (Ed.: W. Krenkel), WILEY-VCH, Weinheim, Germany (2003), p. 95–121.
28. R. J. Diefendorf and R. P. Boisvert, Siliciumcarbid-Composites durch polymere Ausgangsstoffe, in *Proceedings of Verbundwerk 1988*, Demat Exposition Managing, Frankfurt, Germany (1988), p. 13.01–13.37.
29. T. Haug, R. Ostertag and W. Schfer, Fiber Reinforced Ceramics for Aerospace Applications, *Advanced Materials and Structures from Research to Application* (Eds.: J. Brandt, H. Hornfeld, M. Neitzel), SAMPE European Chapter (1992), p. 163–173.
30. A. Mühlratzer, K. Handrick and H. Pfeiffer, *Acta Astronautica*, **42** (1998), p. 533–540.
31. C. C. Evans, A. C. Parmee and R. W. Rainbow, Silicon Treatment of Carbon Fiber-Carbon Composites, in *Proceedings of 4th London Conference on Carbon and Graphite* (1974), p. 231–235.
32. W. B. Hillig, R. L. Mehan, C. R. Morelock, V. I. DeCarlo and W. Laskow, Silicon/Silicon Carbide Composites, *Ceramic Bulletin*, **Vol. 54**, No. 12 (1975).
33. R. Gadow, Die Silizierung von Kohlenstoff, *Doctoral Thesis*, University of Karlsruhe (1986).
34. W. Krenkel, Development of a Cost Efficient Process for the Manufacture of CMC Components, *Doctoral Thesis,* University of Stuttgart, DLR-Forschungsbericht 2000–4 (2000).
35. W. Krenkel, Cost Effective Processing of CMC Composites by Melt Infiltration (LSI-Process), *Ceramic Engineering and Science Proceedings* (Ed: American Ceramic Society), **Vol. 22**, Issue 3 (2001), p. 443–454.
36. W. Krenkel and H. Hald, Liquid Infiltrated C/SiC – An Alternative Material for Hot Space Structures in *Spacecraft Structures and Mechanical Testing, European Space Agency Publications Division*, Paris, France, ESA SP-289 (1989), p. 325.
37. R. Kochendörfer and W. Krenkel, CMC Intake Ramp for Hypersonic Propulsion Systems, *High-Temperature Ceramic-Matrix Composites I: Design, Durability and Performance* (Eds.: A. G. Evans, R. Naslain), Ceramic Transactions, **Vol. 57** (1995), p. 13–22.
38. J. Schulte-Fischedick, A. Zern, J. Mayer, M. Rühle, M. Frieß, W. Krenkel and R. Kochendörfer, The Morphology of Silicon Carbide in C/C-SiC Composites, *Journal of Material Science and Engineering A*, Elsevier Science B.V. (2002), p. 146–152.
39. W. Krenkel, R. Renz and B. Heidenreich, Lightweight and Wear Resistant CMC Brakes, *Ceramic Materials and Components for Engines* (Eds.: J. G. Heinrich, F. Aldinger), WILEY-VCH, Weinheim, Germany (2001), p. 63–67.
40. C. A. Nannetti, A. Borello and D. A. de Pinto, C-Fiber Reinforced Ceramic Matrix Composites by a Combination of CVI, PIP and RB, *High Temperature Ceramic Matrix Composites* (Eds.: W. Krenkel, R. Naslain, H. Schneider), WILEY-VCH, Weinheim, Germany (2001), p. 368–374.
41. M. Frieß and W. Krenkel, Silicon Based Preceramic Polymers in Combination With the LSI-Process in *Proceedings of Materials Week 2000,* International Congress on Advanced Materials their Processes and Applications, Munich, Germany (2002).
42. A. Mühlratzer, Production, Properties and Applications of Ceramic Matrix Composites, *CFI/Ber. DKG 76*, No. 4 (1999), p. 30–35.
43. D. Desnoyer, A. Lacombe and J. M. Rouges, Large Thin Composite Thermo-structural Parts in *Proceedings of Intern. Conference "Spacecraft Structures and Mechanical Testing",* Noordwijk/Netherlands, (1991), ESA/ESTEC.
44. Data sheet from GE Power Systems, Engineering Data of Power Systems Composites, LCC, Newark, Delaware, (2003).
45. W. Krenkel, Designing with C/C-SiC Composites, in *Advances in Ceramic Matrix Composites IX* (Eds.: N.P. Bansal, J.P. Singh, W.K. Kriven, H. Schneider), *Ceram. Trans.*, **153** (2003), p. 103–123.
46. Data sheet from SGL Carbon Group, SIGRASIC 6010 GNJ – Faserverstärkte Keramik für Bremsscheiben (1998).
47. W. Krenkel, C/C-SiC Composites for Hot Structures and Advanced Friction Systems in *Ceram. Eng. Sci. Proc.*, **24** [4] (2003), p. 583–592.

7

Silicon Carbide Fiber-Reinforced Silicon Nitride Composites

Ramakrishna T. Bhatt

U.S. Army Vehicle Technology Directorate
NASA Glenn Research Center
Cleveland, OH 44135
Phone# (216)433-5513
Ramakrishna.T.Bhatt@grc.nasa.gov

ABSTRACT

The fabrication and thermo-mechanical properties of SiC monofilament- and SiC fiber tow-reinforced reaction-bonded silicon nitride matrix composites (SiC/RBSN) and fully dense SiC monofilament-reinforced silicon nitride matrix composites (SiC/Si$_3$N$_4$) have been reviewed. The SiC/RBSN composites display high strength and toughness, and low thermal conductivity in the as-fabricated condition, but are susceptible to internal oxidation and mechanical property degradation in the intermediate temperature range from 400 to 1100°C. The internal oxidation problem can be avoided by applying functionally graded oxidation resistant coatings on external surfaces of the composite. On the other hand, the fully dense SiC/Si$_3$N$_4$ composites fabricated by using SiC monofilaments and pressure assisted sintering methods display moderate strength and limited strain capability beyond matrix cracking stress. Advantages, disadvantages, and potential applications of SiC fiber-reinforced silicon nitride matrix composites fabricated using various fiber types and processing methods are discussed.

I. INTRODUCTION

The realization of improved efficiency for engines used for aero and space propulsion as well as for land-based power generation will depend strongly on advancements made in the upper use temperature and life capability of the structural materials used for the engine hot-section components. Components with improved thermal capability and longer life between maintenance cycles will allow improved system performance by reducing cooling requirements and life-cycle costs. This in turn is expected to reduce fuel consumption, to allow improved thrust-to-weight and performance for space and military aircraft; and to reduce emissions and power costs for the electrical power industry.

Currently, the major thrust for achieving these benefits is by the development of fiber-reinforced ceramic matrix composites (FRCMC). These materials are not only lighter and capable of higher use temperatures than state-of-the-art metallic alloys (~1100°C), but also capable of providing significantly better static and dynamic toughness than monolithic ceramics while maintaining their advantages; namely, high temperature strength, low density, erosion and oxidation resistance. However, FRCMC are more difficult to process than monolithic ceramics. The requirement for stable multifunctional fiber-matrix interfaces for optimized composite mechanical properties combined with the relative instability of many fibers with respect to temperature, atmosphere, and chemical interaction with the matrix places stringent limitations on FRCMC processing temperatures and times. For the last twenty years, a variety of FRCMC such as silicon carbide fiber reinforced silicon nitride, silicon carbide fiber reinforced silicon carbide and alumina fiber reinforced alumina matrix composites have been developed. This review covers only SiC fiber-reinforced silicon nitride matrix composites. The other composite systems will be discussed in other chapters of this hand book.

The interest in development of SiC/Si_3N_4 composites stems from the fact that: (a) SiC and Si_3N_4 are thermodynamically compatible and stable at temperatures to 1700°C, (b) silicon nitride processing temperature can be tailored to avoid fiber degradation, (c) silicon nitride matrix microstructure can be controlled to improve composite properties such as matrix cracking strength and thermal conductivity.

The fabrication of SiC fiber reinforced Si_3N_4-matrix begins by the use of monofilaments, fiber tows, and textile processes (weaving, braiding) to form multi-fiber bundles or tows of ceramic fibers into 2D and 3D fiber architectures or preforms that meet product size and shape requirements. For assuring crack deflection between the fibers and final matrix, a thin fiber coating or interphase material with a mechanically weak microstructure, such as boron nitride (BN) or pyrolytic carbon is then applied to the fiber surfaces by chemically vapor infiltration (CVI). Silicon nitride-based matrices are then formed within the coated preforms by a variety processes: (1) CVI of precursor ammonia and silicon-containing gases that react on the preform surfaces to leave a dense Si_3N_4 product; (2) reaction-bonded silicon nitride (RBSN) where silicon powder in the fiber preform is converted to silicon nitride matrix; (3) polymer infiltration and pyrolysis (PIP) in which a pre-ceramic Si_3N_4-forming polymer is repeatedly infiltrated into open porosity remaining in the composite preform and then pyrolyzed by high-temperature thermal treatments; (4) hot pressing (HP) and (5) hot-isostatic pressing (HIP) in which fiber preform filled with silicon nitride powder and sintering additives is uniaxially or iso-statically pressed at high temperatures.

The pressureless sintering method commonly used for the fabrication of monolithic silicon nitride ceramics cannot be successfully adapted for the fabrication of SiC fiber reinforced silicon nitride matrix composites because of significant strength degradation of currently available fibers at the high temperatures required for sintering, and also because of the problems associated with retardation of densification of the matrix material by the presence of already dense fibers. Fabrication of SiC/Si$_3$N$_4$ composites by CVI was not successful because of the difficulty of maintaining stoichiometry of silicon nitride, and infiltrating into thick sections of the preforms. Also, the textile process for the manufacture of 2-D woven and 3-D braided SiC/Si$_3$N$_4$ has not been investigated. Only SiC monofilament- and SiC fiber tow-reinforced Si$_3$N$_4$ matrix composites fabricated by reaction-bonding, or hot pressing, or hot-isostatic pressing have been reported in the ceramic literature.

II. SiC FIBER-REINFORCED REACTION-BONDED SILICON NITRIDE COMPOSITES

Fabrication of a fiber reinforced reaction-bonded silicon nitride involves heating a fiber-reinforced porous silicon powder preform in pure nitrogen or nitrogen containing gas mixtures at high enough temperature for silicon to react with nitrogen to form silicon nitride. The material produced by the process of converting silicon-to-silicon nitride matrix by gas phase reaction is referred to as reaction-bonded silicon nitride (RBSN). The nitrogen gas diffuses from the outer surfaces of the porous fibrous preform towards its center. A slow, controlled increase in processing temperature can result in reacted material filling the internal pores with no shrinkage of the overall body. In order for the reaction to proceed to completion, a continuous pore structure must be maintained for diffusion of nitrogen. If the number of pores and pore channels is decreasing, then the depth to which the reactions may proceed becomes a limitation. The fiber stability, green density of the fiber preform, particle size and purity of silicon powder, purity of the nitriding gas, and the thickness of the preform are some of the important factors that control the nitriding temperature and time. The greater the density, the less the depth to which the reaction may proceed. The nitrided composites generally contain considerable amount of porosity (20 to 40%). However, the shape, size and distribution of pores in the nitrided composites can be controlled by controlling fabrication variables.

If SiC fiber tows are used as reinforcements, the reaction-bonded silicon nitride process has the advantage of synthesizing SiC/Si$_3$N$_4$ composites in potentially complex and large shapes with near net shape capability, reducing or even eliminating, post-consolidation machining. The SiC/Si$_3$N$_4$ composite fabricated by the RBSN process has several advantages: (a) absence of shrinkage during consolidation, (b) RBSN process yields materials with higher purities than typical dense silicon nitride containing sintering aids, (c) improved purity of RBSN matrix can permit achievement of high temperature strength and creep resistance in composites, (d) RBSN processing temperatures are frequently much lower and processing times much shorter than for conventional pressure assisted sintering, (e) service temperature limits can substantially exceed processing temperatures, (f) non-equilibrium phase chemistries or combinations of phases that are inaccessible to conventional processing routes can be achieved.

Because RBSN employs gaseous reactants, the SiC/RBSN material tends to have higher levels of frequently interconnected, residual porosity than the SiC/Si$_3$N$_4$ composite fabricated by pressure assisted sintering methods. Interconnected residual porosity remains an important issue for two reasons: oxidation and thermal conductivity. Internal oxidation can lead to internal stresses which may cause premature matrix cracking and fiber delamination. Thus, to avoid internal oxidation protective coatings may be necessary for these materials.

Thermal conductivity is also a function of matrix porosity. As the porosity increases, thermal conductivity decreases. Therefore, internal porosity may significantly reduce the thermal conductivity of SiC/RBSN composites. The reduced modulus that intrinsically accompanies uniformly distributed porosity in RBSN matrix facilitates load transfer to reinforcements, and also lowers thermal stresses in properly designed composites.

The process of converting silicon to Si_3N_4 is exothermic and can be difficult to control. Heat management is a significant issue in those reactions that are exothermic because overheating causes loss of microstructural control and phase chemistry of the Si_3N_4 matrix. Even though reactions are normally carried out at relatively low temperatures ($\sim$1350°C), the combination of long reaction times and the reactivity of constituents can pose serious detrimental interactions.

The earliest study on SiC fiber-reinforced RBSN was performed by Lindley and Godfrey [1]. Although fabricated composites showed no strength enhancement over unreinforced RBSN, they reported improved work of fracture for the composites by controlling the fiber/matrix interface. The poor strength properties of the composites were attributed to degradation of the SiC fiber at the fabrication temperature. Interest in SiC/RBSN was rejuvenated in the early 1980's because of the need to develop strong and tough ceramic materials. A feasibility study [2–5] investigating the fabrication of SiC fiber-reinforced RBSN composites using small diameter (<20 μm) SiC fiber tows and large diameter (142 μm) SiC monofilament fibers showed mixed results. The SiC monofilament composites showed some improvement in flexural strength, but the SiC fiber tow-reinforced composites did not. Again instability of the SiC fiber tows at the fabrication temperatures and inadequate bonding between the SiC fibers and RBSN matrix are the primary causes for inferior properties. All the above studies used traditional processing methodology and a nitridation cycle developed for monolithic RBSN for the fabrication of RBSN composites. To overcome the fiber degradation problem, a low temperature, short time nitridation cycle was developed by controlling the particle size, hence surface area, of the silicon powder [6]. Using this nitridation cycle, strong and tough RBSN composites reinforced by monofilament SiC fibers or SiC fiber tows have been fabricated [6–11]. The following sections review the fabrication, properties, and limitations of these materials.

II.1. SiC Monofilament-Reinforced RBSN Composite

II.1.1. Processing

The SiC fiber monofilaments consist of a SiC sheath with an outer diameter of 142 μm surrounding a pyrolitic graphite-coated carbon core with a diameter of 37 μm. On the surface of the SiC fiber contained two layers of carbon-rich surface coating. Each layer is a mixture of amorphous carbon and SiC.

The steps involved in the fabrication of SiC/RBSN composites are shown in Fig. 1. The details of the composite fabrication procedure were described in Reference 6. The starting materials for composite fabrication were SiC fiber mats and silicon powder cloth. The SiC fiber mats were prepared by winding the SiC fibers with desired spacing on a cylindrical drum. The fiber spacing used depended on the desired fiber volume fraction in the composite. The fiber mats were coated with a fugitive polymer binder such as polymethylmethacralate (PMMA) to maintain the fiber spacing.

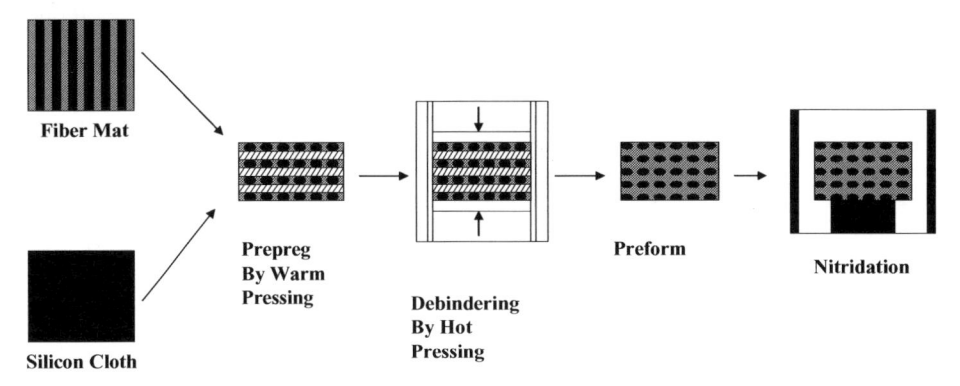

FIGURE 1. Fabrication steps for SiC fiber reinforced reaction-bonded silicon nitride composites.

For silicon cloth preparation, the attrition-milled silicon powder was mixed with a polymer fugitive binder (Teflon) and an organic solvent (a kerosene based solvent-Stoddard), and then rolled to the desired thickness to produce silicon cloth.

For composite preform fabrication, the fiber mat and the silicon cloth were alternately stacked in a metal die and hot pressed in a vacuum furnace at 1000°C and 69 MPa pressure to remove the fugitive polymer binder. The resultant composite preform was nitrided at 1250°C for 10 to 24 h to convert silicon to silicon nitride matrix. The fabrication time varies with the panel size. The RBSN matrix shows ~60% α-Si_3N_4, 32% β-Si_3N_4, and ~8% excess silicon. The nitrided composite typically contains ~25 vol% SiC fibers, 35 to 45 vol% silicon nitride matrix, and 30 to 40 vol% porosity. The average pore volume, relative amounts of open to closed porosity, and the density of the nitrided composites can be manipulated by silicon powder characteristics and consolidation parameters. Generally the higher the preform density the higher will be the nitrided density. The composite fabricated by the above approach shows a uniform distribution of SiC fibers in the RBSN matrix and stable interface coating on the SiC fibers, both of which are essential for attaining high strength and toughness in composites

II.1.2. Properties of Monofilament SiC/RBSN composites

II.1.2a. Physical and Mechanical Properties

The properties of a SiC/RBSN composite depend on the volume fraction of the constituents and the bonding between SiC fibers and the RBSN matrix. The room temperature tensile stress-strain curves for the SiC/RBSN composites and the unreinforced RBSN matrix are shown in Fig. 2. The stress-strain curve for the unreinforced RBSN matrix shows only an initial linear elastic region and no strain capability beyond matrix fracture. In contrast, the room temperature tensile stress-strain curves for the 1-D and 2-D SiC/RBSN composites display three distinct regions: an initial linear elastic region, followed by a non-linear region, and then a second linear region. The non-linearity in the stress-strain curve is due to matrix micro-cracking normal to the loading direction. The strain capability beyond initial matrix fracture is possible because of the weak or frictional bond formed between the fiber and RBSN matrix and the retention of a large fraction of as-produced fiber strength under the

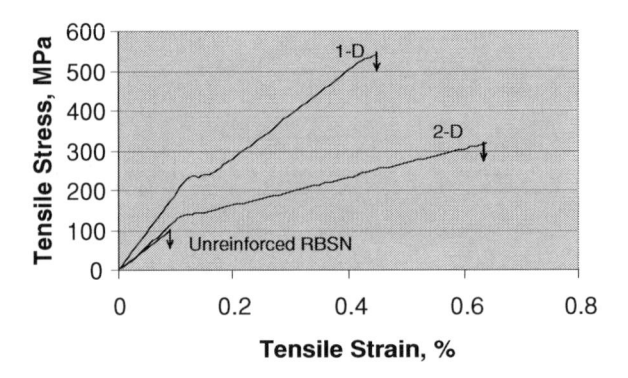

FIGURE 2. Room temperature tensile stress-strain curves for 1-D and 2-D SiC/RBSN composites containing ~24 vol% fibers, and unreinforced RBSN [8, 9]

fabrication conditions. If the fibers were strongly bonded to the matrix, the composite would have behaved similar to that of the unreinforced matrix.

The slope of the initial elastic portion of the stress-strain curve represents Young's modulus, which primarily depends on volume fraction and elastic modulus of the fibers and the matrix. Theoretically, Young's modulus (E_C) of the composite can be estimated using the rule-of-mixtures.

$$E_C = E_f V_f + E_m V_m$$

Where E is the elastic modulus, V is volume fraction of the constituent, and the subscripts f and m refer to the fiber and matrix, respectively. The elastic modulus of the RBSN matrix decreases with increasing porosity. To estimate the elastic modulus of RBSN with known porosity, the following equation may be used [12].

$$E_m = E_o \exp(-3V_P)$$

Where V_P is the total volume fraction of porosity and E_o is the Young's modulus of dense silicon nitride which has a value of 300 GPa [13].

The slope of the second linear line represents the secondary modulus which is primarily controlled by the fibers. If fibers are not broken until the ultimate tensile strength is reached, the secondary modulus should correspond to $\sim E_f V_f$. A secondary modulus value less than $\sim E_f V_f$ indicates the fracture of fibers in the early stages of deformation.

At the inflection point, a single through-the thickness matrix crack forms normal to the loading direction and the crack occurs at the largest matrix flaw. As the loading is increased beyond the inflection point, the pieces of matrix on both sides of the crack load and crack again at the next largest matrix flaw. This process continues until matrix blocks cannot sustain load. The stress at which the first matrix crack forms can be modeled by knowing the volume fractions and elastic moduli of the constituents, interfacial shear strength between the fiber and the matrix, fracture toughness of the matrix, fiber diameter, and nature of residual stress in the composite. Various fracture mechanics-based models can fairly accurately predict the matrix cracking strength.

TABLE I. Room temperature physical and mechanical properties for SiC/RBSN composites [8–10,14].

Property	1-D	2-D
Physical Properties		
Fiber content, %	24–30	24–30
Density, gm/cc	2.2–2.4	2.2–2.4
Porosity, %	30–40	30–40
Flexural Properties[@]		
Elastic Modulus, GPa	182 ± 10	NA
Matrix Cracking Stress, MPa	206 ± 40	NA
Matrix Cracking Strain, %	0.1	NA
Ultimate Strength, MPa	965 ± 138	NA
Ultimate Strain, %	1.2	NA
Tensile Properties		
Elastic Modulus, GPa	186 ± 20	118 ± 15
Matrix Cracking Stress, MPa	227 ± 41	130 ± 30
Matrix Cracking Strain, %	0.13	0.1
Ultimate Tensile Strength, MPa	690 ± 138	316 ± 62
Ultimate Tensile Strain, %	1	0.6
Poisson's Ratio	0.21	NA
Compression Properties		
Elastic Modulus, GPa	180 ± 7	157 ± 5
Ultimate Tensile Strength, MPa	1752 ± 207	722 ± 74
Ultimate Tensile Strain, %	1	0.5
Shear Properties		
Shear Modulus, GPa	31 ± 3	NA
Interfacial Shear Strength[*], MPa	11 ± 7	NA
Interlaminar Shear Strength[#], MPa	40 ± 4	NA
Fracture Toughness, MPa(m)$^{1/2}$	13	NA

[@] 4-point flexural test, specimen size: 3 mm (T) × 4 mm (W) × 50 mm (L) NA-Data not available
[#] Double notch shear test
[*] Fiber push-out method

The ultimate tensile strength of the composite is controlled by the fibers. Therefore, knowing the bundle strength of the fiber, the ultimate tensile strength of the composite, σ_C, can be estimated from the equation,

$$\sigma_C = \sigma_{fB} V_f$$

where σ_{fB} is the fiber bundle strength and V_f is the fiber volume fraction. The tensile properties of unidirectional SiC/RBSN composites are anisotropic, i.e., in the fiber direction the composite is strong, but in the direction transverse to the fibers it is weaker than even the unreinforced matrix. To achieve isotropic properties, multi directional reinforcement is needed, but the penalty of multi directional reinforcement is reduced mechanical properties as illustrated by the tensile stress- strain curve for the 2-D SiC/RBSN composite (Figure 2). Table I summarizes the room temperature mechanical properties of 1-D and 2-D SiC/RBSN composites.

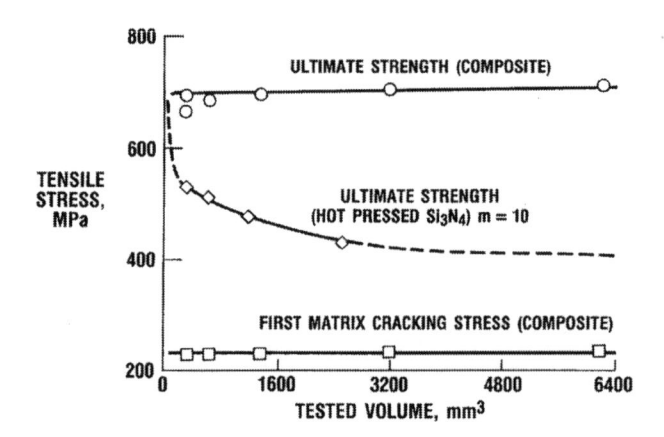

FIGURE 3. Variation of room temperature tensile strength with tested volume for 1-D SiC/RBSN containing ~24 vol% fibers and hot-pressed silicon nitride [8, 15].

The SiC/RBSN composites showed several advantages over monolithic Si_3N_4. The room temperature tensile strength properties of unidirectionally reinforced SiC/RBSN composites are relatively independent of tested volume (Fig. 3); whereas the ultimate strength of monolithic ceramics decreases with increasing volume because of greater probability of finding a strength-limiting flaw with increasing volume [8, 15]. In other words, the strength of a fiber-reinforced ceramic matrix composite does not depend on the largest flaw formed during fabrication or during specimen preparation, but it depends on the complex interaction of local stress fields around the flaws and their growth due to global stresses. This implies that for applications requiring large material volumes, such as, turbine blades or rotors, the strength of SiC/RBSN composites may be better than that of dense, commercially available hot-pressed Si_3N_4. In addition, SiC/RBSN composites display notch insensitive strength behavior and better impact resistance than unreinforced RBSN or dense Si_3N_4 material.

The variation of tensile strength with temperature in air for a 1-D SiC/RBSN composite is shown in Fig. 4 [16]. The elastic modulus and matrix cracking strength decreases slowly with increase in temperature, but the ultimate tensile strength remains relatively constant from 25 to 600°C. Beyond this temperature it decreases due to oxidation of the carbon coating on the SiC fibers and creep effects.

II.1.2b. Thermal Properties

II.1.2b.1. Thermal Expansion Thermal expansion of unidirectional SiC/RBSN composite is mainly a function of constituents' volume fractions and measurement direction relative to the fiber, and is not affected by constituents' porosity. Measurement of linear thermal expansion with temperature in nitrogen for the 1-D SiC/RBSN composites parallel and perpendicular to the fibers indicates a small amount of anisotropy (Fig. 5). This is attributed to small difference in thermal expansion coefficients of SiC fibers (4.2×10^{-6}) and RBSN matrix (3.8×10^{-6}) as well as anisotropic thermal expansion of carbon coating on SiC fibers. In the fiber direction, linear thermal expansion is controlled by the SiC fiber, and in the direction perpendicular to the fiber, it is controlled by the RBSN matrix.

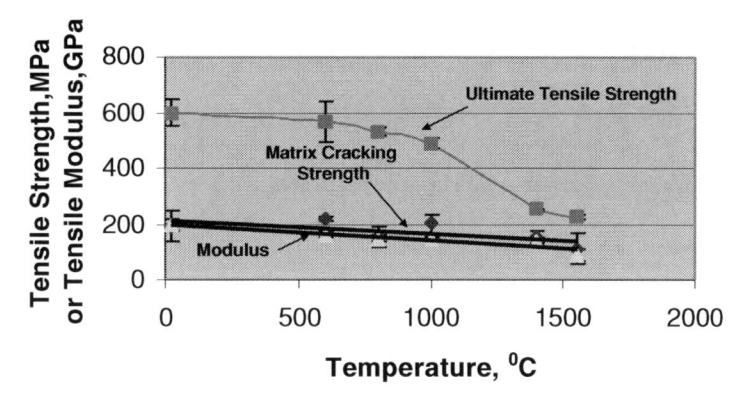

FIGURE 4. Variation of mechanical properties with temperature for 1-D SiC/RBSN composites containing ~24 vol% fibers in tested air [16].

Thermally cycling the 1-D SiC/RBSN composites between 25 and 1400°C in nitrogen had no significant effect on the linear thermal expansion curve transverse to the fibers, but thermal cycling in oxygen caused a positive shift in $((L-L_0)/L_0)$ in the first cycle indicating length (L) increase (Fig. 6). With each additional cycle, a further increase in specimen length was observed, but with each additional cycle, a further increase in specimen length was observed, but the rate of increase in length/cycle consistently decreased. After the third cycle, reproducible thermal expansion curves with no apparent hysteresis were obtained.

Since RBSN is a porous material and most of the porosity is interconnected, exposing it to oxygen results in growth of silica on the geometrical surfaces and the surfaces of pores as well. However, the pore wall oxidation ends quickly with the sealing of the pore channels near the surface because of the 82% increase in volume during conversion of a mole of silicon nitride to a mole of silica. Beyond this stage, oxidation is limited to the geometrical surfaces. During cycling the surface silica layer appears to have cracked due to

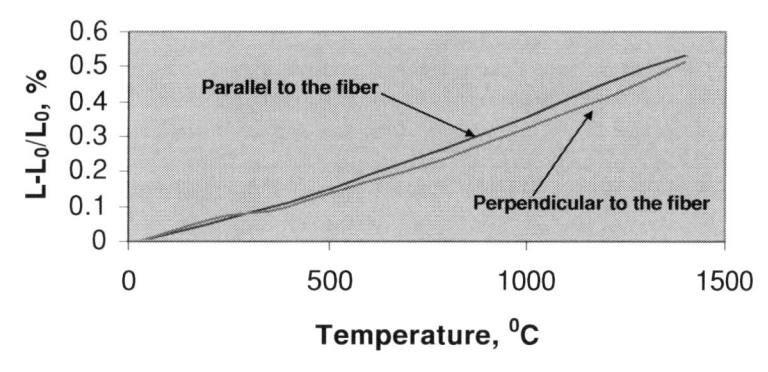

FIGURE 5. Linear thermal expansion curve during heating in nitrogen for 1-D SiC/RBSN composites containing ~24 vol% fibers measured parallel and perpendicular to the fibers [17].

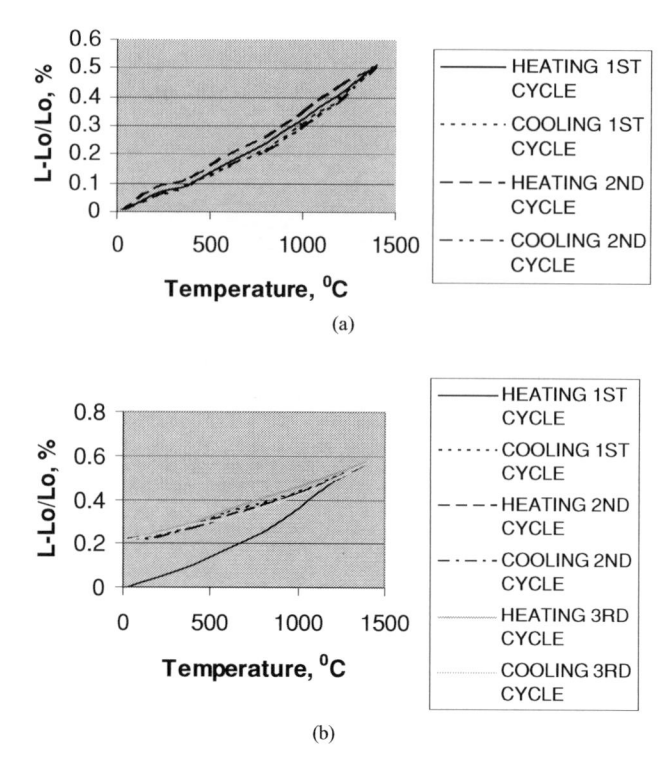

FIGURE 6. Thermal expansion and contraction curves for a 1-D SiC/RBSN composite containing ~24 vol% fibers measured transverse to the fibers in (a) nitrogen, and (b) oxygen showing influence of internal oxidation [17].

crystallization and phase transformation. Growth of silica on the pore walls as well as on the external surfaces accounts for length increase in the SiC/RBSN composites.

Internal oxidation of SiC/RBSN is severe between 800 to 1100°C, and the depth of the oxidation damage zone is directly related to the pore size; the smaller the pore size, the lower is the oxidation induced damage. In addition, internal oxidation is also found to generate tensile residual stresses which affect strength properties of SiC/RBSN.

Thermal cycling of SiC/RBSN in nitrogen had no effect on mechanical properties, but in oxygen environment, composite properties degraded due to internal oxidation (discussed in Section 1.2d) and swelling as shown in Table II.

II.1.2b.2. Thermal Conductivity The influence of temperature on calculated thermal conductivity for 1-D SiC/RBSN composites parallel and perpendicular to the fiber is plotted in Fig. 7. The thermal conductivity (K) was calculated from the equation,

$$K = \rho\, C_P\, \alpha$$

where C_P is the specific heat, α is the thermal diffusivity, and ρ is the density of the material. The specific heat, thermal diffusivity, and the density of the material were measured

TABLE II. Room temperature tensile properties of unidirectional SiC/RBSN composites ($V_f \sim 0.24$) [17]

Room-temperature properties	As-fabricated[a] (not cycled)	After 5 cycles in nitrogen[b] 25 to 1400°C	After 5 cycles in oxygen[b] 25 to 1400°C
Elastic modulus, GPa	178 ± 16	186	173
Tensile strength, MPa			
First matrix	220 ± 24	213	138
Ultimate	680 ± 62	551	138
Strain to failure, %			
First matrix	0.12	0.11	0.08
Ultimate	0.45	0.40	0.08

[a] Average of 5 specimens
[b] Average of 3 specimens

quantities. Thermal conductivity parallel to the fibers is greater than perpendicular to the fibers. Thermal conductivity generally decreases with temperature. In the fiber direction, both the SiC fiber and the RBSN matrix contribute to the thermal conductivity, whereas in the direction perpendicular to the fiber, thermal conductivity is predominantly controlled by the matrix. The significant difference in thermal conductivity parallel and perpendicular to the fiber is attributed to a boundary gap between the SiC fiber and the RBSN matrix. Existence of a boundary gap has been confirmed by experimental methods [18, 19].

The coefficient of linear thermal expansion, specific heat, thermal diffusivity, thermal conductivity for 1-D and 2-D SiC/RBSN at four temperatures in nitrogen measured parallel and perpendicular to the fibers are summarized in Tables III and IV. In general, through-the thickness thermal conductivity value at room temperature for SiC/RBSN composites is low when compared with a value ~ 7 W/m-k for the unreinforced RBSN or with a value of ~ 30 W/m-k for the sintered silicon nitrides [13]. Both weak bonding between the SiC

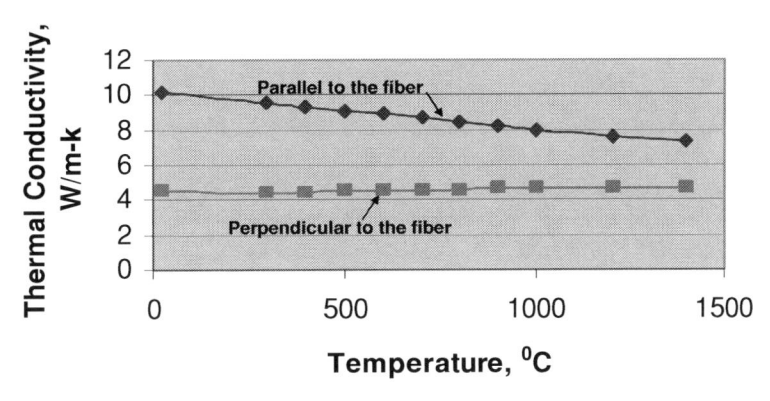

FIGURE 7. Variation of thermal conductivity with temperature for 1-D SiC/RBSN composites containing ~ 24 vol% fibers measured in nitrogen [18, 19].

TABLE III. Thermal property data for 1-D SiC/RBSN composites in nitrogen [18, 19].

Property	Temperature, °C			
	25	600	1000	1400
Thermal Properties				
Thermal Expansion				
In-Plane, $10^{-6}/°C$	NA	3.1	3.6	3.8
Through-the – Thickness, $10^{-6}/°C$	NA	2.9	3.2	3.7
Specific Heat, J/kg-K	675	1250	1312	1330
Thermal Diffusivity				
In-Plane, 10^6 m^2/sec	7	3.3	2.8	2.5
Through-the – Thickness, 10^6 m^2/sec	2.5	1.8	1.8	1.8
Thermal Conductivity				
In-Plane, W/m-K	10	9.0	8.0	7.0
Through-the – Thickness, W/m-K	4.5	4.5	4.5	4.5

fibers and the RBSN matrix, and porosity in the RBSN matrix appear to affect the thermal conductivity in this system.

II.1.2c. Thermal Shock Resistance

The ability to withstand the thermal stresses generated during ignition, flameout, and operating temperature excursions is an important consideration in evaluating potential high temperature materials. Thermally created stresses may initiate delamination and microcracking or cause existing matrix flaws to grow, giving a gradual loss of strength and modulus and eventual loss of component integrity. However, the evaluation of thermal stress resistance is a complex task since performance depends not only on material thermal and mechanical properties, but it is also influenced by heat transfer and geometric factors such as heat transfer coefficient and component size. No reliable testing methods exist to evaluate thermal shock resistance of materials except testing sub-elements in an engine environment. However, the water quench and thermal gradient tests are used to qualitatively evaluate thermal shock performance of potential high temperature materials.

TABLE IV. Thermal property data for 2-D SiC/RBSN composites in nitrogen.

Property	Temperature, °C			
	25	600	1000	1400
Thermal Properties				
Thermal Expansion				
In-Plane, $10^{-6}/°C$	NA	3.3	3.9	3.9
Specific Heat, J/kg-K	680	1160	1240	1270
Thermal Conductivity				
Through-the – Thickness, W/m-K	4.59	4.0	4.0	3.8

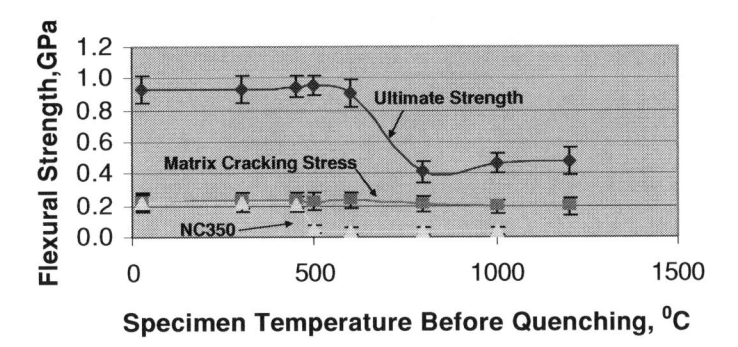

FIGURE 8. Room temperature 4-point flexural strengths for 1-D SiC/RBSN composites containing ~30 vol% fibers and monolithic RBSN (NC350) after quenching [20].

The thermal shock resistance of unidirectionally reinforced SiC/RBSN composites was evaluated using the water quench method. Both room temperature flexural (Fig. 8) and tensile properties (Fig. 9) of 1-D SiC/RBSN composites were measured before and after quenching and compared with the flexural properties of quenched unreinforced RBSN under similar conditions.

When tested under tensile testing mode, the thermally shocked SiC/RBSN composite showed no loss in tensile properties, but under flexural testing mode, the composites showed loss of ultimate flexural strength after quenches from above 600°C. The flexural strength loss behavior for the quenched unreinforced RBSN appeared similar to that for the composite, but for unreinforced RBSN, the strength loss occurred after quenches from temperatures above 425°C which is 175°C less than that observed for the composite. Although thermal shocking did damage to the composite, probably by microcracking of the matrix, it did not affect the tensile properties of the composite. These results indicate that SiC/RBSN composites have better thermal shock resistance than the unreinforced RBSN and thus better toughness or flaw tolerance.

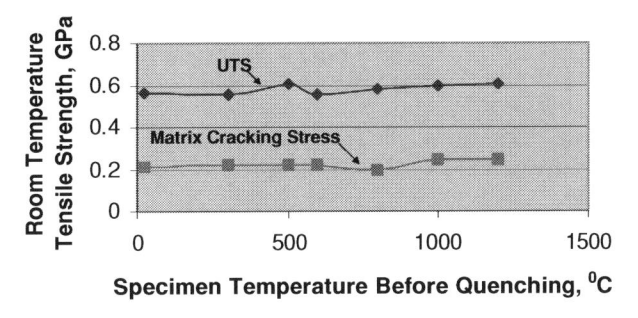

FIGURE 9. Room temperature tensile strengths for 1-D SiC/RBSN composites containing ~30 vol% fibers after quenching [20].

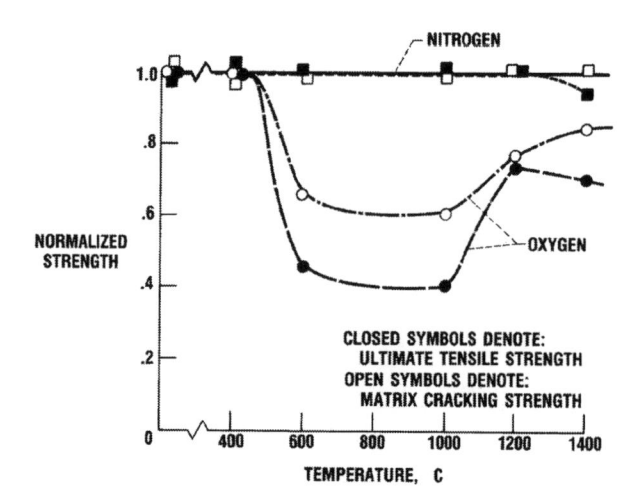

FIGURE 10. Influence of 100 hr exposure in nitrogen and oxygen environments on room temperature tensile properties of 1-D SiC/RBSN composites containing ∼24 vol% fibers [21]. Normalized strength is defined as the ratio of room temperature tensile strength of the environmentally exposed specimens to that of the as-fabricated specimens.

II.1.2d. Environmental Stability

The SiC/RBSN composites are stable in inert environments such as argon, and nitrogen at temperatures to 1400°C for 100h, but in oxygen, oxidation starts as low as 400°C [21]. Oxidation effects on unidirectionally reinforced SiC/RBSN composites areshown in Figure 10. Three oxidation regimes have been identified: oxidation of the fiber/matrix interface, oxidation of the porous Si_3N_4, and intrinsic strength degradation of the SiC fibers. At low temperatures between 400–800°C, where oxidation of porous Si_3N_4 is not kinetically favorable, oxygen could diffuse through the porous matrix and oxidize the carbon coating at the fiber/matrix interface. This reaction decouples the fiber from the matrix, causing loss of load transfer and loss of strength. In the temperature range between 800°–1100°C, where internal oxidation of the porous Si_3N_4 is prevalent, both reactions – formation of silica in the Si_3N_4 pores and oxidation of the fiber/matrix interface – occur simultaneously. Beyond 1100°C, rapid oxidation of the porous Si_3N_4 matrix allows formation of dense silica on the composite surface which reduces permeability of oxygen through the matrix. However, instability of the fiber surface coating and possible fiber degradation still results in some loss in composite strength.

II.1.2e. Creep Properties

Creep resistance is of primary concern in rotating components of a turbine engine. High creep rates can lead to both excessive deformation and uncontrolled stresses. Creep resistance of fiber-reinforced ceramic matrix composites depend on relative creep rates of, stress-relaxation in, and load transfer between constituents. The tensile creep behavior of SiC/RBSN composites containing ∼24 vol% SiC monofilaments was studied in nitrogen at 1300°C at stress levels ranging from 90 to 150 MPa. Under the creep stress conditions the steady state creep rate ranged from 1.2×10^{-9} s^{-1} to 5.1×10^{-8} s^{-1}. At stress levels below

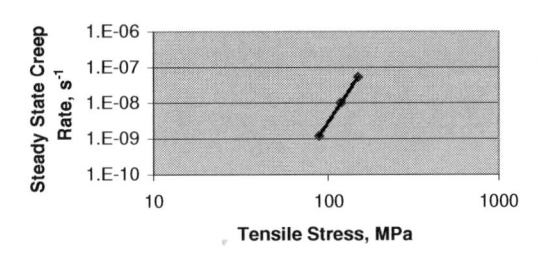

FIGURE 11. Variation of steady state creep rate with applied stress for 1-D SiC/RBSN composites containing ~24 vol% fibers at 1300°C in nitrogen and oxygen [22].

the matrix cracking stress, the RBSN matrix has a lower creep rate than the SiC monofilaments. Therefore, SiC/RBSN composites generally show low creep rate at 1300°C at stress levels up to 120 MPa in inert environments as shown in Fig. 11 [22]. However, because of the high creep resistance of the RBSN matrix, load is shed from SiC monofilaments to the RBSN matrix during creep, leading to progressive increase in matrix stress and a relaxation in fiber stress. This will eventually lead to matrix fracture, and the development of periodic matrix cracks, with load transferred across the cracks by the bridging fibers. This is undesirable from long term creep point of view, since the creep strength of the composite after matrix cracking is now controlled by the rupture strength of the fibers, which must now support the entire creep load across the crack faces. Exposure of the interface and the fibers to oxygen, in a cracked composite will lead to its premature failure.

II.2. Reaction-Bonded Silicon Nitride Systems with Tow Fibers

The SiC/RBSN composites containing SiC monofilaments as discussed earlier cannot be pursued for component manufacturing for three reasons: first, the fact that large diameter fibers cannot be bent to a radius less than 1 mm severely restricted the shape capability for the components; second, machining the components from a block of composite is very expensive and time consuming; third, the SiC/RBSN composites with large diameter fibers did not bridge the matrix cracks effectively. The limitations of large diameter reinforcement can be avoided by employing textile processes (weaving, braiding) to form multi-fiber bundles or tows of ceramic fibers into 2D and 3D fiber architectures. However, fabrication of the 2D and 3D SiC/RBSN has not been fully investigated. Brandt et al. [4] fabricated the 2-D SiC/RBSN matrix composites by infiltrating silicon slurry into a stack of 2-D SiC woven cloth, consolidating the stack by hot-pressing, and then reaction-bonding the hot-pressed stack. The strength and density values of the composite were low, and the composite showed moderate improvement in toughness. Few studies reported fabrication of SiC/RBSN composites using SiC fiber tows and prepreg/hotpressing methods of consolidation [2, 3, and 11]. The fabrication and properties of these composites are discussed below.

II.2.1. Processing
The as-received fiber tows were spread and coated with a layer of BN and then with a layer of SiC; both layers were deposited by chemical vapor deposition. For the fabrication

TABLE V. Physical property data for Hi-Nicalon SiC/RBSN composites [11].

Interface coating	Fiber lay-up	Fiber volume, %	Density, gm/cc	Porosity, %
BN/SiC	0	24 ±1	1.96 ± 0.03	36
BN/SiC	0/90	24 ± 2	1.94 ± 0.01	37

of SiC fiber tow mats filled with silicon power slurry, the BN/SiC coated SiC fiber tow was passed through a series of rollers to spread the tow and then into a tank filled with the silicon slurry. The slurry coated fiber tows were wound on a metal drum at a predetermined spacing to prepare fiber mat. Strips, either all-unidirectional or alternate strips of unidirectional and transverse lay-up, were stacked in a die and pre-pressed at 3.5 MPa at room temperature. The pre-pressed composites were hot pressed at 40 MPa at 800°C for 15 min and then the load was released. Subsequently the temperature was increased to 1200°C and the panel was nitrided for 4 h in flowing nitrogen.

II.2.2. Properties of Tow SiC/RBSN Composites

Room temperature physical property data for as-processed 1-D and 2-D SiC/RBSN tow composites are shown in Table V. The composites contained ~24 vol% SiC fibers and ~36 vol% porosity.

The room temperature tensile stress strain curves for the 1-D and 2-D SiC/RBSN tow composites and that of the unreinforced RBSN areshown in Figure 12. In general, the composite stress-strain curves showed two regions: a linear elastic region followed by a non linear region. At the inflection point where elastic region changes to non linear region a transverse crack formed normal to the loading direction, but this crack propagated partially into the through-the-thickness direction. This damage mechanism is in contrast to that observed for SiC/RBSN composites reinforced by SiC monofilaments in which only through-the-thickness matrix cracks are formed. As the loading is increased, additional transverse cracks are formed along the gage section specimen until final fracture. The fracture surface indicated significant fiber pull out. The room temperature tensile property data for SiC/RBSN tow composites are tabulated in Table VI.

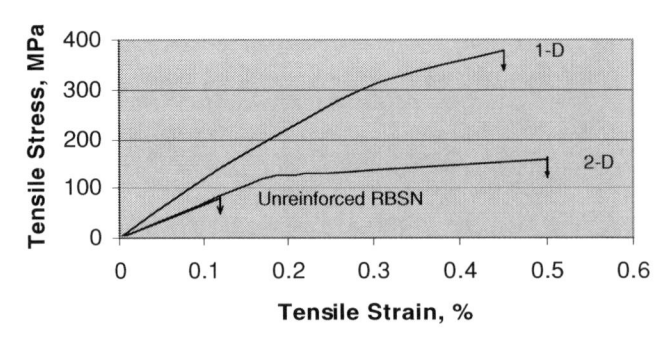

FIGURE 12. Room temperature tensile-stress-strain curves for 1-D and 2-D Hi-Nicalon SiC/RBSN composites containing ~ 24 vol% fibers, and unreinforced RBSN matrix [11].

TABLE VI. Room temperature tensile property data for Hi-Nicalon SiC/RBSN composites [11].

Interface coating	Fiber lay-up	Fiber volume,%	Proportional limit stress, % MPa	Proportional limit strain, %	Elastic modulus, GPa	Ultimate tensile strength, MPa	Ultimate tensile strain, %
BN/SiC	0	24 ± 1	290 ± 21	0.28 ± 0.02	105 ± 1	329 ± 8	0.27 ± 0.13
BN/SiC	0/90	24 ± 2	88 ± 18	0.12 ± 0.03	73 ± 3	133 ± 22	0.35 ± 0.09

II.3. Oxidation Protection Methods for SiC/RBSN Composites

Internal porosity in SiC/RBSN composites can be controlled but cannot be eliminated during processing of the composites. Interconnected internal porosity is a liability for SiC/RBSN composites because of poor oxidation resistance in the intermediate temperature range from 400 to 1100°C and poor thermal cyclic resistance. To avoid the internal oxidation problem several methods such as infiltration of RBSN composites with a silicon nitride-yielding polymer, surface coating the RBSN composites with a layer of CVD SiC or Si_3N_4, or multiple layers of CVD SiC and glass formers have been investigated [23–27]. The SiC/RBSN monofilament composites infiltrated with a silicon nitride-yielding polymer showed better oxidation resistance than uncoated SiC/RBSN composites at temperatures greater than 800°C in the initial stages of oxidation, but after 100 hr exposures, oxidation behavior of polymer infiltrated and uninfiltrated SiC/RBSN composites were the same. During long term exposure at high temperatures, amorphous polymer derived silicon nitride recrystallized, causing shrinkage cracks and porosity which opened the sealed surface for further oxidation. On the other hand, the CVD SiC and glass-former coated SiC/RBSN monofilamentcomposites survived (Fig. 13) both static and cyclic burner rig testing at temperatures to 1600°C for 10 hrs [28]. The carbon core and carbon-rich coating on the monofilament SiC fibers very close to surface coating were still intact. This suggests that externally coated SiC/RBSN composites can be used for high temperature applications provided the coating is not completely cracked in service conditions.

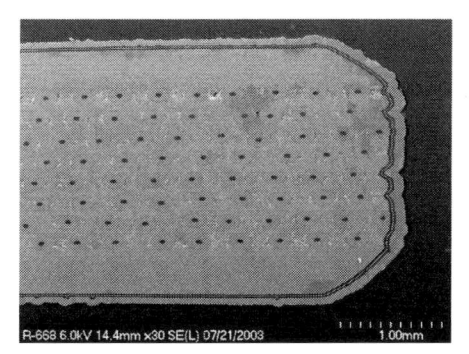

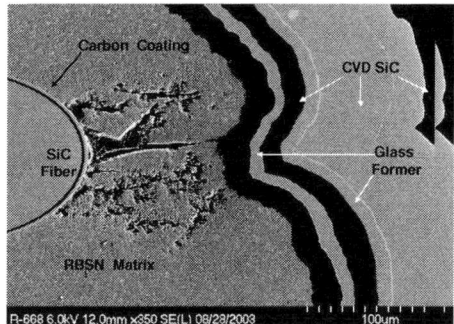

FIGURE 13. Cross sections of a CVD SiC/glass former coated SiC/RBSN monofilament composite after 10 h burner rig testing in air at 1600°C showing stability of carbon core and coating.

II.4. Sub-Element Testing

Oxidative stability of surface coated SiC/RBSN monofilament composites in burner rig testing prompted interest in utilization of this composite for uncooled components for small engine applications. Turbine vanes were machined from blanks of 1-D and 2-D SiC/RBSN composites, and surface coated with a layer of CVD SiC and glass former. Both uncoated and coated vanes were engine tested in at 1315°C for 10 h. The uncoated vanes showed severe damage, but the surface coated vanes survived engine tests with minimal damage [28].

III. DENSE SIC/SILICON NITRIDE COMPOSITES

SiC fiber reinforced silicon nitride powder preforms without any sintering additives are difficult to densify by pressure assisted sintering methods because the volume diffusivity of silicon nitride is not large enough to offset the densification retardation effects of surface diffusion and volatilization phenomena. To promote densification, it is necessary to mix silicon nitride powder with sintering oxide additives while fabricating SiC/Si_3N_4 composites. During pressure assisted sintering, the oxide additive reacts with silica on the surface of silicon nitride powder to form complex silicates. One disadvantage of silicates is that they are less refractory than silicon nitride. Therefore, the high temperature properties of SiC/Si_3N_4 composites are controlled by properties of the silicates, and not by silicon nitride. Two major issues with pressure assisted sintering methods are possible degradation of fibers during processing and limited shape capability.

Rice and coworkers [29] fabricated SiC fiber reinforced Si_3N_4 composites by slip casting and hot pressing methods. The processing scheme included dispersion of an array of SiC fibers in silicon nitride slurry to prepare a green material, and followed by hot pressing the green material at ~1500°C for de-bindering and consolidation. The fabricated composites displayed high fracture toughness, but their strengths were moderate in relation to typical strengths achieved in hot-pressed Si_3N_4. Several factors such as non-uniform distribution of the fibers, degradation of the fibers at the high processing temperatures, and incomplete densifications were suggested as reasons for limited strength. Shetty et al. [30] reported fabrication of dense SiC/Si_3N_4 composites containing monofilament SiC fibers. A combination of slurry coating and filament-winding was used for processing green monotapes of unidirectional silicon carbide monofilament in silicon nitride powder blend which contained 8 wt% Y_2O_3 and 4 wt% Al_2O_3. The monotapes were stacked and hot pressed in a die at 1750°C in N_2 atmosphere at 27 MPa. The 3-point flexural strengths of the resulting composites were significantly lower than that of the unreinforced silicon nitride. Crack-initiation resistance of the composites was comparable or only marginally better than that of the Si_3N_4 matrix, but the crack propagation resistance was significantly high. Strength degradation of fibers from exposure to the processing temperature, residual tensile stress, and filament damage during hot pressing are possible factors responsible for poor strength of the composites. Foulds et al. [31] improving the above processing approaches were able to fabricate fully dense, strong and tough SiC/Si_3N_4. Razzell and Lewis [32] fabricated fully dense SiC/Si_3N_4 composites by post hot-pressing SiC/RBSN composites containing SiC monofilaments and oxide sintering additives. The three-point flexural

SILICON CARBIDE FIBER-REINFORCED SILICON NITRIDE COMPOSITES 167

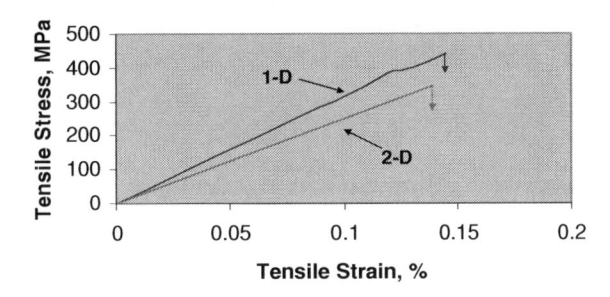

FIGURE 14. Room temperature tensile stress-strain behavior for 1-D and 2-D HP SiC/ Si_3N_4 composites containing ~30 vol% SiC monofilaments [31].

strength of the composite material was similar to that of the unreinforced silicon nitride, but the composites displayed several toughening mechanisms. Nakano et al. [33] reported manufacturing of SiC/Si_3N_4 composites using SiC fiber tows and hot-pressing. Flexural strength and fracture toughness of the composites at room temperature were 702 MPa and 20.2 MPa(m)$^{1/2}$, respectively. The composite degraded when tested at 1400°C. The processing and properties of SiC/Si_3N_4 composites with SiC monofilaments are discussed in the next section.

III.1. Processing

SiC monofilaments were collimated and wound on the outside of a revolving drum, where a resin was used to maintain fiber spacing. The fiber/resin mats were sectioned from the drum surface and used as layers between matrix powders in hot pressing. In an alternate processing method, SiC monofilaments were wound on the wax sheet wrapped drum at the desired spacing, and then, a silicon nitride powder blend-filled polymer slurry was sprayed on the monofilament to produce a flexible fiber/matrix tape. This tape can be cut, laid up, and hot pressed similar to the processing method described above. The drum wrap-spray coat procedure allows for precise fiber spacing and orientation, complete coating of fibers, precise thickness control and scalability to larger structures. The fiber/resin mats were applied sequentially between layers of matrix in a graphite mold. The matrix powder contained 5 wt% Y_2O_3 and 1.25 wt% MgO as densification aids. The preform powder lay up was consolidated at 1700°C under 70 MPa pressure for 1 hr in a vacuum atmosphere. The fabricated composites typically showed a density value of ~3.2 gm/cc, which is the theoretical density for SiC/Si_3N_4 composites containing 30 vol% SiC fibers. Uniform distribution of SiC fiber within the Si_3N_4 matrix was achieved and the carbon coating on the fiber remained intact.

III.2. Physical and Mechanical Properties

The room temperature tensile stress strain behaviors of 1-D and 2-D SiC/ Si_3N_4 monofilament composites showed high matrix cracking stress, but strain capability beyond matrix fracture is limited. Limited fiber pull out was observed on the tensile fracture surfaces [31].

TABLE VII. Physical and mechanical properties for HP SiC/Si$_3$N$_4$ composites [10, 31, 34]

Property	Temperature, °C	
	25	1350–1370
Physical Properties		
Fiber content, %	30	30
Density, gm/cc	3.2	NA
Flexural Properties@		
Elastic Modulus, GPa	319	140
Matrix Cracking Stress, MPa	528	
Matrix Cracking Strain, %	0.16	
Ultimate Strength, MPa	620	366
Ultimate Strain, %	0.28	0.5
Tensile Properties		
Elastic Modulus, GPa	316	260
Matrix Cracking Stress, MPa	348	84
Matrix Cracking Strain, %	0.11	0.03
Ultimate Tensile Strength, MPa	476	290
Ultimate Tensile Strain, %	0.6	0.22
Shear Properties		
Interfacial Shear Strength*, MPa	29 ± 7	NA
Interlaminar Shear Strength#, MPa	59	NA

@ 4-point flexure test, specimen size: 3 mm (T) × 4 mm (W) × 50 mm (L)
# 3-point short beam flexural test
* Fiber push-out method
NA-data not available

Limited property data for HP SCS-6/Si$_3$N$_4$ composites are reported in the literature. The 1-D HP SiC/ Si$_3$N$_4$ composite shows ∼380 MPa and ∼ 450–475 MPa matrix cracking strength and ultimate tensile strength, respectively. A 2-D composite is essentially brittle and shows ∼300 MPa matrix cracking/ultimate tensile strength. Physical and mechanical property data from three studies are summarized in Table VII.

III.3. Creep Properties

The tensile creep behavior of a HP SCS-6/ Si$_3$N$_4$ composite was investigated in air at 1350°C [34, 35]. The unidirectional composite containing 30 vol% SiC monofilaments was creep tested at stress levels at 70, 110, 150, and 190 MPa (Fig. 15). The steady state creep rate ranged from an average of 2.5×10^{-10} s^{-1} at 70 MPa to 5.6×10^{-8} s^{-1} at 150 MPa. The slope of the steady state creep rate vs creep stress curve yields a value of ∼7 which is comparable to the stress exponent of 4 to 6 found for tensile creep of monolithic Si$_3$N$_4$. For a similar stress, the average steady state creep rate of the composite was ∼ four orders of magnitude lower than the tensile creep rate of the HP Si$_3$N$_4$. At low stresses, ∼70 MPa, creep failure was accompanied by extensive fiber pullout and debonding along the fiber/matrix interfaces. The extent of fiber pullout diminished as the creep stress was

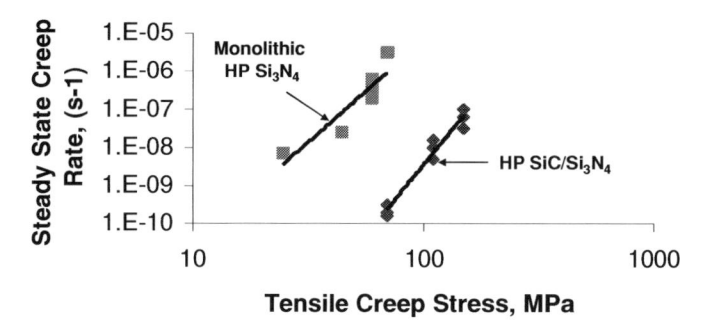

FIGURE 15. Variation of steady-state tensile creep rate with tensile creep stress for monolithic HP-Si$_3$N$_4$ at 1315°C and HP-SiC/Si$_3$N$_4$ composites tested at 1350°C in air [34, 35].

increased. For the specimens crept at stress levels between 70 and 110 MPa, a small fraction of fibers fractured during creep deformation as evidenced by the a small discontinuous strain jumps in the steady state creep regime.

III.4. Impact Resistance

Turbine components, specifically vanes and blades are subjected to impact damage. Therefore, the mechanism and consequences of impact damage on strength properties are also important in design of turbine components. To evaluate ballistic impact resistance of HP-SiC/Si$_3$N$_4$, Foulds et al. [31] impact tested both HP monolithic and 2-D SiC/Si$_3$N$_4$ composite tiles. The composite specimens showed local damage, but the monolithic material fragmented completely

IV. APPLICATIONS

Because of their high specific modulus and strength, toughness, and low thermal conductivity SiC/RBSN composites containing SiC fiber tow reinforcements are potential candidates for uncooled heat engine components such as inter-turbine ducts, combustor liners, nozzle vanes, divergent and convergent flaps, and blades. However, these components must be coated with surface coatings to avoid internal oxidation. Other applications include radome, and exhaust nozzles. On the other hand, uses of SiC/RBSN or fully dense SiC/Si$_3$N$_4$ composites containing SiC monofilaments are limited to simple shaped components such as flame holders, divergent and convergent flaps, and high temperature seals.

V. SUMMARY AND CONCLUDING REMARKS

Various studies indicate that strong, tough, creep and impact resistant SiC/RBSN and SiC/Si$_3$N$_4$ composites can be fabricated. The highlights of these studies are the following:

(1) SiC/RBSN composites containing monofilaments are limited to simple shapes. However, SiC/RBSN shape capability can be improved by using textile processes (weaving, braiding) to form multi-fiber bundles or tows of ceramic fibers into 2-D and 3-D fiber architectures.
(2) Internal pores in SiC/RBSN composites are unavoidable and reduce their oxidation resistance and thermal conductivity. Functionally graded oxidation resistant surface coatings appear to avoid internal oxidation problems for unstressed conditions.
(3) Fully dense SiC/Si$_3$N$_4$ can be fabricated by pressure assisted sintering methods, but the processing methods do not allow complex shape capability. The high temperature fabrication conditions also degrade fiber strength and hence ultimate tensile strength and strain capability. As a result, application potential for dense SiC/Si$_3$N$_4$ composites is very limited.

VI. REFERENCES

[1] M.W. Lindley and D.J. Godfrey, "Silicon Nitride Composites with High Toughness," Nature, 229, 192–193 (1971).
[2] J.W. Lucek, G.A. Rossetti, Jr., and S.D. Hartline, "Stability of Continuous Si-C (-O) Reinforcing Elements in Reaction-Bonded Silicon Nitride Process Environments," pp. 27–38 in Metal Matrix, Carbon, and Ceramic Matrix Composites 1985, NASA CP-2406, Edited by J.D. Buckley, NASA, Washington, D.C., 1985.
[3] N.D. Corbin, G.A. Rossetti, Jr., and S.D. Hartline, "Microstructure/Property Relationships for SiC Filament-Reinforced RBSN," Ceram. Eng. Sci. Proc., 10, [9–10], pp. 1083–1089 (1989).
[4] J. Brandt, K. Rundgren, R. Pompe, H. Swan, C. O'Meara, R. Lundberg, and Pejryd, "SiC Continuous Fiber-Reinforced Si$_3$N$_4$ by Infiltration and Reaction Bonding," Ceram. Eng. Sci. Proc. 13, [9–10], pp. 622–631 (1992).
[5] T.L Starr, D.L. Mohr, W.J. Lackey, and J.A. Hanigofsky, "Continuous Fiber-Reinforced Reaction Sintered Silicon Nitride Composites," Ceram. Eng. Sci. Proc. 14, [9–10], pp. 1125–1132 (1993).
[6] R.T. Bhatt, "Method of Preparing Fiber Reinforced Ceramic Materials," U.S. Pat. No. 4689188, 1987.
[7] R.T. Bhatt, "Mechanical Properties of SiC/RBSN Composites", in "Tailoring Multiphase and Composite Ceramics," (1986), p. 675 edited by R.E. Tressler, G.L. Messing, C.G. Pantano and R.E. Newnham, Plenum Press, New York.
[8] R.T. Bhatt, "The Properties of Silicon Carbide Fiber Reinforced Silicon Nitride Composites," in Whisker- and Fiber-Toughened Ceramics (1988) pp. 199–208 edited by R.A. Bradley, D.E. Clark, D.C. Larsen, and J.O. Stiegler, ASM International, Ohio.
[9] R.T. Bhatt and R.E. Phillips, "Laminate Behavior for SiC Fiber-Reinforced Reaction-Bonded Silicon Nitride Matrix Composites," J. Comp. Techn. Res. *12*, p. 13, 1990.
[10] R.T. Bhatt and J.D. Kiser, "Matrix Density Effects on Mechanical Properties of SiC/RBSN Composites," NASA-TM 103733, 1990.
[11] R.T. Bhatt, "Tensile Properties and Microstructural Characterization of Hi-Nicalon SiC/RBSN Composites," Ceramics International, #26, pp. 535–539, 2000.
[12] A.J. Moulson, "Review: Reaction Bonded Silicon Nitride; Its Formation and Properties," J. Mater. Sci., 14, pp. 1017–1051, 1979.
[13] S. Hamshire, "Engineering Properties of Nitrides," Engineered Materials Handbook, V1, 1987 ASM International, Metals Park, Ohio 44073.
[14] R.T. Bhatt, "Tension, Compression, and Bend Properties for SiC/RBSN Composites," Proc. of the 8th ICCM Conference Vol. III, 23-A, 1991.
[15] G.K. Bansal and W.H. Duckworth, "Effects of Specimen Size on Ceramic Strength," in "Fracture Mechanics of Ceramics" (1978), 3, p. 189 edited by R.C. Bradt, D.P.H. Hasselman and F.F. Lange, Plenum Press, New York.

[16] J.Z. Gykenyesi, "High Temperature Mechanical Characterization of Ceramic Matrix Composites" NASA CR-206611, 1998.
[17] R.T. Bhatt and A.R. Palczer, "Effects of Thermal Cycling on Thermal Expansion and Mechanical Properties of SiC Fiber-Reinforced Silicon Nitride Matrix Composites," J. Mat. Sci. *32*, 1039–1047, 1997.
[18] H. Bhatt, K.Y. Donaldson, D.P.H. Hasselman, and R.T. Bhatt, "Effect of Finite Interfacial Conductance on the Thermal Diffusivity/Conductivity of SiC Fiber Reinforced RBSN," Thermal Conductivity *21*. 1990, ed. C.J. Cremers and H.A. Fines, Plenum Press, New York, NY.
[19] H. Bhatt, K.Y. Donaldson, D.P.H. Hasselman, and R.T. Bhatt, "Role of Interfacial Thermal Barrier in the Effective Thermal Diffusivity / Conductivity of SiC Fiber Reinforced RBSN," J. Am. Ceram. Soc., *73*, [2], 312, 1990.
[20] R.T. Bhatt and R.E. Phillips, "Thermal Effects on the Mechanical Properties of SiC Fiber-Reinforced Reaction-Bonded Silicon Nitride Matrix Composites," J. Mat. Sci., *25*, 3401, 1990.
[21] R.T. Bhatt, "Oxidation Effects on the Mechanical Properties of a SiC Fiber Reinforced Reaction-Bonded Si_3N_4 Matrix Composites," J. Am. Ceram. Soc., *75*, 406, 1992.
[22] G.E. Hilmas, J.W. Holmes, R.T. Bhatt, and J.A. DiCarlo, "Tensile Creep Behavior and Damage Accumulation in a SiC-Fiber/RBSN-Matrix Composite," in Ceramic Transactions Vol. 38 – Advances in Ceramic Matrix Composites, N. Bansal (ed.), ACS, Westerville, OH, pp. 291–304, 1993.
[23] A.P.M. Adriaansen, and H. Gooijer, "Comparative Study of the Oxidation of RBSN and RBSN Coated With CVD Si_3N_4," Euro. Ceram., V3, 1989, pp. 569–574.
[24] O.J. Gregory and M.H. Richman, "Thermal Oxidation of Sputter-Coated Reaction-Bonded Silicon Nitride," J. Amer. Ceram. Soc. 67, [5], pp. 335–340 (1984).
[25] J. Desmaison, N. Roels, and P. Belair, "High-Temperature Oxidation-Protection CVD Coatings for Structural Ceramics: Oxidation Behavior of CVD-Coated Reaction-Bonded Silicon Nitride," Mater. Sci. and Eng., A121 441–447 (1989).
[26] D.S. Fox, "Oxidation Kinetics of Coated SiC/RBSN," 6th Annual HITEMP Review – 1993 NASA CP 19117 pp. 69–1 to 69–3.
[27] D.S. Fox, "Oxidation Protection of Porous Reaction-Bonded Silicon Nitride," J. Mat. Sci. *29* (21), 5693–5698, 1994.
[28] W. Fohey, R.T. Bhatt, and G.Y. Baaklini, "Burner Rig and Engine Test Results on SiC/RBSN Composites," 6th Annual HITEMP Review – 1993 NASA CP 19117 pp. 68–1 to 68–3.
[29] R.W. Rice, P.F. Becher, S.W. Freiman, and W.J. McDonough, "Thermal Structural Ceramic Composites," Ceram. Eng. Sci. Proc., 1, [7–8], pp. 424–443 (1980).
[30] D.K. Shetty, M.R. Pascucci, B.C. Mutsudy, and R.R. Wills, "SiC Monofilament-Reinforced Si_3N_4 Matrix Composites," Ceram. Eng. Sci. Proc. 16, [7–8], pp. 632–645 (1985).
[31] W. Foulds, J.F. Lecostaouec, C. Landry, and S. Dipietro, "Tough Silicon Nitride Matrix Composites Using Textron Silicon Carbide Monofilaments," Ceram. Eng. Sci. Proc. 10, [9–10], pp. 1083–1089 (1989).
[32] A.G. Razzel and M.H. Lewis, "Silicon Carbide/SRBSN Composites," Ceram. Eng. Sci. Proc. 12, [7–8], pp. 1304–1317 (1991).
[33] K. Nakano, S. Kume, K. Sasaki, and H. Saka, "Microstructure and Mechanical Properties of Hi-Nicalon Fiber Reinforced Si_3N_4 Matrix Composites," Ceram. Eng. Sci. Proc. 17, [4], pp. 324–332 (1996).
[34] J.W. Holmes, Y. Parks, and J.W. Jones, "Tensile Creep and Creep Recovery Behavior of a SiC Fiber Si_3N_4 Matrix Composite," J. Am. Ceram. Soc., 76 [5] 1281–1293 (1993).
[35] J.W. Holmes, "Tensile Creep Behavior of a Fiber Reinforced SiC-Si_3N_4 Composite," J. Mat. Sci., *26*, 1808–1814 (1991).

8

MoSi$_2$-Base Composites

Mohan G. Hebsur
National Aeronautics and Space Administration
Glenn Research Center
21000 Brookpark Road
Cleveland, Ohio 44135
Phone: 216–433–3266
Email: mohan.g.hebsur@grc.nasa.gov

ABSTRACT

Addition of 30 to 50 vol% of Si_3N_4 particulate to $MoSi_2$ eliminated its low temperature catastrophic failure, improved room temperature fracture toughness and the creep resistance. The hybrid composite SCS-6/$MoSi_2$-Si_3N_4 did not show any matrix cracking and exhibited excellent mechanical and environmental properties. Hi-Nicalon continuous fiber reinforced $MoSi_2$-Si_3N_4 also showed good strength and toughness. A new $MoSi_2$-base composite containing in-situ whisker-type βSi_3N_4 grains in a $MoSi_2$ matrix is also described.

I. INTRODUCTION

Due to high specific strength and stiffness, and the potential for increased temperature capability, composite materials are attractive for subsonic aircraft and future space propulsion systems. Based on high temperature oxidation behavior, it appears that $MoSi_2$ is one of the few intermetallics to have potential for further development. It also has a higher melting point (2296 K) and lower density (6.1 g/cm^3) than superalloys, and has electrical and thermal conductivity advantages over ceramics.[1] However, the use of $MoSi_2$ has been hindered due to its brittle nature at low temperatures, inadequate creep resistance at high temperatures, accelerated ('pest') oxidation 773 K, and its relatively high coefficient of thermal expansion (CTE) compared to potential reinforcing fibers such as SiC. The CTE mismatch between the fiber and the matrix results in severe matrix cracking during thermal cycling.

In the last 15 years, an extensive amount of work has been carried out in efforts to improve the high temperature properties of $MoSi_2$ by solid solution alloying, discontinuous reinforcement, and fiber reinforcement. Alloying with refractory metals has improved high temperature creep strength. Substantial improvements in strength have also been achieved by adding particulate, platelets or whiskers of SiC,[2] TiB_2, and HfB_2.[3] However, the effects of grain refinement may limit the creep strength of these types of composites. To date, $MoSi_2$ alloyed with W and containing 40 vol% SiC has achieved the creep strength superior to that of the superalloys.[4] The addition of SiC whiskers has also yielded improvements in room temperature toughness. However, it appears that the strength and damage tolerance required for high temperature aerospace applications could only be achieved by reinforcement with high strength continuous fibers.

Maloney and Hecht[5] have done extensive work on the development of continuous fiber reinforced $MoSi_2$-base composites to achieve high temperature creep resistance and room temperature toughness. SiC, sapphire, ductile Mo, and W alloy fibers were investigated. The refractory metal fibers increased both creep strength and fracture toughness, although reaction with the matrix was still a problem. The addition of about 40 vol% of SiC in the form of whiskers and particulate was used to lower the thermal expansion of the $MoSi_2$ base matrix and prevented matrix cracking in the SiC reinforced composites. However, matrix cracking was still observed in an SCS-6 fiber reinforced composite even with the matrix containing up to 40 vol% SiC whiskers. This composite also suffered catastrophic pest attack at 773 K. Sapphire fiber reinforced composites showed no evidence of matrix cracking due to the good thermal expansion match between $MoSi_2$ and Al_2O_3. However, the strong fiber-matrix bond did not provide any toughness improvement.

In earlier work[6] of developing $MoSi_2$ suitable for SiC fiber reinforcement, it was found that the addition of about 30 to 50 vol% of Si_3N_4 particulate to $MoSi_2$ improved the low temperature accelerated oxidation resistance by forming a Si_2ON_2 protective scale and thereby eliminated catastrophic pest failure. The Si_3N_4 addition also improved the high temperature oxidation resistance and compressive strength. More importantly, the Si_3N_4 addition significantly lowered the CTE of the $MoSi_2$ and eliminated matrix cracking in SCS-6 reinforced composites even after thermal cycling.[7] The progress made in developing, processing, and characterizing $MoSi_2$-base composites is reported here.

II. PROCESSING

Several batches containing a mixture of commercially available $MoSi_2$ and either 30 or 50 vol% of Si_3N_4 were mechanically alloyed. The $MoSi_2$-Si_3N_4 powder was consolidated into "matrix-only" plates by vacuum hot pressing followed by hot isostatic pressing (HIP) to achieve full density. Composite plates of various thickness consisting of 6, 12, or 56 plies of 30 vol% SCS-6 fibers having 0, 0/90 and 90° orientations in a $MoSi_2$-Si_3N_4 matrix were prepared by the powder cloth technique and consolidated in the same manner as the material without fibers. This resulted in fully dense material without excessive reaction or damage to the fibers. From the consolidated material, ASTM standard specimens for several tests

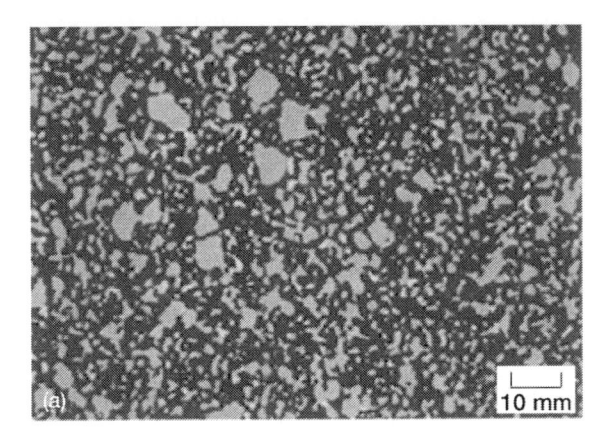

FIGURE 1. SEM micrograph of consolidated MoSi$_2$-50Si$_3$N$_4$. MoSi$_2$ is the light phase.

such as compression, fracture toughness, impact and oxidation were machined by EDM and grinding techniques. Details of the specimen preparation and testing procedures are described elsewhere.

III. PROPERTIES OF MoSi$_2$-BASE COMPOSITES

3.1. Microstructure of As-Fabricated Composite

Figure 1 shows the microstructure of the as-consolidated MoSi$_2$-Si$_3$N$_4$ matrix. As expected from thermodynamic predictions, the Si$_3$N$_4$ particles appeared to be quite stable, with very little or no reaction with the MoSi$_2$ even after exposure at 1773 K. X-ray diffraction of MoSi$_2$-Si$_3$N$_4$ showed only the presence of MoSi$_2$ and a mixture of α and β Si$_3$N$_4$ phases. Even though there is a significant CTE mismatch between MoSi$_2$ and Si$_3$N$_4$, the small particles of Si$_3$N$_4$ prevented thermally induced microcracking. As fabricated SCS-6/MoSi$_2$-Si$_3$N$_4$ composite did not show any matrix cracking. A reaction zone around the fibers was generally 1 micron in thickness and resulted from reaction of the carbon layer to form SiC and Mo$_5$Si$_3$.[6] The CTE of the matrix and composites plotted as a function of temperature are compared with the monolithic constituents in Figure 2. The addition of Si$_3$N$_4$ to MoSi$_2$ has effectively lowered the CTE of the matrix, achieving the desired result of eliminating matrix cracking. Furthermore, no cracks were found in either the matrix or the reaction zone even after 1000 thermal cycles between 1473 and 473 K in vacuum. These results show that the use of Si$_3$N$_4$ was much more effective than SiC.

3.2. Oxidation Behavior

3.2.1. Low Temperature Oxidation

Figure 3 shows the specific weight gain versus number of cycles at 773 K. Both MoSi$_2$-30 Si$_3$N$_4$ and MoSi$_2$-50 Si$_3$N$_4$ show very little weight gain indicating the absence of accelerated oxidation. XRD analysis of both these specimens indicated strong peaks of

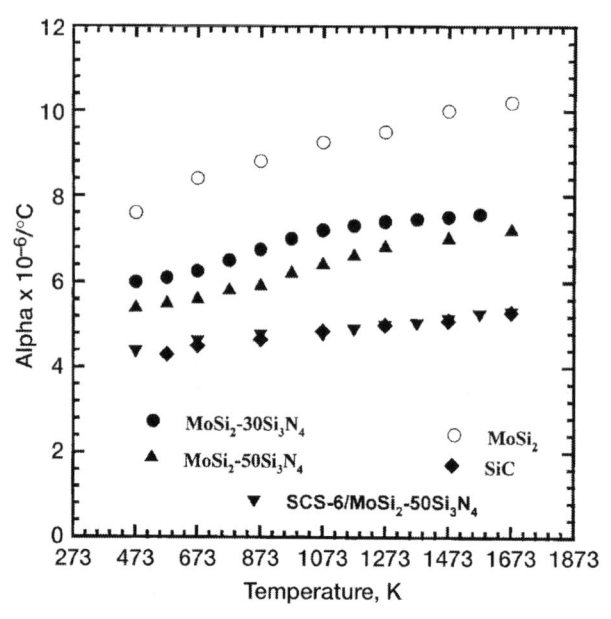

FIGURE 2. Thermal expansion data for several $MoSi_2$-base materials and SiC reinforcing phase.

Si_2ON_2 and the absence of MoO_3. The $MoSi_2$ exhibited accelerated oxidation followed by pesting.

Figure 4 shows the SCS-6/$MoSi_2$ and SCS-6/$MoSi_2$-30Si_3N_4 composites exposed at 773 K. The SCS-6/$MoSi_2$ specimen, which had matrix cracks, was completely disintegrated into powder within 24 cycles, whereas the SCS-6/$MoSi_2$-30Si_3N_4 specimen was intact

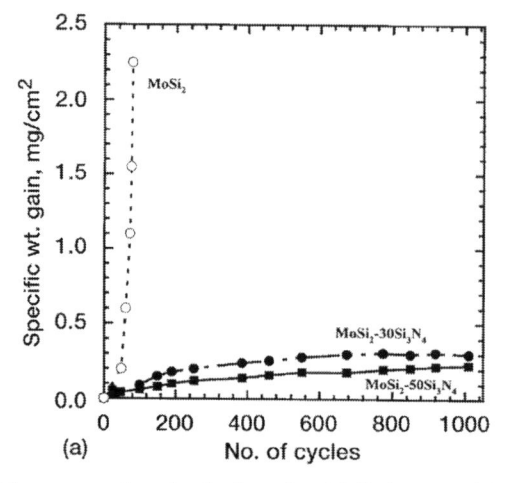

FIGURE 3. Specific weight versus number of cycles for various $MoSi_2$-base materials cyclic oxidized at 773 K in air.

FIGURE 4. SCS-6 fiber reinforced specimens cyclic oxidized at 773 K in air. (a) MoSi$_2$ matrix after 24 cycles. (b) MoSi$_2$-30Si$_3$N$_4$ matrix after 200 cycles.

even after 200 cycles. This is again in strong contrast to previous work,[5] where both SCS-6/MoSi$_2$-40 vol% SiC and Al$_2$O$_3$/MoSi$_2$ composites were reduced to powder after exposure at 773 K. All these observations are consistent with the elimination of pest attack in MoSi$_2$-Si$_3$N$_4$ composites due to a mechanism involving elimination of the accelerated oxidation associated with a non-protective MoO$_3$ oxide scale. The Si$_2$ON$_2$ scale forms rapidly, and is protective even at cracks, pores, and interfaces.

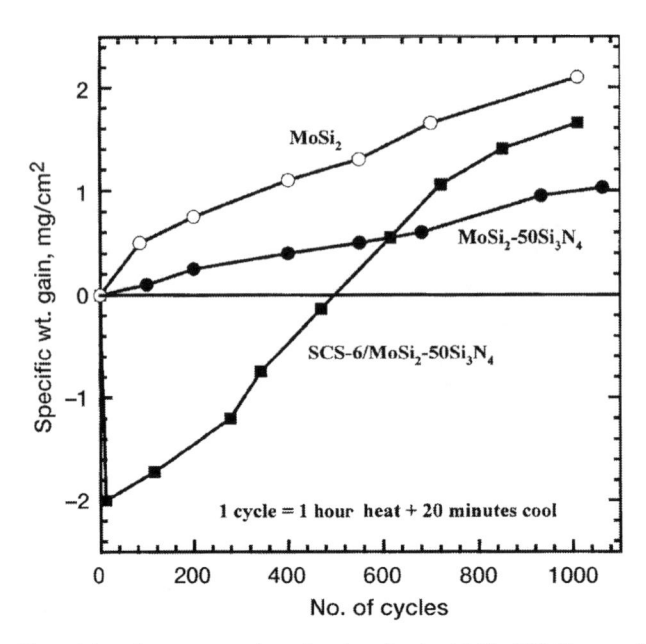

FIGURE 5. Specific weight gain versus number of cycles plot for $MoSi_2$-$50Si_3N_4$ monolithic and SCS-6/$MoSi_2$-$50Si_3N_4$ hybrid composite cyclic oxidized at 1523 K in air.

3.2.2. High Temperature Oxidation

The results of cyclic oxidation tests at 1473 K, which more closely approximates the conditions under which the material would be subjected in a structural application, are shown in Figure 5. The materials in Figure 5 were subjected to 1 hr heating cycles to 1473 K, followed by 20 minute cooling cycles. The $MoSi_2$-$50Si_3N_4$ particulate composite exhibited superior oxidation resistance as compared to $MoSi_2$ alone. The specific weight gain of $MoSi_2$-$50Si_3N_4$ was only about 1 mg/cm^2 in 1000 hr, almost comparable to CVD-SiC, which is considered the best SiO_2 former available. The composite initially lost weight due to oxidation of the carbon on the SCS-6 fiber followed by steady weight gain, less than 2 mg/cm^2 in 1000 hr. XRD of surface oxides on $MoSi_2$-50 Si_3N_4 and hybrid composite indicated strong peaks of α-cristobalite.

3.3. Mechanical Properties of $MoSi_2$-Si_3N_4 Matrix

3.3.1. Compressive Creep

Previous work[6,7] showed that the Si_3N_4 additions substantially increased compressive strength at all temperatures. This has been augmented with additional testing to further characterize this material. Figure 6 shows the results of constant load compression creep tests at 1473 K on $MoSi_2$-$50Si_3N_4$ plotted as second stage creep rate (ε) versus specific stress. For comparison, several materials such as $MoSi_2$, $MoSi_2$-40SiC,[4] and a single crystal Ni-base superalloy are also included. $MoSi_2$-$50Si_3N_4$ is almost five orders of magnitude stronger than binary $MoSi_2$ and comparable to $MoSi_2$-40SiC. This again confirms the previous

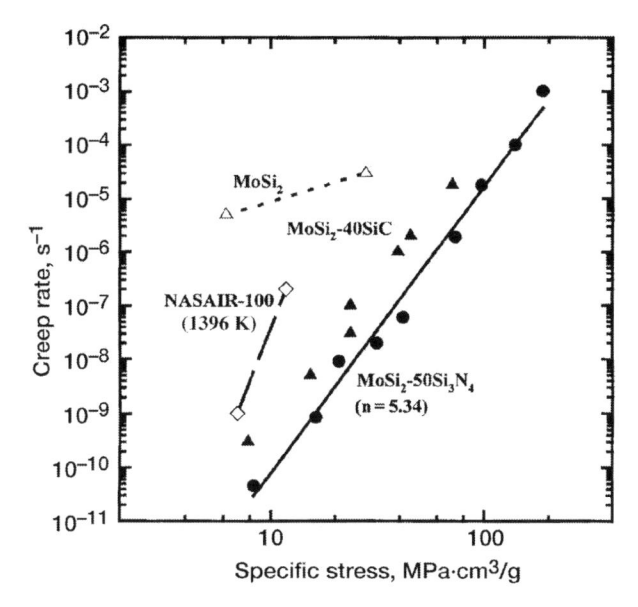

FIGURE 6. Second stage creep rate versus specific stress at 1473 K for $MoSi_2$-$50Si_3N_4$ compared with other materials.

observation of beneficial effects of particulate reinforcement. The derived stress exponent, n = 5.3, and the activation energy of 520 kJ/mol calculated from the temperature dependence of creep rate at constant stress, imply a diffusion controlled dislocation mechanism as the rate controlling mechanism.

3.3.2. Fracture Toughness

The fracture toughness of $MoSi_2$ and $MoSi_2$-Si_3N_4 base materials were measured on chevron notched 4 point bend specimens. Figure 7 shows the fracture toughness of $MoSi_2$-$50Si_3N_4$ as a function of temperature. For comparison, results for two monolithic SiC and Si_3N_4 are also included. The room temperature fracture toughness of both $MoSi_2$-$30Si_3N_4$ and $MoSi_2$-$50Si_3N_4$ matrix was ~5.2 MPa$\sqrt{m}$, which is about twice the value measured on monolithic $MoSi_2$. Figure 8 also shows that fracture toughness of $MoSi_2$-Si_3N_4 increases with temperature, especially beyond 1473 K, which is the BDTT for this material. All of the ceramics maintain the same toughness as temperature is increased.

3.3.3. Erosion

Alman et al.,[9] studied the solid particle erosion behavior of $MoSi_2$-Si_3N_4 at room temperature and at elevated temperature. Alumina particles entrained in a stream of nitrogen gas impacted the target material at a velocity of 40 m.s^{-1}. Impingement angles of either 60, 75 or 90 degrees were used. It was found that the erosion rate for the $MoSi_2$-Si_3N_4 composite measured at room temperature was a maximum at the 90 degree incident angle, which is typical of brittle materials. The erosion rate of the composite at 348 K, increased slightly

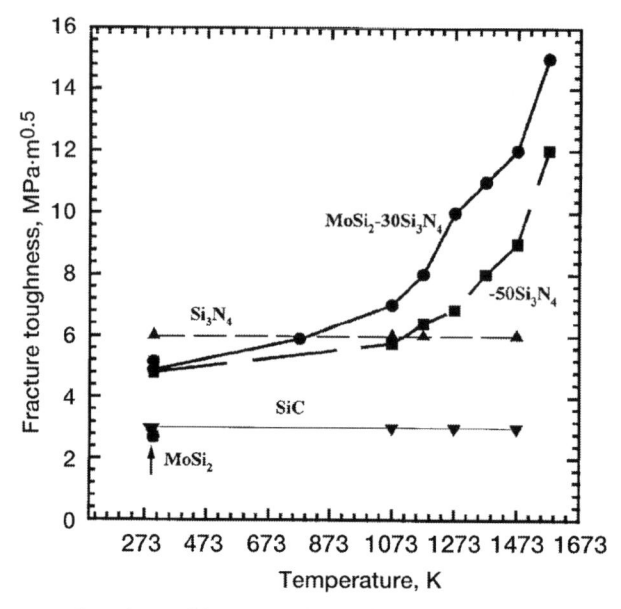

FIGURE 7. Temperature dependence of fracture toughness of MoSi$_2$-base materials compared with ceramic matrices.

with increasing test temperature up to 973 K, i.e. from 4.1 to 4.9 mm^3 g^{-1}. At 1173 K, the measured erosion rate decreased to 2.9 mm^3 g^{-1}. However, at the present time it is unclear if this is a consequence of the MoSi$_2$, phase becoming more ductile at this temperature or is attributable to some other factor. At temperature below 1173 K, the erosion rate of the MoSi$_2$-Si$_3$N$_4$ composite was similar to that of βSi$_3$N$_4$ and lower than that of Stellite-6B (Figure 8).

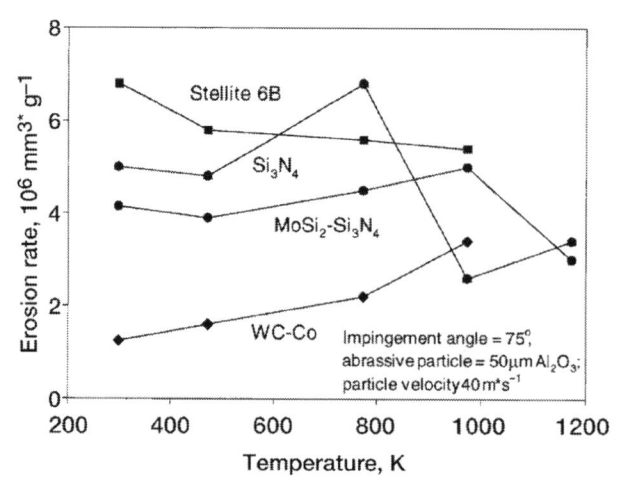

FIGURE 8. Effect of test temperatures on the erosion rate at a fixed impingement angle of 75°.

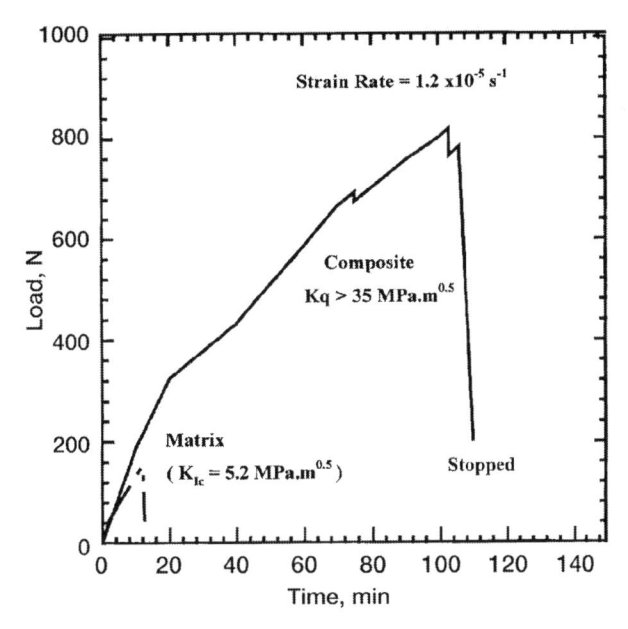

FIGURE 9. Load-time curves from a chevron notched 4 point bend specimens of SCS-6/MoSi$_2$-50Si$_3$N$_4$ and MoSi$_2$-50Si$_3$N$_4$ tested at room temperature.

3.4. Mechanical Properties of SCS-6/MoSi$_2$–Si$_3$N$_4$ Composite

3.4.1. Fracture Toughness

Figure 9 shows the load verses displacement plot for SCS-6/MoSi$_2$-30Si$_3$N$_4$ monolithic chevron notched 4 point bend specimens tested at room temperature. The composite specimen did not break even after testing for 2 hr. The apparent critical stress intensity factor, K_q, calculated from the maximum load data was greater than 35 MPa$\sqrt{m}$, which is 7 times tougher than the monolithic material. The toughness of the hybrid composite also increased with temperature reaching as high as 65 MPa$\sqrt{m}$, at 1673 K in argon atmosphere.

3.4.2. Tensile Behavior

Figure 10(a) shows the room temperature tensile stress strain curve for SCS-6/MoSi$_2$-Si$_3$N$_4$, indicating composite-like behavior; and three distinct regions, an initial linear region, followed by a nonlinear region and a second linear region. The nonlinear region is due to the matrix cracking normal to the loading direction. The second linear region is controlled by fiber strength. Individual SCS-6 fibers were tensile tested at room temperature in the as-received, as-etched and etched-from-composite conditions, and produced average strength values of 3.52 ± 0.8, 3.35 ± 0.6, and 3.4 ± 1 GPa, respectively. Thus, neither etching nor consolidation conditions degraded the strength of the fibers.

Fiber/matrix interfacial properties play an important role in composite mechanical behavior. In the case of this composite system, the carbon layer on SCS-6 provides an appropriate level of bonding that produces adequate strengthening and toughening. The

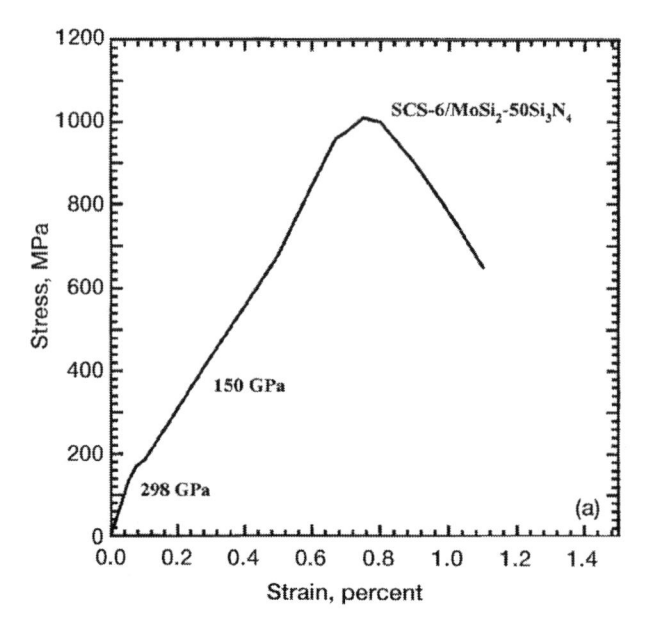

FIGURE 10(a). Room temperature tensile stress-strain curves for 6-ply SCS-6/MoSi$_2$-Si$_3$N$_4$ hybrid composites.

carbon can react with MoSi$_2$ to form SiC and Mo$_5$Si$_3$, although the carbon layer is still retained and the reaction zone thickness is not very large at typical HIP temperatures. The fiber matrix interfacial shear strengths determined from a fiber push out test using thin polished sections produced values near 50 MPa, indicating a week bond between the matrix and the fiber.

It was found that the room temperature ultimate tensile strength and strain to fracture were reduced by only 20 percent in a specimen with exposed fibers preoxidized at 1473 K for 200 hr. High temperature tensile tests were performed in air at temperatures up to 1673 K. Figure 10(b) shows the temperature dependence of ultimate tensile strength, along with the data from competitive materials, such as Ni-base single crystal alloy PWA1480 and SCS-6/RBSN. Ni-base single crystal alloy PWA 1480 exhibits higher tensile strength than both MoSi$_2$-base and RBSN-base composites between room temperature and 1273 K; however, PWA 1480 is almost 3 times denser than both composites, and hence is at a disadvantage on a specific strength basis. Figure 10(b) also shows the tensile strength data for the SCS-6 fibers, re-emphasizing the fiber-dominated behavior of the composites. The MoSi$_2$-base composites also exhibited elastic modulus values of $\sim$290/200 GPa between RT and 1473 K.

Several tensile creep tests were carried out on SCS-6$_{[0]}$/MoSi$_2$-50Si$_3$N$_4$ composite specimens between 1273 and 1473 K in vacuum. Test durations of $\sim$1000 hr were achieved and some idea of long term durability was obtained. Specimens tested at these temperatures exhibited a short primary creep stage and an extended secondary stage. The minimum creep rates ranged from 1.0×10^{-9} to 2.0×10^{-9} s^{-1} at 70 MPa between 1373 and 1473 K.

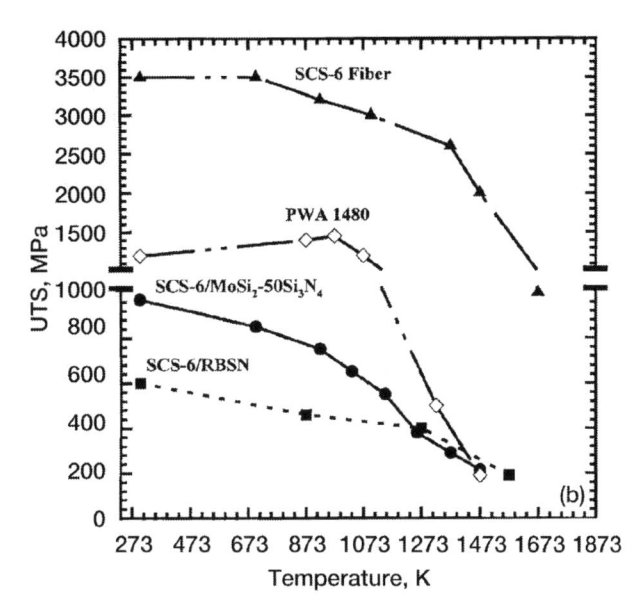

FIGURE 10(b). Temperature dependence of ultimate tensile strength of 6-ply SCS-6/MoSi$_2$-50Si$_3$N$_4$ base composites compared with other materials.

3.4.3. Impact Behavior

The Charpy V-notch (CVN) impact tests were conducted on full size specimens of MoSi$_2$-50Si$_3$N$_4$ matrix and SCS-6$_{[0]}$ and $_{[0/90]}$ oriented /MoSi$_2$-50Si$_3$N$_4$ hybrid composites between liquid nitrogen temperature (77 K) and 1673 K in air. Figure 11(a) shows the force time curves obtained from the instrumented impact tests at room temperature for monolithic MoSi$_2$-50Si$_3$N$_4$, SCS-6$_{[0]}$ and $_{[0/90]}$/MoSi$_2$-50Si$_3$N$_4$ composites. The maximum value of force represents the elastic energy required for crack initiation. The hybrid composite in [0] orientation exhibited the highest peak force values, followed by the cross-plied and finally the monolithic material. At 1673 K, the peak force values for all three materials were higher than their corresponding values at room temperature. The hybrid composite exhibited a gradual, stepwise decrease in load after the peak force was achieved. This indicates substantial energy absorption during crack propagation, and was especially pronounced in the [0] orientation.

Figure 11(b) shows the temperature dependence of CVN energy for MoSi$_2$-base materials compared with other potential materials such as superalloys, and ceramics. The CVN energy for both the monolithic MoSi$_2$-50Si$_3$N$_4$ and the hybrid composites increased with increasing temperature. The fiber reinforcement in [0] orientation increased the impact resistance by 5 times and in [0/90] orientation nearly two times. The CVN energy of SCS-6/MoSi$_2$-50Si$_3$N$_4$ was comparable to the cast superalloy B-1900 but substantially lower than the wrought superalloy Hastelloy X. The CVN energy of MoSi$_2$-50Si$_3$N$_4$ monolithic was comparable to Mo alloys and in-situ toughened Si$_3$N$_4$ (AS-800) and was far superior to NiAl, and monolithic hot pressed Si$_3$N$_4$, and SiC. Unlike MoSi$_2$-50Si$_3$N$_4$ which shows increased CVN energy with temperature (beyond 1273 K), the AS-800 shows a constant

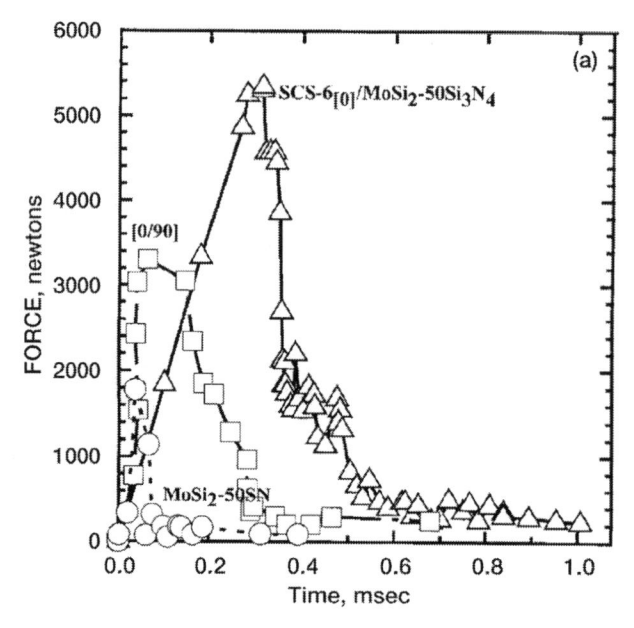

FIGURE 11(a). Force-time curves for $MoSi_2$-Si_3N_4 monolithic and SCS-6/$MoSi_2$-Si_3N_4 hybrid composites obtained from the instrumented Charpy impact tests at room temperature.

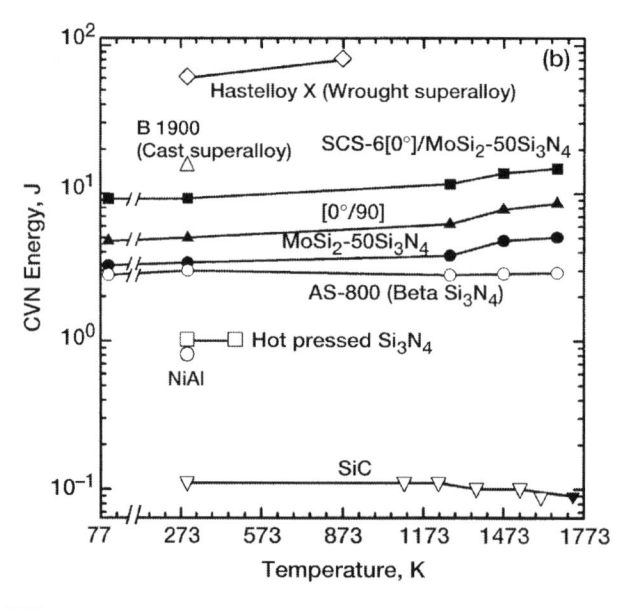

FIGURE 11(b). CVN energy versus temperature plot for $MoSi_2$-base composites compared with other materials.

CVN energy between 77 and 1673 K. The monolithic SiC shows a slight decrease of CVN energy with temperature. This is probably due to the degradation caused by densification aids used with SiC.

3.5. Hi-Nicalon/MoSi$_2$-Si$_3$N$_4$ Composite

Most of the attractive strength and toughness values reported so far were achieved with composites reinforced with SCS-6 fibers. This fiber does not have adequate creep strength at the highest temperatures envisioned for MoSi$_2$ and is too large to be bent around the sharp radii needed to make complex shapes. Finer diameter fibers are preferred on a cost, shape making, creep resistance, and toughness basis. Hi-Nicalon is the best currently available fiber, although Dow Corning's Sylramic® fiber, is also appropriate for this MoSi$_2$-Si$_3$N$_4$ matrix. A transition in effort to Hi-Nicalon fibers was therefore investigated, first using tow fibers that are spread out, wound on a drum and then infiltrated with matrix powder and ultimately woven cloth.

Tape casting was adopted as a powder metallurgy method for composite fabrication. Initially, several casting trials of MoSi$_2$-Si$_3$N$_4$ were carried out to optimize various parameters such as particle size, type and amount of binder and solvent, flow behavior of the slurry, and binder burn-out cycle. A 56 ply composite of SCS-6/MoSi$_2$-Si$_3$N$_4$ was successfully fabricated by tape casting followed by the standard hot press plus HIP consolidation. Composites with small diameter fibers such as SCS-9 (75 μm) and coated Hi-Nicalon (18 to 20 μm) were successfully fabricated.[8]

3.6. Mechanical Properties of Hi-Nicalon/MoSi$_2$-Si$_3$N$_4$ Composite

The influence of fiber diameter and architecture on mechanical properties was investigated by conducting room temperature tensile and fracture toughness tests on specimens of SCS-6, SCS-9 and BN/SiC coated Hi-Nicalon/MoSi$_2$-50Si$_3$N$_4$ hybrid composite. Testing in the [0] direction (longitudinal) produced the highest strength, (700 to 1000 MPa strength and 1.2 percent total strain) to failure. Testing in the [90] direction produced the lowest ultimate tensile strength of only 72 MPa and 0.04 percent strain to failure for SCS-6 reinforced composite. This is not an unexpected result since the fibers cannot bridge matrix cracks in the transverse direction, and cross-plied laminates or woven two or three dimensional architectures are required to achieve more isotropic properties. For example, the Hi-Nicalon reinforced composite exhibited high strength and strain to failure in the 0/90 architecture, about 60 percent of the unidirectional value, Figure 12(a). Figure 12(b) shows that the Hi-Nicalon/MoSi$_2$-Si$_3$N$_4$ in [0/90] direction exhibited higher fracture toughness than the CMC's Hi-Nicalon/SiC and Hi-Nicalon/Si$_3$N$_4$, even though they were tested in the more favorable [0] direction.[10] The CMC's were processed at much higher temperatures, 1873 to 2073 K, causing more fiber degradation than Hi-Nicalon/MoSi$_2$-Si$_3$N$_4$ and therefore exhibited lower toughness. Figure 13 shows the influence of fiber diameter and architecture on flexural stress rupture at 1473 K/210 MPa in air. This figure clearly indicates the limited improvement with large diameter fiber and more than two orders of improvement with fine diameter fiber [0/90 oriented] in stress rupture lives.

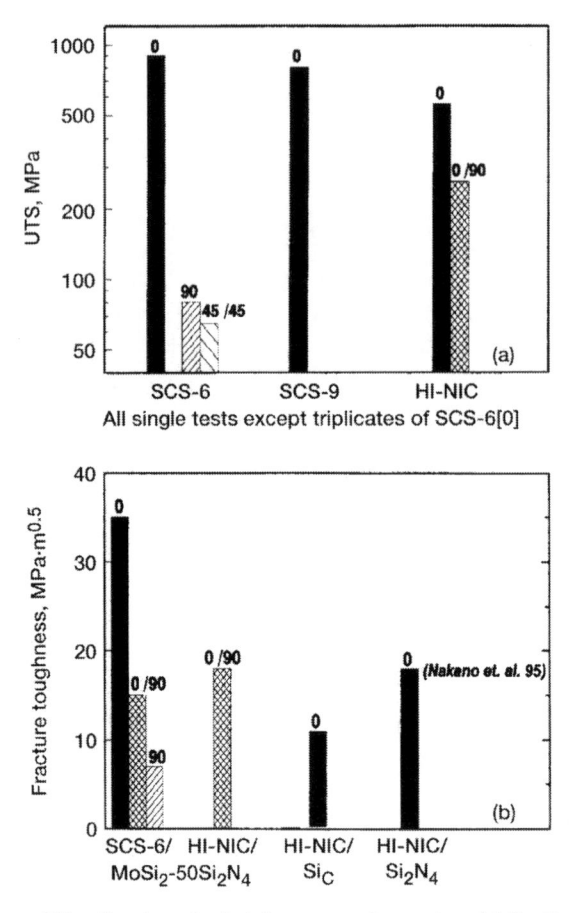

FIGURE 12. Influence of fiber diameter and orientation on room temperature. (a) Tensile strength. (b) Fracture toughness of $MoSi_2$-base hybrid composites.

3.7. 2D Hi-Nicalon/$MoSi_2$-Si_3N_4 Composite by Melt Infiltration

In order to respond to the industrial need of low cost processing of complex shaped $MoSi_2$-base composites, it was decided to initiate the melt infiltration processing of two-dimensional woven Hi-Nicalon preforms. This kind of processing is being used for SiC/SiC composites. The preforms $10 \times 15 \times 0.3$ cm and having a dual layer coating of BN and SiC by chemical vapor infiltration were obtained from DuPont and cut into of 1.2 cm wide × 15 cm long strips. These strips were placed in a Plaster of Paris mold. A water based slurry of $MoSi_2$-Si_3N_4 mixture was vacuum infiltrated into the strips. About 50 vol% infiltration was achieved using this technique. The slurry infiltrated pre-forms were dried and then infiltrated with molten silicon alloy to achieve full density. Preliminary results on microstructure and mechanical property of the melt infiltrated composites are promising.

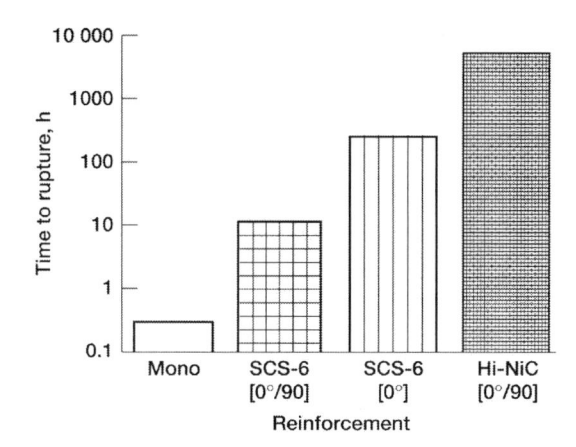

FIGURE 13. Influence of SiC fiber diameter and orientation on flexural stress rupture of $MoSi_2$-$50Si_3N_4$ at 1473 K/210 MPa.

3.8. Applications of SiC/$MoSi_2$-Si_3N_4 Composite

The hybrid composite tested for blade outer airseal component (BOAS) in the aggressive environment of gas turbine engine. Two engine test coupons of the BN/Si_3N_4 coated Hi-Nicalon$_{[0/90]}$ /$MoSi_2$-$50Si_3N_4$ hybrid composite were made according to Pratt and Whitney's design. After machining, the surfaces of these coupons were coated with 2 μm thick by CVD-SiC to protect the exposed fibers from environmental degradation. These coupons were tested in Pratt and Whitney's demonstrator engine, XTC/66/b, to simulate (BOAS) thermal cyclic conditions. One surface of the test coupon was facing the jet fueled flame, which reached approximately 1478 K and produced a thermal gradient of about 873 K between the exposed (front) and unexposed (back) surfaces. This hybrid composite performed significantly better than a SiC whisker reinforced $MoSi_2$, which showed severe cracking within the first few cycles. The hybrid composite was removed after 15 cycles and post-test examination did not reveal any surface or matrix cracking. The other applications of this hybrid composite include exhaust nozzle and combuster components of a gas turbine engine.

IV. $MoSi_2$-βSi_3N_4 COMPOSITE

However, manufacturing of state-of-the-art fiber-reinforced composite is still quite expensive due to the high cost of fibers and fiber coatings. This problem can be alleviated by reinforcing a $MoSi_2$-base alloy with a high volume fraction of randomly oriented, in-situ grown, long whisker-type grains of βSi_3N_4. This engineered microstructure is produced through proper thermomechanical treatment and the use of sintering additives, which promote the growth of long whiskers of βSi_3N_4 in a fully dense $MoSi_2$ matrix. The combination of properties achieved by this new alloy would make it a viable alternative to continuous fiber reinforced composite.

TABLE I. Processing Conditions for the Various Alloys Investigated

Alloy designation	Alloy composition, wt%	Consolidation conditions	Microstructure
MS-60	$MoSi_2$-$35Si_3N_4$-$4Al_2O_3$-$1Y_2O_3$	Hot Press: 1673 K/120 MPa/2-hrs HIP (Ta can): 2173 K/300 MPa/2-hrs	Fully dense βSi_3N_4 with long whisker-type morphology
MS-70	$MoSi_2$-$35Si_3N_4$-$4Al_2O_3$-$1Y_2O_3$	Hot Press: 2073 K/70 MPa/3-hrs HIP (no can necessary): 2173 K/300 MPa/2-hrs	Fully dense βSi_3N_4 with long whisker-type morphology
MS-80	$MoSi_2$-$35Si_3N_4$-$4Al_2O_3$-$1Y_2O_3$ (Processed in 01-HD attritor with significant powder loss)	Hot Press: 1873 K/56 MPa/2-hrs + 1973 K/56 MPa/2-hours HIP (graphoil wrap): 2173 K/280 MPa/2-hrs	Not fully dense βSi_3N_4 with blocky morphology
MS-50	$MoSi_2$-$35Si_3N_4$	Hot Press: 1673 K/120 MPa/2-hours HIP (Ta can): 1773 K/300 MPa/2-hrs	Fully dense βSi_3N_4 with blocky morphology
MS-40	$MoSi_2$-$35Si_3N_4$	Hot Press: 1473 K/200 MPa/2-hrs HIP (Ta can): 1473 K/280 MPa/2-hrs.	Not fully dense Fine grained $MoSi_2$ and blocky βSi_3N_4

4.1. Processing

A series of processing runs, summarized in table I, were made in order to determine the minimum requirements necessary to produce a fully dense material with the desired in-situ toughened microstructure composed of whisker-like Si_3N_4 grains. From the consolidated material, specimens for determining physical, mechanical, and environmental properties were machined using EDM. Details of the processing, specimen design and testing procedures are described elsewhere.[11]

4.2. Microstructure

Figure 14(a) shows SEM back scattered image of a fully dense $MoSi_2$-βSi_3N_4 composite (MS-70). During processing, the original βSi_3N_4 powder particles are transformed into randomly oriented whiskers of βSi_3N_4. These long whiskers are well dispersed throughout the material and appear to be quite stable, with very little or no reaction with the $MoSi_2$, even at 2173 K. Figure 14(b) shows a back-scattered image of $MoSi_2$-βSi_3N_4 (MS-80) with the βSi_3N_4 exhibiting the blocky aggregate type morphology.

4.3. Physical Properties

Density of the in-situ toughened $MoSi_2$-βSi_3N_4 (alloy MS-70) was measured to be 4.57 ± 0.01 g/cm^3, Vickers microhardness was 10.7 ± 0.6 GPa, and average CTE was about 4.0 ppm/K. Figure 15 shows the dynamic Young's Modulus for two batches of $MoSi_2$-βSi_3N_4 (MS-70 and MS-80) as a function of temperature. Young's Modulus for MS-70 decreases

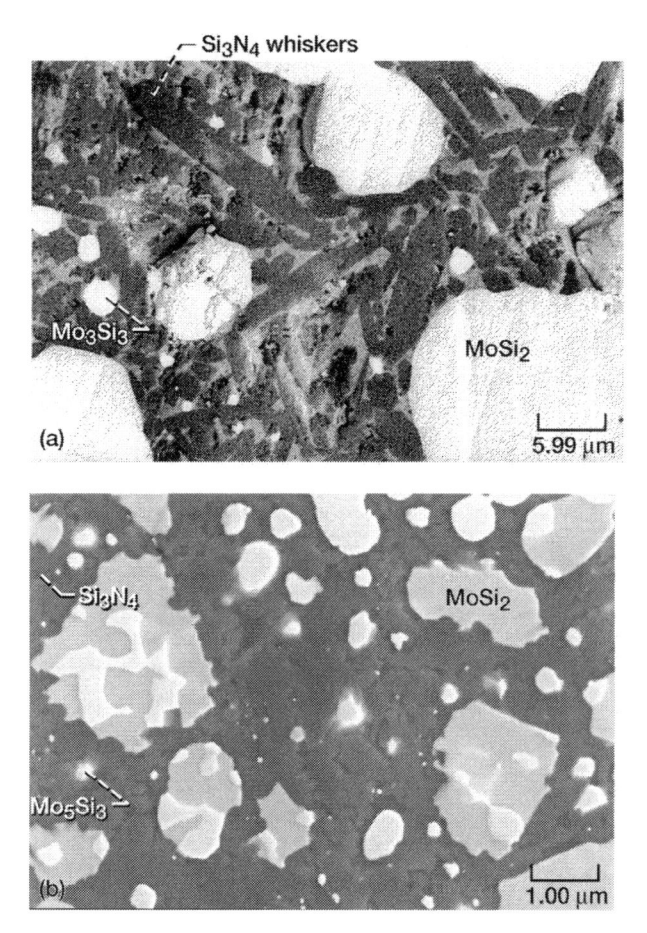

FIGURE 14. SEM back scattered image of $MoSi_2$-βSi_3N_4 (MS-70) showing randomly oriented in-situ grown long whiskers of βSi_3N_4 and large $MoSi_2$ particle size. (b) SEM back scattered image of $MoSi_2$-βSi_3N_4 (MS-80) where the Si_3N_4 has a blocky particulate structure and not a whisker-like morphology.

with increasing temperature but only by about 10 percent over a 1273 K range. As expected, the denser alloy (MS-70) exhibits a higher modulus at all temperatures compared to MS-80, which contained residual porosity after processing.

4.4. Electrical Conductivity

Electrical conductivity is a very important property in determining whether a material can be suitably machined by EDM. Room temperature electrical conductivity of various $MoSi_2$-base materials and Si_3N_4 ceramic were measured on rectangular bars. $MoSi_2$-βSi_3N_4 (MS-70) composite showed more than an order of magnitude higher (20/ohm.com) conductivity for the $MoSi_2$-βSi_3N_4 compared to AS-800 (0.5/ohm.com). Both the $MoSi_2$ and $MoSi_2$-βSi_3N_4 have proven machineable by EDM. In contrast, AS-800 cannot be EDM'ed.

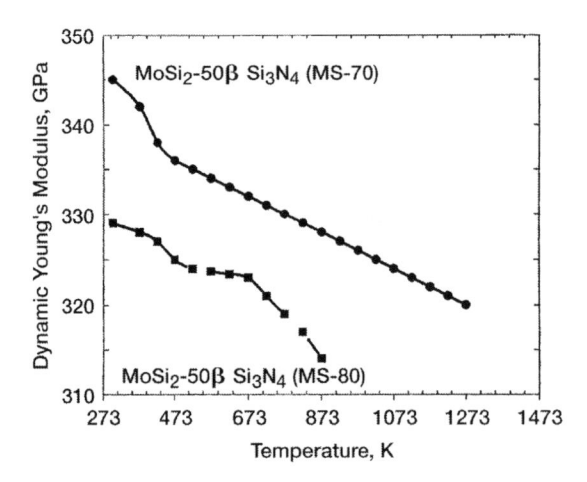

FIGURE 15. Dynamic Young's Modulus versus temperature for $MoSi_2$-βSi_3N_4 materials. MS-70 and MS-80.

4.5. Mechanical Properties

4.5.1. Compression Strength

The compression yield strengths of (MS-80) measured between room temperature and 1673 K are shown in Figure 16. For comparison the compression strengths of $MoSi_2$-βSi_3N_4 (MS-50) are also included. Both materials contain 50 vol% of Si_3N_4 but $MoSi_2$-βSi_3N_4 has higher strength than $MoSi_2$-βSi_3N_4 at all temperatures. Furthermore, the MS-80 material has less than optimum microstructure and it would be anticipated that the in-situ toughened MS-70 or MS-60 material would exhibit even better strength, especially at high temperatures.

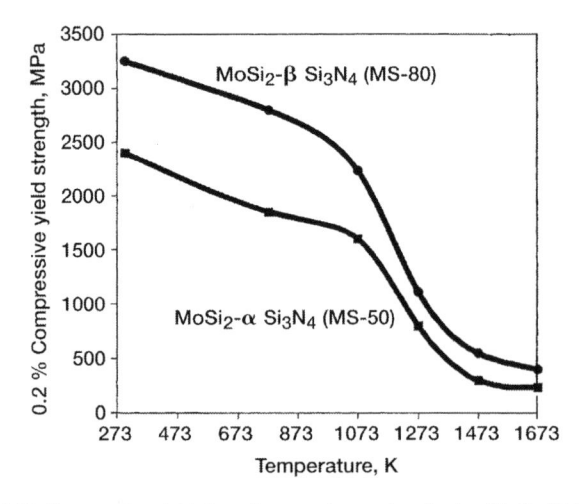

FIGURE 16. Compressive yield strength versus temperature for two $MoSi_2$-Si_3N_4 alloys.

TABLE II. Fracture Toughness of MS-70 Determined at Room Temperature by Four Different Test Methods

Test method	Conditions	Fracture toughness, K_{IC} (MPa$\sqrt{m}$)
Indentation Fracture (IF)	5 indents at 98N and 196 N	$9.4 \pm (0.5)$[b]
Single Edge V-Notched Beam (SEVNB)	Three specimens used; $\alpha = 0.25$[a]	$7.4 \pm (0.4)$[b]
Single Edge Precracked Beam (SEPB) (ASTM C1421)	Three specimens used; $\alpha = 0.4$ to 0.6[a]	$8.9 \pm (0.5)$[b]
Chevron-Notched Beam (CNB) (ASTM C1421)	Average of five specimens	12.2 ± 0.2

[a] Indicates the ratio of precrack size to specimen depth.
[b] Indicates ± 1.0 standard deviation.

4.5.2. Fracture Toughness

A summary of the fracture toughness values for in-situ toughened MoSi$_2$-βSi$_3$N$_4$ (MS-70) determined by four different test methods is presented in table II. The SEVNB method yielded the fracture toughness (7.4 MPa$\sqrt{m}$) and the CNB method resulted in the highest value (12.2 MPa$\sqrt{m}$). A few CNB specimens machined from MS-60 exhibited even higher fracture toughness of 14.5 MPa$\sqrt{m}$. The SEVNB method tends to provide a lower value of fracture toughness presumably due to rising R-curve behavior. It is worth comparing the fracture toughness between MoSi$_2$-βSi$_3$N$_4$ (MS-70) and MoSi$_2$-βSi$_3$N$_4$ (MS-40 and MS-50). The βSi$_3$N$_4$ reinforced materials exhibited K_{Ic} values of 3.5 ± 0.4 MPa$\sqrt{m}$ and 4.9 ± 0.3 MPa$\sqrt{m}$ for MS-40 and MS-50, respectively compared to 8.9 ± 0.5 MPa$\sqrt{m}$ for the in-situ toughened MS-70, as determined by the SEPB method. Therefore, a significant increase in fracture toughness, up to 157 percent by SEPB method, was achieved for MS-70 material through improved material processing and engineered microstructural control. In addition, the fracture toughness of the MS-70 material was greater than that (7.2 ± 0.2 MPa$\sqrt{m}$), of AS-800, an in-situ toughened Si$_3$N$_4$. Figure 17 shows a comparison of the indent crack trajectories generated in the polished surfaces of MS-40, MS-50, and MS-70 materials. Indents were made with a Vickers indenter using an indent load of 98 N. The length of crack eminated from the indent corner was significantly shorter for the MS-70 material, illustrating again that fracture resistance (hence, fracture toughness) was much greater than the MS-40 and MS-50 materials. Also note the more tortuous path of the crack for the MS-70 material.

4.5.3. Flexural Strength and Weibull Behavior

Figure 18 shows a comparison of the four-point flexural strength of MoSi$_2$-βSi$_3$N$_4$ (MS-70), and MoSi$_2$-βSi$_3$N$_4$ (MS-50) and AS-800 as a function of temperature up to 1673 K. AS-800 is stronger by about 10 percent than MoSi$_2$-βSi$_3$N$_4$ (MS-70) at all temperatures. Figure 18 also shows that the in-situ whisker reinforced MoSi$_2$-βSi$_3$N$_4$ (MS-70) is significantly stronger than MoSi$_2$-βSi$_3$N$_4$ (MS-50) at all temperatures. Room temperature four point bend tests were carried out to measure the strength and Weibull modulus of two types MoSi$_2$-βSi$_3$N$_4$ materials, one with the βSi$_3$N$_4$ in a whisker-like morphology (MS-70) and the other with a more characteristic blocky βSi$_3$N$_4$ particulate structure (MS-80).

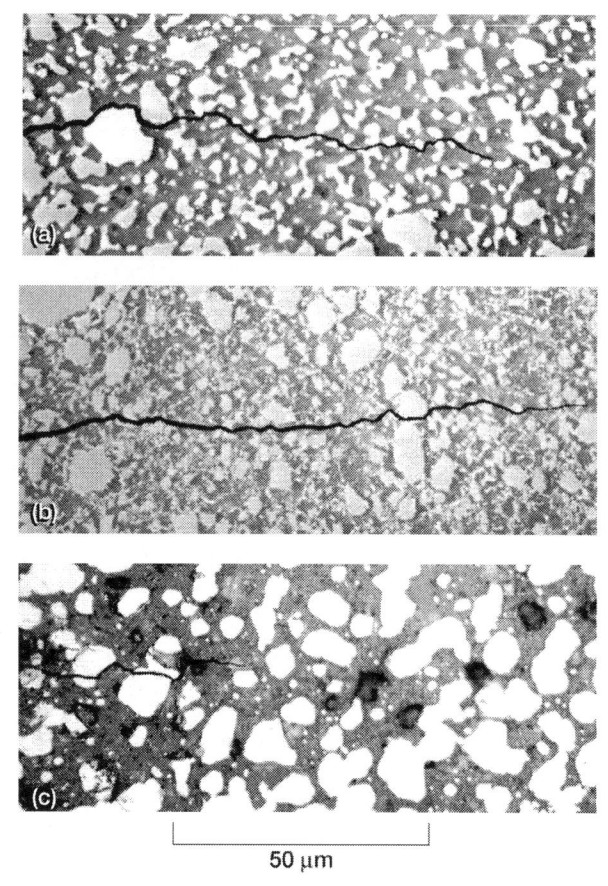

FIGURE 17. Indentation crack length and trajectories for $MoSi_2$-50%, Si_3N_4. (a) MS-40. (b) MS-50. (c) MS-70. An Indent load of 98 N was applied to initiate the cracks.

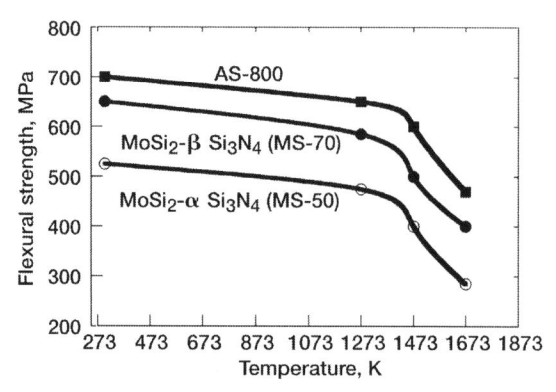

FIGURE 18. Flexural strength as a function of temperature for $MoSi_2$-βSi_3N_4 (MS-70) compared to AS-800 and $MoSi_2$-Si_3N_4 (MS-50).

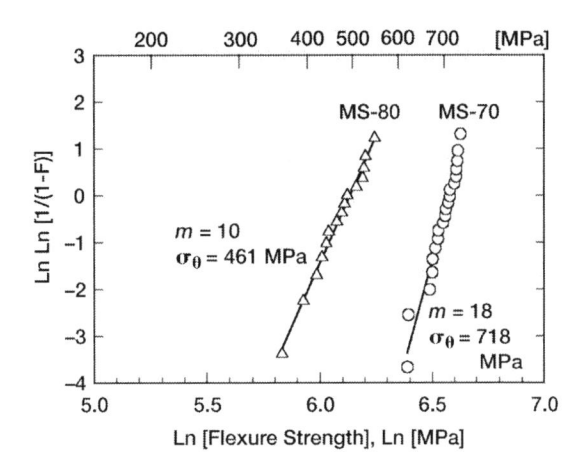

FIGURE 19. Weibull (flexure) strength distribution for $MoSi_2$-βSi_3N_4 (MS-70) and (MS-80) materials.

The resulting Weibull distributions for MS-70 and MS-80 are plotted in Figure 19. The in-situ toughened material (MS-70) exhibits a much higher characteristic strength and Weibull modulus (728 MPa and m = 17) than the MS-80 material (461 MPa and m = 9.8).

The toughening mechanisms operative in the in-situ whisker reinforced $MoSi_2$-βSi_3N_4 are crack deflection, crack bridging, and grain pull-out. Through a combination of these three toughening mechanisms the in-situ whisker reinforced $MoSi_2$-βSi_3N_4 exhibited significant toughness compared to materials of similar composition. In both the $MoSi_2$-βSi_3N_4 (MS-50 and MS-40) and the MS-80 material, the Si_3N_4 did not possess a whisker structure and hence these materials only underwent crack deflection resulting in a lower fracture toughness. Furthermore, because the additional toughening mechanisms, such as bridging and pullout, lead to a rising R-curve behavior, the in-situ whisker reinforced $MoSi_2$-βSi_3N_4 would be expected to have better reliability. Because the results do indicate a higher Weibull modulus and strength for MS-70 compared to other $MoSi_2$-base materials, then either the flaw population in this material is unusually uniform or a rising R-curve behavior is active during fracture. Another consequence of R-curve behavior would be discrepancies in fracture toughness values obtained from the different testing techniques as summarized in table II.

4.6. Environmental Resistance

4.6.1. Low Temperature Oxidation and Pesting

Figure 20 shows specific weight gain versus number of cycles at 773 K for $MoSi_2$-βSi_3N_4 (MS-70) and binary $MoSi_2$. The oxidation behavior of a MoSiB[13] alloy (at.% 84.5Mo-6.5Si-8B-1Hf) is also included for comparison. $MoSi_2$-βSi_3N_4 (MS-70) shows very little weight gain compared to binary $MoSi_2$ and the MoSiB alloy, indicating the absence of accelerated oxidation. In contrast, the binary $MoSi_2$ and MoSiB alloys exhibited accelerated oxidation followed by pesting. XRD analysis of the $MoSi_2$-βSi_3N_4 specimen indicated strong peaks of Si_2ON_2 and the absence of any MoO_3.

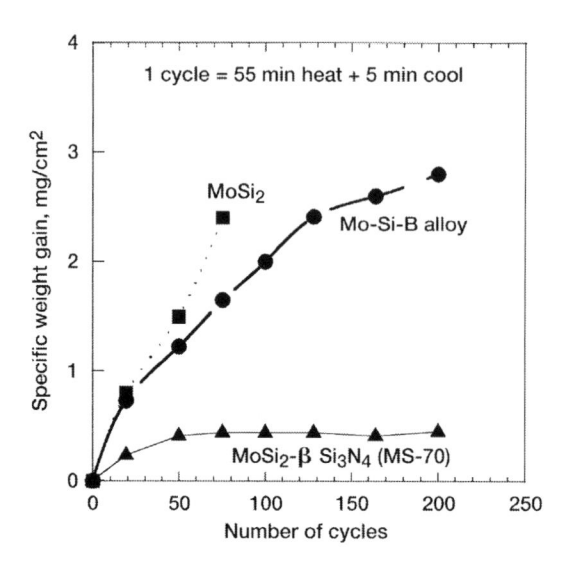

FIGURE 20. Specific weight gain versus number of cycles of $MoSi_2$-$50\beta Si_3N_4$ (MS-70) at 773 K compared with $MoSi_2$ and MoSiB alloy.

4.6.2. High-Temperature Cyclic Oxidation

Cyclic oxidation tests were carried out on rectangular coupons of the in-situ toughened $MoSi_2$-βSi_3N_4 (MS-70) at 1273 and 1623 K in air for about 100 cycles where each cycle consisted of 1 hr heating followed by 20 minutes of cooling. The material exhibited a parabolic oxidation behavior with weight gain of only 0.03 and 0.4 mg/cm^2 at 1273 and 1623 K, respectively. These weight gain values are smaller than those for AS-800 (0.06 and 0.63 at 1273 and 1623 K, respectively). XRD analysis of the oxidized surface of the in-situ toughened $MoSi_2$-βSi_3N_4 indicated strong peaks of α cristobalite.

4.7. Applications of $MoSi_2$-βSi_3N_4 Composite

The mechanical properties of $MoSi_2$-based composites are now adequate for a wide range of industrial and military applications. A few of the potential applications are high lighted below.

4.7.1. Diesel Engine Glow Plug

$MoSi_2$-Si_3N_4 diesel engine glow plug for automotive applications has recently been developed. This glow plug has significant advantages for automotive diesel engines. It possesses a long glow plug lifetime (13 years) in the diesel fuel combustion environments. It also produces higher heating rates than metal glow plugs, allowing for faster starting of the diesel engine by approximately a factor of two. The glow plug is microstructurally tailored $MoSi_2$-Si_3N_4 composite which contains approximately 30 vol% $MoSi_2$ in a Si_3N_4 matrix. It is composed of a core and a sheath both of which have the same phase, but with different $MoSi_2$ morphology. The core has an interconnected $MoSi_2$ phase so is electrically conductive, while the sheath has a non-interconnected $MoSi_2$ phase so it is electrically insulating. Since both the core and sheath are essentially the same composite composition except for phase interconnectivity, there is perfect match of thermal expansion coefficients and material compatibility.

4.7.2. Aircraft Engines

Industrial applications typically require fracture toughness values of 10 MPa$\sqrt{m}$, while aerospace applications require 15 MPa$\sqrt{m}$ or higher. The fracture toughness of $MoSi_2$-βSi_3N_4 is very close to 15 MPa$\sqrt{m}$ making this ideal for blade outer air seal component in an aircraft engine. The low cost of manufacturing near-net shape component and other beneficial properties such as ease of machining by EDM technique of $MoSi_2$-βSi_3N_4 makes this an attractive material in aerospace applications.

V. CONCLUDING REMARKS

A wide spectrum of mechanical and environmental properties have been measured in order to establish feasibility of an $SiC/MoSi_2$-Si_3N_4 hybrid composite. The high impact resistance of the composite is of particular note, as it was a key property of interest for engine applications. Processing issues have also been addressed in order to lower cost and improve shape making capability. The composite system remains competitive with other ceramics as a potential replacement for superalloys. However, the cost of fiber, fiber coating and the processing is a major concern. Therefore a new $MoSi_2$-based composite with an engineered microstructure consisting of large particles of $MoSi_2$ reinforced with long whisker-shaped grains of Si_3N_4 was developed. This in-situ toughened $MoSi_2$-βSi_3N_4 exhibited higher fracture toughness and Weibull modulus than other $MoSi_2$-base materials or a structural Si_3N_4 ceramic. This material also exhibits excellent resistance to intermediate temperature pesting that affects most other Mo-base and $MoSi_2$-base alloys. Given this combination of good fracture resistance and excellent environmental resistance, the in-situ toughened $MoSi_2$-βSi_3N_4 shows great promise as a high temperature structural material. However, the development of reliable processing methods will be needed to broaden interest in this material.

REFERENCES

1. A.K. Vasudevan and J.J. Petrovic, "Key Development in High Temperature Structural Silicides," Mat. Sci. & Eng, **A261,** [1–2] 1–5 (1999).
2. S. Bose, "Engineering Aspect of Creep Deformation of Molybdenum Disilicide," in "High Temperature Silicides," Eds., A.K. Vasudevan and J.J. Petrovic. North Holland, NY, 217–225 (1992).
3. R.M. Aikin, Jr., "High Temperature Mechanical Behavior of Discontinuously Reinforced $MoSi_2$ Composites," Structural Intermetallics, Eds., R. Darolia, J.J. Lewandowski, C.T. Liu, D.B. Miracle, and M.V. Nathal, TMS, Warrendale, PA, 791–798 (1993).
4. K. Sadananda, C.R. Feng, and H. Jones, "Creep of Molybdenum Disilicide" in "High Temperature Silicides." Eds., A.K. Vasudevan and J.J. Petrovic. North Holland, NY, 227–237 (1992).
5. M.J. Maloney and R.J. Hecht, "Development of continuous Fiber-Reinforced $MoSi_2$-base Composites," in High Temperature Silicides, Eds., A.K. Vasudevan and J.J. Petrovic, North Holland, NY, 19–32 (1992).

6. M.G. Hebsur, "Pest resistant and Low CTE $MoSi_2$-Matrix for High Temperature Structural Applications," in "Intermetallic Matrix Composites III," Eds., J.A. Graves, R.R. Bowman, and J.J. Lewandowski, MRS Proc. **350**, [4–6], MRS, Pittsburgh, PA, 177–182 (1994).
7. M.G. Hebsur, "Pest resistant $MoSi_2$ Materials and Method of Making," U.S. Patent, #5,429,997 (1995).
8. M.G. Hebsur, "Development and Characterization of $SiC_{(f)}MoSi_2$-$Si_3N_{4(p)}$ Hybrid Composites," Mater., Sci. & Eng., **A261**, [1–2] 24–37 (1999).
9. D.E. Alman, J.H. Tylczak, J.A. Hawak, and M.G. Hebsur, "Solid Particle Erosion Behavior of an Si_3N_4-$MoSi_2$ Composite at Room and Elevated Temperatures," Mat., Sci., & Eng., **A261,** [1–2] 245–251 (1999).
10. Knakano, K. Sasaki, M. Fujikura, and H. Ichikawa, "SiC-and Si_3N_4-Matrix Composites According to the Hot-Pressing Route" in "High Temperature Ceramic-Matrix Composites II," "Manufacturing and Materials Development," Eds., A.G. Evans and R. Nasalin, American Ceramic Society, Westerville, OH, 215–229 (1995).
11. M.G. Hebsur, S.R. Choi, J.D. Whittenberger, J.A. Salem, and R.D. Noebe, "Development of Tough, Strong and Pest Resistant $MoSi_2$-βSi_3N_4 Composites for High Temperature Structural Applications," in "Structural Intermetallics," Eds., K.J. Hemker, D.M. Dimiduk, H. Clemens, R. Darolia, H. Inui, V.K. Sikka, M. Thomas, and J.D. Whittenberger, TMS, Warrendale, PA, 745–753 (2001).
12. S.R. Choi, and M.G. Hebsur, "Elevated Temperature Slow Crack Growth and Room Temperature Fracture Toughness of $MoSi_2$-$50Si_3N_4$ Composites," Ceram. Eng. Sci. Proc., **19,** [3] 361–369 (1998).
13. D.M. Berczik, "Oxidation Resistant Molybdenum Alloy," U.S. Patent, No. 5,696,150 (1997).

9

Ultra High Temperature Ceramic Composites

Matthew J. Gasch*, Donald T. Ellerby** and Sylvia M. Johnson**

*ELORET – NASA Ames Research Center
M/S 234-1, Moffett Field, CA 94035
650.604.5377, mgasch@mail.arc.nasa.gov
**NASA Ames Research Center
M/S 234-1, Moffett Field, CA 94035
650.604.2811, Donald.T.Ellerby@nasa.gov
650.604.2646, Sylvia.M.Johnson@nasa.gov

ABSTRACT

Ceramic borides, carbides and nitrides are characterized by high melting points, chemical inertness and relatively good oxidation resistance in extreme environments, such as conditions experienced during reentry. This family of ceramic materials has come to be known as Ultra High Temperature Ceramics (UHTCs). Some of the earliest work on UHTCs was conducted by the Air Force in the 1960's and 1970's. Since then, work has continued sporadically and has primarily been funded by NASA, the Navy and the Air Force. This article summarizes some of the early works, with a focus on hafnium diboride and zirconium diboride-based compositions. These works focused on identifying additives, such as SiC, to improve mechanical or thermal properties, and/or to improve oxidation resistance in extreme environments at temperatures greater than 2000°C.

I. INTRODUCTION

Refractory compounds such as ceramic carbides, borides and nitrides are characterized by high melting points, high hardness and good chemical inertness and oxidation resistance.[1-3] These refractory compounds have been broadly termed Ultra High Temperature Ceramics (UHTCs). Some of the earliest and most thorough work, to date, was performed in the 1960's by ManLabs under a research program funded by the Air Force Materials Laboratory (AFML).[1-2] Work on these materials was initiated to meet the need for high temperature materials that would enable the development of maneuverable hypersonic flight vehicles with sharp leading edges.[2] Around the same time NASA was also investigating high temperature materials to meet its own needs.[3] The relatively good oxidation resistance of refractory **diboride** compounds, compared to other refractory intermetallic compounds (i.e. carbides, nitrides and silicides), has focused many research efforts into detailed investigations of transition metal diborides of groups IV and V (Ti, Zr, Hf, Nb, Ta). Of the transition metal diborides, hafnium diboride (HfB_2) and zirconium diboride (ZrB_2) were identified as the most promising candidates for high temperature applications such as nose caps, sharp leading edges, vanes and similar objects for use in high velocity flight or on future generations of reentry vehicles.[1-2,4-6]

Strong covalent bonding is responsible for the high melting points, moduli and hardness of the UHTC family of materials.[6-7] High negative free energies of formation also give UHTCs excellent chemical and thermal stability under many conditions.[6-7] In comparison to carbides and nitrides, the diborides tend to have higher thermal conductivity, which gives them good thermal shock resistance and makes them ideal for many high temperature thermal applications.[7-8]

Realization of the full potential of UHTCs as future high temperature materials requires a thorough knowledge of their properties and behavior under diverse conditions of temperature, environment and stress states.[7] The primary purpose of this article is to provide a historical review of UHTC research carried out thus far and, where available, to provide engineering data on UHTCs. Since the most comprehensive research has been focused on compositions containing HfB_2 or ZrB_2, they will be the primary materials discussed. Results from materials containing HfB_2 or ZrB_2 as the principal component with selected additives designed to enhance one or more of the following: oxidation resistance, mechanical properties and thermal stress resistance will also be presented. When available, data on other UHTC compositions will also be discussed.

II. APPLICATIONS

The need for high temperature materials that can operate with no or limited oxidation or ablation at temperatures greater than 3000K has driven the development of UHTC materials. The potential applications for UHTCs span a wide number of needs arising from future military, industrial and space based projects. Potential industrial applications for UHTCs include use in foundry or refractory processing of materials. Their chemical inertness makes them ideal for molten metal crucibles, thermowell tubes for steel refining and as parts for electrical devices such as heaters and igniters.[9]

The military and aerospace applications for UHTCs range from rocket nozzle inserts and air augmented propulsion system components to leading edges and nose caps for future hypersonic re-entry vehicles.[9-12] Early space vehicle designs, such as the space shuttle, were designed with a large radius, blunt body design to reduce aerothermodynamic heating to maintain moderate temperature limits on all parts of the vehicle. However, the larger the leading edge radii, higher the vehicles drag, which reduces maneuverability and cross range during reentry. Therefore, to improve performance, hypervelocity vehicle concepts have been proposed using slender aerodynamic shapes with sharp leading edges.[13-14] Development of sharp body vehicles increases the lift-to-drag ratio thereby improving the vehicles' reentry cross range. A higher lift to drag ratio also has the potential to improve the overall vehicle system safety in a number of ways. Design of a high L/D vehicle would increase the window during ascent in which a launch could be aborted and the vehicle safely recovered on land, reducing the need for crew to bail out and reducing the possibility of having to ditch the vehicle in the ocean. Secondly, as mentioned previously, the high L/D increases the vehicles cross range during descent from orbit. This provides the vehicle more opportunities to initiate descent while on orbit, and safely land the vehicle at a desirable location. However, the temperature of the leading edge is inversely proportional to the square root of the leading edge radius, i.e. as the leading edge radius decreases the temperature increases.[14] Therefore the successful design of a sharp hypersonic vehicle requires the development of new materials with higher temperature capabilities than the current state-of-the-art materials can provide. Ultra High Temperature Ceramics are a family of materials that are promising candidates for meeting such requirements.[13-14]

III. BONDING AND STRUCTURE OF UHTCs

III.1. Structural Stability of Borides, Carbides and Nitrides

Ceramic borides, carbides and nitrides all have very strong chemical bonds that give them high temperature structural stability.[7,15-17] As a result of the extremely strong bond between carbon atoms, carbides follow the classical definition of a brittle ceramic. They come in three general classes: ionic, covalent and interstitial. None of the ionic carbides have engineering uses because of their extreme brittleness.[15] The two covalent carbides of major importance, SiC and B_4C, both are valuable for their extreme hardness as well as excellent thermal and chemical stability. The largest class of carbides, those of the interstitial type, includes carbides of the metals Hf, Zr, Ti, and Ta. These materials benefit from strong carbon networks and have some of the highest melting points of known materials. They are also known to have high strengths at elevated temperatures.[15-16] From this standpoint, carbides offer a tremendous benefit in many engineering applications. Unfortunately, these materials are hard to fabricate because of their refractoriness, and little has been achieved beyond the laboratory scale.[15-16]

Ceramic nitrides have many of the same properties as carbides and nitrides are also difficult to fabricate, especially in the pure form, due to strong covalent bonding. Silicon nitride and boron nitride are the primary materials in the family of nitrides to be developed for engineering applications.[17]

Then there are the ceramic borides, such as HfB_2 and ZrB_2. These materials also benefit from very strong bonding between boron atoms, although their bonding is not typically as strong as seen in the carbides thus these materials often have melting points below that of the carbides.[7,15] A unique feature of the electronic nature of the boron bonding in these materials results in the borides having high thermal and electrical conductivities, higher than typically found in carbides and nitrides, as well as low coefficients of thermal expansion which combined gives the borides relatively good thermal shock resistance, for ceramics.[7]

TABLE 1. Observed phase stability of transition metal diborides of the AlB_2 structure.

Sc	Ti	V	Cr	Mn	Fe	Co	Ni
Y	Zr	Nb	Mo	Tc	Ru	Rh	Pd
La	Hf	Ta	W	Re	Os	Ir	Pt

- Stable AlB_2 Phase
- Stable at high temp.
- Prepared or detected at high temp.
- AlB_2 phase is metastable

Boride ceramics also appear to have improved oxidation resistance over those of carbides and nitrides, the nature of which is discussed in the following section.

Table 1 lists a number of metallic elements that form binary diboride compounds, with the AlB_2-type structure, shown in Figure 1.[18–19] The AlB_2 structure contains graphite-like layers of boron separated by hexagonal close-packed (h.c.p.) layers of metal atoms. The diborides are comprised of rigid covalent boron lattices, such that the boron atoms have a trigonal prismatic metal environment with three close boron neighbors. The metal atoms coordinate twelve boron atoms, six metallic atoms in the same layer and two metal atoms in the two adjacent layers (top and bottom).[18–19] The boron nets have very strong covalent bonds that hinder an increase in the a_o direction, though no such hindrance occurs in the c_o direction, giving borides the ability to accommodate a wide variety of metals.[18]

Spear calculated three chemical bonding parameters for borides of the AlB_2-type, plotted in Figure 2.[18] The parameter in the top graph is the ratio of the two elliptic axes of the metal atom and provides a measure of the amount of metal atom distortion. The middle

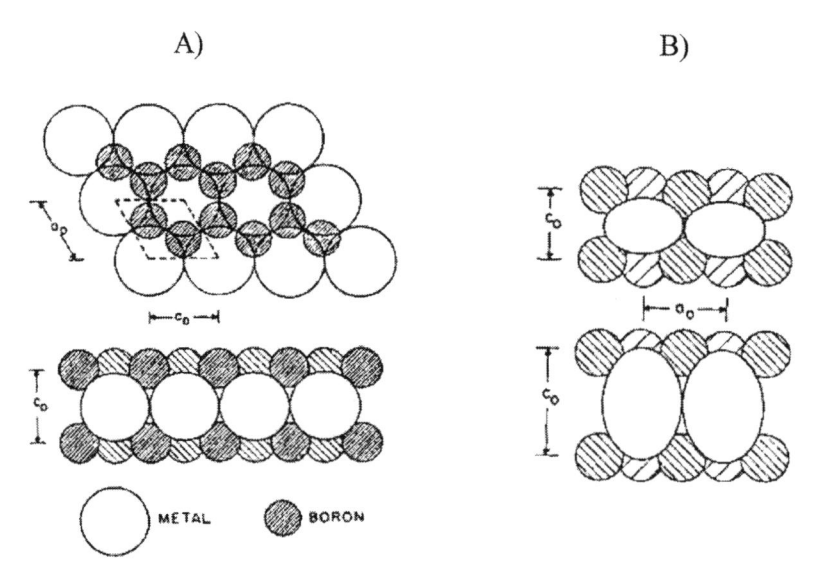

FIGURE 1. A) Atomic projections of the AlB_2-type structure showing top down and side view. B) An illustration of metal atom deformation within the AlB_2-type structure.[18]

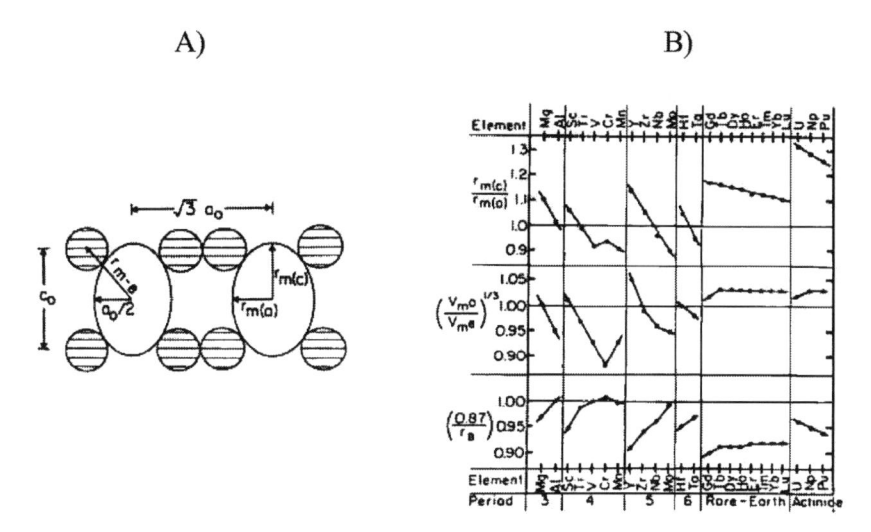

FIGURE 2. A) View of the (110) plane in a diboride structure, illustrating the geometrical crystal parameters as calculated by Spear. B) Table of bonding parameters as calculated from geometrical parameters in A), metals are grouped according to their period in the Periodic Chart.[18]

parameter is a ratio of the volume of the elemental metal atom to the metal atom in the diboride. Ratios less than unity indicate the metal is too small for the space, while ratios greater than one indicate the metal atom is compressed in the structure. Ratios different from unity suggest M-M and possibly M-B bonds will have less than optimum strength. Finally, the parameter on the bottom is a measure of B-B bond strength; it is a ratio of the average minimum free boron radius to that of boron in the diboride. Smaller values are a result of stretched bonds and smaller bond energies.[18] From these calculations, Spear deduced that M-B bonding is likely the leading contributor to the structural integrity of AlB_2 type borides, more so than M-M or B-B bonding. Typically, the less distortion there is to the unit cell, the stronger are the bonds that hold it together. As the degree of bonding increases so does the melting point, modulus and hardness, in the diborides as well as for carbide and nitride ceramics.[7-8,15]

Enthalpies of formation for several boride systems were reported by Samsonov and Vinitiskii.[20] From those results it is clear that the stability of diborides decrease in the order $HfB_2 > TiB_2 > ZrB_2 \gg TaB_2 > NbB_2 > VB_2$, with VB_2 the least stable of the diborides, having the lowest energy of formation. To understand this trend, Guillermet and Grimvall showed that the cohesive properties of these materials could be described by the filling of electron bands, as shown by a plot of density of electron states (bonding or anti-bonding states), N(E), versus the respective energy of the system, Figure 3.[21] Using ZrB_2 as an example, the peaks P_1 and P_2 arise from 2s and 2p electron orbital hybridization with zirconium d orbital electrons. The peaks P_3 and P_4 represent bonding and then anti-bonding states between boron 2p and zirconium 4d electrons.[7] Hybridization thus reduces the strength of B-B bonds but creates strong M-B bonds as predicted by Spear in his analysis. Because the Fermi level, E_F (energy level of the highest filled band at 0K) for ZrB_2 falls between the bonding and anti-bonding peaks, it can be seen that any increase in electron density would reduce bond strength by further filling of the anti-bonding states; similarly, any reduction

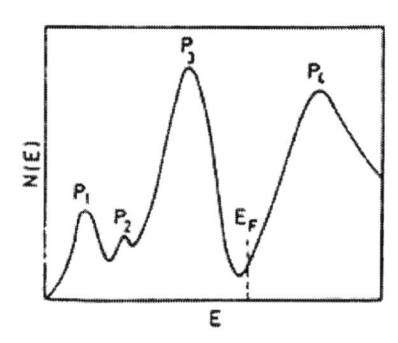

FIGURE 3. Plot of the general features of the electronic density of states, N(E) number of electrons in that state (bonding or anti-bonding) versus energy of the system, for an AlB_2-type compound.[21]

in electron density lowers the number of electrons filling the bonding levels, also reducing bond strength. With those results in mind, the trend in boride stability is clearly shown when looking at a plot of standard entropy, E_S, vs. number of electrons per unit cell, n_e, see Figure 4a.[21] A plot of E_S vs. T_f, melting temperature (Figure 4b) shows the same trends as those illustrated in the plot of E_S vs. n_e.[21] Those materials, with their standard entropy or cohesive energy at a maximum, show the highest stabilities and melting points. Simply put, the materials with a Fermi energy between P3 and P4 have the highest stabilities. Listed in increasing in order of bond strength, those species are materials formed with Ti, Zr and Hf. As the number of electrons, n_e, increases (i.e. V, Nb, Ta) more **anti-bonding** states are filled which decreases the bond strength and the melting temperature of the compound. Similarly, as n_e decreases (i.e. Sc, Y), fewer electrons are available to fill the **bonding** levels and compound stability decreases, also demonstrated by a reduction of the melting temperature, as shown in Figure 4b.[21]

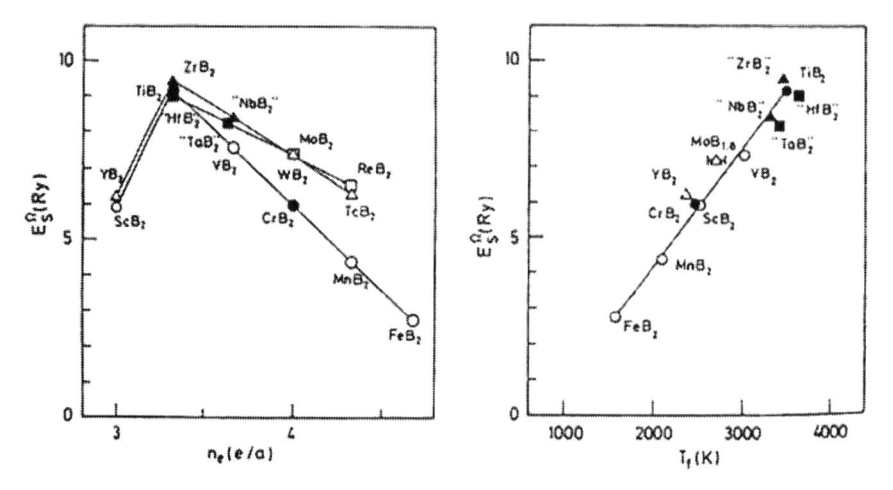

FIGURE 4. Left, plot of characteristic energy, defined from entropy related Debye temperature and plotted vs. the average number of valence electrons. Right, characteristic energy plotted vs. melting temperature.[21]

TABLE 2. Crystal structure, density and melting point of UHTCs.

Material	Crystal Structure	Lattice Parameters			Density	Melting Pt.
		aÅ	bÅ	cÅ	g/cc	°C
HfB_2	Hexagonal	3.142	–	3.476	11.19	3380
HfC	FCC	4.638	4.638	4.638	12.76	3900
HfN	FCC	4.525	4.525	4.525	13.9	3385
ZrB_2	Hexagonal	3.169	–	3.530	6.10	3245
ZrC	FCC	4.693	4.693	4.693	6.56	3400
ZrN	FCC	4.578	4.578	4.578	7.29	2950
TiB_2	Hexagonal	3.030	–	3.230	4.52	3225
TiC	Cubic	4.327	4.327	4.327	4.94	3100
TiN	FCC	4.242	4.242	4.242	5.39	2950
TaB_2	Hexagonal	3.098	–	3.227	12.54	3040
TaC	Cubic	4.455	4.455	4.455	14.50	3800
TaN	Cubic	4.330	4.330	4.330	14.30	2700
SiC	Polymorphic		Various		3.21	2820

Source: Ref 22–26

III.2. Structure

Following the previous description of the atomic structure of boride, carbide and nitride ceramics, Table 2 lists physical crystalline structural differences of a variety of UHTCs along with respective density and melting point.[22–26] Note that density increases with increasing mass of the metal atom. Note also the differences in melting points between materials whereby the carbides typically have the highest melting points, above borides or nitrides of the same metal constituent.

III.3. Thermodynamic Properties

Some of the thermodynamic properties for a few diborides, including HfB_2 and ZrB_2 are listed in Tables 3 and 4. The data in these Tables are from Pankratz et al, who reviewed available data in the early 80's and included only the data they deemed to be reliable.[26] Enthalpies of formation are strongly correlated with Gibbs free energies of formation for borides, because the entropy terms are small. This also means that the free energy is relatively insensitive to temperature.[7]

TABLE 3. Thermodynamic data for UHTCs.

Material	Enthalpy of formation, ΔH_f (kJ/mol)			Gibbs free energy of formation, ΔG_f (kJ/mol)		
	At 298 K	At 1000 K	At 2000 K	At 298 K	At 1000 K	At 2000 K
HfB_2	−335.98	−334.90	–	−332.20	−324.49	–
ZrB_2	−322.14	−326.65	−340.36	−318.16	−306.34	−279.60
TiB_2	−323.84	−326.59	−347.87	−319.69	−308.34	−347.87
TaB_2	−209.20	−209.77	−208.02	−206.53	−200.18	−191.02

In standard state, pure phase, at 0.1 MPa (1 atm), Source: Ref. 26

TABLE 4. Heat capacity for UHTCs.

Material	Heat capacity at constant pressure C_p (J/mol*K)		
	At 298 K	At 1000 K	At 2000 K
HfB_2	49.45	81.67	–
ZrB_2	48.37	71.99	82.66
TiB_2	44.28	76.89	94.54
TaB_2	48.12	76.77	96.71

From the elements, in standard state, pure phase, at 0.1 MPa (1 atm), Source: Ref. 26

IV. PROCESSING

IV.1. Raw Materials

The principal source of elemental zirconium metal is the zirconium silicate mineral, zircon ($ZrSiO_4$). The mineral baddeleyite, a natural form of zirconia (ZrO_2), is a secondary source of zirconium. Zircon is also the primary source of all elemental hafnium, which is contained in zircon at a ratio of about 1 part hafnium to 50 parts zirconium.[27] During diboride production, the respective metal oxide is used as the metal source while boron carbide, crystalline boron or a combination of boron oxide and carbon acts as the boron source. Due to the limited demand for diboride powders of Zr and Hf, the processes for manufacturing these powders have not been refined to the same degree as for other common ceramics, such as SiC or Si_3N_4. Therefore some variation in the powder between lots and vendors has been observed during the manufacturing of these materials. The effect of these variations was briefly investigated by ManLabs, where they found that relatively small variations in the impurity contents within their powders influenced the powder's thermal stability (melting point), the powder's hot pressing characteristics and the final material properties of the condensed billet.[2]

IV.2. Sintering

Fabrication of UHTCs has typically been accomplished by hot pressing in either resistance or induction heated furnaces using graphite dies. Due to the observation of chemical reactions between the graphite dies and the diboride powders, a number of methods have been investigated to produce a diffusion barrier between the powder and the die. ManLabs tested several diffusion barriers including: a) boron nitride wash on a graphite sleeve, b) tungsten foil, c) pyrolytic graphite paper, d) silicon carbide wash and e) pyrolytic graphite paper with an inner BN wash. They found that up to 2100°C the pyrolytic graphite paper with an inner BN wash was most effective, although above 2100°C reaction zones were very apparent, despite the barrier, and approached the width of those found without a diffusion barrier, up to ∼5 mm.[2]

ManLabs also investigated the effects of a number of sintering variables on the final billet density and microstructure, including maximum temperature, pressure and heating schedule. They were not able to detect any changes in the final microstructure due to small fluctuations in heating rate or hold times.[2] ManLabs investigations found that

TABLE 5. List of compositions fabricated by ManLabs.

Sample ID	Sample
I	ZrB_2, no additive
II	HfB_2, no additive
III	HfB_2 with 20v% SiC
IV	HfB_2 with 30v% SiC
V	ZrB_2 with 20v% SiC
VI	HfB_2 with 4v% Hf-Ta alloy
VIII	ZrB_2 with 14v% SiC and 30v% C
X	ZrB_2 with 20v% SiB_6
XIII	ZrB_2 with 50v% C

changes in pressure from 20–40 MPa produced little or no variation in billet density, but they did observe that at pressures below 20 MPa the diboride samples would not densify uniformly.[2]

High-pressure hot pressing technology was also used to fabricate monolithic UHTCs. Hot pressing at temperatures ranging from 1400°C to 1900°C and pressures of 500 MPa to 2 GPa provided a means of preparing fine-grained dense materials without the use of additives. In one study, the influence of pressure on densification was studied by ManLabs on HfB_2 using a fixed pressure of 800 MPa and temperatures from 1400°C to 1900°C with 10 minute holds.[28] Full densification was not achieved at temperatures below 1840°C. Similarly, the influence of pressure was studied at 1790°C with 10 minute holds for a range of pressures from 400 MPa to 1.5 GPa. Full density was achieved at a pressure of 1.5 GPa, while a pressure of 800 MPa yielded a sample with 1.4% porosity, and 400 MPa resulted in a sample with 6.8% porosity.[28]

Pure diborides as well as diborides with several additive systems were investigated in attempts to improve material properties and/or oxidation resistance. The primary ZrB_2 and HfB_2 compositions that ManLabs looked at are listed in Table 5. Due to the high melting temperatures of the diborides, ManLabs encountered considerable difficulty when consolidating samples of pure HfB_2 or ZrB_2. Without the use of high pressure hot pressing techniques they were unable to produce fully dense billets without significant grain growth or cracking, but the high pressure hot pressing significantly limited their maximum billet size.[1-2,28] The additives that were investigated were primarily used to increase oxidation resistance (SiC additions) or thermal stress resistance (C additions, added to reduce the materials elastic modulus) and were not originally considered as potentially advantageous to fabrication. However, ManLabs found that with the intentional additives, billet cracking could be eliminated and dense, fine-grained diboride microstructures were achievable, (see Figure 5).[4,29]

Figure 6 shows the influence of SiC additions on the densification behavior of HfB_2 by comparing the hot pressing characteristics of pure HfB_2 (material II in Table 5) to that of HfB_2-20vol% SiC (material III).[4] The relative densities of these materials hot pressed at equivalent temperatures clearly show that SiC improves HfB_2 densification by reducing the temperature and time required to achieve a comparable density. Similar trends were observed for the ZrB_2 and ZrB_2/SiC materials. A distinguishing feature of the microstructures of the SiC containing materials, both ZrB_2 and HfB_2, was their much smaller grain size, compared

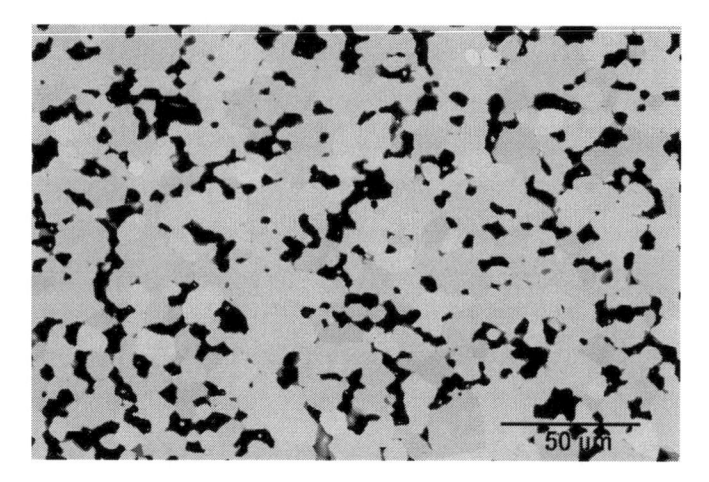

FIGURE 5. HfB$_2$/20v% SiC produced by ManLabs.

to hot pressed samples of pure ZrB$_2$ and HfB$_2$. ManLabs researchers conjectured that the SiC phase promoted densification by restricting the growth of diboride grains, thus promoting diffusional densification as modeled by Nabarro-Herring, which favors small particle sizes.[4]

The presence of chemical impurities was determined by careful microstructural investigation of grain boundaries. Interestingly, ManLabs also found that the SiC additions appear to "getter" impurities within the materials, and that the reduction in impurity phases was a function of the volume percent of SiC. The minimum remaining impurity phases were observed with the addition of at least 10–15v% SiC.[2]

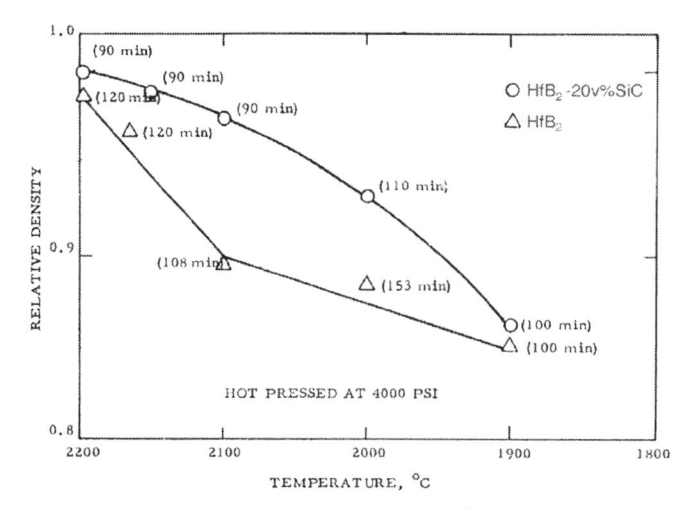

FIGURE 6. Comparison of hot pressing characteristics of pure HfB$_2$ and HfB$_2$-20v% SiC[4].

IV.3. Machining

Due to the hard brittle nature of the UHTC materials, diamond tooling is typically required to machine components. Electron Discharge Machining (EDM) can be used as an alternative to diamond grinding to machine the borides due to the high electrical conductivity of these materials. This is an advantage over some of the carbide, nitride, and oxide materials that typically have higher resistances, which does not allow them to be machined using EDM.

V. PROPERTIES

V.1. Thermal Properties

Typical values for the coefficient of thermal expansion (CTE) and thermal conductivity of some UHTC materials including HfB_2, ZrB_2 and these diborides with SiC are given in Table 6.[5,25,30–32] In general, the CTE's for these materials increase sufficiently with

TABLE 6. Thermal expansion and conductivity of some UHTCs.

Material	Thermal Expansion 10^{-6}/K	Temp. Range °C	Thermal Conductivity W/m*K	Temp. °C	Ref.
HfB_2	6.3	20–1027	105	20	5, 30
	6.8	1027–2027	75	400	5, 30
	7.6	20–2205	70	800	5, 30
HfB_2-20v% SiC	–	–	79	100	31
	–	–	74	500	31
	–	–	62	1000	31
HfC	6.6	20–1500	20	20	5
	–	–	23	400	5
	–	–	30	800	5
HfN	~6.5	20–1000	18	20	5
	–	–	20	400	5
	–	–	22	800	5
ZrB_2	5.9	20–1027	–	–	30
	6.5	1027–2027	–	–	30
	8.3	20–2205	–	–	30
ZrB_2-20v% SiC	5–7.8	400–1600	98.7	100	31
	–	–	84.5	500	31
	–	–	78	1000	31
ZrC	6.7	20–1500	–	–	25
TiB_2	4.6	20–1027	–	–	30
	5.2	1027–2027	–	–	30
	8.6	20–2205	–	–	30
TiC	7.7	20–1500	–	–	25
TaB_2	8.2	20–1027	16.0	20	30
	8.4	1027–2027	16.1	1027	30
	8.4	20–1650	36.2	2027	30
TaC	6.3	20–1500	–	–	25
SiC	1.1	20	114	20	32
	5.0	1000	35.7	1000	32
	5.5	1500	26.3	1500	32

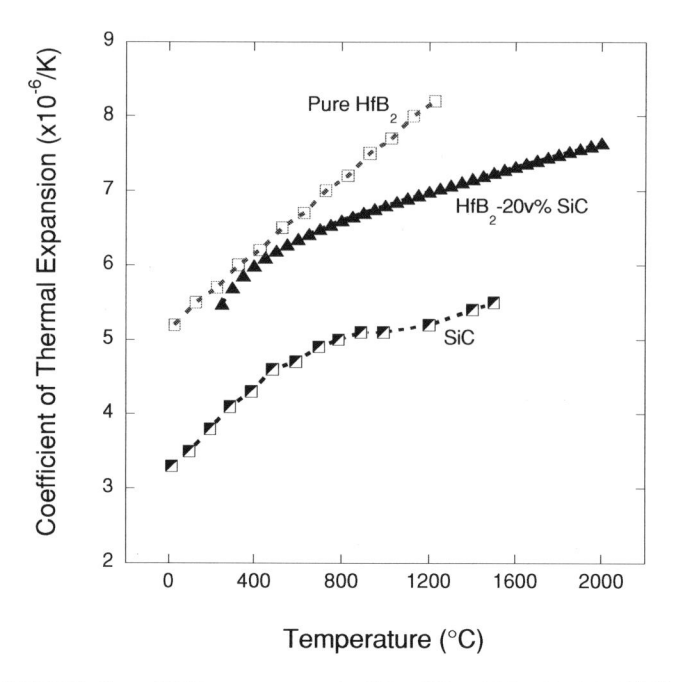

FIGURE 7. Plots of CTE vs. temperature for HfB_2, SiC, and HfB_2-20v% SiC.[31–32]

temperature such that temperature dependent coefficients should be used during design. To illustrate this, a comparison of CTE of pure HfB_2, pure SiC and HfB_2-20v% SiC is shown in Figure 7.[31–32]

As previously indicated, the thermal conductivities of borides are typically high, in comparison to many other ceramics and are a result of both a lattice and an electronic contribution to phonon transport.[7] Figure 8 illustrates the large difference in conductivities of the borides from the nitride and carbide ceramics of hafnium.[5,31–32] Although the conductivities of the borides initially drop with increasing temperature, they eventually even out at levels still above that for the nitrides and carbides. For a leading edge, high thermal conductivity reduces thermal stresses within the material, by reducing the magnitude of the thermal gradients within the part. High thermal conductivity also allows energy to be conducted away from the tip of the leading edge and reradiated out of surfaces of the component with lower heat fluxes, reducing the surface temperature for a given incident heat flux, compared to an insulating material.

The added benefits of high temperature materials with high conductivity can be seen when comparing a sharp UHTC leading edge to a relatively blunt, low conductivity leading edge such as the conventional RCC thermal protection system, used on the shuttle. Because of the low conductivity of conventional leading edge thermal protection system (TPS) materials, the maximum surface temperature is determined by a balance of the incident heat flux with the energy that is re-radiated out of the leading edge, hence the need for materials

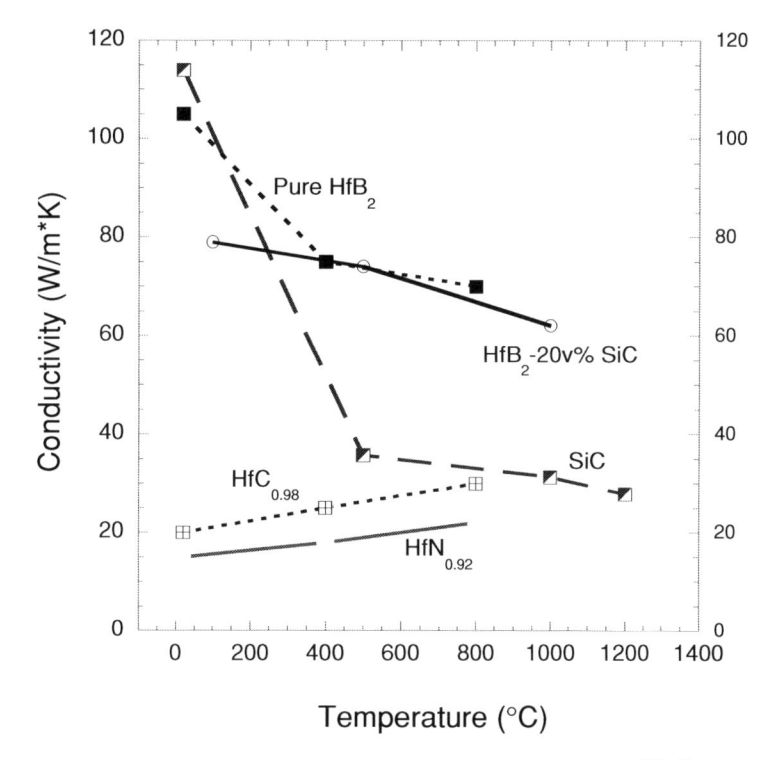

FIGURE 8. Plots of conductivity vs. temperature for several UHTCs[5,31–32]

with as high an emittance as possible. Due to the low thermal conductivity of these materials, there is little or no conduction of energy away from the tip of the leading edge.

Currently the leading edges of operational space vehicles are relatively blunt, due to limits in the temperature capability of the state of the art materials, which are typically based on SiC and limited to ~1600°C, before active oxidation becomes an issue. As the leading edge radius decreases the surface temperature increases. But the blunt leading edges significantly impact the vehicles performance, which effects overall vehicle safety, as mentioned earlier. Therefore, in order to achieve maximum performance, materials are needed that are both capable of withstanding the reentry environment at temperatures greater than 2000°C, and have a high thermal conductivity that will direct more energy away from the tip of the leading edge allowing for even further improvements in vehicle performance, i.e. faster velocities at lower altitudes.

Figure 9 illustrates how the surface temperature for the sharp UHTC leading edges is determined by an energy balance of incident heat flux, re-radiated energy and energy pulled away from the leading edge tip and re-radiated out the sides of the component where the incident heat flux is lower. In this sharp leading edge design, three dimensional heat transfer plays an important role in understanding and achieving the full potential of these materials. Thus the need for highly conducting, yet refractory materials is essential in the design of sharp vehicles.

Blunt Edge

SiC Coated C/C

Low Thermal Conductivity
$\dot{q}_{cond} \approx 0$

$\dot{q}_{conv} \approx \dot{q}_{rad}$

Sharp Edge

UHTC

High Thermal Conductivity

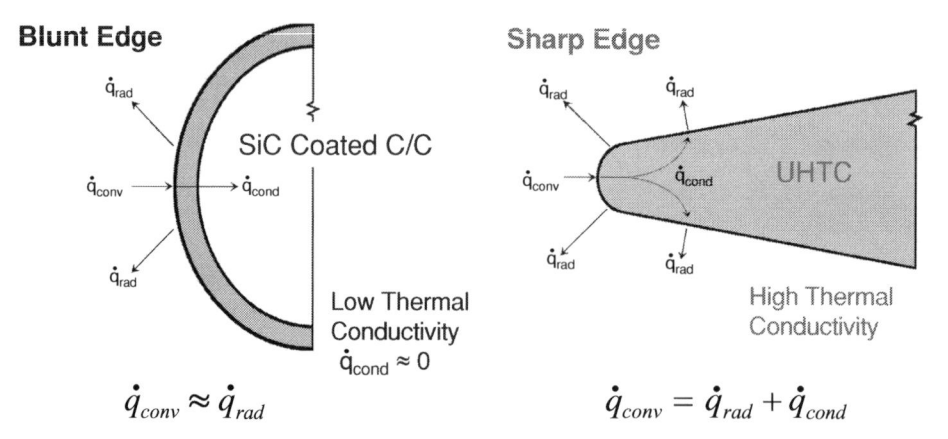

$\dot{q}_{conv} = \dot{q}_{rad} + \dot{q}_{cond}$

FIGURE 9. Comparison of the functionality of conduction for blunt and sharp leading edge designs of a hypersonic space vehicle.[14]

V.2. Mechanical Properties

The mechanical properties of a number of UHTC materials, including HfB_2 and ZrB_2 with and without SiC additions, are given in Table 7. In general, there is a limited amount of property data for the diborides due to the limited number of studies that have been performed to date. Hardness values in these materials are typically relatively high, due to their high degree of covalent bonding. The wide scatter in the data is probably due to differences in processing of these materials resulting in different grain sizes and porosity.[3]

Similarly, the moduli of HfB_2 and ZrB_2 are high due to the high degree of covalent bonding. Typical modulus values for both HfB_2 and ZrB_2 with and without SiC are around 500 GPa at room temperature. As shown in Figure 10, the modulus of HfB_2 and ZrB_2 with and without SiC additions begins to fall off above 800°C.[20,25,32–33]

The flexural strength of HfB_2 and ZrB_2 can also vary a great deal, mainly due to differences in processing and resulting grain size and other processing defects. In general, finer grain size results in higher strengths. Researchers at ManLabs noted that impurity phases could be a source of residual stress within materials, further limiting the useful strengths of HfB_2 and ZrB_2.[2] A plot of strength vs. temperature is shown in Figure 11. Unlike the modulus values at high temperature, strengths increase slightly with increased temperature. However, after 800°C strengths decrease along with modulus. As shown in the plot, ManLabs data indicate that there is another increase in strength above 1400°C; however these results have not been confirmed.

V.3. Electrical Properties

Typical electrical properties for some UHTC-type materials including HfB_2 and ZrB_2 are listed in Table 8.[15,20,25,31] The table shows that the diborides have significantly lower resistivity than the associated carbides. As mentioned before, the low resistivity of the borides allows them to be machined using EDM. The ability to use this technique enables the production of complex shapes at reduced costs compared to traditional diamond machining.

TABLE 7. Mechanical properties of selected UHTCs.

Material	Temp. °C	Young's Modulus GPa	Flexural Strength MPa	Poisson's Ratio	Hardness GPa	Ref.
HfB_2	23	530	480	0.12	21.2–28.4	33
	800	485	570	–	–	33
	1400	300	170	–	–	33
	1800	–	280	–	–	33
HfB_2-20v% SiC	23	540	420	–	–	33
	800	530	380	–	–	33
	1400	410	180	–	–	33
	1800	–	280	–	–	33
HfC	23	352	–	–	26.0	25
ZrB_2	23	500	380	0.11	25.3–28.0	20, 33
	800	480	430	–	–	33
	1400	360	150	–	–	33
	1800	–	200	–	–	33
ZrB_2-20v% SiC	23	540	400	–	–	33
	800	500	450	–	–	33
	1400	430	340	–	–	33
	1800	–	270	–	–	33
ZrC	23	348	–	–	27.0	25
TiB_2	23	551	300–370	0.11	25–33	20
TiC	23	451	–	–	30.0	25
TaB_2	23	257	–	–	19–25	20
TaC	23	285	–	–	18.2	25
SiC	23	415	359	0.16	32	32
	1000	392	397	0.157	8.9	32

V.4. Optical Properties

The optical properties of borides are relatively unexplored.[3] Emission coefficients at a wavelength of 0.65 μm (infrared) on **un**oxidized specimens are generally in the range between 0.5–0.8 at temperatures of 1600°C to 1800°C in vacuum. As shown in Table 9 for unoxidized samples and in Table 10 for oxidized samples, the emittance can vary drastically from material to material and typically changes with temperature.[31] On oxidized samples, the emittance values obtained for HfB_2-20vol% SiC and ZrB_2-20vol% SiC seem to depend strongly on temperature. Above 1900°C these materials exhibit an emittance near 0.60, characteristic of HfO_2 and ZrO_2.[35] For the oxidized samples below 1900°C the emittance is considerably higher than at the higher temperatures. This is probably the result of the formation of a SiO_2 and/or B_2O_3-SiO_2 glass on the surface of the models that would be expected to have a higher emittance. At the higher temperatures, this protective glass layer may have evaporated leaving only HfO_2 or ZrO_2 that would have a lower emittance. In leading edge applications, for reentry vehicles, a high emittance is desirable, as it would reradiate much of the energy from the surface eliminating some of the energy that otherwise the material/component would have to handle, complicating the design and limiting the conditions under which it could operate.

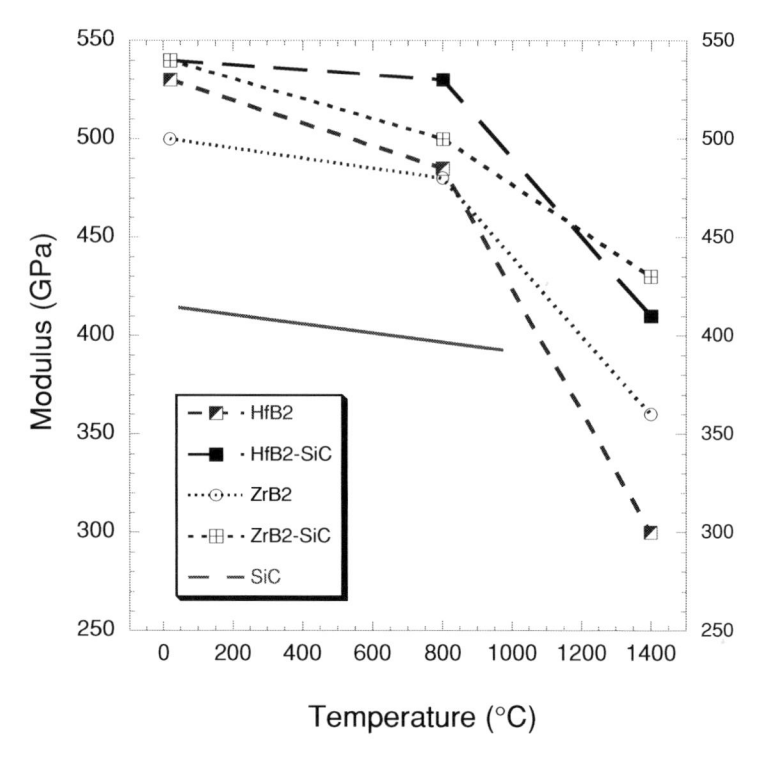

FIGURE 10. Plots of modulus vs. temperature for HfB_2 and ZrB_2 with and without SiC additions.[20,25,32–33]

VI. HIGH TEMPERATURE TESTING

A significant portion of the work conducted by ManLabs was in identifying compositions that would be stable at high temperatures in reentry environments. From investigations of the thermodynamic properties of selected UTHCs, ManLabs determined the diborides to have the following relative stabilities: $HfB_2 > ZrB_2 > TiB_2 > TaB_2 > NbB_2$. Subsequent furnace oxidation studies showed HfB_2 and ZrB_2 to be more oxidation resistant than other diboride materials.[1] Based on these results, ManLabs chose ZrB_2 and HfB_2 from the list of borides, as the best candidates for continued research into high temperature applications.

A comparison of oxidation rates of HfB_2 and ZrB_2 with other high temperature ceramics is shown in Figure 12.[6,35–44] At lower temperatures it is apparent that materials such as SiC and Si_3N_4 have lower oxidation rates than either HfB_2, ZrB_2 or their mixture with SiC. But these Si based materials are only applicable in temperature ranges less than ~1700°C. At higher temperatures the materials are either unstable and dissociate, and/or reach a temperature regime where active oxidation predominates. During active oxidation the protective SiO_2 layer is disrupted by evolution of SiO, exposing the surface below to continued oxidation. Thus, at these ultra high temperatures, materials that are primarily Si based are currently no longer suitable.

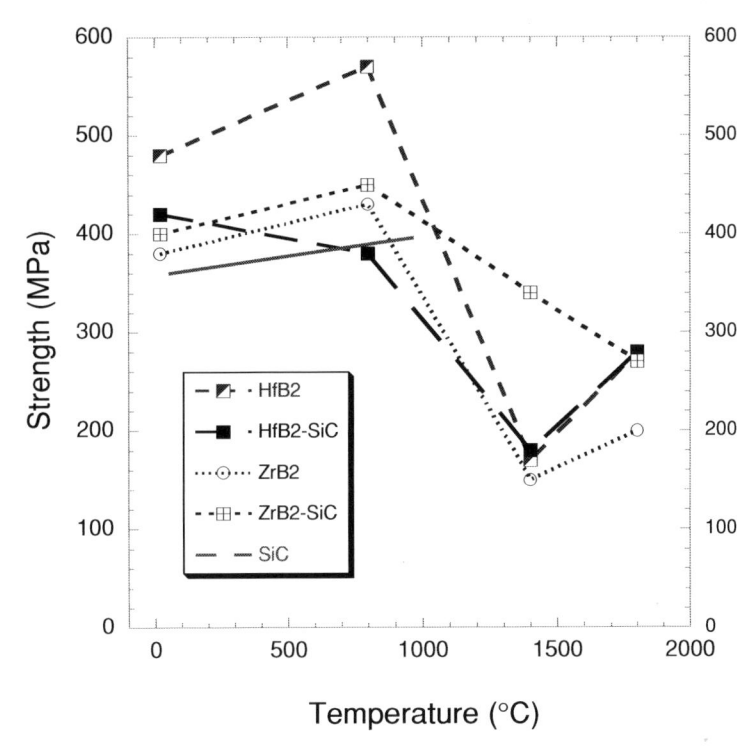

FIGURE 11. Plots of strength vs. temperature for HfB_2 and ZrB_2 with and without SiC additions.[20,25,32–33]

Based on their favorable oxidation resistance, the diborides are one family of materials that show some promise for use in ultra high temperature applications. We see in Figure 12 that HfB_2 has a lower oxidation rate than ZrB_2 and that the HfB_2/SiC mixture has improved oxidation resistance compared to pure HfB_2. Figure 12 also shows that HfB_2 and HfB_2/SiC have lower oxidation rates than do pure HfC.

Oxidation studies by ManLabs also showed that metal rich compositions of HfB_2 or ZrB_2 (i.e. $HfB_{1.9}$) were more oxidation resistant and possessed higher thermal stability than boron rich compositions ($HfB_{2.1}$) of the same diboride.[1]

To determine the effect of SiC content on diboride oxidation, ManLabs conducted a series of furnace oxidation experiments at 1800°, 1950° and 2100°C on HfB_2 specimens with 10v% SiC and ZrB_2 specimens with 5, 10, 15, 35 and 50v% SiC. Samples were heated in argon to the respective test temperature at which time the dry air was introduced for periods of 30–60 minutes, then argon was reintroduced for cool down. They found that 5v% SiC additions provided little improvement in oxidation resistance while samples containing 50v% SiC showed improvements after one hour at 1800°C and 1950°C but were completely oxidized after an hour at 2100°C. They concluded that additions of 35v% SiC to both ZrB_2 and HfB_2 provided the best protection for temperatures up to 2100°C during a furnace test, but that reasonable oxidation protection can be achieved with SiC contents as low as 15v% for ZrB_2 and 10v% for HfB_2.[2,35] A comparison of the oxidation scale thickness as a function

TABLE 8. Electrical properties of some UHTCs.

Material	Electrical Resistivity 10^{-6} Ω*cm	Temp. °C	Ref.
HfB_2	11	25	20
HfB_2-20v% SiC	9.6	20	31
	57.8	1000	31
HfB_2-30v% SiC	20.3	20	31
	82.5	1000	31
HfC	109	25	15
ZrB_2	12.1	20	31
	44.0	1000	31
ZrB_2-20v% SiC	10.2	20	31
	54.5	1000	31
ZrC	63	25	25
TiB_2	9	25	20
TiC	68	25	25
TaB_2	33	25	20
TaC	30	25	25

TABLE 9. Emittance measurements for unoxidized Hf and Zr materials.

Material	Unoxidized Emittance	
	1800°C	2200°C
HfB_2-20v% SiC	0.8	0.58
HfO_2	~0.5	
ZrB_2	0.5	0.4
ZrB_2-20v% SiC	0.6	0.56

Source: Ref 31

TABLE 10. Emittance measurements for oxidized Hf and Zr materials.

Material	Oxidized Emittance	Temp. °C
HfB_2	0.50	1800
HfB_2-20v% SiC	0.90	1500
	0.75	1800
	0.60	2100
ZrB_2	0.57	1800
ZrB_2-20v% SiC	0.9	1500

Source: Ref 48

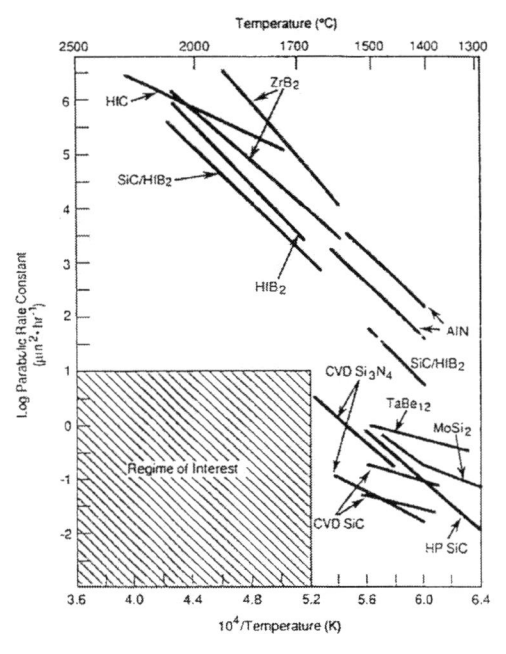

FIGURE 12. Comparison of oxidation rates of several engineering ceramics.[35-44]

of temperature for pure HfB_2, SiC and HfB_2-20v% SiC is shown in Figure 13. This figure clearly shows that at temperatures $>\sim 2100$ K the oxide scale thickness on pure HfB_2 is thinner than that on pure SiC, whose scale thickness is rapidly increasing with temperature above 2050 K. The figure also shows that at all temperatures $>\sim 2050$ K the HfB_2-20% SiC materials have the thinnest oxide layers.

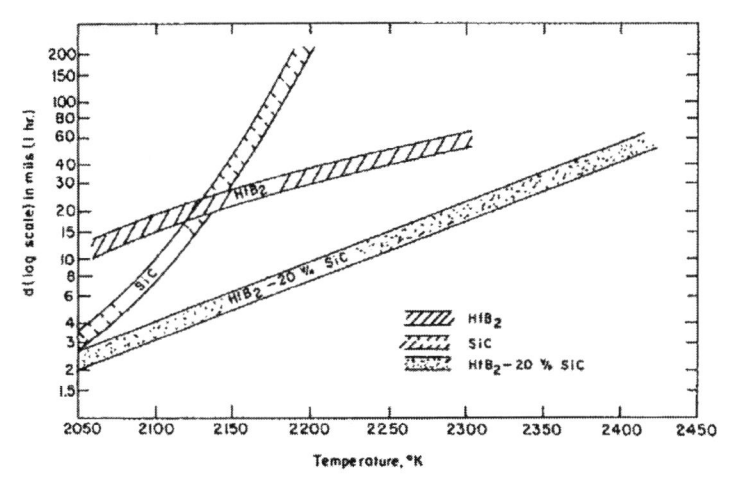

FIGURE 13. Summary comparison of a 1-hour oxidation study of HfB_2, SiC and HfB_2-20v% SiC.[11]

The response of the refractory UHTC materials to high temperature oxidizing conditions imposed by furnace heating has been observed to differ markedly from the behavior observed in arc plasma facilities that provide a simulated reentry environment.[35] Furnace evaluations are normally performed for long times at fixed temperature and slow gas flow with well defined solid/gas-reactant/product chemistry. Arc jet tests, on the other hand, are carried out under high velocity gas flow conditions in which energy flux, rather than temperature, is defined. Furthermore, furnace studies employ air at 1 atmosphere and as diatomic species. But during a typical re-entry profile for a manned space vehicle, the pressures will generally be much less than 1 atmosphere and a significant portion of the gas molecules will be dissociated into highly reactive monatomic species as they cross the bow shock formed during reentry. The resulting monatomic species may recombine at the surface giving up some of their energy to the material and, depending on the catalycity of the substrate, this recombination can add a significant fraction to the overall heating of the articles' surface. Arc jet testing provides the best ground based simulation of the reentry environment, although there are a number of differences between the arc jet environment and the actual reentry environment that must be accounted for when designing an arc jet experiment and when interpreting the data. For example, catalycity can play a more significant role during arc jet testing than in flight because a higher proportion of the air molecules are dissociated in the arc jet than in flight. Because of the differences between static or flowing air oxidation experiments and experiments in the arc jet, correlation of material responses from the two test situations is difficult, if not impossible in many cases, i.e. if material A performs better in the furnace test than material B it does not necessarily hold that the same trend will occur in the arc jet.[46]

VI.1. Furnace Oxidation Testing (Cold Gas/Hot Wall – CG/HW)

In an effort to bridge the gap between furnace oxidation studies and arc jet oxidation tests, ManLabs performed a series of furnace studies on UHTCs under various ranges of time, gas velocity, temperature and pressure.[2,46–50] Furnace test conditions covered temperatures between 500°C and 2500°C, times from five minutes to four hours at air flow rates in two regimes, low velocity between 0.3 and 3 m/s and high velocity, between 3 and 75 m/s. Cyclic exposures were also performed for most of the candidates.

Low velocity CG/HW tests were conducted in a resistively heated tube furnace at temperatures ranging from 500°C to 2300°C. Samples were heated in argon to the respective test temperature at which time the dry air was introduced. Prior to cool down, argon was reintroduced into the furnace. Due to the nature of the heating used in these experiments, there were no appreciable temperature gradients across the sample or across the oxide scales formed during the experiment. Temperature measurement within the furnace was carried out using one-color and two-color pyrometry. (Two color pyrometers were assumed to measure correct temperature and one color pyrometer data was corrected using two color temperatures and a calculated emissivity). Temperatures were also checked by measurements of the melting points of several materials. The tests showed that the hypereutectic carbides HfC+C and ZrC+C oxidize consistently at rates of 1.5–3 mm/hr at temperatures between 650°C and 2300°C.[38] HfB_2 displayed better oxidation resistance than ZrB_2 or the carbides at all temperatures and metal rich compositions of HfB_2 (i.e. $HfB_{1.7-2.0}$) performed better than boride rich compositions. Table 11 summarizes selected ManLabs furnace oxidation

TABLE 11. Furnace oxidation conversion depths after 1-hour exposures in air of some Hf and Zr-based materials.

Material	Conversion Depth (mm)		
	1650°C	1900°C	2200°C
$HfB_{2.1}$	0.3	2.5	2.5
ZrB_2	0.3	2.3	4.8
HfB_2-20v% SiC	0.0	0.8	1.3
$HfB_{2.1}$-20v% SiC	0.0	0.8	1.4
$HfB_{2.1}$-35v% SiC	0.0	0.0	0.3 (2100°C)
HfC+C	1.8	1.8	1.5
ZrC+C	2.0	1.7	2.5

Furnace oxidation depths after 1 hour – Source: Ref 47

results as a function of temperature.[47] The high temperature limit for protective oxide formation for ZrB_2 was found to be 1875°C, and 2050°C for HfB_2. Incorporation of SiC improves the high temperature limits for protective oxide formation for ZrB_2 and HfB_2 to 2000°C and 2100°C respectively. For diboride composites with at least 20 vol% SiC, no oxidation was seen below 1800°C and oxidation rates were reduced to between 0.85 and 1.85 mm/hr at 2100°C.[48] Among all samples with SiC additions, preferential oxidation of the SiC was observed at all temperatures where oxidation of the material was observed (T ≥ 1800°C). This phenomenon leads to the occurrence of three distinct zones in an oxidized specimen: the outer oxide, a porous diboride zone depleted of SiC, and finally, the unaltered base material.[47–48] The preferential oxidation of the SiC may be due to active oxidation of the SiC due to the high temperatures at which the materials were tested. It can be seen that the $HfB_{2.1}$-35v% SiC material appears to have less oxidation conversion than the 20v% material. However, it must be remembered that the addition of SiC was not based solely on oxidation resistance. Optimum mechanical and physical properties, in addition to oxidation resistance, led ManLabs researchers to favor HfB_2-20v% SiC over other compositions.

During high velocity CG/HW tests, samples were heated inductively to temperatures between 1100°C and 2500°C. Temperature measurements of the surface were also obtained by one-color and two-color pyrometry. Use of a two-color pyrometer was again employed to evaluate the spectral emittance of the oxidizing surface. Temperature gradients within the sample were measured by monitoring the temperature at the root of internal holes drilled to within 2.5 mm of the surface.[47]

These tests by ManLabs showed that for materials that form condensed oxide scales, significant temperature gradients develop across the oxide formed during the experiments, on the order of 400°C.[48] These gradients occurred because induction heating only coupled with the substrate and not the oxide. Because the oxide was only heated by conduction from the heated substrate, the oxidized surface was cooler than the base material. The importance of this aspect of material behavior depends on the rate-controlling factor in the oxidation process.[48] Graphites, which form volatile oxidation products showed strong dependence of oxidation on the velocity of the air. But oxidation rates of UHTC materials were insensitive to increased flow rate; rather, they were limited by the minimum temperature of the condensed oxide layer.[48]

An explanation for the insensitivity of oxidation rate on gas flow rate can be found by looking at the influence of the temperature gradients within the oxide scale. In low velocity CG/HW furnace studies, temperature gradients are largely absent, so the measured sample temperature, using a thermocouple, is the same as the surface temperature measured using an optical pyrometer. But during high velocity CG/HW studies, the measured surface temperature is lower than the sample temperature. This occurs because induction heating only couples to the base material, not the oxide. Reduced heat transfer through the oxide thus makes the measured surface temperature less than that of the base material. Since the surface temperatures were used to control most tests, during high velocity testing, the sample temperatures had to be continually increased to maintain a constant surface temperature because as the surface oxide depth increased, the gradient across the surface increased. When low velocity furnace oxidized samples were compared with high velocity oxidized samples with the same *measured surface temperature*, it was found that, for condensed oxide forming materials, the high velocity CG/HW tests and low velocity furnace tests yielded similar oxidation behavior, independent of air flow.[49] Under these conditions, the major oxide phases were condensed, therefore the oxidation rate was primarily driven by the surface temperature of the oxide and not the base material temperature.

VI.2. Arc Jet Testing (Hot Gas/Cold Wall – HG/CW)

The oxidation response of refractory UHTCs, as observed in an arc plasma facility that provides a simulated reentry environment, was carried out by ManLabs using a 10 MW facility. The range of conditions covered stagnation pressures between 0.01 and 1atm at cold wall heat fluxes between 120 and 1400 W/cm^2 for exposure times between 20 and 1,800 seconds with multiple exposure times up to a total run time of 23,000 seconds.[49] Characterization of the test environment was performed prior to each test by measuring stagnation pressure, stagnation enthalpy and cold wall heat flux. These measurements were performed by means of water cooled probes, energy balance measurements and transient and steady state calorimetry. The heat flux calorimeters were of the same geometry as the test models: 13 mm diameter flat face cylinders. Temperature measurement was carried out by means of one-color and two-color pyrometry. One-color pyrometer temperatures were converted to true temperature by employing a suitable emittance value, while the radiated heat flux data were used to compute total normal emittance. A micro-optical pyrometer observed the temperature at the base of a cavity drilled to within 2.5 mm of the heated surface.[49]

A summary of selected ManLabs arc jet results is shown in Table 12.[50] Results from this high velocity CG/HW testing showed that, for oxide forming UHTCs, temperature gradients of 400°C to 800°C can be observed from the surface of the oxide to the substrate. The practical implications of this are substantial, so long as the gradients exist under actual conditions; it indicates that for HG/CW arc jet tests the temperature level experienced by the substrate will be substantially lower than what is seen by the surface (opposite of CG/HW). Thus the oxidation experienced by the substrate should be substantially lower than for the case where gradients are ignored, i.e. furnace studies. Moreover, these gradients appear to exist for long periods of time.[48] Further arc jet testing by ManLabs confirmed that oxide forming samples experienced lower conversion depths than furnace tested materials. For example, a ZrB_2-20v% SiC composite tested for 7200 seconds showed surface temperatures

TABLE 12. Arc jet results for selected Hf and Zr ceramics and composites

Material	Conversion Depth (mm)	Temp. °C	SiC Depletion (mm)	Exposure Time seconds	Experimental Emittance
$HfB_{2.1}$	.5	2500	–	600	0.43
	.3	2500	–	1800	0.38
	.5	2780	–	600	0.45
	1.0	2860	–	1800	0.69
HfB_2-20v% SiC	.6	2580	<26	1800	0.64
	1.1	3030	49	1800	0.70
$HfB_{2.1}$-35v% SiC	.2	2000	3	1800	0.62
	3.1	2570	41	1800	0.43
ZrB_2	1.2	2680	–	600	0.59
ZrB_2-20v% SiC	.7	2530	1	1800	0.61
	2.8	2700	20	1800	0.57
HfC+C	1.2	2450	–	1800	0.39
	1.8	2900	–	1800	0.57
ZrC+C	1.0	2500	–	1800	0.70
	1.4	2770	–	1800	0.68

Source: Ref 50

of 2500°C but with a conversion depth of only 0.66 mm and a HfB_2-20v% SiC composite exposed for a total of 22,500 seconds and a surface temperature of 2670°C, showed measured oxide conversion of only 0.4 mm.[48] SiC depletion depths are reduced similarly; a 0.25 mm SiC depletion depth was observed for HfB_2-20v% SiC furnace tested to a temperature of 1900°C for 30 minutes. The same sample tested in the arc jet with surface temperatures of 2800°C for 30 minutes showed a SiC depletion depth of only 0.25 mm, the same depth as above for the furnace test, even though the surface temperature was 900°C higher in the arc jet.[50]

VII. THERMAL STRESS

As with any monolithic ceramic, UHTCs are inherently susceptibile to tensile stresses generated by temperature gradients within the material, in other words thermal shock.[51–52] However, arc jet testing performed by ManLabs, although limited, did illustrate the favorable thermal shock resistance of Hf and Zr diboride ceramics, as no arc jet models failed during testing due to thermal shock, despite rapid heating and cooling cycles. To further evaluate the levels of thermal stress resistance among compositional and microstructural variations of diboride materials, ManLabs researchers conducted a series of thermal stress tests.[52] Among the materials tested were those mentioned previously (Table 5). Tests were conducted by placing a cylindrical test specimen, 40–50 mm OD × 25 mm ID, around a heating rod, under vacuum. As power to the rod was increased, the UHTC cylinder was heated from the inside out creating a thermal gradient across the cylinder wall. The magnitude of the thermal stresses generated is proportional to the power input into the heating rod. Typically the power is increased at a constant rate until a critical stress is reached and the part fractures.

However, ManLabs researchers experienced considerable difficulty when trying to test these UHTCs to failure using the same configurations generally used for ceramic materials such as refractory oxides and SiC. This test procedure required that the materials being evaluated show linear elastic behavior to the point of failure, i.e. the materials should not show any non-linear or plastic deformation. Instead they found, in these initial tests, that the UHTC materials did not fail. It was surmised that either the conductivity of these materials was too great, such that sufficient thermal gradients could not be developed to induce failure, or the materials were capable of accommodating sufficient non-linear strain so as not to fail.[52]

To achieve fracture in UHTC materials, ManLabs researchers decided to notch all subsequent test specimens. Notches were 6 mm deep × 1.6 mm wide and parallel to the axis of the cylinder, extending inward from the outer surface along the entire length of the specimen. The main disadvantage of using notched specimens was the lack of adequate experimental and analytical data on the shape factor required to compare the experimental results with predicted properties as well as the inability to compare these results with those of previous evaluations. Their results showed that materials with SiC and carbon additions displayed somewhat higher steady state thermal stress resistance than the other compositions. Nonetheless, all the diboride compositions tested showed a level of thermal stress resistance considerably above any other ceramics they had tested.[52]

VIII. RECENT WORK

Research on UHTCs slowed considerably after the work by ManLabs ended, until the early 1990's, when interest in the monolithic UHTC materials was renewed. The high costs of raw materials in addition to the high temperatures and pressures required to hot press UHTC powders has led many new investigations into alternate starting materials and methods of fabricating UHTCs. In addition to conventional methods, researchers are also looking at reactive hot pressing and pressureless sintering by liquid infiltration and reaction.[37,53–54] These new reaction-based processes share the near-net shape and near-net dimension capabilities of gas-phase reaction bonding as well as reduced processing temperatures and times required for solid state sintering.

Basic property evaluation of UHTCs and UHTC composites has resumed at a number of government facilities within NASA and the military, as well as at some universities.[55–59] With this resurgence in basic research, investigations into UHTC carbides and nitrides are seeing added attention as new processing techniques make the fabrication of these materials easier.

NASA Ames began working on UHTCs in the early 1990s, and in 1997 and 2000 conducted two flight experiments, SHARP-B1 and SHARP-B2 (Sharp Hypersonic Aerothermodynamic Research Probes) in collaboration with the Air Force and Sandia National Labs. These experiments briefly exposed the UHTC materials to the actual reentry environment.[13] The SHARP-B1 vehicle tested a HfB_2-SiC nose tip with a 3.5 mm radius. The nose tip was surrounded by a ZrB_2-SiC skirt and was instrumented with internal sensors that measured internal temperatures of 1690°C, which according to models, corresponds to an external temperature above 2760°C. The SHARP-B1 vehicle was not designed to be recovered, and thus post-test characterization of the UHTCs was not possible. The second vehicle, SHARP-B2, was recovered. This test flew 4 segmented strakes on the outside of the reentry vehicle. The strakes were designed to retract within the reentry vehicle at a predetermined altitude after which a parachute was deployed allowing the vehicle and UHTC materials to be recovered. Each UHTC strake was composed of three segments with each segment a different UHTC material (HfB_2 or ZrB_2). The flight experiment was a success, and the materials were recovered, but due to insufficient time for materials development the materials had poor mechanical properties and a number of strake segments failed. NASA is continuing to pursue these materials and has made progress in improving the processing methodologies and resulting material properties.[60]

Aside from the work performed at NASA Ames, recent work by Levine et al., at NASA Glenn, has been conducted to improve the oxidation resistance of monolithic materials and to develop fiber reinforced UHTC materials to improve composite fracture toughness and impart a level of graceful failure in materials.[13–14,55,60]

Methods to improve the environmental durability and oxidation resistance of UHTCs are also being addressed by the Navy. Work by Opeka et al, at the Naval Surface Warfare Center, and others are seeking to provide an understanding of the oxidation of not only HfB_2 and ZrB_2 but of other UHTC materials, such as HfC, ZrC and HfN.[5,56,59] Thus far, their work indicates that the oxidation resistance of HfC and HfN is improved with decreasing C and N contents, when the materials are substoichimetric. Studies indicate that the oxidation of HfC occurs with the formation of a compact protective interlayer of Hf-C-O.[56] In a study of the oxidation of ZrC-ZrB_2-SiC composites, Opeka et al found that ZrB_2 rich composites had the highest oxidation resistance. The oxidation of ZrC, results in the formation of a fine grained oxide that allows diffusion of O_2 to the ZrC surface, providing no oxidation protection to the composite.[5]

IX. CONCLUSIONS

The need for high temperature materials that can maintain mechanical strength and operate with limited oxidation at temperatures >2000°C have driven the development of a variety of ceramics and ceramic composites. These refractory materials have come to be known as Ultra High Temperature Ceramics (UHTCs). The earliest work on UHTCs began in the 1960's by ManLabs under a research program funded by the Air Force Materials Research Lab (AFML). Compared to the other refractory compounds such as carbides and nitrides, ManLabs determined that intermetallic diboride compounds offer the highest degree of oxidation resistance. In comparison to carbides and nitrides, diborides also have high thermal conductivity which gives them good thermal shock resistance. Through further studies, HfB_2 and ZrB_2 were chosen as the most promising candidates for use in high temperature applications.

It was shown that all UHTCs are characterized by very strong bonding which gives them high temperature structural stability. It also makes them very difficult to fabricate and as a result, only a limited amount of research has been invested in these materials despite their many potential benefits. Conventional hot pressing of pure UHTCs typically yields porous materials with poor mechanical properties. But the addition of sintering additives aids in densification yielding materials with strengths measured in excess of 500 MPa. Work by ManLabs demonstrated that SiC additions to UHTC powders aided densification, in addition to increasing oxidation resistance of the composite.

High temperature oxidation testing was conducted in both furnace studies as well as in simulated reentry environments (arc jet). Oxidation rates during furnace tests were found to be much greater than oxidation rates during arc jet testing. This was explained by the formation of significant temperature gradients (between the oxide surface and materials substrate) during arc jet testing, that were otherwise absent during furnace tests. During arc jet testing the gradients maintained the bulk material at a lower temperature than the surface, thus reducing the levels of surface recession. Although ceramic materials are typically of limited use due to their susceptibility to thermal stress, ManLabs testing demonstrated, through repeated arc jet exposures, the favorable thermal shock resistance of diboride UHTCs.

After the ManLabs work, further research on UHTCs slowed until the early 1990's when NASA Ames renewed efforts. Work there resulted in two hypersonic flight experiments, SHARP-B1 and SHARP-B2 that proved the successful application of monolithic high temperature ceramic composites on sharp leading edges. Since then, NASA Glenn has also started conducting research to further improve UHTC oxidation resistance in addition to research into the development of UHTC (fiber) composite materials. Current work by the Navy is also seeking to provide a better understanding of the oxidation of not only HfB_2 and ZrB_2 but HfC, ZrC and HfN as well. Clearly, the potential applications for UHTCs span a wide number of needs arising from future military, industrial and space based projects. While continued work has provided valuable insight on the performance of UHTCs consistent ongoing research efforts are required if future applications are to be realized.

REFERENCES

1. Kaufman, L. and Clougherty, E. V. "Investigation of Boride Compounds for Very High Temperature Applications," RTD-TRD-N63-4096, Part III, ManLabs Inc., Cambridge, MA, (March 1966).
2. Clougherty, E. V, Kalish, D. and Peters, E. T. "Research and Development of Refractory Oxidaton Resistant Diborides," AFML-TR-68-190, ManLabs Inc., Cambridge, MA, (1968).
3. Gangler, J. J. "NASA Research on Refractory Compounds," High Temp. High Press. [3] 487–502 (1971).
4. Clougherty, E. V, Hill, R. J, Rhodes, W. and Peters, E. T. "Research and Development of Refractory Oxidaton Resistant Diborides," AFML-TR-68-190, Part II Vol. II, ManLabs Inc., Cambridge, MA, (1970).
5. Opeka, M. M., Talmy, I. G., Wuchina, E. J., Zaykoski, J. A. and Causey, S. J., "Mechanical, Thermal and Oxidation Properties of Refractory Hafnium and Zirconium Compounds," J. Europ. Ceram. Soc., [**19**] 2405–2414 (1999).
6. Courtright, E. L., Graham, H. C., Katz, A. P. and Kerans, R. J. "Ultra High Temperature Assessment Study – Ceramic Matrix Composites," AFWAL-TR-91-4061, Wright Patterson Air Force Base, Ohio (1992).
7. Cutler, R. A. "Engineering Properties of Borides," in ASTM Engineered Materials Handbook, Vol. 4 – Ceramics and Glasses, Schneider, S. J., technical chairman, p. 787–803 (1991).
8. Guillermet, A. F. and Grimvall, G. "Phase Stability Properties of Transition Metal Diborides," Am. Inst. Phy. Conf. Proc., [**231**] 423–431 (1991).
9. Mroz, C. "Annual Mineral Review; Zirconium Diboride," Am. Ceram. Soc. Bull. [**74**] 165–166 (1995).
10. Upadhya, K., Yang, J. M. and Hoffmann, W.P. "Materials for Ultrahigh Temperature Structural Applications," Am. Ceram. Soc. Bull. [**58**] 51–56 (1997).
11. Monteverde, F., Bellosi, A. and Guicciardi, S. "Processing and Properties of Zirconium Diboride-Based Composites," J. Europ. Ceram. Soc. [**22**] 279–288 (2002).
12. Low, I. M. and McPherson, R. "Fabrication of New Zirconium Boride Ceramics," J. Mat. Sci. Lett. [**8**] 1281–1283 (1989).
13. Kolodziej, P., Salute, J. and Keese, D. L. "First Flight Demonstration of a Sharp Ultra-High Temperature Ceramic Nosetip," NASA TM-112215, (December 1997).

14. Kontinos, D. A., Gee, K. and Prabhu, D. K., "Temperature Constraints at the Sharp Leading Edge of a Crew Transfer Vehicle," AIAA 2001–2886 (June 2001).
15. Shaffer, P. T. B. "Engineering Properties of Carbides," in ASTM Engineered Materials Handbook, Vol. 4 – Ceramics and Glasses, Schneider, S. J., technical chairman, p. 804–811 (1991).
16. Battelle Columbus Laboratories. "Engineering Property Data on Selected Ceramics" Vol. 2: Carbides. Metals and Ceramics Information Center, Battelle Columbus Laboratories, Report MCIC-HB-07-Vol. 2 (1979).
17. Hampshire, S. "Engineering Properties of Nitrides," in ASTM Engineered Materials Handbook, Vol. 4 – Ceramics and Glasses, Schneider, S. J., technical chairman, p. 812–820 (1991).
18. Spear, K. E. "Chemical Bonding in AlB2-Type Borides," J. Less-Common Metals, [47] 195–201 (1976).
19. Burdett, J. K., Canadell, E. and Miller, G. J. "Electronic Structure of Transition Metal Borides with the AlB_2 Structure," J. Am. Chem. Soc., [108] 6561–6568 (1986).
20. Samsonov, G. V. and Vinitskii, I. M. Handbook of Refractory Compounds, Plenum Press (1980).
21. Guillermet, A. F. and Grimvall, G. "Bonding Properties and Vibrational Entropy of Transition Metal MeB_2 (AlB_2) Diborides," J. Less-Common Metals, [169] 257–281 (1991).
22. Jenkins, R. et al., Joint Committee on Powder Diffraction Standards. Powder Diffraction File: from the International Center for Diffraction Data. Swarthmore, PA (1988).
23. Schwetz, K. A., Reinmoth, K. and Lipp, A. Production and Industrial Uses of Refractory Borides, Vol. 3, Radex Rundschau, 568–585 (1981).
24. McColm, I. C. Ceramic Science for Materials Technologists, Leonard Hill, London, 330–343 (1983).
25. Exner, H. E. Int. Metall. Rev., [24] 149–173 (1979).
26. Pankratz, L. B., Stuve, J. M. and Gokcen, N. A. "Thermodynamic Data for Mineral Technology," Buelletin 677, U.S. Bureau of Mines, 98–102 (1984).
27. Hedrick, J. B. "Zirconium and Hafnium," U.S. Geological Survey Minerals Yearbook, 86.2–86.8 (1999).
28. Kalish, D. and Clougherty, E. V. "Fundamental Study of the Sintering Kinetics of Refractory Compound Phases at High Pressure and High Temperature," Contract No. 426200, Summary Report (October 1966).
29. Clougherty, E. V., Pober, R. L. and Kaufman, L., "Synthesis of Oxidation Resistant Metal Diboride Composites," Trans. TMS-AIME, [242] 1077–1082 (1968).
30. Samsonov, G. V. and Serebryakova, T. I. "Classification of Borides," Sov. Powder Metall. Met. Ceram. (English Translation), [17] 116–120 (1978).
31. Clougherty, E. V., Wilkes, K. E. and Tye, R. P. "Research and Development of Refractory Oxidation Resistant Diborides," Part II, Vol. V: Thermal, Physical, Electrical and Optical Properties, AFML-TR-68-190, ManLabs Inc., Cambridge, MA, (1969).
32. Munro, R. G., "Material Properties of a Sintered alpha-SiC," J. Physical and Chemical Reference Data, [26] 1195–1203 (1997).
33. Rhodes, W. H., Clougherty, E. V. and Kalish, D. "Research and Development of Refractory Oxidation Resistant Diborides," Part II, Vol. IV: Mechanical Properties, AFML-TR-68-190, ManLabs Inc., Cambridge, MA, (1970).
34. Lynch, J. F., Ruderer, C. G. and Duckworth, W. H. "Borides," in Engineering Properties of Ceramics. American Ceramic Society – Columbus, OH. p. 5.4.1–6, 5.4.5–6 (1966).
35. Perkins, R., Kaufman, L. and Nesor, H. "Stability Characterization of Refractory Materials Under Velocity Atmospheric Flight Conditions," Experimental Results of High Velocity Cold Gas/Hot Wall Test, Part III Vol. II, AFML-TR-68-84, ManLabs Inc., Cambridge, MA, (1969).
36. Tripp, W. C., Davis, H. H. and Graham, H. C. "Effect of an SiC Addition on the Oxidation of ZrB2," Ceramic Bulletin, [52] 612–616 (1973).
37. Kaufman, L. and Clougherty, E. V. and Berkowitz-Mattuck, J. B. "Oxidation Characterastics of Hafnium and Zirconium Diboride," Trans. TMS-AIME, [239] 458–466 (1967).
38. Berkowitz-Mattuck, J. B., "High Temperature Oxidation – Zirconium and Hafnium Diborides," J. Electrochem. Soc., [113] 908–914 (1966).
39. Strife, J. R. and Sheehan, J. E., "Ceramic Cotings for Carbon-Carbon Composites," Ceramic Bull. [67] 369–374 (1988).
40. Schiroky, G. H., Price, R. J. and Sheehan, J. E., "Oxidation Charcteristics of CVD Silicon Carbide and Silicon Nitride," GA-A18696, General Atomics, La Jolla, CA (1986).
41. Schlichting, J., "Oxygen Transport Through Silica Surface Layers on Silicon-Containing Ceramic Materials," High Temp.-High Press. [14] 717–724 (1982).

42. Laurenko, V. A. and Alexeev, A. F., "Oxidation of Sintered Aluminum Nitride," Ceramic International **[9]** 80 (1983).
43. Chung, S. K. "Fracture Characterization of Armor Ceramics," Am. Ceram. Soc. Bull. **[69]** 358–366 (1990).
44. Lasday, S. B. "Alpha Silicon Carbide Properties Advantageous for Automotive Water Pump Seal Faces Produced at New Facility in W. Germany," Ind. Heat. **[35–39]** (1990).
45. Kittel, C. Introduction to Solid State Physics, John Wiley & Sons (1975).
46. Kaufman, L. and Nesor, H. "Stability Characterization of Refractory Materials Under High Velocity Atmospheric Flight Conditions," Part II Vol. II, AFML-TR-69-84, ManLabs Inc., Cambridge, MA, (1969).
47. Kaufman, L. and Nesor, H. "Stability Characterization of Refractory Materials Under High Velocity Atmospheric Flight Conditions," Part III Vol. I, AFML-TR-69-84, ManLabs Inc., Cambridge, MA, (1969).
48. Kaufman, L. and Nesor, H. "Stability Characterization of Refractory Materials Under High Velocity Atmospheric Flight Conditions," Part III Vol. II, AFML-TR-69-84, ManLabs Inc., Cambridge, MA, (1969).
49. Kaufman, L. and Nesor, H. "Stability Characterization of Refractory Materials Under High Velocity Atmospheric Flight Conditions," Part II Vol. III, AFML-TR-69-84, ManLabs Inc., Cambridge, MA, (1969).
50. Kaufman, L. and Nesor, H. "Stability Characterization of Refractory Materials Under High Velocity Atmospheric Flight Conditions," Part III Vol. III, AFML-TR-69-84, ManLabs Inc., Cambridge, MA, (1970).
51. Kingery, W. D. "Factors Affecting Thermal Stress Resistance of Ceramic Materials," J. Am. Ceram. Soc., **[3]** (1955).
52. Clougherty, E. V., Niesz, D. E. and Mistretta, A. L. "Research and Development of Refractory Oxidation-Resistant Diborides," Thermal Stress Resistance, Part II Vol. VI, AFML-TR-68-190, ManLabs Inc., Cambridge, MA, (1968).
53. Woo, S. K., Kim, C. H. and Kang, E. S. "Fabrication and Microstructural Evaluation of $ZrB_2/ZrC/Zr$ Composites by Liquid Infiltration," J. Mat. Sci. **[2]** 5309–5315 (1994).
54. Zhang, G., Deng, Z., Kondo, N., Yang, J. and Ohji, T. "Reactive Hot Pressing of ZrB_2-Sic Composites," J. Am. Ceram. Soc. **[83]** 2330–2332 (2002).
55. Levine, S., Opila, E., Halbig, M., Kiser, J., Singh, M. and Salem, J., "Evaluation of Ultra-High Temperature Ceramics for Aeropropulsion Use," J. Europ. Ceram. Soc. **[22]** 2757–2767 (2002).
56. Bargeron, C. B., Benson, R. C., Newman, R. W., Jette, A. N. and Phillips, T. E. "Oxidation Mechanisms of Hafnium Carbide and Hafnium Diboride in the Temperature Range 1400–2100C," Johns Hopkins APL Technical Digest, **[14]** 29–35 (1993).
57. Monteverde, F. and Bellosi, A. "Effect of the Addition of Silicon Nitride on Sintering Behavior and Microstructure of Zirconium Diboride," Scripta Materialia, **[46]** 223–228 (2002).
58. Monteverde, F., Bellosi, A and Guicciardi, S. "Processing and Properties of Zirconium Diboride-Based Composites," J. Europ. Ceram. Soc., **[22]** 279–288 (2002).
59. Shimada, S. "A Thermoanalytical Study on the Oxidation of ZrC and HfC Powders with Formation of Carbon," Solid State Ionics, **[149]** 319–326 (2002).
60. Gasch, M., Ellerby, D., Irby, E., Beckman, S., Gusman, M. and Johnson, S., "Processing and Properties of Hafnium Diboride/Silicon Carbide Ultra High Temperature Ceramics," To be published in J. of Materials Science (2003).

Part III

Non-oxide/Oxide Composites

10

SiC Fiber-Reinforced Celsian Composites

Narottam P. Bansal
NASA Glenn Research Center
Cleveland, OH 44135
(216) 433-3855;Narottam.P.Bansal@nasa.gov

ABSTRACT

Celsian is a promising matrix material for fiber-reinforced composites for high temperature structural applications. Processing and fabrication of small diameter multifilament silicon carbide tow reinforced celsian matrix composites are described. Mechanical and microstructural properties of these composites at ambient and elevated temperatures are presented. Effects of high-temperature exposures in air on the mechanical behavior of these composites are also given. The composites show mechanical integrity up to ~1100°C but degrade at higher temperatures in oxidizing atmospheres. A model has been proposed for the degradation of these composites in oxidizing atmospheres at high temperatures.

1. INTRODUCTION

Fiber-reinforced ceramic matrix composites are being developed for high-temperature structural applications in aerospace, energy conservation, power generation, nuclear, petrochemical, and other industries. Glasses and glass-ceramics of various compositions [1, 2] are being investigated as matrix materials for fiber-reinforced composites. Maximum use temperatures of these matrices in composite form are compared in Table 1. Barium aluminosilicate (BAS) glass-ceramic having monoclinic celsian, $BaAl_2Si_2O_8$, as the crystalline phase appears to be the most refractory with a projected maximum use temperature of ~1590°C and a melting point of 1760°C. Celsian shows excellent resistance towards

TABLE 1. Glass and glass-ceramic matrices of interest for fiber-reinforced composites [1–2]. Reprinted with kind permission of Kluwer Academic Publishers

Matrix type	Major constituents	Minor constituents	Major crystalline phases	Maximum use temp. (°C) in composite form
Glasses[a]				
7740 Borosilicate	B_2O_3, SiO_2	Na_2O, Al_2O_3	------------	600
1723 Aluminosilicate	Al_2O_3, MgO, CaO, SiO_2	B_2O_3, BaO	------------	700
7930 High silica	SiO_2	B_2O_3	------------	1150
Glass-ceramics				
LAS-I	Li_2O, Al_2O_3, MgO, SiO_2,	ZnO, ZrO_2, BaO	β-spodumene	1000
LAS-II	Li_2O, Al_2O_3, MgO, SiO_2, Nb_2O_5	ZnO, ZrO_2, BaO	β-spodumene	1100
LAS-III	Li_2O, Al_2O_3, MgO, SiO_2, Nb_2O_5	ZrO_2	β-spodumene	1200
MAS	MgO, Al_2O_3, SiO_2	BaO	Cordierite	1200
BMAS	BaO, MgO, Al_2O_3, SiO_2	--------	Barium osumilite	1250
Ternary mullite	BaO, Al_2O_3, SiO_2	--------	Mullite	~1500
Celsian	BaO, Al_2O_3, SiO_2	--------	Celsian	~1600

[a] 7740, 1723 and 7930 are Corning Glass Works designations.

oxidation and reduction, has good thermal shock resistance because of its low thermal expansion coefficient, and does not undergo any phase transformation up to ~1590°C. It also shows reasonably good resistance against alkali attack. It is chemically compatible with various reinforcement materials such as SiC, Si_3N_4, Al_2O_3, and mullite. It shows thermally stable microwave dielectric properties which makes it useful for applications in electromagnetic windows and radomes, microelectronic packaging, and high voltage condensers. Also it is highly effective [3–4] as an environmental barrier coating (EBC) for the protection of SiC_f/SiC combustor liner in the harsh atmosphere of turbine engines. Therefore, celsian is an attractive matrix material for fiber-reinforced composites for applications as high temperature structural materials in hot sections of gas turbine engines. Celsian composites reinforced with small diameter multifilament tow SiC fibers are described here. Processing and properties of large CVD SiC filament reinforced celsian matrix composites are reported elsewhere [5–11].

2. POLYMORPHS AND CRYSTAL STRUCTURE OF CELSIAN

Celsian (monoclinic space group I2/c) can exist in three different forms: monoclinic, hexagonal and orthorhombic. Hexacelsian (space group P6/mmm) is the high temperature polymorph and is thermodynamically stable above 1590°C, but once formed can metastably exist in the entire temperature range down to room temperature (Fig. 1) [12, 13]. The kinetics of transformation of hexacelsian into monoclinic celsian is very sluggish [14]. Hexacelsian shows a large thermal expansion coefficient of $\sim 8.0 \times 10^{-6}$/°C and undergoes a rapid, reversible structural transformation into the orthorhombic form at ~300°C, accompanied by a large volume change of ~3%. In the presence of hexacelsian,

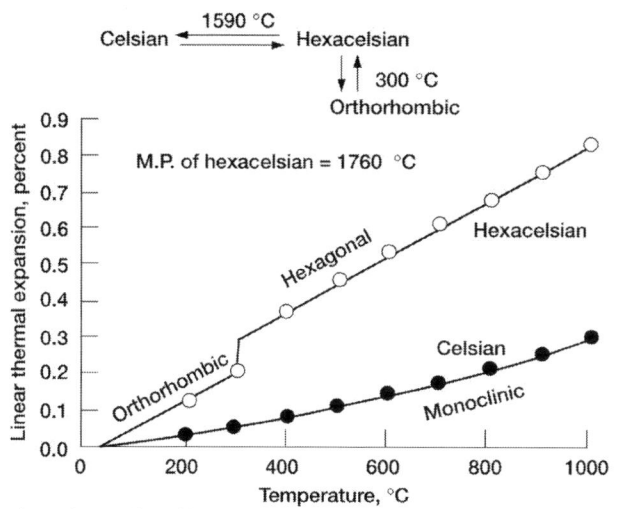

FIGURE 1. Linear thermal expansion of hexacelsian and monoclinic celsian BAS. (Reprinted, from reference 12, with kind permission of Elsevier).

thermal cycling of the composite would result in microcracking of the BAS matrix. Obviously, if BAS is to be successfully used as a matrix material for fiber reinforced composites, the processing conditions need to be designed to completely eliminate the undesirable hexacelsian phase.

The crystal structure of hexacelsian (Fig. 2 and 3a) [15] contains infinite two-dimensional hexagonal sheets, consisting of two layers of silica tetrahedral sharing all four vertices with substitution of Al for Si and charge compensation by Ba between the sheets. It has been suggested [16] that the transformation at 300°C between orthorhombic and hexagonal forms of hexacelsian involves rotation of these sheets with respect to each other. However, according to Takeuchi [16], no significant differences are observed between the powder XRD patterns of the orthorhombic and hexagonal forms of hexacelsian except peak shifts owing to lattice expansion, indicating similar fundamental frameworks of both structures.

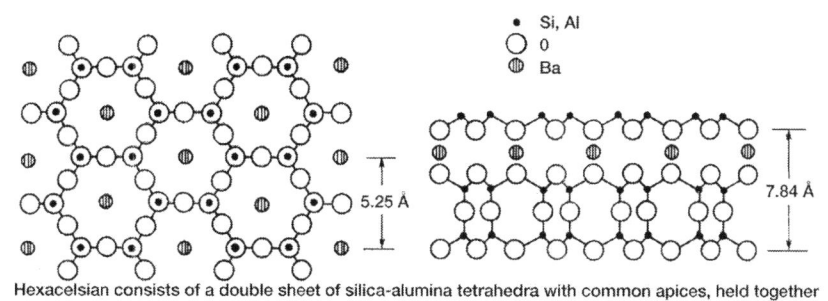

Hexacelsian consists of a double sheet of silica-alumina tetrahedra with common apices, held together by barium ions in 12-fold coordination.

FIGURE 2. Structure of hexagonal $BaAl_2Si_2O_8$. (Reprinted, from reference 15, with permission of The American Ceramic Society).

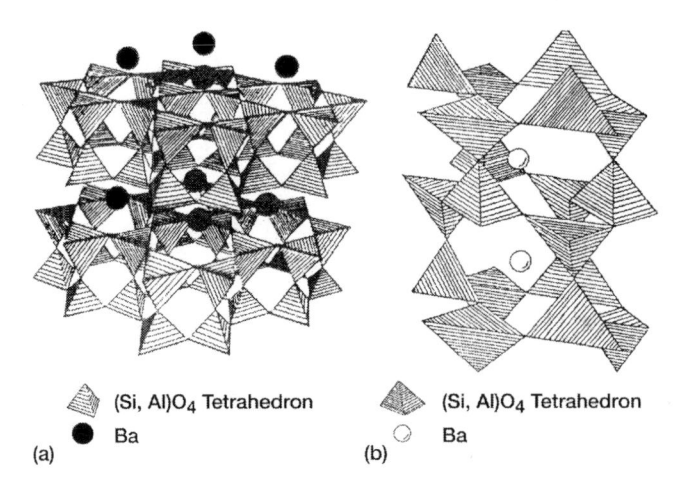

FIGURE 3. Crystal structures of (a) hexagonal and (b) monoclinic celsian $BaAl_2Si_2O_8$. (Reprinted, from reference 15, with permission of The American Ceramic Society).

The celsian structure (Fig. 3b) [17] is similar to the feldspar structure in which all four vertices of the silica tetrahedral are shared, forming a three-dimensional network. As in hexacelsian, Al substitutes for Si with charge compensation by Ba in the larger interstices of the structure. Gay [18] and Newnham and Megaw [17] considered the formation of a superlattice in celsian associated with ordering of the Al-Si atoms.

In celsian crystal structure [19] the Al and Si atoms are statistically distributed over tetrahedral sites, so that each Al tetrahedron is surrounded by four predominantly Si tetrahedral and vice versa. The feature of the hexacelsian structure, however, is the presence of $(Al, Si)O_4$ tetrahedra which share three corners so that a hexagonal sheet results with the remaining apices pointing in the same direction. Two of these sheets share their apical oxygens, forming a double tetrahedral sheet (Fig. 3a). Ba atoms occupy positions between the double sheets with twelve equidistant oxygen neighbors. In celsian, Ba ion has an irregular configuration, so that it has ten oxygen neighbors at several Ba-O distances. Consequently, to transform hexacelsian to celsian would require creation of a three-dimensional network from the two-dimensional sheet structure of hexacelsian as well as rearrangement of the Ba sites. This would entail breaking and reforming of Al-O and Si-O bonds. Some of the physical, mechanical and thermal properties of celsian are given in Table 2 [20–22].

3. FORMATION OF MONOCLINIC CELSIAN

On heat treatment, both hexacelsian and monoclinic celsian phases crystallize in BAS glass [23–25]. Also, hot pressing of barium aluminosilicate (BAS) glass or its composites reinforced with large diameter silicon carbide SCS-6 monofilaments or small diameter multifilament Nicalon or Hi-Nicalon fibers resulted in the crystallization of both hexacelsian and monoclinic celsian phases [12]. On doping BAS with 5 wt.% monoclinic celsian seeds or 10 wt.% strontium aluminosilicate (SAS), only the celsian phase was formed in hot pressed

TABLE 2. Properties of Celsian

Property	Temperature	Value
Melting point, °C	–	1760
Density, g.cm^{-3}	RT	3.39 (monoclinic celsian)
		3.26 (hexacelsian)
CTE, °C^{-1}	20–1000°C	2.29×10^{-6} (monoclinic celsian)
	300–1000°C	8×10^{-6} (hexacelsian)
Flexure modulus, GPa	RT	100 ± 7
	538°C	102 ± 14
	815°C	94 ± 8
	1093°C	87 ± 6
Poisson's ratio	RT	0.32
Flexure (4-point) strength, MPa	RT	115 ± 15
	500°C	120 ± 5
	750°C	100 ± 10
	1000°C	60 ± 5
	1200°C	25 ± 5
	1350°C	25 ± 5
Fracture toughness, MPa.m$^{1/2}$	RT	1.56
Thermal diffusivity, cm^{2}s^{-1}	RT	0.0114
Heat capacity, Jg^{-1}K^{-1}	RT	0.84
	260°C	0.79
	538°C	0.88
	815°C	0.88
	1093°C	0.94
	1372°C	1.02
Thermal conductivity, Wm^{-1}K^{-1}	RT	3.25
	500°C	3.1
	1000°C	3.0
	1400°C	2.9

monolithic specimens. However, in fiber-reinforced composites hot pressed under similar conditions, a small concentration of hexacelsian was still present as hexacelsian nucleates preferentially on the surface and the presence of fibers provides a large surface area. When the additive concentration increased to 10 wt.% celsian seeds or 20 wt.% SAS, celsian was the only phase detected from x-ray diffraction, with complete elimination of hexacelsian, in the hot pressed composites reinforced with SiC fibers. Hence BAS with 20–25% of SAS was used as the matrix for fiber-reinforced composites.

4. CMC PROCESSING

Ceramic grade Nicalon or Hi-Nicalon fiber tows were used as the reinforcement. Fibers having a dual surface layer of BN overcoated with SiC were used. BN acts as a weak, crack deflecting, compliant layer, while the SiC overcoat acts as a barrier to diffusion of boron from BN into the oxide matrix and also prevents diffusion of matrix elements into the fiber.

The matrix of 0.75BaO-0.25SrO-Al$_2$O$_3$-2SiO$_2$ (BSAS) composition was synthesized [20] by a solid-state reaction method from BaCO$_3$, SrCO$_3$, Al$_2$O$_3$, and SiO$_2$ powders.

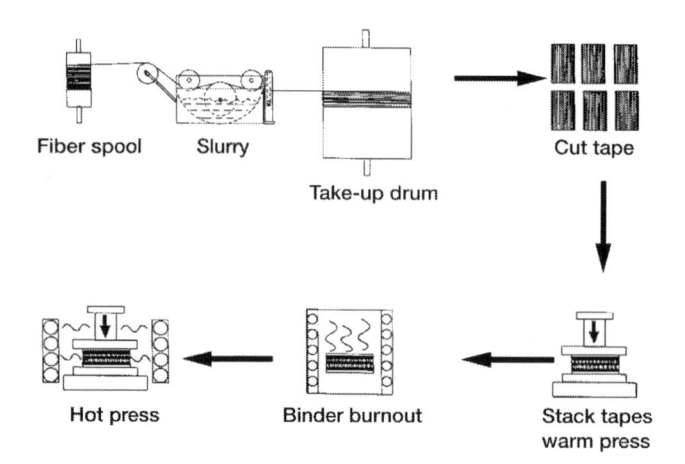

FIGURE 4. Schematic of the set up used for fabrication of small diameter, multifilament fiber tow reinforced ceramic composites. (Reprinted, from reference 26, with kind permission from Elsevier).

The mixed powder was calcined at ~900–910°C for decomposition of the carbonates. The resulting powder consisted of mainly SiO_2 (α-quartz) and $BaAl_2O_4$ phases with small amounts of Ba_2SiO_4, α-Al_2O_3, and $Ba_2Sr_2Al_2O_7$ also present. This powder was made into slurry by dispersing it in methyl ethyl ketone along with organic additives as binder, surfactant, deflocculant and plasticizer followed by ball milling. BSAS glass powder, custom melted by an outside vendor, was also used as precursor to celsian matrix in the fabrication of some composite panels.

Schematic of the set-up used for fabrication of FRC by matrix slurry infiltration is shown in Fig. 4 [26]. The fiber tows are impregnated with the matrix precursor by passing it through the slurry. Excess slurry is squeezed out of the fiber tow before winding it on a rotating drum. After drying, the prepreg tape is cut to size, stacked up in desired orientation, and warm pressed resulting in a "green" composite. The fugitive organics are slowly burned out of the sample in air followed by hot pressing in a graphite die resulting in almost fully dense composite.

X-ray diffraction (Fig. 5) from the surface of a hot pressed composite panel [27] showed only the desired monoclinic celsian. This indicates that the mixed oxide precursor is converted *in situ* into the monoclinic celsian phase during hot pressing of the FRC. The undesired hexacelsian phase was not detected from XRD. However, a small amount of hexacelsian was detected by Raman micro-spectroscopy [28]. The hot pressed composite panel was surface polished and sliced into test bars for mechanical testing.

5. NICALON/CELSIAN COMPOSITES

5.1. Flexure Strength

Flexure strength was measured in four-point bending using an upper span of 0.75″ and a lower span of 2.5″. Flexural properties [29–31] of 0/90° cross-ply Nicalon/BN/SiC/BSAS

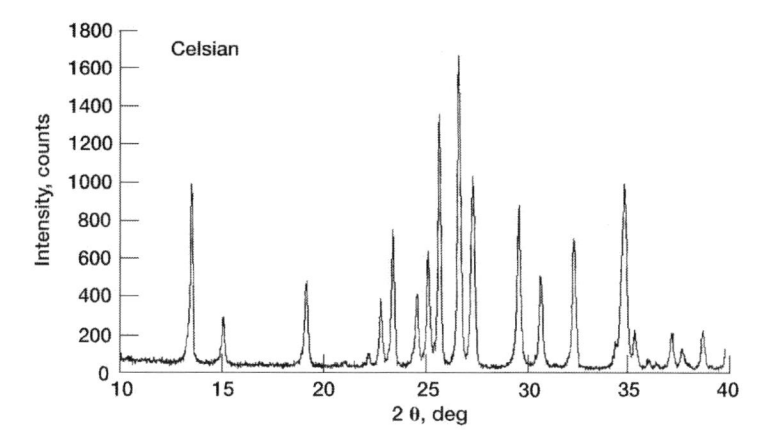

FIGURE 5. X-ray diffraction from surface of a hot pressed Hi-Nicalon/BSAS composite. (Reprinted, from reference 26, with kind permission from Elsevier).

composites, containing 38–40 volume per cent of fibers, at various temperatures in air are shown in Table 3.

5.2. Effects of High-Temperature Annealing in Air

Thermal-oxidative stability of Nicalon/BN/SiC/BSAS composites has also been investigated [29–31]. The 0/90° cross-ply composites, containing ~40 volume per cent of fibers were heat treated in air for 100–500 hrs. at various temperatures between ~550°C to ~1300°C. Glass formation was observed on the surface of specimens annealed at or above 1100°C. The specimens exposed at ~1200°C and ~1300°C were so severely covered with glass, along with the presence of some gas bubbles, that these could not be further tested. The residual flexural strengths of the composites after air-annealing are shown in Fig. 6. The sample annealed at ~550°C showed 40% drop in room temperature strength. This could be due to oxidation of the carbon layer (formed *in situ* during composite processing) at the fiber-matrix interface where the temperature is too low for any residual glassy phase to flow and provide oxidation resistance. However, the specimen exposed at ~1100°C retained

TABLE 3. Mechanical Properties* of 0/90° Nicalon/BN/SiC/BSAS Composites at Various Temperatures in Air; $V_f = 0.38$–0.40

Temperature °C	Yield stress, σ_y MPa	Modulus, E_c GPa	Ultimate strength, σ_u MPa	Failure strain %
25	75 ± 1	92 ± 1	289 ± 9	0.43 ± 0.01
550	85 ± 1	88 ± 5	282 ± 21	0.45 ± 0.04
1100	60 ± 15	80 ± 6	208 ± 1	0.34 ± 0.01
1200	44 ± 5	63 ± 3	187 ± 20	0.39 ± 0.03
1300	32 ± 8	54 ± 3	180 ± 45	0.55 ± 0.10

* Measured in 4-point flexure

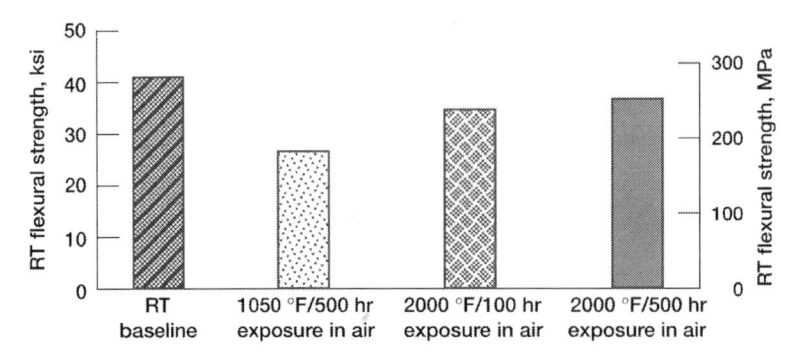

FIGURE 6. Effect of air-annealing at various temperatures on room temperature flexural strength of (0/90) Nicalon/BSAS composites; $V_f = 0.4$.

90% of the baseline strength. At this temperature, the residual glassy phase is less viscous, as seen by gas bubble formation, and may have resulted in the protection of the carbon interface layer from oxidation.

5.3. Tensile Strength

Tensile testing was carried out in air at a displacement rate of 0.02"/minute. Tensile properties of 0/90° cross-ply Nicalon/BN/SiC/BSAS composites, containing 38–40 volume per cent of fibers, at various temperatures are shown [29–31] in Table 4. Above ~1100°C, ultimate strength and the proportional limit fall off fairly rapidly, while modulus decreased by ~40% from ambient temperature to ~1300°C. Nicalon fiber is known to degrade at ~1200°C and degradation rate increases with increase in temperature. Presence of residual glassy phase in the matrix would account for the observed decrease in modulus and increase in failure strain at elevated test temperatures.

Effect of test rate on ultimate tensile strength for 0/90° Nicalon/BSAS composite at 1100°C in air has also been investigated [32]. Results in Fig. 7 exhibit a significant dependency of ultimate strength on test rate. The ultimate strength decreased with decreasing test rate, similar to the behavior observed in many advanced monolithic ceramics (or brittle materials including glass and glass ceramics) at ambient or elevated temperature. The

TABLE 4. Tensile Properties of 0/90° Nicalon/BN/SiC/BSAS Composites at Various Temperatures in Air; $V_f = 0.38–0.40$

Temperature °C	Yield stress, σ_y MPa	Modulus, E_c GPa	Ultimate strength, σ_u MPa	Failure strain %
25	75 ± 1	97 ± 12	178 ± 28	0.43 ± 0.09
550	63 ± 1	87	196	0.64
1100	63 ± 6	71 ± 9	162 ± 16	0.41 ± 0.06
1200	32 ± 9	63 ± 3	112 ± 12	0.50 ± 0.08
1300	29 ± 2	53 ± 3	153 ± 3	0.63 ± 0.06

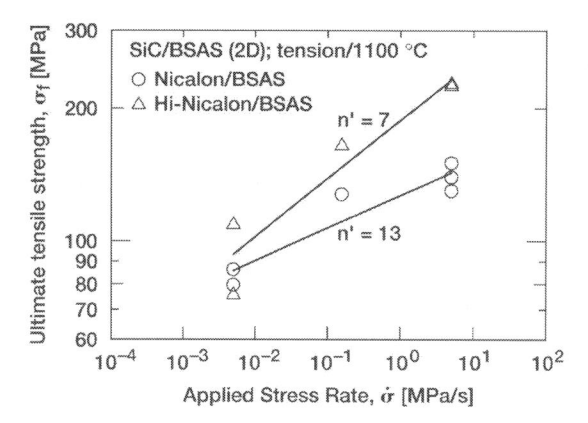

FIGURE 7. Results of ultimate tensile strength as a function of applied stress rate determined for Nicalon/BSAS and Hi-Nicalon/BSAS composites at 1100°C in air. The solid lines represent the best-fit regression based on eq. (1).

strength degradation with decreasing stress rate in brittle materials is known to occur by slow crack growth (delayed failure or fatigue) of an initial crack, and is expressed as:

$$\log \sigma_f = \frac{1}{n+1} \log \dot{\sigma} + \log D \tag{1}$$

where n and D are slow crack growth (SCG) parameters, and σ_f (MPa) and $\dot{\sigma}$ (MPa/s) are strength and stress rate, respectively. These results suggest that care must be exercised when characterizing elevated-temperature ultimate strength of composite materials. This is due to that fact that high-temperature ultimate strength has a relative meaning if a material exhibits rate dependency: the strength simply depends on which test rate one chooses. Therefore, at least two test rates (high and low) are recommended to better characterize the high-temperature ultimate strength behavior of a composite material.

5.4. Shear Strength

Interlaminar shear testing was carried out using the short beam shear technique in 3-point bending at a displacement rate of 0.02"/minute from room temperature to ~1300°C. Results of the interlaminar shear strength [31] for the 0/90° cross-ply Nicalon/BN/SiC/BSAS composites, containing 38–40 volume per cent of fibers are given in Table 5. The failure mode for all the test specimens was shear.

5.5. Tensile Stress-Rupture Strength

To determine if the composite would provide stress oxidative stability after the matrix is microcracked, stepped stress-rupture tests were carried out. The 0/90° cross-ply Nicalon/BN/SiC/BSAS composites, containing ~40 volume% of fibers, were tested in standard dead weight creep machines at 1100°C and 1200°C in air. No testing was done at

TABLE 5. Interlaminar Shear Properties of 0/90°
Nicalon/BN/SiC/BSAS Composites at Various Temperatures
in Air; $V_f = 0.38$–0.40

Temperature, °C	Interlaminar shear stress*, MPa	Failure mode
25	17 ± 2	Shear
1100	10 ± 3	Shear
1200	7 ± 1	Shear
1300	6 ± 1	Shear

* Average of three samples

1300°C because of the presence of residual glass in the matrix. Tests were started at stress level near the proportional limit. Approximately every 50 hours, the stress was increased by 2 ksi (~14MPa) until the sample failed. The results [31] are shown in Fig. 8. The composite survived to a stress level of ~60% of its ultimate tensile strength. Examination of the fracture surface of the specimen tested at ~1100°C/10ksi (~70 MPa) for 8 h revealed some embrittlement.

Results of constant stress (stress-rupture) testing for 0/90° Nicalon/BSAS (for two different batches 'A' and 'B') at 1100°C in air are presented [32] in Fig. 9, where *time to failure* is plotted against *applied stress* in log-log scales. A decrease in time to failure with increasing applied stress, which represents a susceptibility to damage accumulation or delayed failure, is evident for the composite. The mode of fracture in constant stress testing was similar to that in constant stress-rate testing showing some fiber pullout with jagged matrix cracking through the specimen-thickness direction.

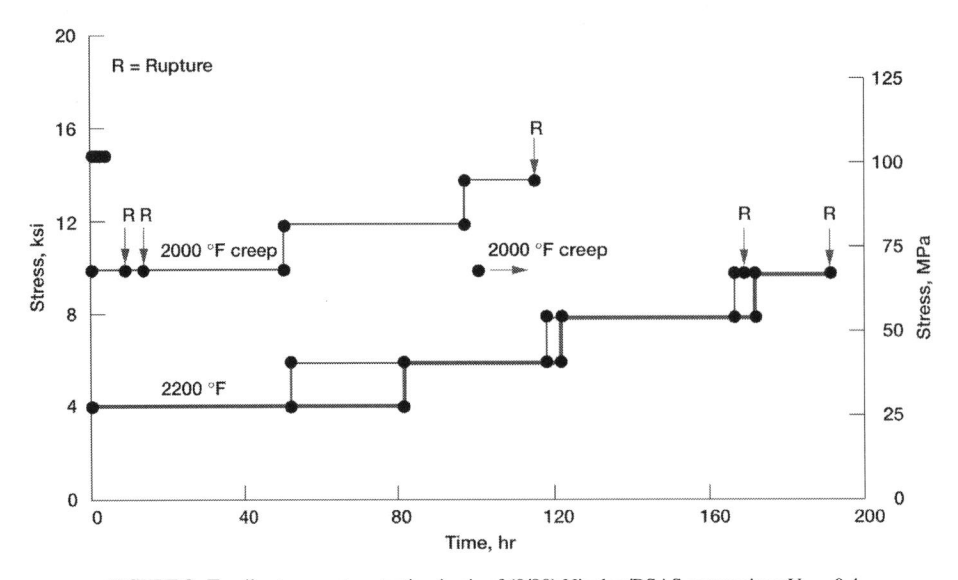

FIGURE 8. Tensile stress rupture testing in air of (0/90) Nicalon/BSAS composites; $V_f = 0.4$

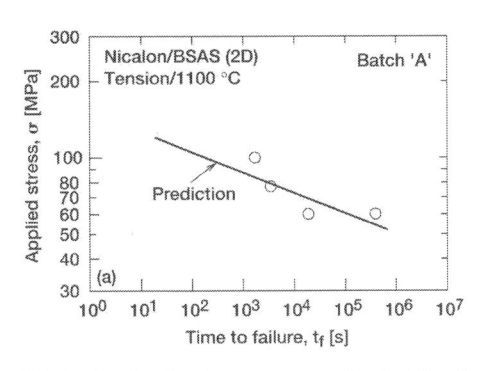

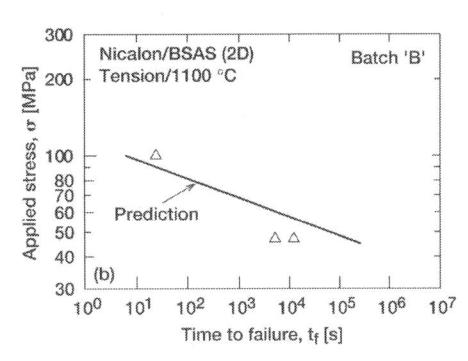

FIGURE 9. Results of constant stress ("static fatigue" or "stress rupture") testing for Nicalon/BSAS composite: (a) Batch 'A' and (b) Batch 'B' tested at 1100°C in air. The solid lines represent the predictions based on the results of constant stress-rate testing (Fig. 7).

6. HI-NICALON/CELSIAN COMPOSITES

SEM micrographs taken from the polished cross-sections of the uncoated and BN/SiC coated Hi-Nicalon fiber-reinforced BSAS composites are shown in Fig. 10. A large variation in fiber diameter within a tow can be seen although the manufacturer reports an average diameter of 14 μm. Uniform fiber distribution, good matrix infiltration within the fiber tow, and high composite density is evident. Occasional pores and second matrix phase, appearing as the bright phase, can also be seen. The amount of this second phase must be small, as it was not detected by XRD. This phase has been tentatively identified [33] as "hyalophane" [(Ba, Sr)AlSi$_3$O$_8$] which is richer in Si and deficient in Al than celsian. The BN coating which is the dark layer is adherent to the fiber, but the SiC coating tends to debond at the BN/SiC interface probably during composite processing.

TEM micrograph from the Hi-Nicalon/BN/SiC/BSAS composite [26] is shown in Fig. 11a. Four distinct layers are seen in the BN coating. The SiC coating was often found to be missing from the BN/matrix interface but was intact in this particular case. Another TEM micrograph [26] is shown in Fig. 11b. The fiber/BN interface and the innermost layer of BN are unchanged. BN has coarsened from outside inwards during composite processing. From EDS, no obvious chemical changes were observed in the BN coating. SiC layer was not affected.

6.1. Flexure Strength

Typical stress-strain curves, recorded in 3-point flexure, for the uncoated and BN/SiC coated fiber-reinforced composites [33–34] are shown in Fig. 12. Also shown for comparison is the stress-strain curve for a hot pressed BSAS monolith. The monolith shows bend strength of 130 MPa, modulus of 98 GPa and fails in brittle mode, as expected. The uncoated fiber-reinforced CMC shows a monolithic-like failure with strength of ∼200 MPa and failure strain of ∼0.1%. This shows that reinforcement of BSAS matrix with uncoated Hi-Nicalon fibers does not yield a strong or tough material. This may indicate a strong bond between the uncoated fibers and the matrix. However, fiber push-in and push-out tests [33, 35]

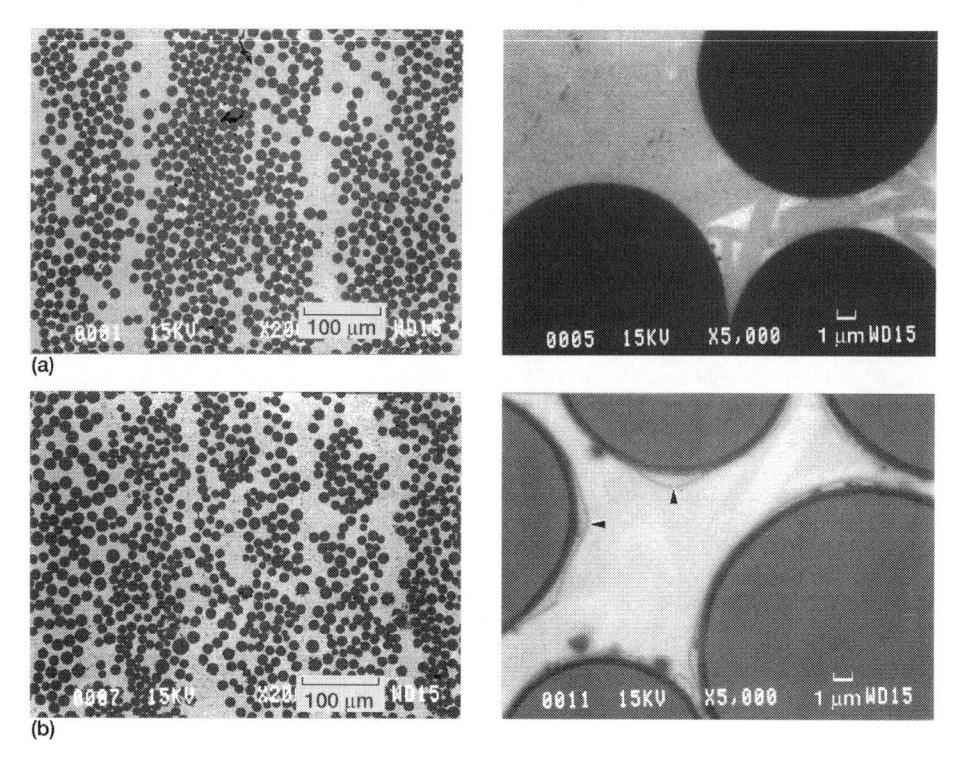

FIGURE 10. SEM micrographs showing polished cross-sections of unidirectional (a) Hi-Nicalon (uncoated)/BSAS and (b) Hi-Nicalon/BN/SiC/BSAS composites. (Reprinted, from reference 33, with kind permission of Materials Research Society).

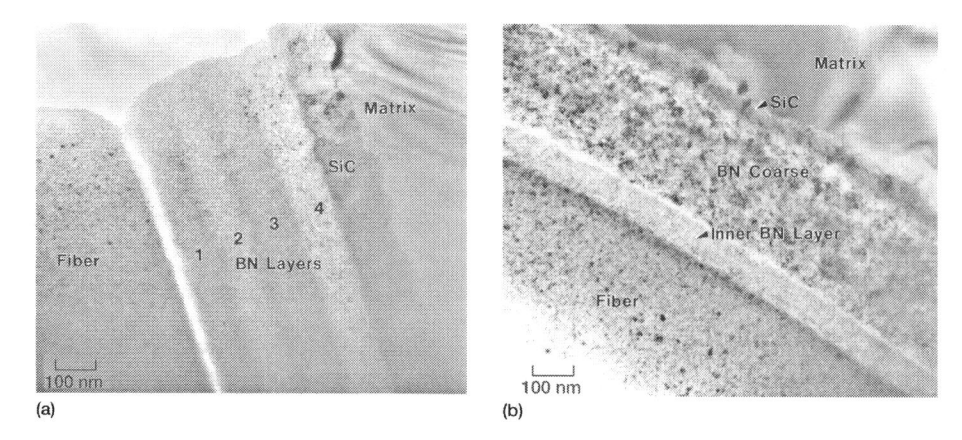

FIGURE 11. TEM bright field image from cross-section of Hi-Nicalon/BN/SiC/BSAS composite showing (a) multiple BN layers and (b) coarsening of the outer three BN layers. The inner BN and the SiC interface layers are unaffected. (Reprinted, from reference 26, with kind permission from Elsevier).

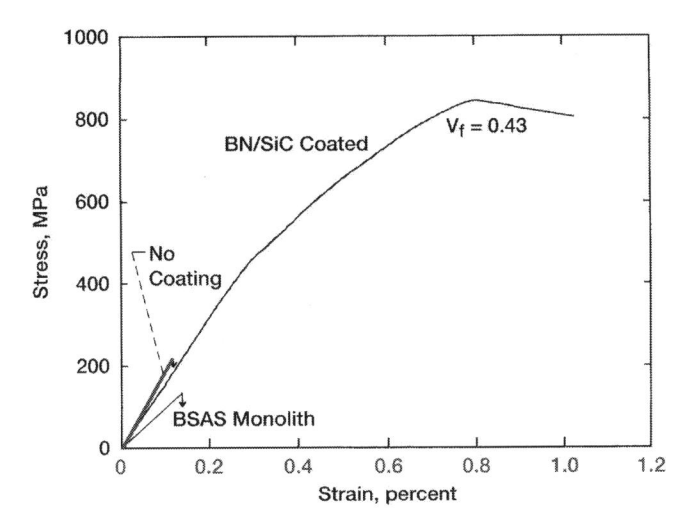

FIGURE 12. Stress-strain curves for unidirectional uncoated and BN/SiC-coated Hi-Nicalon/BSAS composites recorded in 3-point flexure; also shown is the curve for hot pressed monolithic BSAS.

indicated a weak fiber-matrix interface in these CMCs. Low strength of these CMCs was found to be due to mechanical damage [33, 35] to the fiber surface during hot pressing as no compliant layer (such as BN) was present. In contrast, the BN/SiC coated fiber-reinforced CMC shows graceful failure. The first matrix cracking stress was ~400 MPa and ultimate strength greater than 800 MPa. This shows that reinforcement of BSAS matrix with BN/SiC coated Hi-Nicalon fiber results in a strong and tough composite.

6.2. Tensile Strength

Tensile properties of unidirectional BN/SiC coated Hi-Nicalon fiber reinforced celsian matrix composites [36–37] from room temperature to 1200°C in air are shown in Table 6. The value of Young's modulus decreased with increase in test temperature indicating the presence of glassy phase in the matrix. The yield stress decreased from room temperature

TABLE 6. Tensile Properties of Unidirectional Hi-Nicalon/BN/SiC/BSAS Composites at Various Temperatures in Air; $V_f = 0.30$

Temperature °C	Modulus, E_c GPa	Yield stress, σ_y MPa	Yield strain %	Ultimate strength, σ_u MPa	Failure strain %
25	154 ± 2	97 ± 15	0.064 ± 0.09	589 ± 15	0.45 ± 0.01
800	132 ± 3	68 ± 6	0.052 ± 0.06	448 ± 4	0.41 ± 0.03
1000	106 ± 3	70 ± 7	0.064 ± 0.09	480 ± 25	0.40
1100	91 ± 3	92 ± 6	0.10 ± 0.005	400 ± 2	0.47 ± 0.05
1200	73 ± 5	99 ± 40	0.074	370 ± 12	—

to 800°C followed by a slight increase to 1200°C. Ultimate strength decreased with temperature. SEM micrographs of the fracture surfaces of the composite specimens after tensile tests showed extensive fiber pullout. Fracture surfaces of samples tested at 1000°C or higher showed the presence of pores in the matrix which were more predominant in the 1200°C tested specimen. This may be due to oxidation of the Hi-Nicalon fibers producing a gaseous product and also oxidation of the BN coating producing a low melting glassy phase that is partially blown out of the interior of the composite by the escaping gases.

Effect of test rate on ultimate tensile strength of 0/90° Hi-Nicalon/BSAS composite at 1100°C in air has also been investigated [32]. Results in Fig. 7 exhibit a significant dependency of ultimate strength on test rate. The ultimate strength decreased with decreasing test rate, similar to the behavior observed in many advanced monolithic ceramics (or brittle materials including glass and glass ceramics) at ambient or elevated temperature.

6.3. Shear Strength

In-plane and interlaminar shear strength of composites have been measured by the double-notch shear test method at a displacement rate of 2×10^{-3} mm/s. Results for a unidirectional Hi-Nicalon/BN/SiC/BSAS composite containing 42 volume % of fibers from room temperature to 1200°C are shown [38–39] in Fig. 13. The interlaminar shear strength

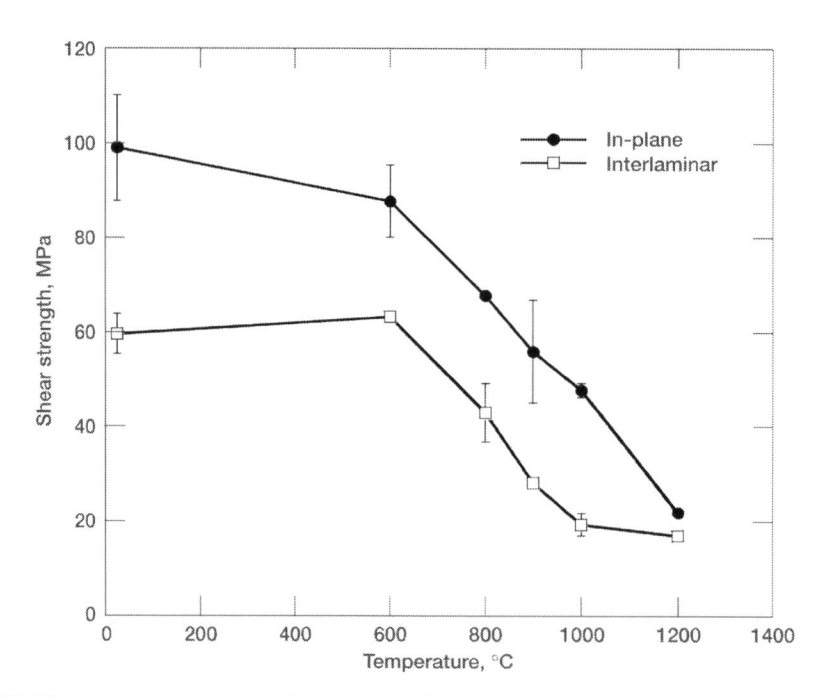

FIGURE 13. Temperature dependence of interlaminar and in-plane shear strength of Hi-Nicalon/BSAS composite in air. (Reprinted, from reference 39, with kind permission from Elsevier).

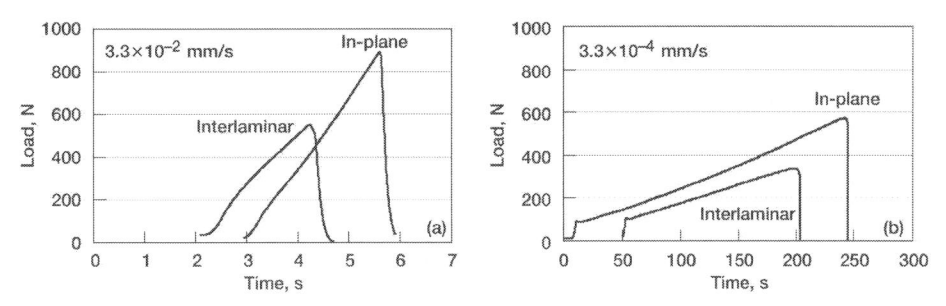

FIGURE 14. Typical load-time curves for Hi-Nicalon/BSAS composite tested in interlaminar and in-plane shear at 1100°C in air; (a) fast test rate of 3.3×10^{-2} mm/s, and (b) slow test rate of 3.3×10^{-4} mm/s.

is lower than the in-plane shear strength at all temperatures. A rapid decrease in strength with temperature is probably due to softening of the residual glassy phase in the matrix.

Test rate dependence of interlaminar and in-plane shear strengths of CMC have also been investigated [40–41] using double notch shear (DNS) test specimens. Typical load versus time curves for two different test rates are shown in Fig. 14. The effects of test rate on shear strengths of a unidirectional Hi-Nicalon™ fiber-reinforced BSAS composite at 1100°C in air are presented in Fig. 15. The composite exhibited a significant effect of test rate on shear strength, regardless of orientation which was either in interlaminar or in in-plane region. Shear-strength degraded by about 50% as test rate decreased from 3.3×10^{-1} mm/s to 3.3×10^{-5} mm/s. The rate dependency of composite's shear strength was very similar to that of ultimate tensile strength at 1100°C observed [32] in a similar composite (2-D Hi-Nicalon/BSAS) in which ultimate tensile strength decreased by about 60% when test rate changed from the highest (5 MPa/s) to the lowest (0.005 MPa/s). A phenomenological,

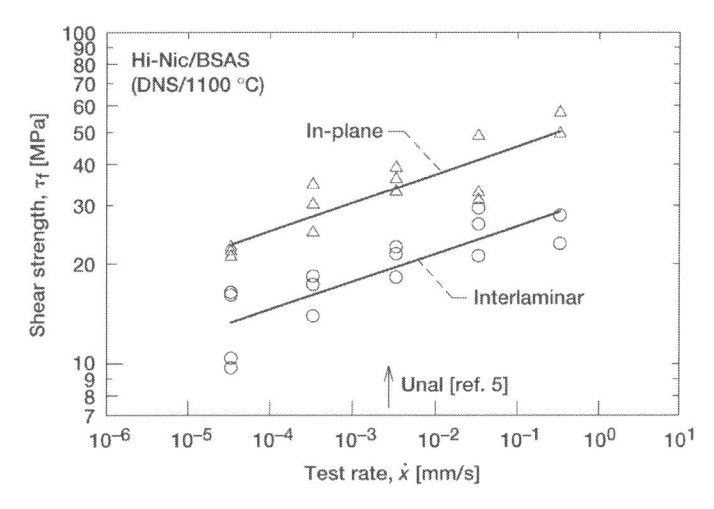

FIGURE 15. Effect of test rate on interlaminar and in-plane shear strengths for Hi-Nicalon/BSAS composite tested at 1100°C in air. The arrow in the 'test-rate' axis indicates the test rate used by Unal and Bansal [38–39].

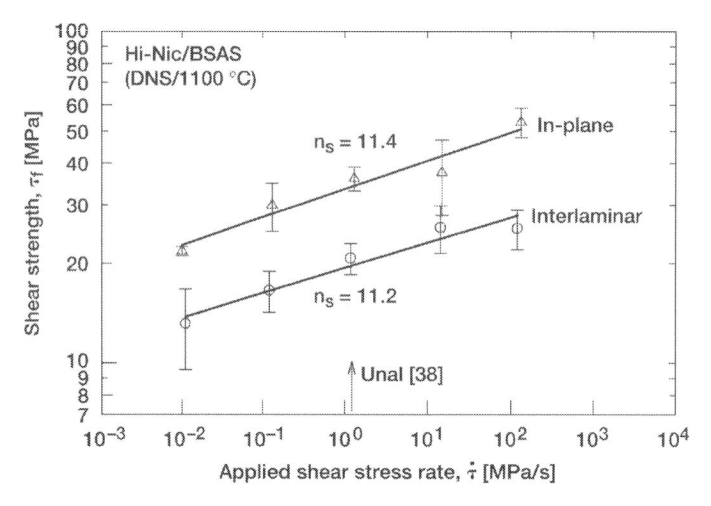

FIGURE 16. Shear strength as a function of applied shear stress rate for Hi-Nicalon/BSAS composite at 1100°C in air, reconstructed from the data in Fig. 15 based on eq. (2). The arrow in the 'stress-rate' axis indicates the stress rate used by Unal and Bansal [38–39].

power-law slow crack growth formulation has been proposed [40–41] to account for the rate dependency of shear strength of the composite.

Results of shear strength (τ_f) as a function of applied shear stress rate ($\dot{\tau}$) based on:

$$\log \tau_f = \frac{1}{n_s + 1} \log \dot{\tau} + \log D_s \qquad (2)$$

with displacement rate being converted to shear stress rate are shown in Fig. 16. Here n_s and D_s are the slow crack growth (SCG) parameters. The values of SCG parameters were determined as $n_s = 11.2$ and $D_s = 19.24$ for interlaminar direction and $n_s = 11.4$ and $D_s = 33.27$ for in-plane direction. It is noteworthy that the value of n_s, a measure of susceptibility to SCG, was same in both directions. The value of SCG parameter n_s ($=11$) in shear also compares with that of n ($=7$) in tension for the similar 2-D Hi-Nicalon/BSAS composite tested at 1100°C in air [32], indicating that the Hi-Nicalon/BSAS composites exhibit significant susceptibility to SCG in both shear and tension. In general, the patterns of SCG for many monolithic advanced ceramics tested at elevated temperatures are well defined in fracture surfaces in terms of its configuration and size. By contrast, the patterns of SCG for CMCs subjected to either in tension or in shear are not obvious and rather obscured by the nature of composite architecture so that it would be very difficult and/or a great challenge to derive a definite evidence of SCG from overall fracture surface examinations. Hence, the study of shear strength degradation with respect to test rate through constant stress-rate testing is a simple, quick and convenient way to check and quantify the degree of SCG in shear for CMCs at elevated temperatures. The residual glassy phase may have been a major cause of SCG in the Hi-Nicalon/BSAS composite under shear. Life prediction diagram (applied shear stress vs. time to failure) in shear for interlaminar and in-plane directions for Hi-Nicalon/BSAS composite [41] at 1100°C in air is shown in Fig. 17.

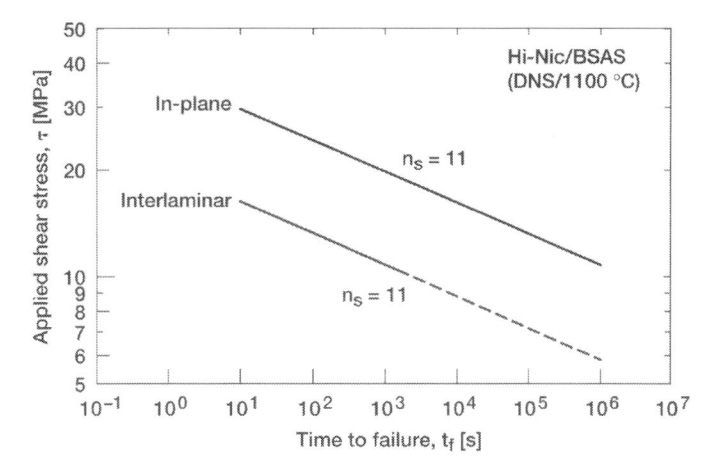

FIGURE 17. Life prediction diagram in shear for Hi-Nicalon/BSAS composite at 1100°C in air in interlaminar and inplane directions under constant stress condition.

6.4. Thermo-oxidative Stability

6.4.1. Effects of High-Temperature Annealing in Air

Thermal-oxidative stability of Hi-Nicalon/BN/SiC/BSAS composites has also been investigated [42–43]. Unidirectional composites were annealed for 100 hrs. in air at various temperatures. Optical photographs showing the physical appearance of the CMC bars before and after annealing at various temperatures are shown in Fig. 18. No changes in physical appearance were observed in samples annealed at 1000°C or lower. However, the specimen annealed at 1100°C was covered with a thin porous white layer that could be easily removed by polishing with a fine emery paper. The samples aged at 1200°C were deformed and

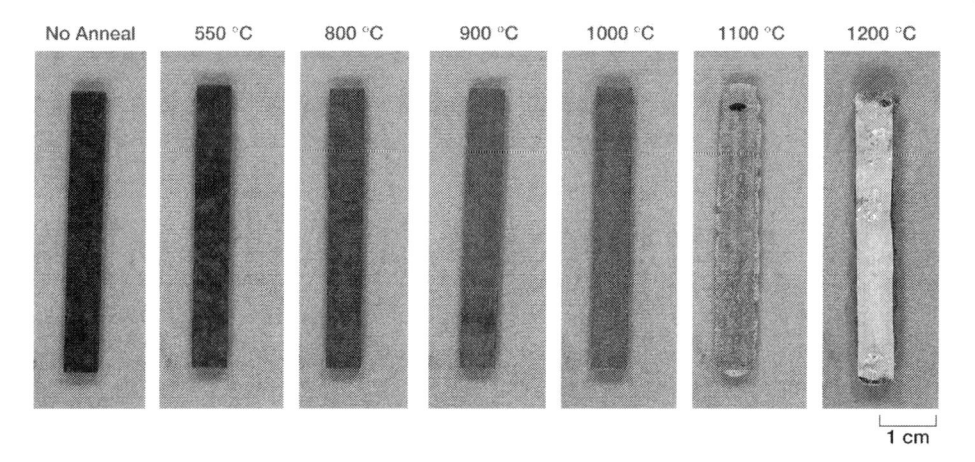

FIGURE 18. Optical photographs showing surface appearance of Hi-Nicalon/BSAS composites after 100 h air-annealing at various temperatures.

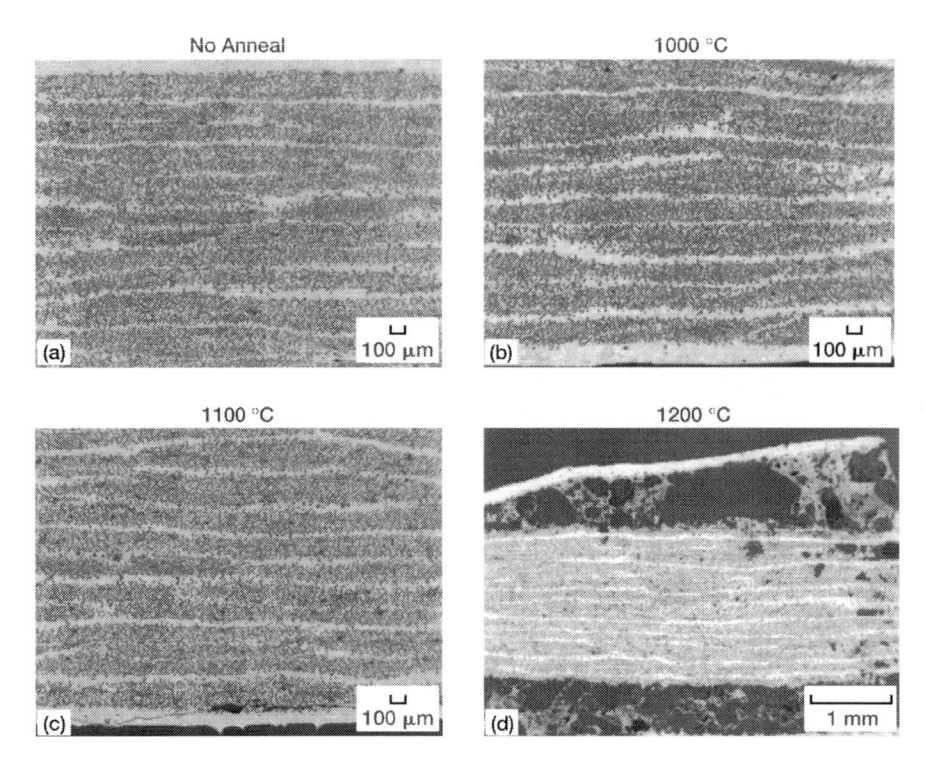

FIGURE 19. SEM micrographs showing polished cross-sections of unidirectional Hi-Nicalon/BSAS composites annealed for 100 h in air at various temperatures; $V_f = 0.32$

developed a thick shiny white glaze on the exposed surfaces. Pores were present in the surface layer. Signs of partial melting and gas bubble formation during heat treatment were also observed. From XRD analysis, both amorphous and Celsian phases were detected in the surface layer. Since such a behavior was not observed [44] in monolithic BSAS material even after heat treatment for 20 h in air at 1500°C, it must be caused by the presence of Hi-Nicalon fibers and/or the BN/SiC coating. In air, BN is probably oxidized to B_2O_3 which reacts with the matrix and/or silica formed from the oxidation of silicon carbide fibers, resulting in low-melting glassy phase which migrates to the sample surface. The bubble formation may be related to the oxidation of Hi-Nicalon fibers producing amorphous silica and CO and CO_2 gases.

The matrix layers on the surface of CMC specimens annealed at 1100 and 1200°C appear to have cracked and delaminated as seen in the SEM micrographs (Fig. 19), taken from the polished cross-sections. A large difference in the coefficients of thermal expansion (CTE) of Hi-Nicalon fiber ($\sim 3.5 \times 10^{-6}$/°C) and the matrix may be responsible for the observed cracking and delamination. This would provide an easy path for the ingression of oxygen to the fiber bundles and accelerate the degradation of fibers from oxidation. No such cracking or delamination was observed in samples annealed at lower temperatures. Higher magnification SEM micrographs from the 1200°C annealed specimens showed the

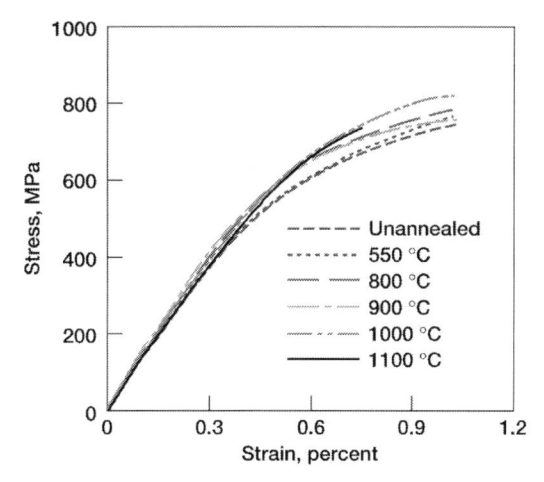

FIGURE 20. Room temperature stress-strain curves recorded in 3-point flexure for unidirectional Hi-Nicalon/BSAS composites annealed for 100 h in air at various temperatures; $V_f = 0.32$.

presence of gas bubbles on its surface. Severe damage was also observed underneath the top debonded layer where damaged broken pieces of the fibers are also present.

Flexure strengths of the annealed CMC specimens were measured at room temperature in 3-point bending. The strain-stress curves of the CMCs annealed at various temperatures are presented in Fig. 20 and the results are summarized in Table 7. Up to the annealing temperature of 1100°C, no net change was observed in the weights of samples. Also, values of Young's modulus (E_c), yield stress (σ_y), yield strain (ε_y), and ultimate strength (σ_u) did not change. However, the specimen annealed at 1200°C, gained 0.43% weight and had appreciably deformed and its strength could not be measured.

6.4.2. Degradation Mechanism at 1200°C

Various steps involved in the degradation of Hi-Nicalon/BN/SiC/BSAS CMC on annealing in air at 1200°C are shown in Fig. 21. During annealing at 1200° C, the surface

TABLE 7. Mechanical Properties* of Unidirectional Hi-Nicalon/BN/SiC/BSAS Composites Annealed at Various Temperatures for 100 h in Air ; $V_f = 0.32$

Annealing temp. (°C)	Density (g/cm³)	Weight change after annealing	E_c, GPa	σ_y, MPa	ε_y, %	σ_u, MPa
------	3.09 ± 0.03	–	137	122	0.091	759
550	3.12	–	145	155	0.108	853
800	3.06	–	150	138	0.096	814
900	3.16	–	151	171	0.114	769
1000	3.04	–	146	134	0.092	819
1100	2.90	–	142	143	0.102	736
1200	Deformed	+0.43%	–	–	–	–

* Measured at room temperature in 3-point flexure

> - Delamination of surface matrix layer in FRC
> - Ingression of oxygen into FRC
> - Oxidation of SiC fibers:
> $$SiC + O_2 \rightarrow SiO_2 + CO + CO_2$$
> - Reaction of SiO_2 and Celsian:
> $$Celsian + SiO_2 \rightarrow \text{Low m.p. ternary phase}$$
> - On cooling: formation of Celsian crystals in glass matrix

FIGURE 21. Proposed degradation mechanism of Hi-Nicalon/BSAS composites at 1200°C in air.

matrix layer delaminates from the composite ply underneath (SEM micrograph of Fig. 19), probably due to the large CTE mismatch between the Hi-Nicalon fibers and the celsian matrix. This facilitates the ingression of oxygen into the composite. This causes the oxidation of SiC fibers, particularly those which have lost the duplex CVD coating (see Fig. 10), resulting in the formation of SiO_2:

$$SiC(s) + O_2(g) \rightarrow SiO_2(s) + CO(g) + CO_2(g) \tag{3}$$

The silica formed reacts with celsian resulting in the formation of a low melting phase. The phase diagram of BaO-Al_2O_3-SiO_2 system (Fig. 22) [45] does show the presence of a ternary phase with a melting point of 1122°C. This phase is richer in SiO_2 but poorer in BaO and Al_2O_3 than celsian. Formation of gaseous by-products CO and CO_2 during reaction (3) results in the evolution of bubbles as observed in SEM micrograph of Fig. 19. On cooling, celsian crystals precipitate from the melt leaving behind a glassy matrix which is richer in SiO_2 and poorer in BaO and Al_2O_3 than celsian. SEM micrograph taken from the surface of CMC annealed at 1200°C showed the presence of elongated crystals in glassy matrix. From qualitative EDS analysis, these crystals were seen to be celsian and the glassy matrix was richer in SiO_2 but poorer in BaO and Al_2O_3 than celsian. The XRD analysis showed only celsian in the as-fabricated sample whereas celsian and an amorphous phase were present in the 1200°C-annealed CMC. Formation of a low-melting glass phase has also been reported [3] in the study of BSAS environmental barrier coating (EBC) on Si-based ceramic substrates such as CVD SiC and SiC_f/SiC composite. On heat treatment, the plasma sprayed BSAS coating reacted with the silica layer, formed from oxidation of the Si-based ceramic substrate, resulting in low melting ternary glass phase which was found to be richer in SiO_2 but poorer in BaO and Al_2O_3 than celsian.

CONCLUDING REMARKS

Hi-Nicalon/Celsian composites are stable up to use temperature of ~1100°C in oxidizing environments and degrade at higher temperatures due to the instability of polymer-derived fibers. The stability of Celsian matrix composites may be extended to higher temperatures by more uniform and stable interface coating(s) and by reinforcement with more advanced silicon carbide fiber (Sylramic) for applications as hot components (combustion liner, air foil, nozzle, etc.) in turbine engines.

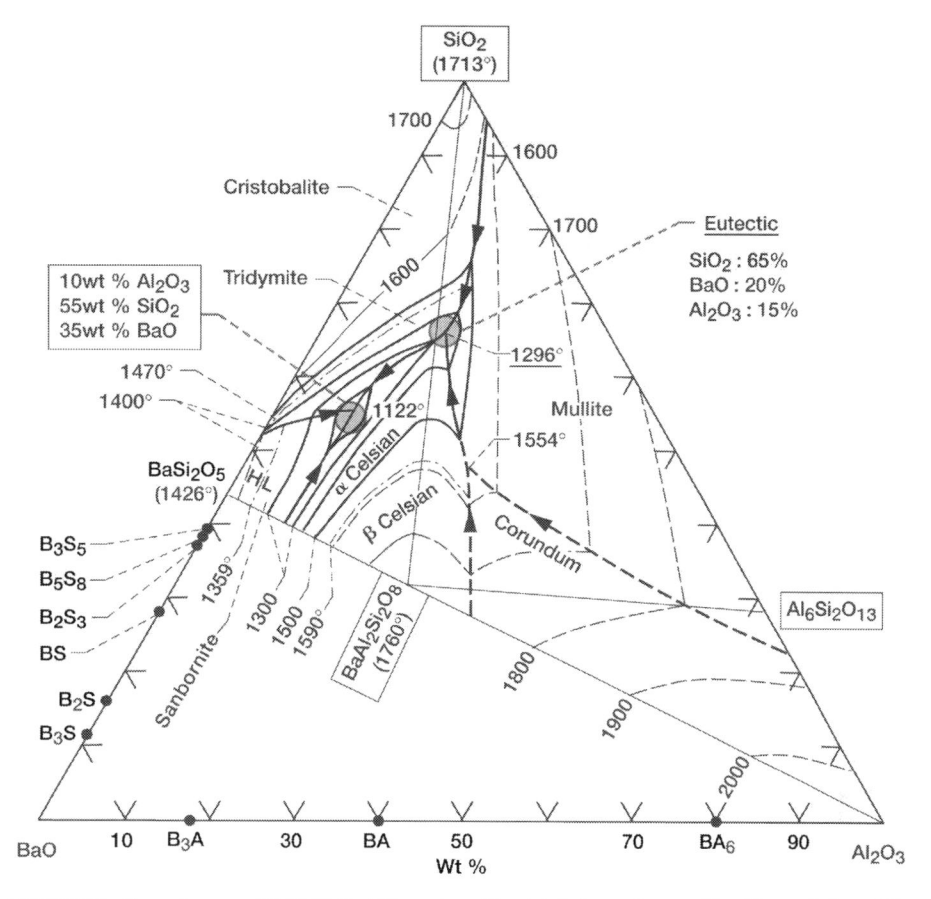

FIGURE 22. Phase diagram of BaO-Al$_2$O$_3$-SiO$_2$ system. (Reprinted, from reference 45, with permission of The American Ceramic Society).

ACKNOWLEDGMENTS

The author would like to thank his collaborators, Dr. Sung Choi, Dr. Jeff Eldridge, Dr. John Gyekenyesi, and Mr. Mark Hyatt of NASA Glenn, Dr. Ozer Unal (Ames Lab.), Prof. Charles Drummond (OSU), Prof. K.P.D. Lagerlof (CWRU), Dr. Rob Dickerson (Los Alamos Lab.), and Dr. Philippe Colomban (CNRS, France) who made valuable contributions to the Celsian matrix composites project over the last many years. John Setlock, Ralph Garlick, Ron Phillips, Bob Angus, and Ralph Pawlik provided valuable technical assistance during composite fabrication and characterization.

REFERENCES

1. K. M. Prewo, Fiber Reinforced Glasses and Glass-Ceramics, in *"Glasses and Glass-Ceramics"*, M. H. Lewis, Ed., Chapman and Hall, New York, NY, 1989; pp. 336–368.

2. J. J. Brennan, Glass and Glass-Ceramic Matrix Composites, in *"Fiber Reinforced Ceramic Composites. Materials, Processing and Technology"*, K. S. Mazdiyasni, Ed., Noyes Publications, Park Ridge, NJ, 1990, pp. 222–259.
3. K. N. Lee, D. S. Fox, J. I. Eldridge, D. Zhu, R. C. Robinson N. P. Bansal, and R. A. Miller, Upper Temperature Limit of Environmental Barrier Coatings Based on Mullite and BSAS, NASA/TM—2002-211372, March 2002. *J. Am. Ceram. Soc.*, **86** (2003).
4. H. E. Eaton, Jr., W. P. Allen, N. S. Jacobson, N. P. Bansal, E. J. Opila, J. L. Smialek, K. N. Lee, I. T. Spitsberg, H. Wang, P. J. Meschter, and K. L. Luthra, Silicon Based Substrate with Environmental/Thermal Barrier Layer, U. S. Patent 6,387,456; May 14, 2002.
5. N. P. Bansal, Ceramic Fiber-Reinforced Glass-Ceramic Matrix Composites, U. S. Patent 5,214,004; May 25, 1993.
6. N. P. Bansal, Method of Producing a Ceramic Fiber-Reinforced Glass-Ceramic Matrix Composite, U. S. Patent 5,281,559; January 25, 1994.
7. N. P. Bansal, Method of Producing a Silicon Carbide Fiber Reinforced Strontium Aluminosilicate Glass-Ceramic Matrix Composite, U. S. Patent # 5,389,321; February 14, 1995.
8. N. P. Bansal, Mechanical Behavior of Silicon Carbide Fiber-Reinforced Strontium Aluminosilicate Glass-Ceramic Composites, *Mater. Sci. Eng. A*, **231** [1–2] 117–127 (1997).
9. N. P. Bansal, CVD SiC (SCS-0) Fiber-Reinforced Strontium Aluminosilicate Glass-Ceramic Composites, *J. Mater. Res.*, **12** [3] 745–753 (1997).
10. N. P. Bansal, Influence of Fiber Volume Fraction on Mechanical Behavior of CVD SiC Fiber/$SrAl_2Si_2O_8$ Glass-Ceramic Matrix Composites, *SAMPE J. Advanced Mater.*, **28** [1] 48–58 (1996).
11. N. P. Bansal, CVD SiC Fiber-Reinforced Barium Aluminosilicate Glass-Ceramic Matrix Composites, *Mater. Sci. Eng. A*, **220** [1–2] 129–139 (1996).
12. N. P. Bansal, Celsian Formation in Fiber-Reinforced Barium Aluminosilicate Glass-Ceramic Matrix Composites, *Mater. Sci. Eng. A*, **342** [1–2] 23–27 (2003).
13. J. S. Moya Corral and G. Verduch, The Solid Solution of Silica in Celsian, *Trans. J. Br. Ceram. Soc.*, **77**, 40–44 (1978).
14. D. Bahat, Kinetic Study on the Hexacelsian-Celsian Phase Transformation, *J. Mater. Sci.*, **5**, 805–810 (1970).
15. B. Yoshiki and K. Matsumoto, High Temperature Modification of Barium Feldspar, *J. Am. Ceram. Soc.*, **34** [9], 283–286 (1951).
16. Y. Takeuchi, A Detailed Investigation of the Structure of Hexagonal $BaAl_2Si_2O_8$ With Reference to its α-beta Inversion, *Min. J. Japan*, **2** [5], 311–332 (1958).
17. R. E. Newnham and H. D. Megaw The Crystal Structure of Celsian (Barium Feldspar), *Acta Cryst.*, **13**, 303–312 (1960).
18. P. Gay, A Note on Celsian, *Acta Cryst.*, **9**, 474 (1956).
19. M. C. Guillem and C. Guillem, Kinetics and Mechanism of Formation of Celsian from Barium Carbonate and Kaolin, *Trans. J. Br. Ceram. Soc.*, **83**, 150–154 (1984).
20. N. P. Bansal, Solid State Synthesis and Properties of Monoclinic Celsian, *J. Mater. Sci.*, **33** [19] 4711–4715 (1998).
21. J. J. Buzniak, K. P. D. Lagerlof, and N. P. Bansal, Hot Pressing and High Temperature Mechanical Properties of $BaAl_2Si_2O_8$ (BAS) and $SrAl_2Si_2O_8$ (SAS), in Advances in Ceramic Matrix Composites (N. P. Bansal, Ed.), Am. Ceram. Soc., Westerville, OH; *Ceram. Trans.*, **38**, 789–801 (1993).
22. D. Zhu, N. P. Bansal, K. N. Lee, and R. A. Miller, Thermal Conductivity of Ceramic Thermal Barrier and Environmental Barrier Coating Materials, NASA/TM-2001-211122, Sept. 2001.
23. 23 N. P. Bansal and M. J. Hyatt, Crystallization Kinetics of Barium Aluminosilicate Glasses, *J. Mater. Res.*, **4** 1257 (1989).
24. M. J. Hyatt and N. P. Bansal, Crystal Growth Kinetics in $BaOAl_2O_3 2SiO_2$ and $SrOAl_2O_3 2SiO_2$ Glasses, *J. Mater. Sci.*, **31** [1] 172–184 (1996).
25. N. P. Bansal and C. H. Drummond III, Kinetics of Hexacelsian to Monoclinic Celsian Phase Transformation in $SrAl_2Si_2O_8$, *J. Am. Ceram. Soc.*, **76** [5] 1321–1324 (1993).
26. N. P. Bansal and J. A. Setlock, Fabrication of Fiber-Reinforced Celsian Matrix Composites, *Comp. Part A: Applied Science and Manufacturing*, **32** [8] 1021–1029 (2001).
27. N. P. Bansal, Strong and Tough Hi-Nicalon Fiber-Reinforced Celsian Matrix Composites, *J. Am. Ceram. Soc.*, **80** [9] 2407–2409 (1997).

28. G. Gouadec, P. Colomban, and N. P. Bansal, Raman Study of Hi-Nicalon Fiber-Reinforced Celsian Composites. Part 1: Distribution and Nanostructure of Different Phases, *J. Am. Ceram. Soc.*, **84** [5] 1129–1135 (2001).
29. N. P. Bansal, P. H. McCluskey, G. D. Linsey, D. Murphy, and G. Levan, Processing and Properties of Nicalon Reinforced Barium Aluminosilicate (BAS) Glass-Ceramic Matrix Composites, in *Ceramic Matrix Composites for Rocket Nozzle, Leading Edge, and Turbine Applications*, Ed. M. M. Opeka, DoD Ceramics Information Analysis Center, West Lafayette, IN; pp. 335–358 (1995).
30. N. P. Bansal, P. H. McCluskey, G. D. Linsey, D. Murphy, and G. Levan, Nicalon Fiber-Reinforced Celsian Glass-Ceramic Composites, in "HITEMP Review 1995 – Advanced High Temperature Engine Materials Technology Program. *Volume III: Turbine Materials —CMC's, Fiber and Interface Issues*", NASA CP 10178, pp. 41–1 to 41–14 (1995).
31. G. D. Linsey, P. McCluskey, D. Murphy, and G. Levan, Processing and Properties of Barium Aluminosilicate Glass-Ceramic Matrix Composites, NASA CR 198369, August 1995.
32. S. R. Choi, N. P. Bansal, and J. P. Gyekenyesi, Ultimate Tensile Strength as a Function of Test Rate for Various Ceramic Matrix Composites at Elevated Temperatures, NASA/TM-2002-211579, June 2002.
33. N. P. Bansal and J. I. Eldridge, Hi-Nicalon Fiber-Reinforced Celsian Matrix Composites: Influence of Interface Modification, *J. Mater. Res.*, **13** [6] 1530–1537 (1998).
34. Ö. Ünal and N. P. Bansal, Temperature Dependency of Strength of a Unidirectional SiC Fiber-Reinforced (Ba, Sr)$Al_2Si_2O_8$ Celsian Composite, in *Advances in Ceramic Matrix Composites IV* (J. P. Singh and N. P. Bansal, Eds.), Am. Ceram. Soc., Westerville, OH; *Ceram. Trans.*, **96**, 135–147 (1999).
35. N. P. Bansal and J. I. Eldridge, Effects of Fiber/Matrix Interface and its Composition on Mechanical Properties of Hi-Nicalon/Celsian Composites, NASA/TM-1999-209057, March 1999; *Proc. ICCM-12 Conference*, Paper No. 147, Paris, France, July 1999; ISBN 2-9514526-2-4.
36. J. Z. Gyekenyesi and N. P. Bansal, High Temperature Mechanical Properties of Hi-Nicalon Fiber-Reinforced Celsian Composites, in *Advances in Ceramic Matrix Composites V* (N. P. Bansal, J. P. Singh, and E. Ustundag Eds.), Am. Ceram. Soc., Westerville, OH; *Ceram. Trans.*, **103**, 291–306 (2000).
37. J. Z. Gyekenyesi and N. P. Bansal, High Temperature Tensile Properties of Unidirectional Hi-Nicalon/Celsian Composites in Air, NASA/TM-2000-210214, July 2000.
38. Ö. Ünal and N. P. Bansal, Interlaminar Shear Strength of a Unidirectional Fiber-Reinforced Celsian Composite by Short-Beam and Double-Notched Shear Tests, *Ceram. Eng. Sci. Proc.*, **22** [3] 585–595 (2001).
39. Ö. Ünal and N. P. Bansal, In-Plane and Interlaminar Shear Strength of Silicon Carbide Fiber-Reinforced Celsian Composite, *Ceram. Int.*, **28** [5] 527–540 (2002).
40. S. R. Choi, N. P. Bansal, and J. P. Gyekenyesi, Rate Dependency of Shear Strength in SiC_f/BSAS Composite at Elevated Temperature, *Ceram. Eng. Sci. Proc.*, **24** [4] 435–441 (2003).
41. S. R. Choi, N. P. Bansal, and J. P. Gyekenyesi, Dependency of Shear Strength on Test Rate in SiC/BSAS Ceramic matrix Composite at Elevated Temperature, NASA/TM-2003-212182, April 2003.
42. N. P. Bansal, Mechanical Properties of SiC Fiber-Reinforced Celsian Composites After High-temperature Exposures in Air, in *Proc. 8th International Conference on Composites Engineering (ICCE-8),* D. Hui, Editor, Tenerife, Spain, August 5–11 (2001); p. 59.
43. N. P. Bansal, Effects of Thermal Ageing in air on Microstructure and Mechanical Properties of Hi-Nicalon Fiber-Reinforced Celsian Composites, unpublished work.
44. N. P. Bansal, M. J. Hyatt, and C. H. Drummond III, Crystallization and Properties of Sr-Ba Aluminosilicate Glass-Ceramic Matrices, *Ceram. Eng. Sci. Proc.*, **12** [7–8] 1222–1234 (1991).
45. E. M. Levin and H. F. McMurdie, Phase Diagram for Ceramists, Vol. III, Fig. 4544, p. 220 (1975); The Am. Ceram. Soc., Westerville, OH.

11

In situ Reinforced Silicon Nitride – Barium Aluminosilicate Composite

Kenneth W. White, Feng Yu and Yi Fang
Department of Mechanical Engineering
University of Houston
Houston, TX 77204-4006
kwwhite@uh.edu
Tel.: (713) 743-4526

Processing and properties of a low cost, high performance *in situ* reinforced silicon nitride – barium aluminosilicate (BAS) composite are presented in this chapter. BAS glass-ceramic serves as an effective liquid-phase-sintering aid, to attain full densification and complete the α–β Si_3N_4 phase transformation. Si_3N_4 whiskers grow in random directions in a near completely crystallized matrix of hexacelsian BAS. Mechanical properties of this composite are comparable with those of conventional hot pressed or HIPed dense silicon nitride ceramics. The in situ reinforced silicon nitride – BAS composite is an ideal candidate for applications requiring a combination of high strength, high toughness, low dielectric constant, low loss tangent, high thermal shock resistance, high thermal stability, low thermal expansion and low cost.

1. INTRODUCTION

The increasing demand for high performance materials in aerospace and other structural applications, where a combination of high temperature strength and resistance to environmental degradation are important, has led to the development of a variety of ceramic materials based on Si_3N_4, Al_2O_3 and SiC[1;2]. Among these ceramics, silicon nitride (Si_3N_4) is one

TABLE 1. Some typical properties of dense Si_3N_4 and three other popular engineering ceramics (after Ziegler, et al.,[4] Lee and Rainforth[16] and Becher and Warwick[72])

	Dense Si_3N_4	Dense Al_2O_3	ZrO_2	SiC
Theoretical density (g/cm^3)	3.2–3.9	3.96–3.98	5.9–6.1	3.08–3.20
Microhardness (GPa)	14–18	12.8–19.3	8–17	21–25
Bend strength at room temperature (MPa)	400–1200	230–600	400–1000	350–640
K_{IC} at room temperature (MPa$\sqrt{m}$)	3.4–11.2	3.8–4.5	6–20	3.0–5.7
Thermal expansion coefficient 0–1000°C (10^{-6}/K)	2–3	6.5–8.9	6.8–10.6	3.7–4.9
Elastic modulus (GPa)	280–320	300–410	170–220	280–450
Thermal shock behavior*, Critical ΔT (°C)	850	225	300	N/A
Decomposition temperature (°C)	1870	N/A	N/A	2100
Melting temperature (°C)	N/A	2000–2050	>2600	N/A

* measured under 22°C water quench and sample thickness is 3 mm

the most promising materials for high temperature engineering applications due to its high strength at high temperature, low specific gravity, high hardness, good thermal shock resistance, and relatively good resistance to oxidation when compared to other high temperature structural materials[3–8].

Table 1 gives some typical properties of dense Si_3N_4. For comparison, properties of Al_2O_3, ZrO_2, and SiC are also included.

Although silicon nitride was discovered nearly a century ago, it is only in the last three decades that its potential as an engineering material has been explored. Significant efforts within both academia and industry to put this ceramic into practical applications, has resulted in many successful applications of silicon nitride, used as cutting tools, grinding media and wear components. Si_3N_4 components have been adopted in commercial applications and include: auto turbochargers, Diesel engines, hydraulic pumps, among others[9–12].

Silicon nitride, however, as with other engineering and electronic ceramics, suffers from relatively poor reliability. In addition, high processing pressures are often required for fabrication of dense silicon nitride, which incurs a manufacturing cost penalty because of the need for specialized equipment and restricts the geometric shapes of components. Therefore, the challenge that ceramists are facing in applying this material in a broader range is mainly three-fold: increasing the reliability to satisfy design criteria for practical applications, improving the reproducibility, and lowering the manufacturing cost. Currently, the most satisfactory approach is to design the microstructure of silicon nitride ceramics with improved resistance to fracture.

To economically fabricate complex-shaped components from dense silicon nitride near-net-shape, pressureless sintering is preferred to minimize machining and tooling costs. However, the inherent sluggish densification of silicon nitride typically requires more sintering additives than pressure-assisted sintering methods. Barium aluminum silicate (BAS) has been shown to serve both as a sintering additive, as well as a matrix material, yielding a dense silicon nitride-based ceramic matrix composite by pressureless sintering, which enables the elimination of costly processing steps.

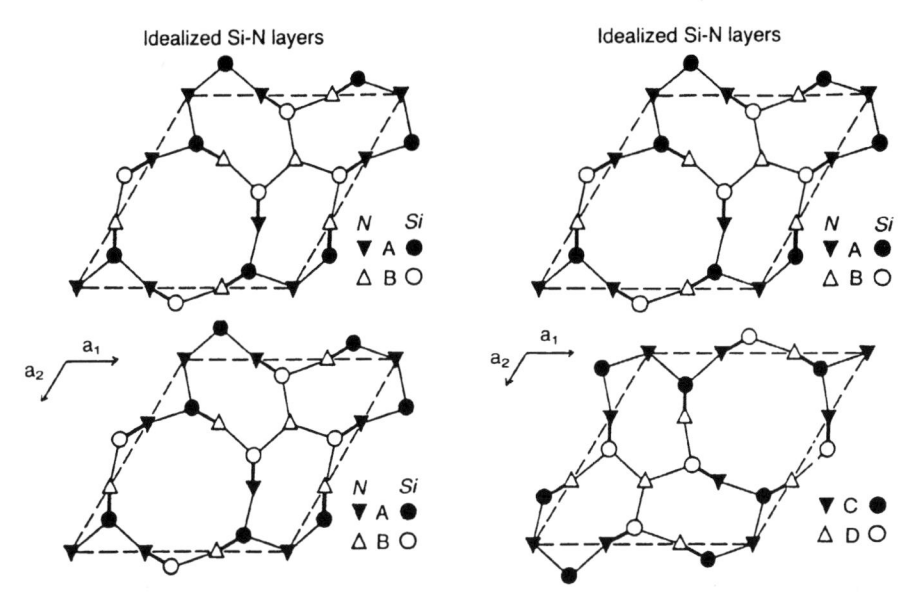

FIGURE 1. (a) The AB and AB layers in the β-Si$_3$N$_4$ and (b) the AB and CD layers in the α-Si$_3$N$_4$ crystal structure[16].

2. COMPOSITE COMPONENTS

2.1. Silicon Nitride

Silicon nitride has two major polymorphs α and β for engineering application, in addition to the amorphous and the cubic polymorph first reported in 1999[13]. Both silicon nitride (Si$_3$N$_4$) and its form in solid solution (SiAlON) have been reviewed in detail[12;14].

Based on powder XRD, Hardie and Jack[15] report that both α and β forms are hexagonal with a c-dimension of α approximately twice as that of the β polymorph. The Si atom in Si$_3$N$_4$ is in a hybridized state of sp^3 configuration producing a tetrahedral arrangement of valence orbital electrons bonding covalently with four N atoms. The bonding between Si-N is about 70% covalent and 30% ionic and the building blocks of both α and β-Si$_3$N$_4$ are SiN$_4$ tetrahedra, similar to the SiO$_4$ tetrahedra of silicate glasses. As presented by Lee and Rainforth[16], both α and β-Si$_3$N$_4$ can be considered as three-dimensional assemblies SiN$_4$ tetrahedra. The tetrahedra share corners in such a way that each N corner is common to three tetrahedra. The structures can be regarded as consisting of layers of Si and N atoms in the sequence ABABA ... or ABCDABCDA ... along the c-axis for the β and α phases respectively (Figure 1). The AB layer is identical in both the α and β phases and the CD layer in the α phase is related to the AB layer by a 180° rotation about an axis perpendicular to c axis so that the c spacing of α-Si$_3$N$_4$ is approximately twice as that of β. Crystallographic data for the silicon nitride polymorphs are summarized in Table 2.

TABLE 2. Crystallographic data for α and β Si_3N_4 polymorphs[16]

Polymorph	Space Group	Lattice Parameters (Å)	JCPDS Card
α-Si_3N_4	P31c	a = 7.7553 c = 5.618	41–360
α-sialon	P31c	Composition dependent	33–261
β-Si_3N_4	$P6_3$ or $P6_3/m$	a = 7.595 c = 2.9023	33–1160
β-sialon	$P6_3$ or $P6_3/m$	Composition dependent	41–1013

2.2. Barium Aluminum Silicate

BAS, based on the composition, $BaO \cdot Al_2O_3 \cdot 2SiO_2$, has been studied for decades, both in its synthetic and in its naturally occurring forms[17–22]. The location of the BAS in the general ternary system BaO-Al_2O_3-SiO_2 is depicted in Figure 2. Four distinct crystalline modifications of $BaAl_2Si_2O_8$ have been recognized. Celsian and paracelsian, both monoclinic, are known as natural minerals of limited occurrence[23]. The remaining two, hexagonal hexacelsian and orthorhombic form, are encountered only in synthetic products[24]. Crystallographic data for the BAS polymorphs are reviewed in Table 3.

The crystal structure of the hexagonal type of BAS was identified by Ito using X-ray methods[17]. The structure of hexacelsian (Figure 3) contains infinite two-dimensional hexagonal sheets consisting of two layers of silica tetrahedra sharing all four vertices. The double sheet of linked oxygen tetrahedra around Si (and Al), with the composition (Si, Al)O_2, is parallel to (0001). Barium atoms occupy the positions between such double sheets and

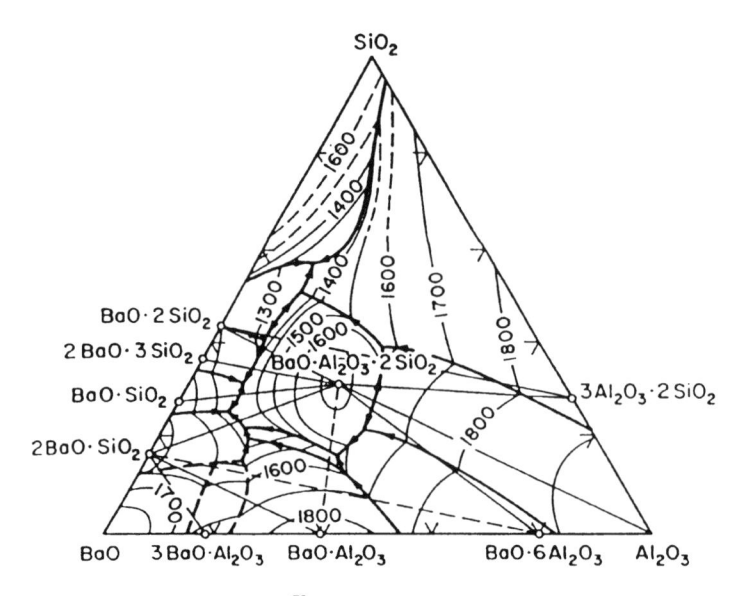

FIGURE 2. BaO–Al_2O_3–SiO_2 phase diagram[73] (Reprinted with permission of The American Ceramic Society, www.ceramics.org. Copyright [1975]. All rights reserved.)

TABLE 3. Crystallographic data for BAS polymorphs

Polymorph	Space Group	Lattice Parameters (Å)	JCPDS Card
Monoclinic Phase (Celsian)	C2/m	a = 8.641, b = 13.047, c = 7.203, β = 115.08°	38–1450
Hexagonal Phase (Hexacelsian)	P6/mm	a = 5.313, c = 7.805	12–726
Orthorhombic Phase	C	a = 5.293, b = 9.168, c = 7.790	12–725

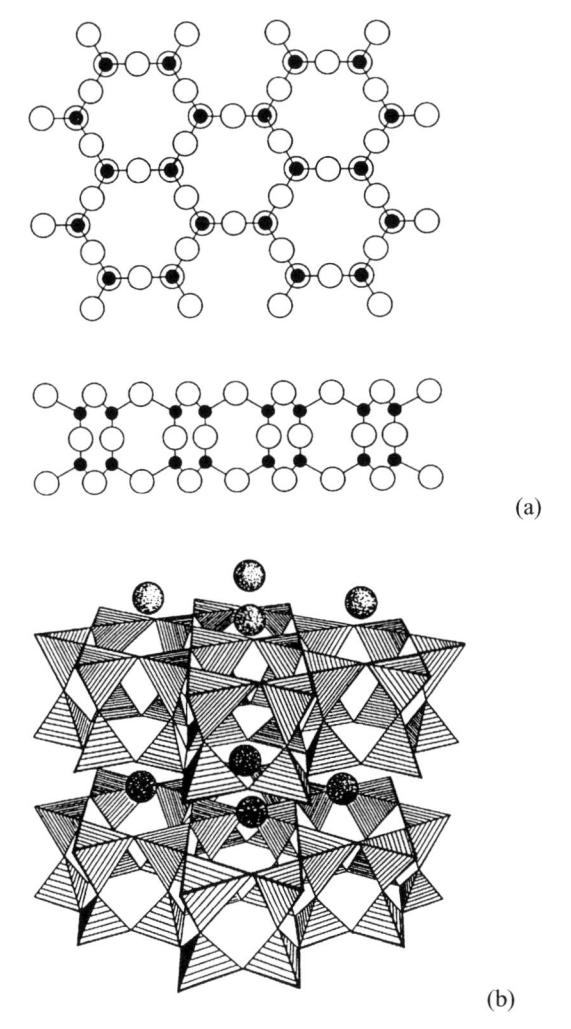

FIGURE 3. (a) Hexacelsian structure. Black spheres are Al or Si atoms and blank circles represent O atoms, and (b) 3-dimensional schematic showing the structure of hexacelsian. Large spheres represent Ba atoms and shadowed tetrahedra represent Si(Al)O$_4$ tetrahedra.[24] (Reprinted with permission of The American Ceramic Society, www.ceramics.org. Copyright [1951]. All rights reserved.)

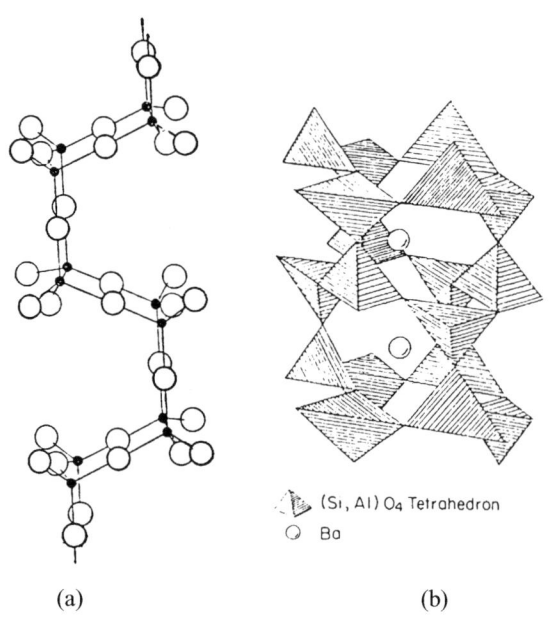

FIGURE 4. (a) The chain consisting of the broken hexagonal rings of silicon oxygen tetrahedra in celsian, (b) 3-dimentional schematic showing the celsian structure[21] (Reprinted with permission of The American Ceramic Society, www.ceramics.org. Copyright [1989]. All rights reserved.)

hold them together to make up the bulk of the structure. For the orthorhombic structure, it has been stated by Takeuchi[25], "No significant differences are observed between the powder XRD patterns of the orthorhombic and hexagonal hexacelsian except peak shifts owing to lattice expansion, indicating that the fundamental frameworks of both structures are the same". The only difference between the orthorhombic and hexagonal structures is a minor adjustment in oxygen atom positions. The structure of monoclinic celsian[26] contains a chain of rings stretched along the a-axis. These rings, each consisting of four linked tetrahedra of oxygen atoms, are superimposed upon each other and linked by common oxygen to form a zigzag chain of the peculiar type (Figure 4). As in hexacelsian, the Al substitutes for Si with charge compensation by Ba in the larger interstices of the structure.

The polymorphism of stoichiometric BAS at atmospheric pressure is represented graphically in Figure 5. Below the melting temperature of 1760°C, BAS first undergoes a solid state phase transformation at 1590°C, where the stable hexagonal BAS (hexacelsian) transforms into the monoclinic phase (celsian), which remains thermodynamically stable through room temperature. However, this hexagonal to monoclinic reconstructive transformation is extremely sluggish, causing hexagonal BAS to persist in a metastable state at temperatures below 1590°C. Guillem, et. al.[27] suggested that it requires creation of a three-dimensional network from the two-dimensional sheet structure of hexacelsian as well as re-arrangement of the Ba sites to convert hexacelsian to celsian. This transformation also entails breaking and reforming Al-O and Si-O bonds. Therefore, there is a kinetic barrier to the nucleation of the celsian phase from hexacelsian. Upon undercooling, the metastable hexacelsian will transform rapidly into an orthorhombic phase at approximately 300°C. This reversible

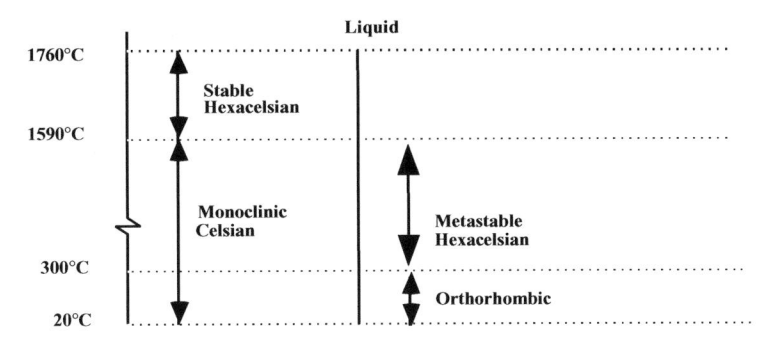

FIGURE 5. Schematic representation of the possible phase transitions in the BAS system under atmospheric conditions[33] (Reprinted with permission of The American Ceramic Society, www.ceramics.org. Copyright [2001]. All rights reserved.)

transformation is accompanied by about a 3% volume change[24;25] that is usually destructive. The persistence of hexacelsian phase below 1590°C, therefore, is generally regarded as an undesirable feature of the BAS ceramic systems.

Because of the destructive character of the phase transformation from hexacelsian to orthorhombic, care should be exercised to avoid hexacelsian. In monolithic BAS, various

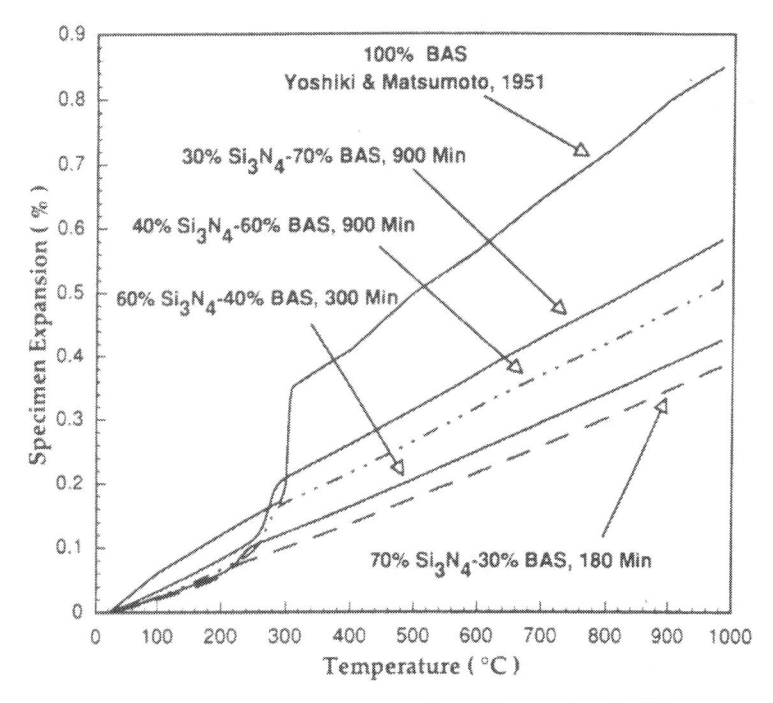

FIGURE 6. Linear expansion as a function of temperature of SN-BAS composite with different amounts of BAS, pure BAS is also included as the reference[31]

methods, such as seeding with celsian particles, adding mineralisers such as Li_2O, to the melt and hot pressing, have been tried with limited success in promoting the formation of celsian[20–22;28;29]. In addition, neither electron diffraction nor XRD analysis has been effective for detecting the hexacelsian to orthorhombic BAS transformation, since only small changes in the O-atom positions within a common framework characterize this transformation[25]. Using high-temperature XRD techniques, Quander, et. al.[30] monitored the phase transformation in a Si_3N_4–BAS composite in situ up to 550°C and observed no significant change in the XRD spectrum with the test temperature. However, their dilatometry studies identified the potential existence of this transformation in Si_3N_4 with a high BAS content[30;31]. As shown in Fig. 6, the dilatometric indication of transformation for the 30% Si_3N_4–70% BAS composite was significant, whereas that for the 70% Si_3N_4–30% BAS composite was almost negligible. The magnitude of the volume change also decreased significantly as the degree of α-Si_3N_4–β-Si_3N_4 phase transformation increased[31]. All these features indicate that the presence of more β-Si_3N_4 whiskers had a tendency to preclude the hexacelsian-to-orthorhombic transformation. Richardson, et. al.[32] conducted thermal fatigue tests on both 30vol%-Si_3N_4 – 70vol%-BAS and 70vol%-Si_3N_4–30vol%-BAS composites, between room temperature and 600°C. They reported no decrease in strength after thermal cycling with the 70-vol%-Si_3N_4 composite, which indicated that either this transformation was suppressed or it was no longer destructive.

3. DEVELOPMENT OF Si_3N_4–BAS COMPOSITE

3.1. Processing

The composite starting powder is prepared by mixing the BAS constituent powders ($BaCO_3$, Al_2O_3 and SiO_2 powders[30;32;33]) with Si_3N_4 powder. Composites with compositions with various ranges, from 90% BAS: 10% Si_3N_4 to 30%BAS: 70% Si_3N_4 have been processed. Focus has been centered on the 30%BAS: 70% Si_3N_4 composition previously developed[31;32], for ease of sinterability and avoiding the adverse BAS phase transformation. The well blended starting powders are pressed into desired shape through conventional pressing methods, such as uniaxial pressing or cold-isostatic-pressing. During pressureless sintering, a Si_3N_4-based powder bed and a Nitrogen atmosphere are necessary to prevent the decomposition of Si_3N_4 when the temperatures are close or above 1870°C. Temperatures higher than 1950°C are avoided during sintering because of unacceptable surface degradation[34].

3.2. Densification and Phase Transformation

Densification is not detected at temperatures below 1200°C since only solid state reactions are observed[35]. It is found that the composite densifies quickly between 1650°C and 1740°C (Figure 7) which is attributed to the formation of a liquid phase. The densification process is almost completed at sintering temperatures below the melting temperature of crystallized BAS due to the constitutive BAS powder used in the process.

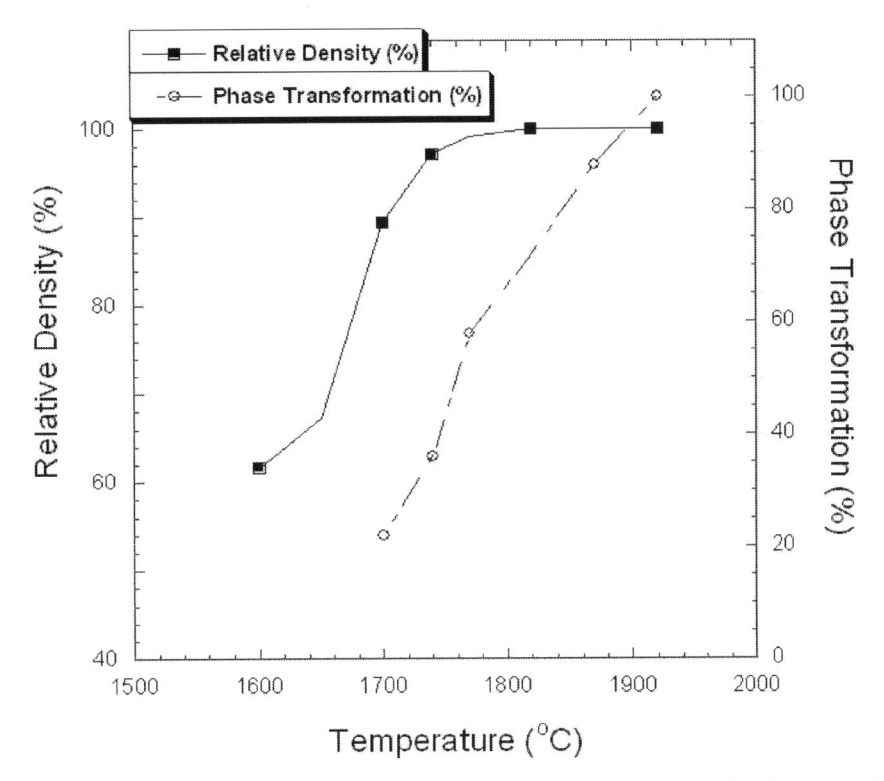

FIGURE 7. Relative density and weight percentage of α to β-Si_3N_4 phase transformation of as-sintered samples, as a function of sintering temperature (with a constant sintering time for 30 min, relative density and β silicon nitride content in green body are ~55% and ~5% respectively)

The α to β-Si_3N_4-phase transformation requires the presence of a liquid phase, through which the reconstructive transformation may occur[36]. For the Si_3N_4-BAS composite, the α-Si_3N_4 dissolves into a solution of liquid BAS[37] and precipitates on the pre-existing β particles as shown in Figure 8. As the heterogeneous nucleation is considered the dominant nucleation mechanism during the α to β-Si_3N_4-phase transformation[6;7], homogeneous nucleation of β-Si_3N_4 grain is not expected, due to large liquid phase presence which limits the local supersaturation of Si and N[38].

The progress of the α to β-Si_3N_4-phase transformation can be monitored through the change of the X-ray peak intensity of both α and β-Si_3N_4. Stronger peaks from β-Si_3N_4 are observed with longer sintering time and higher temperatures. Quantitative analysis of the phase transformation is calculated using the method proposed by Gazzara and Messier[39] which also is presented in Figure 7. Feng, et. al. find that, with a constant sintering time for 30 minutes, the phase transformation becomes detectable at temperatures lower than 1760°C (the melting temperature for BAS), while transformation is accomplished in the samples that are processed at 1950°C. Study is also performed on fixed sintering temperature at 1920°C and different dwelling times. Similar behavior was observed, achieving a relative density of

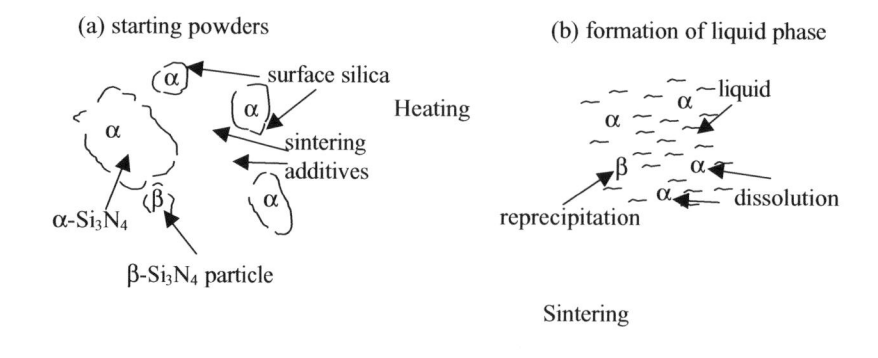

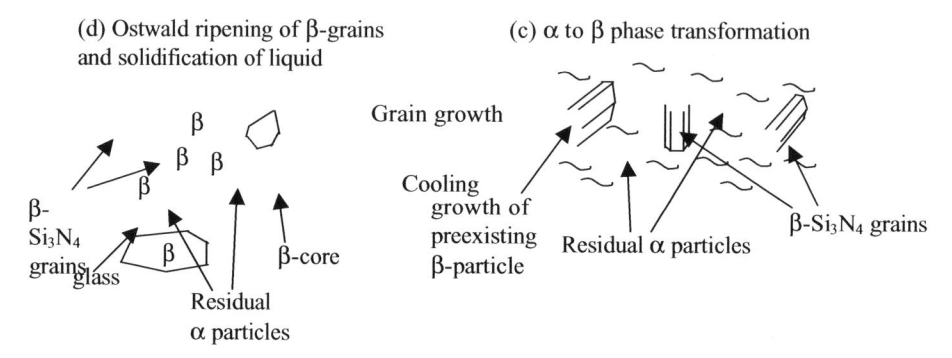

FIGURE 8. Schematic of solution-precipitation mechanism in sintering of Si_3N_4 (after Lee and Rainforth[16])

97%, with ~68% of the α-Si_3N_4 being transformed to β-Si_3N_4 after sintering only 5 min at 1920°C. When the sintering time was extended to 120 min, no residual α-Si_3N_4 could be detected by XRD.

3.3. Microstructure Evolution and Whisker Morphology

The evolution of Si_3N_4-BAS composite is similar to other silicon nitride ceramics, normally comprising four different stages: densification due to the formation of BAS liquid phase, nucleation of the β-Si_3N_4 particles, growth of β grains during the α to β-Si_3N_4 phase transformation, and subsequently β-Si_3N_4 grain growth. This process is shown schematically in Figure 8.

Figure 9 demonstrates the typical microstructures of Si_3N_4–BAS composite with a complete α to β transformation. No visible pores are observed, and the randomly oriented β-Si_3N_4 whiskers in a continuous BAS matrix, is expected to promote isotropic mechanical performance. The average aspect ratio of the whiskers in most sintering conditions described previously are larger than 10. Statistical analysis suggests a normal distribution of whisker width/length for all samples. Grain coarsening is evident with the extension of sintering time (Figure 10), however, the rate of diameter increase is small. Longer sintering times promote growth of all whiskers, which shifts and broadens the entire distribution curve.

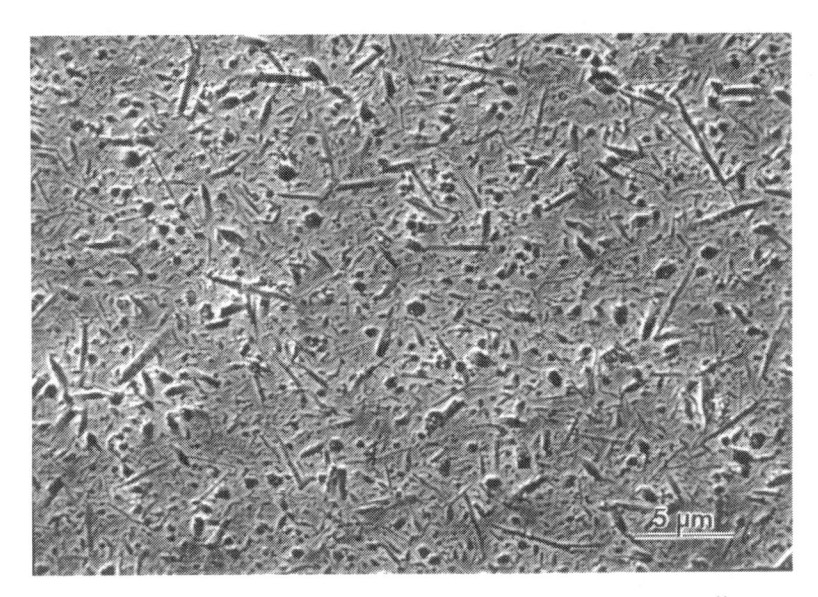

FIGURE 9. SEM micrograph showing the whisker morphology of Si_3N_4-BAS composite[33] (Reprinted with permission of The American Ceramic Society, www.ceramics.org. Copyright [2001]. All rights reserved.)

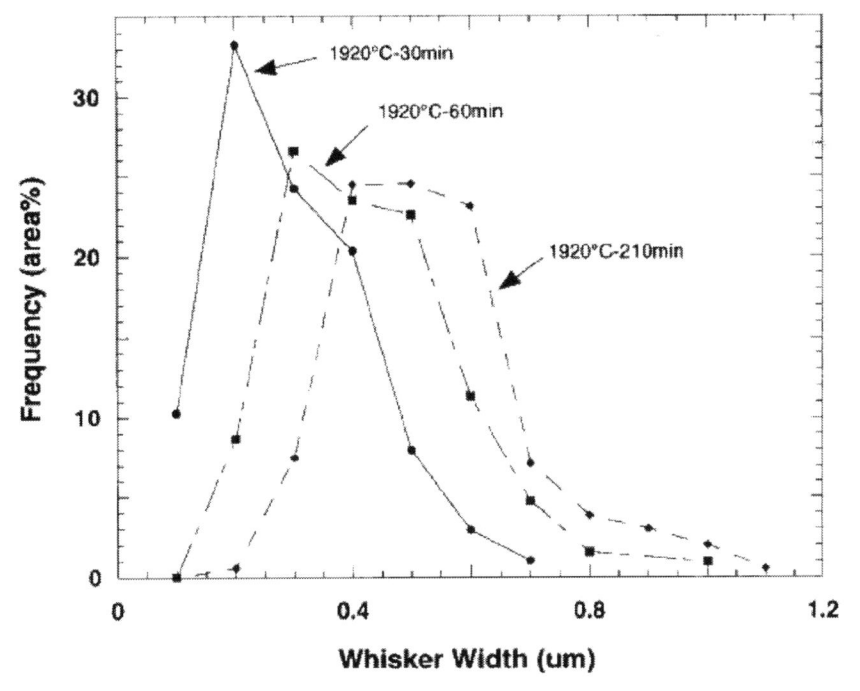

FIGURE 10. Distribution of whisker width for the composites sintered at 1920°C for 30, 60 and 210 min[33] (Reprinted with permission of The American Ceramic Society, www.ceramics.org. Copyright [2001]. All rights reserved.)

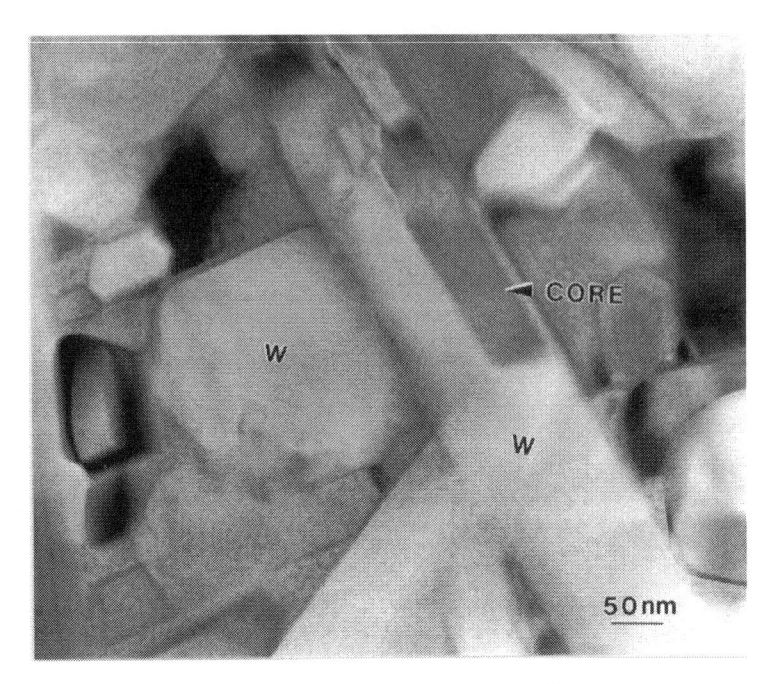

FIGURE 11. TEM micrograph of a $\beta-Si_3N_4$ grain with a core-rim structure[33] (Reprinted with permission of The American Ceramic Society, www.ceramics.org. Copyright [2001]. All rights reserved.)

Several authors[40;41] report that the $\beta-Si_3N_4$ component in the Si_3N_4 – BAS system as $\beta-SiAlON$. The amount of aluminum substitution in the $\beta-SiAlON$ in this system depends on the amount of alumina available and processing conditions. Higher sintering temperatures and longer processing times can encourage the dissolution of aluminum into the $\beta-Si_3N_4$ lattice. TEM observation[33] has revealed the characteristic core-rim structure of a $\beta-Si_3N_4$ grain in this composite as shown in Figure 11, which is believed due to the formation of $\beta-SiAlON$. EDS analysis confirms the difference in aluminum content between the core and the rim, where the virtually aluminum free core is believed to be pre-existing $\beta-Si_3N_4$ nuclei. The aluminum content within the rim structure, resulting from the formation of a SiAlON solid solution, varies with the location and size of the whiskers[42–44].

3.4. Crystallization of BAS Matrix

As noted, the BAS glass-ceramic serves as a liquid phase sintering aid for the composite and remains as a structural matrix. Volume fraction as high as 30% are used, in contrast to sintering aid contents of typically <10% in conventional pressure-sintered silicon nitride ceramic. Therefore, a pronounced influence on the composite properties can be expected. Although, BAS glass is one of the most refractory in its family with a softening point of 925°C[20], a crystallized BAS matrix still would offer additional improvement for the composite high-temperature properties.

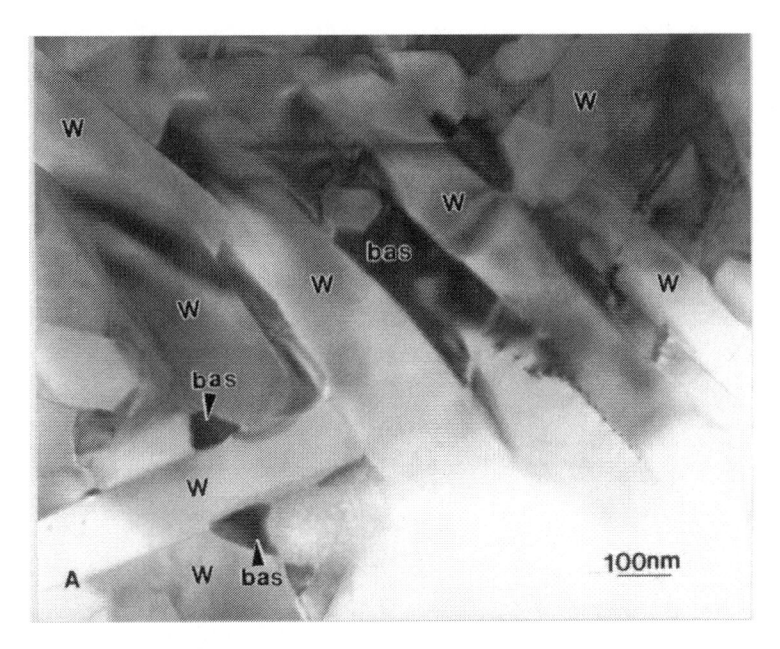

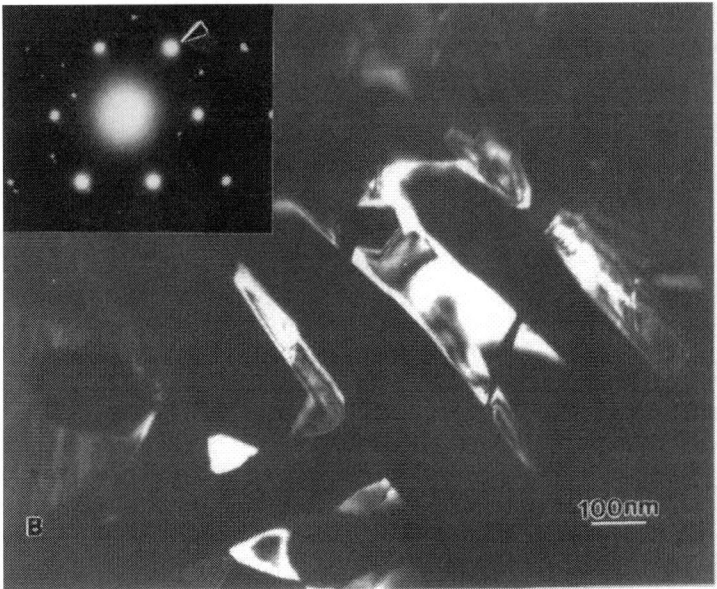

FIGURE 12. (a) Bright field and (b) dark-field TEM images showing the almost complete crystallization of BAS matrix[33] (Reprinted with permission of The American Ceramic Society, www.ceramics.org. Copyright [2001]. All rights reserved.)

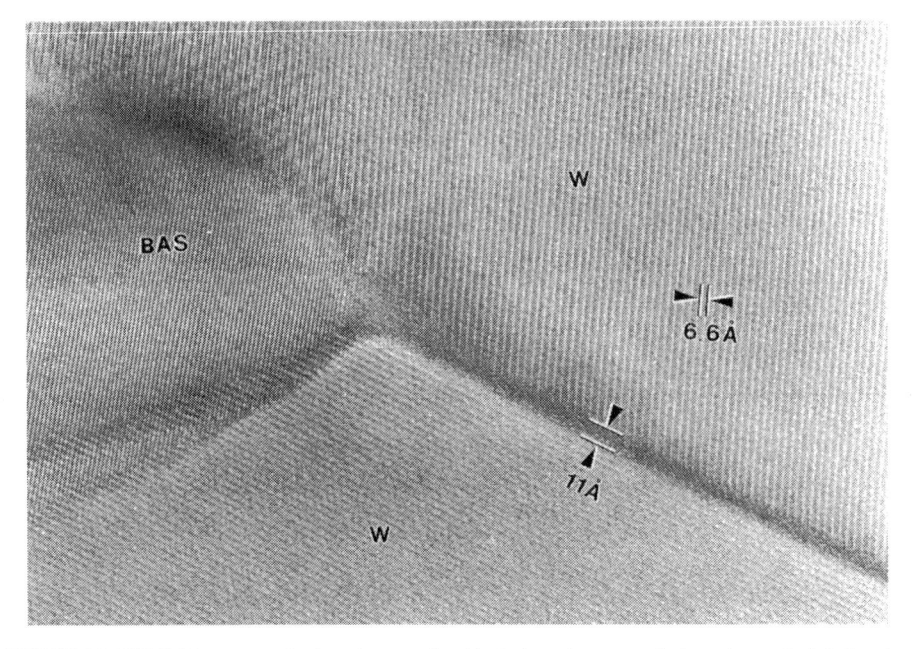

FIGURE 13. HRTEM image reveals the existence of residual glass phase at grain boundary and triple junction area. (W: $\beta-Si_3N_4$ whisker)[33] (Reprinted with permission of The American Ceramic Society, www.ceramics.org. Copyright [2001]. All rights reserved.)

Extensive studies by Bansal and his colleagues have shown the monolithic $BaO \cdot Al_2O_3 \cdot 2SiO_2$ glass always to crystallize into hexacelsian phase first, before the appearance of celsian phase[20;21;45]. When used as a matrix in Si_3N_4-BAS composites, it is shown that the formation of celsian is suppressed and hexacelsian persists, even with only 30 vol% Si_3N_4[30;31]. TEM studies identify nearly complete crystallization of the BAS matrix in all Si_3N_4-BAS samples[33;46]. The bright-field TEM image, Figure 12(a) shows the microstructure of $\beta-Si_3N_4$ whiskers and a BAS matrix of a sample sintered at 1920°C for 120 min. Figure 12(b) is the corresponding dark-field image, obtained using a $[1\bar{1}01]$ diffracted beam from the BAS$[1\bar{2}1\bar{3}]$ diffraction pattern (see diffraction pattern inset). Using HRTEM, some residual glassy phase was observed (Figure 13), predominantly at triple-grain junctions of whiskers and BAS grains. Also, an interlayer of glassy phase (~1 nm thick) persisted between whiskers. No residual glass is found between BAS grains.

4. COMPOSITE PROPERTIES

4.1. Hardness and Elastic Modulus

The modulus of the fully densified composite is measured by various methods including: tensile test, ultrasonic and impulse excitation of vibration. The results of all of these methods

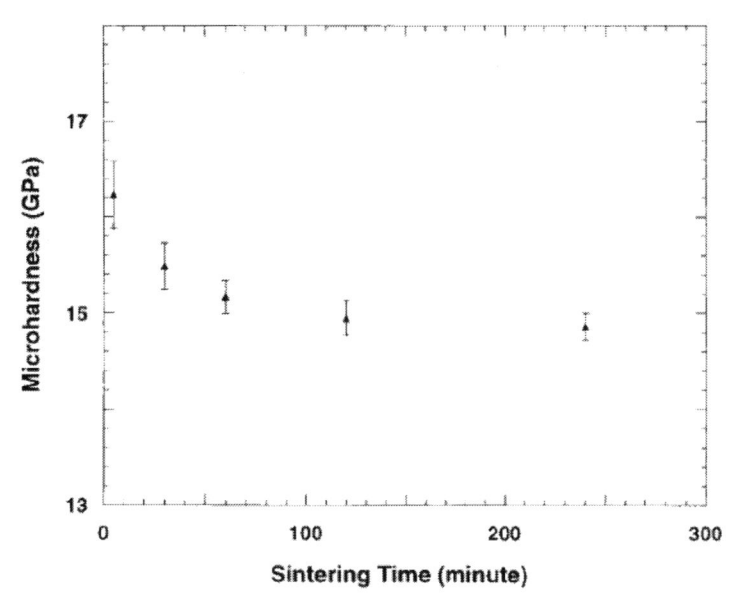

FIGURE 14. Vicker's hardness, as a function of sintering time at 1920°C[33] (Reprinted with permission of The American Ceramic Society, www.ceramics.org. Copyright [2001]. All rights reserved.)

fall between $220 \sim 245$ GPa[32;33]. The average shear modulus is 90 GPa, the bulk modulus is about 156 GPa, and Poisson's ratio is 0.256[32].

The Vicker's hardness, which is in the range of $14.9 \sim 16.2$ GPa, changed with differences in the phase composition and microstructure. This value is somewhat less than that of conventional dense silicon nitride 17–20 GPa, probably due to the 30 vol% BAS content. The decreasing hardness with sintering time at 1920°C (shown in Figure 14) is attributed to the progressive transformation of $\alpha-Si_3N_4$ to the softer β-grains[47]. After phase transformation is complete, the hardness decrease is governed only by changes in grain size.

4.2. Strength and Toughness

It is well known that the needle-like $\beta-Si_3N_4$ grains contribute to the high strength and toughness of the silicon nitride materials[48–51]. With the equiaxed $\alpha-Si_3N_4$ microstructure transforming into a $\beta-Si_3N_4$ structure characterized by elongated grains, the strength, toughness increase significantly. Therefore, the Si_3N_4-BAS composite with high strength and toughness is not practical without a fully α to β Si_3N_4 transformation. Sintering at 1800°C, the strength of a fully densified 70% Si_3N_4–30% BAS is only about 300 MPa when the silicon nitride transformation is at only 40%[30]. Extending the sintering time to 7 hours results a near-complete phase transformation (94%), however, the strength of the composite is still not very interesting (at 490 MPa)[32]. When the sintering temperature is increased to above 1900°C, the composite shows a significant improvement in strength[52]. The composite strength is well above 600 MPa after sintering at 1920°C for only 5 minutes[33] even though

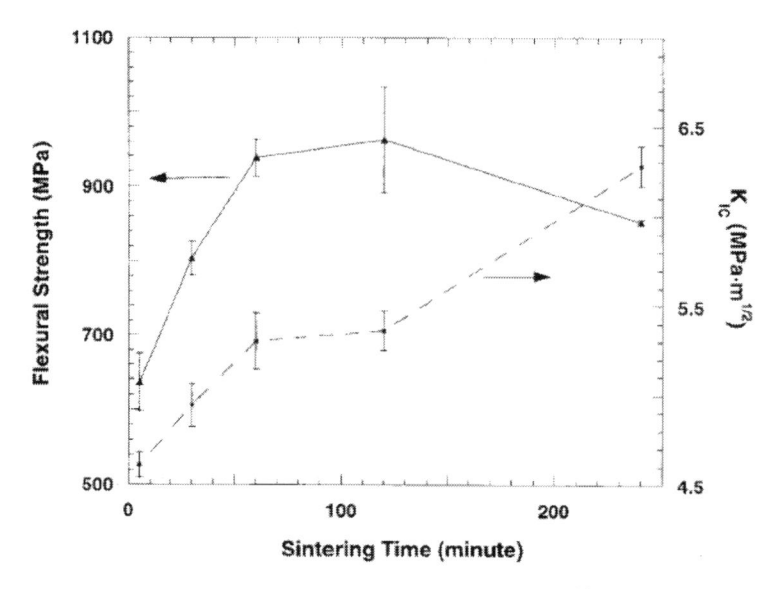

FIGURE 15. Strength and toughness as a function of sintering time at 1920°C[33] (Reprinted with permission of The American Ceramic Society, www.ceramics.org. Copyright [2001]. All rights reserved.)

the silicon nitride transformation is only around 70%[52]. The reason for the higher strength at these processing temperatures is still not clear. At 1920°C, the observed strength increase with the sintering time follows the progress of silicon nitride phase transformation, and peaks at 962 MPa after around 2 hours (Figure 15). Accordingly, fracture toughness of the composite increased from 4.5 to 5.5 MPa·$\sqrt{m}$ with the development of whisker like $\beta-Si_3N_4$ grains. The combination of properties is comparable with that of conventional hot-pressed silicon nitride ceramics. Longer processing time up to 4 hours further increased the toughness to 6.3 MPa·$\sqrt{m}$ due to the coarsening of the microstructure while the flexural strength dropped to 852 MPa for the same reason. Rising fracture resistance behavior is observed in the fully transformed composite with a wake process zone of approximately 200 μm [53].

It has been well established that the whisker-shaped $\beta-Si_3N_4$ grains not only contribute to the high strength but also the toughness[48;49;54-57]. Li and Yamanis[58] showed that silicon nitrides exhibit steady-state toughness close to 10 MPa·m$^{1/2}$ when containing large (>1 μm diameter) elongated β whiskers. Kawashima, et. al.[59] reported increasing fracture toughness of silicon nitride with the diameter of the large elongated grains, consistent with predictions from grain bridging and pullout models[54]. However, with whisker coarsening, a conflict between strength and fracture toughness arises in most silicon nitride ceramics. A bimodal/duplex microstructure of abnormally grown elongated Si_3N_4 grains, surrounded by a fine matrix, is believed to provide the optimum compromise[6;56;60-63]. From both theoretical studies and experimental results,[54;55;63] increasing the reinforcing (elongated Si_3N_4) grain size enhances the contributions from both frictional bridging and pullout. The observed increase in the rising fracture resistance behavior corresponds with the bimodal grain diameter distribution shifts to larger sizes. Also, the toughening effect achieved minimized the strength reduction, although the larger elongated grains are often

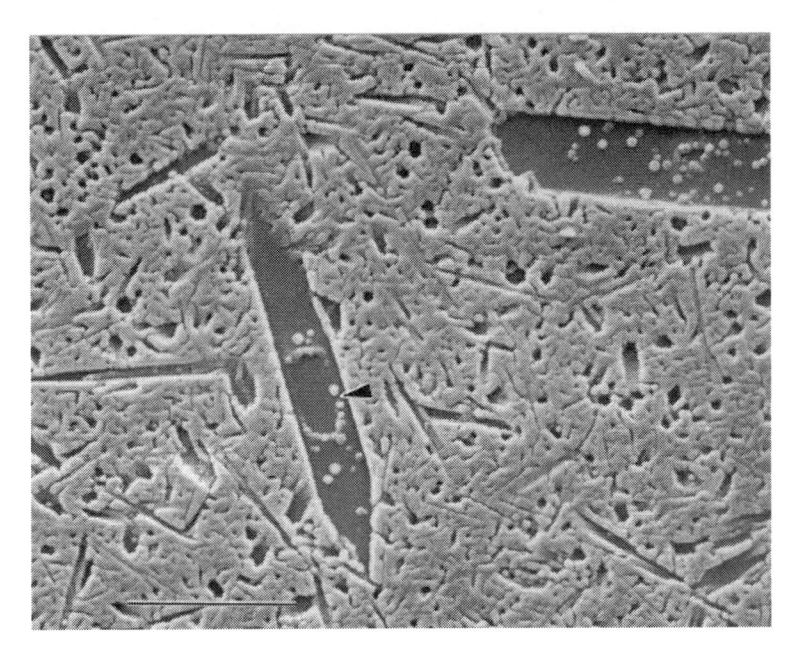

FIGURE 16. typical bimodal microstructure of Si_3N_4-BAS composite from seeding (scale bar 4 μm, the arrow indicates the core area)[38] (Reprinted with permission of The American Ceramic Society, www.ceramics.org. Copyright [2000]. All rights reserved.)

suggested as fracture-critical flaws[55]. Observations support the notion that clusters of large grains, rather than individual large grains, serve as fracture origins[60] and the loss in strength appears to be aggravated by such large grain clusters[63]. Therefore, the key factor in optimizing both the fracture toughness and strength in Si_3N_4 materials is to generation of a distinct bimodal microstructure with the large elongated grains well dispersed in a fine matrix[63].

In the Si_3N_4 – BAS composite, bimodal microstructure is introduced by either seeding[38] or tailoring α-rich Si_3N_4 starting powders[64]. An example of a seeded microstructure is shown in Figure 16. Feng and White reported[64] that bimodal microstructure could be obtained by introducing various fractions of two different Si_3N_4 powders (UBE Industries, E10 and ESP), where the size distribution and initial β phase content of the starting powders is controlled. The characteristic pronounced abnormal grain growth (Table 4) suggests the role of the ESP powders in improved fracture toughness, although no significant changes in strength seem evident. The flexural strength remains unchanged with the initial growth of those larger whiskers at 50E10 and 70E10 samples. Further increase in large whisker amount causes a dramatic strength drop to around 800 MPa[64]. The result indicates that the achievement of both high fracture toughness and high strength in this composite is made possible by encouraging the abnormal grain growth, where the size and content of large grains is controlled as mentioned above. The strength distributions of the 100E10, 70E10 and 50E10 powder samples that have been sintered at 1920°C for 210 min are shown in Figure 17. The 100E10 powder sample has the largest distribution range, from 764 to 1159 MPa, whereas the 50E10 powder sample has the smallest distribution range, ranging from 797 to 1049 MPa.

TABLE 4. Grain size distributions for the Si_3N_4–BAS composite from different ratio of starting silicon nitride powders with their indentation fracture toughness and flexural strength

Sample	Frequency (area%)				Toughness (MPa·$\sqrt{m}$)	Strength (MPa)
	<0.5 μm	0.5–1 μm	1–1.5 μm	>1.5 μm		
100E10	32.6	64.9	2.5	0.0	5.4	960
70E10	23.1	66.9	9.4	0.5	6.5	955
50E10	12.9	69.0	14.1	4.1	6.8	921

All samples were sintered at 1920°C for 210 min; sample designation shows the percentage of the E10 powder in the starting Si_3N_4 powder.

According to several research groups[65–67], an increase of the fracture toughness of a ceramic does not guarantee an increase of Weibull modulus (i.e., the reliability) of the ceramic. Instead Weibull modulus benefits from the strong rising fracture resistance behavior. For example, Hirosaki, *et. al.* showed that even the silicon nitride toughness increased from 8.5 to 10.3 MPa·$m^{1/2}$, the Weibull modulus dropped from 53 to 25. For the Si_3N_4-BAS composite, the Weibull modulus increases from 13.6 for the 100E10 powder sample to 16.3 for the 50E10 powder sample. And the increase of the Weibull modulus is consistent with the stronger rising R-curve observed in these samples with abnormally grown grains[68].

Another important feature that occurs in all silicon nitride ceramics at elevated temperatures is degradation in strength, which generally has been recognized as a result of subcritical crack growth or creep. The residual glassy phase at the grain boundary and the triple junctions of grains can soften when the test temperature exceeds its softening temperature, which results in grain boundary sliding, separation, void formation, and cracking. Both the amount and viscosity of the glassy phase are important to the high-temperature

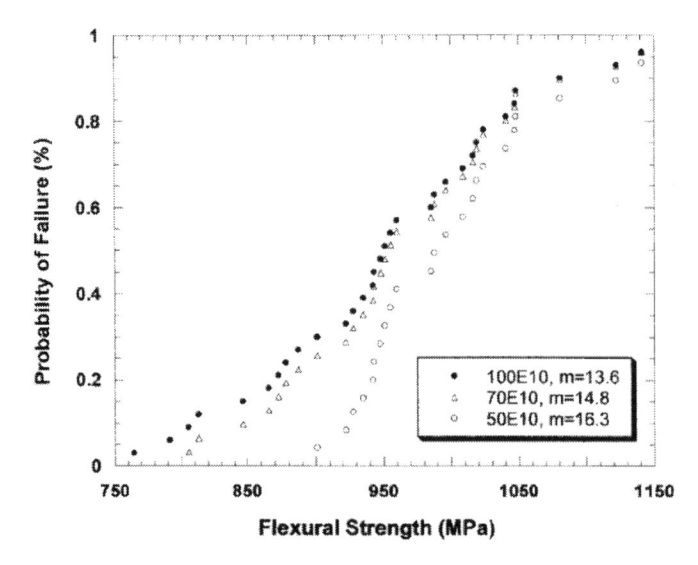

FIGURE 17. Distribution of flexural strength of the 100E10, 70E10 and 50E10 powder samples[64] (Reprinted with permission of The American Ceramic Society, www.ceramics.org. Copyright [2001]. All rights reserved.)

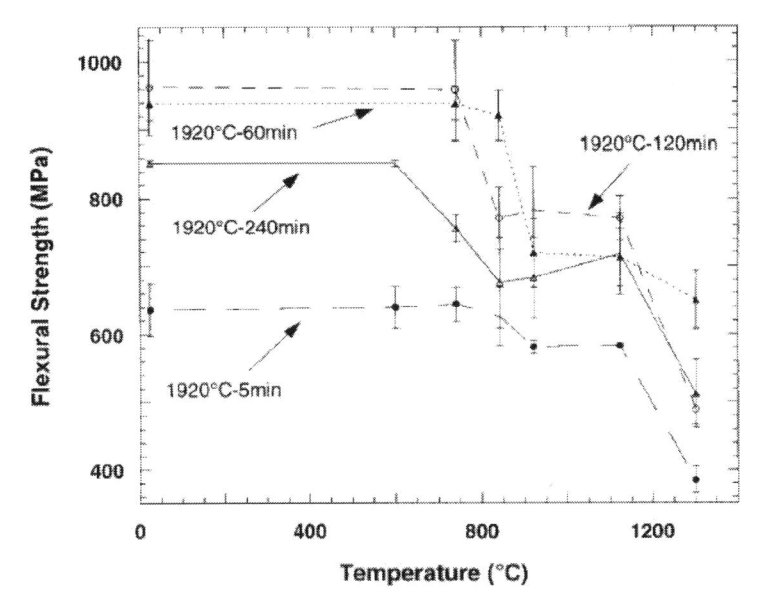

FIGURE 18. Flexural strength of 100E10 sample sintered at 1920°C for 5, 60, 120 and 240 min, as a function of testing temperature.[33] (Reprinted with permission of The American Ceramic Society, www.ceramics.org. Copyright [2001]. All rights reserved.)

performance of silicon nitride ceramics. Figure 18 shows the trend of flexural strength of Si_3N_4-BAS composite sintered at 1920°C for various times. All samples retain room temperature strength to 600°C, where the strength begins to decrease at temperatures between 600 to 920°C. The strength drop in this temperature range is believed due to the residual glassy phase shown in Figure 13. The inconsistency between this temperature range and the softening point of BAS glass implies that the residual glassy phase composition is deviated from the stoichiometric BAS. Also, it is found that longer sintering times, results in earlier strength drops. It further confirms the dissolution of Al_2O_3 into the silicon nitride lattice, as proposed earlier. Within the temperature range of 920 to 1120°C, a characteristic plateau is observed for the entire composite group under study. It is attributed to the crystallized BAS matrix which restricts further softening of the residual glassy phase. At 1300°C, the best sample (sintered at 1920°C, 120 min) supports up to 600 MPa. This value is higher than that of most silicon nitride ceramics with a crystallized matrix that have been processed either with or without pressure.

4.3. Thermal shock resistance

Thermal shock resistance of silicon nitride ceramics is strongly affected by its microstructure, considering the grain morphology, average grain size, phase composition (β content) and the composition and amount of the glassy phase[5]. Maximum thermal shock resistance has been achieved in the dense silicon nitride ceramics with complete α to β phase transformation and an optimized grain boundary phase[5]. Satisfying the above prerequisites, the Si_3N_4-BAS composite shows excellent thermal shock resistance. The low thermal

TABLE 5. Coefficient of Thermal Expansion for
BAS, Si_3N_4 and Si_3N_4-BAS composite[32;69]

Material	CTE ($\times 10^{-6}/°C$)
Celsian BAS	2.29
Hexacelsian BAS	7.99
Si_3N_4	3.31
Si_3N_4-BAS	3.8 ~ 4.0

expansion coefficient (CTE) contributes the Si_3N_4-BAS composite excellent thermal shock resistance. Measurements show, over the range of room temperature to 1400°C, the CTE varies from $2.7 \times 10^{-6}/°C$ to $4.4 \times 10^{-6}/°C$ with an average around $3.8 \sim 4.0 \times 10^{-6}/°C$[32;69]. For comparison, the coefficient of thermal expansion of Si_3N_4, BAS and Si_3N_4-BAS composite are listed in Table 5. In addition, the characteristic rising fracture resistance behavior also improves the thermal shock resistance[6;70]. Cyclic thermal shock test performed between room temperature and 600°C shows that the degradation of flexural strength could not be observed after 200 cycles[71]. Freitag and Richardson[69] also reported that sub-scale radomes fabricated from Si_3N_4-BAS composite survived under repeated cycle over a temperature range of 21°C and 1150°C in rapid succession. All these tests indicate that this composite is suitable for elevated temperature structural applications.

4.4. Tribology Behavior

Fretting tests with ball-on-flat configuration has been performed to investigate the tribological behavior of the Si_3N_4 – BAS composite under the following parameters; stroke: 200 μm; frequency: 20 Hz; load: 10 N; cycles: 100,000; temperature: 25 °C. With a Si_3N_4

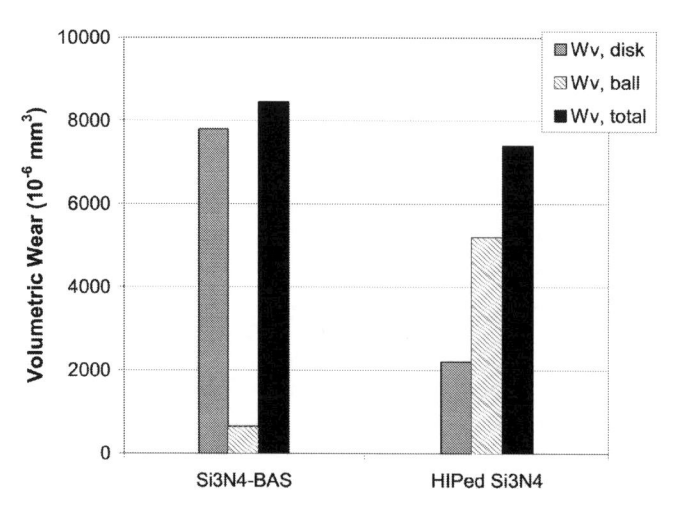

FIGURE 19. Volumetric wear at ball, disk and total wear, determined in Fretting tests with HIPed Si_3N_4 ball against Si_3N_4-BAS and HIPed Si_3N_4 disks for comparison[74]

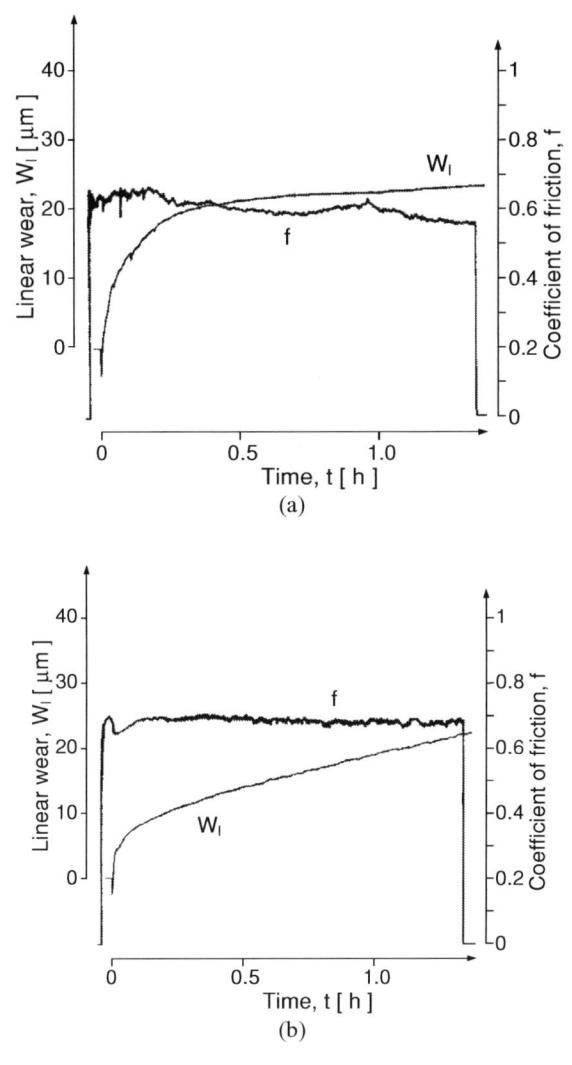

FIGURE 20. Evolution of coefficient of friction (f) and linear wear (W_l) in tests with HIPed Si_3N_4 ball against Si_3N_4-BAS composite in (a) dry air (R.H. 4%), (b) normal (R.H. 50%) and (c) moist air (R.H. 100%)[74]

ball against Si_3N_4–BAS composite and HIPed Si_3N_4 for comparison, it is found that wear of Si_3N_4–BAS composite is much higher than on HIPed Si_3N_4 at 50% relative humidity (R.H.), as shown in Figure 19. The coefficient of friction for the Si_3N_4–BAS composite is around 0.67 which is higher than the value of 0.49 for HIPed silicon nitride. These results are believed due to the low hardness of the BAS composite constituent. By varying the relative humidity, an increase of differential wear rate is found for increasing R.H. and the evolution of the coefficient of friction and linear wear under different R.H. is presented in Figure 20.

4.5. Dielectric Properties

Development of the Si_3N_4-BAS composite combines excellent mechanical properties of silicon nitride with the excellent electrical properties of barium aluminosilicate (BAS). When totally isotropic and randomly reinforced, the electrical performance of composite is greatly enhanced. The dielectric constant of the composite at room temperature is 7.3 and increases linearly to 8.6 at 1400°C, at a constant 35 GHz[69]. The values measured fall between those of hot pressed silicon nitride and BAS. A loss tangent value of 0.0003 has been measured at room temperature[69].

5. SUMMARY

A low cost-high performance silicon nitride-BAS composite can be fabricated via pressureless sintering. In this composite, the BAS glass-ceramic serves as an effective liquid-phase-sintering aid, to attain full densification and complete the α–β Si_3N_4 phase transformation, and remains as a structural matrix that is reinforced by the silicon nitride whiskers. Si_3N_4 whiskers grow in random directions in a near completely crystallized matrix of hexacelsian BAS.

With a room temperature flexural strength well above 900 MPa and toughness close to 7 MPa·m$^{1/2}$, the composite mechanical properties are comparable with conventional hot pressed or HIPed dense silicon nitride ceramics. Combining with its excellent thermomechanical and electrical properties, the in situ reinforced silicon nitride – BAS composite is an ideal candidate for applications requiring a combination of high strength, high toughness, low dielectric constant, low loss tangent, high thermal shock resistance, high thermal stability, low thermal expansion and loss cost. A list of potential applications includes: radomes, microelectronic packaging, ceramic valves, etc.

REFERENCES

1. D.W. Richerson, "Evolution in the U.S. of ceramic technology for turbine engines." *Am. Ceram. Soc. Bull.*, 64[2] 282–86 (1985)
2. R. Raj, "Fundamental research in structural ceramics for service near 2000°C," *J. Am. Ceram. Soc.*, 76[9] 2147–74 (1993)
3. F.F. Lange, "Silicon nitride polyphase systems fabrication, microstructure, and properties," *International Metals Reviews*, 1, 1 (1980)
4. G. Ziegler, J. Heinrich, G. Wotting, "Review Relationships between processing, microstructure and properties of dense and reaction-bonded silicon nitride," *J. Mater. Sci.*, 122 3041–86 (1987)
5. G. Ziegler, "Thermo-mechanical properties of silicon nitride and their dependence on microstructure," *Mater. Sci. Forum* 47 162–203 (1989)
6. M.J. Hoffmann, G. Petzow, "Microstructures design of Si_3N_4 based ceramics," *Mater. Res. Soc. Symp. Proc.*, vol. 287 3–14 (1993)
7. M.J. Hoffmann, G. Petzow, "Tailored microstructures of silicon nitride ceramics," *Pure & Appl. Chem.*, 66[9] 1807–14 (1994)
8. M.J. Hoffmann, "High temperature properties of Si_3N_4 ceramics," *MRS Bulletin*, 10 [2] 28–32 (1995)
9. D.W. Richardson, P.M. Stephan, "Evolution of applications Si_3N_4 based materials," *Mater. Sci. Forum*, 47 282 (1989)
10. M. Savitz, "Commercialization of advanced structural ceramics I," *Am. Ceram. Soc. Bull.*, 78 [1] 53–56 (1999)

11. M. Savitz, "Commercialization of advanced structural ceramics II, patience is a necessity," *Am. Ceram. Soc. Bull.*, 78 [3] 52–56 (1999)
12. F.L. Riley, "Silicon nitride and related materials,"*J. Am. Ceram. Soc.*, 83 [2] 245–65 (2000)
13. A. Zerr, G. Miehe, G. Serghiou, M. Schwarz, E. Kroke, R. Riedel, H. Fuess; P. Kroll; R. Boehler, "Synthesis of cubic silicon nitride," *Nature* 400 340–42 (1999)
14. T. Ekström, M. Nygren, "SiAlON Ceramics,"*J. Am. Ceram. Soc.*, 75 [2], 259–76 (1992)
15. D. Hardie, K.H. Jack "Crystal structures of silicon nitride," *Nature* 180 332–33 (1957)
16. W.E. Lee, W.M. Rainforth, *Ceramic Microsructures property control by processing*, 1st Ed. Chapman & Hall, London, 1994
17. T. Ito, *X-ray studies on polymorphism*. Japan Maruzen Co. Ltd., Tokyo, 1950
18. H.C. Lin, W.R. Foster, "Studies in the system $BaO-Al_2O_3-SiO_2$, I. the polymorphism of celsian," *Am. Mineral*, 53 134–144 (1968)
19. D. Bahat, "Kinetic study on the hexacelsian-celsian phase transformation,"*J. Mater. Sci.*, 5 805–10 (1970)
20. N.P. Bansal, M.J. Hyatt, "Crystallization Kinetics of $BaO-Al_2O_3-SiO_2$ Glasses,"*J. Mater. Res.*, 4 [5] 1257–65 (1989)
21. C.H. Drummond I, W.E. Lee, N.P. Bansal, M.J. Hyatt, "Crystallization of a Barium-Aluminosilicate Glass," *Ceram. Eng. & Sci. Proc.*, 10 [9–10] 1485–502 (1989)
22. C.H. Drummond I, N.P. Bansal. Crystallization behavior and properties of $BaO \cdot Al_2O_3 \cdot 2SiO_2$ Glass Matrices. *Ceram. Eng. & Sci. Proc.*, 11 [7] 1072–86 (1990)
23. W.F. Muller, "On polymorphism of $BaAl_2Si_2O_8$"; pp. 354–60 in *Electron Microscopy in Mineralogy*, Edited by H.R. Wenk, Springer-Verlag, Berlin, 1976
24. B. Yoshiki, K. Matsumoto, "High – temperature modification of barium feldspar,"*J. Am. Ceram. Soc.*, 34 [9] 283–86 (1951)
25. Y. Takeuchi, "A detailed investigation of the structure of hexagonal $BaAl_2Si_2O_8$ with reference to its α–β inversion," *Min. J. Japan*, 2 [5] 311–22 (1958)
26. R.E. Newnham, H.D. Megaw, "The crystal structure of celsian (Barium feldspar)," *Acta Cryst.*, 13 303–12 (1960)
27. M.C. Guillem, C. Guillem, "Kinetics and mechanism of formation of celsian from barium carbonate and kaolin," *Trans. J. Br. Ceram. Soc.*, 83 150–54 (1984)
28. M.J. Hyatt, N.P. Bansal, "Crystallization kinetics of barium and strontium aluminosilicate glasses of feldspar composition," NASA Technical Memorandum, No. 106624, 1994
29. M.J. Hyatt, N.P. Bansal, "Crystal growth kinetics in $BaO-Al_2O_3-2SiO_2$ glasses," *J. Mater. Sci.*, 31 [1] 172–84 (1996)
30. S.W. Quander, A. Bandyopadhyay, P.B. Aswath, "Synthesis and properties of in situ Si_3N_4-reinforced $BaO \cdot Al_2O_3 \cdot 2SiO_2$ ceramic matrix composites,"*J. Mater. Sci.*, 32 2021–29 (1997)
31. A. Bandyopadhyay, P.B. Aswath, W.D. Porter, O.B. Cavin, "The low temperature hexagonal to orthorhombic transformation in Si_3N_4 reinforced BAS matrix composite,"*J. Mater. Res.*, 10 1256–63 (1995)
32. K.K. Richardson, D.W. Freitag, D.L. Hunn, "Barium Aluminosilicate Reinforced In Situ with silicon nitride," *J. Am. Ceram. Soc.*, 78 [10] 2662–68 (1995)
33. F. Yu, N. Nagarajan, Y. Fang, K.W. White, "Microstructural Control of a 70% Silicon Nitride – 30% Barium Aluminum Silicate Self-Reinforced Composite,"*J. Am. Ceram. Soc.*, 84 [1] 13–22 (2001)
34. F. Yu, N. Nagarajan, Y. Fang, K.W. White, "The development of Si_3N_4/BAS ceramic matrix composite," *Ceramic Transactions* 85 381–92 (1998)
35. K.T. Lee, P.B. Aswath, "Synthesis of hexacelsian barium aluminosilicate by a solid-state process," *J. Am. Ceram. Soc.*, 83 [12] 2907–12 (2000)
36. R.J. Brook, D.R. Messier, F.L. Riley, "The α/β silicon nitride phase transformation," *J. Mater. Sci.*, 13 1199–1205 (1978)
37. A. Bandyopadhyay, S.W. Quander, P.B. Aswath, D.W. Freitag, K.K. Richardson, D.L. Hunn, "Kinetics of in-situ α to β Si_3N_4 transformation in a barium aluminosilicate matrix," *Scripta Metall. et Mater.*, 32 [9] 1417–22 (1995)
38. Y. Fang, F. Yu, K.W. White, "Bimodal Microstructure in silicon Nitride – Barium Aluminum Silicate Ceramic-Matrix Composites by Pressureless Sintering," *J. Am. Ceram. Soc.*, 83 [7] 1828–30 (2000)
39. C.P. Gazzara, D.R. Messier, "Determination of Phase Content of Si_3N_4 by X-ray Diffraction Analysis," *Am. Ceram. Soc. Bull.*, 56 [9] 777–780 (1977)

40. H. Pickup, R.J. Brook, "Barium oxide as a sintering aid for silicon nitride," *Br. Ceram. Soc. Proc.*, 39 69–76 (1987)
41. C.J. Hwang, R.A. Newman, "Silicon nitride ceramics with celsian as an additive,"*J. Mater. Sci.*, 31 150–56 (1996)
42. D.A. Bonnell, M. Ruhle, T.Y. Tien, "Redistribution of aluminum ions during processing of SiAlON ceramics," *J. Am. Ceram. Soc.*, 69 [8] 623–27 (1986)
43. N.K. Kim, D.Y. Kim, A. Kranzmann, E. Bischoff, S.-J.L. Kang, "Variation of aluminum concentration in β'-sialon grains formed during liquid-phase sintering of Si_3N_4-Al_2O_3-Nd_2O_3,"*J. Mater. Sci.*, 28 4355–58 (1993)
44. S.-J.L. Kang, S.M. Han, "Grain growth in Si_3N_4-based materials," *MRS Bulletin* 10 [2] 33–37 (1995)
45. M.J. Hyatt, N.P. Bansal, "Crystal growth kinetics in $BaO \cdot Al_2O_3 \cdot 2SiO_2$ and $SrO \cdot Al_2O_3 \cdot 2SiO_2$ glasses," *J. Mater. Sci.*, 31 172–184 (1996)
46. F. Yu, C.R. Ortiz-Longo, K.W. White, D.L. Hunn "The Microstructural Characterization of In Situ Grown Si_3N_4 whisker-Reinforced Barium Aluminum Silicate Ceramic Matrix Composite," *J. Mater. Sci.*, 34 2821–35 (1999)
47. C. Greskovich, G.E. Gazza, "Hardness of dense α and β Si_3N_4ceramics," *J. Mater. Sci. Lett.*, 4 195–96 (1985)
48. F.F. Lange, "Relation between strength, fracture energy and microstructure of hot pressed Si_3N_4," *J. Am. Ceram. Soc.*, 56 [10] 518–22 (1973)
49. F.F. Lange, "Fracture toughness of Si_3N_4 as a function of the initial α phase content," *J. Am. Ceram. Soc.*, 62 [7] 428–30 (1979)
50. G. Himsolt, H. Knoch, H. Huebner, F.W. Kleinlein, "Mechanical properties of hot-pressed silicon nitride with different grain structures,"*J. Am. Ceram. Soc.*, 62 [1] 29–32 (1979)
51. E. Tani, S. Umebayashi, K. Kishi, K. Kobayashi, M. Nishijima, "Gas-pressure sintering of Si_3N_4 with concurrent addition of Al_2O_3 and 5wt% rare earth oxide high fracture toughness Si_3N_4 with fiber like structure," *Am. Ceram. Soc. Bull.*, 65 [9] 1311–15 (1986)
52. F. Yu. "Development of a 70% Si_3N_4-30% barium aluminum silicate self-reinforce ceramic matrix composite by microstructural modification". Ph.D., Dissertation, University of Houston, 1998
53. Y. Fang, F. Yu, K.W. White, "Microstructural influence on the R-curve behavior of a 70% Si_3N_4-30% barium aluminum silicate self-reinforced composite," *J. Mater. Sci.*, 35 2695–99 (2000)
54. P.F. Becher, "Microstructural design of toughened ceramics," *J. Am. Ceram. Soc.*, 74 [2] 255–69 (1991)
55. P.F. Becher, H.T. Lin, S.L. Hwang, M.J. Hoffmann, I.W. Chen, "The influence of microstructure on the mechanical behavior of silicon nitride ceramics," *Mater. Res. Soc. Symp. Proc.* 287 147–59 (1993)
56. P.F. Becher, S.L. Hwang, H.T. Lin, T.N. Tiegs, "Microstructural contributions to the fracture resistance of silicon nitride ceramics," pp. 87–100 in *Tailoring of mechanical properties of Si_3N_4 ceramics*, Edited by M.J. Hoffmann and G. Petzow, Kluwer Academic Publishers, Dordrecht, 1994
57. P.F. Becher, S.L. Hwang, C.H. Hsueh, "Using microstructure to attack the brittle nature of silicon nitride ceramics," *MRS Bulletin* 10 [2] 23–27 (1995)
58. C.W. Li, J. Yamanis, "Super tough silicon nitride with R-curve behavior," *Ceram. Eng. & Sci. Proc.*, 10 [7] 632–45 (1989)
59. T. Kawashima, H. Okamoto, H. Yamamoto, A. Kitamura, "Grain size dependence of the fracture toughness of silicon nitride ceramics,"*J. Ceram. Soc. Jpn.,Int. Ed.*, 99 310–13 (1991)
60. M.J. Hoffmann "Analysis of microstructural development and mechanical properties of Si_3N_4 ceramics," pp. 59–72 in *Tailoring of mechanical properties of Si_3N_4 ceramics*, Edited by M.J. Hoffmann and G. Petzow, Kluwer Academic Publishers, Dordrecht, 1994.
61. W. Dressler, H.J. Kleebe, M.J. Hoffmann, M. Ruhle, G. Petzow, "Model experiments concerning abnormal grain growth in silicon nitride,"*J. Euro. Ceram. Soc.*, 16 3–14 (1996)
62. H. Emoto, M. Mitomo, "Control and characterization of abnormally grown grains in silicon nitride ceramics," *J. Euro. Ceram. Soc.*, 17 797–804 (1997)
63. P.F. Becher, E.Y. Sun, K.P. Plucknett, K.B. Alexander, C.H. Hsueh, H.T. Lin, S.B. Waters, C.G. Westmoreland, E.S. Kang, K. Hirao, M.E. Brito, "Microstructural design of silicon nitride with improved fracture toughness I, effects of grain shape size,"*J. Am. Ceram. Soc.*, 81 [11] 2821–30 (1998)
64. F. Yu, K.W. White, "Relationship between Microstructure and Mechanical Performance of a 70% Silicon Nitride – 30% Barium Aluminum Silicate Self-Reinforced Ceramic Composite," *J. Am. Ceram. Soc.*, 84 [1] 5–12 (2001)

65. K. Kendall, N. McN, S.R. Tan, J.D. Birchall, "Influence of toughness on Weibull modulus of ceramic bending strength," *J. Mater. Res.*, 1 [1] 120–23 (1986)
66. R.F. Cook, D.R. Clarke, "Fracture stability, R-curves and strength variability," *Acta Mater.*, 36 [3] 555–62 (1988)
67. D.K. Shetty, J.S. Wang, "Crack stability and strength distribution of ceramics that exhibit rising crack-growth-resistance (R-Curve) behavior," *J. Am. Ceram. Soc.*, 72 1158–62 (1989)
68. Y. Fang, F. Yu, K.W. White, "Microstructural Modification to Improve Mechanical Properties of a 70% Si_3N_4 – 30% BAS Self-Reinforced Ceramic Composite," *J. Mater. Sci.*, 37 4411–17 (2002)
69. D.W. Freitag, K.K. Richardson, "BAS reinforced in-situ with silicon nitride," U.S. Patent, No. 5,358,912 (1994).
70. M.J. Hoffmann, G.A. Schneider, G. Petzow, "The potential of Si_3N_4 for thermal shock applications," pp. 49–58 in *Thermal shock and thermal fatigue behavior of advanced ceramics*, Edited by. G.A. Schneider and G. Petzow, Kluwer Academic Publishers, Dordrecht, 1993
71. F. Yu, H. Fang, K. Ravi-Chandar, K.W. White, "Thermal fatigue test of Si_3N_4-BAS composite," The American Ceramic Society 102nd Annual Meeting and Exposition. Paper No: B2P-013-00, St. Louis, Missouri, 2000
72. P.F. Becher, W.H. Warwick, "Factors influencing the thermal shock behavior of ceramics," pp. 37–48 in *Thermal shock and thermal fatigue behavior of advanced ceramics*, Edited by G.A. Schneider and G. Petzow, Kluwer Academic Publishers, Dordrecht, 1993
73. E.M. Levin, H.F. McMurdie. *Phase diagrams for ceramists*. American Ceramic Society, Columbus, OH, 1975
74. F. Yu, K.W. White, "Fretting test with Si_3N_4/BAS and Si_3N_4/SAS against Si_3N_4," unpublished work (1999)

12

Silicon Carbide and Oxide Fiber Reinforced Alumina Matrix Composites Fabricated Via Directed Metal Oxidation

Ali S. Fareed

Power Systems Composites, LLC, Newark, DE
e-mail: ali.fareed@ps.ge.com
Tel.: (302) 631-1309

ABSTRACT

Directed metal oxidation was used to fabricate a variety of fiber-reinforced ceramic matrix composites. Reinforcements included silicon carbide and oxide based fibers. SiC_f/Al_2O_3 components successfully completed engine and rig tests with operating temperatures between 1000 and 1400°C. Replacement of ceramic grade (CG) Nicalon™ fibers with reduced oxygen containing Hi-Nicalon™ fibers resulted in enhanced properties, particularly residual strengths following elevated temperature exposure. Oxide based fibers offer the advantage of a closer coefficient of thermal expansion match with the alumina matrix. Aluminum oxide and mullite based fibers were evaluated as reinforcements. The oxide-oxide composites exhibited a significant drop in strength as test temperatures approached 1000°C. In contrast, composites reinforced with silicon carbide fibers retained over 80% of their room temperature strength and fracture toughness at temperatures as high as 1200°C in air.

1. INTRODUCTION

The refractory nature and low density of ceramic materials make them an ideal choice for a host of high temperature applications such as turbine engines, aerospace, chemical process equipment and heat exchangers. It has been well established over the last decade that the inherently low fracture toughness of ceramics can be significantly improved by incorporating continuous fibers as a reinforcement phase. Hence, there has been an increasing amount of interest in these reinforced materials.

One approach for fabricating fiber reinforced ceramic matrix composites is the directed oxidation of metals, a process first introduced by Lanxide Corporation [1, 2] and later used successfully to produce turbine engine and aerospace components. Rights to the "DIMOX" technology, as it was identified, were ultimately acquired by Power Systems Composites, L.L.C., a subsidiary of the Power Systems business of the General Electric Company.

The DIMOX^{TM} directed metal oxidation process offers the potential to overcome many of the limitations of conventional ceramic processing technologies such as sintering and hot pressing. The absence of densification shrinkage coupled with no displacement or damage to fiber preforms during matrix infiltration, results in net or near net shape processing. Hence, machining costs are minimized. Other attractive features include a low cost matrix infiltration process, thermal stability of the matrix and the ability to incorporate a broad range of fiber architectures and fiber types. To date, the major development effort for fiber-reinforced CMC fabrication via directed metal oxidation has been with an aluminum oxide matrix. A limitation of these composites is their low matrix thermal conductivity.

The versatility of the process is manifest in the demonstrated fabrication of a variety of matrices with different types of reinforcements.

A variety of CMC systems have been developed and fabricated in the past. These have included aluminum oxide, aluminum nitride and silicon nitride matrix composites [3–5]. The reinforcement has predominantly consisted of silicon carbide based fibers. Oxide based fibers have also been evaluated over the years as and when they have become available. This chapter reviews the development effort of silicon carbide reinforced aluminum oxide matrix composites fabricated via directed metal oxidation and compares them with those reinforced with oxide fibers.

2. COMPOSITE FABRICATION

The generic process for fabrication of fiber-reinforced aluminum oxide matrix composites by directed metal oxidation includes preforming, fiber-matrix interface coating, matrix growth and removal of residual aluminum. A flow chart with the various processing steps is shown in Fig. 1.

Typically, the fabrication of composite plates begins by stacking layers of woven fabric (e.g. 8 harness satin weave) with alternate plies being rotated by 90° to achieve a more balanced lay-up. Advanced performing techniques such as braiding and 3-D weaving can be used for complex shaped components. Following preform fabrication, a BN/SiC duplex coating [6] is applied on the fibers by chemical vapor infiltration (CVI). The inner BN layer provides a weak fiber-matrix interface, which is essential in order to achieve the desired composite toughness and non-catastrophic failure. The outer coating of SiC protects the underlying BN layer and fiber from oxidation and reaction with molten aluminum during the subsequent matrix growth processing step. The coated fiber preform is brought into contact with a molten aluminum alloy. At temperatures in the 900 to 1000°C range, the aluminum reacts with oxygen from the ambient air to form a three-dimensionally interconnected

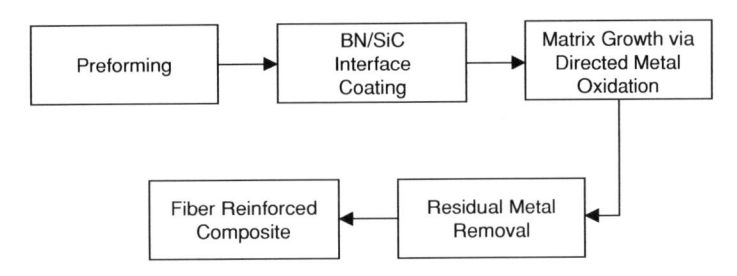

FIGURE 1. Fabrication sequence for 2-D Nicalon^{TM}/Al$_2$O$_3$ composites.

matrix of aluminum oxide that grows into the fiber preform. An interconnected network of microscopic metal channels that are at most a few microns in diameter is present in the matrix phase. The molten alloy wicks from the aluminum alloy reservoir through these microchannels to the growth front, where it reacts to continue the matrix growth process. A gas permeable barrier layer is applied to one or more preform surfaces, which terminates the oxidation reaction when the growth front comes in contact with it. Hence, net or near net shape fabrication can be achieved. The residual aluminum present in the matrix has a low melting temperature (~600°C) relative to that required for typical turbine engine applications (>1000°C). Molten aluminum is highly reactive and has the potential of reacting with adjoining metallic components. Hence, the aluminum metal in the matrix is removed following matrix growth. This removal is carried out by a wicking process, which is facilitated by the fact that the residual aluminum is predominantly in the form of a three-dimensional interconnected network. The removal of residual aluminum results in the formation of fine, uniformly distributed microchannels of porosity. This metal removal has no adverse effect on the strength and toughness of the fiber-reinforced composite [4]. Overall porosity levels are typically 5–10 vol.%. Fiber volume fractions range from 0.30 to 0.35.

3. SiC FIBER REINFORCED ALUMINUM OXIDE MATRIX COMPOSITES

3.1. Role of the BN/SiC Fiber-Matrix Interface

The fiber-matrix interface plays a critical role in the functioning of a fiber-reinforced CMC. A key requirement to obtaining strong and tough CMCs is a weak fiber-matrix interface that allows fibers to debond from and slide relative to the matrix. The use of carbon or boron nitride debond coatings has been reported for CMCs with silicon carbide [7] and glass-ceramic [8] matrices. For composites fabricated by directed metal oxidation, thin layers of carbon or BN would rapidly oxidize during matrix densification. Hence, a novel, duplex BN/SiC coating concept was developed in order to provide and maintain the desired weak interface. The rationale behind choosing this coating architecture is twofold. The interface provides the necessary weak fiber-matrix bond and protects fibers from molten aluminum and oxidation during matrix densification.

Individual thicknesses of BN and SiC coating layers can play a pivotal role in the resulting mechanical properties of the composite [6]. In general, optimal SiC coating thicknesses fall in the 2–4 μm range. Higher SiC coating thicknesses can lead to canning (i.e., sealing) of the preform surfaces, which inhibits transport of oxygen or metal to the growth front during the matrix growth process. This in turn causes porosity due to incomplete matrix infiltration. Low SiC thicknesses, of about a micron or less lead to inadequate fiber and BN coating protection during the growth process, thus resulting in decreased strengths. BN coating thicknesses that are too low (<0.2 μm) do not provide the necessary debonding. No additional benefit was observed as the BN coating thickness was increased above 0.5 μm. Thicknesses above a micron tend to make the composite more susceptible to interlaminar shear type failures, a mode of failure which is not typically observed. Hence, the optimum BN coating thickness range is between 0.2 and 0.5 μm.

3.2. Microstructures of 2-D Nicalon^{TM}/Al$_2$O$_3$ Composites

Representative low and high magnification optical micrographs of polished cross-sections of SiC$_f$/Al$_2$O$_3$ composites are shown in Figs. 2 and 3, respectively. Uniform matrix infiltration is observed within as well as between fiber bundles, which is typical of fiber-reinforced composites fabricated via directed metal oxidation. A point to note is the ability of the process to fill up spaces as large as 200 μm in width (Fig. 2) as well as gaps of the order of a micron or less (Fig. 3). A small amount of porosity is evident within some of the fiber bundles, which is attributed to sealing or canning of individual groups of fibers in close contact during the CVI coating process. Hence, there is no access for matrix growth into

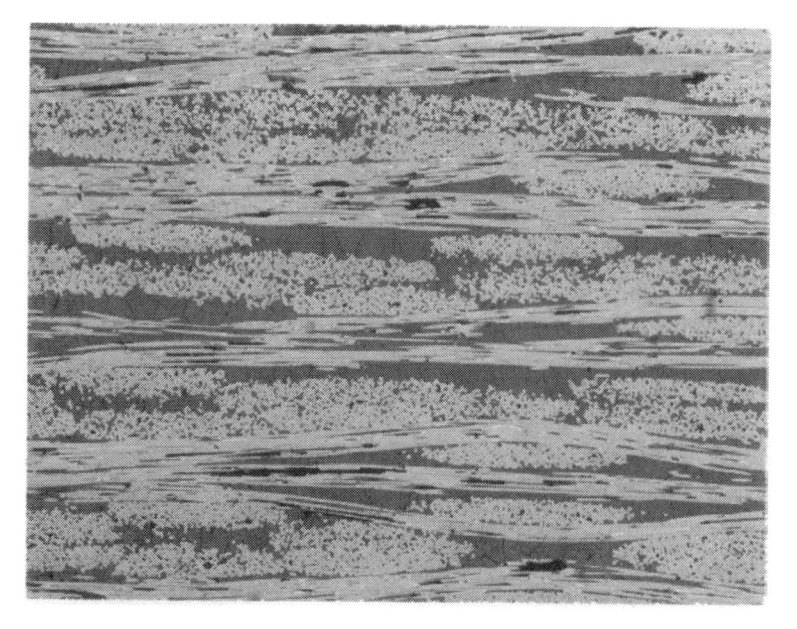

FIGURE 2. Microstructure of a 2-D Nicalon^{TM}/Al$_2$O$_3$ composite exhibiting matrix infiltration within fiber bundles and between fabric plies. The dark regions associated with the horizontal fibers result from fiber pull-out during polishing and are not porosity.

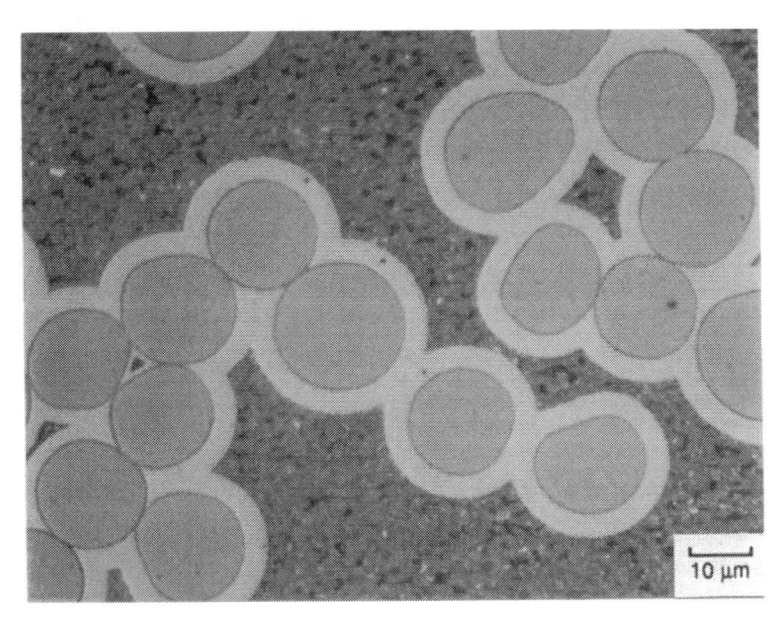

FIGURE 3. Microstructure showing Nicalon™ fibers with uniform BN/SiC duplex coatings in an alumina matrix. The fine distributed microporosity present in the matrix is a result of residual aluminum metal removal. The micrograph also demonstrates the ability of the matrix to grow through regions of micron or submicron dimensions.

these regions. The fine, uniformly distributed microchannels of matrix porosity observed at the higher magnification, in Fig. 3, is a result of the removal of interconnected residual aluminum following the directed metal oxidation process. Uniform BN/SiC duplex coatings are observed around the fibers.

A coefficient of thermal expansion (CTE) mismatch is present between the Nicalon™ fibers ($4.0 \times 10^{-6}/°C$ between 0 and 900°C) and aluminum oxide matrix ($8 \times 10^{-6}/°C$ between RT and 1000°C). This mismatch leads to the development of residual tensile stresses in the matrix on cool down from matrix growth temperatures. In 2-D woven composites, relatively larger areas of unreinforced matrix are likely to occur between fabric plies. Hence, for 2-D Nicalon™/Al_2O_3 composites, these unreinforced regions are likely to develop residual tensile stresses. In some cases, the residual tensile stresses are high enough to cause matrix microcracking. The presence of these microcracks is apparent in Fig. 2.

3.3. 2-D Nicalon™/Al_2O_3 Composite Properties

The composite properties reviewed in this section are for 2-D woven Nicalon™/Al_2O_3 composites consisting of approximately 35 vol.% of fibers [6, 9–13]. All samples were cut from ground plates and tested in the as-machined condition with no externally applied oxidation protection coating. Elevated temperature testing was conducted in an air atmosphere.

TABLE I. Physical and Thermal Properties for 2-D
Nicalon™/Al$_2$O$_3$ Composites

Density	2.8 g/cm^3
Young's Modulus	160 GPa
In-Plane CTE (25-1200°C)	5.5×10^{-6}/°C
Through Thickness Thermal Conductivity	
@25°C	8.7 W/mK
@1000°C	5.7 W/mK
@1200°C	5.5 W/mK

Physical and thermal properties for CG Nicalon™/Al$_2$O$_3$ composites are summarized in Table I. Four point flexural strengths and fracture toughness at different temperatures are listed in Table II. All tests were done in air following a hold at temperature for 15 minutes. Room temperature flexural strengths of 460 MPa were retained in full at 1200°C. Fracture toughness exhibited 80% retention up to 1200°C in air.

Fractured surfaces exhibit a significant amount of fiber pull-out at room temperature and 1200°C, as shown in Figs. 4 and 5, respectively. This pull-out is indicative of a weak fiber-matrix interface. Debonding typically occurs at the BN/fiber interface (Fig. 4b). Hence, fiber surface roughness can play a significant role in determining the debonding characteristics of these composites.

Samples measuring 50 × 6 × 3 mm were thermally shocked by quenching from 1000 or 1200°C into room temperature water. Flexural strengths measured before and after thermal shock are shown in Table III. The relatively low composite thermal conductivity (Table I) does not adversely affect thermal shock behavior. Repeated thermal shocks from 1000°C to room temperature water for five cycles results in only a modest decrease (15%) in room temperature strength.

As there are no fibers aligned in the "through-thickness" direction of 2-D fiber-reinforced composites, a possible mode of failure is by shear between the fabric plies. Hence, interlaminar shear strength is an important property. Tests were conducted using the ASTM D2344-84 short beam method. The average interlaminar shear strength from five samples was measured to be 62 MPa at room temperature.

Tensile test results (Fig. 6) show retention of the room temperature UTS of 260 MPa up to 1200°C in air, which is in agreement with the four point flexural strength data (Table II).

TABLE II. 2-D Nicalon™/Al$_2$O$_3$ Mechanical Properties

Test Temperature (C)	Flexural Strength* (MPa)	Fracture Toughness** (MPa(m)$^{1/2}$)
Room Temperature	461 (28 (8)	28 (5 (3)
1200 (C	488 (22 (12)	23 (3 (3)
1300 (C	400 (12 (4)	19 (3 (3)
1400 (C	340 (11 (4)	16 (4 (3)

* MIL-STD-1942A
** chevron notch[14]

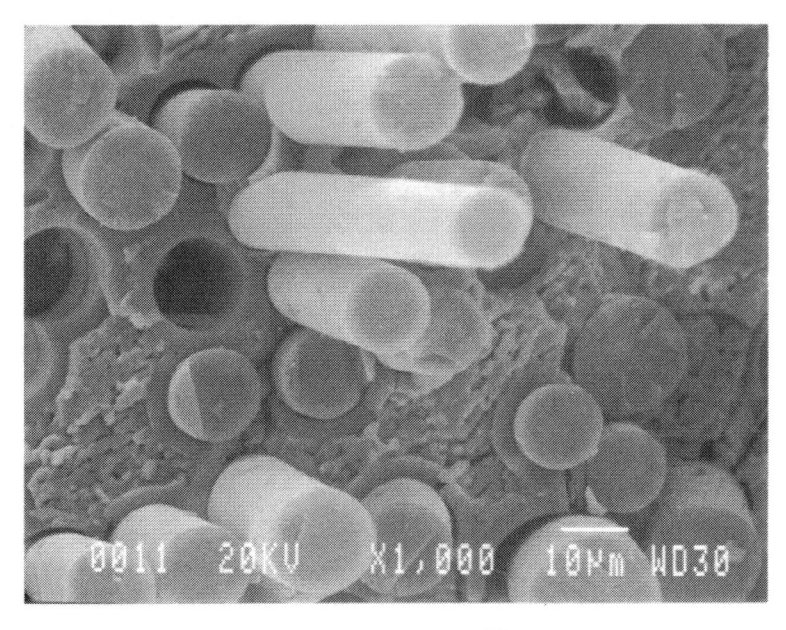

FIGURE 4a. Room temperature fracture surface of a 2-D Nicalon™/Al$_2$O$_3$ composite showing fiber pull-out.

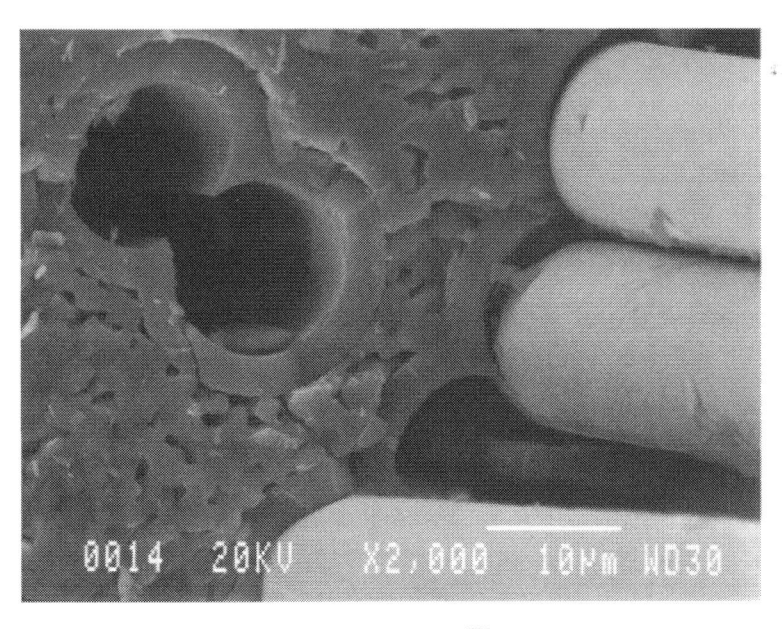

FIGURE 4b. Room temperature fracture surface of a 2-D Nicalon™/Al$_2$O$_3$ composite showing debonding at the BN-fiber interface.

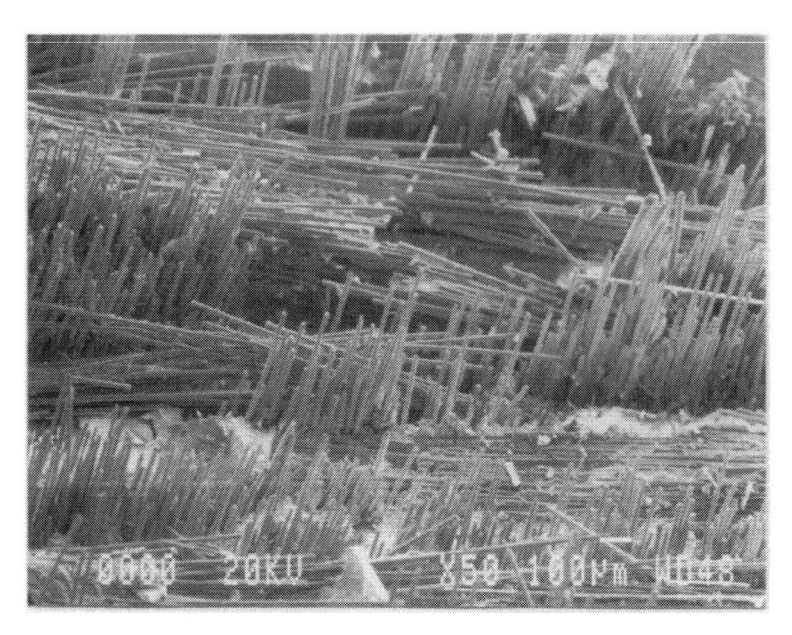

FIGURE 5. Fracture surface at 1200°C showing Nicalon™ fiber pull-out.

A typical room temperature tensile stress-strain curve (Fig. 7) shows a proportional limit of 60 MPa. The change in slope of the stress-strain curve at this point is an indication of matrix microcracking and microcrack propagation. At elevated temperatures, these microcracks can provide a path for oxygen ingress, potentially leading to fiber or coating degradation. Hence, stress rupture tests were conducted with applied tensile stresses above 60 MPa, in order to evaluate the impact of composite performance above the proportional limit. The tests consisted of loading the samples in tension and holding isothermally at elevated temperatures in air. In some tests, the stress level was increased periodically until failure occurred, as shown in Fig. 8. All samples tested were in the as-machined condition, with no externally applied oxidation protection coatings. The data shows that these Nicalon™ fiber reinforced alumina matrix composites can withstand sustained exposures well over 1000 h at 1100 and 1200°C with an applied stress which exceeds the minimum required to cause matrix microcrack opening. A lifetime of over 6500 h has been demonstrated without

TABLE III. Thermal Shock Results for 2-D Nicalon™/Al_2O_3 Composites

Exposure Temperature (°C)	# of Thermal Shock Cycles	Flexural Strength (MPa)	% Strength Retained
None	—	385	—
1000	1	357	93%
1000	5	328	85%
1200	1	322	84%

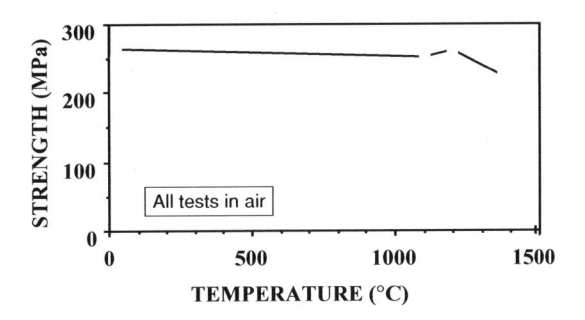

FIGURE 6. 2-D Nicalon™/Al$_2$O$_3$ tensile strength as a function of temperature (data generated by the University of Michigan for Williams International).

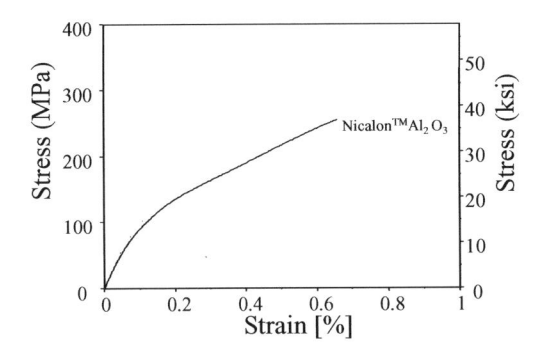

FIGURE 7. Tensile stress-strain curve at room temperature for a 2-D Nicalon™/Al$_2$O$_3$ composite.

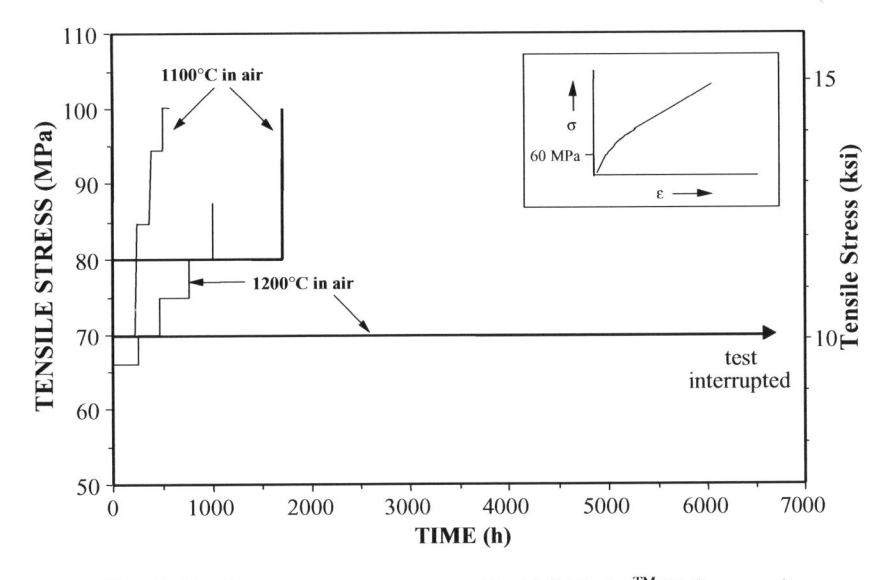

FIGURE 8. Tensile stress rupture tests on as-machined 2-D Nicalon™/Al$_2$O$_3$ composites.

failure at 1200°C in air under a constant stress of 70 MPa. This composite survivability at levels above the proportional limit is attributed to internal oxidation protection mechanisms. Partial oxidation of the BN/SiC layer result in glass formation that seal small microcracks (Fig. 9a&b), thus preserving composite integrity. However, once the microcracks are large enough (i.e., at a stress level of 90–100 MPa), composite degradation occurs as a result of fiber and/or coating oxidation. This degradation leads to a brittle failure of composites with minimal fiber pull-out, as observed in the fracture surface of a sample subjected to a stress rupture test at 1100°C in air (Fig. 10). In this test, the applied stresses were sufficient to cause significant microcrack opening.

In order to further simulate turbine engine conditions, a temperature cycle was superimposed during elevated temperature exposure under an applied tensile stress. Each thermal cycle consisted of a hold for 1 h at 1100°C in air followed by a cool down to 600°C in 45 minutes. After a hold at 600°C for 1 h, the temperature was ramped back up to 1100°C in 45 minutes. A strain versus time plot for this test is shown in Fig. 11. After a few initial thermal cycles with an applied tensile stress of 12 MPa, the stress was increased to 50 MPa. Following 115 thermal cycles at 50 MPa, the stress level was raised to 70 MPa. Sample failure occurred after an additional 22 cycles at 70 MPa. In all, the test was run for over 500 h with a total of 142 thermal cycles. An enlargement of the 70 MPa region in Fig. 11 shows the oscillation of the sample strain per thermal cycle as measured by a noninvasive laser strain measurement system. The increase in strain observed with increasing number of cycles for a given stress is attributed to continuously increasing matrix microcracking. A second test was conducted under the same thermal cycle conditions but with a constant applied stress of 50 MPa. The test was run for 1000 h in air with a total of 282 thermal cycles being completed prior to sample failure.

Low cycle tension-tension fatigue tests were conducted using a trapezoidal waveform (Figs. 12. and 13) and a stress ratio of 0.1. The S-N plot at 1100°C in air (Fig. 12) shows an endurance limit of 100 MPa. The post LCF tensile strength at 1100°C in air following run-out was 212 MPa, retention of 88% of the as-fabricated UTS at 1100°C. Fractured surfaces of this post LCF tested specimen exhibit extensive fiber pull-out (Fig. 14), which is further indication of interface coating and fiber integrity preservation during fatigue testing for 100,000 cycles at 1100°C in air. Additionally, tension-tension fatigue tests were also carried out as a function of temperature using a maximum stress level of 120 MPa. The results are shown in Fig. 13. There are no deleterious effects occurring at intermediate temperatures of 600 and 800°C.

3.4. 2-D Hi-Nicalon^{TM}/Al$_2$O$_3$ Composite Properties

Residual tensile strengths following exposure to 800 and 1100°C for 1000 h in air are shown in Fig. 15 for CG Nicalon^{TM}/Al$_2$O$_3$ composites. A substantial decrease in UTS is observed following the 1100°C exposure. The percentage drop is greater for residual strengths measured at room temperature relative to those measured at 1100°C. However, replacement of CG with Hi-Nicalon fibers resulted in a substantial improvement in strength retention following exposure to 1100°C in air for periods of 500 and 1000 h (Fig. 16). Hi-Nicalon fibers have significantly lower oxygen content of 0.5% relative to the 12% present in the CG grade [15]. The increased oxygen content in CG Nicalon coupled with the inherent

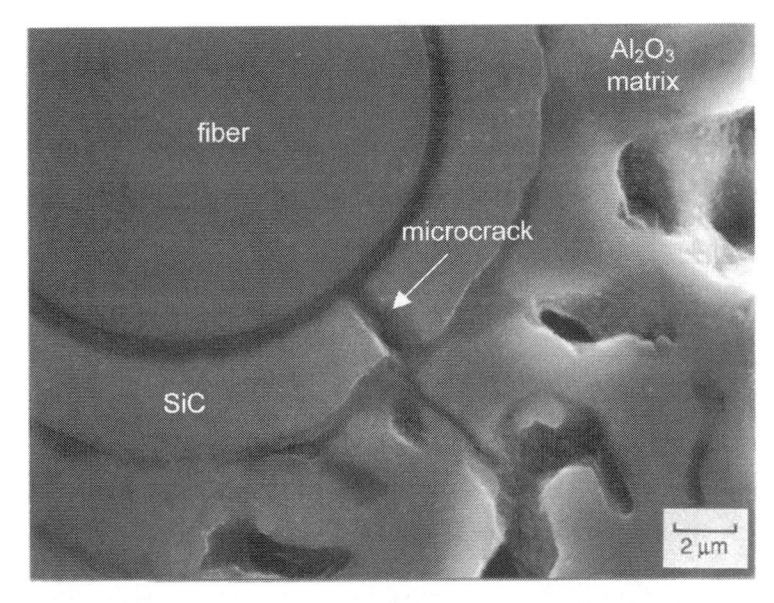

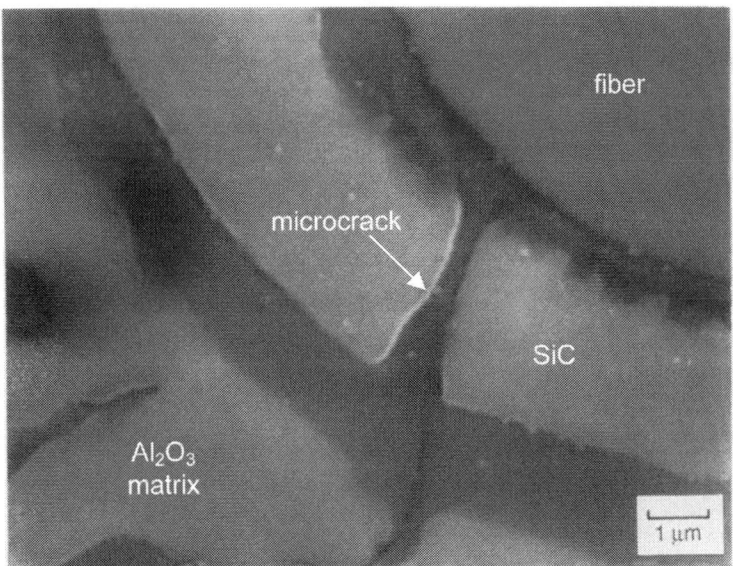

FIGURE 9. Microstructures of a 2-D Nicalon™/Al$_2$O$_3$ composite showing microcrack sealing following elevated temperature exposure.

FIGURE 10. Fracture surface of a 2-D Nicalon™/Al$_2$O$_3$ composite following stress rupture testing at 1100°C in air.

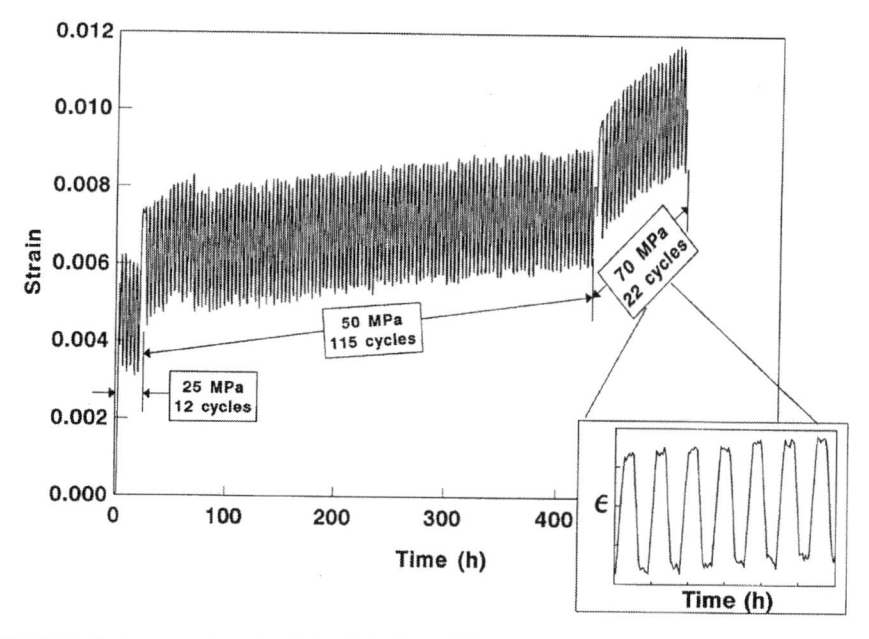

FIGURE 11. Strain versus time plotted for a 2-D Nicalon™/Al$_2$O$_3$ composite during thermal cycling between 1100 and 600°C in air with a simultaneously applied tensile stress.

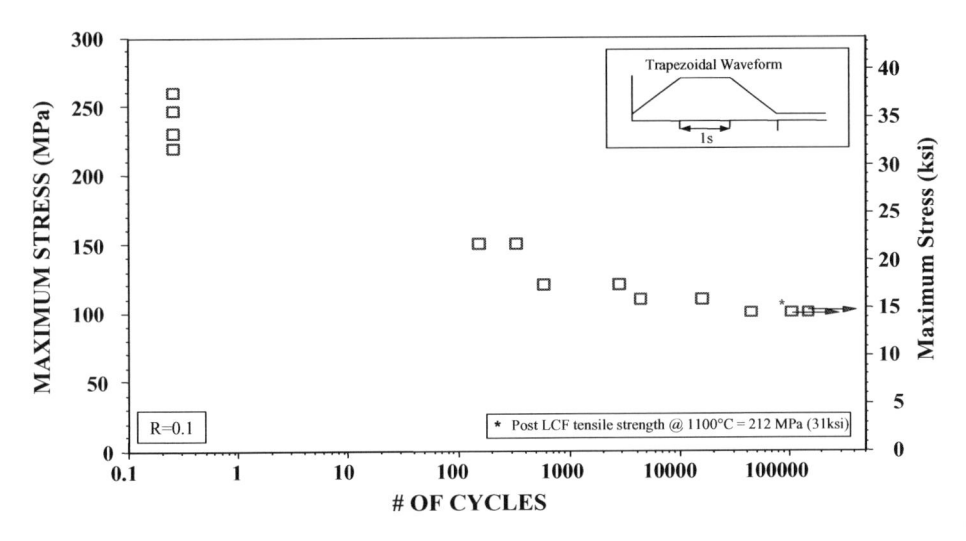

FIGURE 12. S-N plot for 2-D Nicalon™/Al$_2$O$_3$ composites describing low cycle tension-tension fatigue test results obtained at 1100°C in air.

microcracked matrix of Nicalon™/Al$_2$O$_3$ composites result in a more rapid degradation of mechanical properties. It is on the basis of these tests that a shift was made from CG Nicalon™ to a reduced oxygen content Hi-Nicalon fiber.

Tensile stress-strain curves for 2-D Hi-Nicalon™/Al$_2$O$_3$ composites are shown in Fig. 17. Tensile test fracture surfaces for these composites were similar to that

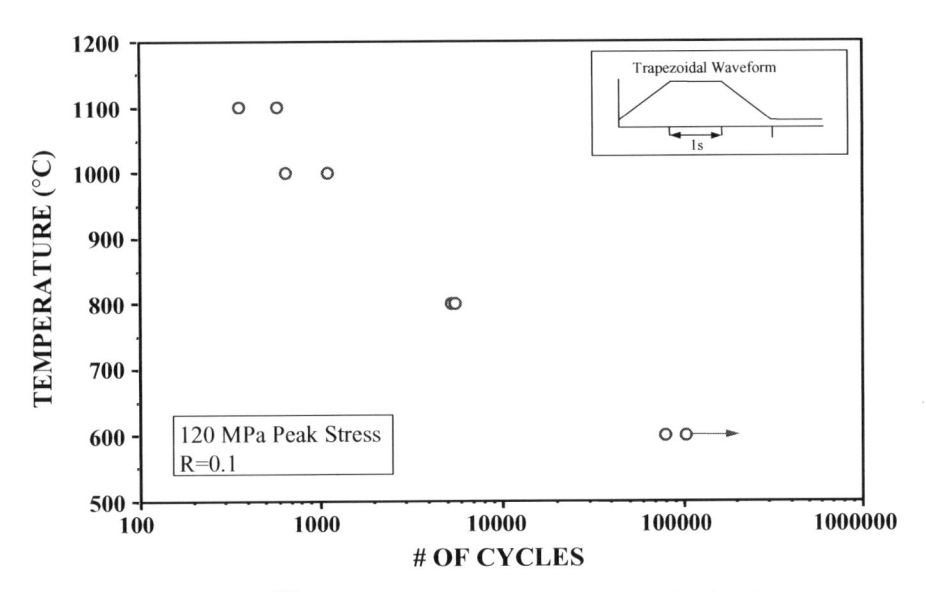

FIGURE 13. 2-D Nicalon™/Al$_2$O$_3$ low cycle tension-tension fatigue as a function of temperature.

FIGURE 14. 2-D Nicalon™/Al$_2$O$_3$ fracture surface of a specimen tested at 1100°C in air following 100,000 tension-tension fatigue cycles at 1100°C.

observed for Nicalon™/Al$_2$O$_3$ composites. Fiber pull-out was substantial (Fig. 18a) with debonding occurring at the fiber-BN interface (Fig. 18b). Further development of the Hi-Nicalon™/Al$_2$O$_3$ composite system was significantly reduced by the introduction of silicon carbide fiber reinforced silicon carbide matrix composites by the melt infiltration process [16]. These melt infiltrated composites utilize the duplex BN/SiC coatings invented for

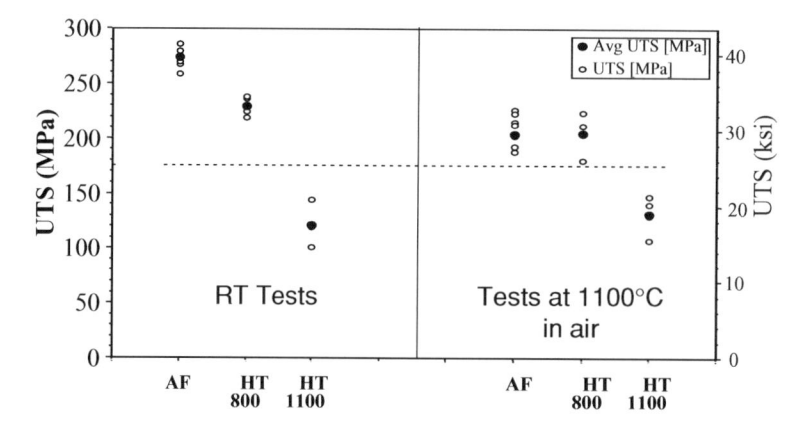

FIGURE 15. 2-D Nicalon™/Al$_2$O$_3$ residual tensile strength following heat-treatment at 800 and 1100°C in air for 1000 h.

SILICON CARBIDE AND OXIDE FIBER REINFORCED ALUMINA MATRIX COMPOSITES 291

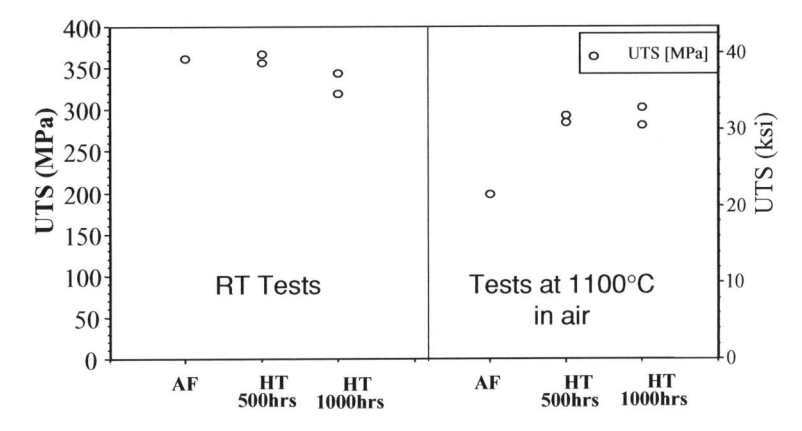

FIGURE 16. 2-D Hi-Nicalon^{TM}/Al$_2$O$_3$ residual tensile strength following heat-treatment at 1100°C in air for 500 and 1000 h.

Nicalon^{TM}/Al$_2$O$_3$. They offer the advantage of a denser, higher thermal conductivity and microcrack-free matrix. Furthermore, the residual silicon metal present in the matrix of melt infiltrated composites has a higher melting temperature (1410°C) relative to the aluminum (623°C) in the matrix of Nicalon^{TM}/Al$_2$O$_3$ composites. Hence, there is no need for a metal removal processing step in SiC$_f$/SiC melt infiltrated composites. From a manufacturing perspective, the reduction of a process step can have significant implications with regard to lowering cost and improving overall process yield. Aluminum oxide matrices fabricated via directed metal oxidation offer a more suitable match for reinforcement with oxide fibers. A preliminary evaluation of various oxide fibers used as reinforcement is discussed later on in this chapter.

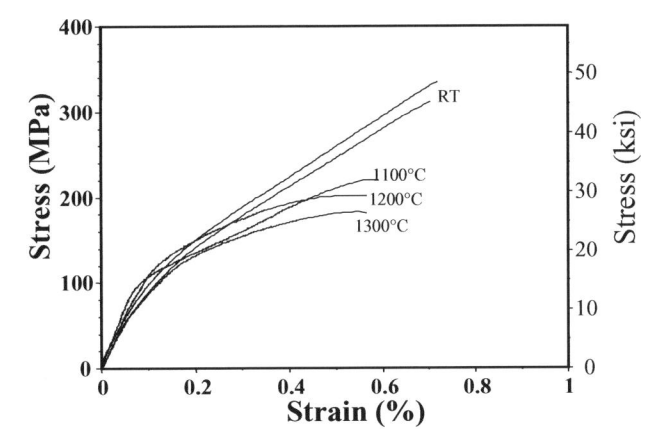

FIGURE 17. Tensile stress-strain curves at various temperatures for 2-D Hi-Nicalon^{TM}/Al$_2$O$_3$ composites.

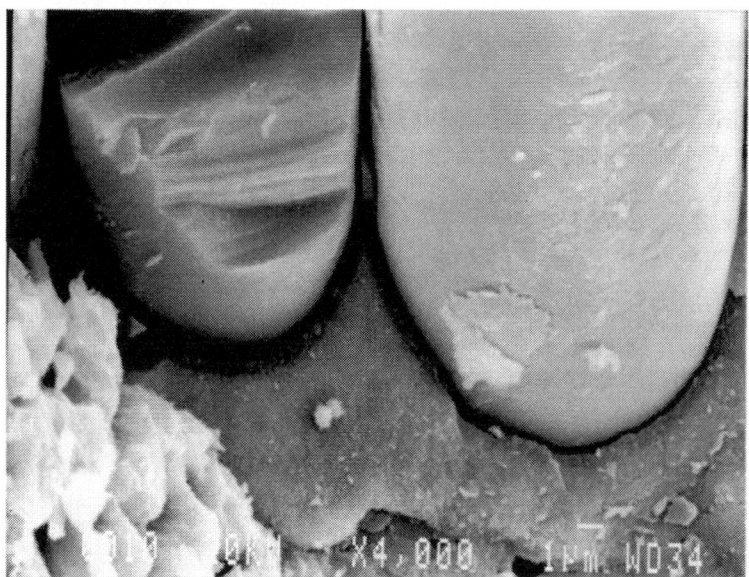

FIGURE 18. Room temperature fracture surfaces for 2-D Hi-Nicalon™/Al$_2$O$_3$ composites showing (a) fiber pull-out and (b) debonding at the fiber-BN interface.

FIGURE 19. Nicalon™/Al$_2$O$_3$ components fabricated using the directed metal oxidation process.

3.5. Turbine Engine Test Results

Several Nicalon™/Al$_2$O$_3$ turbine engine components were fabricated and engine or rig tested by various engine manufacturers. Some of these components fabricated are shown in Fig. 19 . The only machining required for most of these components is along the edges and for the holes. A typical microstructure of a curved section of a component (Fig. 20) shows that the dense microstructure observed in flat panels (Fig. 2) is maintained in components.

Prototype combustor components were fabricated for Williams International for use in expendable engine testing, under an ARPA and US Air Force Wright Laboratory/Materials Directorate funded program. A representative component is shown in the lower left corner of Fig. 19. This combustor components are 25 cm in diameter. Flexural strength specimens were machined from one of the combustor components and compared with those from flat plates supplied to Williams International. Room temperature three point flexural strengths of 8 specimens sectioned from the combustor averaged 350 MPa, whereas four point flexural strengths from the flat plate averaged (8 specimens) 370 MPa. The similarity in strengths confirms that composite fabrication in flat plates are carried over very well to shaped components using directed metal oxidation.

Two sets of combustor components were successfully engine tested for 10 h each by Williams International. The engine tests consisted of operating at temperatures mostly between 1175 and 1400°C with over 5 cold starts in each test. The components experienced peak temperatures as high as 1500°C.

Flameholders for man-rated engine applications were successfully run for up to 87 h in full scale engine tests involving 2500 cycles between 250 and 1000°C. Rig tests representing

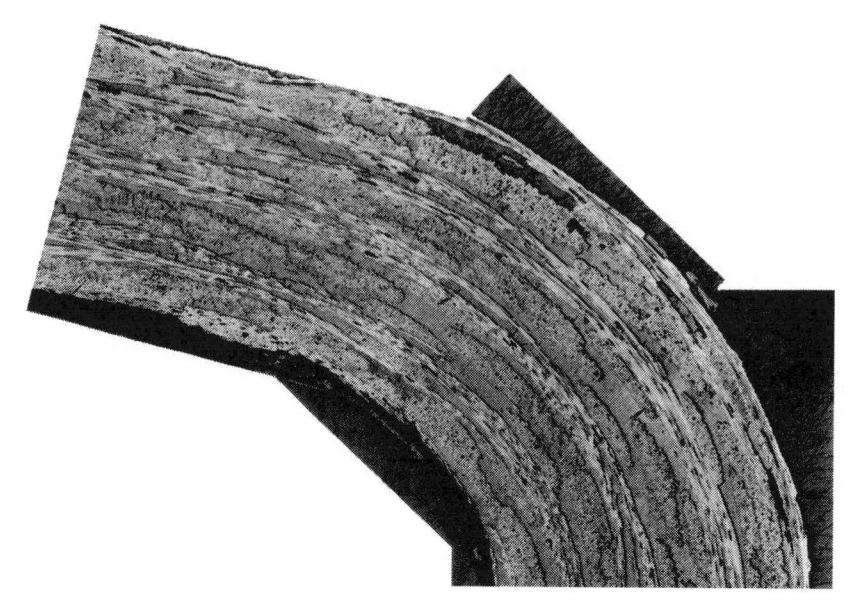

FIGURE 20. Microstructure of a curved section from a 2-D Nicalon™/Al$_2$O$_3$ component.

actual afterburner operation involved higher temperature cycling between 1000 and 1300°C. These were conducted for a total of 150 afterburner cycles with no visible component damage or deterioration.

4. OXIDE FIBER REINFORCED ALUMINUM OXIDE MATRIX COMPOSITES

Various oxide fibers have been used as a reinforcement to fabricate aluminum oxide matrix composites via directed metal oxidation. The procedure used is identical to the one shown for fabricating Nicalon™/Al$_2$O$_3$ composites in Fig. 1. A listing of oxide fibers evaluated is provided in Table IV, along with selected fiber properties [17–19]. While all the composite systems fabricated were evaluated microstructurally, mechanical properties were measured on composites reinforced with PRD-166, Altex™, Almax, Nextel™ 610 and Nextel™ 720 fibers. Among the oxide/oxide systems fabricated to date, a relatively larger effort has been devoted towards the development of composites reinforced with high purity alumina Nextel™ 610 fibers.

4.1. Microstructure of Oxide/Oxide Composites

A representative low magnification optical micrograph of a polished cross-section of Nextel™ 610/Al$_2$O$_3$ is shown in Fig. 21. Uniform matrix infiltration is observed within as well as between fiber bundles, similar to that observed for 2-D Nicalon™/Al$_2$O$_3$ composites in Fig. 2. The interaction of interface coatings with a microcrack introduced via micro hardness indentation on an Almax/Al$_2$O$_3$ composite is shown in Fig. 22. Crack deflection

TABLE IV. Selected Properties of Oxide Fibers Used to Fabricate CMCs Via Directed Metal Oxidation [17–19]

Fiber & Manufacturer	Chemical Composition	RT σ (GPa)	RT E (GPa)	CTE ($10^{-6}/°C$)	Diameter (μm)	Density (g/cm^3)
FP DuPont	>99% Al_2O_3	1.5	390	6.8	20	3.97
PRD-166 DuPont	80% Al_2O_3, 20% ZrO_2	2	380	9	20	4.2
Altex™ Sumitomo	85% Al_2O_3, 15% SiO_2	1.7	190	8.8	15–17	3.25
Nextel™ 440 3M	70% Al_2O_3, 28% SiO_2, 2% B_2O_3	2	190	5	10–12	3.05
Saphikon Saphikon	single crystal Al_2O_3	2.0–3.4	410	9	125–150	3.97
Almax Mitsui Mining	99.50% Al_2O_3	1.8	320	7.1	10	3.6
Nextel™ 610 3M	>99% Al_2O_3, 0.4–0.7% Fe_2O_3, 0.2–0.3% SiO_2	1.6–3.0	360–390	7.9	10–12	3.7–3.8
Nextel™ 720 3M	85% Al_2O_3, 15% SiO_2	2	260	6	12	3.4

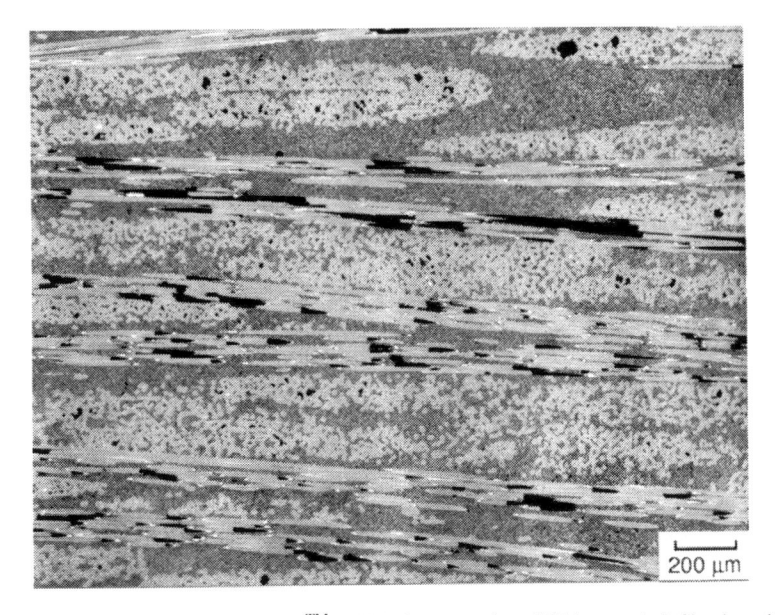

FIGURE 21. Microstructure of a 2-D Nextel™ 610/Al_2O_3 composite exhibiting matrix infiltration within fiber bundles and between fabric plies. The dark regions associated with the horizontal fibers result from fiber pull-out during polishing and are not porosity.

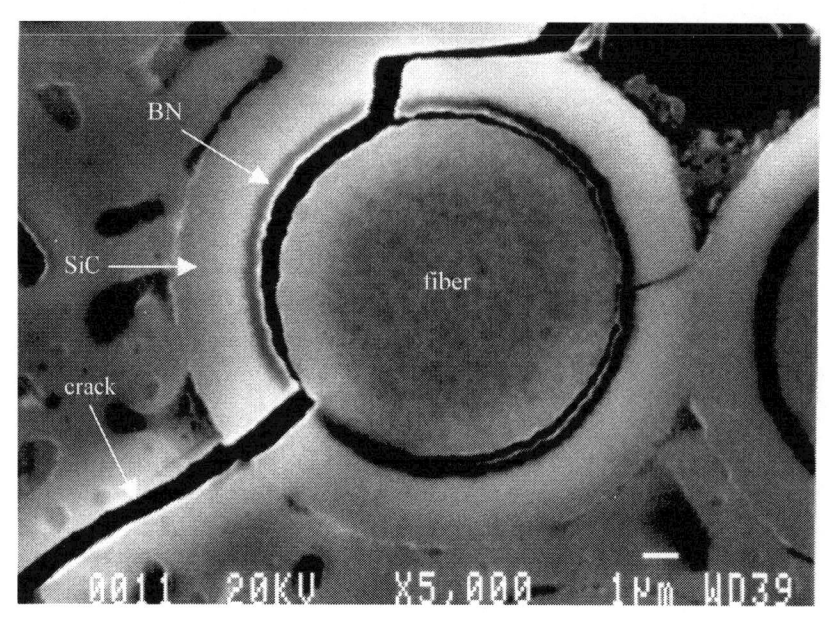

FIGURE 22. Microstructure showing crack deflection at the BN-fiber interface for Almax/Al$_2$O$_3$ composites.

occurs at the BN/fiber interface, which is typical for all the oxide/oxide composites evaluated and similar to the SiC$_f$/Al$_2$O$_3$ system discussed earlier (Figs. 4b and 18b).

4.2. Mechanical Properties of Oxide/Oxide Composites

All mechanical property testing was conducted on specimens with as-machined surfaces. No external coatings were applied. Elevated temperature tests were conducted in air. Four-point flexural strengths were measured per MIL-STD-1942A while the Chevron notch technique [14] was used to obtain composite fracture toughness. Tensile stress strain curves were generated using dogbone shaped specimens and a clip gage extensometer.

The variation of four point flexural strength as a function of temperature for composites reinforced with Almax, Nextel^{TM} 610 and Nextel^{TM} 720 fibers is shown in Fig. 23. Each data point corresponds to an average of a minimum of three tests. All three oxide/oxide composites exhibit a significant decrease in strength with increasing test temperatures. Approximately 50% of the room temperature strength is lost at 1000°C in air. In contrast, a standard Nicalon^{TM}/Al$_2$O$_3$ composite with similar fiber loading and fiber architecture exhibits only a minor loss in flexural strength between room temperature and 1200°C in air (Fig. 23). Early vintage alumina composites, reinforced with PRD-166 and Altex^{TM} aluminum oxide based fibers, also exhibited a decreasing trend of flexural strength with increasing test temperature, similar to the oxide/oxide composites discussed above.

Almax and Nextel^{TM} 610 fiber-reinforced composite samples were exposed to 1000°C in air for 100 h and then tested in four point flexure. At least 80% of the as-fabricated

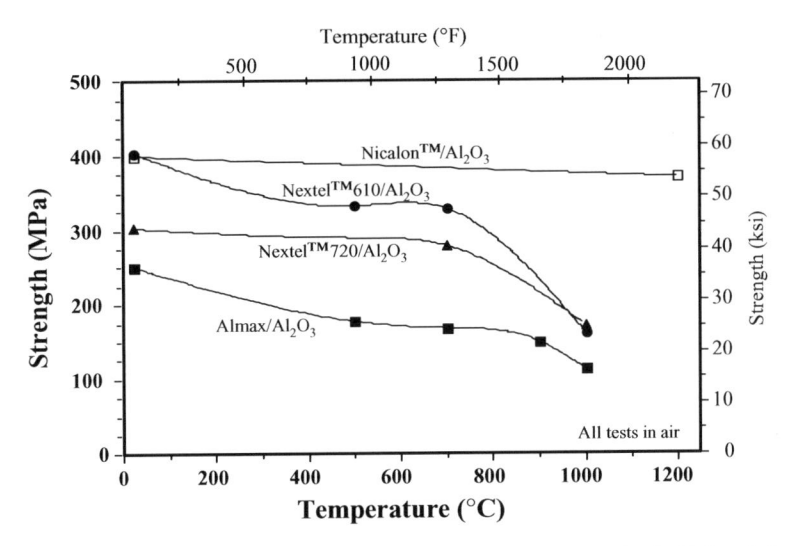

FIGURE 23. Four point flexural strength as a function of temperature for various CMCs fabricated via directed metal oxidation.

composite strength is retained following the heat-treatment (Table V). Hence, mechanical property testing at elevated temperatures appears to be significantly more damaging to the oxide/oxide composites when compared to testing at room temperature following a heat-treatment.

Tensile properties for composites reinforced with Almax, Nextel™ 610 and Nextel™ 720 fibers are summarized in Table VI. The trend of strongest to weakest (viz., Nextel™ 610/Al_2O_3 to Almax/Al_2O_3) is similar to that observed for room temperature four point flexural strengths (Fig. 23) and for individual fiber strengths reported in the literature (Table IV). Stress strain curves for the three oxide/oxide composite systems are shown in Fig. 24 and compared to that of Nicalon™/Al_2O_3. The change in reinforcement from Nicalon™ to oxide based fibers results in a significant (>50%) decrease in strain to failure. The Almax and Nextel™ 720 fiber-reinforced composites also exhibit a substantial decrease in ultimate tensile strength (UTS) relative to the Nicalon™/Al_2O_3 material.

A 40% drop in tensile strength is observed for the Nextel™ 720/Al_2O_3 composite between room temperature and 1000°C in air (Table VI). A similar decrease was observed in the four point flexural strength data (Fig. 23). Tensile stress strain curves at the two

TABLE V. Four Point Flexural Strengths Prior to and Following Composite Exposure at 1000°C for 100 h in Air.

Composite System	Flex. Strength @ RT (MPa)		Flex. Strength @ 1000°C (MPa)	
	As-Fabricated	Heat-Treated	As-Fabricated	Heat-Treated
Almax/Al_2O_3	251	209	113	108
Nextel™ 610/Al_2O_3	404	329	162	129

TABLE VI. Tensile Properties of Oxide/Oxide CMCs Fabricated Via Directed Metal Oxidation

	Test Temp.	UTS* (MPa)	Failure Strain* (%)	Modulus* (GPa)
Almax/Al$_2$O$_3$	23°C	130 ± 12 (5)	0.11 ± 0.02 (5)	155 ± 10 (5)
Nextel^{TM} 610/Al$_2$O$_3$	23°C	230 ± 30 (7)	0.27 ± 0.06 (7)	170 ± 13 (7)
Nextel^{TM} 720/Al$_2$O$_3$	23°C	189 ± 4 (5)	0.21 ± 0.01 (5)	152 ± 7 (5)
Nextel^{TM} 720/Al$_2$O$_3$	1000°C	108 ± 7 (3)	0.09 ± 0.01 (3)	152 ± 7 (3)

* Average ± standard deviation (# tested)

temperatures are compared in Fig. 25. On the basis of the low (0.09%) failure strain at a 1000°C, it is highly improbable that Nextel^{TM} 720/Al$_2$O$_3$ will be used as a high temperature structural material in turbine engine applications.

Fracture toughness was measured for composites reinforced with the aluminum oxide fibers. A decreasing trend in toughness was observed with increasing temperature, similar to the strength data. Nextel^{TM} 610/Al$_2$O$_3$ composites exhibited a decrease in fracture toughness from 17 MPa$\sqrt{m}$ at room temperature to 9 MPa$\sqrt{m}$ at 1000°C in air. Fracture toughness for the Almax/Al$_2$O$_3$ composite ranged from 19 MPa$\sqrt{m}$ at room temperature to 6.6 MPa$\sqrt{m}$ at 700°C to 3.6 MPa$\sqrt{m}$ at 1000°C in air.

Hence, for all of the oxide/oxide composites tested, a similar trend was observed of mechanical property variation as a function of test temperature. There is a significant drop in flexural strength, tensile strength and fracture toughness of the composites between room temperature and 1000°C.

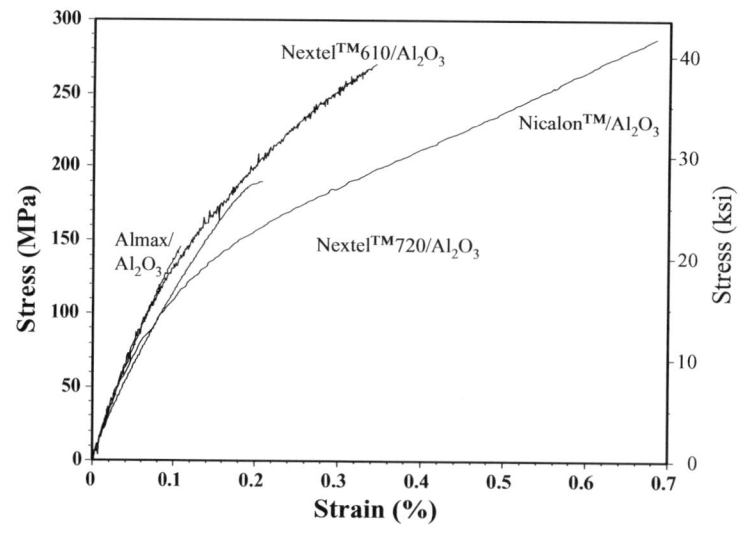

FIGURE 24. Room temperature tensile stress-strain curves for various CMCs fabricated via directed metal oxidation.

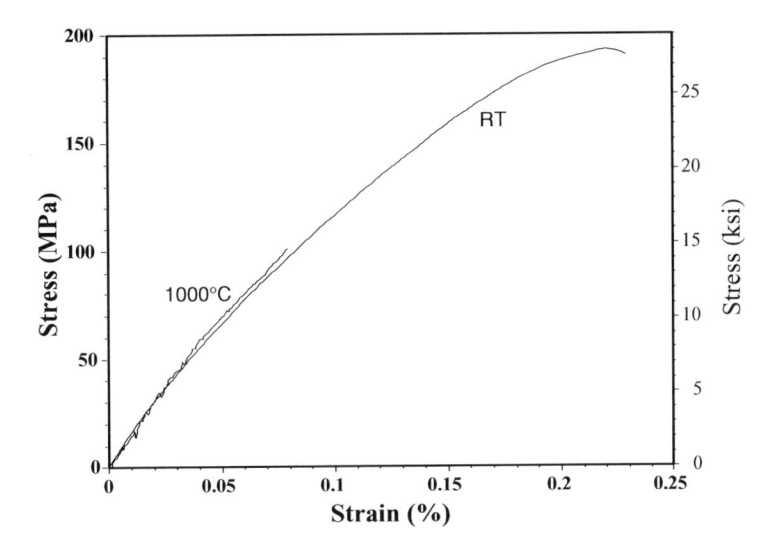

FIGURE 25. Tensile stress strain curves at room temperature and 1000°C in air for a Nextel™ 720/Al$_2$O$_3$ composite.

4.3. Fracture Surface Analysis of Oxide/Oxide Composites

Although a 55% drop in flexural strength was observed for Almax/Al$_2$O$_3$ composites between room temperature and 1000°C in air (Fig. 23), no significant difference was observed in the degree of fiber pull-out at these two temperatures (Figs. 26 and 27). Fiber-matrix debonding is observed at all test temperatures, which is an indication that the BN coating is functioning as a weak interface throughout the test temperature range. Debonding occurs at the fiber-BN interface, as shown by the arrow in Fig. 28.

Further evaluation of the individual fiber fracture surfaces following composite flexural testing revealed a distinct change in fiber fracture morphology from transgranular at room temperature to intergranular at 1000°C (Fig. 29). The intergranular fiber fracture implies weak or low melting grain boundary phases at 1000°C, which would result in a significant drop in fiber strength. Hence, a substantial decrease in composite strength is observed between room temperature and 1000°C in air. Fiber fracture surfaces at 700°C showed a mixed fiber failure mode, with both transgranular and intergranular fractures being observed.

The average size of the alumina grains in Almax is between 0.5 and 1.0 μm (Fig. 29b). Sub-micron sized pores are observed in the room temperature fiber fracture surfaces (Fig. 29a). These pores are not evident in the intergranular fiber fracture surfaces at 1000°C (Fig. 29b) which implies that the porosity in Almax fibers is located within grains and not at the grain boundaries. The relatively low room temperature strengths of the high purity alumina Almax fiber-reinforced composites compared to those reinforced with Nextel™ 610 alumina fibers (Figs. 23 and 24) can be attributed to the porosity present in Almax. The low tensile strain to failure and relatively short fiber pull-out lengths observed in fracture

FIGURE 26. Almax/Al$_2$O$_3$ composite fracture surface at RT.

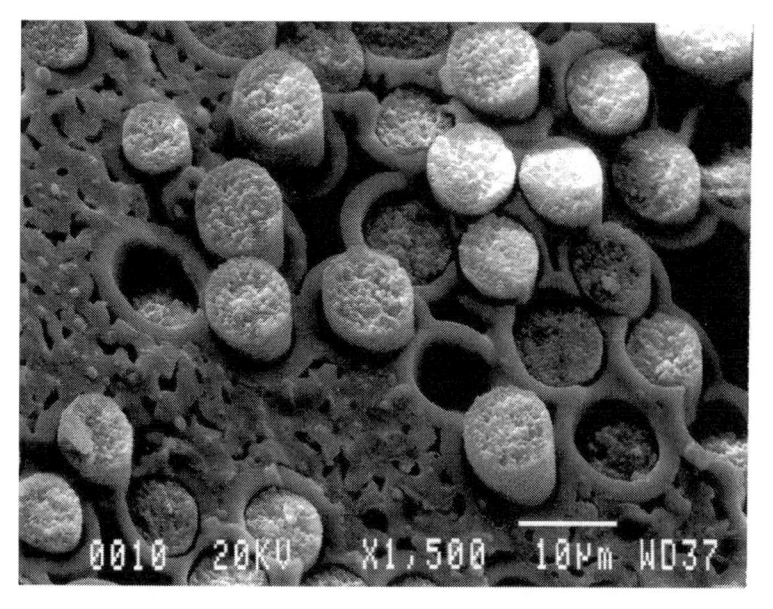

FIGURE 27. Almax/Al$_2$O$_3$ composite fracture surface at 1000°C.

FIGURE 28. Almax/Al$_2$O$_3$ composite fracture surface at 700°C.

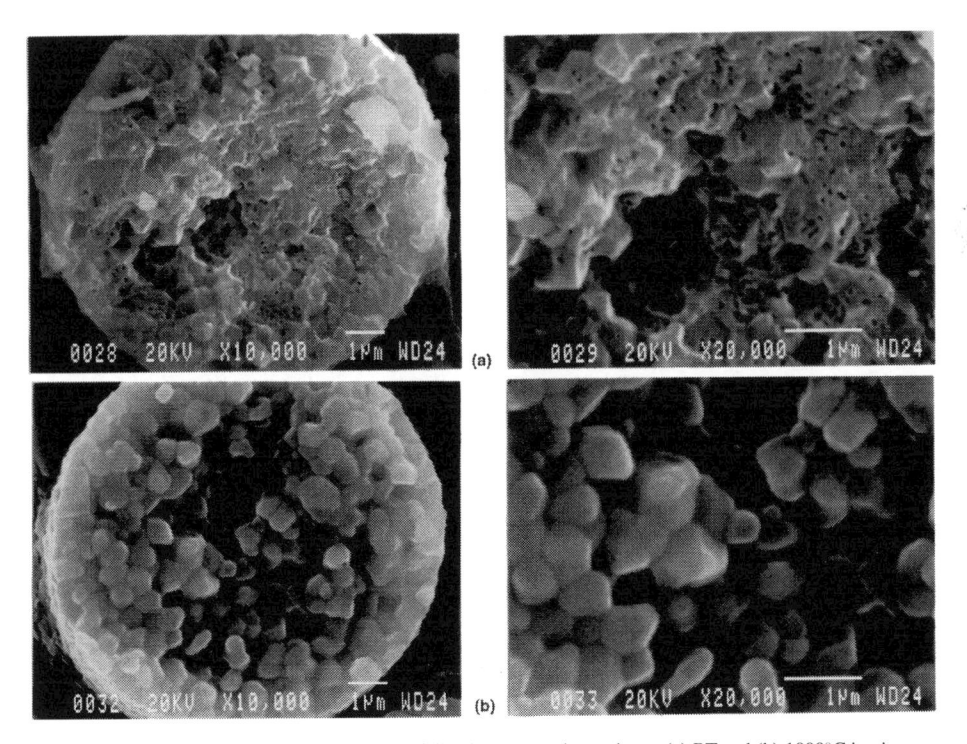

FIGURE 29. Almax fiber fracture surface following composite testing at (a) RT and (b) 1000°C in air.

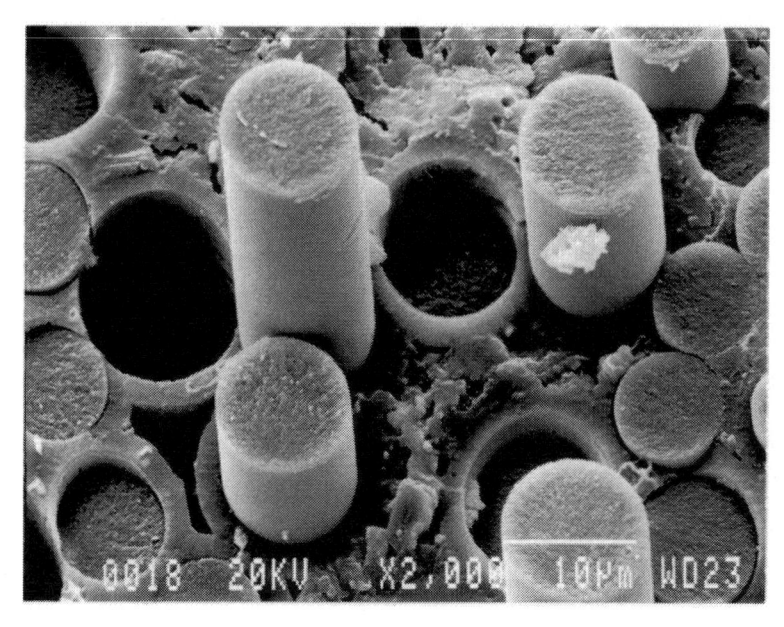

FIGURE 30. Nextel™ 610/Al$_2$O$_3$ composite fracture surface at 1000°C.

surfaces of Almax/Al$_2$O$_3$ composites is also associated with weak, porous fibers. Another potential contributor to the short fiber pull-out lengths observed is the relatively rough surface of Almax fibers (Fig. 22), which is a result of the processing approach used for fiber manufacture, viz., sintering of fine alumina particles. The rough fiber surface can result in mechanical interlocking, which inhibits to some extent, fiber-matrix sliding and reduces the degree of fiber pull-out.

Fracture surfaces for Nextel™ 610/Al$_2$O$_3$ composites showed longer pull-out lengths relative to the Almax/Al$_2$O$_3$ composites which was attributed to denser and hence, stronger fibers and a smoother fiber surface. Nextel™ 610 fiber pull-out following flexural strength testing at 1000°C is shown in Fig. 30, and is an indication of adequate debonding functionality of the BN coating at this temperature. As with the Almax/Al$_2$O$_3$ composite, debonding occurs at the Nextel™ 610 fiber-BN coating interface. An evaluation of the individual fiber fracture surfaces also exhibits a change in fracture morphology for Nextel™ 610 fibers from transgranular at room temperature to intergranular at 1000°C [20], which would account for the loss in composite strength as a function of increasing temperature (Fig. 23). Fiber fracture surfaces indicate dense fibers with an alumina grain size of ~0.2 μm [20]. Overall, the temperature dependence of strength and fracture behavior of composites reinforced with the high purity alumina fibers, Nextel™ 610 and Almax, are similar. The higher room temperature strength of Nextel™ 610/Al$_2$O$_3$ (>60%) is attributed to the higher strength of Nextel™ 610 fibers, predominantly because they are denser and have a smaller grain size relative to Almax. A room temperature tensile fracture surface of Nextel™ 610/Al$_2$O$_3$

FIGURE 31. Nextel™ 610/Al$_2$O$_3$ composite fracture surface following tensile testing at room temperature.

(Fig. 31) exhibits a significant amount of fiber pull-out, which is fairly similar to that observed in CG as well as Hi-Nicalon™/Al$_2$O$_3$ composites (Fig. 18).

5. SUMMARY

Directed metal oxidation is a versatile process capable of fabricating aluminum oxide matrix composites with a variety of fiber reinforcements. The major effort to date for these composites has been focused on silicon carbide fiber reinforcements. Long lifetimes of over 6000 h have been demonstrated at 1200°C in air with an applied stress of 70 MPa. Engine and rig tests were successfully completed using SiC$_f$/Al$_2$O$_3$ components at temperatures between 1000 and 1400°C. Compared to ceramic grade Nicalon™, the lower oxygen containing Hi-Nicalon™ fibers proved to be a superior reinforcement with respect to retained mechanical properties following exposure to temperatures of 1100°C in air. In comparison, composites fabricated using oxide fiber reinforcements exhibited reduced strengths and toughnesses. The rapid strength loss as a function of increasing temperature for the oxide/oxide system is attributed to a change in fiber fracture morphology from transgranular at room temperature to intergranular at 1000°C. The intergranular mode of failure suggests the presence of weak grain boundary phases at the elevated temperatures, which would contribute towards lowering fiber strength. The results of this study reinforce the critical need for a small diameter oxide fiber with high temperature thermomechanical stability. Directed metal oxidation is an ideal matrix formation process for such an oxide fiber as and when one becomes commercially viable.

ACKNOWLEDGMENTS

A portion of the technical development work leading to the results described was supported by the Defense Advanced Research Projects Agency through the Office of Naval Research. Oxide fiber composites were partially funded by the Department of Energy as part of the CFCC program. Additional funding was provided by the DuPont Company and Lanxide Corporation in support of the DuPont Lanxide Composites Inc. joint venture.

The author respectfully acknowledges the many stimulating discussions held with various colleagues at Lanxide, DuPont Lanxide Composites and Power Systems Composites, LLC. Williams International and the University of Michigan generated some of the data presented. The permission to use these data from all parties concerned is gratefully acknowledged.

APPENDIX

Nicalon^{TM} is a trademark of NCK Nippon Carbon Co., Ltd
Nextel^{TM} is a trademark of 3M Co.
Altex^{TM} is a trademark of Sumitomo Alumina Chemical Co.

REFERENCES

1. M.S. Newkirk, A.W. Urquhart, H.R. Zwicker, and E. Breval, "Formation of Lanxide^{TM} Ceramic Composite Materials," J. Mater. Res., 1 [1] 81–89 (1986).
2. M.S. Newkirk, H.D. Lesher, D.R. White, C.R. Kennedy, A.W. Urquhart, and T.D. Claar, "Preparation of Lanxide^{TM} Ceramic Matrix Composites: Matrix Formation by the Directed Oxidation of Molten Metals," Ceram. Eng. Sci. Proc., 8 [7–8] 879–85 (1987).
3. A.W. Urquhart, "Novel Reinforced Ceramics and Metals – A Review of Lanxide's Composite Technologies," Mater. Sci. and Eng. A144 75–82 (1991).
4. A.S. Fareed, B. Sonuparlak, C.T. Lee, A.J. Fortini, G.H. Schiroky, "Mechanical Properties of 2–D Nicalon^{TM} Fiber Reinforced LANXIDE^{TM} Aluminum Oxide and Aluminum Nitride Matrix Composites," Ceram. Eng. Sci. Proc., 11 [7–8] 782–94 (1990).
5. W.B. Johnson, "Reinforced Si_3N_4 Matrix Composites Formed by the Directed Metal Oxidation Process," Ceram. Eng. & Sci. Proc. 13 [9–10] 573–580 (1992).
6. A.S. Fareed, G.H. Schiroky, and C.R. Kennedy, "Development of BN/SiC Fiber Coatings for Fiber-Reinforced Alumina Matrix Composites Fabricated by Directed Metal Oxidation," Ceram. Eng. Sci. Proc., 14 [9–10] 794–801(1993).
7. A.J. Caputo, W.J. Lackey, and D.P. Stinton, "Development of a New, Faster Process for the Fabrication of Ceramic Fiber-Reinforced Ceramic Composites by Chemical Vapor Infiltration," Ceram. Eng. Sci. Proc. 6 [7–8] 694–706 (1985).
8. R.L. Lehman, "Glass-Ceramic-Matrix Fiber Composites," in: Flight-Vehicle Materials, Structures and Dynamics – Assessment and Future Directions, Vol. 3, S.R. Levine, ed., ASME, New York, 77–92 (1992).
9. A.S. Fareed, "Ceramic Matrix Composite Fabrication and Processing: Directed Metal Oxidation," Handbook on Continuous Fiber Reinforced Ceramic Matrix Composites, ed. R.L. Lehman, S.K. El-Rahaiby Jr and J.B. Wachtman, Jr., CIAC/ACers, 301–324 (1995).
10. A.S. Fareed, D.J. Landini, T.A. Johnson, A.N. Patel, P.A. Craig, "High Temperature Ceramic Matrix Composites by the Directed Metal Oxidation Process," Proc. High Temperature Composites Clinic, SME, (1992).

11. A.S. Fareed, B. Sonuparlak, P.A. Craig, and J.E. Garnier, "Effect of Sustained High Temperature Exposure on the Mechanical Properties of Nicalon^{TM}/Al$_2$O$_3$ Composites," Ceram. Eng. & Science Proc. 13 [9–10] 573–580 (1992).
12. G.H. Schiroky, A.S. Fareed, B. Sonuparlak, C.T. Lee, and B. Sorenson, "Fabrication and Properties of Fiber-Reinforced Ceramic Composites Made by Directed Metal Oxidation," pp. 151–163, in: Flight-Vehicle Materials, Structures and Dynamics – Assessment and Future Directions, Vol. 3, S.R. Levine Jr, ed., ASME, New York, 1992.
13. C.A. Andersson, P. Barron-Antolin, G.H. Schiroky, and A.S. Fareed, "Properties of Fiber-reinforced Lanxide^{TM} Alumina Matrix Composites," pp. 209–15 in: ASM Proc. Int. Conf. Whisker- and Fiber-toughened Ceram., 1988.
14. D.G. Munz, J.L. Shannon, and R.T. Bubsey, "Fracture Toughness Calculations from Maximum Load in Four Point Bend Tests of Chevron Notch Specimens," Int. J. of Fracture 16 R137–R141 (1980).
15. M. Takeda, "Mechanical and Structural Analysis of Silicon Carbide Fiber Hi-Nicalon Types," Ceramic Engineering and Science Proceedings 17(4–5) 35–42 (1996).
16. D.J. Landini, A.S. Fareed, H. Wang, P.A. Craig Jr and S. Hemstad, "Ceramic Matrix Composites Development at GE Power Systems Composites, LLC," submitted for publication in Ceramic Gas Turbine Design and Test Experience – Progress in Ceramic Gas Turbine Development, Vol. 2, ed. D. Richerson, M. Ferber and M. Van Roode, ASME Press.
17. J.K. Weddell, "Continuous Ceramic Fibers," J. Text. Inst., 81[4] 333–359 (1990).
18. T.L. Tomkins, "Ceramic Oxide Fibers: Building Blocks for New Applications," Ceramic Industry, April (1995).
19. Ceramic Fibers and Coatings – Advanced Materials for the Twenty-first Century, Publication NMAB-494, National Academy Press, Washington D.C., 22 (1998).
20. A.S. Fareed, and G.H. Schiroky, "Microstructure and Properties of Nextel^{TM} 610 Fiber Reinforced Ceramic and Metal Matrix Composites," Ceram. Eng. Sci. Proc., **15** [4] 344–352(1994).

13

SiC Whisker Reinforced Alumina

Terry Tiegs

Oak Ridge National Laboratory
Oak Ridge, TN 37831-6087
Telephone: 865-574-5173; Email: tiegstn@ornl.gov

ABSTRACT

SiC whisker reinforced alumina composites exhibit significant improvements in mechanical properties, such as strength and fracture toughness. These composites are typically densified by pressure-assisted sintering (i.e. hot-pressing) with SiC whisker contents ranging from 10 to 30 vol.%. Cutting tools for high nickel alloys are the major application, but other wear and structural uses are also being developed.

1. INTRODUCTION

As a class of ceramic materials, SiC whisker-reinforced ceramic composites were developed for potential structural applications because of the significant improvements in the mechanical properties these materials offered as compared to the monolithic materials. The incorporation of SiC whiskers into alumina ceramics resulted in increases in strength, fracture toughness, thermal conductivity, thermal shock resistance and high temperature creep resistance. These discoveries initiated several years of intense study into this class of composites.

SiC whiskers were initially developed in the early 1960s. However, their use was originally to be applied to the reinforcement of metal matrices, such as aluminum. These metal matrix composites were only a small commercial success, mainly because of the high cost of the whiskers. The first application of whisker reinforcement to ceramics, and in particular alumina, did not occur until the 1980's. For the purposes of this article, SiC whiskers will be defined to be acicular or needle-like shaped, discontinuous, nearly single crystals,

0.1 to 5 μm in diameter and having lengths greater than 5 μm. This differentiates them from other fibers materials that are polycrystalline, amorphous or consist of multiple coatings. It also limits the dimensional range to relatively small sizes, but not small enough to be in the nanoscale range. Much of the early research and published data prior to the mid-1970s employed the vapor-liquid-solid (VLS) mechanism to grow small quantities of SiC whiskers [1,2]. However, at that time there was limited markets for these materials and the costs were prohibitively high since they were produced in such small quantities. Since that time, methods were developed to mass-produce whiskers at an acceptable cost. These mass-production methods, discussed mainly in the patent literature, used carbothermic reduction reactions of low-cost silica and carbon precursors, such as rice hulls, to produce industrial-scale quantities of whiskers at competitive costs. Thus, SiC whiskers have become economically viable reinforcing agents in applications of large-scale, high-volume components.

With the development of some composite materials and the commercial success of some of these products, the demand for whiskers has increased. Numerous companies experimented with different whisker growth processes over the years and several patents have been issued in the area. At the present time, commercial sale of whiskers is limited to a few companies in the world with SiC whiskers the most predominant. In addition, whiskers of TiC, TiN, Al_2O_3, mullite, Si_3N_4, and B_4C have also been produced and studied. However, SiC remains the whisker of most interest and commercial success.

2. TOUGHENING BEHAVIOR OF WHISKER-REINFORCED COMPOSITES

Analysis of the toughening behavior responsible in the SiC whisker reinforced alumina composites showed that crack-whisker interaction resulting in crack bridging, whisker pull-out and crack deflection are the major toughening mechanisms. For these mechanisms to operate, debonding along the alumina matrix-whisker interface (often associated with crack deflection) must occur during crack propagation and allow the whiskers to bridge the crack in its wake. Examination of the fracture surfaces of whisker-reinforced composites reveals they are microscopically rough with whiskers readily evident as shown in Figs. 1 and 2.

Micro-mechanical modeling and available experimental evidence indicates that the composite toughness, K_{IC} (composite), can be described as the sum of the matrix toughness, K_{IC} (matrix), and a contribution due to whisker toughening, ΔK_{IC} (whisker reinforcement). In other words,

$$K_{IC} \text{ (composite)} = K_{IC} \text{ (matrix)} + \Delta K_{IC} \text{ (whisker reinforcement)} \quad (1)$$

In one analysis [3], a relationship was derived for the increase in fracture toughness due to whisker reinforcement:

$$\Delta K_{Ic}(W.R.) = \sigma_f \cdot \sqrt{\frac{V_f \cdot r}{B \cdot (1-v^2)} \cdot \frac{E_c}{E_w} \cdot \frac{\gamma_m}{\gamma_i}} \quad (2)$$

where,

ΔK_{Ic} (W. R.) = increase in fracture toughness due to whisker reinforcement
σ_f = fracture strength of whiskers
V_f = volume fraction of whiskers

SIC WHISKER REINFORCED ALUMINA

FIGURE 1. Fracture surface of SiC whisker reinforced alumina showing very rough surface and tortuous crack propagation.

r = whisker radius
v = Poisson's ratio for whiskers
E = Young's modulus for composite (c) and whiskers (w)
γ = fracture energy for matrix (m) and matrix-whisker interface (i)
B = constant that depends on the bridging stress profile (approximately 6 for alumina-SiC whisker composites)

In a similar fashion from a second analysis [4], the following was used to relate whisker reinforcement to toughening:

$$\Delta K_{Ic}(W.R.) \sim f\,d\,S^2/E + 4\mathbf{L}_i\,f(d/R)/(1-f) \tag{3}$$

where,

ΔK_{Ic} (W.R.) = increase in fracture toughness due to whisker reinforcement
f = volume fraction of whiskers
d = whisker debond length
S = strength of whiskers
E = Young's modulus of composite
R = whisker radius

In both analyses of toughening behavior, increases in toughness for whisker-reinforced composites are dependent on the following parameters: (1) whisker strength, (2) volume fraction of whiskers, (3) elastic modulus of the composite and whisker, (4) whisker diameter, and (5) interfacial fracture energies.

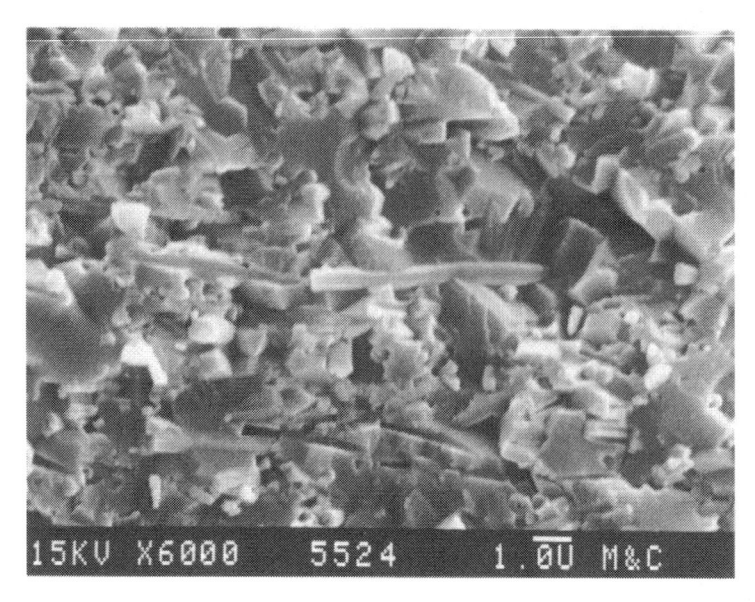

FIGURE 2. Fracture surface of SiC whisker reinforced alumina. SiC whiskers clearly observed on exposed surfaces.

As can be deduced from the discussion above, the characteristics of the whiskers, such as diameter and strength, have a direct effect on the toughening behavior and mechanical behavior. The surface chemistry of the whiskers influences the nature of the interface bond between the whiskers and the matrix, which is addressed in the γ_m/γ_i term in Eq. 2. It has been found that whiskers from various manufacturers have different characteristics, which affect their performance as reinforcements.

3. SiC WHISKER CHARACTERISTICS

SiC whiskers can vary in their physical characteristics depending on the fabrication techniques used in manufacturing. In some cases lot-to-lot variations have been observed from the same manufacturer. While some of the differences appear minor, many have significant influence on the behavior of whiskers in alumina matrix composites. The variability is a result of differences in the SiC growth processes employed by the various manufacturers. Parameters that affect the growth processes include such things as types of raw materials, catalysts, atmospheres, fabrication temperatures, heating rates, gas flow behavior, and reactor geometry [1]. In addition, from the same process, fluctuations in starting material chemistry and normal fabrication conditions can influence the whisker growth process.

A general summary of typical physical characteristics of selected SiC whiskers is given in Table 1. The performance of the whiskers in a number of applications is dependent in part on the diameter and aspect ratio. For example, the toughening behavior in alumina matrix composites is dependent on the ability of the whiskers to bridge propagating cracks. This

TABLE 1. Summary of typical physical characteristics and bulk chemistry of selected SiC whiskers.

	American Matrix	ARCO/ACMC	Tokai Carbon	Tateho
Ave. Diameter	1.3	0.6	0.5	0.4
Approx. Mean Aspect Ratio	—	30 : 1	25 : 1	50 : 1
Impurities (ppm)				
B	>1000	3	—	3
Ca	400	400	20	400
Fe	200	200	50	200
K	100	100	3	20
Mg	100	100	3	30
Mn	3	300	5	30
Na	100	100	50	5
O	2.9%	1.3%	N. D.	1.0%

ability, in turn, is dependent on the whisker diameter and debond length. Scanning electron microscopy (SEM) reveals a wide variety of morphologies from the different SiC whisker manufacturers as shown in Figs. 3 and 4. The surface roughness has been observed to vary considerably from relatively smooth to very faceted surfaces.

The chemistry of the whiskers is important since it affects many areas in fabrication and final composite properties. The bulk chemistry influences the stability of the whiskers at high temperatures and in reactive environments. The surface chemistry can change the interface properties between the whisker and the matrix and thereby change the debonding behavior

FIGURE 3. Appearance of SiC whiskers produced by Tateho Chemical Co.

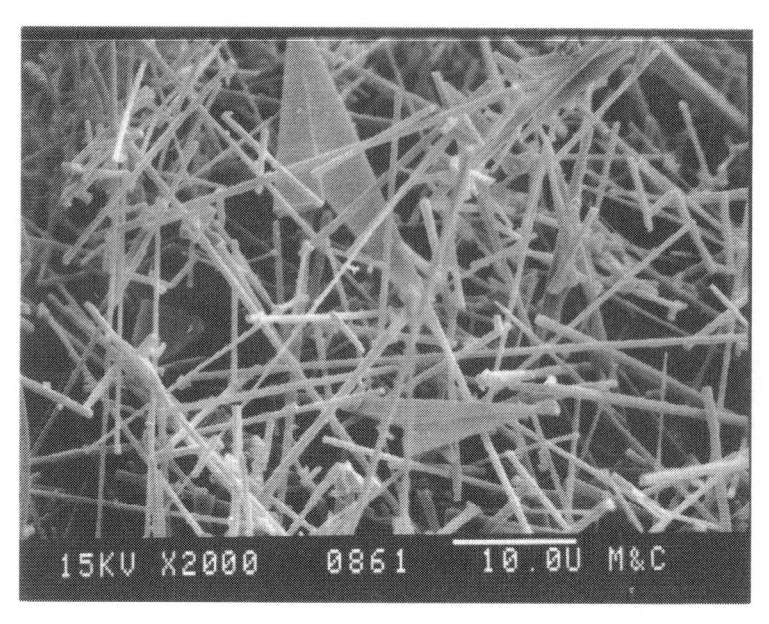

FIGURE 4. Appearance of SiC whiskers produced by Advanced Composites Materials Corp.

and consequently the fracture toughness. The surface chemistry also determines what, if any, chemical species can be incorporated into the matrix phase during high temperature processing and alter its properties.

The bulk chemistry of SiC whiskers varies, and for the most part like the physical characteristics just described, depends on the starting materials and/or the catalysts used in the whisker growth processes. A general comparison of bulk chemistries for selected SiC whiskers is also shown in Table 1. The differences can be directly related to the precursors, catalysts and fabrication techniques used in manufacture of the SiC whiskers [1]. For example, American Matrix Inc. (AMI and now part of Advanced Refractory Technology, Inc.) reportedly used boric acid as a catalyst in the growth process and consequently, high concentrations remain in the whiskers. In both the Advanced Composites Materials Corp. (ACMC) and Tateho processes, rice-hulls are used as precursor materials. As a result, in those materials, high calcium contents attributable to the rice hulls are readily observed in the final products. A further example points to the high Co content of the Tokai Carbon whiskers where the Co is reportedly added for use as a catalyst. The distribution of the impurities occurs both on external whisker surfaces and within internal inclusions. The oxygen contents are predominantly attributable to silica and depend mainly on post-growth oxidation treatments given the whiskers to remove any residual carbon. Post-oxidation treatments, such as HF acid leaching, can be used to reduce the oxygen content.

In addition to the bulk chemistry, the SiC whisker surface chemistry is also very important in alumina matrix composites. This is because it directly influences the nature of the interface bond between the alumina matrix and the SiC whisker. The properties of the interface are important for the toughening mechanisms operable in the composite to work. A wide range of surface chemistries have been observed from the various whisker sources. Based on x-ray photon spectroscopy (XPS or ESCA), SiC whiskers can be classified into three broad general categories [5]. The first category is high in surface oxygen and XPS results show strong Si-O bonding on the surface typical of silica. These types of surfaces maybe a result of the whisker growth process or may be attributed to a post-growth oxidation treatment. Such treatments are commonly given to whiskers derived from carbothermal reduction process where excess carbon is used. The second general category exhibits predominantly Si-C bonding with a relative "clean" SiC chemistry. These are whiskers where an oxidation step may have been followed by an HF acid treatment to remove any excess SiO_2. The last category also has strong Si-C bonding, like the "clean" surfaces, but has excess surface carbon and strong C-C bonding is also observed.

Because whiskers are single crystals with very ordered structures, they typically can have very high tensile strengths. This makes them good reinforcement materials for composites. The theoretical strength of perfect flaw-free crystals is on the order of one-tenth of the elastic modulus. Thus, SiC whiskers should have theoretical strengths >40 GPa. However, the highest reported values are only up to 20 GPa and most are in the range of 4–11 GPa [6]. The elastic moduli of SiC whiskers have been reported to be ~550 GPa and the fracture toughness was determined to be 3.23 MPa$\sqrt{m}$ [7].

The strength limiting flaws in selected SiC whiskers were calculated to be in the size range of 0.1–0.7 μm, which corresponds to defects observed in many commercial whiskers [8]. Defects of this size range include the core regions, which are an accumulation of smaller inclusions and also voids located within some whisker types. Another defect decreasing whisker strength is excessive surface roughness, which is usually associated with crystallographic phase changes and results in stress concentrations. Extensive investigations by transmission electron microscopy (TEM) of the flaws in SiC whiskers derived from rice-hulls have been described in previous papers. The defects were described as small, partially crystalline inclusions (approximately 0.01 μm diameter) containing calcium, manganese and oxygen concentrated in a whisker core region. Similar observations have been made on whiskers not derived from rice-hulls.

The ideal SiC whisker for reinforcement of ceramic composites would have several characteristics that would also depend on the matrix phase. In every case, the ideal whisker would have no internal structural imperfections, be relatively smooth and possess high strength. For alumina matrices where the difference in thermal expansion is great between the matrix and the whisker, the whisker diameter would be in the range of 1.5–2.0 μm [9]. For alumina-SiC composites, it was determined that the presence of excess surface carbon on the whiskers resulted in a weak interfacial bond and produced materials with high toughness and strength [9].

4. FABRICATION OF WHISKER COMPOSITES

The whiskers are normally received from commercial sources in an agglomerated form and appear to be spray-dried balls. Deagglomeration and dispersion of the SiC whiskers has been found to be extremely important in the fabrication of composites with good mechanical

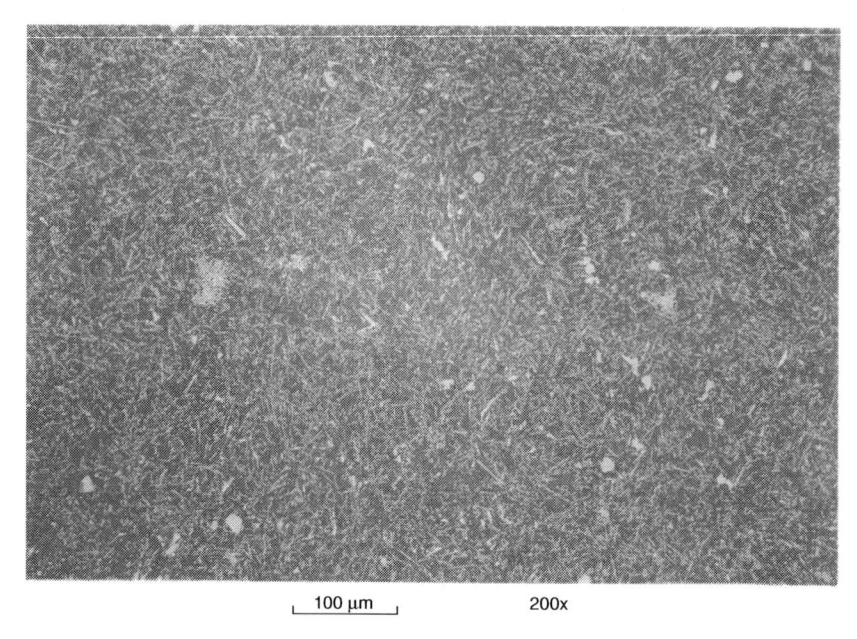

FIGURE 5. Microstructure of SiC whisker reinforced alumina. In addition to SiC whiskers, some particulates are also observed.

properties. Numerous methods have been used to disperse the SiC whiskers and mix them with the matrix powders. These include ultrasonic homogenization, high shear mixing, ball milling, and turbomilling [5]. If the whiskers remain as agglomerates, these remain in the final composite as low density regions and will degrade the final mechanical properties. Dispersion of whiskers can also be important for achieving good fracture toughness. In one report, an Al_2O_3-40 wt% SiC whisker composite showed $K_{Ic} = 8.1$ MPa$\sqrt{m}$ with a homogeneous distribution of whiskers [10]. When the whiskers were poorly dispersed and agglomerated the $K_{Ic} = 6.2$ MPa$\sqrt{m}$. Typical microstructures of SiC whisker reinforced alumina with good whisker dispersion are shown in Figs. 5 and 6.

In contrast to fiber-reinforced composites that require hand lay-up techniques, whisker composites are also amenable to more conventional processing techniques normally employed in the ceramic industry [5]. After mixing the whiskers with matrix powders, 'green' articles can be formed by standard procedures such as dry-pressing, slip-casting, extrusion and injection molding. The process of mixing whiskers (with high aspect ratios) and essentially equiaxed particles (alumina powder) has been treated theoretically. The analysis shows that to obtain composites with high densities low aspect ratio whiskers are needed [11]. For example, at an aspect ratio of 50, the calculated packing density of whiskers is only 10%. But as the aspect ratio is lowered to 10, the calculated packing density is increased to 50%. The analyses also suggest the use of matrix powders with diameters less than the diameter of the whiskers for optimum packing of the mixtures. Computer models of percolation theory calculate that the critical fraction to form a network of whiskers is inversely

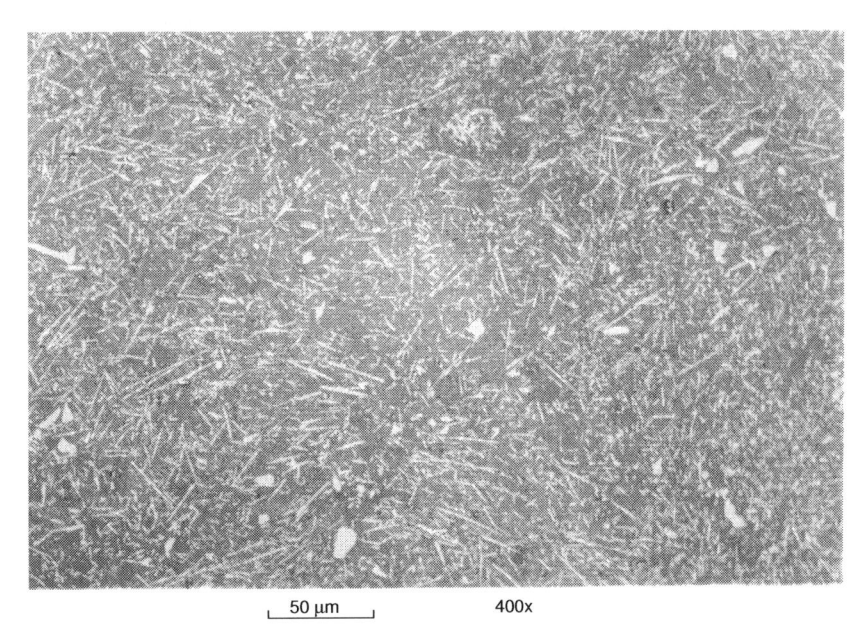

FIGURE 6. Microstructure of SiC whisker reinforced alumina.

dependent on the aspect ratio. For example, with an aspect ratio of 7, the critical fraction is 27 vol.%, which agrees well with experimental evidence. As whiskers are randomly packed, they tend to form semi-rigid arrays that inhibit further consolidation.

During dry-pressing, orientation of the whiskers perpendicular to the pressing axis has been observed. This results in anisotropic shrinkage of the whisker composites relative to the original uniaxial pressing direction during densification. For example, in an alumina-10 vol.% SiC whisker compact, typical sintering shrinkages were 17–25% in the direction parallel and 10–15% in the direction perpendicular to the original uniaxial pressing direction [12].

When whiskers are added to a typical ceramic slurry, rheology changes are dependent on the whisker content. At low whisker content of ~5 vol.%, viscosities have actually been observed to decrease due to the improved packing densities. As whisker contents approach 10–15 vol.%, increases in viscosity are observed in alumina matrix composite slurries [13]. During slip-casting the whiskers can become oriented with preferential alignment parallel to the mold surface. Like the effects observed in dry-pressing, this orientation results in anisotropic shrinkage during sintering. In one report, a maximum linear shrinkage of 21% was observed perpendicular to the whisker plane, but only 7% parallel to the whisker plane. Green densities for the slip-cast composites is generally higher than for dry-pressed mixtures at comparable whisker loadings. In one study of alumina-SiC, the suspensions were optimized and green densities of 69% T. D. were obtained at a whisker loading of 15 vol.% [13].

Green forming of whisker composites can also be done by extrusion, injection molding, and tape casting. In all these cases, orientation of the whiskers is pronounced. In the case of extruded alumina-SiC tubes, the orientation of the whiskers resulted in sintering axial shrinkages of 12% along the length of the tube and 19% radial shrinkage in the diametral dimension. For injection molding, severe distortion of a complex shaped article can occur during densification of these composites. Some of these problems can be minimized by using higher solids loading in the injection molding mixture and by altering the mold gating pattern.

Densification methods fall into two general categories: (1) pressureless-sintering and (2) pressure-assisted [5]. Pressureless sintering is the most economical and is capable of producing complex shaped parts. However, whisker loadings are limited where high sintered densities can be achieved. The pressure-assisted techniques include hot-pressing and hot-isostatic-pressing (HIP). These methods produce high densities at high whisker loadings.

It has been well documented that sintering of alumina-SiC whisker composites is severely inhibited because the whiskers interfere with particle rearrangement and composite shrinkage. This is because as shrinkage occurs, the whiskers form a semi-rigid network that inhibits further densification. In other words, the force to overcome the network are the greater than the forces driving the sintering process. So, as the whisker volume content increases, the ability to sinter the composites to high densities decreases. To promote sintering, a number of techniques are effective, such as lowering the whisker aspect ratio, adjusting the matrix particle size to optimize green density, and the use of a liquid phase sintering aid to encourage rearrangement. Reduction of the aspect ratio improves densification by increasing particle-whisker packing for higher green densities. The lower aspect ratio also enhances the ability of the whiskers to rearrange themselves during sintering. Increasing the amount of liquid phases present during sintering also improves densification by aiding whisker rearrangement. For practical purposes, the whisker contents of pressureless-sintered bodies are generally limited to <20 vol.%.

Hot-pressing overcomes the problem of the formation of a whisker network by using the simultaneous application of high temperature and pressure. The technique was used almost exclusively in the early work on whisker-reinforced composites. With this method, whisker loadings as high as 60% have been achieved while maintaining densities >95%. Hot-pressing, in its simplest form, involves charging of the dried composite powder mixture into a die (typically graphite), followed by the application of pressure and heating of the die. The level of pressure used is dictated by the strength of the die material used. Typically, pressures of 14–69 MPa (2–10 ksi) are used with graphite dies. Heating can be supplied by external heating elements or by induction heating of the graphite die itself. During pressing, orientation of the whiskers perpendicular to the hot-press direction occurs. Hot-isostatic-pressing (HIP) has been used successfully by numerous studies to densify composites with high whisker loadings. It uses green parts fabricated by any of the methods described above, including complex shaped articles. The process involves encapsulation of the green (or partially sintered) parts with either glass or a refractory metal covering. The encapsulated parts are then heated in a furnace with the simultaneous application of gas pressure. Pressures as high as 207 MPa (30 ksi) have been used. After densification, the glass or metal covering is removed. In addition to the more conventional processing described above, whisker-reinforced composites have been fabricated by several other techniques. These include microwave sintering and plasma spraying.

5. POSSIBLE HEALTH EFFECTS OF SiC WHISKERS

As with all materials that have been recently developed, the health risks associated with SiC whiskers are not well known; however, because their sizes and shapes are similar to that of asbestos, therefore they are considered hazardous. Airborne dispersion and subsequent inhalation are the most serious health hazards. However, with proper workplace handling requirements and procedures, large quantities of SiC whiskers can be safely processed. No release of whiskers has been observed from dense ceramic matrix composites during fracture or wear processes. Material Safety Data Sheets (MSDS) are packaged with all products containing loose whiskers and the directions in the data sheets should be followed. At the present time, the American Society for Testing of Materials (ASTM) has developed procedures and handling practice standards for SiC whiskers [14].

6. PROPERTIES OF SiC WHISKER-REINFORCED ALUMINA

As mentioned before, the development of SiC whisker reinforced alumina was promoted because the mechanical property improvements observed with the incorporation of SiC whiskers into ceramic matrices were unprecedented. For example, the fracture toughness of alumina was increased from ~3.0 MPa$\sqrt{m}$ to 8.5 MPa$\sqrt{m}$ with the addition of 20 v/o whiskers. The increases in toughness for various whisker contents are shown in Fig. 7. The data is summarized from a number of reported studies and so reflect some of the variations in raw materials and processing methods. The results also reflect the maximum toughening

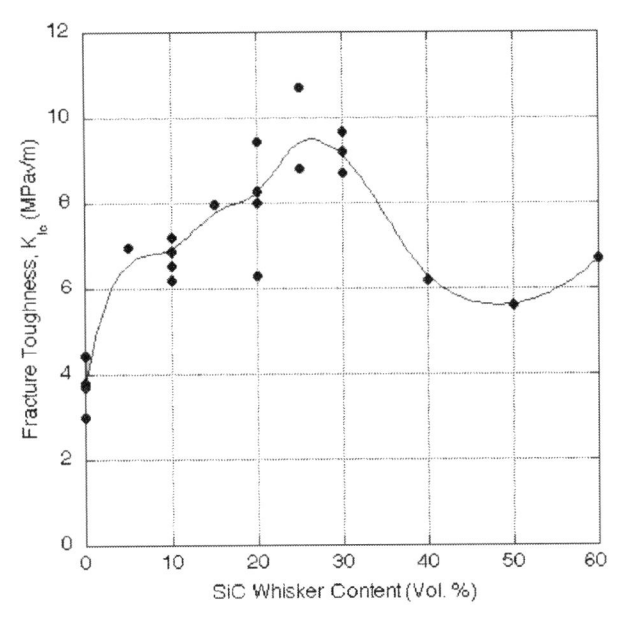

FIGURE 7. Fracture toughness as a function of SiC whisker content in alumina.

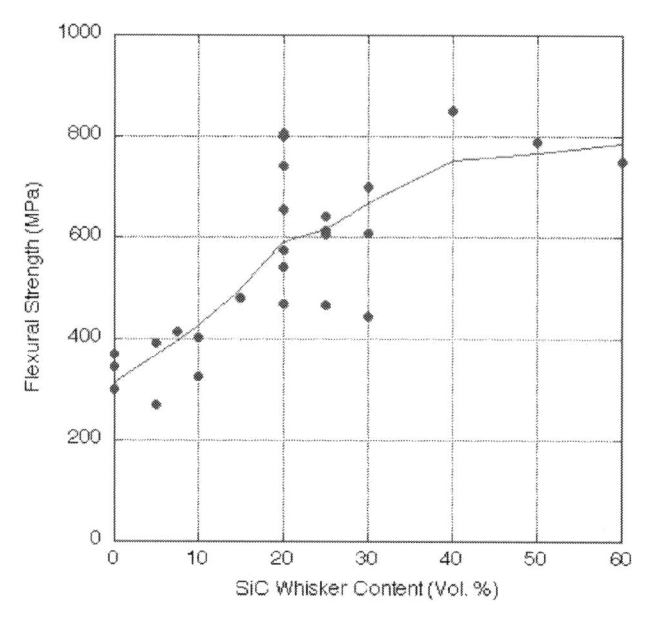

FIGURE 8. Flexural strength as a function of SiC whisker content in alumina.

effects observed for the composites. Because of the orientation effects on the SiC whiskers during processing, the fracture toughness also exhibits anisotropy. For hot-pressed alumina with 20 vol.% SiC whiskers, the toughness (K_{Ic}) was observed to be 8.7 MPa$\sqrt{}$m when the crack propagation was perpendicular to the hot-press axis (and consequently perpendicular to many of the whisker axes) [15]. In contrast, when the toughness was measured parallel to the hot-press axis with crack propagation along the length of the whiskers, it was observed to be 5.5 MPa$\sqrt{}$m. Qualitatively, the toughening effects can be observed by the very rough fracture surfaces associated with SiC whisker reinforced alumina as shown previously in Figs. 1 and 2. The toughness is essentially constant up to temperatures of 1100°C [16]. At higher temperatures, increased toughness is observed, but this is associated with creep crack nucleation and propagation.

The increased toughness was accompanied by increased fracture strengths of 700–800 MPa versus ≤400 MPa in unreinforced alumina. The strength improvements are summarized in Fig. 8. As before, the data is summarized from a number of reports. Just as importantly, the strength improvements are retained to elevated temperatures as shown in Fig. 9, unlike some other toughened ceramic systems. The Weibull modulus of an alumina-20 vol.% SiC whisker composite has been reported to be in the range of 10–13 [9].

The resistance to slow crack growth was shown to markedly increase for an alumina-20 vol.% SiC whisker composite, as compared to unreinforced material [17]. This means that much higher stress intensities and stresses can be applied to these composites while still avoiding time-dependent failure.

Thermal shock testing of an alumina-20 vol.% SiC whisker composite showed no decrease in flexural strength with temperature differences up to 900°C [18]. Alumina, on

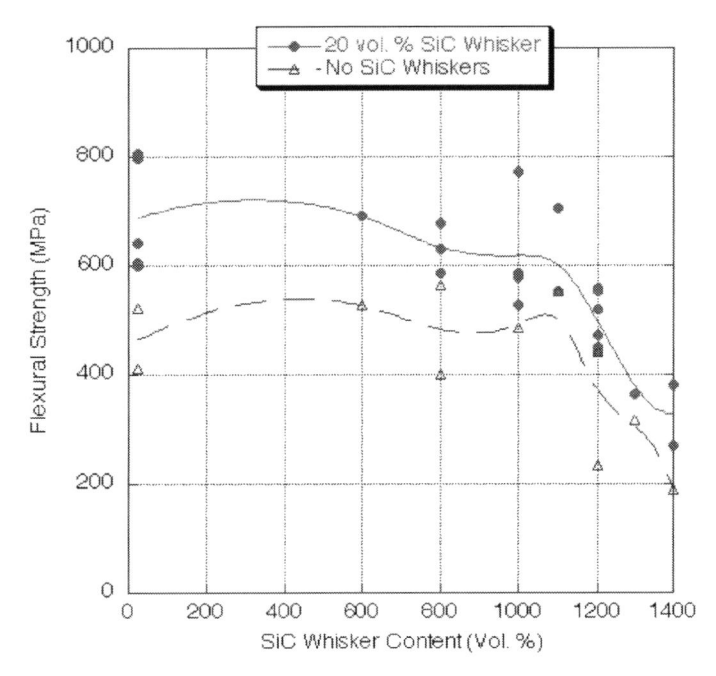

FIGURE 9. Flexural strength as a function of temperature for an alumina-20 vol.% SiC whisker composite and an unreinforced alumina.

the other hand, normally shows a significant decrease in flexural strength with temperature changes of >400°C. The improvement is a result of interaction between the SiC whiskers and thermal-shock induced cracks in the matrix, which prevents coalescence of the cracks into critical flaws. Creep measurements demonstrated that SiC whisker additions to alumina significantly inhibit high-temperature creep as compared to the monolithic material. For example, a 20 vol.% loading of whiskers in an alumina matrix reduced the creep rate by up to two orders of magnitude at temperatures up to 1300°C [16].

Because SiC and alumina have significantly different properties, the composites have physical properties, which are a combination of both [19]. The addition of SiC whiskers to an alumina matrix increases the hardness and thermal conductivity as shown in Figs. 10 and 11, respectively. Conversely, the electrical resistivity and thermal expansion are observed to decrease with SiC whisker additions as shown in Figs. 12 and 13, respectively. Other detailed property summaries on SiC whisker reinforced alumina can be found in Reference 20.

7. APPLICATIONS OF SiC WHISKER-REINFORCED ALUMINA

The initial development of SiC whisker reinforced alumina was done in the early 1980's and the first commercial product was introduced in 1985. The remarkably fast development of SiC whisker-reinforced alumina from initial laboratory samples to a competitive

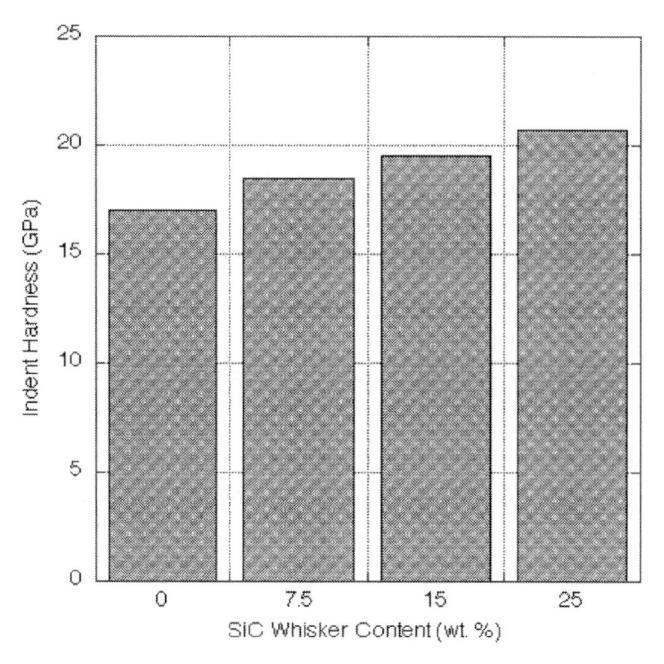

FIGURE 10. Indent hardness as a function of SiC whisker content in alumina.

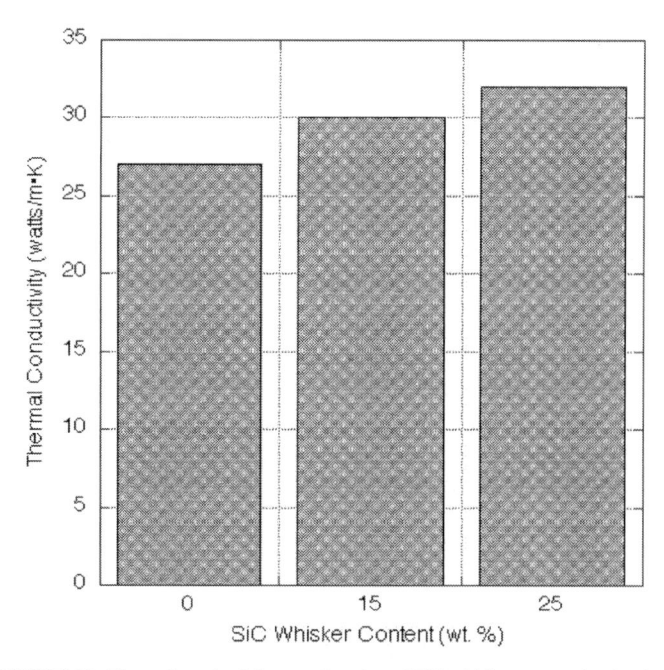

FIGURE 11. Thermal conductivity as a function of SiC whisker content in alumina.

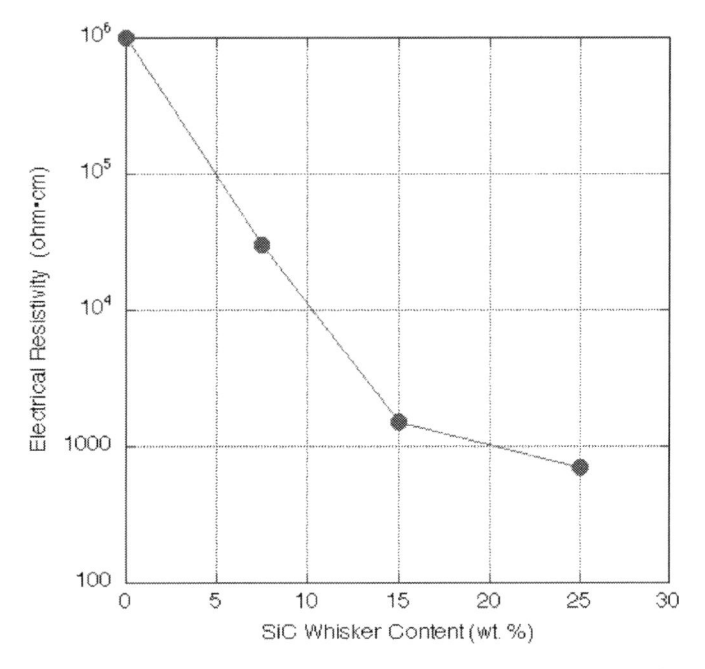

FIGURE 12. Electrical resistivity as a function of SiC whisker content in alumina.

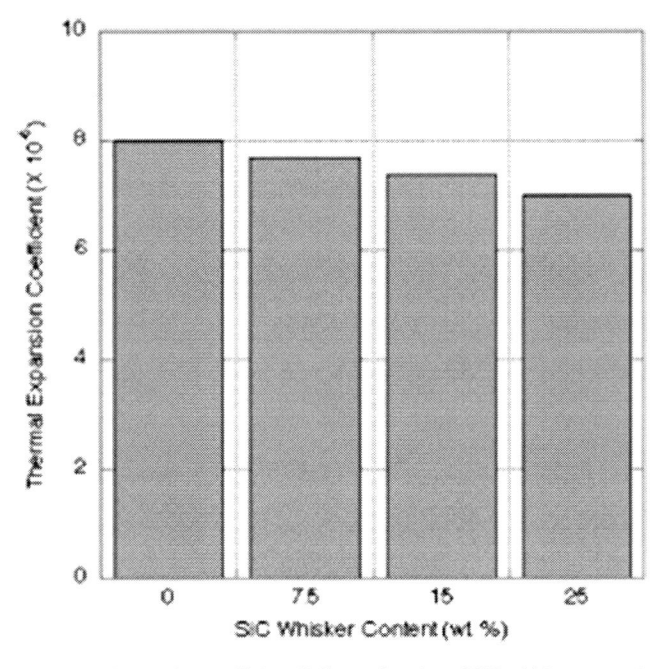

FIGURE 13. Thermal expansion coefficient ($/°C$) as a function of SiC whisker content in alumina.

commercial product took under just three years. The short development time was possible because the materials found a niche market in the cutting of nickel-based alloys where they had exceptional performance.

At the present time, whisker-reinforced alumina is being fabricated and marketed extensively as a cutting tool for high nickel alloys [21]. It was found that the addition of SiC whiskers to alumina not only improved the strength and fracture toughness, but did so without compromising the hot hardness of the matrix. This combination revolutionized machining of high-nickel alloys that are used in the jet engine industry. The SiC whisker-reinforced alumina enabled a ten-fold increase in metal removal rates at increased speeds. For example, in one reported case history, changing from a conventional tool to a SiC whisker-alumina one reduced a 5-hour machining operation of inconel to 20 minutes. Improvements in machining of cast iron and steel products have also been observed.

Because of the excellent wear resistance of whisker-reinforced ceramic composites, near-term applications include wear parts, such as pump seal rings, grit-blast nozzles, aluminum can tooling and dies for metal extrusion and wire pulling. Testing of ceramics under nonlubricating conditions showed that the wear coefficient for whisker-reinforced alumina was several orders of magnitude lower than for comparable monolithic ceramic materials. Research on the joining of ceramics has shown that alumina-SiC whisker composites can be joined to itself and to metals using titanium containing brazing filler metals.

8. CONCLUSIONS

SiC whisker reinforced alumina has significantly improved mechanical properties compared to monolithic alumina ceramics. The composites can be fabricated by several methods using typical conventional powder processing techniques. Cutting tools are the main application for the composites at the present time.

REFERENCES

1. T. N. Tiegs and S. C. Weaver, "Whisker and Platelet Synthesis Processes," pp. 411–432 in Carbide, Nitride and Boride Materials Synthesis and Processing, Chapman and Hall, New York, NY (1997).
2. G. A. Bootsma, W. F. Kippenberg and G. Verspui "Growth of SiC Whiskers in the System SiO_2-C-H_2 Nucleated by Iron," *J. Crystal Growth* 11, 297–309 (1971).
3. P. F. Becher, C. H. Hsueh, P. Angelini and T. N. Tiegs,"Toughening Behavior in Whisker Reinforced Ceramic Matrix Composites," *J. Am. Ceram. Soc.*, 71 [12]1050–61(1988).
4. A. G. Evans, "Perspective on the Development of High Toughness Ceramics," *J. Am. Ceram. Soc.*, 73 [2] 187–206 (1990).
5. T. N. Tiegs and K. J. Bowman, "Fabrication of Particulate-, Platelet- and Whisker-Reinforced Ceramic Matrix Composites," pp. 91–138 in Handbook on Discontinuously Reinforced Ceramic Matrix Composites, *Am. Ceram. Soc.*, Westerville, OH (1995).
6. T. N. Tiegs, L. F. Allard, P. F. Becher and M. K. Ferber, "Identification and Development of Optimum Silicon Carbide Whiskers for Silicon Nitride," pp. 167–172 in Proceedings of the Twenty-Seventh Automotive Technology Development Contractors' Coordination Meeting P-230, Society of Automotive Engineers, Warrendale, PA. (1990).
7. J. J. Petrovic, J. V. Milewski, D. L. Rohr and F. D. Gac, "Tensile Mechanical Properties of SiC Whiskers," *J. Mater. Sci.* 20, 1167–1177 (1985).

8. S. R. Nutt, "Defects in Silicon Carbide Whiskers, *J. Am. Ceram. Soc.*, 67 [6] 428–431 (1985).
9. T. N. Tiegs, "Structural and Physical Properties of Ceramic Matrix Composites," pp. 225–273 in Handbook on Discontinuously Reinforced Ceramic Matrix Composites, *Am. Ceram. Soc.*, Westerville, OH (1995).
10. S. Iio, M. Watanabe, M. Matsubara, and Y. Matsuo, "Mechanical Properties of Alumina/Silicon Carbide Whisker Composites," *J. Am. Ceram. Soc.*, 72 [10] 1880–18884 (1989).
11. J.V. Milewski, "Efficient Use of Whiskers in the Reinforcement of Ceramics," *Adv. Ceram. Mater.*, 1 [1] 36–41 (1986).
12. T. N. Tiegs and P. F. Becher, "Sintered Al_2O_3-SiC Composites," *Am. Ceram. Soc. Bull.* 66(2) 339–342 (1987).
13. M. D. Sacks, H. W. Lee, and O. E. Rojas, "Suspension Processing of Al_2O_3-SiC Whisker Composites," *J. Am. Ceram. Soc.*, 71 [5] 370–379 (1988).
14. "Standard Practice for Handling Silicon Carbide Whiskers," ASTM Designation E 1437-91 (Philadelphia: ASTM 1991).
15. P. F. Becher and G. C. Wei, "Toughening Behavior in SiC-Whisker-Reinforced Alumina," *J. Am.Ceram. Soc.*, 67 [12]C-267-C-269 (1984).
16. P. F. Becher and T. N. Tiegs, "Temperature Dependence of Strengthening by Whisker Reinforcement: SiC Whisker-Reinforced Alumina in Air," *Advanced Ceram. Mater.* 3 [2] 148–53 (1988).
17. Becher, P. F., Tiegs, T. N., Ogle, J. C., and Warwick, W. H., "Toughening of Ceramics by Whisker Reinforcement," pp. 61–73 in *Fracture Mechanics of Ceramics, Vol. 7*, Plenum Press, New York, N.Y. (1986).
18. T. N. Tiegs and P. F. Becher, "Thermal Shock Behavior of an Alumina-SiC Whisker Composite", *Am. Ceram. Soc. Comm.*, 70 [5]C-109-111(1987).
19. B. J. Wrona, J. F. Rhodes, and W. M. Rogers, "Silicon Carbide Whisker Reinforced Alumina," *Ceram. Eng. Sci. Proc.*, 13 [9–10] 653–660 (1992).
20. C. X. Campbell and S. K. El-Rahaiby, "Databook on Mechanical and thermophysical Properties of Whisker-Reinforced Ceramic Matrix Composites," *Am. Ceram. Soc.*, Westerville, OH (1995).
21. B. J. Wrona, "Design Methodology and Applications For Ceramic Matrix Composites," pp. 275–307 in Handbook on Discontinuously Reinforced Ceramic Matrix Composites, *Am. Ceram. Soc.*, Westerville, OH (1995).

14

Mullite-SiC Whisker and Mullite-ZrO$_2$-SiC Whisker Composites

Robert Ruh
Universal Technology Corporation
Beavercreek, OHIO 45432-2600

ABSTRACT

The effect of a transformation-toughened ZrO$_2$ phase, and the incorporation of SiC whiskers on the microstructure and physical, mechanical and electrical properties were investigated. High-purity mullite-SiC whisker composites and mullite-ZrO$_2$-SiC whisker composites were fabricated by hot pressing, using a matrix prepared by the alkoxide process. Varying amounts of ZrO$_2$ stabilization were achieved by varying amounts of Y$_2$O$_3$ and MgO additions. Composites and matrices were characterized by microstructural studies, thermal expansion, thermal diffusivity, room temperature strength (as fabricated and after thermal exposure), elevated temperature strength, high temperature crack growth, deformation and creep, fracture toughness, thermal shock, and permittivity, permeability, impedance and resistivity.

I. INTRODUCTION

High Purity mullite has excellent potential for high temperature applications because of its excellent stability, low thermal expansion and good high temperature strength and creep resistance. Its crystal structure is orthorhombic, space group Pbam, and its lattice parameters are $a = 0.7526$ nm, $b = 0.7682$ nm and $c = 0.2878$ nm.[1] Cameron[2] has shown that the a parameter is directly related to the Al:Si ratio.

Mullite is identified by the formula $3Al_2O_3 \cdot 2SiO_2$, but does exist over a small composition range. Phase relationships in the vicinity of mullite are very sluggish, and considerable research has been devoted to their clarification, often with conflicting results. Prochazka and Klug[3] determined that mullite exists over the 72 to 75% composition range at 1600°C, changing to 74 to 76% at 1800°C, and incongruent melting at 1890°C. If compositions are not within this range, mullite plus an Al_2O_3-SiO_2 phase, or mullite plus Al_2O_3, are present.

While mullite has good overall properties, improvements in strength and fracture toughness would significantly enhance its potential. Some research has been directed toward this goal. For mullite, room temperature flexural strengths of 130 to 360 MPa and fracture toughness of 2.4 to 2.8 MPa $\cdot$ m$^{1/2}$ have been reported.[1,4–6] With the incorporation of a transformation-toughened ZrO_2 phase, room temperature flexural strengths of 270 to 400 MPa and fracture toughness of 3.2 to 5.2 MPa $\cdot$ m$^{1/2}$ were obtained.[7–11] The incorporation of 20 vol% SiC whiskers[12] resulted in a flexural strength of 438 MPa and a fracture toughness of 4.6 MPa $\cdot$ m$^{1/2}$. The incorporation of SiC whiskers and a ZrO_2 phase demonstrated further improvement in properties.[13,14] For pure mullite, strengths of 244 MPa and fracture toughness of 2.8 MPa $\cdot$ m$^{1/2}$ were obtained, which increased to 452 MPa and 4.4 MPa $\cdot$ m$^{1/2}$ for 20 vol% SiC whiskers and 10 vol% ZrO_2.

With the improved properties demonstrated for this composite, a comprehensive study was initiated to systematically investigate this composite system. Materials were prepared in situ by the alkoxide process to exploit the full potential of this promising material. A range of SiC whisker additions and ZrO_2 with different kinds and amounts of stabilizer were employed. An original study was accomplished at the Air Force Research Laboratory,[15] which was complemented by a number of cooperative studies by experts in different disciplines. In the final analysis impressive properties were demonstrated, but some disappointing results were also obtained.

II. EXPERIMENTAL PROCEDURE AND MATERIALS CHARACTERIZATION

(1) Compositions and Rationale

Matrix materials used in this study were prepared by the alkoxide process. The SiC whiskers were designated Silar SC-9 from Arco Chemical Co., Greer, SC. A considerable number of compositions were prepared, but only those discussed in this paper, along with a short designation, are given in Table I. Each composition was made as a separate batch. Compositions are given in weight percent unless otherwise noted. The ZrO_2 phase was partially stabilized with 0.86, 1.5 or 3.0% Y_2O_3. According to the ZrO_2-Y_2O_3 phase diagram,[16] the 0.86% Y_2O_3 composition is in the tetragonal region down to 500°C, and then near the monoclinic/monoclinic plus cubic phase boundary at lower temperatures. The 1.5% Y_2O_3 composition is within the tetragonal plus cubic region to 500°C, and then in the monoclinic plus cubic region at low temperatures. The 3.0% Y_2O_3 composition is well into the tetragonal plus cubic region to 500°C, and in the monoclinic plus cubic region at lower temperatures. For the ZrO_2-MgO materials, MgO contents of 1.0 and 2.2% were used. According to the ZrO_2-MgO phase diagram,[17] the 1.0% MgO composition is in the tetragonal plus cubic region above 1400°C, the tetragonal plus MgO region from 1400° to 1100°C, and in the

TABLE I. Compositions and Their Designations

Designation	Composition
M	mullite
MS	M-30% SiC whiskers
MZ	M-29.5% ZrO_2
MZS	MZ-30% SiC whiskers
MZ0.9Y	M-31.5% ZrO_2-0.86% Y_2O_3
MZ0.9YS	MZ0.9Y-30% SiC whiskers
MZ1.5Y	M-29.5% ZrO_2-1.5% Y_2O_3
MZ1.5YS	MZ1.5Y-30% SiC whiskers
MZ1.0M	M-38.5% ZrO_2-1.0% MgO
MZ1.0MS	MZ1.0M-30% SiC whiskers
MZ2.2M	M-32.5% ZrO_2-2.2% MgO
MZ2.2MS	MZ2.2M-30% SiC whiskers

monoclinic plus MgO region below 1100°C. The 2.2% MgO composition is in the cubic plus MgO region above 1400°C, the tetragonal plus MgO region from 1400° to 1100°C, and in the monoclinic plus MgO region below 1100°C.

(2) Preparation of Suspensions for Matrix

Hydroxide suspensions of mullite with partially stabilized zirconia were prepared from metal alkoxide starting materials. Zirconium isoamyloxide and yttrium isopropoxide were mixed in the proper proportions for the partially stabilized zirconia. Aluminum isopropoxide and silicon ethoxide were added in the proper proportions for mullite. These solutions were refluxed for 4 to 8 h and then hydrolyzed to a pH of 0.5 using HNO_3. The solutions were then neutralized to a pH of 8 using NH_4OH. The resulting suspensions then contained the proper amount of each component as a hydroxide to form the mullite partially-stabilized zirconia matrix materials. With MgO as a stabilizer, procedures were identical to the above, except that MgO ethoxide was used instead of yttrium isopropoxide. Batch sizes were 25 to 100 g of final material.

Batches of mullite without the zirconia phase were also prepared by the base-hydrolyzed procedure just described. In addition, mullite batches were prepared by an acid-hydrolyzed procedure. In this case, hydrolysis to a pH of 0.5 was accomplished by the addition of mixed metal alkoxides to a small amount of NH_4F in water, followed by neutralization with concentrated NH_4OH to a pH of 8.5 to 9. Batch sizes for the base-hydrolyzed and acid-hydrolyzed mullite were 40 to 60 g.

(3) Preparation of Composites

The hydroxide suspensions were mixed with the SiC whiskers on the weight basis of the final fired composition of the mullite-zirconia or mullite matrix. Mixing was accomplished in a food blender for 12 m. The slurry-whisker mixtures were then cast into dishes and vacuum dried at 75°C for 24 h. The dried mixtures were powdered with a B_4C mortar and pestle. Powders were calcined to 700°C in air to drive off the volatile species. Other methods could include evaporating the slurry-whisker mixtures to dryness using a rotary evaporator, or spray drying the homogeneous powder mixtures. For samples without whiskers, the same procedures were used.

The calcined sample powders were vacuum hot pressed in graphite dies lined with graphite foil in a commercial hot press. Samples 2.5 cm in diameter × 0.2 to 0.5 cm thick were fabricated. Firing temperatures and soak times varied for the different batches. While a firing temperature of 1700°C was needed to densify mullite to a near theoretical value, the mullite-ZrO_2-Y_2O_3 composites densified at 1500°C, and the mullite-ZrO_2-MgO composites densified at 1400°C. The soak times were 20 m. Samples were heated to 1100°C under a pressure of 6.9 MPa and held at this temperature until a vacuum of ~200 torr was attained (~30 m), then a pressure of 34.5 MPa was applied and the temperature raised to the soak temperature. Pressure was not released after soaking, but was allowed to decrease during cooling. Theoretical densities were calculated by the rule of mixtures; the densities of the fired samples were determined from weight and dimensions, and by Archimedes' Technique.

The impurity contents of the fired samples were determined by emission spectrographic analysis, and detailed data are presented in Reference 18. In summary, for the mullite batches, impurity levels were less than 1000 ppm, with W, Zr and Na as the major impurities. For the mullite-ZrO_2 batches, total impurity contents were 1000 to 1500 ppm, with Fe, W, Ni and Ca as the major impurities. It is believed that most of the impurities were introduced during the powder processing. The SiC whiskers contained ~7000 ppm impurities, with Ca, Al, Mg, Fe and Cr as the major contributors.

The mullite compositions were established chemically by determining the amount of Al_2O_3 and SiO_2. Some batches were stoichiometric, some were slightly Al_2O_3-rich and some were slightly SiO_2-rich. The amount of Zr, Y and Mg were determined chemically for the mullite-ZrO_2 batches, and values converted to ZrO_2, Y_2O_3 and MgO. Detailed data are given in Reference 18.

(4) Phase and Microstructural Analysis

The phases present in experimental samples were determined by x-ray diffraction analysis. The phases identified were mullite, monoclinic ZrO_2, tetragonal ZrO_2 and α-SiC. The amount of monoclinic and tetragonal ZrO_2 was also determined and correlated with the type and amount of stabilizer. Detailed data are given in Reference 15. The existence of metastable tetragonal ZrO_2 was confirmed, thus confirming transformation toughening.

Microstructural analysis was accomplished on mullite and mullite-ZrO_2 compositions using SEM and EDX.[15] A stoichiometric single-phase mullite sample thermally etched at 1400°C for 1 h in seen in Fig. 1. A considerable variation in grain size from <1 to >10 μm is seen. If a composition was Al_2O_3-rich, an Al_2O_3 phase was seen, while if a composition was SiO_2-rich, a fine amorphous $Al_2O_3 \cdot SiO_2$ phase was seen. A mullite-36.1% ZrO_2 sample thermally etched at 1300°C for 1 h is given in Fig. 2. In this figure the ZrO_2 is the light phase and has a particle size of 0.2 to 0.8 μm. Variation in particle size was seen for compositions with either Y_2O_3 or MgO additive. More extensive microstructural characterization has been

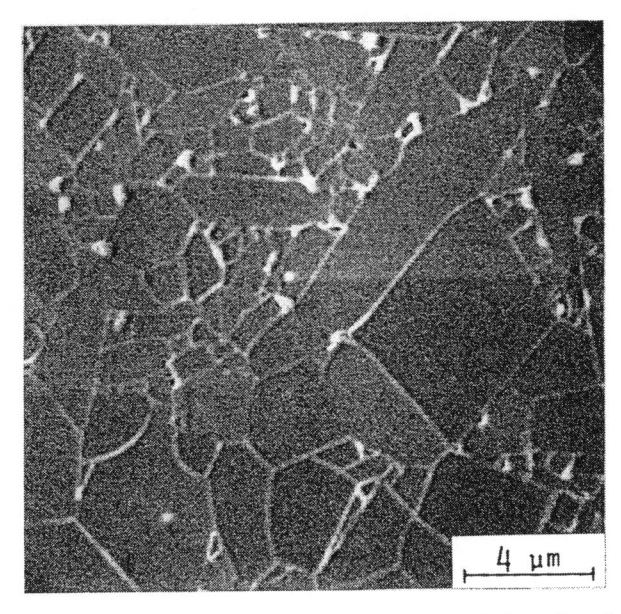

FIGURE 1. SEM micrograph of a stoichiometric single-phase mullite sample thermally etched in air at 1400°C for 1 h. Published with permission of the J. Am. Ceram. Soc. (15)

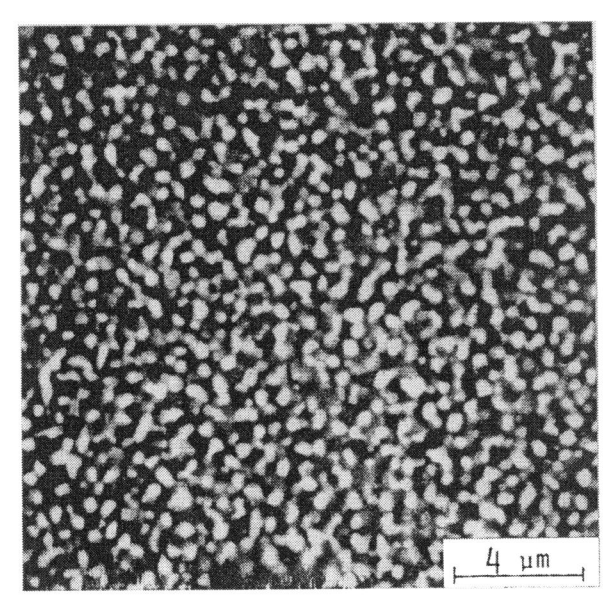

FIGURE 2. SEM micrograph of a mullite-36.1% ZrO_2 sample thermally etched in air at 1300°C for 1 h. Published with permission of the J. Am. Ceram. Soc. (15)

given[15,19] as well as phase and microstructural evolution.[20,21] Modes and mechanisms of oxidation of these composites has also been studied.[22,23]

III. PHYSICAL PROPERTIES

(1) Thermal Expansion

The linear thermal expansion of mullite-ZrO_2 compositions with and without SiC whiskers was determined using an automatic single push rod dilatometer.[24] Results for the room temperature to 500°C temperature range were 4.02×10^{-6}/°C for mullite, 3.51×10^{-6}/°C for mullite-30% SiC whiskers, 5.10×10^{-6}/°C for mullite-24% ZrO_2 and 4.15×10^{-6}/°C for mullite-24% ZrO_2-30% SiC whiskers. For the room temperature to 1200°C range,[15] values of 6.25 to 7.15×10^{-6}/°C were found for mullite-ZrO_2-Y_2O_3 or MgO without SiC whiskers, and 5.65 to 6.25×10^{-6}/°C for those composites with SiC whiskers. Data for a mullite-32.5% ZrO_2-2.2% MgO-20% SiC whiskers are seen in Fig. 3. There may be an indication of the monoclinic-tetragonal transformation, but it is not well defined. Other investigators have seen the transformation in thermal expansion measurements.[13]

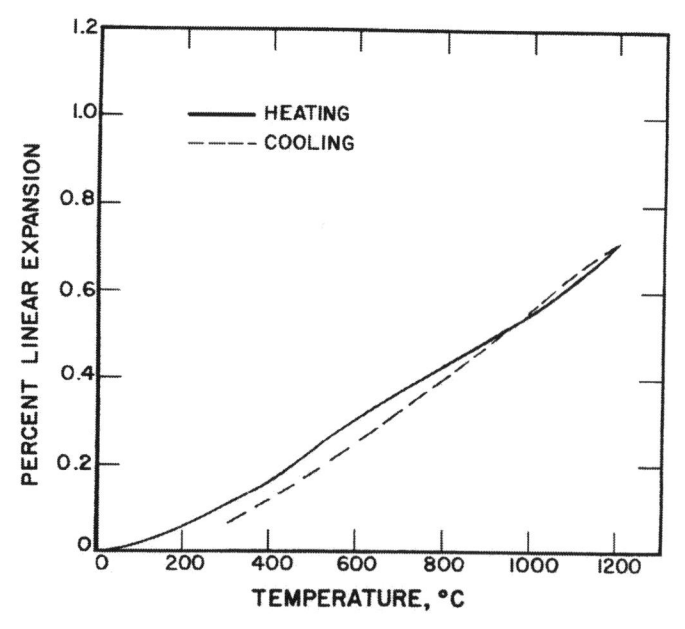

FIGURE 3. Linear thermal expansion of mullite-32.4% ZrO_2-2.2% MgO-20% SiC whiskers. Published with permission of the J. Am. Ceram. Soc. (15)

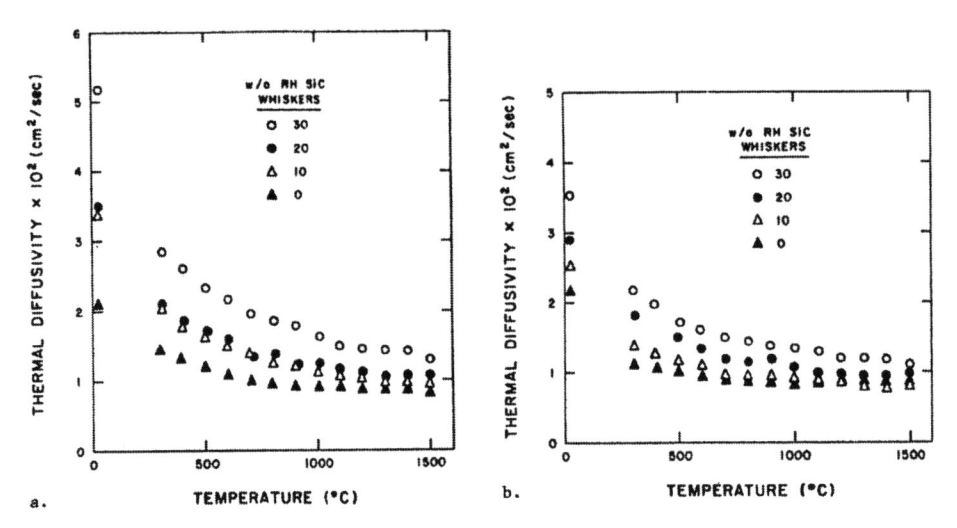

FIGURE 4. Thermal diffusivity versus temperature for SiC-whisker-reinforced mullite for a range of whisker contents. Measurements are (a) perpendicular and (b) parallel to the hot-pressing direction. Published with permission of the J. Am. Ceram. Soc. (25)

(2) Thermal Diffusivity and Specific Heat

The thermal diffusivity of mullite with 0 to 30% SiC whiskers was determined from room temperature to 1500°C.[25] Data are presented in Fig. 4 for measurements made perpendicular to the hot-pressing direction. It is seen that the SiC whiskers increase the thermal diffusivity at room temperature, but values decrease for all compositions with increasing temperature. Similar trends are seen for values measured parallel to the hot-pressing direction, but values were lower. Lower values were expected because the whiskers orient perpendicularly to the hot-pressing direction during hot pressing.

The thermal diffusivity of mullite-ZrO_2 with and without 30% SiC whiskers was determined at room and elevated temperatures, as seen in Fig. 5.[26] At room temperature thermal diffusivity values for the matrix material show little anisotrophy. These values generally increase with increasing monoclinic ZrO_2 phase, consistent with higher values measured for the monoclinic phase.[27] Thermal diffusivity values for mullite fall midway in the range of values for the mullite-ZrO_2 matrices used in this study.

The addition of SiC whiskers generally results in a significant increase in thermal diffusivity values over those measured for the unreinforced matrix. Anisotrophy in thermal diffusivity values also increases with increasing SiC whisker content. The same general trends were found for SiC whisker additions to mullite.[25]

The specific heat of mullite-36% ZrO_2 and mullite-31.5% ZrO_2-0.86% Y_2O_3 and their respective composites are given in Fig. 6. Values for all samples increase up to 700°C, the maximum temperature of measurement. Results are somewhat lower than those obtained for mullite and mullite-30% SiC whiskers, which had very little difference over the same temperature range. The specific heat of both matrices and their composites is comparable

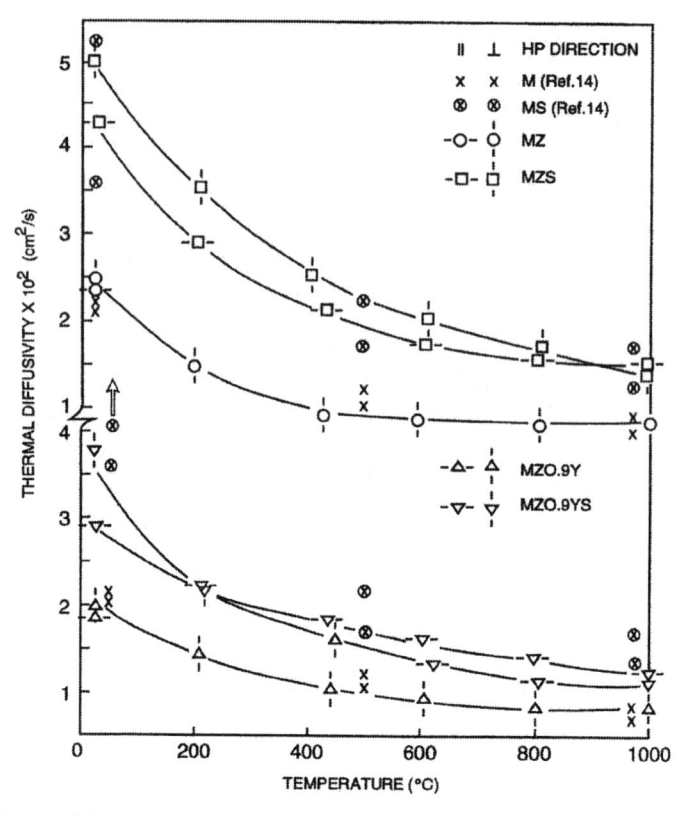

FIGURE 5. Thermal diffusivity parallel and perpendicular to the hot-pressing direction for mullite-ZrO$_2$-SiC whisker composites, as a function of temperature. Published with permission of the J. Am. Ceram. Soc. (26)

at low temperatures, and with increasing temperatures, except for the mullite-31.5% ZrO$_2$-0.86% Y$_2$O$_3$, which increases at a lower rate. The thermal conductivity values were found to follow the same trends as the thermal diffusivity values.

IV. MECHANICAL PROPERTIES

(1) Room Temperature Flexural Strength As Fabricated

The room temperature four-point flexural strength of three batches of mullite with different amounts of SiC whisker additions is given in Fig. 7.[15] Batch 15 was Al$_2$O$_3$-rich and contained Al$_2$O$_3$ crystals of significant size in the microstructure. These crystals acted as flaws and decreased the strength. Batches 16 and 17 were slightly SiO$_2$-rich and had 0.5 to 2 μm amorphous particles uniformly distributed throughout the mullite phase. This microstructure was conducive to high strength. The addition of SiC whiskers increased the strength of the composites in proportion to the amount present, up to 30%. At 40%

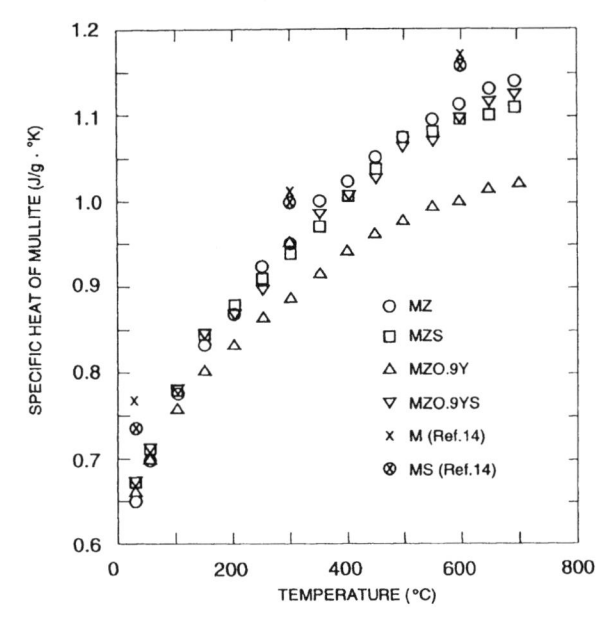

FIGURE 6. Specific heat of mullite-ZrO_2-SiC whisker composites, as a function of temperature. Published with permission of the J. Am. Ceram. Soc. (26)

SiC whiskers the strength decreased for batch 15, and this behavior was seen by other investigators.

The room temperature four-point flexural strength of mullite-ZrO_2-Y_2O_3 composites with different amounts of Y_2O_3 and 0 to 30% SiC whisker additions is given in Fig. 8.[15] The effect of different amounts of Y_2O_3 is discussed in Reference 15. Silicon carbide whisker additions to all compositions increased the strength dramatically and the rate of increase was about the same for all Y_2O_3 compositions. A composition with 30% SiC powder was also tested, and its strength was very similar to the matrix material, as seen in Fig. 8.

Data were also generated on mullite-ZrO_2-MgO composites with different amounts of MgO and 0 to 30% SiC whisker additions. Results were similar to those obtained for mullite-ZrO_2-Y_2O_3 composites.[15]

(2) Flexural Strength After Thermal Exposure

Flexural strength changes for as-fabricated samples versus samples exposed for 260 h at 1000°C or 1200°C in air are given in Fig. 9.[28] Values are given as a percentage of as hot-pressed strength. No significant change in strength was seen for mullite-SiC whisker composites at 1000° or 1200°C. In contrast, the mullite-ZrO_2 samples lost 36% of their strength after the 1000°C exposure and 67% after the 1200°C exposure. The strength of the mullite-ZrO_2-SiC whisker samples did not degrade after 1000°C exposure, but degraded nearly 30% after 1200°C exposure. The mullite-ZrO_2-MgO samples with 1% MgO degraded

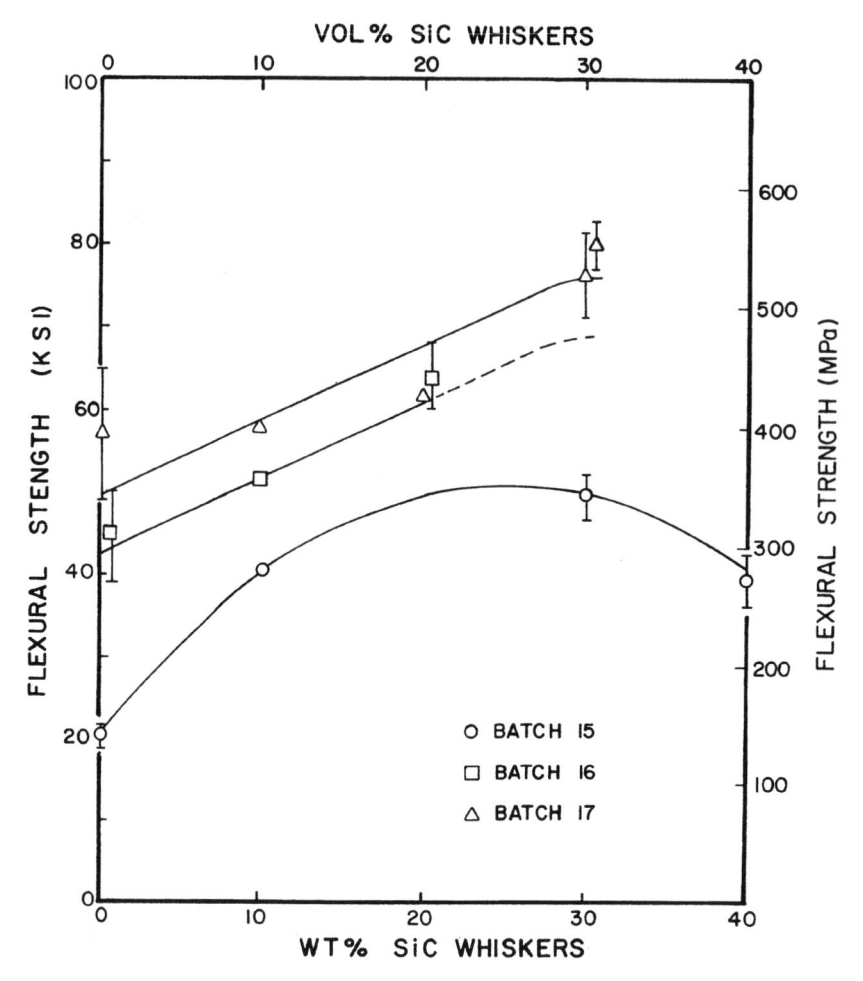

FIGURE 7. Room temperature four-point flexural strength of mullite-SiC whisker composites versus SiC whisker content. Published with permission of the J. Am. Ceram. Soc. (15)

12% at 1000° and 35% at 1200°C. With 30% SiC whisker additions, degradation was 17% at 1000° and 39% at 1200°C. For the mullite-ZrO_2-MgO compositions with 2.2% MgO, no degradation was seen at 1000°C, but a 22% strength decrease was seen at 1200°C. With 30% SiC whiskers, degradation was 17% at 1000° and 11% at 1200°C. All mullite-ZrO_2-Y_2O_3 compositions had large decreases in strength (30 to 55%) after 1000°C exposure, and large decreases (25 to 60%) after 1200°C exposure. An exception was the 1.5% Y_2O_3 composition, which did not degrade in strength. This exception is not understood. For the mullite-ZrO_2-Y_2O_3-SiC whisker composites, extensive degradation was seen after the 1000°C exposure

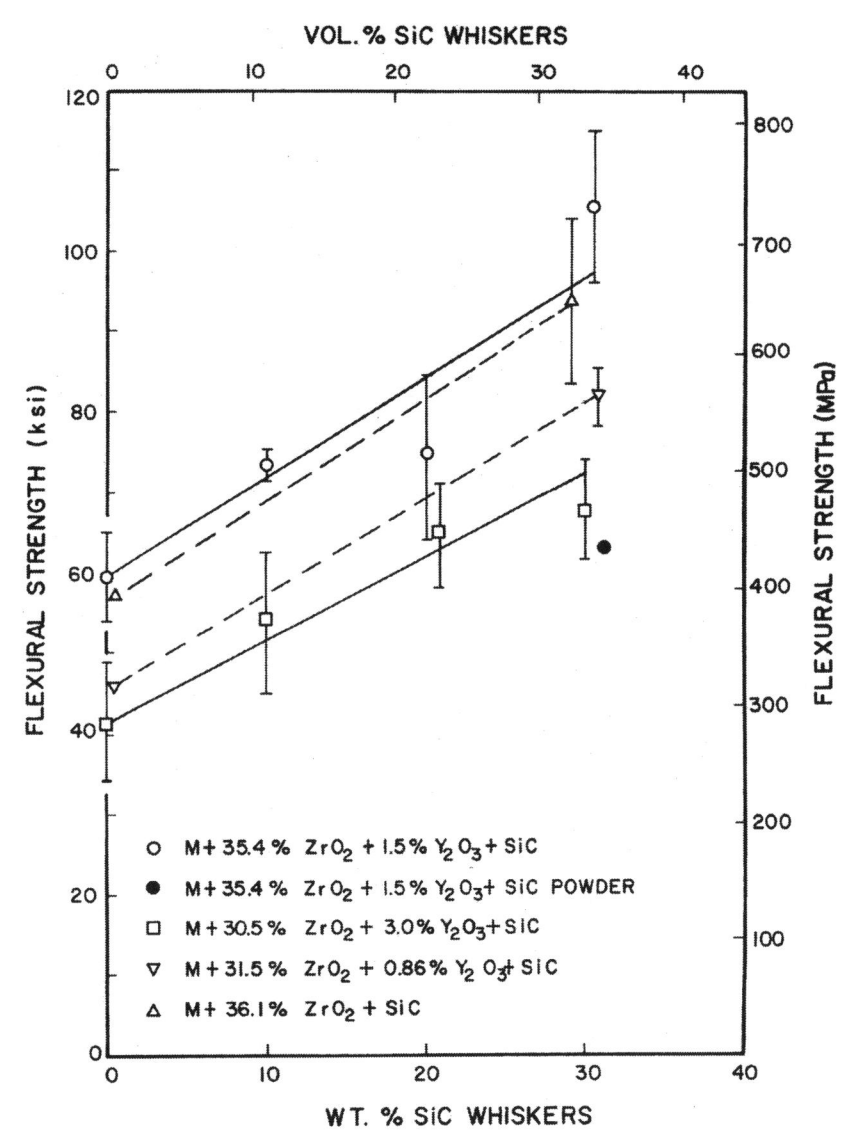

FIGURE 8. Room temperature four-point flexural strength of mullite-ZrO_2-Y_2O_3-SiC whisker composites versus SiC whisker content. Published with permission of the J. Am. Ceram. Soc. (15)

(40 to 80%), and somewhat lesser degradation after the 1200°C exposure (30 to 60%). In summary, only the mullite-SiC whisker samples did not degrade after exposure at 1000°C and 1200°C. All other compositions with and without SiC whiskers degraded in varying amounts. Compositions with MgO additions degraded the least, while compositions with Y_2O_3 additions degraded the most.

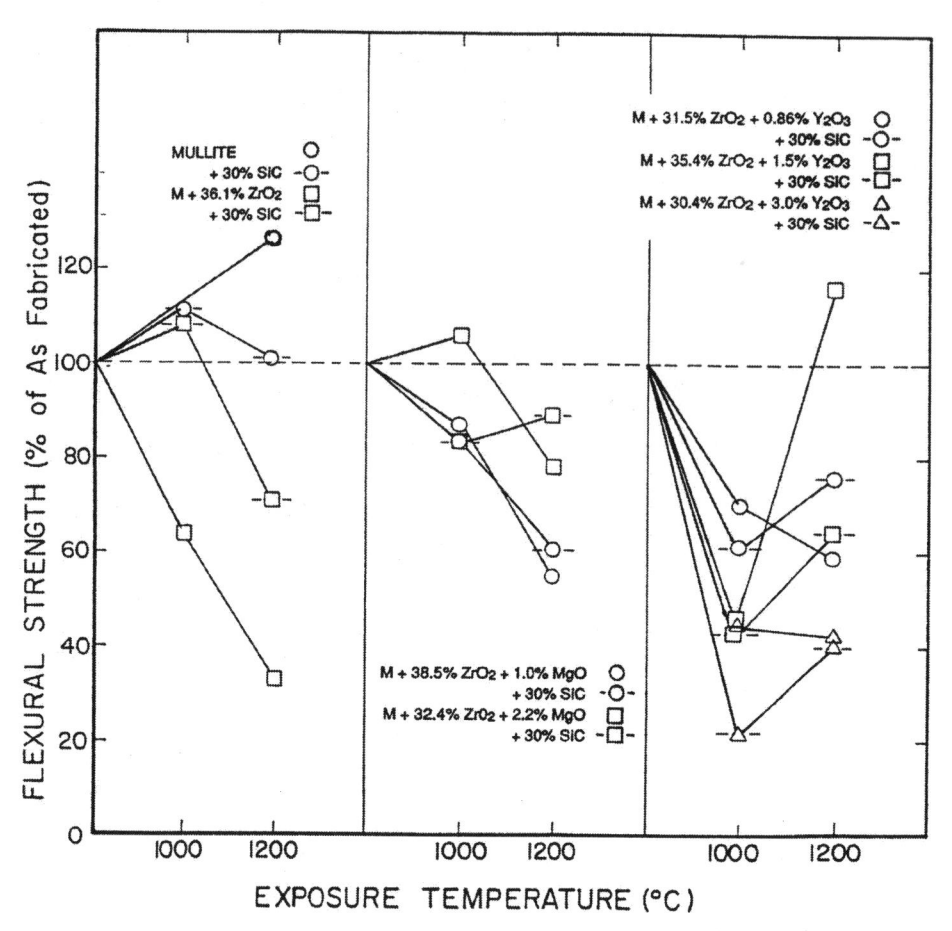

FIGURE 9. Changes in flexural strength for as-fabricated samples versus those exposed for 260 h. at 1000° or 1200°C.

Surface x-ray diffraction analysis data on samples before and after thermal exposure revealed little change in surface composition. This is surprising in light of the large degradation of strength seen. Weight change measurements on samples before and after exposure revealed little change for all except those with mullite-ZrO_2-Y_2O_3-SiC whiskers, and these had weight gains of 2 to 10%.

(3) Elevated Temperature Strength

Limited flexural strength measurements were made at 1250°C on all compositions.[29] Samples were tested in four-point bending and vacuum. A comparison of room temperature and high temperature strength data is given in Fig. 10. It is seen that the strengths of mullite

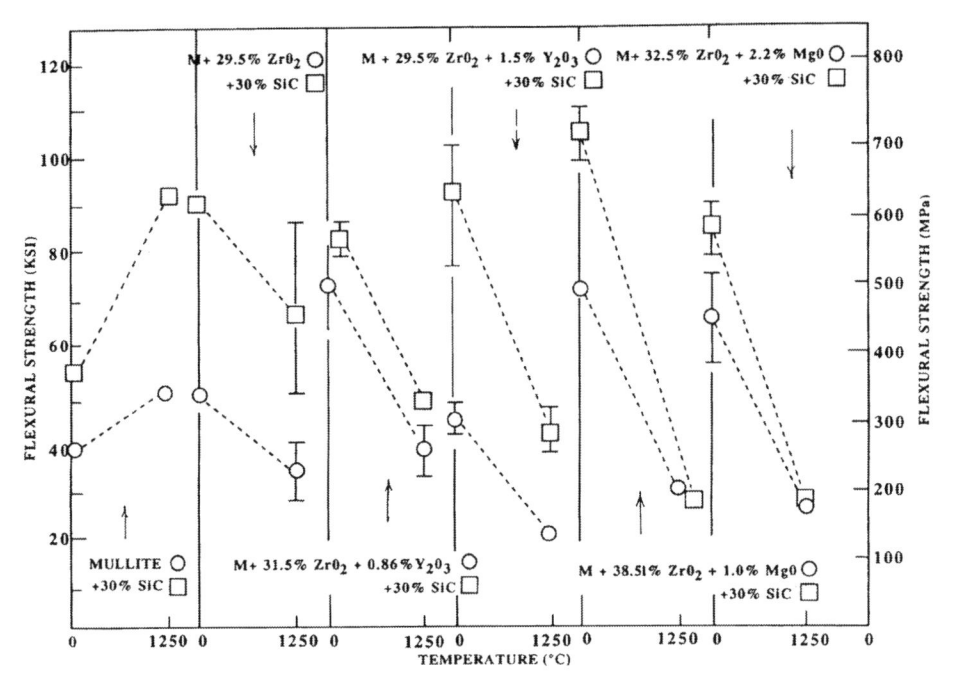

FIGURE 10. Comparison of room temperature and 1250°C flexural strengths for mullite-SiC and mullite-ZrO_2-SiC whisker composites.

and mullite-SiC whisker composites actually increase at high temperatures. However, the strengths of all other compositions decrease sharply at high temperatures. The compositions with MgO stabilizer are particularly weak, and there is no benefit from the SiC whiskers. For compositions with Y_2O_3 stabilizer, the strength degradation is less severe, and those samples with SiC whisker additions are stronger than those without. While the strength of samples with ZrO_2 and no stabilizer still degrade at high temperatures, degradation is the least severe.

Microstructural studies on the above samples[30] determined that the addition of Y_2O_3 or MgO caused a transition from a transcrystalline to intercrystalline fracture. This transition was due to softening of the glassy phase, which resulted in slow crack growth at low stress levels.

(4) High Temperature Crack Growth

High temperature crack growth behavior under creep conditions was studied in SiC whisker-reinforced mullite.[31] Four-point flexurural specimens were prepared for crack growth experiments by introducing Vickers indentation-induced cracks on the tensile surface. Samples were tested in creep in air at a nominal surface tensile stress at 150 MPa and a temperature of 1400°C. Crack growth and nucleation, and extent of creep-strain were

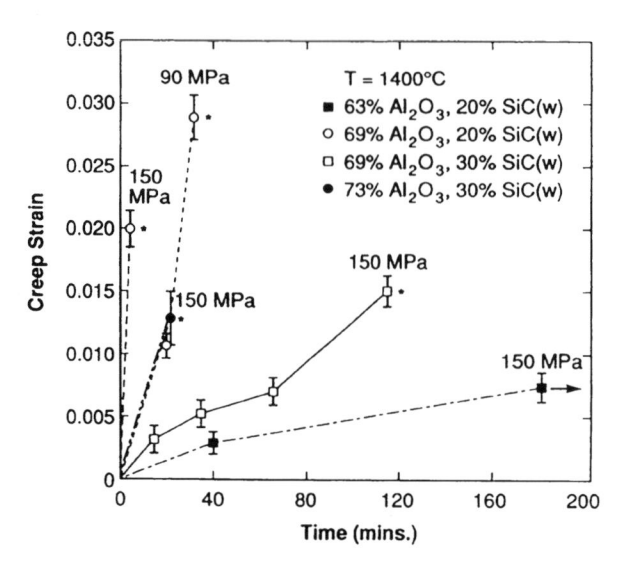

FIGURE 11. Permanent tensile surface creep strain in mullite-SiC whisker composites versus time. Star indicates that the specimen failed. Published with permission of the J. Am. Ceram. Soc. (31)

monitored by a successive interruption technique. A strong linear correlation was observed between the crack growth rate and the creep strain rate. Considerable variation in creep strain was found with time depending on mullite composition and SiC whisker content, as seen in Fig. 11. General conclusions were (1) the higher the stress, the greater the creep rate, (2) the higher the SiC whisker content, the greater the resistance to creep deformation, (3) the neutral axis shifted toward the compressive side of the specimen, implying that the material creeps faster in tension than in compression, (4) no evidence of crack bridging was found, since the exposed SiC whiskers were consumed in the high temperature oxidation reaction, (5) the crack velocity was higher for higher creep rates, and (6) both creep and crack growth rates decreased with time.

(5) High Temperature Deformation

In order to gain a better understanding of high temperature deformation mechanisms in multiphase materials, Parthasarathy, et al.,[32] compared the high temperature deformation of mullite-38.5% ZrO_2-1% MgO matrix to that of mullite-38.5% ZrO_2-1% MgO-30% SiC whisker composite. High temperature deformation was studied in vacuum in compression at constant stress rates of 10^{-6} to 5×10^{-4}/s in the temperature range of 1300° to 1400°C. Samples were oriented with the compression axis perpendicular to the hot-pressing direction. Microstructures are complex, but some conclusions can be drawn by comparing the creep behavior of these composites with literature data on the individual constituents, mullite and zirconia. The SiC whiskers will essentially be rigid under the test conditions.

The microstructures were studied for phase stability, grain size, interphase connectivity, boundary phases and dislocation structures. The examination of a matrix sample tested at

1395°C revealed that the mullite grains were 0.95 to 1 μm in size, and their aspect ratio was ~2. The monoclinic-ZrO_2 grains were 0.8 μm and the tetragonal ZrO_2 grains were mostly intergranular and averaged 0.1 μm in size. These grains were larger than the undeformed material, so apparently some grain growth occurred during creep.

For the composite sample there was little change in the grain size or microstructure after deformation. The monoclinic zirconia grains were 0.2 to 0.32 μm in size; the tetragonal ZrO_2 grains were much smaller, with grain sizes in the 45 to 70 nm range. The tetragonal ZrO_2 tended to be intragranular in mullite, while the monoclinic ZrO_2 was intergranular. The mullite grains had an average size of 0.67 to 0.8 μm. A glass phase was found to wet all SiC/ZrO_2 and SiC/mullite boundaries, but mullite/mullite grain boundaries were free of glassy phase. The mullite/ZrO_2 boundaries were sometimes partially wetted and sometimes free of glassy phase.

A stress exponent of ~1.5 was found for the matrix material for all test temperatures. The first order assumption is that the creep rate is determined by the stronger continuous phase, which is mullite, since ZrO_2 is significantly weaker and discontinuous.

A comparison of the SiC whisker reinforced composite and matrix material revealed that the whisker reinforcement does not improve creep resistance at high stress. However, at lower stress there is a significant improvement.

(6) High Temperature Creep

The compressive creep rate of mullite and mullite-31.7% ZrO_2-30% SiC whiskers was measured at a stress of 50 MPa in the 1290° to 1481°C temperature range.[33] The mullite sample was slightly SiO_2-rich (71.6% Al_2O_3), which generally results in higher strength than stoichiometric or especially Al_2O_3-rich mullite.[15] Strain was calculated as $\Delta l/l_0$, where l_0 is the initial sample length and Δl is the change in length. The creep rate was taken as the parameter 'c' in the Garofalo equation:[34]

$$\epsilon = b_1\{1 - b_2 \exp(-b_3 t)\} + ct$$

where ϵ is strain, b_1, b_2 and b_3 are parameters, c is the steady state creep rate and t is time. Results are presented in Fig. 12, which also includes some date from Parthasarathy, et al.,[32] and some data from Hynes.[35] The creep rate of the mullite is similar to that found by Hynes at high temperatures, but is greater at lower temperatures. The mullite-31.7% ZrO_2-30% SiC whisker sample has a creep rate an order of magnitude lower than mullite, and about two orders of magnitude lower than that found by Parthasarathy, et al., for the mullite-38.5% ZrO_2-1% MgO-30% SiC whisker composite. The large difference in creep rates for the two composites is consistent with large differences in their 1250°C flexural strengths. This is probably attributed to the presence or absence of MgO.

(7) Fracture Toughness

The fracture toughness values of mullite and various composites was determined using an indentation-fracture technique.[36] Vickers diamond-pyramid measurements were made on polished specimens perpendicular to the hot-pressing direction using a 3 kg indentation load. This load produced cracks and hardness impressions of reasonable dimensions.

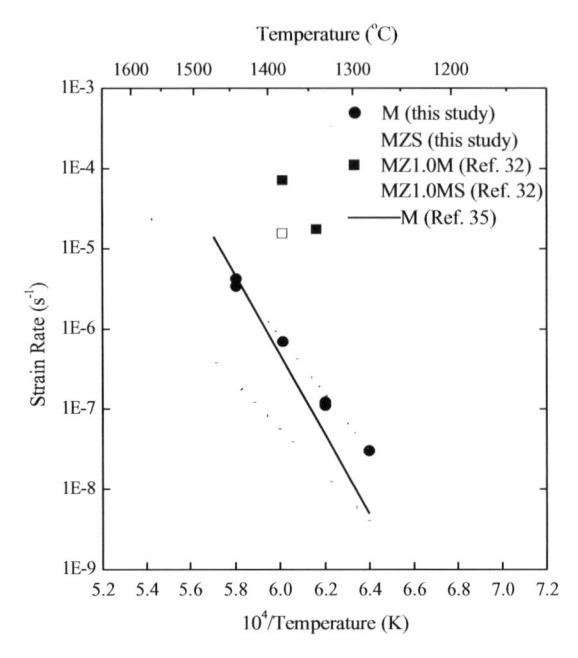

FIGURE 12. Strain rate versus reciprocal temperature for mullite, mullite-ZrO_2-SiC whisker, mullite-ZrO_2-MgO, and mullite-ZrO_2-MgO-SiC whisker. All data are taken at or normalized to 50 MPa. Published with permission of Am. Ceram. Soc. (33)

The variation of fracture toughness versus SiC whisker content for different compositions is shown in Fig. 13, where significant increases in toughness are seen for increasing whisker content. For mullite, values increase from 2.3 MPa·m$^{1/2}$ to 4.7 MPa·m$^{1/2}$ for mullite-30% SiC whiskers. Fig. 14 illustrates the microstructure of the latter composite with a hardness indent. There is a good distribution of SiC whiskers and the grain size is smaller than mullite. (Fig. 1) The mullite-ZrO_2-MgO matrix has a fracture toughness of 3.8 MPa·m$^{1/2}$, demonstrating a significant increase over mullite. For this matrix with 30% SiC whiskers, a value of 5.3 MPa·m$^{1/2}$ was obtained, demonstrating a further increase in fracture toughness. Similar results were seen for the mullite-35.4% ZrO_2-1.5% Y_2O_3 with a value of 4.0 for this matrix and 6.2 for its composite.

In summary, the incorporation of SiC whiskers in mullite increases the toughness 100%. The incorporation of ZrO_2 particles in mullite also increases the toughness 100%. The incorporation of SiC whiskers and a ZrO_2 phase in mullite further increases the toughness 130 to 170% compared to mullite.

(8) Thermal Shock

The thermal shock resistance of mullite-SiC whisker composites and mullite-ZrO_2-SiC whisker composites was evaluated by Kelly, et al.[24] using samples prepared as described by Ruh and Mazdiyasni.[15] Thermal expansion, flexural strength and Young's modulus were

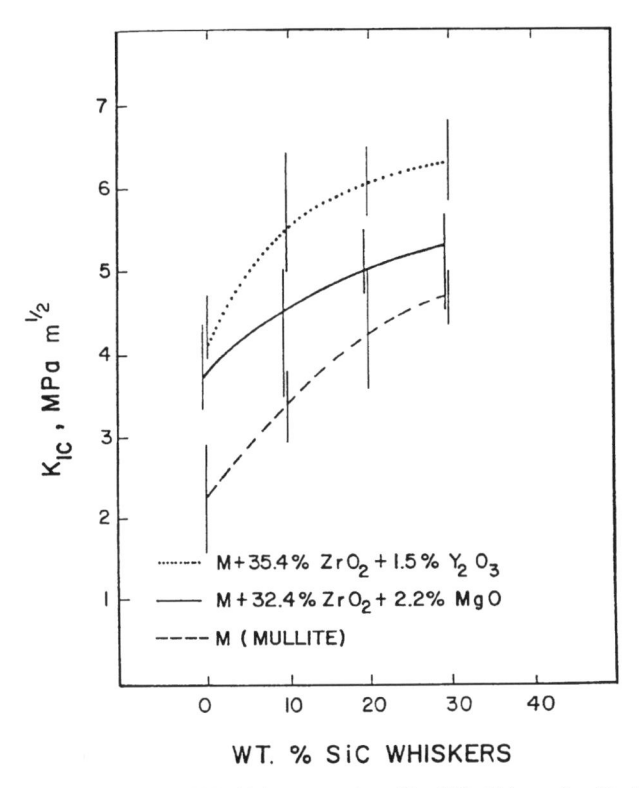

FIGURE 13. Fracture toughness versus SiC whisker content in mullite-SiC whisker and mullite-ZrO_2-SiC whisker composites. Published with permission of the J. Am. Ceram. Soc. (15)

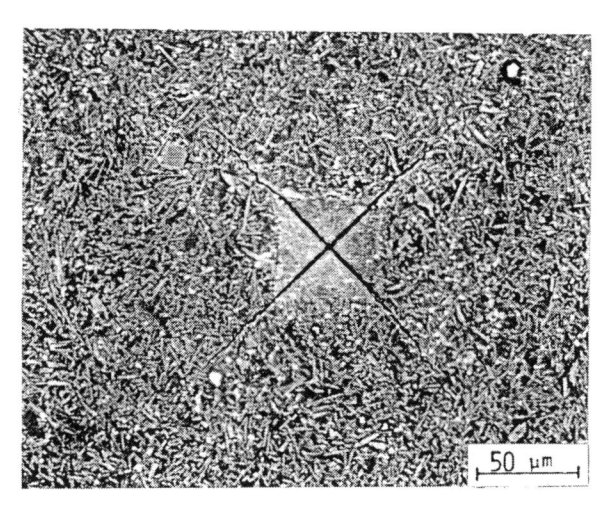

FIGURE 14. SEM micrograph of mullite-30% SiC whisker sample with a microhardness indent. Published with permission of the J. Am. Ceram. Soc. (15)

determined experimentally. Using these data and values of Poisson's ratio from the literature, R-parameters were calculated from the formula:

$$R = \frac{\sigma_F(1-\nu)}{E\alpha} = \frac{\Delta T\max}{S}$$

Where R = resistance to fracture initiation
σ_F = fracture stress
ν = Poisson's ratio
E = Young's modulus
α = coefficient of thermal expansion
$\Delta T\max$ = critical quench temperature difference
S = shape factor

For mullite, mullite-30% SiC, mullite-36.1% ZrO_2 and mullite-36.1% ZrO_2-30% SiC, the R values were 195°, 285°, 224° and 381°C. Thermal shock was measured using a water quench test and microscopically determining the temperature at which cracking occurred. Values of ΔT_{max} for the above materials were 277°, 381°, 303° and 381°C, respectively. Calculating the shape factors S, one obtains values between 1.33 and 1.42, which is good agreement. The R" and R'" parameters were also calculated, the former using the thermal diffusivity data of Russell, et al.[26] The thermal shock resistance of Al_2O_3-20% SiC whiskers was evaluated by Tiegs and Becher[14] by thermal shocking flexure specimens and then measuring residual strength. Whereas Al_2O_3 showed a significant decrease in flexural strength for a temperature change of >400°C, the composites showed practically no strength decrease for temperature differences of up to 900°C. Why such differences were obtained for the two similar SiC whisker composites is not understood, although test methods differed.

V. ELECTRICAL PROPERTIES

(1) Permittivity and Permeability at High Frequencies

Real and imaginary permittivity values were determined on mullite-SiC whisker composites and mullite-ZrO_2 composites.[37] A two-port coaxial measurement system and network analyzer were used, and measurements made on samples perpendicular to the hot-pressing direction over the 2 to 18 GHz frequency range. Results are given in Fig. 15. (Data on spinel-SiC whisker composites are also included.) Values of 6.6 to 6.9 were found for mullite. As the SiC whisker content increased, the real permittivity values increased to 16 to 22 for the 30% SiC whisker composites. Imaginary permittivity values for mullite were 0, and 10 for the 30% SiC whisker composites. For mullite-ZrO_2 composites the real permittivity values were 9.3 and imaginary values were 0 to 0.2. Real and imaginary permeability values for all the materials were 1 and 0 respectively.

(2) Permittivity, Impedance and Resistivity at Low Frequencies

The electrical behavior of hot-pressed mullite composites reinforced with 10, 20 and 30% SiC whiskers was evaluated in the 100 Hz to 10 MHz frequency range.[38] Samples

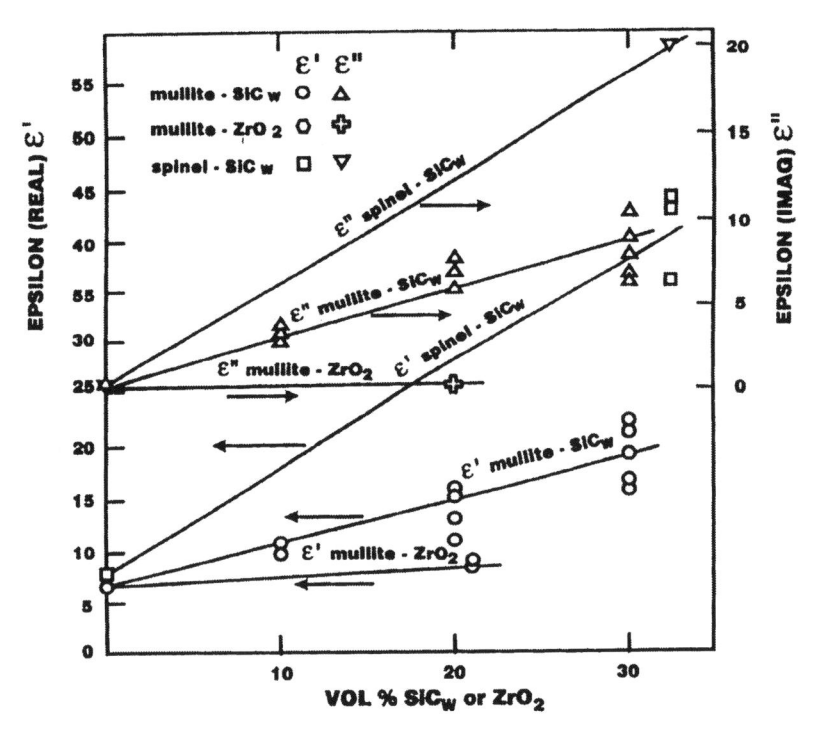

FIGURE 15. Real and imaginary permittivity values for mullite composites. Published with permission of the J. Am. Ceram. Soc. (37)

were hot pressed using alkoxide-derived materials. This resulted in the SiC whiskers being aligned in planes perpendicular to the hot-pressing direction, and randomly oriented within the planes. For samples measured parallel to the hot-pressing direction, the addition of SiC whiskers dramatically increased the dielectric constant of the composite, especially at the lower frequencies. In addition to the volume fraction dependence, these measurements also revealed whisker-orientation dependence, and the increase in dielectric constant was much greater for samples measured perpendicular versus parallel to the hot-pressing direction.

The complex impedance spectra revealed that the resistance or real part of the impedance decreased as the SiC whisker content increased. A SiC whisker orientation effect was also found, with lower impedance being observed for samples measured perpendicular versus parallel to the hot-pressing direction.

Direct current resistivity measurements revealed that the data are sensitive to the volume fraction of SiC whiskers, but no clear distinction can be made between measurements perpendicular versus parallel to the hot-pressing direction. The percolation threshold for SiC whiskers in mullite occurs before the 10% SiC whisker composition.

VI. CONCLUSION

A comprehensive study and characterization has been accomplished on the mullite-SiC whisker and mullite-ZrO_2-SiC whisker composite systems. From the basic research standpoint, the viability of alkoxide fabrication was demonstrated, which made possible the optimum investigation of these model systems. This work showed that significant property improvements were possible with the incorporation of SiC whiskers and a transformation-toughened ZrO_2 phase. High room temperature strength and fracture toughness were demonstrated, as well as improved creep and thermal shock resistance. The effect of SiC whiskers on the thermal diffusivity, specific heat and electrical properties was also demonstrated.

From a practical standpoint, the mullite-SiC whisker composites also demonstrated high strength after thermal exposure and high strength at elevated temperatures. In contrast, the mullite-ZrO_2-SiC whisker composites, with either Y_2O_3 or MgO as a stabilizer, had poor strength after thermal exposure and poor high temperature strength. Thus, for high temperature applications, the mullite-SiC whisker composites are promising and the mullite-ZrO_2-SiC whisker composites are not.

ACKNOWLEDGMENTS

I am very indebted to K. S. Mazdiyasni for the suggestion to investigate the mullite-ZrO_2-SiC whisker system, and for the preparation of the alkoxide batches. I gratefully acknowledge Professor Avigdor Zangvil and his students for stimulating discussions and outstanding microstructural characterization of these composites. Also gratefully acknowledged are the technical experts in different disciplines, whose studies contributed significantly to the understanding of these complex materials. I would like to acknowledge Drs. Henry Graham and Ronald Kerans, and the Air Force team for their encouragement, support and contributions. Finally, I am very grateful to my wife, Sally, for keying, editing and proofreading the manuscript. Without all of the above, this study would not have been possible.

REFERENCES

1. K. S. Mazdiyasni and L. M. Brown, "The Synthesis and Mechanical Properties of Stoichiometric Aluminum Silicate,"*J. Am. Ceram. Soc.*, **55** [11] 548–92 (1972).
2. W. E. Cameron, "Composition and Cell Dimensions of Mullite," *Amer. Ceram. Soc. Bull.*, **56** [11] 1003–11 (1977).
3. S. Prochazka and F. J. Klug, "Infrared-Transparent Mullite Ceramic,"*J. Am. Ceram. Soc.*, **66** [12] 874–80 (1983).
4. R. A. Penty, "Pressure-Centered Kinetics and Mechanical Properties of High Purity Fine-Grained Mullite"; Ph.D. Dissertation. Lehigh University, Bethlehem, PA, 1972.
5. B. L. Metcalfe and J. H. Sant, "The Synthesis, Microstructure, and Physical Properties of High-Purity Mullite," *Trans. J. Br. Ceram. Soc.*, **74** [6] 193–201 (1975).
6. T.-I. Mah and K. S. Mazdiyasni, "Mechanical Properties of Mullite,"*J. Am. Ceram. Soc.*, **66** [10] 699–703 (1983).
7. N. Claussen and J. Jahn, "Mechanical Properties of Sintered, In situ-Reacted Mullite-Zirconia Composites," *J. Am. Ceram. Soc.*, **63** [3–4] 228–29 (1980).

8. J. S. Moya and M. I. Oscendi, "Microstructure and Mechanical Properties of Mullite/ ZrO_2 Composites," *J. Mater. Sci.*, **19** [8] 2909–14 (1984).
9. P. Boch and J. P. Giry, "Preparation and Properties of Reaction-Sintered Mullite-ZrO_2 Ceramics," *Mater. Sci. Eng.*, **71**, 39–48 (1985).
10. C. Baudin De La Lastra, C. Leblud, A. Leriche, F. Cambier, and M. R. Anseau, "K_{IC} Calculations for Some Mullite-Zirconia Composites Prepared by Reaction Sintering,"*J. Mater. Sci. Lett.*, **4** [9] 1099–101 (1985).
11. G. Orange, G. Fantozzi, F. Cambier, C. Leblud, M. R. Anseau, and A. Leriche, "High Temperature Mechanical Properties of Reaction-Sintered Mullite/Zirconia and Mullite-Alumina/Zirconia Composites,"*J. Mater. Sci.*, **20** [7] 2533–40 (1985).
12. G. C. Wei and P. F. Becher, "Development of SiC-Whisker-Reinforced Ceramics," *Am. Ceram. Soc. Bull.*, **64** [2] 298–304 (1985).
13. N. Claussen and G. Petzow, "Whisker-Reinforced Zirconia-Toughened Ceramics"; pp. 649–62 in Tailoring Multiphase and Composite Ceramics, Vol. 20. Edited by R. E. Tressler, G. L. Messing, C. G. Pantano, and R. E. Newnham. Plenum Press, New York, 1986.
14. T. N. Tiegs and P. F. Becher, "Whisker-Reinforced Ceramic Composites"; pp. 639–47 in Tailoring Multiphase and Composite Ceramics, Vol. 20. Edited by R. E. Tressler, G. L. Messing, C. G. Pantano, and R. E. Newnham. Plenum Press, New York, 1986.
15. R. Ruh, K. S. Mazdiyasni, and M. G. Mendiratta, "Mechanical and Microstructural Characterization of Mullite and Mullite-SiC-Whisker and ZrO_2-Toughened-Mullite-SiC-Whisker Composites,"*J. Am. Ceram. Soc.*, **71** [6] 503–12 (1988).
16. R. Ruh, K. S. Mazdiyasni, P. G. Valentine, and H. O. Bielstein, "Phase Relationships in the System ZrO_2-Y_2O_3 at Low Y_2O_3 Contents,"*J. Am. Ceram. Soc.*, **67** [9] C-190-C-192 (1984).
17. V. S. Stubican; private communication.
18. R. Ruh and K. S. Mazdiyasni, "Fabrication of Mullite-SiC-Whisker Composites and Mullite-Partially Stabilized ZrO_2-Whisker Composites"; pp. 205–15 in Metal Matrix, Carbon, and Ceramic Matrix Composites, NASA Conference Publication 2406. Langley Research Center, Hampton, VA, 1985.
19. A. Zangvil, C.-C. Lin, and R. Ruh, "Microstructural Studies in Alkoxide-Derived Mullite/Zirconia/Silicon Carbide-Whisker-Composites,"*J. Am. Ceram. Soc.*, **75** [5] 1254–65 (1992).
20. C.-C. Lin, A. Zangvil, and R. Ruh, "Phase and Microstructural Evolution in Alkoxide-Derived Mullite MgO Partially-Stabilized Zirconia,"*J. Am. Ceram. Soc.*, **78** [5] 1361–71 (1995).
21. C.-C. Lin, A. Zangvil, and R. Ruh, "Phase Evolution in Silicon Carbide-Whisker-Reinforced Mullite/Zirconia Composites during Long-Term Oxidation at 1000° to 1350°C,"*J. Am. Ceram. Soc.*, **83** [7] 1797–803 (2000).
22. C.-C. Lin, A. Zangvil, and R. Ruh, "Modes of Oxidation in SiC-Reinforced Mullite/ZrO_2 Composites: Oxidation VS Depth Behavior," *Acta Mater.*, **47** [6] 1977–86 (1999).
23. C.-C. Lin, A. Zangvil, and R. Ruh, "Microscopic Mechanisms of Oxidation in SiC-Whisker-Reinforced Mullite/ ZrO_2 Matrix Composites,"*J. Am. Ceram. Soc.*, **82** [10] 2833–40 (1999).
24. W. H. Kelly, A. N. Palazotto, R. Ruh, J. K. Heuer, and A. Zangvil, "Thermal Shock Resistance of Mullite and Mullite-ZrO_2-SiC-Whisker Composites," *Ceram. Eng. Sci. Proc.*, **18** [3] 195–203 (1997).
25. L. M. Russell, L. F. Johnson, D. P. H. Hasselman, and R. Ruh, "Thermal Conductivity/ Diffusivity of Silicon Carbide Whisker Reinforced Mullite,"*J. Am. Ceram. Soc.*, **70** [10] C-226-C-229 (1987).
26. L. M. Russell, K. Y. Donaldson, D. P. H. Hasselman, R. Ruh, and J. W. Adams, "Thermal Diffusivity/Conductivity and Specific Heat of Mullite-Zirconia-Silicon Carbide Whisker Composites,"*J. Am. Ceram. Soc.*, **79** [10] 2767–70 (1996).
27. D. P. H. Hasselman, L. F. Johnson, L . D. Bentsen, R. Syed, H. L. Lee, and M. V. Swain," Thermal Diffusivity and Conductivity of Dense Polycrystalline ZrO_2 Ceramics: A Survey," *Am. Ceram. Soc. Bull.*, **66** [5] 799–806 (1987).
28. R. Ruh; unpublished data.
29. R. Ruh; unpublished data.
30. Y. Xu, A. Zangvil, and R. Ruh, "High Temperature Fracture Mechanisms in Alkoxide-Derived Mullite/ZrO_2/SiC-Whisker Composites," *Ceram. Eng. Sci. Proc.*, **14** [9–10] 982–90 (1993).
31. J. E. Ritter, K. Jakus, M. H. Godin, and R. Ruh, "Comparison of High-Temperature Crack Growth in SiC-Whisker-Reinforced Mullite and Si_3N_4,"*J. Am. Ceram. Soc.*, **75** [7] 1760–66 (1992).
32. T. A. Parthasarathy, R . S. Hay, and R. Ruh, "High-Temperature Deformation of SiC-Whisker-Reinforced MgO-PSZ/Mullite Composites,"*J. Am. Ceram. Soc.*, **79** [2] 475–83 (1996).

33. A. Hynes, R. H. Doremus, R. Ruh, Y. Xu, and A. Zangvil, "Creep of Mullite Composites Containing Zirconia and Silicon Carbide Whiskers"; pp. 509–21 in Advances in Ceramic Matrix Composites III, Ceramic Transactions, Vol. 74. Edited by N. P. Bansal and J. P. Singh. The American Ceramic Society, Westerville, OH, 1996.
34. F. Garofalo, *Fundamentals of Creep and Creep-rupture in Metals.* MacMillan, NY, 1965.
35. A. Hynes, "A Study of High Temperature Creep of Polycrystalline Mullite"; Ph.D. Dissertation. Rensselaer Polytechnic Institute, Troy, NY, 1993.
36. A. G. Evans and E. A. Charles, "Fracture Toughness Determinations by Indentation," *J. Am. Ceram. Soc.*, **59** [7–8] 371–72 (1976).
37. R. Ruh and H. Chizever, "Permittivity and Permeability of Mullite-SiC-Whisker and Spinel-SiC-Whisker Composites," *J. Am. Ceram. Soc.*, **81** [4] 1069–70 (1998).
38. R. A. Gerhardt and R. Ruh, "Volume Fraction and Whisker Orientation Dependence of the Electrical Properties of SiC-Whisker-Reinforced Mullite Composites," *J. Am. Ceram. Soc.*, **84** [10] 2328–34 (2001).

15

Nextel™ 312/Silicon Oxycarbide Ceramic Composites

Stephen T. Gonczy* and John G. Sikonia**

*Gateway Materials Technology, Inc.
221 S. Emerson, Mt. Prospect, IL, 847-870-1621
Gatewaymt@aol.com
**Sikonia Consulting
20423 Bullblock Rd., Bend, OR, 541-312-4328
sikonia@earthlink.net

Fiber reinforced ceramic matrix composites (CMCs) are under active consideration for large, complex high temperature structural components in aerospace and automotive applications. The Blackglas™ resin system (a low cost polymer-derived ceramic [PDC] technology) was combined with the Nextel™ 312 ceramic fiber (with a boron nitride interface layer) to produce a silicon oxycarbide CMC system that was extensively characterized for mechanical, thermal, and electronic properties and oxidation, creep rupture, and fatigue. A gas turbine tailcone was fabricated and showed excellent performance in a 1500-hour engine test.

1. INTRODUCTION

Continuous fiber-reinforced ceramic matrix composites (CMCs) offer many potential advantages in high temperature applications – tough, high strain ($>0.2\%$) failure; weight savings over metals at temperatures where polymers cannot survive; and thermal and corrosion resistance at conditions where metals fail.[1,2,3] The most promising near term opportunity for ceramic matrix composites is in hot secondary structural components in high performance jet engines (See Figure 1). High performance brakes are another near-term CMC application of interest for the aerospace industry, as well as for land transportation. In the longer term CMCs have potential applications in the automotive industry as cylinder liners

FIGURE 1. Pratt & Whitney F-119 Turbofan Engine with Exhaust Divertors (for the F22 Raptor)

and heads, pistons, and exhaust manifolds.[4] A third area of potential CMC application is in power and combustion applications, where CMCs offer higher performance temperatures and greater reliability for combustors cans, environmental filters, and cogeneration heat-exchangers.[5]

But CMCs will be commercially successful only when they are produced cost-effectively.[6] Polymer-derived ceramic (PDC) technology is one of the most promising low cost fabrication methods for ceramic matrix composites, particularly for large, complex shapes.[7] In PDC technology, a silicon-based polymer (siloxane, carbosilane, silazane, etc) with fiber or particle reinforcement is shaped and cured in the polymer condition and then pyrolyzed in a controlled atmosphere to form a stable silicon-based ceramic, such as silicon carbide, silicon nitride, silicon oxycarbide, or silicon oxynitride.

These silicon-based polymers can be formed using conventional polymer composite fabrication techniques, such as lamination, resin-transfer molding, infiltration, filament winding, compression molding, etc. In a pyrolysis heat-treatment the polymer converts to the silicon ceramic, losing some weight and increasing in density during the transition. Pyrolysis is done in a neutral atmosphere (nitrogen, argon, vacuum, etc.) at temperatures from 800°C to 1300°C. PDC technology is directly applicable to ceramic matrix composites, because the fiber reinforced ceramic composites can be formed into complex shapes in the polymer stage and then converted to a ceramic in a near-net shape condition.

This chapter will describe the processing and properties of an oxide fiber reinforced ceramic matrix composite with a silicon oxycarbide matrix based on a PDC technology, introduced by AlliedSignal (now Honeywell International) under the trademark of Blackglas™ ceramic.[8] The oxide fiber in this CMC system is the Nextel 312™ fiber (3M, Inc.) that has been treated to form a boron nitride surface coating. The information that follows was primarily developed from Low Cost Ceramic Matrix Composites (LC^3) program funded by DARPA from 1991–1997.

2. BLACKGLAS POLYMER DERIVED CERAMIC TECHNOLOGY

2.1. Blackglas Polymer and Silicon Oxycarbide Ceramic

Silicon oxycarbide (SiOC) ceramic (trade name = Blackglas ceramic) contains unusually high levels (20–30% by weight) of carbon that are fully incorporated into the molecular structure.[9,10] The introduction of carbon into silica glass was historically done by impregnating porous glass with a concentrated solution of an organic compound and subsequent firing in a reducing or neutral atmosphere.[11] The resulting carbon-containing glass is regarded as a composite containing discrete carbon particles in a silica matrix. The Blackglas amorphous SiOC, however, is formed through the pyrolysis of a thermosetting polymeric precursor, in which carbon atoms are an initial part of the polymer matrix and are incorporated into the silica matrix during pyrolysis. The ceramic has a higher softening point than fused silica, and greater resistance to devitrification than previously known carbon-silicon dioxide glass.[12]

The low-molecular weight Blackglas polymer precursor is cross-linked through hydrosilation of vinyl groups on the siloxanes in the presence of a thermally activated catalyst, forming a 3D network polymer. In this proprietary chemistry there are no gaseous byproducts during the curing process. The exothermic curing reaction in air at 50 to 150°C yields a clear polymer with the heat evolution of about 600 J/gm. In components with large mass or thick cross-section some care must be used to control the temperature during the curing process.

The cured polymer is converted to an amorphous ceramic by heating in an inert environment (nitrogen, argon, or vacuum) to temperatures ranging from 800 to 1200°C. During pyrolysis the solid goes through a high surface area, high pore volume intermediate stage before sintering to a low surface area (<1 m^2/g) ceramic. The pyrolyzed ceramic has a density of 2.2 g/cm^3 and an approximate composition of 47 wt% Si, 27 wt% O, 25 wt% C, and 0.5 wt% H. The ceramic (char) yield is 80 to 85 wt% (based on polymer). The final product has Si in the following configurations based on NMR analysis[13]: C_4Si, C_3SiO, C_2SiO_2, $CsiO_3$, and SiO_4. Carbon bonding in the final product is in the forms of carbide and graphite.

The Blackglas PDC system has a number of processing advantages compared to competitive PDC resins. The low viscosity (10 cp) Blackglas resin cures and handles like epoxy, is insensitive to moisture, and has a controllable cure cycle through catalyst content and temperature ramping.

2.2. Nextel 312 Fiber with a Boron Nitride Coating

Nextel 312 fiber is manufactured by 3M and is composed of alumina, boria, and silica. It is available in tow, fabric and woven tape configurations. This material retains strength and flexibility with little shrinkage up to 1200°C. The properties of the Nextel 312 fiber are: density = 2.70 g/cm^3, tensile strength = 1720 MPa (250 ksi), elastic modulus = 150 GPa (22 MSI), unstressed continuous use temperature = 1204°C, short-term use temperature = 1426°C, melt temperature = 1800°C.

Ceramic matrix composites require a fiber interface coating to prevent fiber-matrix bonding and to produce crack deflection and fibrous fracture in the composite. Work at the

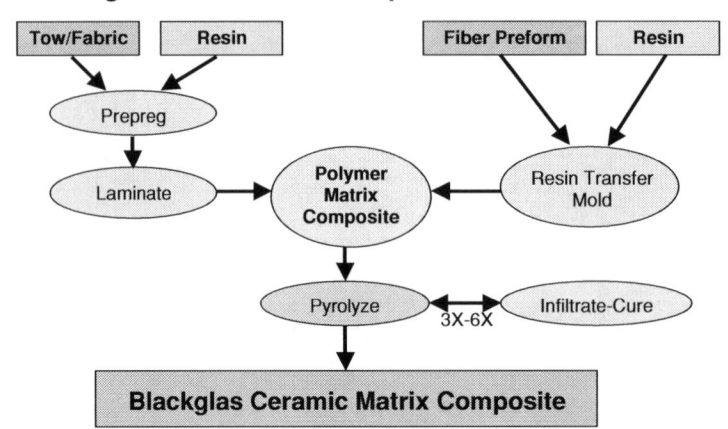

Blackglas Ceramic Fiber Composite Fabrication Process

FIGURE 2. Schematic of Composite Fabrication Process Using Blackglas Resin and Ceramic Fibers

Naval Research Laboratory showed that boron nitride was an effective interface layer.[14] Scientists at Boeing discovered that when Nextel 312 fiber is heated to high temperatures in the presence of ammonia and hydrogen, some of the boria near the surface is reduced and converted to form a thin (<0.1 micron) boron nitride coating.[15,16] AlliedSignal scientists also did work in this area.[17] A detailed research investigation of this process was conducted at the University of Illinois at Chicago in 1997.[18] This process can be applied to all forms of heat cleaned Nextel 312 material including tow, fabric, and woven or braided performs. The BN coating acts as a debonding interface layer between the fiber and the matrix.[19]

2.3. Composite Fabrication

With the Blackglas resin, ceramic composites are made in a multi-stage process, analogous to the method in which carbon-carbon composites are fabricated using resin technology (See Figure 2).[20] First, a polymer composite is made through a forming and curing process, such as auto-clave lamination or resin transfer molding. Then the polymer composite is pyrolyzed to convert the polymer matrix to a ceramic. During pyrolysis, the matrix fraction of the composite develops open porosity, which is then filled by multiple reinfiltration/cure/pyrolysis cycles. The number of reinfiltration cycles required depends on the char yield, the penetration, and the pore structure for a given preceramic polymer system.

Blackglas technology is particularly applicable to forming CMC composites, with exceptional ease and flexibility in processing. Blackglas materials resins offer the following advantages – complex shapes can be made in the polymer stage by standard lamination techniques or resin transfer molding; there is no vapor evolution during the cure cycle; excellent near net shape is achieved; and the part can be machined in the polymer state.

The Nextel 312 (BN) fiber reinforcement should be designed to achieve a fiber volume of 40–50 per cent. Initial curing of the Blackglas polymer is typically done under vacuum at temperatures in the range of 50°C to 150°C.[20]

Blackglas™ ceramic is produced by the pyrolysis of the proprietary thermosetting siloxane resin.[21] During pyrolysis under nitrogen at maximum temperatures between 800°C and 1200°C, the cured Blackglas polymer loses 15% and 20% wt% in weight and increases in density from 1.16 to 2.2 g/cm^3 as it transforms to the silicon oxycarbide ceramic form with the release of methane, C_2 hydrocarbons, C_3 hydrocarbons, and hydrogen. The pyrolysis cycle extends over periods of 6–36 hours with temperature ramp rates ranging from 1 to 5°C/minute depending on the size and thickness of the part. For larger, and especially thicker composites, an extended cycle will reduce the possibility of delamination caused by pyrolysis gas evolution.

Blackglas composites are densified through repeated infiltration/cure/pyrolysis procedures. This procedure fills the porosity formed during pyrolysis of the cured polymer. The normal procedures for re-infiltration of Blackglas composites call for a vacuum purge followed by a resin infiltration. The infiltrated parts are allowed to drain and cured at 50°C for a period of 30 minutes to three hours. After curing, the infiltrated composite is pyrolyzed using the same time-temperature cycle used in the first pyrolysis. This process is typically repeated until the weight gain per cycle is less than 1% and the final open porosity is less than 2%. Studies with the Nextel 312 (BN) and other fibers have shown that maximum, as-prepared strength is achieved with four or five infiltration cycles.[22] However, there are benefits in extending this process. Strength retention after oxidation is enhanced with additional infiltrations.

3. FUNCTIONAL PROPERTIES

In long-life aerospace structural applications, the primary performance requirements for CMCs fall into two categories – initial mechanical properties and long-term mechanical property retention at high operating temperatures. As part of the DARPA Program – Low Cost Ceramic Matrix Composites (LC^3), a well-defined silicon oxycarbide (SiOC) Nextel 312 boron nitride (BN) composite system was extensively tested for mechanical properties, high temperature durability, environmental effects, thermal properties, and electrical properties. The data developed in that study are representative of the range and reproducibility of the properties achievable with this composite system.[23]

3.1. Specimen Preparation

The SiOC matrix was prepared using the Blackglas 493-type preceramic resin. The Nextel 312 fabric was an AF-10 5H satin weave (600 denier tows, 275 g/m^2 areal weight). Nextel 312 is the lowest cost ceramic fiber with acceptable strength and high temperature capability that is commercially available. The boron nitride interface layer on the Nextel 312 fibers was produced by the previously described high temperature nitridation process.

The SiOC-Nextel 312 BN 2-D composites were fabricated as an 8-ply balanced lay-up using a six-step lamination and multiple pyrolysis/infiltration method:

1. Prepreg the nitrided Nextel 312 fabric using a Blackglas 493C resin.
2. Cut and stack prepreg layers and laminate 8 ply composites in a 0/90 symmetrical lay-up using a 150°C final cure.

3. Cut test bar blanks in the polymer condition to appropriate widths and lengths.
4. Pyrolyze the bars in flowing nitrogen at 1000°C with a 1 hour hold and a 2°C/min heat-up and a free furnace cool.
5. Densify the two sets of bars with 5X and 8X reinfiltrations, respectively, using a vacuum infiltration/curing/ pyrolysis cycle.
6. Water jet cut the final width, length and gage sections on the tensile, compression and double notch (DN) shear bars to insure parallel edges.

3.2. Mechanical Tests

All mechanical tests were done at ambient temperatures (20°C). Tensile, shear, and compression tests were done using a stress rate of 25 MPa/s. Load and extension data were recorded by computer data collection.[24]

Tensile tests were done on dog-bone shaped tensile coupons (with dimensions shown in Figure 3) with tapered glass epoxy tabs on the grip sections. The tensile strain was monitored

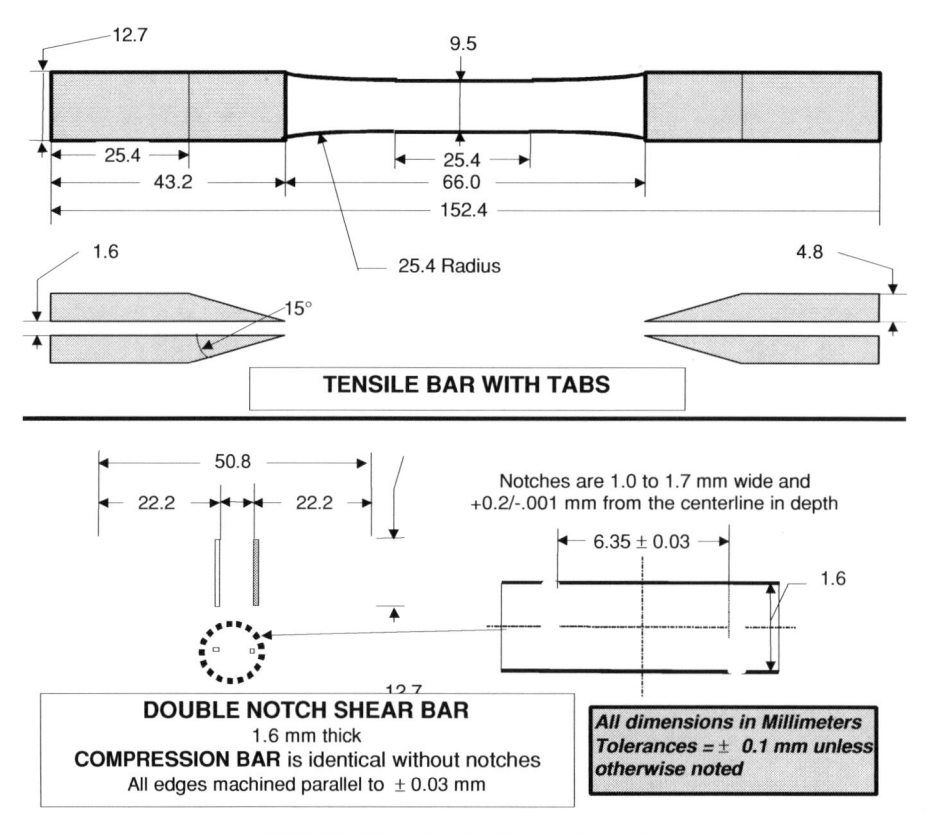

FIGURE 3. Mechanical Test Coupon Geometries

with a single strain gage attached to the woven specimen face and two contact extensometers with a gage length of 25.4 mm.

Elastic modulus values were calculated from a mathematical least squares fit on that linear portion of the tensile stress-strain curves that produced a correlation factor of 0.99. Proportional limit stress and strain values were measured at the upper limit of the elastic modulus defined section of the stress-strain curves.

The shear strengths of the composites were measured in a double notch shear geometry, loaded in compression with an anti-buckling fixture. Compression testing was done on test coupons with dimensions as shown in Figure 3, using an anti-buckling fixture.

All fractured samples were examined visually to identify the mode of failure – tension, compression, shear. Selected samples were analyzed by SEM to characterize the degree of fiber pull-out and the failure mode.

3.3. Oxidation Exposure Testing

Test specimens were mechanically tested in the as-prepared condition and after a 600°C 1000-hour oxidation exposure in flowing air. Test bar sets were weighed at 24, 100, 500, and 1000 hours. Partial sample sets were removed for mechanical testing at the required time intervals.[25]

3.4. Specimen Characterization – As-Prepared and After 600°C Oxidation

As-Prepared Properties – The composites test bars had a nominal fiber volume fraction of ≈46%, based on composite thickness (1.6 mm), ply-count (8), fabric areal weight (250 g/m^2) after de-sizing, and fiber density (2.7 g/cm^3). The theoretical density of the composites at 46 vol% is 2.41 g/cm^3 based on a density of 2.2 g/cm^3 for Blackglas ceramic and 2.7 g/cm^3 for Nextel 312 fiber. The Archimedes density measurements on the 5X and 8X samples were:

Composite Condition	Archimedes. Density (g/cc)	Open Porosity (%)
5X Infil. Samples	2.44	2.1%
8X Infil. Samples	2.35	1.1%

Single test bars were removed and measured by Archimedes density methods after 25, 100, 500, and 1000 hours oxidation at 600°C. The data are given in Table 1. For both test bar sets, the changes in density and porosity after oxidation are relatively small.

TABLE 1. 600°C Oxidation Effects on Density and Porosity of SiOC-Nextel 312 BN 2-D Composites

Condition	As-Prep	600°C/25 hr	600°C/100 hr	600°C/500 hr	600°C/1000 hr
5X-Density (g/cm^3)	2.44	2.45	2.46	2.49	2.43
5X-% Open Porosity	2.1%	2.6%	3.2%	4.1%	4.1%
8X-Density (g/cm^3)	2.35	2.38	2.43	2.41	2.42
8X-% Open Porosity	1.1%	1.4%	1.7%	2.0%	3.6%

600C Oxidation Weight Loss for SiOC-Nextel 312 BN 2-D Composites

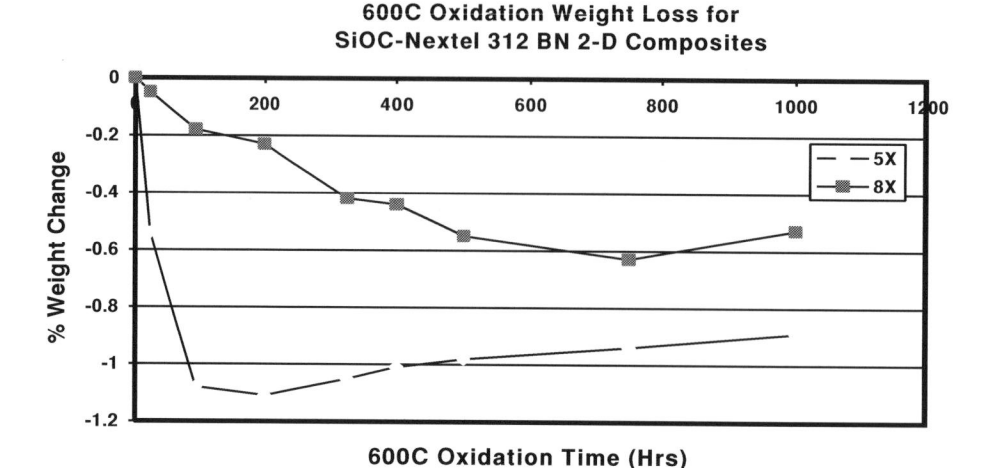

FIGURE 4. Weight Change with 600°C Oxidation Exposure for SiOC Nextel 312 BN 2-D Composites

The percent weight changes during oxidation (Figure 4) for the two sets are again small, producing losses on the order of less than 1 wt% out to the 1000 hour mark. However, the 5X bars show a much more rapid 1.1% loss (within 100 hours) followed by a slow but steady weight increase. The 8X bars show a slower and a reduced weight loss with time reaching a maximum weight loss of 0.6% at 500 hours, followed by a slight weight gain at the 1000-hour mark. It is hypothesized that the initial weight loss is that of "free carbon". Additional slow weight increases may occur with the loss of carbon from the oxycarbide and the oxidation of free silicon bonds.

3.5. Mechanical Properties – As-Prepared and After 600°C 1000-hour Oxidation Exposure

Tensile Properties – Table 2 gives values for the tensile properties as a function of the as-prepared condition and the five 600°C oxidation times.[25] Typical tensile stress strain curves are shown in Figure 5 for the 5X and 8X as-prepared and post oxidation composites.

In the as-prepared condition, additional infiltrations and higher density do produce a difference in the tensile properties, as shown both in Table 2 and in the tensile stress-strain curves of Figure 5. The additional infiltrations produce a higher modulus for the composite, as would be expected through the reduction of porosity. The reduction in tensile strength and tensile strain, however, is unexpected. If a higher density matrix produces a more brittle failure (lower strain and lower stress), it implies that the matrix porosity has a contribution to the composite toughness and non-brittle failure. Given the limited number of tensile test bars, the small differences between the 5X and 8X mean values for the proportional limit stress and strain are probably not significant.

The effects of oxidation on the ambient temperature tensile properties of the composites are shown in Figures 6, 7, and 8. The effects of 600°C oxidation on tensile properties out to 1000 hours are gradual and not very dramatic. Figure 6 shows the effect of 600°C oxidation

TABLE 2. Tensile Properties of SiOC Nextel 312 BN 2-D Composites
As-Prepared and with 600°C Oxidation Exposure

Condition/Property	600°C Oxidation Time (hr)				
	As-Prep	24	100	500	1000
5X Infiltrated – # of Bars	6	2	2	3	3
Ult. Tensile Strength (MPa) Mean/S.D.	150 ± 7	103 ± 40	133 ± 5	141 ± 8	105 ± 18
Ult. Tensile Strain (%) Mean/S.D.	0.25 ± 0.03	0.29 ± 0.14	0.37 ± 0.02	0.32 ± 0.04	0.24 ± 0.06
Elastic Modulus (GPa) Mean/S.D.	88.7 ± 3.7	44.1 ± 0.2	57.6 ± 2.7	56.7 ± 4.1	54.2 ± 2.1
Prop. Limit Stress (MPa) Mean/S.D.	81.3 ± 6.6	58.5 ± 6.9	23.9 ± 2.3	88.8 ± 10.6	30.5 ± 3.6
Prop. Limit Strain (%) Mean/S.D.	0.09 ± 0.09	0.13 ± .02	0.04 ± 0.00	0.16 ± 0.02	0.06 ± 0.01
8X Infiltrated – # of Bars	4	2	2	3	2
Ult. Tensile Strength (MPa) Mean/S.D.	128 ± 11	99 ± 22	144 ± 4	163 ± 6	115 ± 17
Ult. Tensile Strain (%) Mean/S.D.	0.17 ± 0.04	0.18 ± 0.04	0.27 ± 0.02	0.31 ± 0.03	0.20 ± 0.04
Elastic Modulus (GPa) Mean/S.D.	102.5 ± 7.1	76.6 ± 11.8	77.7 ± 0.82	67.1 ± 3.7	73.5 ± 3.7
Prop. Limit Stress (MPa) Mean/S.D.	97.7 ± 2.2	70.6 ± 8.2	76.9 ± 0.8	49.2 ± 2.0	15.0 ± 0.2
Prop. Limit Strain (%) Mean/S.D.	0.10 ± 0.07	0.09 ± 0.00	0.10 ± 0.00	0.07 ± 0.01	0.02 ± 0.00

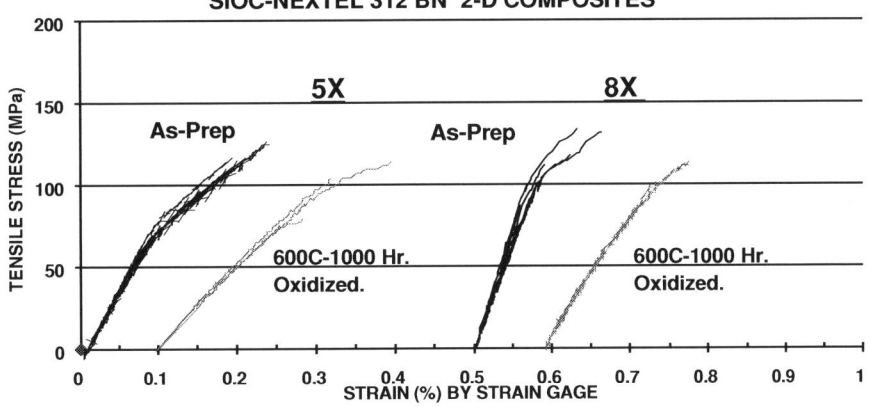

FIGURE 5. Tensile Stress-Strain Curves for As-Prepared and Post 600°C-1000 Hr. Oxidation SiOC-Nextel 312 BN 2-D Composites

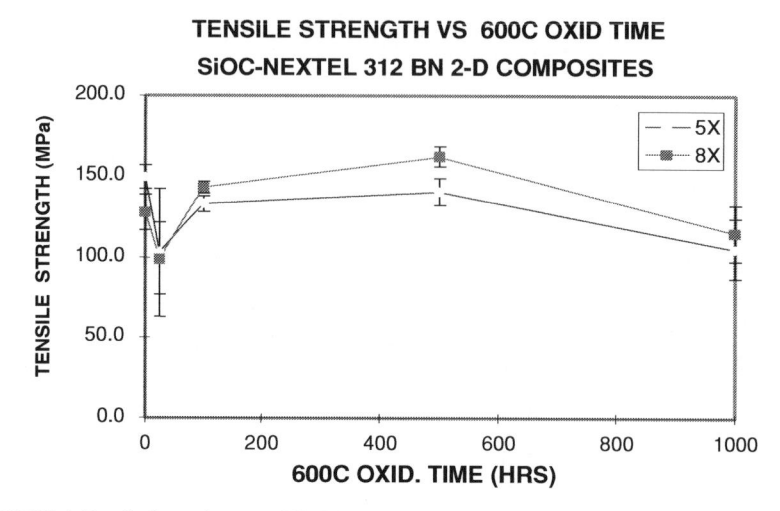

FIGURE 6. Tensile Strength versus 600°C Oxidation Time for SiOC Nextel 312 BN 2-D Composites

time on tensile strength. With the initial 24-hour oxidation there is a 20–30% loss in tensile strength for both sets of composites. But that loss is recovered with oxidation at 100 and 500 hours. Only with 1000 hours of oxidation is there additional loss in tensile strength, dropping to values of about 110 MPa, as compared to the initial strengths of 130 to 150 MPa.

The effect of oxidation time on tensile strain is shown in Figure 7. For the 5X specimens, the tensile strain increases as a function of oxidation time to 100 hours and then falls. But

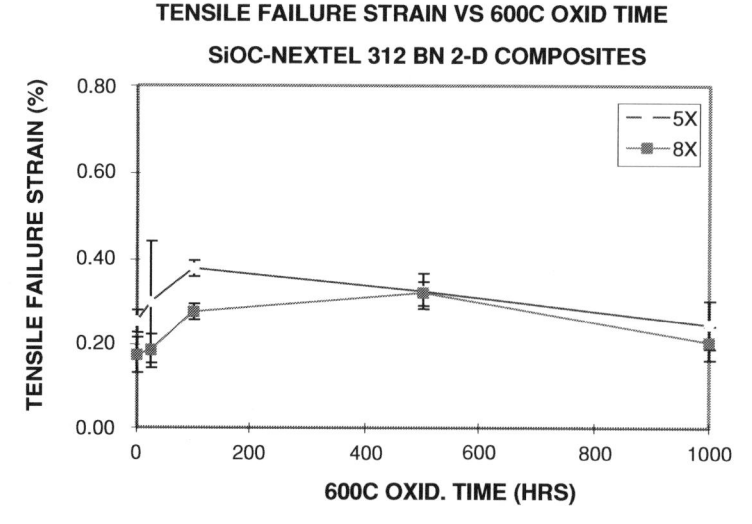

FIGURE 7. Tensile Failure Strain versus 600°C Oxidation Time for SiOC Nextel 312 BN 2-D Composites

TENSILE MODULUS VS 600C OXID TIME
SiOC-NEXTEL 312 BN 2-D COMPOSITES

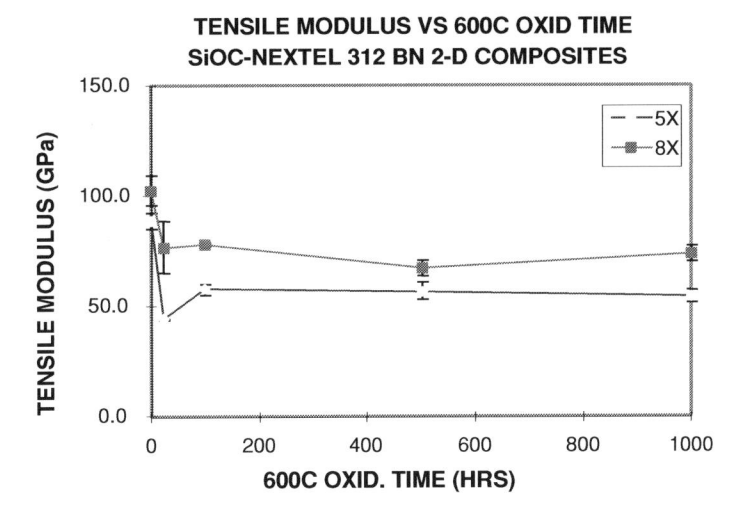

FIGURE 8. Tensile Elastic Modulus versus 600°C Oxidation Time for SiOC Nextel 312 BN 2-D Composites

even with that decrease, the tensile strain is still at the same level (even at 1000 hr) seen in the as-prepared condition. For the 8X specimens, the tensile failure strain increases with oxidation time to 500 hours and then falls at 1000 hours of oxidation. These results suggest that even after 1000 hours of oxidation, the composites may not have reached an oxidatively stable state, with respect to tensile failure strain. Again, the strain level after 1000 hours is approximately equal to the original tensile strain. In both cases the drop off in strain at the 1000 hour level does raise the question of whether there are additional property losses with longer term oxidation

The effect of oxidation time on the tensile modulus is shown in Figure 8. The 8X specimens have a higher modulus than the 5X specimens for all oxidation times. The tensile modulus values decreased after 24 hours of oxidation for both 5X and 8X specimens. The tensile modulus then remained fairly constant for both the 5X and 8X specimens through 1000 hours of oxidation.

The stress-strain curves for the 600°C, 1000-hour oxidized specimens in Figure 5 show that with extended 600°C oxidation the differences in the as-prepared properties of the 5X and 8X coupons are reduced, so that the two systems have similar tensile properties. However, the 8X sample still maintains a slightly higher modulus.

The crack plane in tension remained generally perpendicular to the direction of the applied load, without any deviation into the interlaminar plies. Oxidation of the composites did not change this general trend. SEM was performed on the tensile fracture surfaces to characterize the nature of the tensile failure. Micrographs of the fracture surfaces of the 5X and 8X as-prepared composites show a relatively rough fracture surface, but with very limited fiber pull-out. This corresponds well with the relatively low tensile failure strains of 0.2% seen in both of these composites.

SEM analysis of the tensile fracture surface for the 5X and 8X bars after the 600°C, 1000-hour, oxidation shows that the fracture surfaces for the oxidized test bars are rougher

TABLE 3. Double Notch Shear Strength of SiOC Nextel 312 BN 2-D Composites with 600°C Oxidation Exposure Effects

Condition/Property	600°C Oxidation Time				
	As-Prep	24 hr	100 hr	500 hr	1000 hr
5X Infiltrated – # of Bars	12	3	2	3	9
Ult. Shear Strength (MPa) Mean/S.D.	18.0 ± 6.1	17.6 ± 1.7	15.4 ± 3.2	13.5 ± 4.3	11.6 ± 2.4
8X Infiltrated – # of Bars	11	3	2	3	9
Ult. Shear Strength (MPa) Mean/S.D.	25.5 ± 7.3	26.9 ± 2.3	30.0 ± 1.2	14.3 ± .5	9.2 ± 2.6

with a greater degree of fiber pull-out. This corresponds with the increase in tensile failure strain after oxidation.

The effects of 600°C oxidation on the proportional stress and strain values are not clear in this study. For both the 5X and 8X test sets, the proportional limit stress and strain values in the as-prepared condition are within the expected range – 80–100 MPa stress and 0.1% strain. The 8X stress value is higher than the 5X value. However with 600°C oxidation, the stress and strain values shift higher and lower with time. The 5X values for stress and strain vary considerably with no clear trend. This may be a measurement artifact, based on the use of the least squares fit mathematics to define the linear range.

The 8X proportional limit values show a clearer trend with time, dropping gradually to a value of 15 MPa and 0.02% at the 1000-hour point. However, the limits of the least squares fit approach need to be considered here also. Overall, more work needs to be done on evaluating the oxidation effects on the proportional limit values.

Shear Properties – Double notch shear testing is preferred to short-beam shear, because of the multiple stress condition for the short beam shear geometry. In the double notch shear (DNS) test, the compression load translates into an interlaminar shear stress along the midplane connecting the two notches in the composite (See Figure 3). Table 3 lists the ultimate double notch shear strengths for the two sets of composite test bars.

Although the mean DN shear strength values for the as-prepared 5X and 8X composites differ by about 50% (18 MPa and 25 MPa), the standard deviations for the two sets of numbers are large enough, that the difference may not be statistically valid.

Table 3 and Figure 9 present the double notch shear (DN) strength results as a function of oxidation time for the 5X and 8X composites. With 600°C oxidation, the 5X test bars show a gradual decrease in shear strength, dropping from 18 MPa to about 12 MPa. On the other hand, the 8X samples maintain a uniform shear strength of 25 MPa out to the 100 hour mark; but then at the 500 hour mark the shear strength has dropped to 14 MPa. At 1000 hours the shear strength is close to 10 MPa.

This sudden drop in shear strength for the 8X bars may indicate that an oxidation front reached the center plane of the composite between the 100-hour and 500-hour mark. In the DN geometry, the shear stresses are developed just at the center laminar plane. If the oxidation front doesn't reach that plane, then no change in shear response would be expected. This difference between the 5X and the 8X bars may indicate a difference in oxygen diffusivity between the two infiltration/density conditions.

DOUBLE NOTCH SHEAR VS 600C OXID TIME SiOC-NEXTEL 312 BN 2-D COMPOSITES

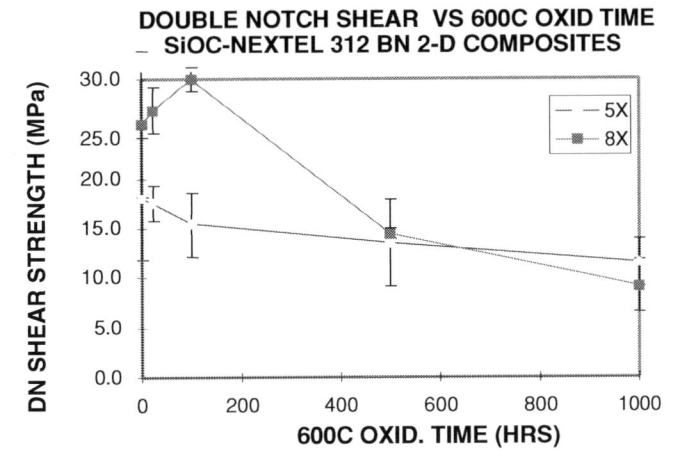

FIGURE 9. Double Notch Shear Strength versus 600°C Oxidation Time for SiOC Nextel 312 BN 2-D Composites

Two types of fracture path were seen at the root of the notch. The type of fracture path depended on whether the bottom of the saw notch coincided with an interlaminar plane or whether it lined up with a fabric ply. The schematic diagram (Figure 10) shows the two types of shear failure cases, which depend on how the cut notches were located with respect to the interlaminar fracture plane. These two types of notch/fabric plane geometries may be a partial explanation for some of the relatively large shear stress values found in the data set. It is hypothesized that the high values were for test bars where the notch root ended in a fabric plane and required higher loads to produce failure. SEM analysis of the DNS fracture surface consistently showed a rough fracture with the fibrous structure observed along the interlaminar fracture surface.

Compression Properties – Table 4 lists the mean values for the ultimate compression strength for the 5X and 8X sets of SiOC-Nextel 312 composite test bars. The small increase in as-prepared compression strength for the 8X bars compared to the 5X test bars is less than expected. Given the higher density and lower porosity of the 8X bars, the compression strength would be expected to be significantly higher compared to the 5X bars.

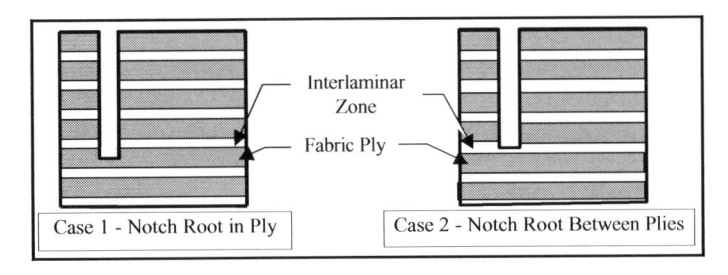

FIGURE 10. Schematic of Shear Notch Geometry with Fabric Plies

TABLE 4. Compression Strength of SiOC Nextel 312 BN 2-D Composites with 600°C Oxidation Exposure Effects

Condition/Property	600°C Oxidation Time (hr)				
	As-Prep	24	100	500	1000
5X Infiltrated – # of Bars	10	3	3	3	9
Ult. Compression Strength (MPa) Mean/S.D.	216 ± 48	126 ± 16	92 ± 14	65 ± 7	63 ± 6
8X Infiltrated – # of Bars	9	3	3	3	6
Ult. Compression Strength (MPa) Mean/S.D.	237 ± 39	266 ± 14	147 ± 37	120 ± 28	110 ± 13

Table 4 and Figure 11 present the compression strength results as a function of oxidation time for the 5X and 8X composites. With 600 °C oxidation the 5X samples lose compression strength very rapidly dropping from 210 MPa to 130 MPa after 24 hours. The compression strength at 500 and 1000 hours of oxidation is approximately 65 MPa. It does appear that the compression strength has stabilized after the 500 hour exposure.

The change in compression strength for the 8X samples is similar, but the drop in strength takes place more slowly and the final value is higher. At 1000 hours the compressive strength is still 110 MPa. Diffusion effects are the most likely explanation for the reduced rate of compressive strength loss, compared to the 5X condition.

The most likely cause of the compression strength loss is oxidative change of the matrix material, in which the loss of carbon and formation of silicon dioxide bonds disrupts the physical continuity of the matrix, markedly reducing its inherent strength. Since the compressive properties are dominated by the matrix strength properties, the matrix changes produced by long term oxidation dramatically reduce (70% for 5X and 50% for 8X after

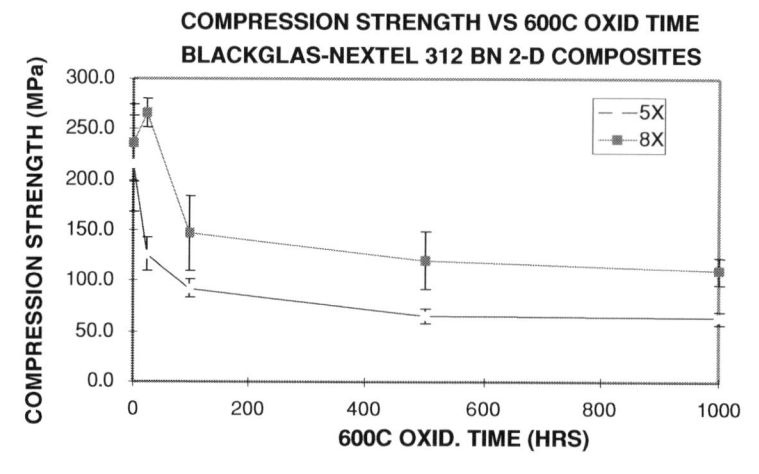

FIGURE 11. Compression Strength versus 600°C Oxidation Time for SiOC Nextel 312 BN 2-D Composites

TABLE 5. Oxidation Temperature Effects on the 3-Point Flexure Properties of SiOC Nextel 312 BN 2-D Composites

	Hours	5X Infiltration			8X Infiltration		
		Stress, MPa	Strain, %	Mod, GPa	Stress, MPa	Strain, %	Mod, GPa
As-Prep	0	202.4	0.31	77.1	215.7	0.33	76.8
500°C Ox	24	254.7	0.39	77.3	223.5	0.34	76.2
	100	235.7	0.49	58.8	244.2	0.39	73.6
	500	251.4	0.56	57.3	275.4	0.48	71.0
	1000	241.8	0.55	54.9	270.4	0.51	64.1
	2000	189.8	0.46	50.8	226.7	0.50	55.4
	4000	172.4	0.43	50.4	216.0	0.46	58.8
600°C Ox	24	244.6	0.44	65.2	218.5	0.36	70.8
	100	242.7	0.48	59.3	240.1	0.45	65.3
	500	212.9	0.5	52.3	229.8	0.50	56.4
	1000	193.8	0.47	49.3	205.5	0.45	55.4
	2000	148.2	0.38	45.9	132.6	0.39	46.6
	4000	79.8	0.17	51.1	75.9	0.14	55.9
700°C Ox	24	241.9	0.44	63.7	222.3	0.35	74.5
	100	215.6	0.45	55.6	234.0	0.42	66.6
	500	111.9	0.26	46.6	132.8	0.29	57.6
	1000	84.3	0.18	49.2	87.1	0.19	53.8
	2000	74.7	0.17	47.1	74.0	0.15	52.9
	4000	73.7	0.16	47.6	73.4	0.13	56.6

500 hours) the composite compression strength. This is in marked contrast to the oxidation effects on tensile properties (losses of 30% for 5X and 10% for 8X after 1000 hours), where tensile properties are dominated by the fiber properties.

3.6. 4000-Hour Oxidation Exposure Effects on Flexure Strength

The first oxidation study on the SiOC-Nextel 312 composites was done at 600°C out to 1000 hours, because of the previously observed oxidation susceptibility at temperatures below 1000°C. A follow-on exposure study was then done at 500°C, 600°C, and 700°C to assess the effects of temperature compared to the previously observed 600°C oxidation.[25] The 4000-hour exposure study used flexure test specimens, because of the ease of preparation and testing. Specimens were prepared using two levels of infiltration: 5X and 8X.

The samples were exposed to dry, flowing air at 500°C, 600°C, and 700°C for periods out to 4000 hours. Five samples from each set were retained for "as-prepared" mechanical evaluation and at least five samples of each set were removed from the furnaces at intervals of 24, 100, 500, 1000, 2000, and 4000 hours. All samples were tested in three-point flexure at room temperature, using an outer loading span of 25.4 mm and a loading rate of 25 MPa/s.

The results of this study are presented in tabular form in Table 5. Flexure strength and flexure failure strain are plotted for the three oxidation temperatures in Figures 12, 13, 14, and 15.

Oxidation Temperature Effects on Flexure Strength of SiOC N312 BN 2-D Composites (5X Infiltrations)

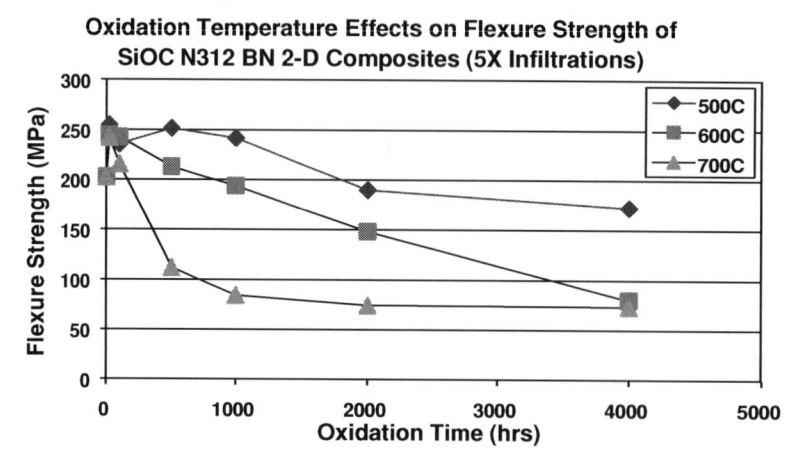

FIGURE 12. Oxidation Temperature Effects on the Flexure Strength of 5X SiOC Nextel 312 BN 2-D Composites

The major conclusions from the long-term oxidation temperature study are as follows:

- Higher oxidation temperatures have a significant effect on the flexural properties.
  - After 500°C oxidation for 4000 hours, the composites still had very nearly all their as-prepared strength, but with 10–20% increased strain capability. The elastic modulus was reduced by 15–20%.
  - At 700°C, the composites reached a baseline level of about 30% of the as-prepared strength and 50% of as-prepared strain after about 1000 hours. The elastic modulus was reduced by 20–35%. Little additional change in properties was observed between 1000 and 4000 hours.
  - The 600°C exposure results are in the middle of the two extremes. After reaching a maximum strength after about 100 hours of exposure, all the mechanical properties decreased to the 700°C baseline level between 2000 and 4000 hours.

Oxidation Temperature Effects on Flexure Strength of SiOC N312 BN 2-D Composites (8X Infiltrations)

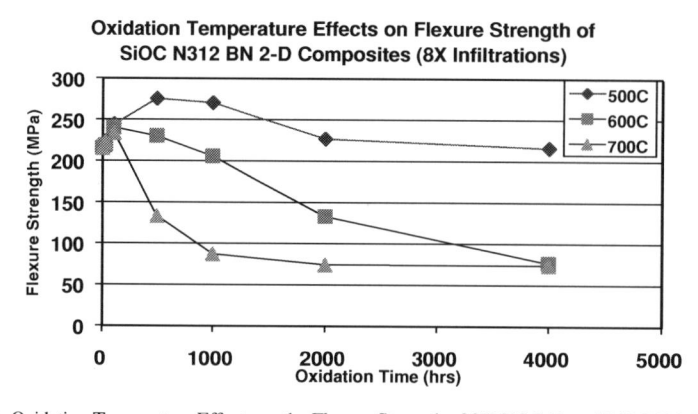

FIGURE 13. Oxidation Temperature Effects on the Flexure Strength of 8X SiOC Nextel 312 BN 2-D Composites

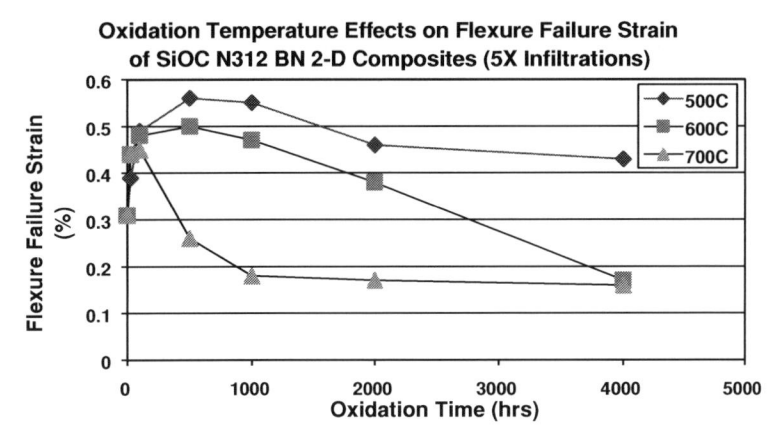

FIGURE 14. Oxidation Temperature Effects on the Flexure Failure Strain of 5X SiOC Nextel 312 BN 2-D Composites

- Using eight infiltrations rather than five results in improved strength retention at 500°C and equal performance at the higher temperatures.
- As the samples become oxidized, the mechanical failure shifted from the tensile to the compressive side of the flexural test specimens.

3.7. Creep Rupture Assessment at 566°C (1050°F)

Creep rupture tests are used to measure the long-term response of a material to a continuously applied stress at a given temperature. The ideal material should be able to support significant stresses for extended periods of time without accumulated permanent strain or breakage. The SiOC-Nextel 312 BN 2-D composites were tested in limited stress

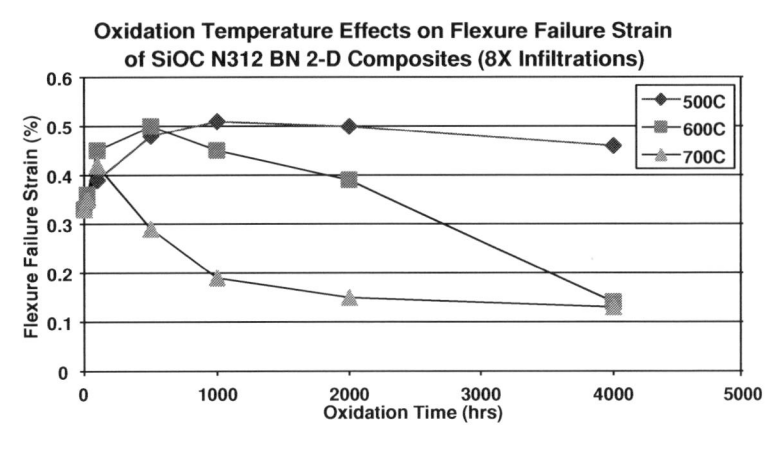

FIGURE 15. Oxidation Temperature Effects on the Flexure Failure Strain of 8X SiOC Nextel 312 BN 2-D Composites

TABLE 6. Tensile Creep Rupture Tests at 566°C/1050°F of SiOC Nextel 312 BN 2-D Composites

ID #	Infiltration Number	Test Condition	Applied Stress MPa (ksi)	Steady-State Strain Rate, sec^{-1}	Final Strain, %	Creep/Stress Rupture Lifetime, hrs
A3	5X	As-Prep	55 (8)	NA		110→
A4	5X	As-Prep	55 (8)	5.5×10^{-10}	0.24	101→
A7	5X	As-Prep	69 (10)	5.5×10^{-10}	0.23	100→
A1	5X	Oxidized	83 (12)	1.1×10^{-8}	0.21	25.4
A8	5X	Oxidized	96 (14)	5.5×10^{-10}	0.14	24
B3	8X	As-Prep	55 (8)	NA	NA	104→
B4	8X	As-Prep	55 (8)	1.5×10^{-9}	0.14	136^d
B10	8X	As-Prep	55 (8)	1.1×10^{-9}	0.17	102→
B7	8X	Oxidized	69 (10)	1.5×10^{-9}	0.29	133→
B6	8X	Oxidized	83 (12)	9.3×10^{-10}	0.18	76
B1	8X	Oxidized	96 (14)	3.9×10^{-9}	0.10	17.5

Strain rates measured over the last 10 hours of the test.
→ – Test stopped without sample breakage or failure.
d – Test continued by loading up to 96 MPa. Specimen failed after 4.3 hours.

creep rupture for the purpose of initial evaluation. The composites in the study were from the same lots as those used in the mechanical property study described earlier, using both five and eight infiltration cycles.[26]

An edge-loaded tensile specimen was used for the creep tests. The sample geometry is based on test methods developed and provided by John Holmes at the University of Michigan. The overall specimen length is 111.7 mm with a top and bottom width of 14.0 mm. The gage section is 33.1 mm long and 6.3 mm wide with an as-prepared thickness of 1.6 mm. The shoulders have an 8° taper with a 152.4 mm radius into the gage section. The tensile creep apparatus consisted of a clamshell high temperature furnace, a self-aligning load train, a laser extensometer, a data acquisition and control system, and computer. The laser extensometer measured the creep strain by scanning the distance between two silicon carbide flags on the specimen's gage section. The sensitivity of the measurement of flag movement is +/− 2 microns.

All the creep tests were performed under constant load to determine the creep rates, maximum strains and times to failure. Tests were performed at 566°C (1050°F) and at stresses of 55, 69, 83, and 96 MPa (8, 10, 12, and 14 ksi). Four types of specimens were tested. Composites with five and eight infiltrations were examined. Sample specimens infiltrated 5 times were given specimen designations beginning with "A" while those infiltrated eight times were designated with "B". Specimens were tested after a 600°C oxidation exposure in flowing air for 100 hours. The results of this work are summarized in Table 6.

A representative stress-strain creep curve for a 5X CMC specimen is shown in Figure 16. In analyzing these limited data, it was concluded that extrapolating the estimated creep rates to longer times and lower and high stress levels could not be done with accuracy. However, as a means of comparing this composite system with others, the data can be used to estimate the stress level to exceed a time-before-failure of 100 hours. These data indicate that a stress level of 11 ksi will just exceed the 100-hour target.

Creep deformation in 2-D ceramic composites is a complex phenomenon with deformation possible in the matrix, the woven fabric, the fibers, and the interface layers. Analysis of the data and of the fracture surfaces in these creep tests does not clearly indicate what

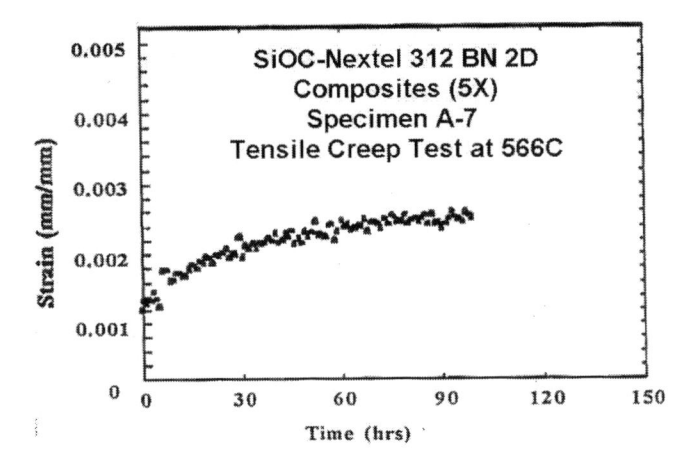

FIGURE 16. Tensile Creep Rupture Curve (566°C) for SiOC Nextel 312 BN 2-D Composites

the dominant creep mechanism is at these temperatures. However, corollary work indicates that the Nextel 312 fibers do have measurable creep rates on the order of 10^{-10} s^{-1} at 600°C at these stresses.

3.8. Low Cycle Fatigue Assessment at 566°C (1050°F)

Cyclic fatigue is another long-term durability requirement for ceramic matrix composites. As part of the LC^3 project, a limited assessment was made of the isothermal fatigue performance of the SiOC Nextel 312 BN 2-D composites.

The isothermal tensile fatigue tests were performed at 566°C (1050°F) in air. The specimen geometry was the same edge loaded tensile geometry used in the creep tests. The test cycle involved linearly increasing the tensile stress from the minimum value (14 MPa) to the maximum value over a one second period. The sample was held at the maximum stress for one second and then the stress was linearly reduced to the minimum value over an additional one second period.

The specimens used in this fatigue study had five infiltrations. Three tests were done in three conditions: as-prepared at 69 MPa stress and 41 MPa on specimens pre-oxidized at 600°C in flowing air for 100 and 900 hours. The preoxidation was done to determine if extended oxidation had a marked on effect the fatigue properties. The results of this testing are given in Table 7.

TABLE 7. Tensile Cyclic Fatigue Tests at 566°C/1050°F of SiOC Nextel 312 BN 2-D Composites

Sample ID	Oxidation Time, hr	Max. Stress, MPa	Failure Time, hr	Number of Cycles	Residual Strength, ksi
139-4	100	69 MPa	11.6	13,880	NA
139-2	100	41 MPa	>100	>120,000	14.2
139-8	900	41 MPa	>100	>120,000	13.5

The residual tensile strength of the composite aged for 900 hours at 600°C plus the 100 hours at 566 °C during the testing was about 90% of that of the sample that underwent static aging. This very limited testing indicated the need for more extensive and detailed work across a range of temperatures and stresses to define the full range of fatigue properties of the composite.

3.9. Thermal Properties

Baseline thermal properties of the SiOC-Nextel 312 BN 2-D composites have been measured.[27]

Thermal Diffusivity – The thermal diffusivity [$D = k/(\rho c_p)$] of SiOC-N312 BN 2-D composites was determined by the laser flash method in which a laser is used as a heat source and the thermal pulse transmission speed is measured in the desired orientation. Thermal diffusivity measurements were made both in-plane and through-plane of the 2-D composite. The specimen size was $9 \times 9 \times 2$ mm square. The thermal diffusivity was calculated from solution of the diffusion equation for heat flow with the known boundary conditions. Details of this procedure are found in ASTM Standard Test Method E37.05 (Thermal Diffusivity by the Flash Method).

The temperature dependencies of the thermal diffusivity for in-plane and thru-plane directions are shown in Figure 17 as a function of temperature up to 600°C. These results indicate that the heat flow in-plane is significantly greater than heat flow thru-plane (perpendicular to the fabric plane) throughout most of the temperature range investigated.

Thermal Heat Capacity – The heat capacity of SiOC-N312 BN 2-D composites was measured by differential scanning calorimetry (DSC). In this test a sample of dimensions $4.24 \times 4.24 \times 1$ mm is placed in a calibrated heating chamber along with a known heat capacity standard, and the chamber is heated at a fixed heating rate. The temperature difference between the standard and the composite is recorded, and the heat capacity is calculated from the measured temperature difference, the heat capacity of the standard, and the calibration constraints for the system.

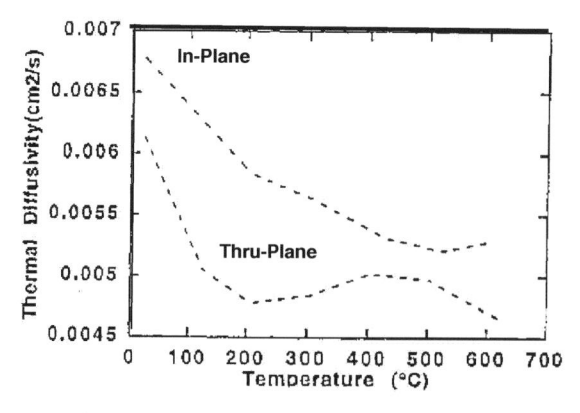

FIGURE 17. Thermal Diffusivity (In-Plane and Thru-Plane) of SiOC Nextel 312 BN 2-D Composites

TABLE 8. Heat Capacity of SiOC Nextel 312 BN 2-D Composites

Temperature (°C)	20°C	100°C	200°C	300°C	400°C	500°C	600°C
Heat Capacity (J/g°-C)	0.79	0.91	1.01	1.11	1.18	1.21	1.24

The temperature dependence of the heat capacity is shown in Table 8. These results indicate that the heat capacity is a smoothly increasing function of temperature and gradually reaches a constant value of 1.25 J/g-°C at 600°C. For reference, the heat capacity of the Nextel 312 at 500°C is 1.04 J/g-°C.

Thermal Conductivity – The thermal conductivity (k) of SiOC-N312 BN 2-D composites can be computed from the thermal diffusivity (D), density (ρ), and heat capacity (C_p):

$$k = D^* \rho^* C_p$$

Calculated values for the in-plane and thru-plane thermal conductivity from 20°C to 600°C are shown in Figure 18. The thermal conductivity ranges from 1.2 to 1.6 W/m-K, depending on temperature and anisotropy.

Thermal Expansion – Thermal expansion was measured using a horizontal pushrod dilatometer with alumina fixtures under flowing nitrogen. Specimens of approximately 10 mm in length were used. The temperature was measured with a platinum thermocouple located approximately 8 mm from the sample. Heating rates of 5°C/min were used to heat the sample from room temperature to 600°C. The sample was cooled at 10°C/min.

Figure 19 shows the expansion data for both in-plane and through-plane directions. For the in-plane case, two regions were observed: Region 1 is from room temperature to 200°C, and Region 2 is from 200 to 600°C. In Region 1, the expansion coefficient is approximately 3.6×10^{-6}/°C while Region 2 indicates about 4.5×10^{-6}/°C. For the through-plane direction the average expansion coefficient is about 4.3×10^{-6}/°C. These results indicate that the thermal expansion is not strongly temperature dependent and is not strongly

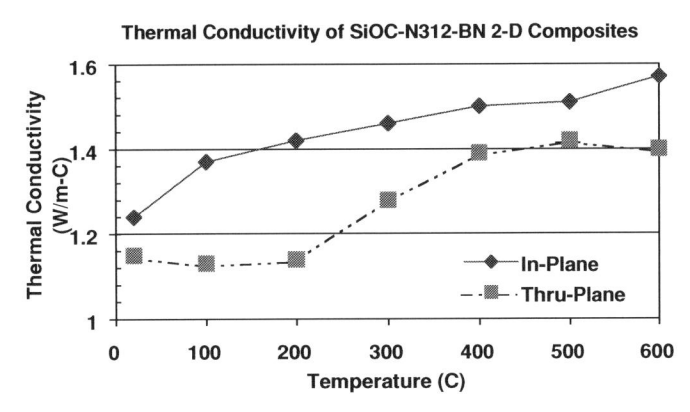

FIGURE 18. Thermal Conductivity (In-Plane and Thru-Plane) of SiOC Nextel 312 BN 2-D Composites

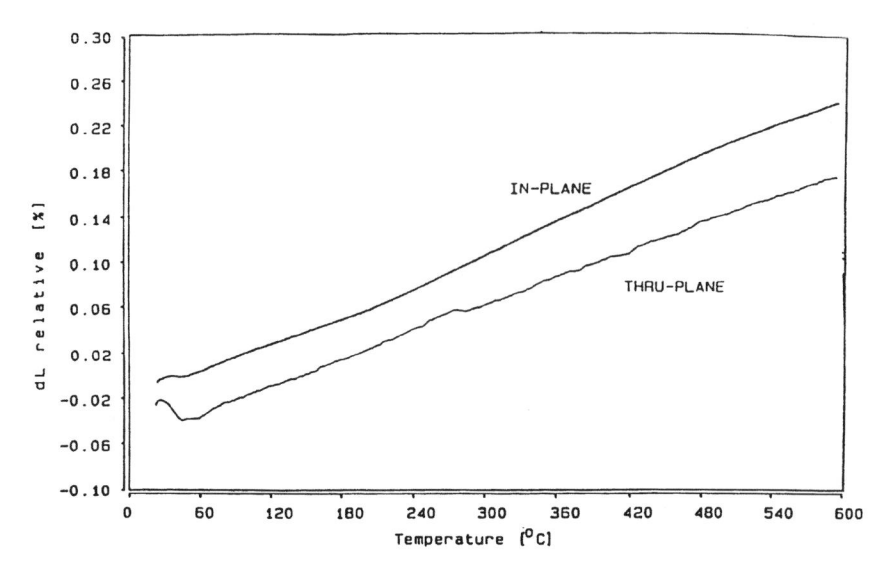

FIGURE 19. Thermal Expansion (In-Plane and Thru-Plane) of SiOC Nextel 312 BN 2-D Composites

influenced by fiber anisotropy. For comparison, the thermal expansion coefficient reported by 3M for the Nextel 312 fiber in the temperature range 25 to 500°C is $3.0 \times 10^{-6}/°C$.

3.10. Dielectric Properties

Ceramic matrix composites are under active consideration for low-observable military applications, where dielectric properties are a key performance factor. The frequency range of interest is the 8–12 GHz microwave range. The composite system must have low electrical conductivity. The SiOC-Nextel 312 system meets that requirement (as compared to ceramic composites made with conductive silicon carbide fibers).

Two screening studies have been done on the dielectric properties of an SiOC-Nextel 312 BN composites. In the first study in 1992 by Lockheed Missiles, a frequency range of 8 to 12 GHz was used and the properties were measured by either a Horn-Lens dielectrometer or a Fabry-Perot resonator system.[28] All measurements were taken at room temperature. The composites were prepared from eight plies of AF-14 plain weave fabric. Four infiltration cycles were used for densification. The total porosity of the composites was about 11%. The data generated for the Nextel 312 system are shown in Figure 20. Note that as an additional processing variable, two test specimens were heated in air (at 450°C and at 750°C) to remove any residual carbon.

The data show that there is a significant oxidation effect on both the dielectric constant and the loss tangent of residual carbon content in the composites. However, for the Nextel 312 (BN) system, heat treatment at temperatures above 450°C is definitely **not** recommended, because of degradation of the boron nitride interface at those temperatures.

In 1994, the Wright Laboratory of the USAF collaborated with AlliedSignal (now Honeywell) to measure the dielectric properties of a SiOC Nextel 312 BN woven 3-D

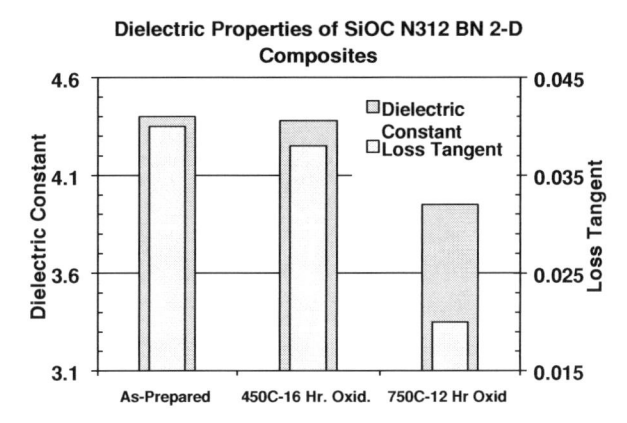

FIGURE 20. Dielectric Properties of SiOC Nextel 312 BN 2-D Composites

composite.[29] In this case the frequency range investigated was also 8–12 GHz, but the test temperature ranged from room temperature to 1000°C. The Nextel 312 3D preforms were woven by Techniweave, Inc. using their angle interlock architecture. Seven infiltration/pyrolysis cycles were used to densify the composites. The Archimedes density of the composites was 2.35 g/cm^3 and the open porosity was about 2%.

The dielectric constant and loss tangent for these composites are shown as a function of temperature in Figure 21. The dielectric constant of the SiOC-N312 was about 4.5. This value can be reduced somewhat with a mild oxidation. The dielectric constant is fairly constant over the temperature range from ambient to 1000°C. At the frequencies investigated, the dielectric loss tangent was about 0.03. This measurement can also be reduced somewhat by mild oxidation. The loss tangent remains constant until the temperature increases above 700°C and then increases to about 0.06 at 1000°C.

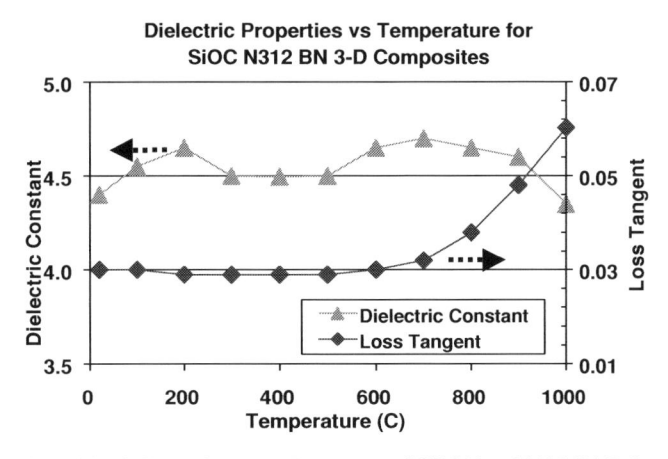

FIGURE 21. Dielectric Properties versus Temperature of SiOC Nextel 312 BN 3-D Composites

4. APPLICATIONS AND COMPONENT TESTING

Since its introduction in 1990 by AlliedSignal, Blackglas resin-based fiber reinforced composites have been considered for many aircraft turbine engine applications: secondary hot structures (400 to 1000°C), hot stealth components (where a low dielectric constant is important), and hot structures including rotors, stators, and combustors. Of these, secondary hot structures including those impinged with jet engine exhaust have been shown to be the best match of mechanical properties and production cost. In the near term, the primary opportunity for ceramic matrix composites is in hot secondary structural aerospace components with a requirement for long life at operational temperatures from 500°C to 1000°C.

4.1. Jet Engine Tailcone

The culmination of the Blackglas/Nextel 312 (BN) application testing effort was done under Agreement MDA972-93-2-0007 between the Defense Advanced Projects Agency (DARPA) and Northrop Grumman Corporation. An 18-inch diameter tailcone was fabricated for the Allison Engine AE2100 gas turbine, turboprop engine.[30] The CMC tailcone was designed as a weight saving feature and at 1.27 kg achieved a 30% weight reduction over the baseline metal design. (See Figure 22)

The tailcone was subjected to an accelerated multi-step mission durability test that produced tailcone measured maximum temperatures of 566°C (1050 °F) with a 28 kPa

FIGURE 22. SiOC Nextel 312 BN Tailcone Installed for Allison Engine Test (after 1500 hour test exposure)

delta pressure difference. During the testing, hundreds of operating cycles were completed. Because this was the first, long-term testing of the part in an authentic engine environment, there were concerns about the effects of acoustic fatigue and creep. The data show that after 1545 hours of operation, there was no evidence of external stress or degradation. The interior of the tailcone showed no distress. Internal flaws caused from the wrinkled preform did not increase in size or depth. The only damage to the tailcone was some loss of material and delamination at the attachment bolt slot caused by the loss of gasket material and high engine vibration.

4.2. High Performance Brake Rotors and Pads

High performance brakes are another near-term CMC application of interest to both the aerospace industry and the high performance automotive community. The interest is driven by the weight penalty of metal brake rotors and pads and the performance limitations of carbon-carbon materials. Work done by Northrop Grumman[31] showed that Blackglas-Nextel 312 composite brakes components had significant advantages over metal and carbon-carbon materials – extended brake life with lower wear, superior stopping power with no brake fade, tailorable and stable friction coefficient, lower production cost than carbon-carbon, and lower weight than metal. These CMC brake components were tested as prototypes for a range of different brake applications – racing motorcycles, military trucks, fighter aircraft, and heavy trucks. (See Figure 23).

FIGURE 23. Temrok™ [Blackglas-Nextel 312 (BN)] Ceramic Composite Brake Discs and Pads from Northrop Grumman

5. CONCLUSIONS

The successful commercialization of ceramic matrix composites requires a cost-effective production process that is amenable to the full range of shapes and sizes required for aerospace, automotive, and industrial power applications. Polymer derived ceramic technology offers a production route that combines low temperature polymer formability of polymer with the high temperature durability of ceramics. Blackglas polymer technology from Honeywell was combined with boron nitride coated Nextel 312 fabric from 3M to produce a 2-D laminated ceramic matrix composite systems with low-cost starting materials and a flexible polymer-based fabrication route. The SiOC Nextel 312 BN 2-D composites were characterized for mechanical strength, thermal properties, dielectric performance, and oxidation and corrosion resistance at temperatures of 500°–700°C out to 4000 hours. The SiOC Nextel 312 BN system showed moderate strength with fibrous fracture and 600°C 1000 hour mechanical durability. The composite system was successfully tested as a tail cone prototype in a jet engine test for over 1500 hours of cycle time. The CMC system has potential applications in secondary structural components that require 600°C durability.

REFERENCES

1. J. J. Mecholsky Jr., "Engineering Research Needs of Advanced Ceramics and Ceramic Matrix Composites," *Bulletin of the American Ceramic Society*, 68 (2) (1989), pp. 367–375.
2. K. W. Prewo, "Fiber Reinforced Ceramics: New Opportunities for Composite Materials," *Bulletin of the American Ceramic Society*, 68 (2) (1989), pp. 395–400.
3. J. R. Strife, J. J. Brennan, and K. M. Prewo, "Status of Continuous Fiber-Reinforced Ceramic Matrix Composite Processing Technology," *Ceramic Engineering & Science Proc.*, 11 (7) (1990), pp. 871–919.
4. Automotive Consulting Group, An Assessment of the Benefits of Ceramics in Automotive and Truck Engines, Prepared for DOE Oak Ridge National Laboratory, December 1993.
5. RCG/Hagler, Bailly, Inc., "Industrial Environmental Market Opportunities for Continuous-Fiber Ceramic Composites", DOE/OR Report # 950 prepared for the Office of Industrial Technologies, U.S. DOE, December 1990.
6. S. T. Gonczy, "Continuous Fiber-Reinforced Ceramic Composites and Blackglas™ Technology – Challenges and Opportunities," *Proceedings – High Temperature High Performance Materials for Rocket Engines and Space Applications*, Minerals, Metals, and Materials Society (Warrendale, PA: 1995), pp. 121–128.
7. D. R. Petrak, "Polymer-Derived Ceramics," *Engineered Materials Handbook, Vol. 4 Ceramics and Glasses* (Metals Park, OH: ASM International, 1991), pp. 223–226.
8. G. M. Renlund., and S. Prochazka, "Silicon Oxycarbide Glasses: Part II. Structure and Properties," *J. Mat. Res.*, 6 (12) (1991), p. 2723.
9. R. Y. Leung, S. T. Gonczy, M.S. Shum "Carbon-Containing Black Glass Monoliths," U.S. Patent 5,242,866, Aug 93.
10. R. Y. Leung, S. T. Gonczy, "Process for Preparing Black Glass Using Cyclosiloxane Precursors," U.S. Patent 5,231,059, Jul 93.
11. D. C. Larsen, J. W. Adams, H. H. Nakumara, Y. Harada, "SiC-Reinforced Black Glass Matrix Composites," Report AFWAL-TR-88-43005, Materials Laboratory, Air Force Wright Aeronautical Laboratories, January 1989.
12. S. Gonczy, R. Leung, J. Sikonia, "Near Net Shape Formability, Fibrous Fracture, and Low Dielectric Properties in Glass Matrix Composites Reinforced with Continuous Fibers", DARPA Workshop on Ceramic Matrix Composites, Alexandria, VA, August 1, 1991.
13. M. Meador, F. Hurwitz, S. T. Gonczy, "NMR Study of Redistribution Reactions in Blackglas™ and Their Influence on Oxidative Stability", *Ceramic Engineering & Science Proceedings*, 20th Conference on Composites & Advanced Ceramics, 7–11 January 1996, 17 (3), p. 394.

14. R. W. Rice, "BN Coating of Ceramic Fibers for Ceramic Fiber Composites," U.S. Patent 4,642,271, Feb 1987.
15. F. H. Simpson, J. Verzemnieks, Barrier Coated Ceramic Fiber and Coating Method, U.S. Patent 4,605,588, Aug 86.
16. F. H. Simpson, J. Verzemnieks, Boron Coated Ceramic Fibers and Coating Method, U.S. Patent 4,948,662, Aug 90.
17. S. Campbell, "Silicon Carboxide Composite Reinforced with Ceramic Fibers Having a Surface Enriched in Boron Nitride", U.S. Patent 5,955,194, Sept 99.
18. S. Campbell, "Formation of Boron Nitride on Surfaces of Aluminoborosilicate Ceramic", Ph.D. Thesis, University of Chicago at Illinois, 1997.
19. N. R. Khasgiwale, E. P. Butler, L. Tsakalakos, D. A. Hensley, W. R. Cannon, S. T. Gonczy, and S. C. Danforth, "Characterization of BN Rich Layer on Ammonia treated Nextel 312 fibers," *Mater. Res. Soc. Symp. Proc.*, Vol. 365 (19950), pp. 389–99.
20. R. Y. Leung, S. T. Gonczy, G. T. Stranford, C. E. Southern, & D. M. Lipkin, "Near-Net Shape Formability and Fibrous Fracture in Glass Matrix Composites Reinforced with Continuous Ceramic Fibers," *NASA Conference Publication 3097, Proceedings of the 14th Conference on Metal Matrix, Carbon, and Ceramic Matrix Composites,* (Jan 17–19,1990), p. 147.
21. R. Y. Leung, W. D. Porter, "Curing and Pyrolysis of Blackglas(TM) Resins and Composites", *Ceramic Engineering & Science Proceedings,* 20th Conference on Composites & Advanced Ceramics, 7–11 January 1996, Vol. 17 (3).
22. S. T. Gonczy, P. D. Dubois, "Flexural Properties of a 2-D Blackglas Nicalon Composite as a Function of Processing and Porosity," *Proceedings – Materials Challenge: Diversification and the Future,* Vol. 40-I, Society for the Advancement of Materials and Process Engineering (Covina, CA: 1995), pp. 446–456.
23. Technical Reports 8, 25, and 46, Low Cost Ceramic Composites (LC3), DARPA Contract MDA 972-93-2-007, 1994–1997.
24. E. Butler, S. Danforth, W. Cannon, and S. Gonczy, "Oxidation Effects at 600°C on the Mechanical Properties of a Blackglas^{TM}/Nextel 312 2-D composite with a Boron Nitride Interface Layer," DARPA Program – Low Cost Ceramic Composites, MDA 972-93-2-007, Technical Report LC^3, #46.
25. S. S. Campbell, S. T. Gonczy, M. McNallan, A. Cox, "Performance of Blackglas^{TM} Composites in 4000-Hour Oxidation Study," *Ceramic Engineering & Science Proceedings*, 20th Conference on Composites & Advanced Ceramics, 7–11 January 1996, Vol. 17 (3), p. 411.
26. K. Vaidyanathan, W. Cannon, S. Danforth, and A. Tobin, "Creep Resistance of Nextel^{TM} 312(BN) Blackglas Composites", *Ceramic Engineering & Science Proceedings*, 19th Conference on Composites & Advanced Ceramics, 8–12 January 1996, Vol. 16(4).
27. A. Tobin, "Component Material Design Data for BN-Nextel 312/Blackglas Composites" DARPA Low Cost Ceramic Composites (LC3), DARPA Contract MDA 972-93–2-007. Technical Report LC^3 15.
28. R. D. Yasukawa, D. G. Polensky, R. Y. Leung, S. T. Gonczy, J. G. Sikonia, "Blackglas Ceramic Matrix Composites with Dielectric Properties and Fibrous Fracture," Presentation at 2d Annual Ceramic Composites and Structures Meeting, USAF, West Palm Beach, FL, Oct. 20–22, 1992.
29. S. T. Gonczy, M. Kearns, "3-D Blackglas-Nextel 312 Composites with Boron Nitride Interface Coatings – Oxidation Stability and Dielectric Properties," Presentation at Aeromat 94 Meeting, Anaheim, CA Jun 6–9, 1994.
30. Durell Wildman, LC3 Program Tailcone Testing Results, Allison Advanced Development Company, private communication, August 18, 2003.
31. Product Literature on Temrok^{TM} Ceramic Matrix Composites from Northrop Grumman Advanced Systems and Technology, Pico Rivera, CA, 1997.

Part IV

Oxide/Oxide Composites

16

Oxide-Oxide composites

Kristin A. Keller*, George Jefferson** and Ronald J. Kerans
Air Force Research Laboratory
Materials and Manufacturing Directorate, AFRL/MLLN
Wright-Patterson AFB, OH 45433-7817
*UES Inc., 4401 Dayton-Xenia Rd., Dayton, OH 45432
**National Research Council, 500 Fifth St., NW, Washington, D.C. 20001
K. K. - phone: 937/656-4072; Kristin.Keller@wpafb.af.mil
G. J. - phone: 937/255-1307; George.Jefferson@wpafb.af.mil
R. K. - phone: 937/255-9823; Ronald.Kerans@wpafb.af.mil

ABSTRACT

The need for high temperature, oxidation-resistant materials has driven research into oxide-oxide composites. These materials require a crack deflecting mechanism to prevent brittle failure; both interface coatings and porous matrices have proven successful in providing this function. Composites containing interface coatings are still in the research stage, while porous matrix materials are currently more advanced. A review of porous matrix composite properties is given, along with composite processing information. This chapter provides an overview of oxide-oxide composite technology, potential applications, current research issues, and what might be expected in future derivatives. This review will center on high-temperature ($>1000°C$) oxide-oxide composites.

I. INTRODUCTION

Advancing aerospace technology is driving a need for structural materials with ever increasing thermal capabilities. For example, all types of engines benefit thermodynamically when materials permit operation at higher combustion temperatures and/or with reduced cooling requirements. Likewise future space/reentry vehicle designs will benefit greatly from improved thermal protection systems with the mechanical integrity to serve structural functions. Fiber reinforced ceramics (terminology includes "ceramic matrix composites" {CMC}, "continuous fiber ceramic composites" {CFCC}, and "ceramic fiber-matrix composites" {CFMC}) offer thermal performance beyond the service limits of high temperature metals, with damage tolerance superior to monolithic ceramics [1]. The most advanced, current ceramic composites are based on silicon carbide fibers and are very robust materials, but they are ultimately limited in lifetime by oxidation and are subject to more rapid degradation in combustion environments [2, 3]. Superior environmental stability, albeit with some sacrifice in mechanical performance, can be achieved by using oxides for all constituents: fibers, matrices and interface-control coatings. In addition, oxide fibers and some oxide composites also have a significant cost advantage over SiC-based composites. Ultimately, the two types of composites have different strong points and each will be preferable for particular uses. Potential applications for oxide-oxide composites will be discussed briefly in section II.

The basic requirement for a tough composite comprising brittle constituents, whatever their chemistry, is that cracks that initiate in the matrix do not propagate into fibers but bypass them by deflecting into fiber-matrix debonding cracks. This requires that something in the fiber-matrix interfacial region – an interface, a fiber coating, or the near-fiber region of the matrix – be sufficiently weak to fail before the fiber fails. Most available oxide composites rely solely on the latter case, where matrix porosity allows for matrix fracture in the interface region before causing a stress concentration on the fiber sufficient to fracture it [4]. Organic matrix and carbon-carbon composites are the prototypes for this behavior. Essentially all of the currently well-developed oxide composites utilize a porous matrix, however, interface control using oxide coatings has been demonstrated and looks very promising. These issues will be discussed in more detail in sections III and IV.

The superior environmental stability of the oxide composites generally comes at the expense of somewhat reduced mechanical properties when compared to non-oxide ceramic composites. This is particularly true of creep resistance. Since this is the life limiting property in many high temperature applications, it is of particular concern. Further, oxides do not have the very low thermal expansivities of silicon-based ceramics, and their thermal conductivities are low. Consequently, thermal strains are larger than in non-oxide composites. To some degree, lower elastic moduli moderate the resulting stresses, but typically strengths are lower also. Hence, the design of components to be used in thermal gradients and transients will need to pay special attention to these issues. However, there is significant room for improvement in all constituents and improved properties can be expected in future materials. Mechanical properties will be discussed and design issues briefly considered in later sections (V, VII).

There are myriad issues associated with the processing of ceramic composites. The task of infiltrating and consolidating ceramics in the presence of constraining fibers without also damaging the fibers is problematic by its very nature. This topic will be surveyed in section VI.

The objective of this chapter is to provide an overview of oxide-oxide composite technology, potential applications, current research issues, and what might be expected in future derivatives. These topics will be reviewed with a focus on high-temperature ($>1000°C$) oxide-oxide composites. Glass and glass-ceramic matrices and/or fibers will not be considered in this chapter. The reader will find additional insight from other chapters in this volume, along with reviews of oxide CMCs given in references [5–7].

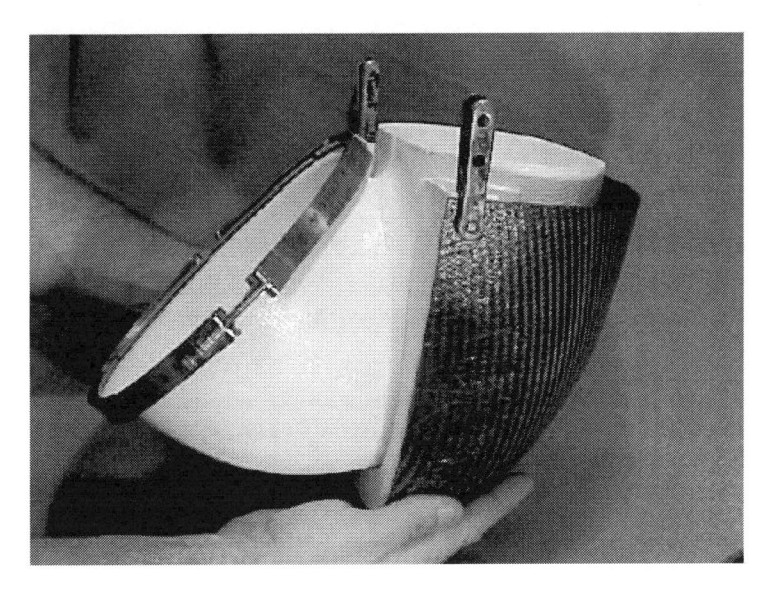

FIGURE 1. A lightweight helicopter exhaust duct for insulating and protecting the surrounding structure. (Courtesy COI Composites Inc., Materials Research and Design, Inc., & U.S. Navy)

II. APPLICATIONS

Virtually all forms of heat engines can benefit from the use of oxide-oxide composites [8–11]. Exhaust components are probably the simplest and lowest risk, whether they are tubes for a reciprocating engine or nozzle flaps for the augmentor of a fighter aircraft turbine engine [12]. An example of a light weight exhaust duct used as a means of insulating and protecting the surrounding structure is shown in Figure 1 [13].

Moving forward in a turbine engine, various other components appear to be viable replacement candidates, including the seal/casing shroud around the turbines [14], the combustor, hot ducts and the stationary vanes. Rotating components seem less likely candidates given the modest strengths of current technology oxide composites. There have been several research efforts aimed at evaluating oxide-oxide composites for static applications [12, 15]. The limited results indicate that the strength and fatigue resistance of these materials make them attractive candidates; however, in at least one instance, fretting wear by adjacent metal parts was problematic [15]. The use of oxide CMCs in combustors will allow higher operating temperatures, reduced cooling and greater design flexibility to lower NO_x emissions and increase efficiencies.

It is expected that the highest efficiency reciprocating engines will have no cooling systems, and will, in fact, be insulated, and will therefore require high temperature capable materials in all the major components. Particularly for lightweight engines, oxide composites seem worth considering for the cylinder head, block, pistons and valves, as well as the exhaust system.

There are also numerous semi-structural applications that may be good applications for all-oxide composites [16, 17]. Thermal Protection Systems (TPS) are combinations of elements – typically including insulation, an outer skin and some provision for attachment – that protect underlying structure from exposure to high temperature and hot gases. An example of such a system is the silica-based composites currently used as thermal blankets for the space shuttles (maximum temperature ~650°C) [18]. The durability of such systems is enhanced if the outer surface is robust against abuse such as impact damage and erosion. Recent research has indicated that Nextel™ 440/monazite composites allow for over a 50% increase in the maximum temperature capability of the current TPS system (SiO_2-based), while also being much more durable [19]. Of course, the optimal situation would be for the hot surface to also act as a structural piece and actually carry loads. Ceramic composites offer the best route to that end, but considerable development is required.

It seems unlikely that polycrystalline oxide fibers will ever approach the use temperatures of SiC-based or C fibers ($>\sim 1500°C$). Hence, it can be anticipated that even in long-term use, materials selection for refractory composites will balance the high temperature properties of non-oxide composites versus the superior enviro-thermal durability of oxide composites.

III. COMPOSITE DESIGN FOR TOUGH BEHAVIOR

A tough ceramic composite requires crack deflection at or near the fiber/matrix interface. There are two basic mechanisms utilized for designing tough ceramic matrix composites, one being the use of a weak engineered interface, which typically involves a fiber coating and the second being the use of a sufficiently weak (i.e. porous) matrix. It is important to note that 'sufficiently weak' in this context means relative to the fiber in the finished composite. This is important because virtually all fabrication processes have some effect on the strength of the fiber. Ultimately crack behavior is the result of a competition between fracture in the fiber, coating (if present) and matrix. The discussion of this section borrows from reference [20], which provides a comprehensive treatment.

3.1. Interface Control

The prototypical approach to enabling tough behavior in ceramic composites is the use of an engineered weak fiber-matrix interface (Figure 2a). In this case, as a matrix crack approaches a coated fiber, the crack is deflected either at the matrix/coating interface, within the coating layer or at the coating/fiber interface. The fiber ultimately fractures at a point away from the crack plane and begins to pull out from the matrix upon continued loading of the composite. The pull-out of the fiber requires some amount of load to overcome the roughness and frictional effects. A typical fracture surface showing fiber pull-out is shown in Figure 2b. For comparison, an example of a brittle fracture surface, with virtually no fiber pull-out, is shown in Figure 2c.

The optimal interface properties, such as coating thickness, frictional stresses, debond lengths, etc., are dependent upon the particular composite constituents and must be evaluated on an individual basis. There are some general guidelines associated with choosing an interface coating with appropriate properties; one of the most commonly used requirements is that the ratio of the fracture energies of the interface and the fiber be less than ~0.25 (He

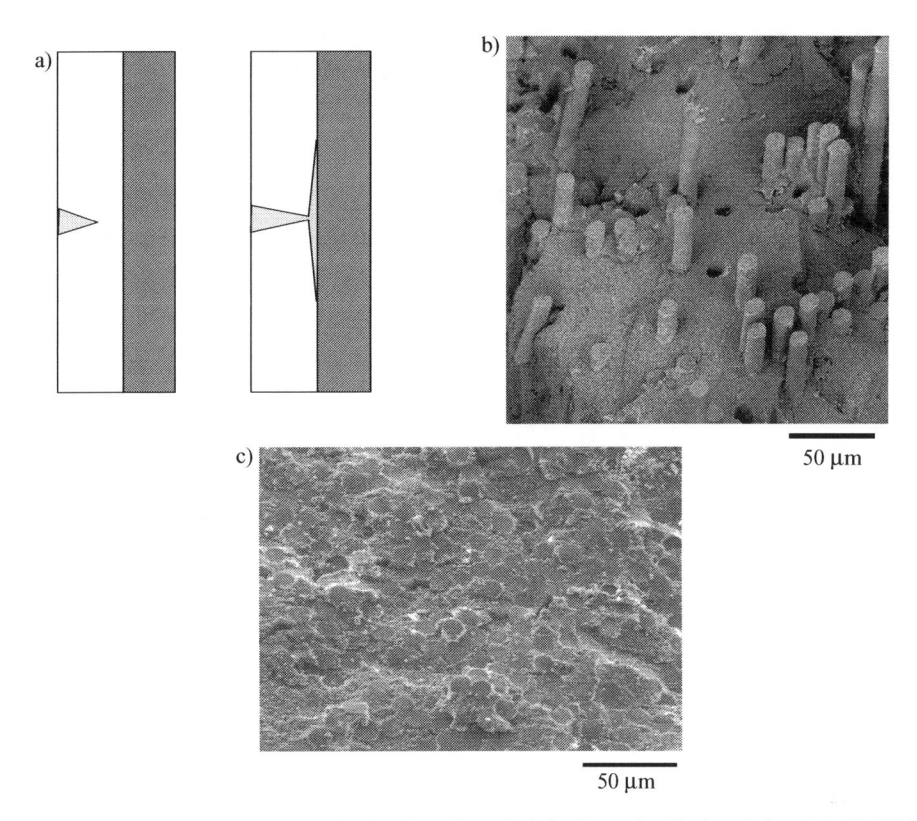

FIGURE 2. a) Schematic showing toughening through crack deflection at the fiber/matrix interface, b) SEM micrograph of fracture surface of Nextel 610/monazite/alumina composite tested at 1200°C showing fiber pull-out and c) fracture surface of Nextel 610/alumina composite tested under the same conditions. Absence of crack deflection mechanism led to brittle failure in c).

and Hutchinson criteria) [21, 22]. A thorough treatment of this topic is beyond the scope of this chapter and the reader is referred to references [20, 23, 24] for a more in-depth discussion.

The design and evaluation of oxide fiber-coating substitutes for controlled interface properties is a multi-faceted problem. Carbon (C) and boron nitride (BN) have unusual combinations of properties that make them ideal choices for fiber coatings, except for oxidation resistance. Our accumulated understanding of composites based on C and BN coatings provides a good foundation, but composites based on oxide coatings can be expected to behave in very different ways. For example, oxide coatings are less compliant; hence the optimum coating thicknesses are likely to be substantially different [25]. Even with thicker coatings, the low levels of friction achievable with C are probably unattainable. This will put an even higher priority on strong fibers and the retention of that strength throughout processing and service in order to attain sufficiently long debond lengths for high levels of distributed damage.

Despite the significant challenges, several coatings, LaPO$_4$ (monazite) and CaWO$_4$ (scheelite), have been definitively demonstrated to provide not only the crack deflection function, but, in the former case, substantial improvement in alumina-based composite properties [26, 27]. For example, the use of monazite coatings in alumina reinforced alumina composites was found to increase the composite life as compared to a coating-less control composite by a factor of several hundred [27]. Such dramatic improvements are rare in structural materials and portend significant improvements in future composite systems.

3.2. Porous Matrices

The commonly used alternative to an engineered fiber-matrix interlayer is a relatively weak matrix [28–42]. It was observed that one type of weak interface is created by using a highly porous oxide fiber coating [43]. By extension, fabricating the composite such that the entire matrix is porous results in effective crack deflection and toughening [4].

In porous-matrix composites, a matrix crack is deflected within the matrix in the region of the fiber/matrix interface. There may be strong bonding between fibers and matrix particles; however, it is postulated that cracks will generally meander to connect pores and hence arrive at fibers in pores, imposing lesser stress concentration on the fibers than in the dense matrix system. In any case, the matrix is unable to impose sufficient stress concentrations to fracture the fibers until the fiber stress is quite high. Fiber failure occurs in a statistical manner at various points/planes within the fiber tow and final failure occurs when the differing crack planes unite to form a continual crack across the sample and comminution of the matrix has occurred [4]. A macroscopic comparison of a) porous matrix composite behavior and b) brittle composite behavior is shown in Figure 3.

One should note that the terms "porous-matrix composite" and "dense-matrix composite" are often used to differentiate between families of composite behavior and are used loosely with regard to the actual density of the material. Current technology composites of the porous-matrix type are not fully sintered and have approximately 35–50% matrix porosity/25–40% total porosity (see Table 2 for examples and references). (The composite consists of a volume fraction of fibers, f. The remainder of the volume, $(1-f)$ is considered to be matrix. The term matrix porosity (p_m) refers to the pore or void fraction of the matrix, so that, as the fibers are assumed to be fully dense, the composite or "total" porosity is $p_c = p_m(1-f)$. For example if $f = 0.4$ a 30% matrix porosity would correspond to 18% total porosity.) Matrices tend to develop microcracks during drying and/or sintering due to the constraint induced by the fibers. A typical microstructure of a porous-matrix composite is shown in Figure 4. The weak matrix necessary for crack deflection compromises matrix-dominated properties to some degree, e.g. shear properties, and transverse and interlaminar properties in 2-D lay-ups.

The mechanical behavior of porous-matrix composites has been modeled, but the compromises between matrix properties and toughness are not yet fully quantified. Initial estimates indicated that a matrix porosity of ~30% is needed for crack deflection within a porous matrix [4, 43]. Subsequent efforts have been aimed at better quantifying the transition between "porous" and "dense" CMC failure behavior [44]; however, further investigation is essential, especially considering the temperature dependent nature of the porous microstructure.

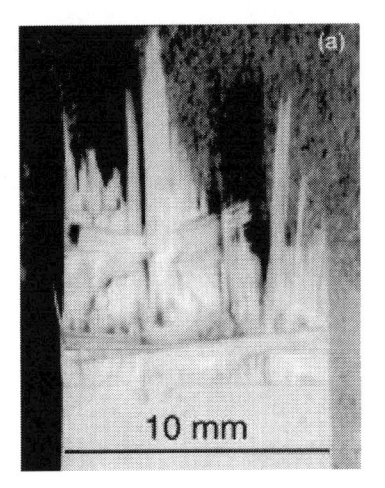

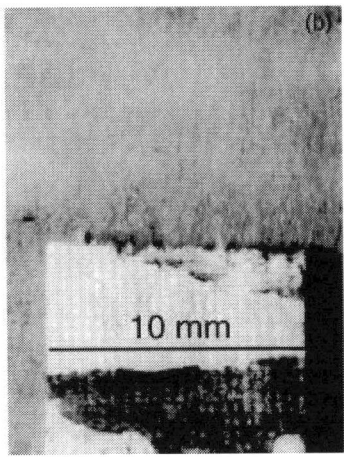

FIGURE 3. Low magnification views showing dramatic difference between a) a fibrous failure normally seen with porous matrix composites and b) a brittle failure [30]. Reprinted with permission of The American Ceramic Society, www.ceramics.org.copyright 2003. All rights reserved.

While mechanistic details remain a subject of investigation, the time (t)-temperature (T) use limits of porous-matrix composites seem ultimately to be limited by matrix sintering. At high temperature, the matrix continues to sinter gaining strength (and stiffness) until it is no longer effective in deflecting cracks around fibers. The current generation of polycrystalline fibers have t-T capabilities similar to the matrix materials, so the matrix-related limits are somewhat masked. While it is not coincidental that the matrix and fiber limits are similar, there are apparent paths to improved fibers that can be pursued when it is clear that the fibers are the limiting factor. Moreover, recent results on composites using fiber coatings for interface control indicate that greater improvements than anticipated can be attained even with current fibers. This topic will be discussed in further detail in the next section.

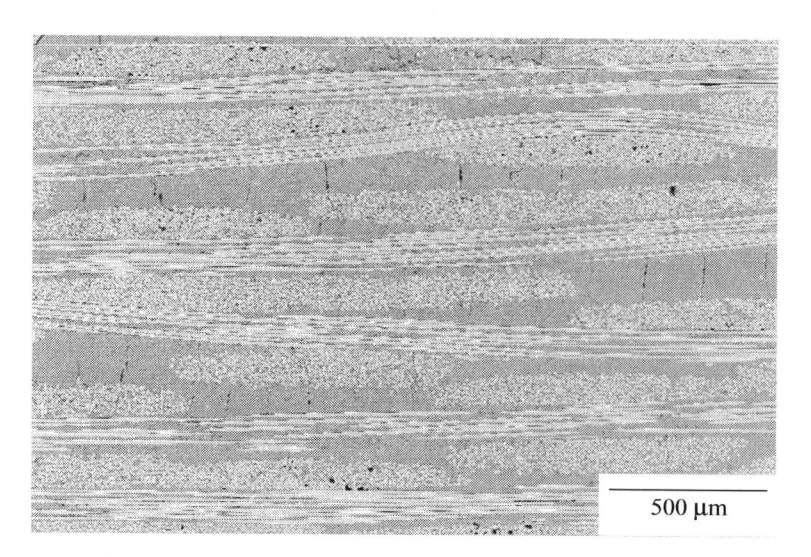

FIGURE 4. Typical microstructure of porous matrix composite (Nextel 720/mullite-alumina, UCSB material)

Finally, the terminology of "weak matrix" in referring to porous-matrix composites may be somewhat controversial, with some justification. A porous ceramic is certainly weak compared to a fully dense material, but fully dense matrices in continuous-fiber composites are a rarity, and probably an unachievable ideal for most of the more refractory materials and viable processes that do not damage fibers. Hence the difference in matrix porosity and the corresponding debit in composite properties associated with an optimum porous matrix compared to a maximum-achievable-density matrix may be small, and there may be a broad variety of applications for porous-matrix, coating-less composites. Nevertheless, it is expected that ultimately, oxidation resistant fiber-matrix interface coatings, improved composites and process designs will yield significantly better properties, higher use temperatures, and improved environmental resistance for these materials.

IV. COMPOSITE SYSTEMS

While there are abundant reasons for academic consideration of porous-matrix oxide-oxide composites, much of the practical interest, and opportunity to pursue them, is owed to two factors: the economical availability of numerous types of fibers, and the absence of the necessity to coat fibers. Both of these factors significantly affect the ease and cost of experimentation, and, more importantly, the ultimate cost of the materials. An additional significant advantage is that in many cases fabrication can be very similar to that employed for polymeric composites. The majority of this section is therefore focused on porous matrix composites, since these materials are the most mature. Research on composites containing interface coatings is also addressed in this section *(4.3)*.

4.1. Fibers

For oxide-oxide composites, the fine diameter (~10–12 μm) Nextel™¥ fibers are typically the reinforcement of choice. There are a limited number of other available oxide fibers, as discussed in a previous chapter; however, the Nextel fibers are the most mature and widely used. Of these, Nextel 610 (Alumina, Al_2O_3) and 720 (Mullite ($3Al_2O_3$-$2SiO_2$)-Al_2O_3) fibers are used for high temperature applications (>1000°C), while the 312 and 440 fibers (Al_2O_3-SiO_2-B_2O_3) are generally used for lower temperature (<1000°C) applications, such as thermal insulation. Nextel 550 fiber contains only γ-Al_2O_3 and amorphous SiO_2 and its use temperature is limited by the onset of crystallization in the fiber (<1200°C). The presence of a glassy phase in the Nextel 312 and 440 fibers dramatically affects their creep properties, which, in turn, limits their maximum use temperature [45].

For the higher temperature fibers, Nextel 610 has a higher tensile strength than the 720, while the 720 has superior creep resistance at high temperature. The research level Nextel 650 fiber (Al_2O_3-YSZ (yttria-stabilized zirconia)) displays intermediate creep resistance, between the 610 and 720 fibers, without the presence of silica. 3M has also presented information on an improved Nextel 720 fiber that displays a slight shrinkage at 1200°C under creep conditions, versus the elongation seen in the current Nextel 720 fiber [46]. Selected properties of the Nextel fibers are given in Table 1, while a more comprehensive treatment is given in the chapter on oxide fibers.

Sapphire (single crystal alumina) fibers* have also been utilized in oxide-oxide composites (see, for example, [47, 48]). Sapphire fibers generally have a substantially larger diameter (>100 μm) and are useable at higher temperatures than the polycrystalline Nextel fibers [49], although an intermediate temperature strength drop and susceptibility to slow crack growth have been reported for sapphire [50]. These single crystal fibers are grown primarily by the EFG (Edge-defined Film-Fed Growth) process versus the sol-gel process used for the Nextel fibers [51]. The EFG process leads to significantly higher costs, which are generally prohibitive for practical application. Further, these fibers are also not as easily handleable as the fine diameter tows, since they cannot be woven into fabrics or easily wound due to the large diameter. An alternative method involving "internal crystallization" of sapphire fibers has been investigated and initial reports indicate progress [52, 53].

There are several oxide fiber types in the developmental stage, including polycrystalline YAG ($Y_3Al_5O_{12}$ – yttrium aluminum garnet) fibers as well as various eutectic (ex: [Al_2O_3-YAG]) compositions. YAG is arguably the most creep resistant oxide known and therefore a good material choice [54]. Fine diameter YAG fibers have been produced [55–57]. Eutectic compositions can also provide improved strength and creep resistance to higher temperatures, depending upon the chosen constituents. However, current eutectic fibers suffer from the same limitations as the sapphire fibers, due to their larger (>100 micron) size (see, for example, references [58–60]). Preliminary work has also been conducted on the production of mullite single crystal fibers through the "internal crystallization" method [61].

In terms of fiber architecture, the most mature composites utilize Nextel fibers in the woven form and usually in an 8 harness-satin weave, although other weaving patterns are available (i.e. plain weave). The woven fabric enables easy handling of the fiber and is formable to more complex shapes. Alternately, fiber tows can be utilized by forming

---

¥ 3M corporation, St. Paul, MN
* Major supplier is Saphikon, Inc., Milford, NH

TABLE 1. Constituent properties.

	Composition, wt%			Strength, MPa			Modulus GPa	Dens g/cc	dia. μm	thermal limits, °C		expansion ppm/°C
	Al_2O_3	SiO_2	other	filament[d]	roving[e]	tow[d]				creep[f]	strength[g]	
fibers												
Nextel 312	63	25[b]	B_2O_3	1700	850	–	150	2.7	10–12	600	850	3.0
Nextel 440	70	28[b]	B_2O_3	2000	1100	–	190	3.1	10–12	900	1050	5.3
Nextel 550	73	27[b]	–	2000	1100	–	193	3.0	10–12	1000	1150	5.3
Nextel 610	>99	<0.3	Fe_2O_3	3100	1130	1600	380	3.9	10–12	1000	1225	8.0
Nextel 650	89	–	ZrO_2, Y_2O_3	2550	820	–	358	4.1	10–12	1100	1050	8.0
Nextel 720	85	15[c]	–	2100	540	800	260	3.4	10–12	1200	1350	6.0
sapphire[a]	100	0	–	3100	–	2250	435	3.8	70–250	–	–	–
matrix constituents (fully dense properties)												
alumina	100	0	0	–	–	–	380	3.96	–	–	–	8.1
mullite	75.5	24[c]	<.5	–	–	–	220	3.16	–	–	–	5.0

[a] Saphikon inc., c-axis single-crystal; [b] partly amorphous silica; [c] in crystalline mullite phase; [d] 25 mm gage length; [e] 150 mm gage length; [f] 1% strain after 1000 h @70 MPa, single filament; [g] 90% tow strength retention after 1.5 minute exposure; Refs: fibers [45, 49, 51, 55, 88, 154, 155, 210, 211]; mullite / alumina: [4, 66, 67, 212]

TABLE 2. Physical properties of selected woven cloth CMC laminates

Designation	Source	Fiber	Matrix	Fiber Volume, %	Matrix Porosity, %	Composite Porosity, %	Density g/cc
COI-312/AS	COI[a]	Nextel 312	Aluminosilicate	48	42	22	2.30
COI-550/AS	COI	Nextel 550	Aluminosilicate	36	39	25	2.41
COI-610/AS	COI	Nextel 610	Aluminosilicate	51	51	25	2.82
GE-610/GEN-IV	GE[b]	Nextel 610	Aluminosilicate	30	35	25	2.90
UCSB-610/M	UCSB[c]	Nextel 610	Mullite-Alumina	40	40	24	–
UCSB-720/M	UCSB	Nextel 720	Mullite-Alumina	40	29–35[d]	17–21	–
COI-720/AS	COI	Nextel 720	Aluminosilicate	48	48	25	2.60
COI-720/A-1	COI	Nextel 720	Alumina	46	46	25	2.71

[a] COI Ceramics Inc., San Diego, CA; [b] General Electric corporation, Cincinnati, OH; [c] University of California, Santa Barbara; [d] variation in properties with porosity adjustment examined; Refs [30, 44, 75, 79, 140, 141, 143]

tapes of the fibers and then stacking them for consolidation into a green body. These tapes can be formed through filament winding [62] or tape casting [63], although the latter is more applicable to larger diameter fibers (~100 μm) than to the small diameter (10–12 μm) Nextel fibers. In a few cases, 3-D woven oxide-oxide composites have been produced and evaluated, although these have been primarily with the lower temperature Nextel 312 and 440 fibers [64, 65]. The startup costs associated with the manufacturing of 3-D performs will necessitate either substantial production volumes or significant cost penalties.

4.2. Porous Matrix Systems

The aforementioned fibers are combined with various matrix materials to produce oxide-oxide CMCs. Coatingless, porous-matrix CMCs based on fabric reinforcement are the most fully developed and examples of these materials are given in Table 2. General Electric's (GE) GEN-IV was one of the earliest porous matrix, all-oxide fiber composites, while COI Ceramics (COI[♣]) is currently the most prominent commercial supplier of oxide-oxide composites. The matrix materials are primarily aluminosilicate (ex. manufacturers: GE[#], COI), alumina-mullite (University of California, Santa Barbara [UCSB]), and/or alumina (COI).

For the aluminosilicate matrices, the main constituent is alumina, which is held together by a silica binder. The silica may be present as a highly porous filler between the alumina particles, as seen in the GE-based materials, or it may be a continuous SiO_2 film on the alumina particles, as seen in the COI-based materials (examples shown in the TEM images of Figure 5). The porosity in the former case is much finer than the porosity seen in the COI material.

Alternately, an alumina-mullite matrix was developed in an effort to improve long-term thermal stability of porous-matrix CMCs. This matrix utilizes a bimodal mixture of mullite (~1 μm) and alumina (~0.2 μm) powders [4]. The mullite exhibits relatively slow sintering that inhibits shrinkage and maintains the overall porous structure, while the fine alumina sinters more readily, cementing the mullite particles and providing strength to the porous matrix structure [66, 67]. Zirconia has also been studied as the primary phase, with mullite used to stabilize the matrix [68].

[♣] COI Ceramics Inc., San Diego, CA
[#] General Electric Corp., Cincinnati, OH.

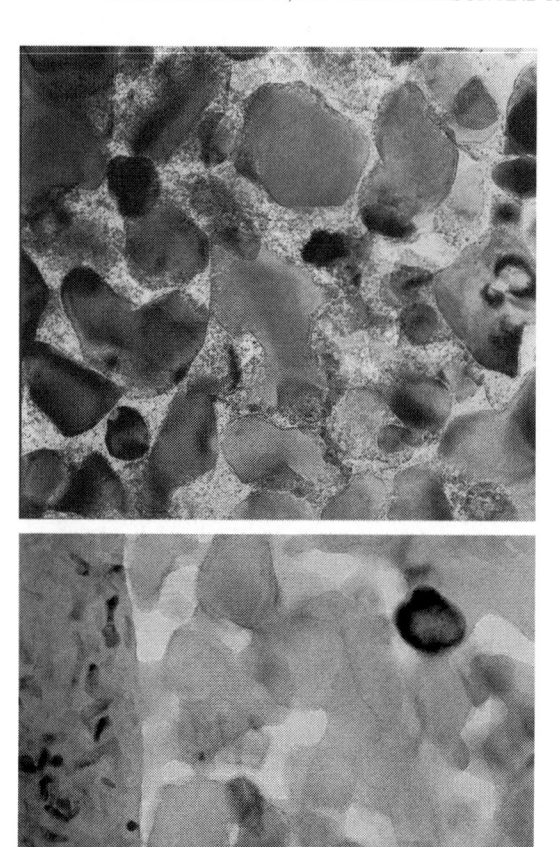

FIGURE 5. TEM shots of aluminosilicate matrices from a) GE and b) COI. Silica is present as a porous network between alumina particles in a) and as a particle coating in b) [20]. Reprinted with permission of The American Ceramic Society, www.ceramics.org. Copyright 2002. All rights reserved.

Another relatively mature porous-matrix composite, WHIPOX (Wound Highly Porous Oxide Ceramic Composite) has been developed; this material is the subject of a preceding chapter in this Handbook and will therefore not be addressed in this text. These materials utilize either Nextel 720 or 610 fibers filament wound into a highly porous mullite or alumina matrix [69, 70].

Both single-phase mullite and two-phase mullite-alumina mixtures can be termed "silica-free". However, due to the high silica activity in mullite and because it may in some instances contain some residual glassy phase material, it may be subject to degradation in environments that are detrimental to silica [71]. For example, vapor-deposited mullite coatings exhibited a vaporization of the SiO_2 in a high pressure, high temperature

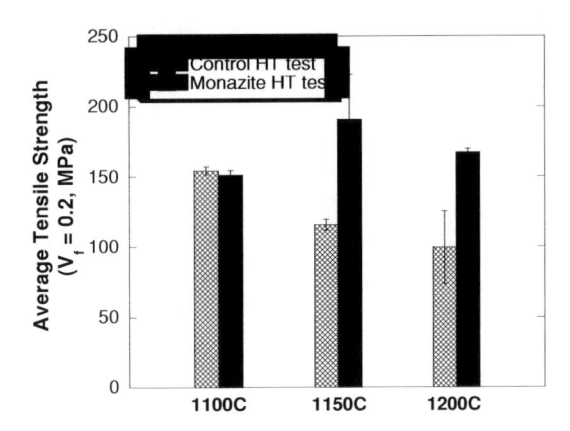

FIGURE 6. Tensile test results of Nextel 610/monazite/alumina composites versus control samples (Nextel 610/alumina) tested at high temperature. Samples sintered for five hours at 1100–1200°C; sintering temperature ≡ test temperature (10 min. soak @ test temperature.)

H_2O environment, leading to cracking and coating recession [72, 73]. Conversely, CVD mullite coatings, characterized as high purity and fully crystalline, appeared to be stable in similar environments [74].

For high temperature applications, there is an effort toward eliminating all silica in the matrix (i.e. even in mullite form) due to concerns over degradation in high temperature water and fuel rich environments [45, 75, 76]. Further, the use of SiO_2 in industrial products may meet with future difficulty, due to the concern surrounding the possibility of silicosis in workers [45, 77, 78]. Therefore, matrices such as all-alumina and alumina-YAG have been investigated for high-temperature use in oxide-oxide composites. COI has utilized the bimodal concept to increase the thermal stability of an all alumina matrix [79]. In this case, colloidal alumina is used as the filler for the larger alumina powder. In work on unidirectional porous-matrix composites based on alumina-YAG matrices reinforced with Nextel 610 fiber, the addition of even a small amount of YAG or Y^+ was shown to significantly slow the sintering of fine (~0.2 μm) alumina in the composite [80, 81]. Previous work on alumina monoliths showed similar results with the addition of Y^+ [82]. Higher volume fractions of YAG (~50%) have been used in a bimodal morphology similar to mullite-alumina, where the YAG phase is much larger (~1–2 μm) than the finer alumina phase (~0.2 μm) [81].

4.3. Composites with Interface Coatings

Porous-matrix, coating-less composites have distinct advantages, but require certain compromises. The most important of these compromises, at least in the long-range view, are in matrix-dominated mechanical properties and ultimate use temperature. Both of these points have been discussed earlier, but to reiterate briefly, when the matrix strength exceeds a critical level, the crack deflection function ceases to operate. Moreover, apparently this level can be exceeded by over-sintering without any total densification, that is, even though the matrix porosity is maintained, crack deflection is lost. Consequently, achieving the best balance of properties and highest temperature capabilities will require oxidation resistant, crack

FIGURE 7. SEM micrograph showing Nextel 610/monazite/alumina fracture surface. Monazite coatings can be seen on the pulled-out fibers and in the fiber troughs.

deflecting coatings and the densest possible matrices. (The following discussion borrows heavily from reference [20].)

Interpretation of results on interface control is complicated by several factors; hence studies on this topic should be read with several caveats in mind. As discussed elsewhere in this chapter, crack deflection is a competition between fracture of the interface/coating and fracture of the fiber and many processes are aggressive to fibers. It is very difficult to distinguish between degradation of a crack deflection mechanism and degradation of the fibers [83]. On the other hand, due to the problematic nature of making high-density matrices, it can be equally difficult to distinguish between a successful coating and the porous-matrix debonding mechanism.

Additionally, research on interface control has been performed on composites with an extremely wide range of fiber fractions. They range from single fibers with very little matrix (high volume fraction), to single or multiple tow minicomposites, unidirectional and 0°/90° fabric composites (moderate volume fractions, ~20–50%), to hand-assembled mono-filament reinforced composites (very low fiber volume fractions). Since the tensile strengths of composites are highly dependent upon the volume fraction of fiber in the loading direction, this needs to be considered when making comparisons.

Finally, these factors make it especially important to carefully choose appropriate control specimens (with no fiber coatings) and control tests to properly evaluate the effect of the coatings on the composite behavior. Unfortunately, this is not often done. All of these issues should be taken into consideration when comparing strengths of composites and the success of interface control mechanisms.

4.3.1. Weak Oxide-Oxide Phase Boundaries/Weak Oxides

Monazites and xenotimes are two classes of weak oxide coatings that satisfy the criteria for crack deflection for the currently available fibers [84, 85]. Of these, monazite ($LaPO_4$) has been the most widely studied.

Monazite ($LaPO_4$) is desirable as an interfacial material due to its high melting point (>2000°C) and its high temperature stability with oxides such as alumina [86]. This material also bonds weakly to oxides, particularly alumina. Morgan and Marshall first demonstrated that monazite is a promising weak interface material by testing monazite-coated sapphire fiber hot pressed in an alumina matrix [87]. Cracks formed from indentation testing did not penetrate into the fiber; instead the crack was deflected at the monazite/fiber interface. This demonstration provided a catalyst for concentrated research on monazite coatings.

Since that time, they and other researchers have examined the dense sapphire/ monazite/alumina system, but with low fiber volume fractions (5–10 vol.%) [88–92]. In these works, indentation testing, push-out testing, and 3-pt. bend testing have been utilized to gain information about the strength of the fiber/matrix interface. Similarly, preliminary push-out testing completed on dense monazite-coated Nextel fibers in a dense (>90%) oxide matrix indicated that these coatings did provide the weak interface function in a dense oxide matrix. [93] (See, for example, references [94–98] for information on push-out testing.)

Monazite coatings were also deposited onto sapphire, single crystal mullite, and two types of directionally solidified fibers (Al_2O_3-YAG, Al_2O_3-YSZ) and consolidated into alumina matrices to evaluate the effect of fiber roughness on the interfacial properties. Results of push-out testing indicated that the coated fibers could be pushed in all cases, indicating weak interfaces. The monazite coating was deformed during push-out in relation to the roughness of the interface, which further validates the effectiveness of monazite as a weak interfacial layer [99, 100].

The production of composites containing higher volume fractions of monazite-coated fibers has been somewhat limited and a summary of the relevant work is listed in Table 3. These composites were subjected to mechanical testing, either tensile or flexural, to evaluate the effectiveness of the coatings.

As discussed earlier in this article, discrete monazite coatings were shown to provide increased strengths and use temperatures for Nextel 610/(porous) alumina composites, as compared to similar composites containing no interfacial coatings (reference [27], Table 3). In these composites, an initial strength loss (~28%) was seen after thermal exposure at 1200°C for 100 h; conversely, control samples containing uncoated fiber displayed >70% strength losses after only 5 h at 1200°C. Although these composites displayed an initial strength loss with short-term thermal exposure at 1200°C, the remaining composite strength (>60%) was retained for up to 1000 h at 1200°C [27]. High temperature tensile test results for Nextel 610/monazite/alumina composites also indicated no difference between samples tested at room temperature or at temperatures up to 1200°C (Figure 6) [101]. The fracture surfaces of all of the monazite-containing specimens revealed coated fibers that had indeed pulled out from the alumina matrix, with portions of monazite coatings remaining on the pulled-out fibers and in the fiber troughs (Figure 7). In all cases, control samples containing uncoated fibers showed brittle behavior, with no fiber pull-out.

Monazite coatings were also evaluated in Nextel 312 and 610/aluminosilicate matrix composites [102]. In this case, the monazite coating was deposited using an aqueous slurry (for 312 fabric) or solution (for 610 fabric). Composites containing either fiber type displayed

TABLE 3. Oxide-Oxide Composites with Weak Oxide (ABO$_4$) Interface Coatings

Fiber	Interface Coating	Matrix	Fiber Arch.	V_f	%Comp. Density	Test Type	Test Condition[#]	UTS (MPa)	%ε	Ref.
N610	LaPO$_4$	Al$_2$O$_3$	0°	~20%*	~70–75%	Tensile	RT	198 ± 12	~0.1	27
							RT, after HT @ 1200°C/100 h	143 ± 7		27
N610	Uncoated control	Al$_2$O$_3$	0°	~20%*	~70–75%		RT	45 ± 20	~0.03	27
N610	LaPO$_4$	Al$_2$O$_3$	0°	~20%*	~70–75%	Tensile (10 min. soak @ T)	RT	168.4 ± 10.5	~0.1	101
							1200°C	167.2 ± 2.5		101
N610	Uncoated control	Al$_2$O$_3$	0°	~20%*	~70–75%		RT	93.15 ± 6.4	~0.05	101
							1200°C	99.6 ± 26		101
N610	LaPO$_4$	AS**	8HS	–	–	Tensile	RT	~140	~0.23	102
N610	Uncoated control	AS**	8HS	–	–	Tensile	RT	~210	~0.45	102
N610	In-situ LaPO$_4$	Al$_2$O$_3$-LaPO$_4$	8HS	40%	~80%	Tensile	RT	200–250	–	103,105
N610	CaWO$_4$	Al$_2$O$_3$-CaWO$_4$	0°	–	~80%	4-pt. bend (1h soak @ T)	RT	~350	0.25	26,84
N720	NdPO$_4$	Mullite + 5 wt.% ZrO$_2$	8HS	35%	~86%		RT	235 ± 32	–	107
							1000°C	234 ± 17		
							1200°C	233 ± 33		
							1300°C	230 ± 41		

[#] RT ≡ Room Temperature test; HT ≡ High Temperature test
* Composite strengths normalized to a V_f = 20%. Actual V_f ranged from 18–35%, although most ~18–25%
** Aluminosilicate matrix from COI

lower strengths than control samples (no fiber coating), due presumably to strength degradation of the fiber from the coating process. The stoichiometry of the coating precursor has been shown to be critical in the retention of fiber strength during coating [103]. However, washing of the coating precursor solution has been effective for retaining the fiber strength [104].

Alternately, an in-situ monazite coating was developed by Davis et al., for the production of porous matrix Nextel 610-based composites [103, 105]. The in-situ monazite coating was formed by mixing the monazite precursor in with the alumina matrix powder during composite fabrication (~10–20 vol.% monazite). The final product displayed higher damage tolerance and a reduced notch sensitivity compared to similar porous matrix composites without the monazite phase. This process is beneficial in that it precludes an individual coating step for the fiber. Either monazite alone or monazite mixed with a refractory oxide is machinable, which further enhances its use [106]. The choice of a discrete monazite coating or a monazite-containing matrix provides alternatives to composite users. For example, a monazite-containing matrix may be too soft for a specific application so that a composite containing a discrete monazite coating may be required.

Scheelite ($CaWO_4$) coatings were also found to provide a weak interface in Nextel 610/alumina + $CaWO_4$ composites [26, 84]. Goettler, et al., reported that relatively dense (>80% dense) unidirectional composites were produced with tensile strengths approaching 350 MPa and strains-to-failure on the order of 0.25%. Debonding at the fiber/matrix interface was evident through microstructural evaluation (SEM, TEM). As with monazite, the stoichiometry of the scheelite coating precursor was found to be critical in maintaining the strength of the fibers. If precise control is kept over the precursor material, scheelite coatings may be very useful in oxide-oxide CMCs.

$NdPO_4$ has also been examined as a coating material [85, 107]. Kaya et al., produced Nextel 720/$NdPO_4$/mullite composites and the resultant samples were tested using 4-pt. flexural tests at high temperature. The composites were reasonably dense (~86%), with some porosity within the fiber tows. The strengths of these materials were ~235 MPa at room temperature; this strength was retained up to 1300°C. The strength decreased <15% after 300 thermal cycles to 1150°C. The results indicated that $NdPO_4$ is also a viable coating candidate.

Although limited, research into composites containing weak oxide interfaces is on-going and will likely continue until a commercial product is developed and the full advantages of these coatings are realized.

4.3.2. Porous coatings

Porous matrices have been discussed in this chapter as a mechanism for providing toughness to a CMC. A porous coating can provide a similar function, with the porosity localized around the fiber within a coating. Various porous coatings have been examined, including zircon [108], ZrO_2 [48], and rare-earth aluminates [109].

Research centered on composites containing porous coatings was conducted as part of a project focused on developing and producing oxide-oxide composites for combustor applications [48]. Sapphire fiber reinforcements (~30 vol.%) were used to reach the high temperature goals (~1400°C) of the project. Porous zirconia (ZrO_2) coatings were formed on the fibers using a mixed carbon + ZrO_2 slurry; the carbon was removed by oxidation after incorporation of the coated fibers in an alumina matrix. Flat combustor tiles were fabricated and exposed in a combustor test rig for times up to 1.5 h. Examination of the tested samples showed microcracking of the matrix, but the fibers remained intact and held the matrix material together. These initial results indicated that oxide-oxide composites are indeed promising for combustor applications; however, the cost and difficulty associated with producing composites containing sapphire may limit its use.

4.3.3. Fugitive Coatings

The term fugitive is used for coatings that are (or can be) removed after composite fabrication. Various materials have been examined for this function, such as molybdenum [110], but carbon is the obvious choice, based on cost and commercial availability. Carbon coatings can be incorporated into oxide-oxide composites and then removed through oxidation either prior to or during use, leaving a gap at the fiber/matrix interface. Carbon coatings were unsuccessful in SiC-based composites, due to intermediate temperature embrittlement [111–113]. In this case, the carbon coating would begin to oxidize at low temperature ($\sim$500°C), leaving a gap between the matrix and fiber. The fibers would oxidize as the carbon was removed, leading to silica at the interface and subsequently strong bonding between the matrix and fiber.

With oxides, however, the carbon does not serve the same environmental protective function and can be removed without concern for fiber degradation. A review of this concept is given in reference [114]. Fugitive carbon coatings were shown to provide a weak interface for sapphire-reinforced YAG and for Nextel 720/CAS (calcium aluminosilicate, glass-ceramic) composites. Push-out testing was utilized for the sapphire/fugitive carbon/YAG composites, while tensile testing, in the 0° and the ±45° directions, was used with the Nextel 720/fugitive carbon/CAS composites. In the latter case, higher strengths were found for the fugitive carbon-containing samples (92 ± 42 MPa) versus control samples with no interfacial coating (22 ± 3 MPa). The thickness of the carbon was found to be highly important, as also indicated by other researchers for SiC-based composites [115].

Two-layer coatings of fugitive carbon + zirconia (ZrO_2) have also been shown to provide a weak interface in a sapphire/alumina composite [116]. In this work, the carbon coating alone did not improve the composite properties; however, it is possible that the carbon coating was not thick enough or was discontinuous. In any case, the fugitive layer in combination with the ZrO_2 was successful, providing flexural strengths of $\sim$600 MPa after oxidation of the carbon. Other work on duplex fugitive carbon + ZrO_2 coatings can be found in [117, 118]. Fugitive carbon coatings were also examined for the small diameter Nextel 610 fibers in a mullite matrix [119]. These coatings were not successful due to incomplete coverage of the fibers. These examples illustrate the need for strict control over the fugitive coating characteristics.

Molybdenum has been examined as an alternate fugitive coating, due to its transformation to MoO_3 at low pO_2 and the subsequent sublimation of MoO_3 at $\sim$1155°C. The results of the published works indicate that molybdenum does not provide significantly improved properties over fugitive carbon coatings and can tend to be discontinuous.

4.3.4. Other coatings

Research has been conducted on other interfacial coating concepts; however, these will only be briefly mentioned in this text. Reference [20] gives an in-depth description of these classes: easy-cleaving coatings, including phyllosilicates [120, 121], hexaluminates [122] and layered perovskites [123], ductile coatings such as platinum [124, 125], segregation weakened interfaces [126], multilayered coatings [127], and reactive coatings [47, 128, 129].

Aside from these classes, various other interface coatings have been proposed and studied. For example, a commercial product, "Soft-Cera", is reportedly available from Mitsui-Mining Materials Corp. It consists of Almax fabric (Al_2O_3) coated with a ZrO_2 coating, incorporated in an alumina porous matrix [130, 131]. Initial reports indicate relatively low strengths (50–60 MPa, $V_f \sim 22\%$–$V_f \sim 11\%$ loading direction) and little effect of gage length on the composite strength. Similarly, work by Sudre indicated that ZrO_2 coatings did not provide the same level of strengthening as porous ZrO_2 [116]. Other coating materials that have been investigated for CMCs include Al_2TiO_5 [132], YAG [132], mullite [132, 133], CeO_2 [134], AlN [135, 136], TiO_2 [137], ZrO_2-SiO_2 [138] and diborides [139].

V. PROPERTIES

The physical properties of composites are dependent upon the constituent properties and the details of their assembly. Because target service conditions for ceramic composites are in thermal and environmental regimes where fibers, matrices and interfaces evolve microstructurally, and potentially react and degrade, it is important to develop an understanding of the time-temperature dependence and environmental performance of the materials. However, as these are emerging materials, mostly under active development, property characterization is limited, and it is not unusual to encounter significant variations in important properties resulting from small refinements in the processing. The best possible understanding of the dependence of time-temperature performance on microstructural details, such as matrix composition and porosity, must be developed so that the limited available data may be meaningfully interpreted. Most of the available data has been generated on the porous-matrix oxide-oxide composites produced by General Electric, COI Ceramics and UCSB. These will be the primary materials considered in this section.

This section is organized as follows: the basic physical characteristics are introduced, followed by discussions of uniaxial room temperature mechanical properties, the effect of long term thermal exposure, creep behavior, notch sensitivity, and fatigue performance. The significantly different off-axis, matrix-dominated properties are described in the final section.

5.1. Basic Physical Characteristics

A number of characteristics are common to most fibrous oxide-oxide composites. As discussed elsewhere, whether intentional or not, nearly all have a large matrix void volume (30–50%). Furthermore, all of the significantly characterized materials depend on that void volume for crack deflection and toughness. While we frequently refer to these as porous-matrix, the void volume often includes shrinkage cracks and micro-cracks, rather than literal pores alone. The relative amounts of these forms of porosity can be expected to significantly affect properties, but the topic is essentially uninvestigated. In any case, the matrices are typically very compliant and very weak compared to the fibers. The upshot is that the fiber-dominated properties are very much so, and the matrix-dominated properties are quite

TABLE 4. Thermal and Electrical characteristics of all oxide CFCC's

	GE-610/GEN-IV	COI-720/A-1
thermal expansion coef.[a] @ RT, ppm/C	4.5	3.5
thermal expansion coef.[a] @ 1000 C, ppm/C	8.2	6.2
thermal conductivity[a], RT, W/mK	4.5	4.25
thermal conductivity[a], 1000 C, W/mK	2.5	2.3
specific heat @ RT, W-s/gK	–	0.75
specfic heat @ 1000 C, W-s/gK	–	1.30
dielectric constant (loss tangent)	–	5.74 (<.1)

[a] Principal fiber direction; Refs [12, 75]

modest. The stiffness and strength in loading directions that put fibers in tension are nearly those of the loaded fibers. The properties in unreinforced directions are those of porous, and sometimes poorly consolidated, oxides.

Selected composite materials and their basic physical properties are summarized in Table 2. The GE GEN-IV all-oxide fiber composites have a relatively low volume fraction (30%) of Nextel™ 610 fiber and a matrix composed of alumina and silica with approximately 35% porosity [12, 30, 37]. COI produces a number of aluminosilicate matrix materials with a selection of different Nextel™ oxide fibers, but with higher fiber loading and matrix porosity than the GE materials [140, 141]. Two examples of nominally silica-free porous composites (Table 2) with improved thermal stabilities are the mullite-alumina formulation developed at UCSB [4, 44, 142, 143] and the all-alumina COI N720/A-1 [75, 79, 141, 144].

Matrix porosity on the order of 40% leads to overall porosity near 25%, resulting in the relatively low composite densities shown in Table 2. Specific surface area measurements of the GEN-IV composite show $30\,m^2/g$. This is 100 times greater than dense matrix CMC's and is indicative of fine interconnected porosity [30]. Such high matrix porosity is undesirable where hermetic fluid containment is required and also leads to concerns with use in erosive environments [76].

The thermal and dielectric properties of several composite systems are shown in Table 4. Note that the thermal expansion, conductivity, and specific heat all vary by approximately a factor of two over the range of temperature from 23°C to 1000°C.

5.2. Room temperature uniaxial mechanical properties

The as-fabricated room temperature tensile properties of the composites are summarized in Table 5. It is important to recognize that, due to their high porosity, the matrices in all of these systems have elastic moduli and strengths an order of magnitude less than those of the fibers [145], hence the composites are highly orthotropic in most manifestations, e.g. laminates [143, 146]. See [147] for an extensive review of the mechanics of brittle matrix composites. The properties in Table 5 represent the response to loading in the fiber direction. The fiber direction orientation, denoted as 0°/90°, is the most favorable orientation; the strength and stiffness in any other orientation, denoted as 'off-axis', is substantially lower.

TABLE 5. Mechanical Properties–Room Temperature 0/90 Fast Fracture Data

designation	Tension					Flexure			Compression		
	E GPa	v12	prop.limit MPa	UTS, MPa	% ε	E GPa	strength MPa	% ε	E GPa	strength MPa	% ε
COI-312/AS	31.0	–	–	125	0.42	48.3	159	0.33	–	–	–
COI-550/AS	40.0	–	–	148	0.43	57.9	168	0.29	44.1	157	0.35
COI-610/AS	124	–	–	366	0.35	141	352	0.35	141	270	0.24
GE-610/GEN-IV	70	–	100	205	0.33	–	–	–	–	–	–
UCSB-610/M	95	0.05	–	215	0.22	–	255	0.29	–	–	–
UCSB-720/M	60	–	80	150	–	–	215	–	–	–	–
COI-720/AS	75.6	0.09	–	180/220[a]	0.30	98.6	216	0.22	80.0	186	0.22
COI-720/A-1	75.1	–	50	177	0.31	90.9	218	0.23	–	253	–

[a] Improved generation material recently reported by the manufacturer; Refs [30, 44, 75, 140–143, 146]

Due to the extreme difference between the matrix and fiber properties, the in-plane modulus is dominated by the fiber modulus and volume fraction. For example, the GE GEN-IV material has a modulus of 70 GPa. With 15% fiber in the loading direction and a fiber elastic modulus of 380 GPa, the stiffness due to the fibers alone is expected to be 57 GPa. The UCSB Nextel 610-based material with 33% higher fiber content likewise has roughly a 35% higher modulus. The small contribution of the matrix makes it particularly difficult to infer the elastic properties of the matrix from the composite test results [143]. As a result, and because the porous matrix materials are difficult to produce without any fiber reinforcement, their constitutive properties are not yet well characterized [145].

Figure 8 shows a typical 0°/90° orientation stress-strain plot. The response is initially linear and is somewhat stiffer than would be expected due to the fibers alone, however it becomes nonlinear at relatively low strain. This behavior is consistent with the matrix contribution to the modulus decreasing with stress due to cracking [39]. The total matrix contribution to load carrying is sufficiently small that a proportional limit can be difficult to

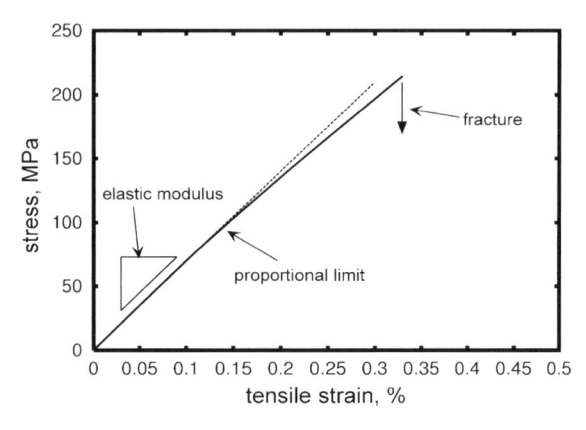

FIGURE 8. Typical CMC tensile stress strain curve.

identify. However, this inelastic behavior contributes to the composite's ability to tolerate stress concentrations, as will be discussed subsequently. As matrix damage develops fully, the tangent modulus decreases to nearly the value that would be expected by considering the fiber contribution alone [147].

Unlike the tensile modulus, the ultimate tensile strength can be substantially impacted by the matrix formulation. In the best case, as the ultimate strength of the composite is approached, damage accumulates in the matrix, without damaging the fibers, until the matrix is not load bearing at all. Failure thus occurs as individual fibers reach their breaking load. The failure characteristics of individual fibers are determined by the statistical distribution of flaws within the fiber. As the weakest fibers fail, their load is transferred to the remaining fibers. If the matrix or fiber interface is sufficiently weak, the failure of one fiber does not result in a localized load transfer to its immediate neighbors, but rather the load redistribution is to a greater number of fibers in a less localized region of the composite. The idealized case of completely independent fibers, as in an untwisted tow, is known as global load sharing (GLS) [148].

In a good quality composite many fibers fail some distance away from the fracture surface. In dense-matrix composites with a weak fiber/matrix interface, they continue to offer some load bearing capability due to frictional sliding as they pull out of the matrix [20], and the fracture surfaces feature significant bare fiber, along with distinct holes resulting from the fiber pull-out (see, for example, figure 2b). Typical fracture surfaces of porous-matrix composites also feature considerable lengths of bare fiber (Figure 3a). While this feature is often called fiber pull-out, by final fracture the matrix in the fracture region has been thoroughly reduced to rubble and has fallen apart, leaving bare fibers behind, so that the pull-out resistance is minimal. Hence, while there is some contribution of friction in porous-matrix composites it is probably a relatively small effect.

In any case, the ultimate strength of the composite approaches the tow strength of the constituent fiber multiplied by the loaded fiber fraction to a reasonable approximation. For example the COI 610/AS composite has 25% fibers in the loading direction with a bundle strength of 1600 MPa, so the composite strength of 366 GPa is within 10% of the value of 400 GPa calculated based on the fiber bundle strength. See [149] for discussion of the relation between fiber filament strength and tow strength. In the opposite extreme the matrix offers no fiber damage protection mechanism and fracture of the composite is observed to occur on a single plane with no fiber pull out. The composite strength in that case is that of a poor quality monolithic ceramic.

Table 5 also shows flexure strengths for comparison; often, these are the only strength values reported for ceramic composites. Flexure tests can be very useful for several reasons: they better mimic the actual loading conditions in many applications, and they sometimes more clearly reveal high damage tolerance. With that said, the values reported should be viewed with caution. The strength value derived from a standard 3 or 4 point flex test is the maximum tensile stress calculated, assuming the material response to be linear. A typical composite displays, or quickly develops, nonlinearity and tension-compression asymmetry due to processing cracks or damage introduced on the tensile side of the bar during the test. Consequently the stress distribution differs from simple beam theory, so that the 'flexure strength' calculated is not exactly the maximum stress borne by the material. Flexure strengths are generally higher than tensile strengths derived from uniaxial tension tests by as much as 20–30%. Even when efforts are made to properly account for the material

TABLE 6. UCSB 720, properties variation with porosity

	Precursor Treatment Cycles		
	$N=2$	$N=4$	$N=7$
matrix porosity, volume %	35	33	31
flex strength, MPa	215	175	112
notched flex strength, MPa	185	160	110
ILS, short beam shear, MPa	10	12	9

Ref [44]

nonlinearity, the flexure test still results in a higher strength value than a simple tension test. This is interpreted as a scale dependence, i.e. a somewhat higher stress is allowable if it is localized over a small volume. Such volume dependence is a consequence of the statistical nature of the fracture of brittle materials, and may play an important role in the notch tolerant behavior of ceramic composites [148, 150, 151].

The strengths of the composites in compression are also listed in Table 5. It is notable that the compressive strength is comparable to the tensile strength. This is in contrast with monolithic ceramics where the compressive strength can be substantially higher than the tensile strength. In the fibrous composite case, compressive failure occurs by fiber buckling, and thus the strength is limited because of the low strength and stiffness of the matrix [152, 153]. The effect of matrix porosity has been examined extensively for the UCSB 720/Mullite-alumina composites. The baseline matrix consists of an 80/20 volume percent mixture of mullite and alumina particles, with approximately 35% matrix porosity by volume. The matrix is then infiltrated with an alumina precursor solution, resulting in a small density increase with each infiltration [44]. Table 6 shows the strength variation with the number of precursor infiltrations, N. As the matrix density increases, the fiber-dominated flexure strength decreases and the fracture mechanism changes from fibrous to monolithic, with only a 2–4% change in matrix density. This shift in character is presumed to be driven not only by the increasing strength of the bond between the matrix and fiber, but also by the increasing elastic modulus of the matrix [147]. The changing elastic response of the matrix is sufficiently small, however, that the elastic response of the composite is not significantly affected. As noted by the authors, fiber degradation resulting from the numerous infiltration and heat treatment cycles could play a significant role. The real possibility of progressive fiber degradation seriously complicates interpretation of results of this kind. As discussed above, crack deflection is a competition between matrix fracture and fiber fracture. It is very difficult to distinguish between a decrease in fiber strength and an increase in matrix/coating strength.

In contrast to the fiber dominated in-plane strength, Table 6 shows that the interlaminar shear strength of the composite improves, to a point, as the matrix density is increased. Hence there is a trade off to be considered when tailoring the matrix design [44]. Perhaps more importantly, however, recognizing this strong sensitivity of room temperature fracture behavior to matrix properties provides general insight into the behavior of the porous composites under thermal exposure, to be discussed in following sections.

5.3. Long term thermal exposure

The composite service temperature/time limit for porous-matrix oxide composites is determined by the combination of two inherent temperature/time limits; degradation of the fibers themselves and degradation of the crack-deflection mechanism via matrix evolution. The details of the degradation of the crack deflection mechanism are not fully understood and probably vary somewhat from system to system. Nevertheless, it is some combination of densification of the matrix, sintering and strengthening within the matrix without significant densification, and sintering of the matrix to the fiber. At some point, the crack deflection mechanism ceases to operate and the properties in all loading directions become those of a poor-quality monolithic ceramic. This is particularly a problem for the alumino-silicate matrix systems [141]. The COI 720/AS composite strength drops dramatically when exposed to temperatures above 1000°C for 1000 hours (Figure 9). At 1200°C the degradation occurs quickly, well within 100 hours. The alumino-silicate matrix therefore may be serviceable for short term exposures as high as 1000°C [141]. Likewise the GE GEN-IV material suffers only 15% strength degradation for short term exposures up to 1100°C [30].

The UCSB matrix is composed primarily of a network of relatively coarse mullite. Mullite is resistant to sintering up to 1300°C, which increases the temperature at which shrinkage and degradation would be problematic to above the degradation temperature of the fiber. The alumina in the matrix serves to bond the mullite particles together, but because the mullite forms a continuous network, there is no global shrinkage for long duration high temperature exposure [4, 142]. The efficacy of this concept is shown in Figure 9 as the mullite-alumina composite strength remains essentially unchanged following 1000 hour heat treatments up to 1200°C. The COI all-alumina matrix performs comparably well, by employing a high initial matrix porosity as well as a heterogeneous alumina particle formulation to enable long-term stability against densification.

As discussed previously, an alumina-YAG matrix behaves in a manner similar to the UCSB material ($d_{YAG} \sim 12$ μm, $d_{alumina} \sim 0.2$ μm) and has shown long-term stability against

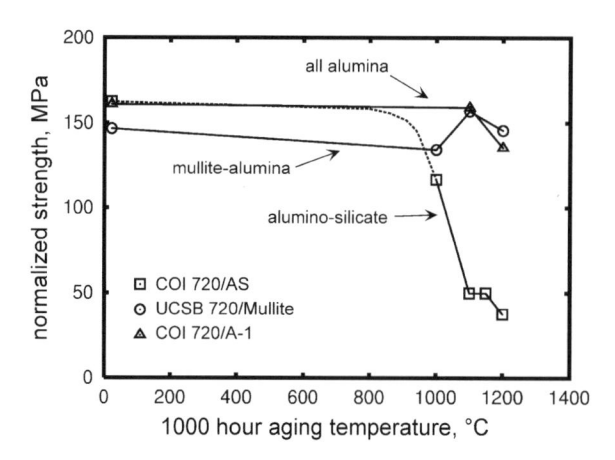

FIGURE 9. Long duration – high temperature exposure strength retention in porous-matrix all oxide composites. Stresses are normalized to 40% fiber fraction. The dashed line indicates that the AS matrix is believed to be stable to ~800C. References [75, 141, 142].

5.4. Creep behavior

Fiber selection is a compromise between strength and creep resistance. The fine grain size necessary for the very high strengths of quality crystalline fibers also makes them prone to creep. Fibers such as Nextel 312, 440 and 550 have especially low creep resistance due to the viscous behavior of the glass-phase silica, while the fully crystalline 610, 650, and 720 fibers offer improved creep characteristics. At 1100°C and a stress of 100 MPa, Nextel 610 fibers creep at a rate of $\dot{\varepsilon} \sim 10^{-10}/s$, due to its fine grain size. Under the same conditions, the creep rate of Nextel 720 is four orders of magnitude lower with $\dot{\varepsilon} \sim 10^{-6}/s$, owing to its coarser grained mullite morphology. Nextel 650 fibers offer intermediate creep resistance without the presence of any silica or mullite, and also offers improved environmental resistance over Nextel 720 [55, 154, 155]. The creep characteristics of the fibers translate directly to the creep behavior of the composites in the principle fiber directions. The deformation rate of both fibers and composites has a stress dependence given by,

$$\dot{\varepsilon} = A\sigma^n \tag{1}$$

where the creep exponent n is approximately 3 for both Nextel 610 and 720 fibers and is somewhat lower for Nextel 650 [55]. Similar stress exponents are found for porous-matrix CMC's. For example, the GE GEN-IV 610 based composite shows an exponent of 3.06. The strain rate at 1000°C under loads from 75–135 MPa is nearly constant until failure; there is no primary or tertiary creep. It is believed that the creep deformation is dominated by the creep behavior of the fibers, and that there is no environmental degradation associated with the matrix cracking. The 610/aluminosilicate composite creep rate is $\sim 2.7 \cdot 10^{-8}/s$ at 1000°C and 75 MPa stress, and retains a residual strength of 205 MPa after 100 hours under those conditions. At 1100°C, the creep rate jumps two orders of magnitude and the specimen ruptures in less than one hour, clearly limiting this system to 1000°C applications even under moderate load [30, 156, 157].

The Nextel 720 based composites show substantially improved creep performance in correlation with the superior fiber properties. A creep limit of 145 MPa, which is 80% of the ultimate strength, was identified by John et al. [158] as the level where the COI 720/AS composite could survive 100 hours at 1100°C. The linear creep rate is $\sim 1 \times 10^{-8}/s$ under these conditions, however the creep response at this high load is significantly greater during the initial 10 hours before decreasing to a steady rate [159].

The reader is referred to [160, 161] for extensive reviews of creep of monolithic ceramics and ceramic composites. Models for creep of ceramic matrix composites are given in [161–163].

5.5. Notch sensitivity and toughness

Notch sensitivity is evaluated by considering a tensile specimen with a stress concentrating feature such as a hole. For an isotropic elastic tensile specimen with a circular hole of one half of the section width, as an example, linear elasticity theory predicts a maximum stress at the edge of the hole in the plate to be 2.15 times the average section stress

[164], while for the 0°/90° layup orthotropic materials under consideration here the concentration is somewhat higher, having a value of ~2.5 [165]. A notch sensitive material will fail as a result of that peak stress, hence the hole effectively reduces the strength of the part by a factor of more than 2 compared with the ultimate strength measured in simple tension.

The inelastic behavior of ceramic composites that results from the formation of microcracks and fiber-matrix debonding at high stress levels serves to substantially reduce the peak stress, thereby improving the material's notch sensitivity and damage tolerance [143]. This peak stress reduction is notably independent of the size of the stress concentrating feature, i.e. geometrically self-similar specimens would be expected to fail at the same stress regardless of their absolute size. However, experiments show that the notch sensitivity of fibrous composites is strongly affected by size scale – notch sensitivity generally improves with smaller defect sizes. The UCSB 610/M composite shows a 30% effective 0°/90° tensile strength reduction due to the presence of a 10 mm diameter hole, but less than a 10% reduction due to a 3 mm hole; a highly notch sensitive material would show a 65% strength reduction in both cases. The 610 notched tensile specimens exhibit a stable progressive failure following fracture initiation, thus permitting the calculation of a fracture energy of 6000 J/m^2. Notably, this is an order of magnitude higher than the fracture energy typical of monolithic ceramics [143]. The UCSB 720 composite shows a fracture energy of about 3000 J/m^2 [44]. The COI 720 material is likewise shown to be notch insensitive, with negligible strength reduction in the presence of sharp or semi-circular edge notches in small size tensile specimens [166], however significant notch sensitivity develops at temperatures above 1100°C [167].

The scale dependence on strength can be explained by utilizing the *point stress* failure criterion, that is used with polymer matrix composites [168, 169], as well as by considering the statistics of fiber strength distribution [148]. These models suggest a characteristic length scale of ~0.5 mm over which the stress must exceed the strength of the material in order for failure to initiate.

5.6. Fatigue

The aluminosilicate matrix materials, GE GEN-IV and COI 720/AS show excellent room temperature fatigue resistance, surviving 10^5 cycles at approximately 85% of their room temperature strength. This fatigue resistance is retained with increasing temperature throughout the regime in which the matrix is microstructurally stable. The fatigue resistance severely declines at the same temperature where long-term strength retention is lost due to matrix sintering and coarsening [156, 170]. For example the COI 720/AS survives only 10^4 cycles at 60% of its ultimate strength at 1200°C [79]. In a series of high temperature fatigue tests where the peak loads were sustained for times exceeding 10 seconds per cycle, it was found that the composite strain paralleled the static creep behavior; this indicates that time-at-temperature-and-stress is more significant than the cyclic loading [156].

Silica-free COI 720/A composites have shown similar fatigue resistance, with a fatigue strength for 10^5 tension cycles at room temperature essentially equal to their static ultimate strength and retaining 85% of their strength at 1200°C [79]. Note that these tests were performed on an early generation of the material with relatively poor room temperature strengths. Fatigue behavior of the improved material has not been reported.

TABLE 7. Anisotropic elasticity and thermal
properties of COI 720/AS.

$E_{11} = E_{22}$	80 GPa
E_{33}	50[a] GPa
ν_{12}	0.09
$\nu_{13} = \nu_{23}$	0.12
G_{12}	17[b] GPa
E^{45}	50 GPa
$G_{13} = G_{23}$	40[c] GPa
thermal expansion, in plane	6 ppm/K
thermal expansion, through thickness	7 ppm/K
thermal conductivity in plane	2 W/mK
thermal conductivity through thickness	1.6 W/mK

[a] Ultrasonic estimate; [b] calculated; [c] model based estimate; Ref [146]

5.7. Off Axis Properties

As noted, these composites are highly orthotropic. Although there is a small difference between the warp and fill directions of the fiber fabric itself [45], the composites for which data have been reported were fabricated with the warp directions of alternating layers of cloth at 0° and ±90°; the resulting laminate has two essentially identical principal fiber directions orientations. Although it is well understood that the off-axis properties are strongly influenced by the porous matrix properties – the strength and stiffness in any load direction not aligned with the fibers directions is substantially lower – there is relatively little published test data for off-axis properties. The most complete set of test data for a single material is that of the COI 720/AS, as shown in Table 7 [146]. Subscripted quantities in the table are the elastic tensor components in the coordinate system aligned with the fiber directions, i.e. 1, 2 refer to the in-plane principle fiber directions with 3 being the through thickness direction. The elastic moduli in the through-thickness direction (E_{33}) and in the ±45° in-plane orientation (E^{45}) are nearly equal and approximately 60% of the 0°/90° value. The 0°/90° and ±45° moduli, along with the value of 0.09 for the principle direction Poisson's ratio, have been used to compute the shear modulus values in Table 7 according to [164],

$$G_{12} = \frac{E_{11}}{2(1 + \nu_{12} + 2(E_{11}/E^{45} - 1))}. \qquad (2)$$

The UCSB Nextel™ 720 reinforced material with its lower stiffness mullite matrix has a ±45° tensile elastic modulus of only 23 MPa, or less than 40% of its fiber direction stiffness [142].

The thermal expansion coefficient and thermal conductivity of the COI 720 material are orthotropic, but with lower anisotropy than the elastic properties, as Table 7 indicates. The thermo-elastic properties are reasonably computed by micromechanical models that incorporate both porosity and microcracking in the matrix [146].

The anisotropy of the composite strength is even greater than the anisotropy of the thermo-elastic properties (Table 8). The ultimate strength of specimens loaded in a ±45° tensile configuration is approximately 20% of the fiber-direction strength. Failure

TABLE 8. Off-axis strength.

	GE GEN-IV	COI 610/AS	COI 720/AS	UCSB 720 (N = 2)
0/90 strength, MPa	205	366	195	150
±45° Strength, MPa	54	—	—	28
in plane shear, MPa	27[a]	48.3[b]	31.0[b]	—
inter laminar shear, MPa	–	15.2	11.7	10
trans thickness, MPa	7.1	—	2.7	—

[a] ±45° test; [b] Iosipescu (V-Notch) shear test; Refs [30, 44, 140, 142, 172]

in these specimens typically occurs in shear bands at 45 degrees, i.e. parallel to one set of fibers and perpendicular to the other, hence the resolved shear stress on that plane may be interpreted as the shear strength of the material [30]. However, there is also a tensile stress component on the 45° plane, so that the measured apparent shear strength is typically somewhat lower than the strength measured in pure shear [151, 171]. The aforementioned UCSB study of matrix density found that the ±45° tensile fracture mode transitions from matrix failure and fiber scissoring to tensile fiber failure in the higher density, stronger, matrices [142]. (Again, some caution in interpretation is warranted as this result could be a consequence of fiber degradation.) Interlaminar shear and transthickness tensile strengths are even lower, as the failure occurs by delamination and is purely a matrix failure [143, 172].

Off-axis properties are difficult to interpret and additional test techniques and dedicated modeling will be necessary to clarify the design implications. Notably, strengths derived from ±45° testing show a strong dependence on specimen geometry, owing to the fibers ability to rotate and stiffen the response of wider specimens following the initial shear failure of the matrix [173]. This is just one of many issues that complicate characterization of composite failure under multiaxial stress states [174, 175]. Hence, the need to measure basic properties and develop a thorough understanding of materials design for actual application requirements is reinforced.

VI. PROCESSING METHODS

The processing options for fibrous composites are seriously complicated by the presence of the fibers. While ceramic monoliths can be sintered relatively easily to near full density, the constraint introduced by the fibers inhibits the densification [176, 177]. Furthermore, temperatures, pressures and chemistries are limited by the need to avoid damaging the fibers. The problem of creating a dense matrix is, therefore, one of avoiding microstructural and morphological changes in one refractory constituent while affecting large changes in another nearly equally refractory constituent. This situation is further aggravated by the introduction of 3-D reinforcements. 2-D cloth or tape layups have their external dimensions constrained by the fibers in 2 directions, but have some latitude for shrinkage in the third. The external dimensions of a 3-D composite are essentially completely determined by the fiber preform. Processing options are confined to filling that space and most processes are inherently unable to fill it completely. Since some porosity is virtually unavoidable, the composite designers' options are restricted to deciding how best to distribute the void volume. For example, uniformly distributed fine, closed porosity is probably preferable to large cracks. There has been relatively little research devoted to this issue and it is expected to eventually become a fertile area. While the constraint provided by the fibers is a significant hindrance to obtaining dense matrices, it makes crack-deflecting porous matrices easy to achieve and maintain to relatively high temperatures.

The processing of a composite typically involves the following separate steps: 1) coating fibers, 2) arranging fibers in the desired architecture, 3) infiltrating the matrix precursor, 4) consolidating the matrix, 5) final machining. This section is arranged in the same order, with the following exceptions. Steps 2 and 5 are outside the scope of this article and are addressed only incidentally. Also, in some processes, step 2 can precede step 1, and step 3 can precede step 2 and, for porous-matrix composites, step 1 is skipped entirely. The methods outlined below have been used for both porous-matrix and dense matrix composite production.

6.1. Processing of Interface Coatings

Coatings have been deposited onto ceramic fibers using a variety of methods, including solution/sol/slurry precursors [178–180], chemical vapor deposition (CVD) [118, 181–183] and electrophoretic deposition (EPD) [184]. (A review of current coating technology is given in reference [20].) The latter two methods are also used for complete matrix infiltration and will be discussed further in the next sections; note that only minor modifications are needed for coating fibers. The choice of coating process is often dictated by the material being deposited. For example, the CVD process has been very successful for the production of carbon (C) and boron nitride (BN) coatings (see, for example, [185–188]); however, there is difficulty in applying coatings of crystalline high temperature oxides, such as alumina, and especially mixed oxides. EPD has been utilized for the deposition of oxide coatings on fibers [184, 189]. This process, however, does not currently allow for discrete coatings to be applied on fabrics.

All of the aforementioned coating processes are useful for coating single filaments, such as sapphire, since they are generally easier to coat versus fiber tows. There is a basic difficulty in producing discrete coatings on fibers within a tow and further, on tows within a fabric. As the coating thickness increases, the propensity toward coating bridging increases and ultimately a crust can form around the fiber tow (as with the EPD process). The deposition of a discrete fabric coating (i.e. each fiber individually coated without excessive coating bridging), aside from CVD processing, has yet to be realized. Oxide coatings have been deposited onto fabrics by dip-coating (immersing the fabric within a sol, followed by drying), electrophoretic deposition and by coating the tows prior to weaving. The latter case is in the trial stage and has not been utilized in a composite [190].

At the current time, liquid precursor (solution/sol/slurry) coating techniques have been the most successful for the deposition of oxides, such as monazite ($LaPO_4$) and porous oxides (carbon-oxide mixtures) onto oxide fiber tows [178, 179, 191, 192]. A typical fiber coater is shown in Figure 10a [193]. In the coating process, a lighter, immiscible liquid is floated on the surface of the coating precursor. The immiscible liquid is used to remove excess sol from the coated tow and it allows for the coating of individual filaments with

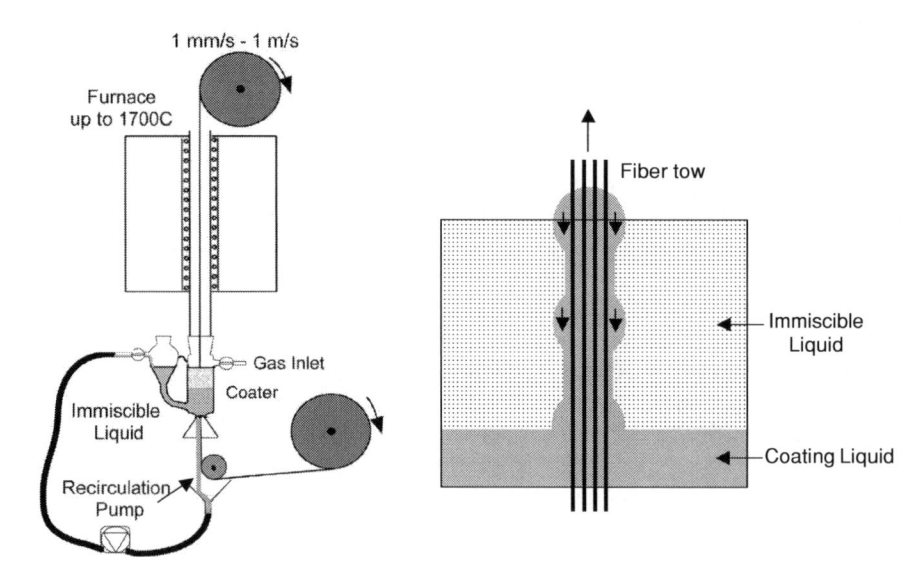

FIGURE 10. a) Fiber coating apparatus [192] and b) immiscible liquid solution. Reprinted with permission of The American Ceramic Society, www.ceramics.org. Copyright 2003. All rights reserved.

minimal bridging (Figure 10b). As discussed previously, the sol chemistry must be carefully controlled to avoid fiber degradation.

Coated fibers are generally characterized in terms of coating uniformity, continuity, thickness and composition. Scanning electron microscopy can be used for the first three types of characterization, while X-ray powder diffraction and/or transmission electron microscopy (+ EDS/EELS) can provide compositional information [194]. The residual porosity within the coating is also considered, particularly if a hermetic coating is needed for environmental protection of the fiber.

6.2. Matrix Infiltration

For processing oxide-oxide composites, either oxide fabric or fiber tapes can be used as reinforcement. The woven fabric can be utilized directly using the techniques outlined in this section. Single filaments and fiber roving/tows, however, must be incorporated into a handleable form, i.e. a tape. This is most commonly achieved through filament winding of the fiber onto a mandrel [62]. The tow can be passed through a ceramic slurry during this process to infiltrate the matrix material within the tow. The resultant tape has some level of green strength, which enables handling of the fiber tows. An alternate method is to first filament wind the fiber and then tape cast an oxide layer within/on the fiber; this process is practical for the larger (>100 μm) diameter fibers. Hand lay-up of fibers can be used, although this is labor intensive and used only with the larger diameter (>100 μm) fibers. 3-D preforms generally do not require additional forming prior to the composite processing.

The oxide slurries used in the various processes outlined here are critical to the successful fabrication of CMCs. Dispersed powder slurries are most commonly used and the formulation of the slurry is extremely important for attaining the most homogeneous green body and the optimal final product. A review of particle-particle interactions and the stabilization of slurries are given in reference [195].

6.2.1. Pressure Infiltration

This processing technique is used primarily with fabric-based samples, although it can be used for fiber tow samples [195]. In this process the fabric is first desized; that is, the organic coating (sizing) applied to the yarn to improve handling is removed by dissolving chemically or by oxidation in an air furnace (heat cleaning). Layers of cloth are then stacked into a mold in the desired organization. It is important to constrain the fabric during the infiltration process to avoid "floating" of the fabric, which would lead to undesirably thick matrix regions between layers and consequently a low fiber volume fraction. Depending upon the individual set-up, the mold can either be pressurized itself or the mold can be placed within a pressure vessel.

Prior to pressurization, a disperse slurry is poured into the mold. The amount of slurry is dictated by the volume fraction of solids and the optimal green matrix density, which is typically ~65% of full density (corresponding to close-packing of powders). The slurry is produced by dispersing the oxide powder within a solvent, generally water. A powder-specific dispersant is added to ensure the stability of the slurry. Without stabilization, sedimentation of the slurry during the infiltration process could occur, leading to matrix gradients within the sample. This can be a particular problem with bi-modal (or higher) powder size distributions and with increasing variations between the particle sizes. Pressure is applied to the unit to force the slurry through the fabric preform. The fabric essentially acts as a filter, building the matrix through the thickness of the sample. Vacuum may also be applied to aid in the filtration.

Once the slurry has filtered through the sample, the green body is removed from the unit. Care must be taken during this step, depending upon the amount of pressure applied to the sample. If excessive pressure is used, there can be a "springback" action of the green body upon release of the pressure, as discussed by Lange et al. [195]. Proper control over processing parameters can reduce this effect. The green body can then be dried and pressureless sintered to form the final porous oxide-oxide composite. Subsequent infiltrations with liquid precursors, such as aqueous colloidal sols, can increase the matrix density significantly, although a dense matrix is difficult to form in this manner. The density is increased rapidly during the first few infiltrations; however, the effect is greatly diminished as the interior channels begin to fill with material and/or the outer pores fill up, preventing the flow of liquid into the interior of the body. Moreover, volatiles must escape during each infiltration cycle, which requires a minimum level of connected porosity.

6.2.2. Pre-preg processing

Many of the processing techniques used in polymer matrix composite fabrication have been adapted for use in producing oxide-oxide composites. For example, COI produces their Nextel 720/aluminosilicate composites by submersing the desized fabric within a slurry ("pre-impregnating") to form a matrix-infiltrated fabric known as a prepreg. (Alternately, desized fiber tows can be filament wound to form a tape.) These prepregs are then stacked on to tooling (Figure 11) or within a mold and consolidated using vacuum bagging and warm (<150°C) temperatures, followed by pressureless sintering at high temperature

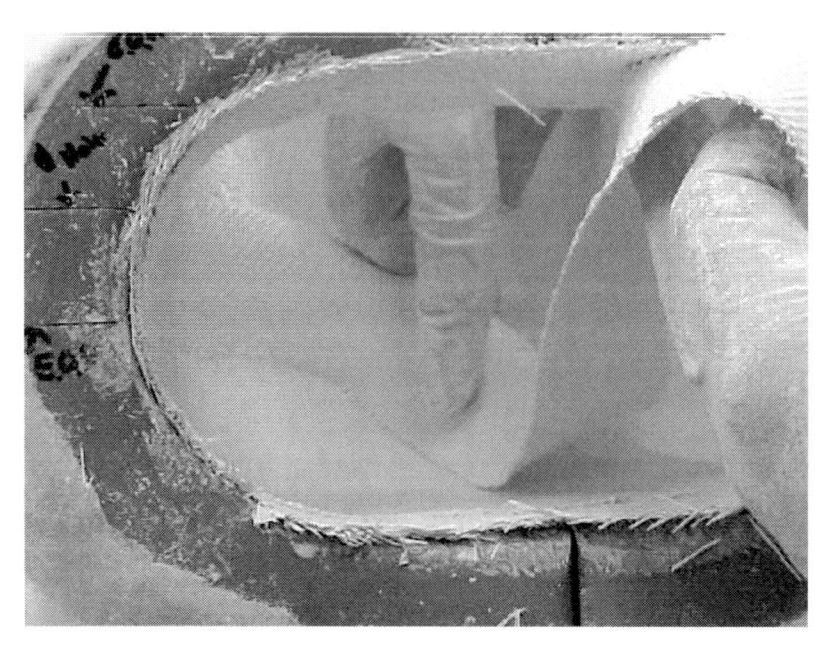

FIGURE 11. Laying of prepreg segments into a mold to produce the exhaust duct of Figure 1. (Courtesy COI Composites Inc., Materials Research and Design, Inc., & U.S. Navy).

(1000–1200°C) [141]. This is akin to the stacking of polymer-impregnated tapes into a mold and then autoclaving, which involves pulling a vacuum on the material, applying exterior gas pressure in the autoclave and heating to cure the polymer resin. In the case of oxide-oxide composites, polymerization generally does not occur and the warm temperatures are used to drive off any water or remaining solvents. The primary advantage of the pre-pregging process is the ability to form near-net shape, complex parts with relatively inexpensive tooling.

6.2.3. Electrophoretic Deposition (EPD)

EPD is a process that utilizes the behavior of charged particles under an applied electric field to infiltrate ceramic fibers/fabric (for further information on this topic, see references [196–198]). In this method, colloidal sols containing fine charged particles are subjected to an electric field, which drives the charged particles toward a deposition electrode. Oxides are insulators, which makes the application of an electric current very difficult. A non-conducting oxide fabric/tape, therefore, must either be coated with a conductive coating or be held against a conductive material to allow for infiltration of the colloidal particles. A strong field must also be applied, which gives rise to concerns over the stability of the EPD coating solution under such conditions.

The process is initially quite rapid, however, the rate decreases as the thickness of the infiltrated layer increases. The use of EPD allows for reasonable infiltration within the fabric tow. Single layer fabric mats have been infiltrated with this process and then taken through the pressure infiltration process [199]. EPD is generally limited in terms of the size and thickness of the material that can be formed. Thinner layers, such as a single fabric layer, can be infiltrated reasonably well; however, the process is not amenable to thicker samples, due to the decreased deposition rate with increasing distance from the electrode. Results of work completed at the University of Birmingham [200] indicated that ~81%–86% composite densities could be attained by using EPD combined with pressure filtration.

6.3. Consolidation

6.3.1. Pressureless Sintering

Green composites attained through prepregging, pressure infiltration, etc. can be sintered at high temperature in air to form a porous-matrix CMC with virtually no external pressure on the material. For these materials, the goal in the sintering step is not necessarily to densify the matrix, but to strengthen the necks between the particles to provide strength to the matrix. Significant densification of the porous matrix would result in the further opening of microcracks within the matrix. Basic discussions of particle sintering can be found in references [201] and [177].

6.3.2. Hot Pressing

Hot pressing is a conventional method for the production of ceramic monoliths and for some SiC-reinforced composites [51, 63]. For composites, this process involves loading a graphite die with the material to be consolidated, either as a prepreg or as individual components (i.e. fibers placed by hand within a powder). The die is then heated to high temperature (>1000°C, machine capability typically >2000°C) in an inert atmosphere. High(er) pressures are applied to the material during heating or at temperature to facilitate the consolidation of the powder. The die is then cooled for removal. High densities (>90%) can be attained with this method, particularly for glass and glass-ceramic matrices, which soften and flow during heating. Additives can also be added to the starting material to enhance densification.

There are several drawbacks associated with this process for the production of oxide-oxide composites with no glassy phases. With non-glassy (i.e.-non-silica containing) matrices, the temperatures needed for complete consolidation of the matrix typically exceed the limitations of the polycrystalline fibers, leading to possible fiber damage. Fiber damage can also occur through the high applied pressures involved in the process. This a particular problem for hot-pressing woven fabrics due to the cross-over points within the weave. Another drawback of this method is the shape and size limitations of the products. Flat panels are generally produced due to the uniaxial pressure and the sample size is limited by the hot press itself. With increasing sample size, the hot press must become larger in order to apply the needed high pressures. Further, complex, intricate shapes cannot be easily formed with this method due to the uniaxial consolidation.

6.3.3. Hot Isostatic Pressing

Hot isostatic pressing (HIP) is similar in nature to hot pressing (high T, P), but it utilizes gas pressure for consolidation versus the uniaxial pressure applied in hot pressing [202]. This allows for much higher pressures on the sample. As with other processes, prepreg tapes are formed through filament winding, tape casting, or through slurry infiltration into a woven fabric. The tapes are stacked in the desired configuration within a rigid mold. Research has shown that such a mold is needed to prevent distortion of the final product due to the anisotropy introduced by the reinforcing fibers [202]. Once the mold is loaded, it is sealed and then heated and pressurized using an inert gas. Sealing the mold prevents gas intrusion within the sample. There is little constraint on the shape complexity of the composite since the gas pressure is the same for all surfaces and directions (i.e. isotropic). The primary drawback for this process is the high cost relative to pressureless sintering techniques. Another area of concern is the mold or canning material, which can introduce contaminants into the product. This can be a particular problem due to the porous nature of the prepregs.

6.3.4. Chemical Vapor Deposition/Infiltration

Chemical vapor deposition/infiltration (CVD/CVI) has been used extensively in the forming of SiC/SiC and C/C composites [203, 204]; however, it has not been utilized to any great degree to form oxide-oxide composites. This process involves forming a preform through stacking filament wound tapes, stacking woven fabric or utilizing a 3-D woven structure. The preform is placed within the deposition chamber, and the desired product (oxide) is deposited on and within the heated preform from gaseous precursors. The deposition process is slow (up to days depending upon sample size) and is further hindered due to the filling of the outer portions of the preform, which inhibits infiltration within the preform. The reader is referred to reference [203] for a more in-depth discussion of this topic.

CVI SiC/SiC composites generally contain ~15% matrix porosity (~90–94% composite density), which is above the ~30% porosity estimated for matrix porosity enabled toughening, but it is still somewhat porous. The process itself holds promise for producing dense oxide-oxide composites; however, there are several problems associated with the process. Forming multi-component oxide matrices, such as mullite ($3Al_2O_3$-$2SiO_2$) or YAG ($3Y_2O_3$-$5Al_2O_3$), can be extremely difficult. Single component oxides may also be difficult to form, due to either a lack of proper precursor materials or to the stability of the deposited material. As stated, the deposition process is also slow, which adversely affects the cost. Due to the time constraints, CVD techniques are more practical for producing fiber coatings, rather than full composites [182].

6.4. Metal Oxidation Processing

6.4.1. Directed Metal Oxidation

This method is commonly referred to as the DIMOX (directed metal oxidation) process [205] and it has been primarily utilized with SiC-based reinforcements. In this technique, a composite preform is coated with a barrier material on five sides (assuming a rectangular shape) and the sixth side is held adjacent to a molten metal, generally aluminum. Oxidation of the metal occurs and the resultant oxide advances into the preform, filling voids. By adding dopants to the aluminum, such as magnesium, the formation of a protective oxide scale can be avoided, which enables the continual oxidation of the metal. Additional additives may be needed to ensure wetting of the fiber by the molten aluminum, while fiber coatings may be needed to prevent damaging interaction between the fiber and the molten metal. During this process, channels are formed within the matrix that allow for wicking of the metal through capillary action to the oxidation front. The resultant composite contains a matrix of aluminum + alumina; the majority of the aluminum can be removed through oxidation. A small amount of metal ($<5\%$) is typically left in the composite and this amount can be reasonably controlled.

Similarly, melt infiltration involves infiltrating a ceramic preform with a molten metal [206, 207]. In this case, the metal is melted and then is forced into a composite preform using pressure, vacuum wetting agents, etc. The fiber (or fabric) preform can be initially infiltrated with an oxide material to increase the density and provide strength to the preform. The wetting of the preform with the metal is critical and special care should be taken to properly prepare the sample. After infiltration, the metal is oxidized, which generally results in a pore-filling volume expansion.

6.4.2. Reaction Bonding

The reaction-bonding process has been increasingly investigated over the last decade, particularly due to its cost-effectiveness. It has been primarily applied to the formation of monoliths, but it has been successfully demonstrated with sapphire fiber reinforcements [125, 208] and with Nextel 610 fibers [209]. It is similar in nature to DIMOX; however, in this process, all of the metal can be oxidized, leaving no residual metallic phase. This process involves combining a metal powder, such as aluminum, with an oxide powder, such as alumina. The powders are attrition milled together to obtain the optimal size and shape of the metal powders, which enables complete oxidation of the metal without excess hydroxide production. The milling parameters are crucial to the final product. Undermilling results in larger metal particles that may not oxidize completely, whereas overmilling produces hydroxides on the surface of the powder which may not be removed until very high temperature, resulting in bloating due to gas formation.

The use of metal in forming a green composite allows for increased formability, since the metals can generally be plastically deformed around the fibers. Upon heat treatment in air, the metal oxidizes, and depending on the specific metal, will result in a given volume change. Ideally, volume expansion will fill porosity present in the sample, resulting in a higher density composite. The oxidation of the metal must be fully understood, however, to obtain the optimal results. Metastable or low temperature phases, formed upon oxidation, may convert upon exposure to higher temperatures to less desirable phases that lead to a lower density than predicted. Multi-component oxides, such as mullite, have been made with this process. As mentioned previously, there has been little application of this process to composite fabrication; however, it appears to merit further exploration.

VII. ENABLING THE NEXT GENERATION OF OXIDE-OXIDE COMPOSITES

The full potential of oxide-oxide composites has yet to be realized. There have been two major obstacles: higher temperature-capable fibers and oxidation resistant interface control that allows higher-strength matrices. The motivation to attack either of them vigorously is dependent upon expectations for the other. Improved fibers are not much use without effective interface control, for example. As discussed in an earlier section, recent advances in interface/coating technology look promising, at least for some composite chemistries. There has also been progress on improved fibers; however, it is likely that any concerted effort on fiber development will await the evolution of fiber coating technology to routine practice in current technology composites. The key issue then, will be whether there is sufficient payoff for the operating temperatures of current technology composites to justify the expense associated with the adoption of fiber coatings.

A positive note is that there are clear approaches to the development of higher temperature, fine diameter oxide fibers. For example, YAG is the most creep resistant oxide reported [54] and therefore a good candidate for a fiber, and mullite and zircon both have good creep resistance and low thermal expansivities. There are also microstructural approaches to improving creep resistance. It is hoped that improving prospects for all of the other constituents for higher temperature composites will motivate vigorous research on improved fibers.

The matrix-dominated properties of foreseeable ceramic composites are modest. Design of structural components will require careful allowance for these properties and careful placement of fibers in all loading directions. This implies a high degree of sophistication in component design and in many cases, designing and producing weavable 3-D fiber architectures. Processing composites with the additional fiber constraint implied by complex architectures will require a corresponding enhancement of processing sophistication. Ultimately these improvements will need to be accomplished without substantial cost penalties. That, in turn implies that these infrastructural needs may require significant commercial markets to provide the production volumes necessary to make them economically viable.

The ultimate high-temperature materials for long-term use are dense matrix, single-crystal reinforced oxide-oxide composites. They are the sole possessors of the potential to achieve all the key properties: environmental resistance, creep resistance and toughness. However, practical economical viability will await a breakthrough process for making single-crystal fibers, or an application so compelling that cost is a secondary consideration.

Finally, tough oxide-oxide composites useful at high temperatures, in some cases to $\sim 1200°C$, have been produced. Some of these are commercially available. The optimum use temperature of oxide-oxide composites is expected to increase with further research and development. There has been significant progress and the current limiting factor in performance is the availability of improved oxide fibers that will enable higher temperature ($>1300°C$), creep resistant oxide composites. However, insertion into many practical applications will also require increased sophistication in design, producing component-compatible fiber architectures and highly repeatable, economical processing.

ACKNOWLEDGMENTS

This work was performed while one of the authors (G. J.) held a National Research Council Research Associateship Award at the Air Force Research Laboratory Materials and Manufacturing Directorate. This work was performed by one of the authors (K. K.) under Air Force Contract #F33615-01-C-5214.

REFERENCES

1. B. O. Elfstrom, The Role of Advanced Materials in Aircraft Engines, *Nouvelle Revue D'Aeronautique et D'Astronautique*, [2] 81–85 (1998).
2. R. C. Robinson and J. L. Smialek, SiC Recession Caused by SiO2 Scale Volatility under Combustion Conditions: I, Experimental Results and Empirical Model, *J. Am. Ceram. Soc.*, **82** [7] 1817–1825 (1999).
3. N. S. Jacobson, Corrosion of Silicon-Based Ceramics in Combustion Environments, *J. Am. Ceram. Soc.*, **76** [1] 3–28 (1993).
4. C. G. Levi, F. W. Zok, J. Y. Yang, M. Mattoni, and J. P. A. Lofvander, Microstructural design of stable porous matrices for all-oxide ceramic composites., *Z. Metallkd.*, **90** [12] 1037–1047 (1999).
5. D. B. Marshall and J. B. Davis, Ceramics for future power generation technology: fiber reinforced oxide composites, *Current Opin. in Solid State and Mater. Sci.*, **5** 283–289 (2001).
6. A. G. Evans, D. B. Marshall, F. Zok, and C. Levi, Recent advances in oxide-oxide composite technology, *Advanced Composite Materials*, **8** [1] 17–23 (1999).
7. R. J. Kerans, R. S. Hay, and T. A. Parthasarathy, Structural Ceramic Composites, *Current Opinion in Solid State and Materials Science*, **4** 445–451 (1999).
8. D. Lewis III, Future Opportunities and Critical Needs for Advanced Ceramics and Ceramic Matrix Composites in Aerospace Applications, *Ceram. Eng. Sci. Proc.*, **21** [3] 3–14 (2000).
9. H. Kaya, The application of ceramic-matrix composites to the automotive ceramic gas turbine, *Comp. Sci. and Tech.*, **59** 861–872 (1999).
10. H. Ohnabe, S. Masaki, M. Onozuka, K. Miyahara, and T. Sasa, Potential application of ceramic matrix composites to aero-engine components, *Comp. Part A*, **30** 489–496 (1999).
11. Y. Liang and S. P. Dutta, Application trend in advanced ceramic technologies, *Technovation*, **21** 61–65 (2001).
12. R. John, L. P. Zawada, and J. L. Kroupa, Stresses due to temperature gradients in ceramic-matrix-composite aerospace components, *J. Am. Ceram. Soc.*, **82** [1] 161–168 (1999).
13. B. Jurf, J. Paretti, et. al., Fabrication and Testing of Insulated CMC Exhaust Pipe. Presented at 14th Advanced Aerospace Materials and Processes Conference. Dayton, OH, June 12, 2003.
14. J. A. Morrison and K. M. Krauth, Design and Analysis of a CMC Turbine Blade Tip Seal for a Land-Based Power Turbine, *Ceram. Eng. Sci. Proc.*, **19** [4] 249–256 (1998).
15. J. M. Staehler and L. Z. Zawada, Performance of four ceramic-matrix composite divergent flap inserts following ground testing on an F110 turbofan engine, *J. Am. Ceram. Soc.*, **83** [7] 1727–1738 (2000).
16. T. J. McMahon, Advanced Hot Gas Filter Development, *Ceram. Eng. Sci. Proc.*, **21** [3] 47–56 (2000).
17. M. G. Holmquist, T. C. Radsick, O. H. Sudre, and F. F. Lange, Fabrication and testing of all-oxide CFCC tubes, *Comp. Part A*, **34** 163–170 (2003).
18. L. J. Korb, C. A. Morant, R. M. Calland, and C. S. Thatcher, The shuttle orbiter thermal protection system, *Am. Ceram. Soc. Bull.*, **60** [11] 1188 (1981).
19. J. B. Davis, D. B. Marshall, K. S. Oka, R. M. Housley, and P. E. D. Morgan, Ceramic composites for thermal protection systems, *Comp. Part A*, **30** 483–488 (1999).
20. R. J. Kerans, R. S. Hay, T. A. Parthasarathy, and M. K. Cinibulk, Interface design for oxidation-resistant ceramic composites, *J. Am Ceram. Soc.*, **85** [11] 2599–2632 (2002).
21. M.-Y. He, A. G. Evans, and J. W. Hutchinson, Crack deflection at an interface between dissimilar elastic materials: role of residual stresses, *Int. J. Solids Struct.*, **31** [24] 3443–3455 (1994).
22. M.-Y. He and J. W. Hutchinson, Crack deflection at an interface between dissimilar elastic materials, *Int. J. Solids Struct.*, **25** [9] 1053–1067 (1989).
23. R. J. Kerans and T. A. Parthasarathy, Crack Deflection in Ceramic Composites and Fiber Coating Design Criteria, *Composites Part A*, **30A** [4] 521–24 (1999).
24. K. T. Faber, Ceramic Composite Interfaces: Properties and Design, *Annu. Rev. Mater. Sci.*, **27** 499–524 (1997).
25. R. J. Kerans, The Role of Coating Compliance and Fiber/Matrix Interfacial Topography on Debonding in Ceramic Composites, *Scr. Met. Mat.*, **32** [4] 505–509 (1995).
26. R. W. Goettler, S. Sambasivan, V. Dravid, and S. Kim, Interfaces in Oxide Fiber – Oxide Matrix Ceramic Composites, in *Computer Aided Design of High Temperature Materials*, A. Pechenik, R. Kalia, and P. Vashishta, Eds., Oxford University Press, (1999). p. 333–349.
27. K. A. Keller, T.-I. Mah, T. A. Parthasarathy, E. E. Boakye, P. Mogilevsky, and M. K. Cinibulk Effectiveness of monazite coatings in oxide/oxide composites after long-term exposure at high temperature, *J. Am. Ceram. Soc.*, **86** [2] 325–332 (2003).

28. G. H. Cullum, Sol-Gel Processing of Ceramic Composites, in *Ceramics and Ceramic Matrix Composites*, S. R. Levine, Ed., American Society of Mechanical Engineers, New York, (1992). p. 139–150.
29. G. H. Cullum, Ceramic Matrix Composite Fabrication and Processing: Sol-Gel Infiltration, in *The Handbook on Continuous Fiber-Reinforced Ceramic Matrix Composites*, R. Lehman, S. El-Rahaiby, and J. Wachtman, Eds., Purdue Research Foundation, (1995).
30. L. P. Zawada, R. S. Hay, S. S. Lee, and J. Staehler, Characterization and high-temperature mechanical behavior of an oxide/oxide composite, *J. Am. Ceram. Soc.*, **86** [6] 981–990 (2003).
31. T. J. Dunyak, D. R. Chang, and M. L. Millard. Thermal Aging Effects in Oxide/Oxide Ceramic Matrix Composites, NASA. Conference Publication, [No. 3235, pt. 2] pp. 675–89 (1994).
32. T. J. Lu, Crack Branching in All-Oxide Ceramic Composites, *J. Am. Ceram. Soc.*, **79** [1] 266–74 (1996).
33. W.-C. Tu, F. F. Lange, and A. G. Evans, Concept for a Damage-Tolerant Ceramic Composite with "Strong" Interfaces, *J. Am. Ceram. Soc.*, **79** [2] 417–24 (1996).
34. A. G. Hegedus, Ceramic Bodies of Controlled Porosity and Process for Making Same. U.S. Patent No. 5,017,522, May 21, 1991. Hexcel Corporation: USA.
35. M. G. Harrison, M. L. Millard, and A. Szweda, Consolidated Member and Method and Preform for Making. U.S. Patent No. 5,306,554, April 26, 1994. General Electric Corporation, USA.
36. A. Szweda, M. L. Millard, and M. G. Harrison, Fiber Reinforced Ceramic Composite Member, U.S. Patent no. 5,488,017 Jan 30, 1996. General Electric Co.: USA.
37. A. Szweda, M. L. Millard, and M. G. Harrison, Fiber reinforced ceramic matrix composite member and method for making U.S. Patent No. 5,601,674. Feb 11, 1997. General Electric Company, USA.
38. L. P. Zawada, Longitudinal and Transthickness Tensile Behavior of Several Oxide/Oxide Composites, *Cer. Eng. Sci. Proc.*, **19** [3] 327–339 (1998).
39. C. G. Levi, J. Y. Yang, B. J. Dalgleish, F. W. Zok, and A. G. Evans, Processing and Performance of an All-Oxide Ceramic Composite, *J. Am. Ceram. Soc.*, **81** [8] 2077–2086 (1998).
40. F. F. Lange, W. C. Tu, and A. G. Evans, Processing of Damage Tolerant, Oxidation Resistant Ceramic-Matrix Composites, *Mater. Sci. Eng.*, **A195** 145–150 (1995).
41. B. N. Cox and F. W. Zok, Advances in Ceramic Composites Reinforced by Continuous Fibers, *Current Opinion in Solid State & Materials Science*, **1** [5] 666–73 (1996).
42. A. G. Evans, The mechanical properties of reinforced ceramic, metal and intermetallic matrix composites, *Mater. Sci. Eng., A*, **143** [1–2] 63–76 (1991).
43. A. G. Evans, F. W. Zok, and J. B. Davis, The Role of Interfaces in Fiber-Reinforced Brittle Matrix Composites, *Compos. Sci. Technol.*, **42** 3–24 (1991).
44. M. A. Mattoni, J. Y. Yang, C. G. Levi, and F. W. Zok, Effects of matrix porosity on the mechanical properties of a porous-matrix, all-oxide ceramic composite". *J. Am. Ceram. Soc.*, **84** [11] 2594–2602 (2001).
45. 3M Nextel Ceramic Textiles Technical Notebook. 3M Ceramic Fibers and Textiles, St Paul MN. (2001).
46. D. M. Wilson. High Temperature Oxide Fibers. *105th Annual Meeting of the American Ceramic Society*. Nashville, TN, April 29, 2003.
47. M. Koopman, S. Duncan, K. K. Chawla, and C. Coffin, Processing and characterization of barium zirconate coated alumina fibers/alumina matrix composites, *Comp. Part A*, **32** 1039–1044 (2001).
48. M. Holmquist, R. Lundberg, O. Sudre, A. G. Razzell, L. Molliex, J. Benoit, and J. Adlerborn, Alumina/alumina composite with a porous zirconia interphase – Processing, properties and component testing, *J. Eur. Ceram. Soc.*, **20** 599–606 (2000).
49. Properties and Benefits of Sapphire: A Quick Reference Guide. Saphikon, Inc., Milford, NH.(2003).
50. S. A. Newcomb and R. E. Tressler, Slow Crack Growth in Sapphire Fibers at 800C to 1500C, *J. Am. Ceram. Soc.*, **76** [10] 2505–2512 (1993).
51. D. W. Richerson, Ceramic Matrix Composites, in *Composites Engineering Handbook*, P. K. Mallick Ed., Marcel Dekker, New York (1997). p. 983–1038.
52. S. T. Mileiko, V. I. Kazmin, V. M. Kiiko, and A. M. Rudnev, Oxide/oxide composites produced by the internal crystallization method, *Comp. Sci. and Tech.*, **57** 1363–1367 (1997).
53. A. A. Kolchin, V. M. Kiiko, N. S. Sarkissyan, and S. T. Mileiko, Oxide/oxide composites with fibres produced by internal crystallisation, *Comp. Sci. and Tech.*, **61** 1079–1082 (2001).
54. G. S. Corman, Creep of Yttrium Aluminum Garnet Single Crystals, *J. Mat. Sci. Let.*, **12** 379–382 (1993).
55. D. M. Wilson and L. R. Visser, High performance oxide fibers for metal and ceramic composites, *Comp. Part A*, **32** 1143–1153 (2001).

56. G. N. Morscher, K. C. Chen, and K. S. Mazdiyasni, Creep Resistance of Developmental Polycrystalline Yttrium-Aluminum Garnet Fibers, *Cer. Eng. Sci. Proc.*, **15** [4] 181–188 (1994).
57. B. H. King and J. W. Halloran, Polycrystalline Yttrium Aluminum Garnet Fibers from Colloidal Sols, *J. Am. Ceram. Soc.*, **78** [8] 2141–2148 (1995).
58. T. Mah, T. A. Parthasarathy, D. Petry, and L. E. Matson, Processing, Structure and Properties of Alumina-YAG Eutectic Fibers, *Ceram. Eng. & Sci. Proc.*, **14** [7–8] 622–638 (1993).
59. S. C. Farmer and A. Sayir, Tensile strength and microstructure of Al_2O_3-ZrO_2 hypo-eutectic fibers, *Eng. Fract. Mech.*, **69** 1015–1024 (2002).
60. J. M. Yang and X. Q. Zhu, Thermo-Mechanical Stability of Directionally Solidified Al_2O_3-ZrO_2 (Y_2O_3) Eutectic Fibers, *Scripta Mater.*, **36** [9] 961–966 (1997).
61. C. H. Ruscher, S. T. Mileiko, and H. Schneider, Mullite single crystal fibres produced by the internal crystallization method (ICM), *J. Eur. Ceram. Soc.*, **23** 3113–3117 (2003).
62. S. T. Peters and Y. M. Tarnopol'skii, Filament Winding, in *Composites Engineering Handbook*, P. K. Mallick Ed., Marcel Dekker, New York (1997). p. 515–548.
63. A. R. Bhatti and P. M. Farries, Preparation of Long-fiber-reinforced Dense Glass and Ceramic Matrix Composites, in *Carbon/carbon, Cement and Ceramic Matrix Composites*, R. Warren, Ed., Elsevier Science Ltd., Oxford, (2000). p. 645–667.
64. E. H. Moore, C. A. Folsom, K. A. Keller, and T. Mah, 3D Composite Fabrication through Matrix Slurry Pressure Infiltration, *Ceram. Eng. Sci. Proc.*, **15** [4] 113–120 (1994).
65. S. M. Sim and R. J. Kerans, Slurry Infiltration of 3-D Woven Composites, *Ceram. Eng. Sci. Proc.*, **13** 632–641 (1992).
66. A. I. Kingdon, R. F. Davis, and M. M. Thackeray, Engineering properties of multicomponent and multiphase oxides, in *Engineered Materials Handbook, Vol. 4, Ceramics and Glasses*, ASM international, Metals City, OH (1991). p. 758–774.
67. M. Miyayama, K. Koumoto, and H. Yanagida, Engineering properties of single oxides, in *Engineered Materials Handbook, Vol. 4, Ceramics and Glasses*, ASM international, Metals City, OH (1991). p. 748–757.
68. J. J. Haslam, K. E. Berroth, and F. F. Lange, Processing and properties of an all-oxide composite with a porous matrix, *J. Eur. Ceram. Soc.*, **20** 607–618 (2000).
69. M. Schmucker, A. Grafmuller, and H. Schneider, Mesostructure of WHIPOX all oxide CMCs, *Comp. Part A*, **34** 613–622 (2003).
70. H. Schneider, J. Goring, B. Kanka, and M. Schmucker, WHIPOX: a new oxide fibre/oxide matrix composite for high-temperature applications, *Keram. Zeit.*, **53** [9] 788–791 (2001).
71. A. Souto and F. Guitian, Purification of mullite by reduction and volatilization of impurities, *J. Am. Ceram. Soc.*, **82** [10] 2660–2664 (1999).
72. N. S. Jacobson, E. J. Opila, and K. N. Lee, Oxidation and corrosion of ceramics and ceramic matrix composites, *Curr. Opin. Solid State Mater. Sci.*, **5** 301–309 (2001).
73. K. N. Lee, N. S. Jacobson, and R. A. Miller, Refractory oxide coatings on SiC ceramics, *MRS Bull.*, **19** [10] 35–38 (1994).
74. J. A. Haynes, M. J. Lance, K. M. Cooley, M. K. Ferber, R. A. Lowden, and D. P. Stinton, CVD mullite coatings in high-temperature, high-pressure air-H_2O, *J. Am. Ceram. Soc.*, **83** [3] 657–659 (2000).
75. S. C. Butner, personal communication.
76. L. P. Zawada, J. Staehler, and S. G. Steel, Consequence of intermittent exposure to moisture and salt fog on the high-temperature fatigue durability of several ceramic matrix composites, *J. Am. Ceram. Soc.*, **86** [8] 1282–1291 (2003).
77. J. T. Kretchik, Regulatory Forecast, *Chemical Health and Safety*, **10** [4] 36 (2003).
78. K. D. Rosenman, M. J. Reilly, and C. Rice, Silicosis among foundry workers: Implication for the need to revise the OSHA standard, *Occ. Health & Indust. Med.*, **36** [1] 40 (1997).
79. S. G. Steel, L. P. Zawada, and S. Mall, Fatigue behavior of a Nextel(tm) 720/alumina (N720/A) composite at room temperature, *Ceram. Eng. Sci. Proc.*, **22** [3] 695–702 (2001).
80. M. K. Cinibulk, K. Keller, T. Mah, and T. A. Parthasarathy, Nextel 610 and 650 Fiber Reinforced Porous Alumina-YAG Matrix Composites, *Ceram. Eng. Sci. Proc.*, **22** [3] 677–686 (2001).
81. M. K. Cinibulk, K. A. Keller, T. Mah, and T. A. Parthasarathy, Nextel 610 Fiber-Reinforced Alumina-YAG Porous Matrix Composites, *Ceram. Eng. Sci. Proc.*, **23** [3] 629–636 (2002).
82. J. D. French, J. Zhao, M. P. Harmer, H. M. Chan, and G. A. Miller, Creep of Duplex Microstructures, *J. Am. Ceram. Soc.*, **77** [11] 2857–2865 (1994).

83. N. P. Bansal and J. I. Eldridge, Effects of Interface Modification on Mechanical Behavior of Hi-Nicalon Fiber-Reinforced Celsian Matrix Composites, *Ceram. Eng. Sci. Proc.*, **18** [3] 379–389 (1997).
84. R. W. Goettler, S. Sambasivan, and V. P. Dravid, Isotropic Complex Oxides as Fiber Coatings for Oxide-Oxide CFCC, *Ceram. Eng. Sci. Proc.*, **18** [3] 279–286 (1997).
85. M. H. Lewis, A. Tye, and et al., Development of Interfaces in Oxide Matrix Composites, *Key Eng. Mater.*, **164–165** 351–356 (1999).
86. P. E. D. Morgan, D. B. Marshall, and R. M. Housley, High Temperature Stability of Monazite-Alumina Composites, *Mat. Sci. Eng.*, **A195** 215–222 (1995).
87. P. E. D. Morgan and D. B. Marshall, Ceramic Composites of Monazite and Alumina, *J. Am. Cer. Soc.*, **78** [6] 1553–63 (1995).
88. T. A. Parthasarathy, E. Boakye, M. K. Cinibulk, and M. D. Petry, Fabrication and Testing of Oxide/Oxide Microcomposites with Monazite and Hibonite as Interlayers, *J. Am. Ceram. Soc.*, **82** [12] 3575–3583 (1999).
89. K. K. Chawla, H. Liu, J. Janczak-Rusch, and S. Sambasivan, Microstructure and properties of monazite ($LaPO_4$) coated saphikon fiber/alumina matrix composites, *J. Eur. Ceram. Soc.*, **20** 551–559 (2000).
90. D. H. Kuo and W. M. Kriven, Chemical stability, microstructure and mechanical behavior of $LaPO_4$-containing ceramics, *Mat. Sci. Eng.*, **A210** 123–134 (1996).
91. R. L. Callender and A. R. Barron, Formation and evaluation of highly uniform aluminate interface coatings for sapphire fiber reinforced ceramic matrix composites (FRCMCs) using carboxylate-alumoxane nanoparticles, *J. Mater. Sci.*, **36** 4977–4987 (2001).
92. D.-H. Kuo, W. M. Kriven, and T. J. Mackin, Control of Interfacial Properties through Fiber Coatings: Monazite Coatings in Oxide-Oxide Composites, *J. Am. Ceram. Soc.*, **80** [12] 2987–2996 (1997).
93. K. A. Keller, T. Mah, E. E. Boakye, T. A. Parthasarathy, and P. Mogilevsky. Evaluation of Dense Monazite Fiber-Coatings in Oxide-Oxide Minicomposites. Presented at the *27th Annual Cocoa Beach Conference & Exposition on Advanced Ceramics & Composites*. Cocoa Beach, FL, January 29, 2003, 2003.
94. D. B. Marshall and W. C. Oliver, Measurement of Interfacial mechanical Properties in Fiber-Reinforced Ceramic Composites, *J. Am. Ceram. Soc.*, **70** [8] 542–48 (1987).
95. R. J. Kerans and T. A. Parthasarathy, Theoretical Analysis of the Fiber Pullout and Pushout Tests, *J. Am. Ceram. Soc.*, **74** [7] 1585–1596 (1991).
96. T. A. Parthasarathy, P. D. Jero, and R. J. Kerans, Extraction of Interface Properties from the Fiber Push-out Test, *Scripta Metall. et Mater.*, **25** [11] 2457–2462 (1991).
97. T. A. Parthasarathy, D. B. Marshall, and R. J. Kerans, Analysis of the Effect of Interfacial Roughness on Fiber Debonding and Sliding in Brittle Matrix Composites, *Acta Met.*, **42** [11] 3773–3784 (1994).
98. P. D. Jero, R. J. Kerans, and T. A. Parthasarathy, Effect of Interfacial Roughness on the Frictional Stress Measured Using Pushout Tests, *J. Am. Cer. Soc.*, **74** [11] 2793–2801 (1991).
99. J. B. Davis, R. S. Hay, D. B. Marshall, P. E. D. Morgan, and A. Sayir, Influence of Interfacial Roughness on Fiber Sliding in Oxide Composites with La-Monazite Interphases., *J. Am. Ceram. Soc.*, **86** [2] (2003).
100. R. S. Hay, Monazite and Scheelite Deformation Mechanisms, *Cer. Eng. Sci. Proc.*, **21** [4] 203–218 (2000).
101. K. A. Keller, T. Mah, E. E. Boakye, T. A. Parthasarathy, and P. Mogilevsky. Evaluation of Nextel 610/Monazite/Alumina Composites at High Temperature. Presented at the *105th Annual Meeting of The American Ceramic Society*. Nashville, TN, April 30, 2003.
102. A. Cazzato, M. Colby, D. Daws, J. Davis, P. Morgan, J. Porter, S. Butner, and B. Jurf, Monazite Interface Coatings in Polymer and Sol-Gel derived Ceramic Matrix Composites, *Ceram. Eng. Sci. Proc.*, **18** [3] 269–278 (1997).
103. J. B. Davis, D. B. Marshall, and P. E. D. Morgan, Monazite Containing Oxide-Oxide Composites, *J. Eur. Ceram. Soc.*, **19** 2421–2426 (1999).
104. E. E. Boakye, R. S. Hay, P. Mogilevsky, and L. M. Douglas, Monazite Coatings on Fibers: II, Coating Without Strength Degradation, *J. Am. Ceram. Soc*, **84** [12] 2793–2801 (2001).
105. J. B. Davis, D. B. Marshall, and P. E. D. Morgan, Oxide Composites of Al_2O_3 and $LaPO_4$, *J. Eur. Ceram. Soc.*, **19** 2421–2426 (1999).
106. J. B. Davis, D. B. Marshall, R. M. Housley, and P. E. D. Morgan, Machinable Ceramics Containing Rare-Earth Phosphates, *J. Am. Ceram. Soc.*, **81** [8] 2169–2175 (1998).
107. C. Kaya, E. G. Butler, A. Selcuk, A. R. Boccaccini, and M. H. Lewis, Mullite (Nextel 720) fibre-reinforced mullite matrix composites exhibiting favourable thermomechancial properties, *J. Eur. Ceram. Soc.*, **22** 2333–2342 (2002).

108. E. Boakye, R. S. Hay, M. D. Petry, and T. A. Parthasarathy, Sol-Gel Synthesis of Zircon-Carbon Precursors and Coatings of Nextel 720 Fiber Tows, *Cer. Eng. Sci. Proc. A*, **20** [3] 165–172 (1999).
109. M. K. Cinibulk, T. A. Parthasarathy, K. A. Keller, and T. Mah, Porous Rare-Earth Aluminate Fiber Coatings for Oxide-Oxide Composites, *Ceram. Eng. Sci. Proc.*, **21** [4] 219–228 (2000).
110. J. B. Davis, J. P. A. Lofvander, A. G. Evans, E. Bischoff, and M. L. Emiliani, Fiber Coating Concepts for Brittle Matrix Composites, *J. Am. Cer. Soc.*, **76** [5] 1249–1257 (1993).
111. G. N. Morscher, D. R. Bryant, and R. E. Tressler, Environmental Durability of BN-Based Interphases (for SiCf-SiCm Composites) in H_2O-Containing Atmospheres at Intermediate Temperatures, *Ceram. Eng. Sci. Proc.*, **18** [3] 525 (1997).
112. E. Lara-Curzio, P. F. Tortorelli, and K. L. More, Stress-Rupture of Nicalon(TM)/SiC at Intermediate Temperatures, *Ceram. Eng. Sci. Proc.*, **18** [4] 209–219 (1997).
113. A. G. Evans, F. W. Zok, R. M. McMeeking, and Z. Z. Du, Models of High Temperature, Environmentally Assisted Embrittlement in Ceramic-Matrix Composites, *J. Am. Ceram. Soc.*, **79** [9] 2345–52 (1996).
114. K. A. Keller, T. Mah, C. Cooke, and T. A. Parthasarathy, Fugitive Interfacial Carbon Coatings for Oxide/Oxide Composites, *J. Am. Ceram. Soc.*, **83** [2] 329–36 (2000).
115. E. Lara-Curzio, M. Ferber, and R. A. Lowden, Effect of fiber coating thickness on the interfacial properties of a continuous fiber ceramic matrix composite, *Cer. Eng. Sci. Proc.*, **15** [5] 989–1003 (1994).
116. O. Sudre, A. G. Razzell, L. Molliex, and M. Holmquist, Alumina Single-Crystal Fibre Reinforced Alumina Matrix for Combustor Tiles, *Ceram. Eng. Sci. Proc.*, **19** [4] 273–280 (1998).
117. B. Saruhan, M. Schmucker, M. Bartsch, H. Schneider, K. Nubian, and G. Wahl, Effect of interphase characteristics on long-term durability of oxide-based fibre-reinforced composites, *Comp. Part A*, **32** 1095–1103 (2001).
118. K. Nubian, B. Saruhan, B. Kanka, M. Schmucker, H. Schneider, and G. Wahl, Chemical vapor deposition of ZrO_2 and C/ZrO_2 on mullite fibers for interfaces in mullite/aluminosilicate fiber-reinforced composites, *J. Eur. Ceram. Soc.*, **20** 537–544 (2000).
119. P. W. M. Peters, B. Daniels, F. Clemens, and W. C. Vogel, Mechanical characterization of mullite-based ceramic matrix composites at test temperatures up to 1200 C, *J. Eur. Ceram. Soc.*, **20** 531–535 (2000).
120. K. Chyung and S. B. Dawes, Fluoromica Coated Nicalon Fiber Reinforced Glass-Ceramic Composites, *Mat. Sci. Eng.*, **A162** 27–33 (1993).
121. G. Demazeau, New Synthetic Mica-like Materials for Controlling Fracture in Ceramic Matrix Composites, *Mat. Tech.*, **10** 43–58 (1995).
122. M. K. Cinibulk and R. S. Hay, Textured Magnetoplumbite Fiber-Matrix Interphase Derived from Sol-Gel Fiber Coatings, *J. Am. Ceram. Soc.*, **79** [5] 1233–1246 (1996).
123. G. Fair, M. Shemkunas, W. T. Petuskey, and S. Sambasivan, Layered Perovskites as 'Soft-ceramics', *J. Eur. Cer. Soc.*, **19** 2437–2447 (1999).
124. M. H. Jaskowiak, W. H. Philipp, L. C. Vetch, and J. B. Hurst, Platinum Interfacial Coatings for Sapphire/Al_2O_3 Composites, *Cer. Eng. Sci. Proc.*, **13** 589–598 (1992).
125. J. Wendorff, R. Janssen, and N. Claussen, Platinum as a Weak Interphase for Fiber-Reinforced Oxide-Matrix Composites, *J. Am. Ceram. Soc.*, **81** [10] 2738–2740 (1998).
126. T. A. Parthasarathy, Effect of Segregation-Induced Interface Weakening on Fiber Pullout in a Sapphire/YAG CMC, *Air Force Research Laboratory*, [unpublished] (1994).
127. M. G. Jenkins and S. S. Kohles, High-Temperature Performance and Retained Strength of an Oxide-Oxide Continuous Fiber Ceramic Composite, *Ceram. Eng. Sci. Proc.*, **19** [3] 317–325 (1998).
128. R. S. Hay, The Use of Solid-State Reactions with Volume Loss to Engineer Stress and Porosity into the Fiber-Matrix Interface of a Ceramic Composite, *Acta Met.*, **43** [9] 3333–3348 (1995).
129. Z. Chen, S. Duncan, K. K. Chawla, M. Koopman, and G. M. Janowski, Characterization of interfacial reaction products in alumina fiber/barium zirconate coating/alumina matrix composite, *Mat. Char.*, **48** 305–314 (2002).
130. T. Mamiya, H. Kakisawa, W. H. Liu, S. J. Zhu, and Y. Kagawa, Tensile damage evolution and notch sensitivity of Al_2O_3 fiber-ZrO_2 matrix minicomposite-reinforced Al_2O_3 matrix composites, *Mat. Sci. Eng. A*, **A325** 405–413 (2002).
131. P. Chivavibul and M. Enoki, Effect of gage length on tensile strength and failure strain of woven fabric Al_2O_3 fiber-ZrO_2 minicomposite-reinforced Al_2O_3 matrix composite, *J. Mater. Sci. Lett.*, **22** 495–498 (2003).
132. L. Pejryd, R. Lundberg, and E. Butler, Ceramic Composite, Particularly for Use at Temperatures Above 1400 Degrees Celsius, U.S. Patent No. 5,567,518, Oct. 22, 1996. Volvo Aero Corporation: USA.

133. S. Shanmugham, D. P. Stinton, F. Rebillat, A. Bleier, T. M. Besmann, E. Lara-Curzio, and P. K. Liaw, Oxidation-Resistant Interfacial Coatings for Continuous Fiber Ceramic Composites, *Cer. Eng. Sci. Proc.*, **16** [4] 389–399 (1995).
134. J. L. Stempin and D. R. Wexell, Fiber Reinforced Ceramic Matrix Composites Exhibiting Improved High-Temperature Strength. U.S. Patent No. 5,422,319, June 6, 1995. Corning, Inc.: USA.
135. S.-K. Lau and C. H. McMurty, Aluminum Nitride-Coated Silicon Carbide Fiber. U.S. Patent No. 5,484,655, Jan. 16, 1996. The Carborundum Company: USA.
136. L. E. Carpenter, Single and Multilayer Coatings Containing Aluminum Nitride. U.S. Patent No. 5,183,684, Feb. 2, 1993. Dow Corning Corporation: USA.
137. L. C. Lev and A. S. Argon, Development of Oxide Coatings for Matching Oxide Fiber-Oxide Matrix Composites, *Cer. Eng. Sci. Proc.*, **15** [5] 743–752 (1994).
138. W. Y. Lee, E. Lara-Curzio, and K. L. More, Multilayered Oxide Interphase Concept for Ceramic-Matrix Composites, *J. Am. Ceram. Soc.*, **81** [3] 717–720 (1998).
139. M. K. Brun, R. A. Giddings, and S. Prochazka, Silicon Carbide Composite with Metal Boride Coated Fiber Reinforcement. U.S. Patent No. 5,316,851, May 31, 1994. General Electric Comp.: USA.
140. Oxide-Oxide CMC Data Sheets. COI Ceramics, Inc., San Diego, CA. (2003).
141. R. A. Jurf and S. C. Butner, Advances in all-oxide CMC, *J. Eng. Gas Turbines Power*, **122** 202–205 (2000).
142. E. A. V. Carelli, H. Fujita, J. Y. Yang, and F. W. Zok, Effects of thermal aging on the mechanical properties of a porous-matrix ceramic composite, *J. Am. Ceram. Soc.*, **85** [3] 595–602 (2002).
143. J. A. Heathcote, X. Gong, J. Y. Yang, U. Ramamurty, and F. W. Zok, In-plane mechanical properties of an all-oxide ceramic composite, *J. Am. Ceram. Soc.*, **82** [10] 2721–2730 (1999).
144. S. Butner and J. Pierce. Processing and properties of a Nextel 720 alumina composite fabricated by composite wet lay-up. *Presented at the 26th Annual Conference on Composites, Materials, and Structures*. Cocoa Beach Florida, Jan 28–Feb 1, 2002.
145. H. Fugita, G. Jefferson, R. M. McMeeking, and F. W. Zok, Mullite/alumina mixtures for use as porous matrices in oxide fiber composites, *J. Am. Ceram. Soc.*, **87** [2] 261–267 (2004).
146. G. P. Tandon, D. J. Buchanan, N. J. Pagano, and R. John, Analytical and experimental characterization of thermo-mechanical properties of a damaged woven oxide-oxide composite, *Ceram. Eng. Sci. Proc.*, **22** [3] 687–694 (2001).
147. A. G. Evans and F. W. Zok, The physics and mechanics of fibre-reinforced brittle matrix composites, *J. Mater. Sci.*, **29** [15] 3857–3896 (1994).
148. J. C. McNulty and F. W. Zok, Application of weakest-link fracture statistics to fiber-reinforced ceramic-matrix composites, J. Am. Ceram. Soc., **80** [6] 1535–1543 (1997).
149. U. Ramamurty, F. W. Zok, F. A. Leckie, and H. E. Deve, Strength variability in alumina fiber reinforced aluminum matrix composites, *Acta Mater.*, **45** [11] 4603–4613 (1997).
150. N. A. Weil and I. M. Daniel, Analysis of fracture probabilities in nonuniformly stressed brittle materials, *J. Am. Ceram. Soc.*, **47** [6] 268–274 (1964).
151. R. Pipes, R. Blake, J. Gillespie, and L. Carlsson, *Delaware Composites Design Encyclopedia, Vol 6, Test Methods*. Technomic, Lancaster, Pa (1990).
152. N. K. Naik and R. S. Kumar, Compressive strength of unidirectional composites : evaluation and comparison of prediction models, *Compos. Struct.*, **46** [3] 299–308 (1999).
153. S. Y. Hsu, T. J. Volger, and S. Kyriakides, Compressive strength predictions for fiber composites, *J. Appl. Mech.*, **65** [1] 7–16 (1998).
154. D. M. Wilson, S. L. Lieder, and D. C. Lueneburg, Microstructure and High Temperature Properties of Nextel 720 Fibers, *Ceram. Eng. Sci. Proc.*, **16** [5] 1005–1014 (1995).
155. D. M. Wilson, D. C. Lueneburg, and S. L. Lieder, High Temperature Properties of Nextel 610 and Alumina-Based Nanocomposite Fibers, *Ceram. Eng. Sci. Proc.*, **14** [7–8] 609–621 (1993).
156. L. P. Zawada and S. S. Lee, The effect of hold times on the fatigue behavior of an oxide/oxide ceramic composite, in *Thermal and Mechanical Test Methods and Behavior of Continuous-Fiber Ceramic Composites*, M. G. Jenkins, et al., Eds., American Society for Testing and Materials, West Conshohocken, PA, (1997).
157. L. P. Zawada and S. S. Lee, Evaluation of Four CMC's for Aerospace Turbine Engine Divergent Flaps and Seals, *Ceram. Eng. and Sci. Proc.*, **16** [4] 337–339 (1995).
158. R. John, D. J. Buchanan, and L. P. Zawada, Notch-sensitivity of a woven oxide/oxide ceramic matrix composite, in *Mechanical, Thermal and Environmental Testing and Performance of Ceramic Composites and*

Components, M. G. Jenkins, E. Lara-Curzio, and S. T. Gonczy, Eds., American Society for Testing and Materials, West Conshohocken, PA (2000). p. 172–181.
159. R. John, D. J. Buchanan, and L. P. Zawada, Creep deformation and rupture behavior of a notched oxide/oxide Nextel720/AS composite, *Ceram. Eng. Sci. Proc.*, **21** [3] 567–574 (2000).
160. W. R. Cannon and T. G. Langdon, Review creep of ceramics: part 1 mechanical characteristics, *J. Mater. Sci.*, **18** [1] 1–50 (1983).
161. S. V. Nair and J. L. Bassani, Macro- and micromechanics of elevated temperature crack growth in ceramic composites, in *High Temperature Mechanical Behavior of Ceramic Composites*, S.V. Nair and K. Jakus, Eds., Butterworth-Heinemann, Boston (1995). p. 437–470.
162. J. R. Zuiker, A model for the creep response of oxide-oxide ceramic matrix composites, in *Thermal and Mechanical Test Methods and Behavior of Continuous-Fiber Ceramic Composites*, M. G. Jenkins, et al., Eds., American Society for Testing and Materials, West Conshohocken, PA, (1997).
163. R. M., McMeeking, Models for the creep of ceramic matrix composite materials, in *High Temperature Mechanical Behavior of Ceramic Composites*, S. V. Nair and K. Jakus, Eds., Butterworth-Heinemann, Boston, (1995). p. 409–436.
164. S. P. Timoshenko and J. N. Goodier, *Theory of Elasticity*. McGraw-Hill, New York (1970).
165. S. C. Tan, *Stress Concentrations in Laminated Composites*. Technomic, Lancaster, PA (1994).
166. D. J. Buchanan, V. A. Kramb, R. John, and L. P. Zawada, Effect of small effusion holes on creep rupture behavior of oxide/oxide Nextel™720/AS composite, *Ceram. Eng. Sci. Proc.*, **22** [3] 659–666 (2001).
167. R. John, D. J. Buchanan, V. A. Kramb, and L. P. Zawada, Creep rupture behavior of oxide/oxide Nextel720/AS and MI SiC/SiC composite with effusion holes, *Ceram. Eng. Sci. Proc.*, **23** [3] 617–628 (2002).
168. J. M. Whitney and R. J. Nuismer, Stress fracture criteria for laminated composites containing stress concentrations, *J. Compos. Mater.*, **8** [3] 253–265 (1974).
169. J. C. McNulty, F. W. Zok, G. M. Genin, and A. G. Evans, Notch-sensitivity of fiber-reinforced ceramic-matrix composites: effects of inelastic straining and volume dependent strength, *J. Am. Ceram. Soc.*, **82** [5] 1217–1228 (1999).
170. S. G. Steel, Monotonic and Fatigue Loading Behavior of an Oxide/Oxide Ceramic Matrix Composite, Masters Thesis, Air Force Institute of Technology, Wright Patterson Air Force Base, OH. (2000).
171. M. Kumosa, G. Odegard, D. Armentrout, L. Kumosa, K. Searles, and J. K. Sutter, Comparison of the +/−45 tensile and Iosipescu shear tests for woven fabric composite materials, *J. Compos. Technol. Res.*, **24** [1] 3–16 (2002).
172. L. P. Zawada and K. E. Goecke, Testing methodology for measuring transthickness tensile strength for ceramic matrix composites, in *Mechanical, Thermal and Environmental Testing and Performance of Ceramic Composites and Components*, M. G. Jenkins, E. Lara-Curzio, and S. T. Gonczy, Eds., American Society for Testing and Materials, West Conshohocken, PA (2000). p. 62–85.
173. G. Odegard, D. Armentrout, K. Searles, L. Kumosa, J. K. Sutter, and M. Kumosa, Failure analysis of +/−45 off-axis woven fabric composite specimens, *J. Compos. Technol. Res.*, **23** [3] 205–224 (2001).
174. S. W. Tsai and E. M. Wu, A general theory of strength for anisotropic materials, *J. Compos. Mater.*, **5** 58–80 (1971).
175. L. J. Hart-Smith, The role of biaxial stresses in discriminating between meaningful and illusory composite failure theories, *Compos. Struct.*, **25** 3–20 (1993).
176. D. C. C. Lam and F. F. Lange, Microstructual Observations on Constrained Densification of Alumina Powder Containing a Periodic Array of Sapphire Fibers, *J. Am. Ceram. Soc.*, **77** [7] 1976–1978 (1994).
177. M. N. Rahaman, *Ceramic Processing and Sintering*. 1st ed. Marcel-Dekker, Inc., New York (1995). 770.
178. D. B. Gundel, P. J. Taylor, and F. E. Wawner, Fabrication of Thin Oxide Coatings on Ceramic Fibres by a Sol-Gel Technique, *J. Mater. Sci.*, **29** 1795–1800 (1994).
179. R. S. Hay and E. E. Hermes, Sol-Gel Coatings on Continuous Ceramic Fibers, *Cer. Eng. Sci. Proc.*, **11** [9–10] 1526–1532 (1990).
180. H. Dislich and E. Hussmann, Amorphous and Crystalline Dip Coatings Obtained from Organometallic Solutions: Procedures, Chemical Processes, and Products, *Thin Solid Films*, **77** 129–139 (1981).
181. P. D. Jero, F. Rebillat, D. J. Kent, and J. G. Jones, Crystallization of Lanthanum Hexaluminate from MOCVD Precursors, *Ceram. Eng. Sci. Proc.*, **19** [3] 359–360 (1998).
182. J. A. Haynes, K. M. Cooley, D. P. Stinton, and R. A. Lowden, Corrosion-resistant CVD mullite coatings for Si_3N_4, *Ceram. Eng. Sci. Proc.*, **20** [4] 355–362 (1999).

183. P. V. Chayka, Liquid MOCVD Precursors and Their Application to Fiber Interface Coatings, *Ceram. Eng. Sci. Proc.*, **18** [3] 287–294 (1997).
184. P. W. Brown, Electrophoretic Deposition of Mullite in a Continuous Fashion Utilizing Non-Aqueous Polymeric Sols, in *Ceramic Transactions, Vol. 56*, K. V. Logan, Editor, American Ceramic Society: Columbus, OH (1995), p. 369–376.
185. R. W. Rice, BN Coating of Ceramic Fibers for Ceramic Fiber Composites, U.S. Patent No. 4,642,271, Feb. 10, 1987. USA.
186. R. Naslain, O. Dugne, A. Guette, J. Sevely, C. R. Brosse, J. P. Rocher, and J. Cotteret, Boron Nitride Interphase in Ceramic Matrix Composites, *J. Am. Cer. Soc.*, **74** [10] 2482–2488 (1991).
187. C. J. Griffin and R. R. Kieschke, CVD Processing of Fiber Coatings for CMCs, *Cer. Eng. Sci. Proc.*, **16** [4] 425–432 (1995).
188. M. H. Lewis, Interfaces in Ceramic Matrix Composites, in *Carbon/Carbon, Cement and Ceramic Matrix Composites*, R. Warren, Ed., Elsevier, Oxford, (2000). p. 289–322.
189. T. J. Illston, C. B. Ponton, P. M. Marquis, and E. G. Butler, Electrophoretic Deposition of Silica/Alumina Colloids for the Manufacture of CMC's, *Cer. Eng. Sci. Proc.*, **15** [5] 1052–1059 (1994).
190. E. E. Boakye, T. Mah, C. M. Cooke, and K. A. Keller, Initial Assessment of the Weavability of Fiber Tows Coated with Monazite, *J. Am. Ceram. Soc.*, (submitted).
191. R. S. Hay, M. D. Petry, K. A. Keller, M. K. Cinibulk, and J. R. Welch, Carbon and Oxide Coatings on Continuous Ceramic Fibers, in *Ceramic Matrix Composites – Advanced High Temperature Structural Materials*, R. A. Lowden, et al., Eds., Materials Research Society (1995). p. 377–382.
192. E. Boakye, R. S. Hay, and M. D. Petry, Continuous Coating of Oxide Fiber Tows Using Liquid Precursors: Monazite Coatings on Nextel 720, *J. Am. Ceram. Soc.*, **82** [9] 2321–2331 (1999).
193. R. S. Hay and E. E. Hermes, Coating Apparatus for Continuous Fibers, U.S. Patent No. 5,217,533, June 8, 1993. USA.
194. R. S. Hay, J. R. Welch, and M. K. Cinibulk, TEM Specimen Preparation and Characterization of Ceramic Coatings on Fiber Tows, *Thin Solid Films*, **308–309** 389–392 (1997).
195. F. F. Lange, C. G. Levi, and F. W. Zok, Processing Fiber Reinforced Ceramics with Porous Matrices, in *Carbon/carbon, Cement and Ceramic Matrix Composites*, R. Warren, Ed., Elsevier Science Ltd., Oxford, (2000). p. 427–447.
196. P. Nicholson, P. Sarkar, and X. Huang, Electrophoretic Depositon and its Use to Synthesize ZrO_2/Al_2O_3 Micro-laminate Ceramic/Ceramic Composites, *J. Mat. Sci.*, **29** 6274–6278 (1993).
197. A. R. Boccaccini, C. Kaya, and K. K. Chawla, Use of electrophoretic deposition in the processing of fibre reinforced ceramic and glass matrix composites: a review, *Comp. Part A*, **32** 997–1006 (2001).
198. S. Kooner, W. S. Westby, C. M. A. Watson, and P. M. Farries, Processing of Nextel 720/mullite composition composite using electrophoretic deposition, *J. Eur. Ceram. Soc.*, **20** 631–638 (2000).
199. C. Kaya, X. Gu, I. A. H. Al-Dawery, and E. G. Butler, Microstructural development of woven mullite fibre-reinforced mullite ceramic matrix composites by infiltration processing, *Sci. & Tech. of Adv. Mater.*, **3** 35–44 (2002).
200. I. A. H. Al-Dawery and E. G. Butler, Fabrication of high-temperature resistant oxide ceramic matrix composites, *Comp. Part A*, **32** 1007–1012 (2001).
201. W. D. Kingery, H. K. Bowen, and D. R. Uhlmann, *Introduction to Ceramics*. A Wiley Interscience Publication. John Wlley & Sons, New York (1976). 1032.
202. H. T. Larker and R. Lundberg, Near Net Shape Production of Monolithic and Composite High Temperature Ceramics by Hot Isostatic Pressing (HIP), *J. Eur. Ceram. Soc.*, **19** 2367–2373 (1999).
203. F. Langlais, Chemical Vapor Infiltration Processing of Ceramic Matrix Composites, in *Carbon/Carbon, Cement, and Ceramic Matrix Composites*, R. Warren, Ed., Elsevier, Oxford, (2000). p. 611–644.
204. J. J. Brennan, Interfacial Studies of Chemical-Vapor Infiltrated Ceramic Matrix Composites, *Mat. Sci. Eng.*, **A126** 203–223 (1990).
205. M. S. Newkirk, A. W. Urquhart, H. R. Zwicker, and E. Breval, Formation of Lanxide Ceramic Composite Materials, *J. Mater. Sci.*, [1] 81–89 (1987).
206. W. B. Hillig, Making Ceramic Composites by Melt Infiltration, *Am. Cer. Soc. Bull.*, **73** 56–62 (1995).
207. W. B. Hillig and H. C. McGuigan, An exploratory study of producing non-silicate all-oxide composites by melt infiltration, *Mat. Sci. Eng.*, **A196** 183–190 (1995).
208. J. Wendorff, R. Janssen, and N. Claussen, Model experiments on pure oxide composites, *Mat. Sci. Eng.*, **A250** 186–193 (1998).

209. K. A. Keller, T. Mah, E. E. Boakye, and T. A. Parthasarathy, Gel-Casting and Reaction Bonding of Oxide-Oxide Minicomposites with Monazite Interphase, *Ceram. Eng. Sci. Proc.*, **21** [3] 525–534 (2000).
210. M. K. Cinibulk, T. A. Parthasarathy, K. A. Keller, and T. Mah, Porous yttrium aluminum garnet fiber coatings for oxide composites, *J. Am. Ceram. Soc.*, **85** [11] 2703–2710 (2002).
211. T. A. Parthasarathy, E. Boakye, K. A. Keller, and R. S. Hay, Evaluation of Porous Zirconia Silica and Monazite Coatings using Nextel 720 Fiber Reinforced Blackglas Minicomposites, *J. Am. Ceram. Soc.*, **84** [7] 1526–1532 (2001).
212. M. D. Petry and T.-I. Mah, Effect of thermal exposures on the strengths of Nextel 550 and 720 filaments, *J. Am. Ceram. Soc.*, **82** [10] 2810–2807 (1999).

17

WHIPOX All Oxide Ceramic Matrix Composites

Martin Schmücker and Hartmut Schneider
German Aerospace Center (DLR)
Institute of Materials Research
Linder Höhe, D-51147 Köln
Phone x2203-6012462, E-mail martin.schmuecker@dlr.de (M.S.)
Phone x2203-6012430, E-mail hartmut.schneider@dlr.d (H.S.)

1. ABSTRACT

WHIPOX is a fiber-reinforced porous matrix CMC developed at the German Aerospace Center (DLR). The composite consists of commercial alumina or alumino silicate fibers and mullite, mullite/alumina, or alumina matrices. WHIPOX manufacturing is based on a continuous fiber bundle infiltration and winding process. After shaping in the moist stage, green bodies of WHIPOX are sintered pressureless in air. WHIPOX CMCs display quasi-ductile deformation and excellent thermal shock resistance. Depending on fiber and matrix compositions, interlaminate shear strength values of 10 MPa and Young's moduli of $\approx$150 GPa can be obtained. Thermal conductivity perpendicular to fiber orientation is about 1W/mK. The simple winding and sintering techniques and the fact that no fiber coating is required makes WHIPOX an inexpensive CMC material.

2. INTRODUCTION

Monolithic oxide ceramics are problematic as structural materials due to their inherent brittleness. Therefore many efforts have been made during the last two decades to overcome the problem of brittleness by reinforcing oxide matrices. To achieve damage-tolerant and favorable failure behavior reinforcing components such as ZrO_2 particles (e.g. [1]), whiskers

TABLE 1. Requirements of hot-gas conducting structures

	Monolithic non-oxide ceramics	Monolithic oxide ceramics	Non-oxide CMC	Oxide CMC
Sufficient strength at elevated temperature	+	o	+	o
Oxidation resistance	−	+	−	+
Thermal shock resistance	−	−	+	+
Graceful failure behavior	−	−	+	+

+ good,
o suitable,
− insufficient

or chopped fibers (e.g. [2]), and continuous fibers were employed, the latter being most promising (e.g. [3]). Although both constituents of fiber-reinforced ceramics, i.e. ceramic fibers and ceramic matrices, are brittle, the composite displays quasi-ductile deformation behavior, due to mechanisms such as crack deflection, crack bridging or fiber pull-out [4]. A premise for these mechanisms to work is the relatively weak bonding between fibers and the matrix. To achieve a weak fiber/matrix bonding either suitable fiber coatings have to be employed (cleavable, porous, low toughness interphases) [5] or, in an alternative approach, a highly porous matrix can be used [6].

In Table 1 the property profile required for hot gas conducting structures of jet propulsion engines or gas turbines are compiled and it is sketched schematically how much the different groups of materials (oxide, non-oxide; monolithic, CMC) may fulfill the demands. It turns out that oxide/oxide CMCs have the potential to make the best compromise. Thus, oxide CMC materials are a major program focus of the German Aerospace Center (DLR). As conventional CMCs consisting of coated fibers within a fairly dense matrix were found to be a less promising approach due to complex and expensive fabrication procedure, major research activities turned to the "Porous Matrix Concept" since the late 1990's. The idea behind this concept is to tailor the matrix strength in a way that stress transfer into the fibers is not sufficient to fracture them. Porous matrix composites are attractive from two reasons: on the one hand, favorable mechanical properties can be achieved, on the other hand, manufacturing is relatively simple and inexpensive since both, costly fiber coating and elaborate matrix densification techniques, are not required. The porous oxide/oxide CMC developed at DLR is called "WHIPOX" which is an acronym of wound highly porous oxide/oxide CMCs. Figure 1 gives an overview of the microstructure of WHIPOX in comparison to a CMC consisting of a dense matrix and coated fibers.

3. FABRICATION OF WHIPOX ALL OXIDE CMCs

In contrast to the fabric infiltration techniques used by various manufacturers WHIPOX is produced by a continuous winding process [7]. The matrix is derived from a commercial pseudoboehmite/amorphous silica precursor with overall compositions

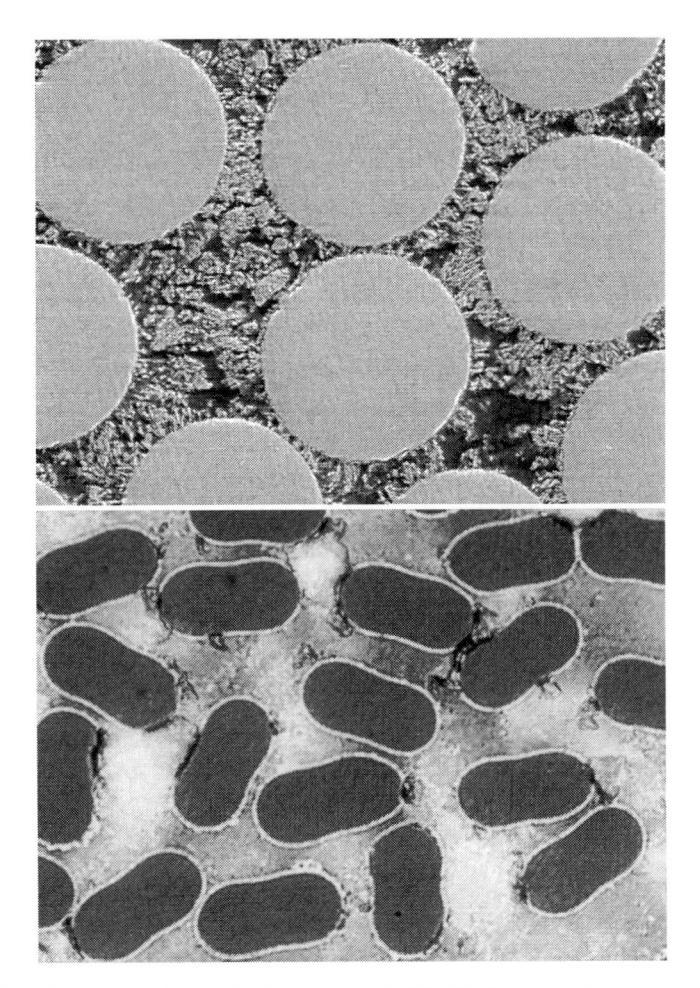

FIGURE 1. Scanning electron micrographs of a porous matrix CMC (top) in comparison to a conventional CMC with dense mullite matrix obtained by hot-pressing and BN fiber coating (bottom).

ranging between 68 and 95 wt.% Al_2O_3 (Figure 2). The winding equipment developed at DLR and the corresponding processing route are shown in Figure 3. In a first step the sizing of the fibers (Nextel 610 or 720, respectively) are burned-off in a tube furnace. Fiber bundles then are spread apart mechanically and are infiltrated with the water-based matrix slurry. The infiltrated rovings are passed through a furnace to stabilize the matrix and are wound in 1D or 2D orientation on plastic mandrels in a computer-controlled process. Green bodies are removed from the mandrel in the moist stage which allows stacking, forming or joining of the prepregs. Once in their final shapes, the green bodies are pressureless sintered in air at $\approx 1300°C$. Fiber contents of WHIPOX CMCs range between 25 and 50 vol.%. Examples of WHIPOX components are shown in Figure 4.

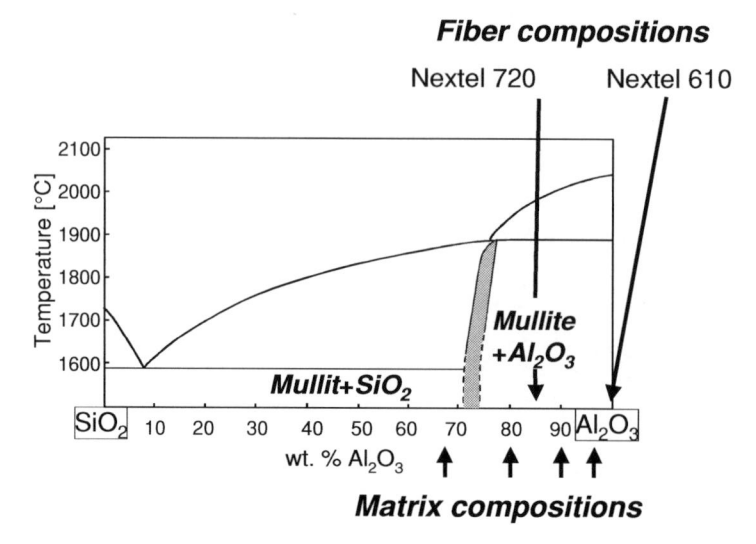

FIGURE 2. The system SiO_2-Al_2O_3 (after Klug et al. [12]). Fiber and matrix compositions used for WHIPOX CMCs are indicated by arrows.

Considerable work has been performed to reduce the matrix porosity in order to enhance shear strength and off-axis properties, yet without the risk of material embrittlement. The matrix porosity can be adjusted by multiple infiltration of the as-sintered components e.g. with $AlCl_3$ solutions. Soaking the infiltrated specimens in ammonia induces the conversion to $Al(OH)_3$ which transforms to alumina after calcination at 1200°C. The re-infiltration

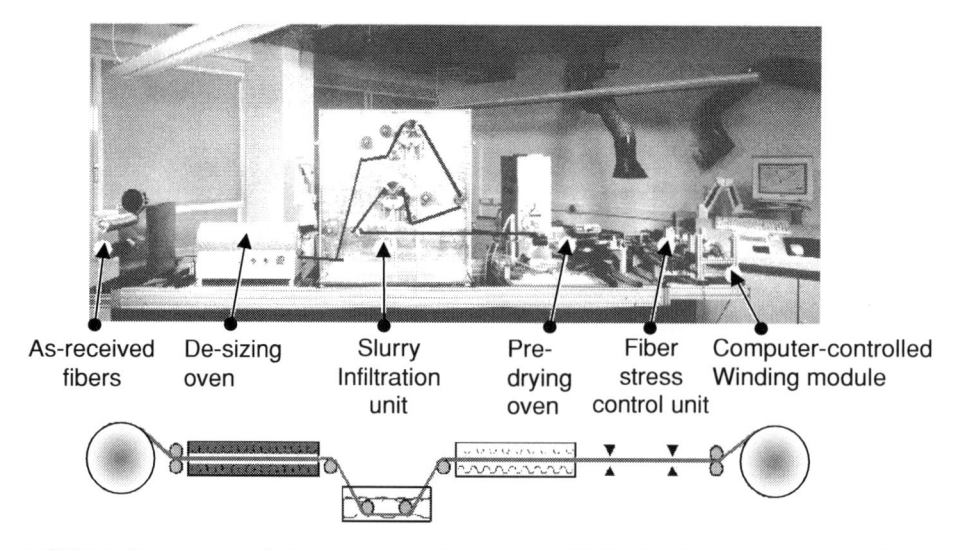

FIGURE 3. Computer-controlled processing route of porous matrix CMCs using the winding equipment of DLR.

WHIPOX ALL OXIDE CERAMIC MATRIX COMPOSITES 427

Wound structures

Shaped and joint structures

Structures after machining

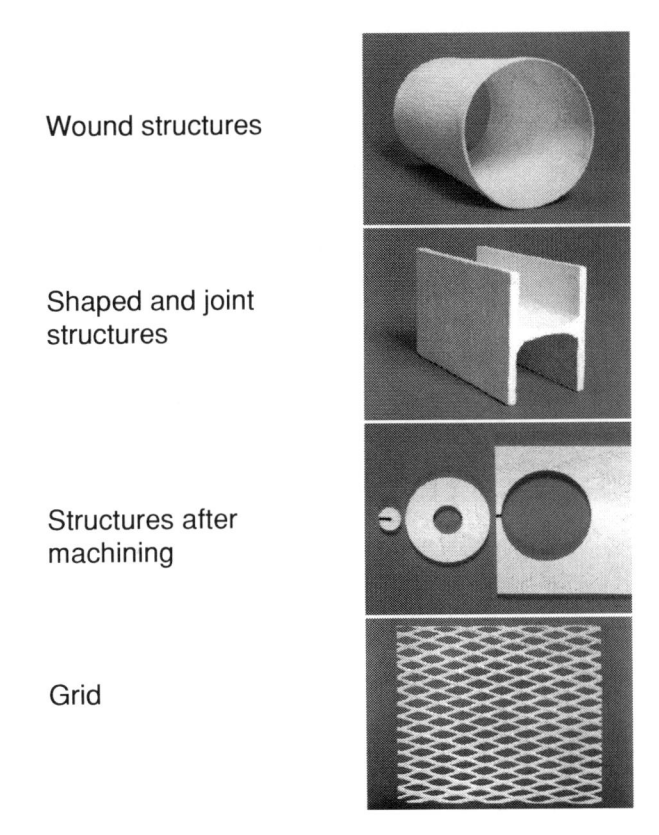

Grid

FIGURE 4. Examples of WHIPOX components fabricated versus the DLR winding process.

on WHIPOX CMCs, however, shows that it is difficult to obtain homogeneous CMCs since the infiltration process is more effective in the outer areas of a sample than it is in its core [8].

4. MESOSTRUCTURE AND MECHANICAL PROPERTIES

Ideally, both, fiber distribution and (sub)micro-scale matrix porosity should be uniform in WHIPOX-CMCs. In reality, however, a laminate type structure exists, with fiber-rich and matrix rich areas. Interlaminate matrix-rich areas of mesoscopic scale are most significant flaws in WHIPOX CMCs as they contain pores up to 100 μm in diameter. Typically, the matrix agglomerations occur at the crossing points of individual fiber bundles as scetched in Figure 5. These kind of defects move across the material at the same angle as the rovings are wound. A statistical method to determine matrix agglomerations in WHIPOX CMCs was recently developed by Schmücker et al. [9]. For that purpose thin sections are analyzed by transmitted light microscopy. The technique utilizes the light conductivity and opacity of fibers and matrix, respectively. Three-dimensional plots of the matrix agglomerations are

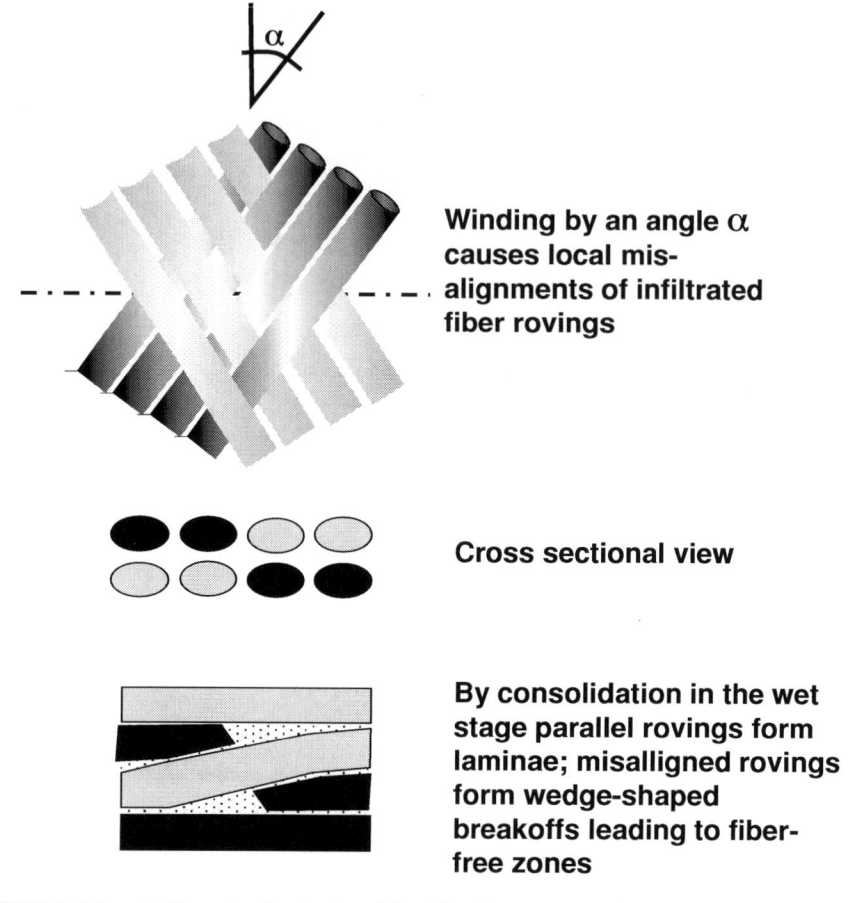

FIGURE 5. Schematic illustration showing the origin of fiber-free areas caused by an angle α between winding direction and the fibers' direction.

obtained by tomographic methods using more than 25 individual slices for each sample. Data analyses actually reveal that matrix agglomerations are not distributed randomly but are concentrated between fiber laminates.

It turned out that a close correlation exists between the degree of interlaminate matrix agglomeration and interlaminate shear strength. The most pronounced interlaminate matrix enrichment acts as "weakest link" whereas the total amount of matrix agglomeration is of minor influence.

Due to the non-isotropic structure of the oxide/oxide composites there are different failure mechanisms in WHIPOX CMCs as illustrated in Figure 6 [10]. Depending on the direction of the applied load, matrix-controlled delamination or fiber-controlled fracture can occur. Beyond, intralaminate debonding and miscellaneous failure mechanisms may emerge. Characteristic values strongly depend on materials, fiber orientation, and manufacturing conditions, especially the fiber/matrix ratio, matrix porosity and sintering temperature. Data

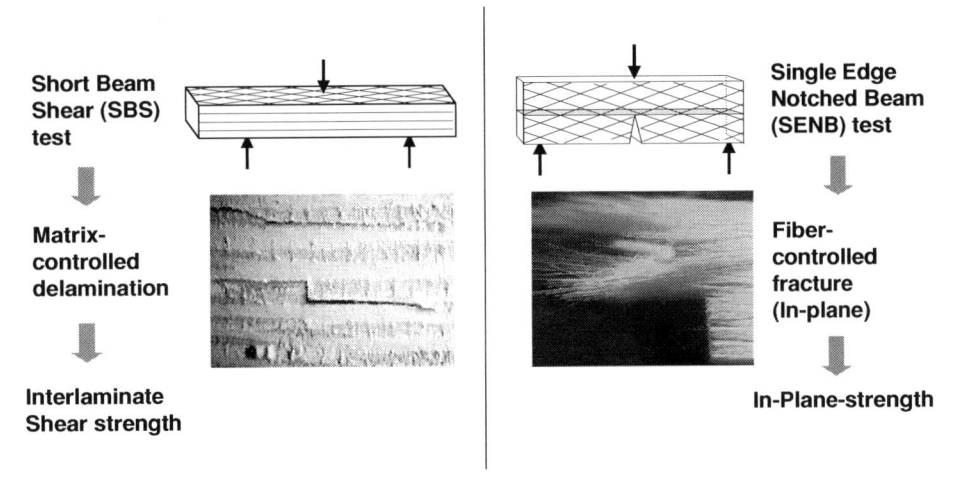

FIGURE 6. Fracture mechanisms in porous matrix CMCs as a response of the non-isotropic microstructures; left: matrix-controlled delamination; right: fiber-controlled fracture.

obtained for WHIPOX CMCs using different fibers and matrix compositions are summarized in Table 2. All WHIPOX qualities display non-brittle fracture behaviour. This is supported by single edge notched beam (SENB) tests (Figure 6) which reveal notch insensitivity of WHIPOX.

The favorable fracture behaviour of WHIPOX CMCs corresponds to their excellent thermal shock and thermal fatigue properties. Thermal shock resistance can be explained by the fact that local stresses dissipate over very short distances and hence thermally induced stress cause only local damage. The excellent thermal shock and thermal fatigue behaviour was demonstrated by local heating of a WHIPOX plate using a solar furnace or an oxyacetylene torch (Figure 7). Despite of the extreme thermal gradients no macroscopic deterioration is detectable even after multiple heating/quenching cycles [10].

The cyclic fatigue behavior at room temperature has been studied for Nextel 720 fiber/ alumino silicate matrix WHIPOX systems by Göring et al. [10]. For this purpose Young's moduli have to be collected as a function of cycles by applying various loads. Figure 8a shows a shear stress/deformation diagram used to determine the "yield shear strength" τ_{el} of the material. As long as cyclic fatigue experiments are carried out in the elastic range,

TABLE 2. In-plane strength, interlaminate shear strength (ILSS) and Young's moduli for WHIPOX CMCs with different fibers and matrix compositions at room temperature

Property	Nextel 610 + Al_2O_3-Matrix	Nextel 720 + Al_2O_3-Matrix	Nextel 720 + mullite Matrix
In-Plane-strength [MPa]	280 ± 20	–	130
ILSS [MPa]	8.0 ± 2,5	9 ± 2	3.6
Young's Modulus [GPa]	140 ± 14	110 ± 11	–

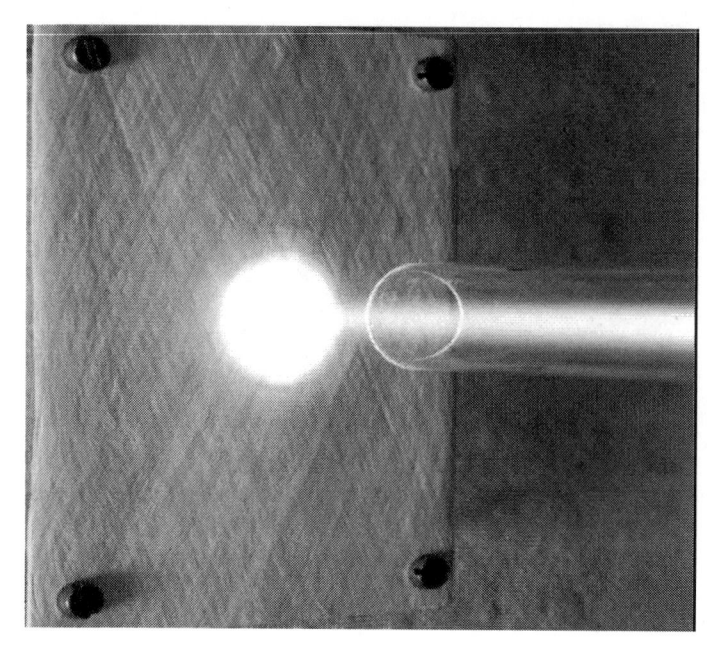

FIGURE 7. Hot spot test for WHIPOX CMCs using an oxyacetylene torch.

$\tau_{max}/\tau_{el} \leq 1$, no degradation can be detected even after 10^4 cycles. On the other hand, if loads greater than the yield strength are applied, a continuous loss in stiffness is observed, indicating fatigue degradation due to an increasing number of defects, such as microcracks and local fiber/matrix debonding (Figure 8b).

5. EFFECTS OF THERMAL AGING

Besides fiber properties, the high temperature use of porous matrix oxide/oxide CMCs depends essentially on the stability of the matrix microstructure against reactions and densification. The influence of matrix composition on microstructural stability and fracture behavior after short-term firing (1 h) was investigated in detail for WHIPOX oxide/oxide CMCs, consisting of Nextel 720 fibers and an aluminosilicate matrix (68wt.% Al_2O_3, 32% SiO_2) corresponding to phase composition of mullite plus minor amounts of silica-rich glass, or, alternatively with an alumino-rich matrix with 95wt.% Al_2O_3, 5% SiO_2 (α-alumina plus minor amounts of mullite). It turned out that the silica-rich matrix reacts with the (alumina-rich) Nextel 720 fibers, thus leading to a new generation of mullite formed at the fibers' outer peripheries (Figure 9) [11]. The silica-rich matrices of these heat-treated materials show significant pore agglomeration and densification effects. Thus, due to the poor matrix stability the fracture behavior of these composites becomes completely brittle. The alumina-rich matrix, on the other hand, is thermodynamically stable in the presence of the Nextel 720 fiber: Only minor pore agglomeration and particle coarsening is observed upon firing at

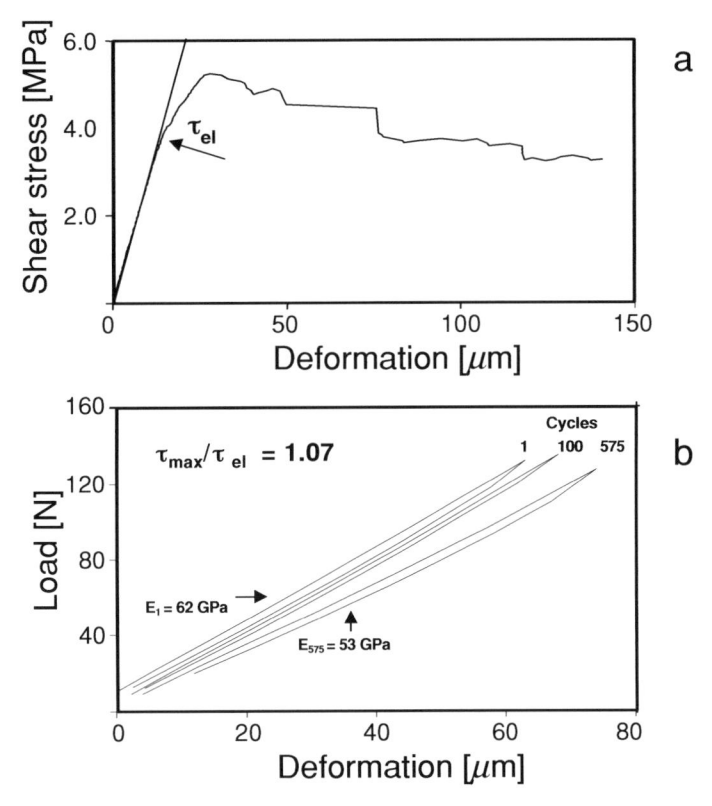

FIGURE 8. Short beam shear test to determine the yield shear strength (τ_{el} of WHIPOX CMC (a) and cyclic fatigue test showing degradation if the load exceeds (τ_{el} (b).

1500 °C. This suggests damage-tolerant fracture behavior of WHIPOX CMCs with alumina-rich matrix, even if they were heat-treated at 1600°C (Figure 10).

6. OTHER PROPERTIES

Various material properties (e. g. thermal diffusivity, air permeability) have been determined for WHIPOX CMCs. Reliable data are important for potential applications such as thermal insulators, filters or burners. Thermal conductivity perpendicular to fiber orientation is about 1W/mK. Closer inspection reveals lower conductivity if a mullite matrix is employed instead of alumina. Thermal conductivity in fiber direction, on the other hand, is about three times higher as perpendicular to the fiber direction, reflecting the non-isotropic structure of the composite (Figure 11).

The air permeability of WHIPOX CMCs strongly depends on processing parameters, matrix composition, and on the composite architecture. According to Figure 12 flow rates range between 10 and 100 l/cm^2 h using overpressures of 250 mbar. It can be anticipated that

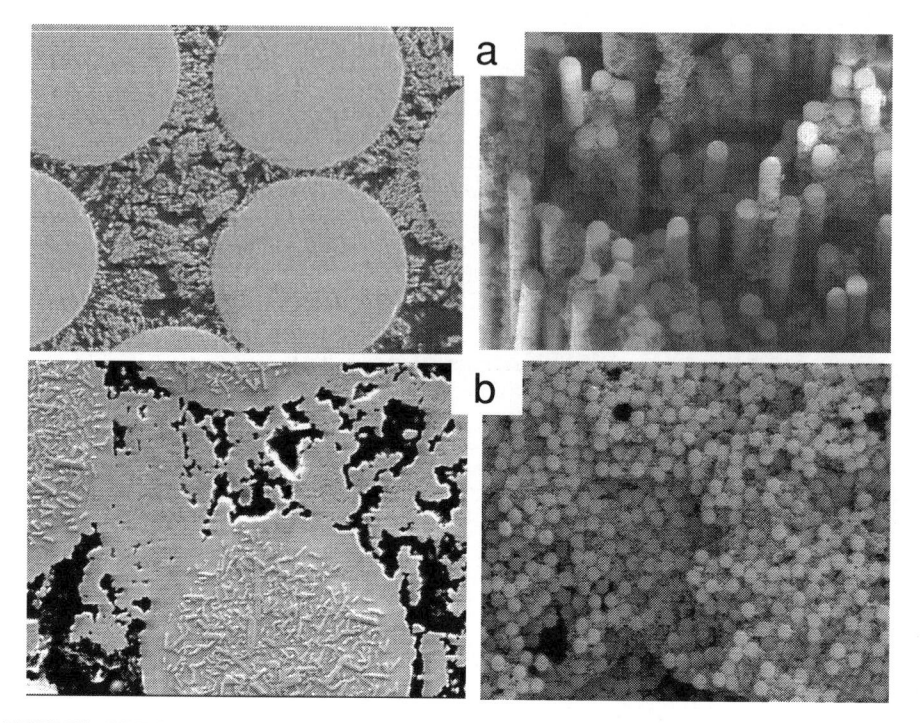

FIGURE 9. WHIPOX CMC with mullite plus SiO$_2$ matrix; a: as-prepared, b: 1600°C, 1 h; left: polished sections; right: fracture surfaces. Note that upon firing the matrix shows pore agglomeration, particle coarsening and reaction with the Nextel 720 fibers, thus leading to catastrophic embrittlement of the composite.

air flow controlling mesopores, typically occurring in regions of matrix agglomerations, are most pronounced, if stiffer 3000 DEN fiber rovings are used instead of more flexible 1500 DEN rovings. Furthermore these data suggest higher air flow if a winding angle of ±45° (2D-orientation) is used instead of ±15° fiber orientation. These findings can be rationalized by higher amounts of fiber cross-over regions occurring in composites of 2D fiber texture and corresponding porous matrix agglomerations. Virtually airtight porous matrix CMCs, on the other hand, may be obtained by subsequent sealing of the surface or by application of external coatings. In preliminary experiments alumina-based WHIPOX CMCs were coated with alumina by plasma spraying. It is anticipated that the coatings do not only affect the air permeability but will increase the erosion resistance of porous matrix CMCs significantly.

7. CONCLUDING REMARKS

Long-term damage tolerant ceramic materials being stable at high temperature, high gas pressures and velocities, and severe chemical environments are needed e.g. for efficient future gas turbine engines. Recently, an oxide fiber/ oxide matrix material which aims to achieve these requirements has been developed at the German Aerospace Center (DLR). The

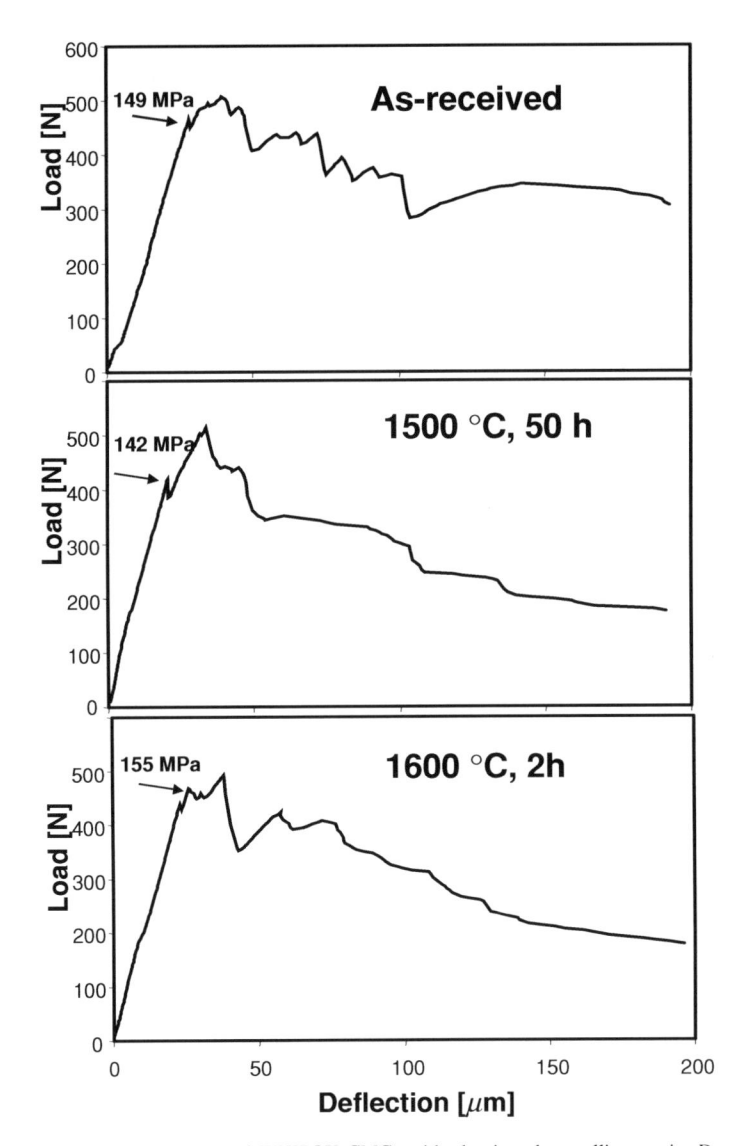

FIGURE 10. Load/deflection curves of WHIPOX CMCs with alumina plus mullite matrix. Damage-tolerant fracture behavior upon firing at 1500°C and 1600°C is indicated.

material is composed of alumino silicate or alumina fibers which toughen mullite or alumina matrices. The matrix of these composites are highly porous, allowing damage tolerance without any specific fiber/matrix interphase. The oxide fiber/oxide matrix composites are produced by a simple winding process, and, thus have been designated as WHIPOX (Wound Highly Porous Oxide CMC). The fabrication technique of WHIPOX allows a variety of sizes and shapes of components by forming, joining and machining. The easy fabrication

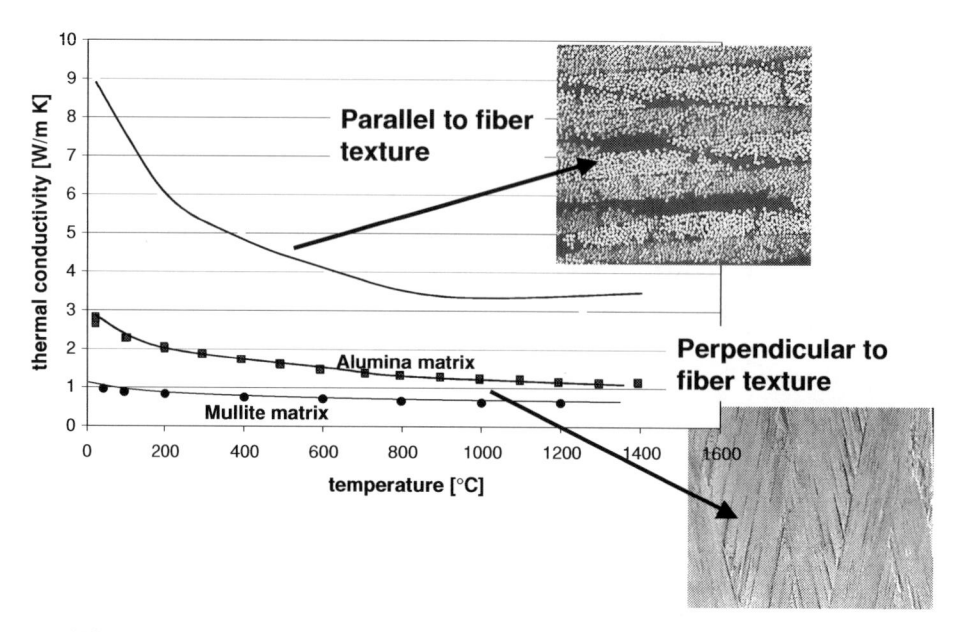

FIGURE 11. Thermal conductivity of WHIPOX CMCs parallel and perpendicular to the fiber texture.

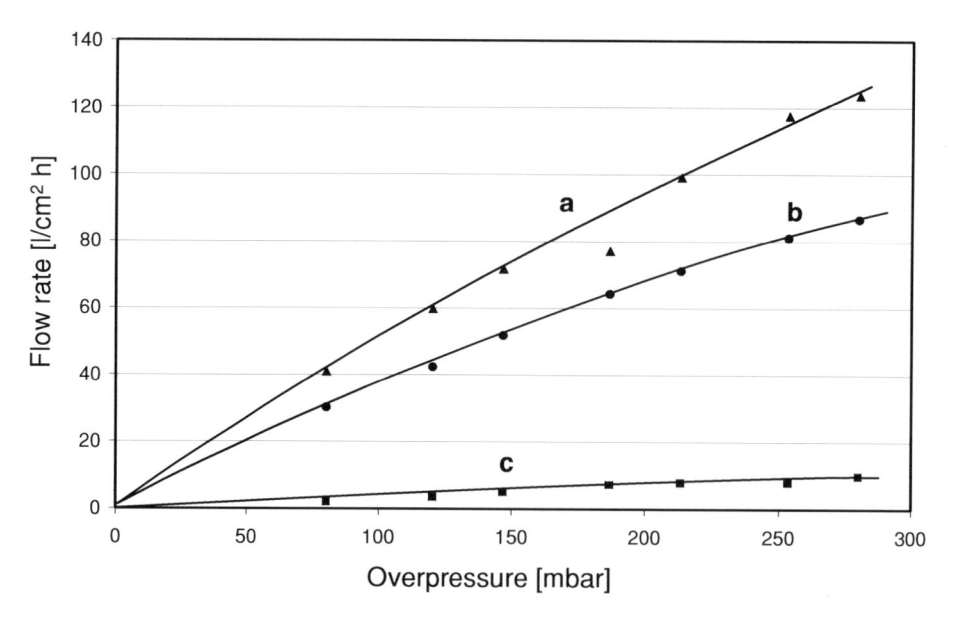

FIGURE 12. Air permeability of different WHIPOX CMCs consisting of Nextel 610 fibers and Al_2O_3/mullite matrix;
a: 3000 DEN fiber roving; fiber orientation ±45°
b: 3000 DEN fiber roving; fiber orientation ±15°
c: 1500 DEN fiber roving; fiber orientation ±15°

of WHIPOX with a simple sintering step in air, makes it a low cost material, although with mechanical and thermal properties suitable for many high-tech applications. The stability of WHIPOX against temperature, erosion and chemical corrosion can be extended by external protection coatings. Thus WHIPOX has an excellent potential for a variety of applications in severe high-temperature environments.

REFERENCES

[1] T. Koyama, S. Hayashi, A. Yasumori, K. Okada, M. Schmücker, H. Schneider, Microstructure and mechanical properties of mullite/zirconia composites prepared from alumina and zircon under various firing conditions, *J. Europ.Ceram. Soc.* **16** 231–237 (1996).
[2] Y. Hirata, S. Matsushita, Y. Ishihara, H. Katsuki, Colloidal processing and mechanical properties of whisker-reinforced mullite matrix composites, *J. Am. Ceram. Soc.* **74** 2438–2442 (1991).
[3] K. K. Chawla, *Composite materials, Science and Engineering, 2nd ed.,* Springer, New York (1998) pp. 212–251.
[4] K. K. Chawla, *Ceramic matrix composites 2nd ed.,* Kluwer Academic Publishers, Boston (2003) pp. 291–354.
[5] R. J. Kerans, R. S. Hay, T. A. Parthasarathy M. K. Cinibulk, Interface design for oxidation-resistant ceramic composites, *J. Amer. Ceram. Soc.* **85** 2599–2632 (2002).
[6] W. C. Tu, F. F. Lange, A. G. Evans, Concept for a damage-tolerant ceramic composite with "strong" interfaces, *J. Amer. Ceram. Soc.* **79** 417–424 (1996).
[7] H. Schneider, M. Schmücker, J. Göring, B. Kanka, J. She, P. Mechnich, Porous alumino silicate fiber/mullite matrix composites: Fabrication and properties, in "*Innovative Processing and Synthesis of Ceramics, Glasses, and Composites IV*" (N. P. Bansal and J. P. Singh, Editors), *Ceram. Trans.* **115**, 415–434 (2000), Am. Ceram. Soc., Westerville, OH.
[8] J. She, P. Mechnich, H. Schneider, M. Schmücker, B. Kanka, Effect of cyclic infiltrations on microstructure and mechanical behavior of porous mullite/mullite composites, *Mat. Sci. Engg.* **A325** 19–24 (2002).
[9] M. Schmücker, A. Grafmüller, H. Schneider, Mesostructure of Whipox all oxide CMCs, *Composites* **A 34** 613–622 (2003).
[10] J. Göring, S. Hackemann, H. Schneider, Oxid/Oxid-Verbundwerkstoffe: Herstellung, Eigenschaften und Anwendungen (in German), in: *Keramische Verbundwerkstoffe,*W. Krenkel (ed.), Wiley-VCh, Weinheim (2003), pp. 123–148.
[11] M. Schmücker, B. Kanka, H. Schneider, Temperature-induced fiber/matrix interactions in porous alumino silicate ceramic matrix composites, *J. Europ. Ceram. Soc.* **20** 2491–2497 (2000).
[12] F. J. Klug, S. Prochazka, R. H. Doremus, Alumina-silica phase diagram in the mullite region, J. Am. Ceram. Soc. 70 (1987) 750–759.

18

Alumina-Reinforced Zirconia Composites

Sung R. Choi* and Narottam P. Bansal**

*NASA Senior Resident Principal Scientist
Ohio Aerospace Institute, Cleveland, OH.
(216) 433-8366
Sung.R.Choi@grc.nasa.gov
**NASA Glenn Research Center
21000 Brookpark Road
Cleveland, OH 44135, USA
(216) 433-3855
Narottam.P.Bansal@nasa.gov

Alumina-reinforced zirconia composites, containing 0 to 30 mol% alumina, were fabricated by hot pressing 10 mol% yttria-stabilized zirconia (10-YSZ) and two different forms of alumina – particulates and platelets. Major mechanical and physical properties, including flexure strength, fracture toughness, slow crack growth, elastic modulus, density, Vickers microhardness, thermal conductivity, and microstructures were determined as a function of alumina content at 25 and 1000°C. Resistance to slow crack growth at 1000°C in air was greater for 30 mol% platelet composite than for 30 mol% particulate composites. Thermal conductivity increased with alumina content.

1. INTRODUCTION

Solid oxide fuel cells (SOFC) are currently being developed for various applications in the automobile, power generation, aeronautic, and other industries. More recently, NASA has explored the possibility of using SOFCs for aero-propulsion under its Zero Carbon Dioxide Emission Technology (ZCET) Project in the Aerospace Propulsion and Power

Program. The SOFC has high-energy conservation efficiency since it converts chemical energy directly into electrical energy. The SOFC is an all solid-state energy conversion device that produces electricity by electrochemical combination of a fuel cell with an oxidant at elevated temperature. The major components of a SOFC are the electrolyte, the anode, the cathode, and the interconnect. The two porous electrodes, anode and cathode, are separated by a fully dense solid electrolyte. Currently, yttria-stabilized zirconia (YSZ) is the most commonly used electrolyte in SOFC because of its high oxygen ion conductivity, stability in both oxidizing and reducing environments, availability, and low cost [1].

In solid oxide fuel cells, YSZ is used in the form of polycrystalline thin films or layers. YSZ must be fabricated in the form of fully dense layers for use as a solid electrolyte in SOFC. Similar to other ceramics, YSZ is brittle and susceptible to fracture due to the existence of flaws, which are introduced during fabrication and use of the SOFC. In addition, the properties of YSZ such as low thermal conductivity and relatively high thermal-expansion coefficient make this material thermal-shock sensitive. Fracture in the solid oxide electrolyte will allow the fuel and oxidant to come in contact with each other resulting in reduced cell efficiency or in some cases malfunction of the SOFC. Therefore, from a structural reliability/life point of view, YSZ solid electrolyte requires high fracture toughness, high strength, and enhanced resistance to slow crack growth at operating temperature (around 1000°C).

To improve the strength and fracture toughness of the electrolyte, the 10 mol% yttria-stabilized zirconia (10-YSZ) was reinforced with 5 to 30 mol% of alumina particulates and platelets. Flexure strength and fracture toughness, determined at 25 and 1000°C in air, together with elastic modulus, density, microhardness, thermal conductivity and other properties of these composites are presented in this chapter. Slow crack growth required for component design and life prediction, evaluated in flexure at 1000°C in air using dynamic fatigue testing for 0 mol% (10-YSZ matrix) and 30 mol% of alumina particulate and platelet composites, is also presented.

2. COMPOSITES PROCESSING

The starting materials used were 10-mol% yttria fully stabilized zirconia powder (HSY-10, average particle size 0.41 μm, specific surface area 5.0 m^2/g) from Daiichi Kigenso Kagaku Kogyo Co., Japan, alumina powder [2, 3] (high purity BAILALOX CR-30, 99.99% purity, average particle size 0.05 μm, specific area 25 m^2/g) from Baikowski International Corporation, Charlotte, NC, and alpha alumina hexagonal platelets (Pyrofine Plat Grade T2) [4–6] from Elf Atochem, France. Appropriate quantities of alumina and zirconia powders were slurry mixed in acetone and mixed for ∼24 h using zirconia media. Acetone was then evaporated and the powder dried in an electric oven. The resulting powder was loaded into a graphite die and hot pressed at 1500°C in vacuum under 30 MPa pressure into 152 mm × 152 mm billets using a hot press. Grafoil was used as spacers between the specimen and the punches. Various hot pressing cycles were tried in order to optimize the hot pressing parameters that would result in dense and crack free ceramic samples. YSZ/alumina composites containing 0 to 30 mol% alumina particulates and platelets were fabricated. Various steps involved during processing of the composite billets are illustrated in Figure 1 [3].

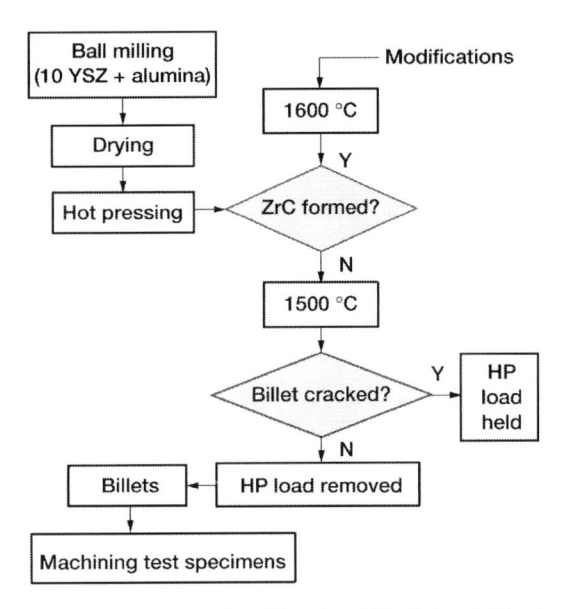

FIGURE 1. Processing flow diagram for 10 mol% yttria-stabilized zirconia/alumina composites [3].

3. TEST SPECIMEN PREPARATION

The hot pressed billets were machined into flexure test specimens with nominal depth, width and length of 3.0 mm × 4.0 mm × 50 mm, respectively, in accordance with ASTM test standard C 1161 [7]. Machining direction was longitudinal along the 50 mm-length direction. It should be noted that unlike transformation-toughened (from *tetragonal* to *monoclinic*) zirconias, the *cubic* ytrria-stabilized zirconia is very unlikely to induce transformation-associated residual stresses on the surfaces of test specimens due to machining. The sharp edges of test specimens were chamfered to reduce any spurious premature failure emanating from those sharp edges.

4. MICROSTRUCTURAL ANALYSIS AND DENSITY

X-ray diffraction (XRD) patterns were recorded at room temperature using a step scan procedure ($0.02°/2\theta$ step, time per step 0.5 or 1 s) on a Philips ADP-3600 automated diffractometer equipped with a crystal monochromator employing Cu K_α radiation [3]. Microstructural analysis was carried out using SEM and TEM, and limited fractographic analysis was performed optically to examine fracture origins and their nature. Typical SEM micrographs for polished cross-sections -planes normal to hot pressing direction- of 20 mol% particulate and platelet composites are shown in Figure 2. The dark areas represent alumina particulates or platelets while the light areas indicate the 10-YSZ matrix, as analyzed through SEM/EDS [2, 3]. The alumina phases were uniformly dispersed within the major continuous 10-YSZ phase. Typical XRD spectra, showing presence of cubic YSZ and α-alumina

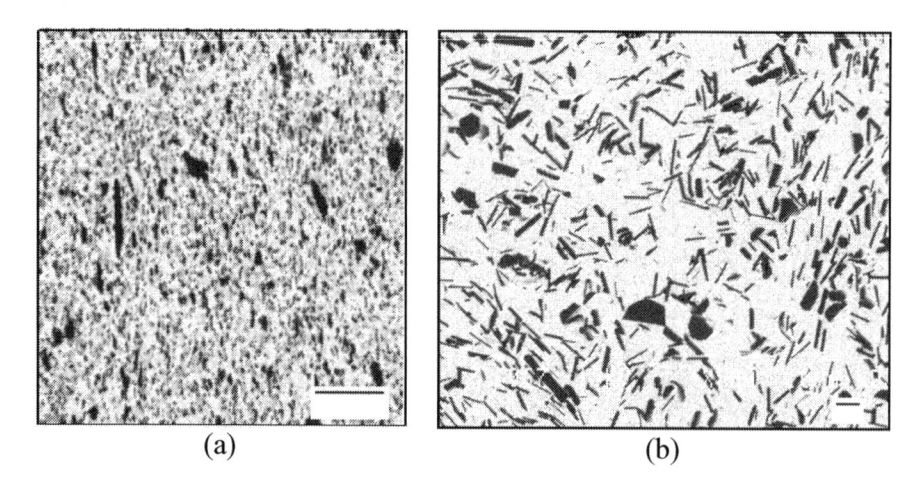

FIGURE 2. SEM micrographs showing polished cross-sections of 10-YSZ/20 mol% alumina particulate and platelet composites: (a) particulate; (b) platelet. Bars = 10 μm.

phases, are presented in Figure 3. Similar XRD results were observed for the platelet composites. TEM micrographs and dot maps for the particulate composites indicated that an average equiaxed grain size was about less than 1.0 μm for either YSZ matrix or alumina and that grain boundaries and triple junctions were clean, an indication of absence of amorphous phase [3]. No appreciable deformation or microcracks of adjacent grains that might occur due to thermoelastic mismatches between YSZ and alumina were observed in the composites [3].

Density was measured with a bulk mass/volume method using the same flexure specimens that were used in elastic modulus measurements. A total of five specimens were used for each of alumina contents. Figure 4 depicts density as a function of alumina content for both particulate and platelet composites [3, 4]. Density decreased linearly with increasing alumina content, yielding good agreement with the prediction based on the rule of mixture. The difference in density between particulate and platelet composites was negligible.

5. MECHANICAL PROPERTIES

5.1. Flexure Strength

Flexure strength was determined at 25 and 1000°C in air using a SiC four-point flexure fixture with 20 mm-inner and 40 mm-outer spans in conjunction with an electromechanical test frame (Model 8562, Instron, Canton, MA). A fast test rate of 50 MPa/s was applied in load control to minimize slow crack growth effect of the materials.[†] For a given temperature,

---

[†] Elevated temperature strength of many advanced ceramics depends on test rate due to slow crack growth during testing. Its has shown that elevated-temperature strength of advanced ceramics increases with increasing test rate and converges to ambient-temperature strength (or inert strength) at "ultra"-fast test rates $\geq 10^5$ MPa/s [8]. Although the test rate of 50 MPa/s used in this work was not sufficient to obtain an appropriate "inert" strength whereby no slow crack growth occurs, the test rate of 50 MPa/s was chosen to determine the conventional, so-called "fast-fracture" strength of the material in which a test rate of around 30–100 MPa/s is typically employed.

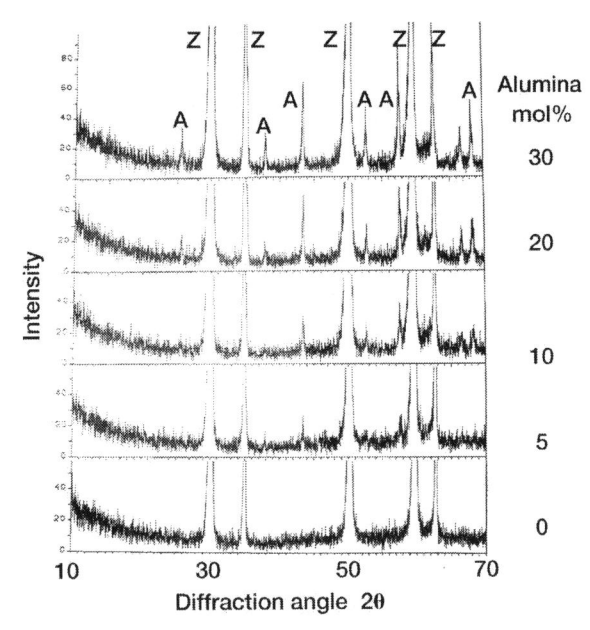

FIGURE 3. X-ray diffraction patterns for 10-YSZ/alumina particulate composites. "Z" and "A" indicate *cubic* YSZ and α-alumina, respectively [2, 3]. Reprinted with permission of The American Ceramic Society, www.ceramics.org. Copyright [2002].

a total of 10 test specimens were tested for each of alumina contents. Testing was followed in accordance with ASTM test standards C1161 and C 1211 [9].

The results of strength testing at 25°C are shown in Figure 5. Strength of the particulate composites increased with increasing alumina content, while strength of the platelet composites remained almost unchanged with increasing alumina content, except at 5 mol%. Hence, increase in strength was much more significant in the particulate composites than in

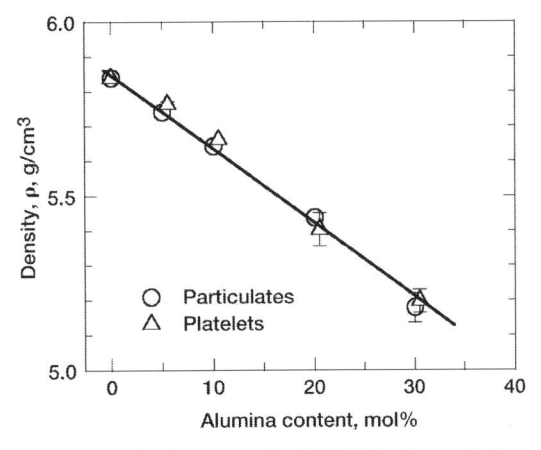

FIGURE 4. Density as a function of alumina content for 10-YSZ/alumina particulate and platelet composites. Error bars indicate ±1.0 standard deviations. The line indicates the prediction based on the rule of mixture [3, 4].

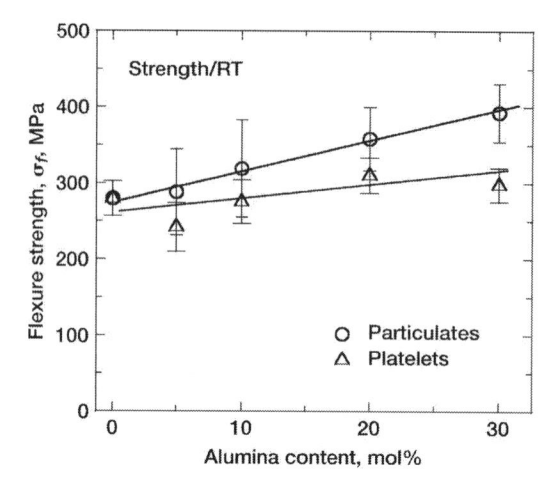

FIGURE 5. Flexure strength of as a function of alumina content for 10-YSZ/alumina particulate and platelet composites at ambient temperature in air [2, 4]. Error bars indicate ±1.0 standard deviations.

the platelet composites. For a given alumina content, strength of the particulate composites was 15 to 30% greater than the platelet composite. Particularly, the strength of 30 mol% alumina particulate composite was 40% greater than the 10-YSZ (matrix) strength. Fracture originated from surface defects, associated with pores and machining damage in many cases. Typical examples of fracture surfaces of specimens showing surface-flaw-associated failure are shown in Figure 6.

The results of elevated-temperature strength testing for both particulate and platelet composites are presented in Figure 7 as a function of alumina content. Like the ambient–temperature counterpart, elevated-temperature strength of the platelet composites did not exhibit any significant dependency on alumina content. Strength of the platelet composites was greater at 5 and 10 mol% but lower at 20 and 30 mol% than the particulate composite counterpart. Strength of the particulate composites with respect to the 10-YSZ strength

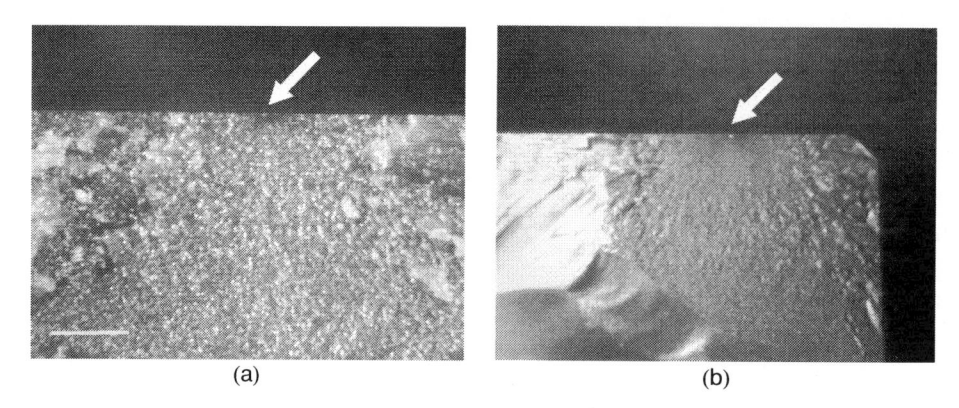

FIGURE 6. Typical examples of fracture surfaces showing fracture origins for (a) 0 mol% (10-YSZ); (b) 30 mol% alumina particulate composite [2]. Bar = 500 μm. Reprinted with permission of The American Ceramic Society, www.ceramics.org. Copyright [2002].

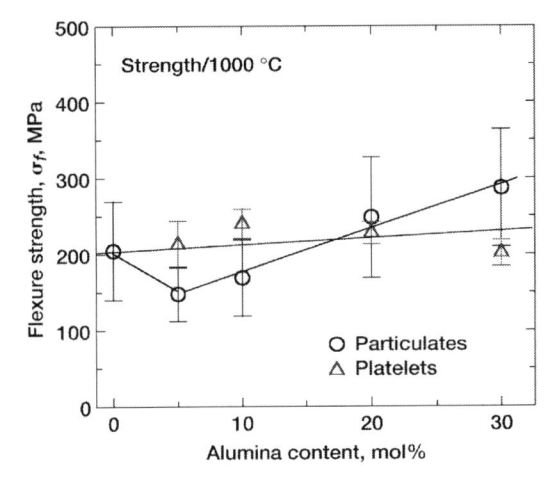

FIGURE 7. Flexure strength as a function of alumina content for 10-YSZ/alumina particulate and platelet composites at 1000°C in air [2, 5, 6]. Error bars indicate ±1.0 standard deviations.

decreased initially at 5 mol% and increased thereafter with increasing alumina content, reaching a maximum at 30 mol%. The strength of the 30 mol% alumina particulate composites was 40% greater than the 10-YSZ strength. Also, the strength of 30 mol% alumina particulate composite was 30% greater than the respective platelet composite strength. The overall elevated-temperature strength of the two composite systems for a given alumina content was lower (20–50% and 10–30%, respectively, for the particulate and platelet composites) than their ambient-temperature counterpart, attributed to strength degradation at elevated temperature presumably by slow crack growth phenomenon. Individual strength data of both composites at 25 and 1000°C are summarized in Tables 1 and 2.

TABLE 1. Summary of major mechanical and physical properties of 10-YSZ reinforced with alumina particulates [2–6, 30]

Properties	Alumina mol%	0	5	10	20	30
Flexure strength, σ_f (MPa)	RT	279.6(23.1)[a]	287.9(56.7)	318.8(64.1)	358.1(41.5)	353.6(89.1)
	1000°C	204.5(64.5)	147.8(35.7)	169.0(49.7)	248.1(79.3)	287.0(77.0)
Fracture toughness, K_{IC} (MPa√m)	RT	1.6(0.1)	2.1(0.3)	1.7(0.2)	2.3(0.1)	2.6(0.3)
	1000°C	1.7(0.2)	2.3(0.2)	1.5(0.2)	2.3(0.2)	2.6(0.3)
RT elastic modulus, E (GPa)		219(2)	225(2)	233(1)	250(1)	262(2)
Density, ρ (g/cm^3)		5.839(0.008)	5.740(0.012)	5.642(0.007)	5.437(0.005)	5.178(0.042)
RT Vickers hardness, H (GPa)		14.1(0.6)	14.5(0.2)	14.9(0.1)	15.7(0.3)	15.4(0.4)
Thermal conductivity at 1000°C, k (W/m-K)		2.2	2.5	2.7	3.0	3.3

[a] The numbers in parentheses indicate ±1.0 standard deviation.

TABLE 2. Summary of Major mechanical and physical properties of 10-YSZ reinforced with alumina platelets [2–6]

Properties	Alumina mol%	0	5	10	20	30
Flexure strength, σ_f (MPa)	RT	279.6(23.1)[a]	241.9(32.3)	275.4(28.9)	310.1(22.9)	297.9(22.2)
	1000°C	204.5(64.5)	212.8(30.6)	239.7(19.0)	227.6(14.5)	201.1(17.1)
Fracture toughness, K_{IC} (MPa$\sqrt{m}$)	RT	1.6(0.1)	1.7(0.1)	1.8(0.1)	2.2(0.1)	2.6(0.1)
	1000°C	1.7(0.2)	2.6(0.5)	2.7(0.4)	2.9(0.1)	3.0(0.1)
RT elastic modulus, E (GPa)		219(2)	226(2)	234(1)	245(1)	259(2)
Density, ρ (g/cm^3)		5.839(0.008)	5.764(0.008)	5.661(0.002)	5.403(0.048)	5.197(0.033)
Vickers hardness, H (GPa)		14.1(0.6)	14.2(0.6)	14.2(0.7)	13.6(0.6)	11.5(0.6)

[a] The numbers in parentheses indicate ±1.0 standard deviation.

The strength scatter at 1000°C was less significant in the platelet composites than in the particulate composites. Weibull modulus (m), estimated despite a limited number (10) of specimens tested at each alumina content, was in the range of $m = 8$ to 20 and $m = 5$ to 10, respectively, for the platelet and the particulate composites. Fracture originated mainly from surface flaws for both composites. However, platelets, particularly at 1000°C, were frequently associated as primary failure origins, so that alumina platelets might have acted as strength-controlling flaws rather than as strengthening media, typical of many platelets-reinforced composites, resulting in less scatter in strength or higher Weibull modulus. The fact that the 30 mol% particulate composite at both temperatures exhibited improved strength over the platelet composite is indicative of an effective approach to the reinforcement of 10-YSZ.

It has been shown that some other zirconia/alumina composites exhibited a strength decrease with increasing alumina content, in part as a result of internal (tensile) residual stresses by the CTE mismatch between zirconia and alumina particulate or platelets [10,11]. Based on the results of strength increase with increasing alumina content particularly for the 30 mol% particulate composite, as seen in Figure 5, it can be stated that the alumina particulates used in this work might not have interacted with the matrix to produce residual stresses by CTE mismatch sufficient enough to degrade composite strength.

5.2. Fracture Toughness

Fracture toughness of the particulate and platelet composites, using the flexure test specimens, was determined at 25 and 1000°C in air using the single edge v-notched beam (SEVNB) method. This method utilizes a razor blade with diamond paste to introduce a final sharp root radius by tapering a saw notch [12]. A starter straight-through notch 0.6 mm deep and 0.026 mm wide was made on the 3-mm wide face of each test specimen. The v-notched specimens with a final notch depth of 0.9 mm and its root radius of about 10 μm were fractured in a SiC four-point flexure fixture with 20 mm-inner and 40 mm-outer spans using the electromechanical testing machine at an actuator speed of 0.5 mm/min. A total of

FIGURE 8. A typical example of a sharp v-notch produced in a single edge V-notched beam (SEVNB) specimen used in fracture toughness testing [2]. Bar = 50 μm. Reprinted with permission of The American Ceramic Society, www.ceramics.org. Copyright [2002].

five specimens were tested for each composite at each test temperature. A typical example of v-notched fracture toughness test specimen is shown in Figure 8. It should be mentioned that an attempt was made to utilize a more convenient fracture toughness technique, single edge precracked beam (SEPB) method (ASTM C 1421 [13]) at both RT and 1000°C. However, this technique was found to be ineffective at 1000°C due to crack healing, which resulted in a considerably high 'apparent' value of fracture toughness, as observed in other ceramics such as silicon nitride and alumina [14]. The SEVNB method, however, may underestimate fracture toughness at ambient temperature (compared to other methods such as SEPB and/or chevron notch) if a material exhibits strong R-curve behavior, since the method tends to give a starting value of fracture toughness on the corresponding R-curve [12 (b)]. The fracture toughness K_{IC} was calculated based on the formula by Srawley and Gross [15].

$$K_{IC} = \frac{P_f(L_o - L_i)}{BW^{3/2}} \frac{3\alpha^{1/2}}{2(1-\alpha)^{3/2}} f(\alpha) \quad (1)$$

where P_f, L_o, L_i, B, W are fracture load, outer span, inner span, specimen width, and specimen depth, respectively, $\alpha = a/W$ with a being precrack size, and $f(\alpha)$ is expressed

$$f(\alpha) = 1.9887 - 1.326\alpha - \frac{\alpha(1-\alpha)(3.49 - 0.68\alpha + 1.35\alpha^2)}{(1+\alpha)^2}$$

A summary of the results of fracture toughness testing at 25°C in air is presented in Figure 9, in which fracture toughness determined by the SEVNB method is plotted as a function of alumina content for both particulate and platelet composites. Similar to the trend in ambient-temperature flexure strength, fracture toughness increased with increasing alumina content. Fracture toughness increased significantly by 65 and 62%, respectively, for the particulate and platelet composites when alumina content increased from 0 to 30 mol%. It is noted that unlike the ambient-temperature flexure strength the difference in fracture toughness between the particulate and platelet composites was negligible. It has been observed that an incompatibility is generally operative for many advanced ceramics between strength and fracture toughness in such a manner that one property increases while the other decreases. However, this was not the case for these two types of composite systems,

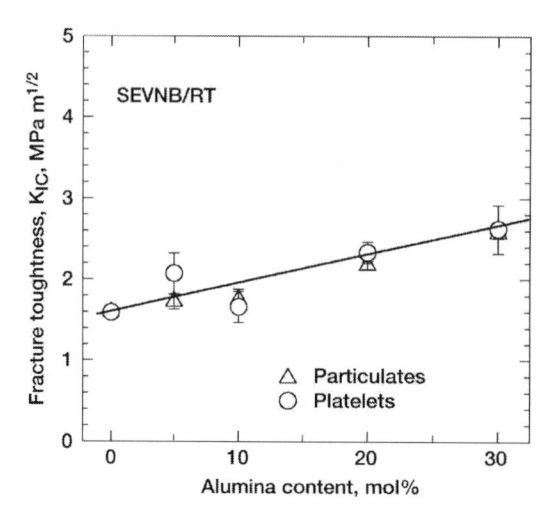

FIGURE 9. Fracture toughness as a function of alumina content for 10-YSZ/alumina particulate and platelet composites, determined by the SEVNB method at room temperature [2, 4]. Error bars indicate ±1.0 standard.

resulting in not only strength increase but also fracture-toughness increase with increasing alumina content.

The indent crack trajectories of both 0% and 30 mol% particulate composites were characterized such that the *straight* path and greater COD (crack opening displacement) of a crack was typified for 10-YSZ (0 mol% composite); whereas, the *tortuous* path around alumina grains and less COD was exemplified for the 30 mol% particulate or platelet composite [2]. More enhanced crack interactions with alumina grains with increasing alumina content is thus believed to be responsible for the increased fracture toughness for both composite systems. A notion that platelets would be more efficient in enhancing fracture toughness than particulates was not applicable in these composite systems at ambient temperature. Note that the *cubic* YSZ is not a stress-induced, transformation toughened ceramic. Therefore, the increased fracture toughness with increasing alumina content would be a logical reason for the increased flexure strength observed from both composite systems, since flaw sizes of the both composites seemed to be narrowly distributed. Typical responses to Vickers indentation are shown in Figure 10 for the particular composites. Note the size of indent cracks decreasing with increasing alumina content, indicative of increasing resultant fracture toughness.

A summary of fracture toughness results determined at 1000°C is presented in Figure 11. The overall fracture toughness of both particulate and platelet composites increased with increasing alumina content. Fracture toughness increased significantly by 50% and 74%, respectively, for the particulate and platelet composites when alumina content increased from 0 to 30 mol%. Unlike the ambient-temperature fracture toughness, the difference in fracture toughness at 1000°C, in general, was distinct between the particulate and platelet composites. This indicates that toughening by addition of platelets to 10-YSZ matrix was more effective at 1000°C than at 25°C. Individual fracture toughness data for both composites at 25 and 1000°C are summarized in Tables 1 and 2.

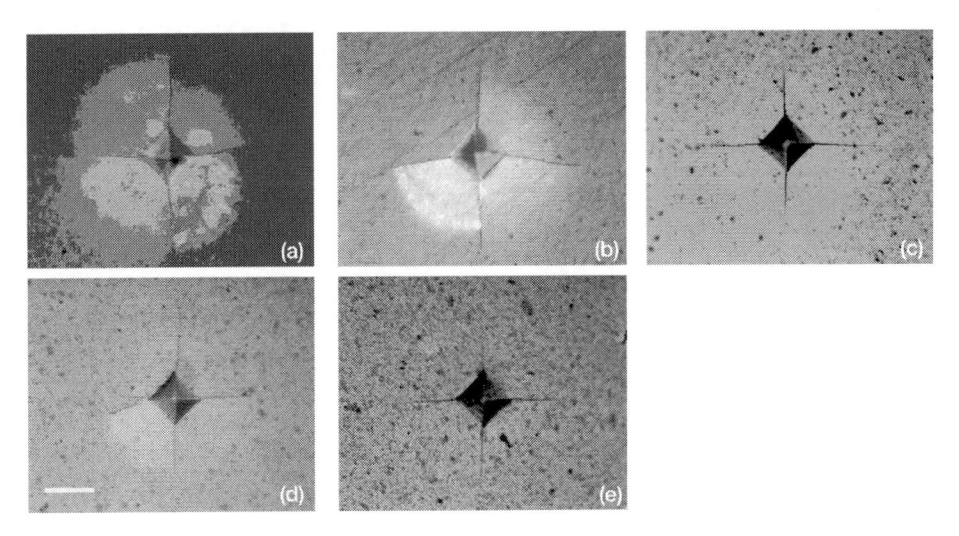

FIGURE 10. Response to Vickers indentation of 10-YSZ/alumina particulate composites at ambient temperature under an indent load of 10 N (a) 0 mol%; (b) 5 mol%; (c) 10 mol%; (d) 20 mol%; (e) 30 mol%. Bar = 30 μm.

5.3. Slow Crack Growth (Dynamic Fatigue)

Slow crack growth (SCG) behavior of some chosen composites was determined at 1000°C in air using dynamic fatigue (or called 'constant stress-rate') testing in accordance with ASTM test method C 1465 [16]. Three different composites including 0 mol% (10-YSZ), 30 mol% alumina particulate and 30 mol% alumina platelet composites were

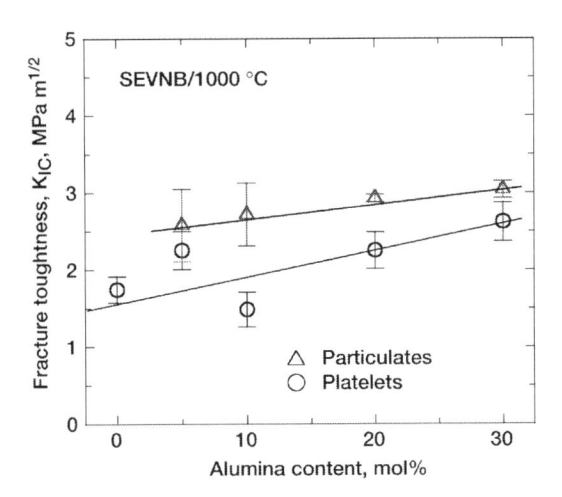

FIGURE 11. Fracture toughness as a function of alumina content for 10-YSZ/alumina particulate and platelet composites at 1000°C in air [2, 5, 6]. Flexure toughness of 10-YSZ/alumina particulate composites [2, 5, 6] is included for comparison. Error bars indicate ±1.0 standard deviations. The lines represent the best fit.

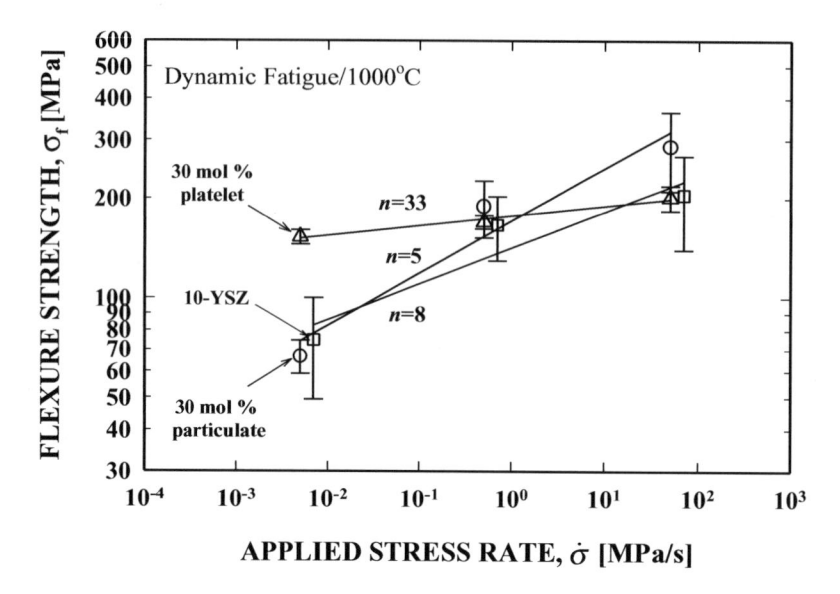

FIGURE 12. Results of dynamic fatigue testing at 1000°C in air for 10-YSZ, 10-YSZ/30 mol% alumina particulate composite, and 10-YSZ/30 mol% alumina platelet composite. The solid lines represent the best-fit line. Slow crack growth parameter n is included.

selected based on the results of optimum strength and fracture toughness properties exhibited by 30 mol% particulate and platelet composites. Dynamic fatigue testing was performed in flexure using flexure test specimens at three different test rates of 50, 0.5 and 0.005 MPa/s for a given composite under load control of the electromechanical test frame. Note that the flexure strength data obtained at 50 MPa/s from flexure strength testing were used as one set for the dynamic fatigue data. A SiC four-point flexure fixture with 20/40 mm spans was used. Typically, a total of 7 to 10 test specimens were used at each test rate for a given composite.

The results of slow crack growth ('dynamic fatigue' or 'constant stress-rate') testing for the three different composites are shown in Figure 12, where fracture stress of each composite was plotted as a function of applied stress rate. The decrease in fracture stress with decreasing stress rate, which represents susceptibility to slow crack growth, was evident for all three composites with its degree of strength degradation with decreasing stress rate being dependent on material. The basic underlying formulation of slow crack growth for many advanced monolithic ceramics and composites (reinforced with particulates, platelets or whiskers) at elevated temperatures follows the following power-law form [17]

$$v = A [K_I/K_{IC}]^n \qquad (2)$$

where v, K_I and K_{IC} are crack velocity, mode I stress intensity factor, and mode I fracture toughness, respectively. A and n are material/environment dependent SCG parameters. The responsible mechanism of SCG at elevated temperatures has been known as grain boundary sliding [18–21]. In case of dynamic fatigue loading, a constant stress rate ($\dot{\sigma}$) is applied to a test specimen until the test specimen fails. The corresponding fracture stress (σ_f)

TABLE 3. Summary of slow crack growth parameters determined for three different composites by dynamic fatigue testing in flexure at 1000°C in air

Materials	Slow crack growth parameters		Correlation coefficient, r^2
	n	D	
10-YSZ (matrix)	8.2	142	0.8968
10-YSZ/30 mol% alumina particulate composite	5.3	171	0.9420
10-YSZ/30 mol% alumina platelet composite	32.6	177	0.9820

can be derived from Eq. (2) and related stress intensity factor with some mathematical manipulations to give [16, 22]

$$\log \sigma_f = \frac{1}{n+1} \log \dot{\sigma} + \log D \qquad (3)$$

where D is another SCG parameter associated with A, n, K_{IC}, inert strength, and crack geometry factor. The SCG parameters n and D can be determined from the slope and intercept by a linear regression analysis when log (*fracture stress*) is plotted as a function of log (*applied stress rate*). Equation (3) is the basis commonly used in dynamic fatigue testing, which has been adopted to determine SCG parameters of advanced ceramics in ASTM test standards at both ambient and elevated temperatures as well [16, 23].

The results shown in Figure 12 were plotted according to Eq. (3) with units of MPa for σ_f and MPa/s for $\dot{\sigma}$. The SCG parameter n was found to be $n = 8$, 5 and 33, respectively, for 0 mol% (10-YSZ), 30 mol% particulate and 30 mol% platelet composites. The SCG parameters n and D are also summarized in Table 3, together with correlation coefficients of regression. Both 0 mol% and 30 mol% particulate composites exhibited very high susceptibility to SCG with a relatively low SCG parameter $n = 6-8$, and the 30 mol% platelet composite exhibited intermediate susceptibility with $n = 33$.[‡] Hence, the 30 mol% platelet composite exhibited greater resistance to SCG as compared with both 0 mol% and 30 mol% particulate composites. The addition of 30 mol% alumina platelets into 10-YSZ matrix resulted in increased resistance to grain boundary sliding, while the addition of 30 mol% fine alumina particulates would not have had any positive effect on reducing or minimizing grain boundary sliding. Note that the value of SCG parameter n determined at 1000°C in air for typical aluminas with 96 to 99% purity is in a range of $n = 7-13$ [20, 24]. Significant improvement in SCG resistance was obtained by the composite approach with 10-YSZ reinforced with 30 mol% alumina platelets, in which each of YSZ and alumina exhibits a significantly high SCG susceptibility ($n = 5-13$) if they are used individually in monolithic form.

A simplified life prediction diagram can give a better interpretation of SCG behavior among the three composites, which was constructed in Figure 13 under the same specimen's

---

[‡] Note that susceptibility to slow crack growth is typically categorized in advanced ceramics such that SCG susceptibility is very high for $n < 20$, intermediate for $n = 30-50$, and very low for $n > 50$.

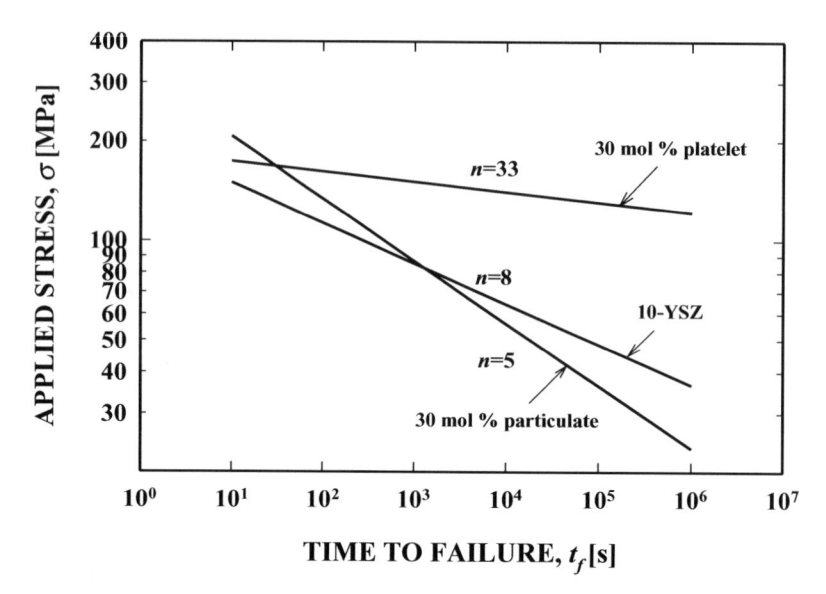

FIGURE 13. Life prediction diagram constructed from the dynamic fatigue results at 1000°C for 10-YSZ, 10-YSZ/30 mol% alumina particulate composite, and 10-YSZ/30 mol% alumina platelet composite. The prediction represents at a failure probability of 50%.

geometrical and dimensional configurations that were used in dynamic fatigue, based on the following relation

$$t_f = \left[\frac{D^{n+1}}{n+1}\right] \sigma^{-n} \qquad (4)$$

where t_f and σ are time to failure (in sec) and constant applied stress (in MPa), respectively. Of course, the prediction is valid when the same failure mechanism(s) is operative, irrespective of loading condition, either dynamic or static. As can be seen from the figure, lifetime is greatest for the 30 mol% platelet composite, and is least for the 10-YSZ or the 30 mol% particulate composite. As a consequence, the 30 mol% platelet composite would be a most reasonable choice among the three materials in conjunction with component life. A detailed life prediction and reliability of actual, complex fuel cell components can be made using analytical (finite element modeling) and reliability tool such as *CARES/Life* integrated computer code [25].

5.4. Elastic Modulus

Elastic modulus of both particulate and platelet composites was determined from 25 to 1000°C as a function of alumina content by the impulse excitation of vibration method, ASTM C 1259 [26] using the flexure specimen configuration. One flexure specimen was used at each of alumina contents for a given composite. A total of five specimens were additionally used at each alumina content to evaluate ambient-temperature elastic modulus of the two composites.

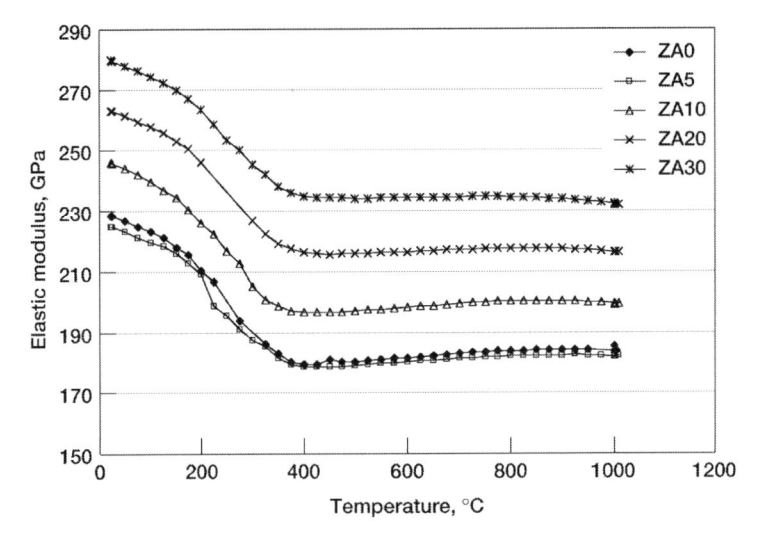

FIGURE 14. Elastic modulus as a function of temperature for 10-YSZ/alumina particulate composites with different alumina contents, determined by the impulse excitation method. ZA0 to ZA30 in the figure indicate respective alumina mol% (e.g., ZA0 = 0 mol%).

Elastic modulus as a function of temperature for the particulate composites are shown in Figure 14. Elastic modulus decreased with increasing temperature up to 400°C and then remained almost unchanged up to 1000°C, thereby forming a unique transition around 400°C, regardless of alumina content. Although not presented here, it was also found that elastic modulus of the platelet composites was very close to that of the platelet composites. A similar transition at 400°C was also observed in 6.5-YSZ by Adams et al. [27]. Figure 15 shows elastic modulus as a function of alumina content for both particulate and platelet composites, determined at ambient temperature. Elastic modulus increased linearly with increasing alumina content for both composites, with little difference in elastic modulus between the two composite systems for a given alumina mol%. The prediction based on the rule of mixture was in good agreement with the experimental data. Individual elastic modulus data at ambient temperature for both composites are summarized in Tables 1 and 2.

5.5. Microhardness

Microhardness of both particulate and platelet composites was evaluated at ambient temperature with a Vickers microhardness indenter with an indent load of 9.8 N using five indents for each composite in accordance with ASTM C 1327 [28]. Figure 16 shows the results of Vickers microhardness measurements for both composites. Microhardness increased linearly with increasing alumina content for the particulate composites up to 20 mol% alumina and then leveled off above 20 mol%. Microhardness of the platelet composites remained almost unchanged up to 10 mol% and then decreased appreciably at 30 mol%, resulting in a significant difference in hardness at 30 mol% between the two composites. Individual microhardness data for both composites are also summarized in Tables 1 and 2.

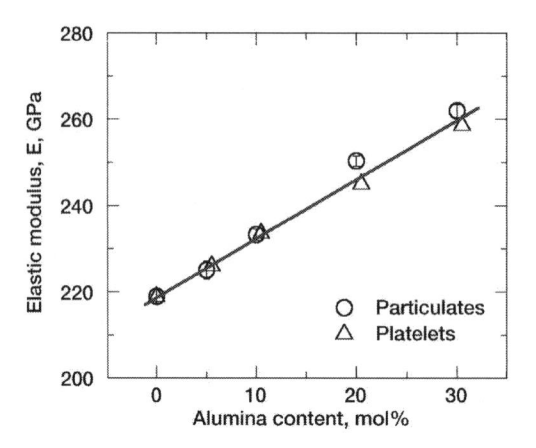

FIGURE 15. Elastic modulus as a function of alumina content for 10-YSZ/alumina particulate and platelet composites, determined by the impulse excitation method at ambient temperature [3, 4]. Error bars indicate ±1.0 standard. The line indicates the prediction based on the rule of mixture.

5.6. Thermal Fatigue

Thermal cycling/fatigue testing was conducted for the 30 mol% platelet composite by applying a total of 10 thermal cycles of heating (1000°C) and cooling (200°C) in air using five flexure specimens [4]. The rate of heating and cooling was about 10°C/min and 20°C/min, respectively. These flexure specimens were then fractured in four-point flexure at ambient temperature to determine their corresponding residual flexure strength. This testing was conducted to better understand the effect of CTE mismatch on flexure strength, possibly resulting in strength degradation due to thermal fatigue associated with residual stresses and/or microcracks induced by CTE mismatch between 10-YSZ matrix and alumina grains.

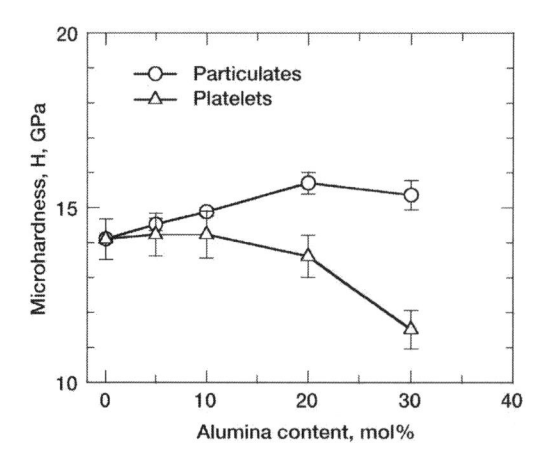

FIGURE 16. Vickers microhardness as a function of alumina contents for 10-YSZ/alumina particulate and platelet composites. Error bars indicate ±1.0 standard deviations.

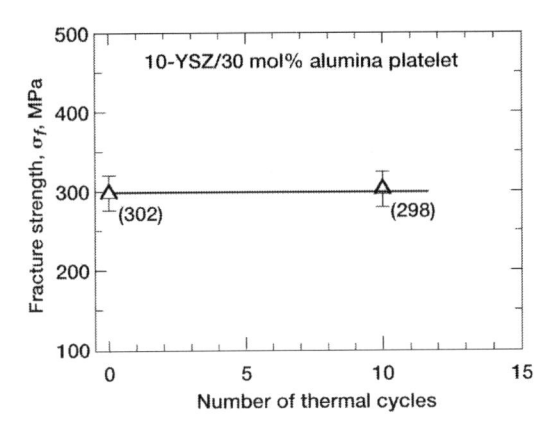

FIGURE 17. Flexure strength as a function of number of thermal cycles (between 200 and 1000°C) for 10-YSZ/30 mol% alumina platelet composite [4]. The numbers in parentheses indicate average strength.

The results of thermal cycling/fatigue tests are shown in Figure 17. As can be seen in the figure, there was no difference in strength between 0 (regular strength test) and 10 thermal cycles, indicating that repeated thermal cycling did not show any significant effect on strength degradation for the composite material. In other words, internal residual stresses and/or microcracks due to CTE mismatch between zirconia and alumina grains possibly occurring in thermal fatigue were negligible to affect residual flexure strength of the composite material of interest. Hence, it is concluded that CTE mismatch would not have been operative sufficient enough to degrade strength of the composites.

6. THERMAL CONDUCTIVITY

One inch diameter hot pressed discs of the particulate composites were used for thermal conductivity measurements. Thermal conductivity testing was carried out using a 3.0 kW CO_2 laser (wavelength 10.6 μm) high-heat flux rig. The general test approach has been described elsewhere [29]. In this steady-state laser heat flux test method, the specimen surface was heated by a laser beam, and backside air-cooling was used to maintain the desired temperature. A uniform laser heat flux was obtained over the 23.9 mm diameter aperture region of the specimen surface by using an integrating ZnSe lens combined with the specimen rotation. Platinum wire flat coils (wire diameter 0.38 mm) were used to form thin air gaps between the top aluminum aperture plate and stainless-steel back plate to minimize the specimen heat losses through the fixture.

Figure 18 shows results of thermal conductivities determined for the particulate composites, as a function of temperature [30]. Thermal conductivity increased with increasing alumina content. This is expected, as the thermal conductivity of alumina measured at 1000°C is much higher (= 6.9 W/m-K) [30] than that of 10-YSZ (= 2 W/m-K). Increase in thermal conductivity with alumina additions is more significant at lower temperatures than at higher temperatures. Thermal conductivity of the composites containing 0, 5, and 10-mol% alumina exhibited a slight change with temperature. However, those containing

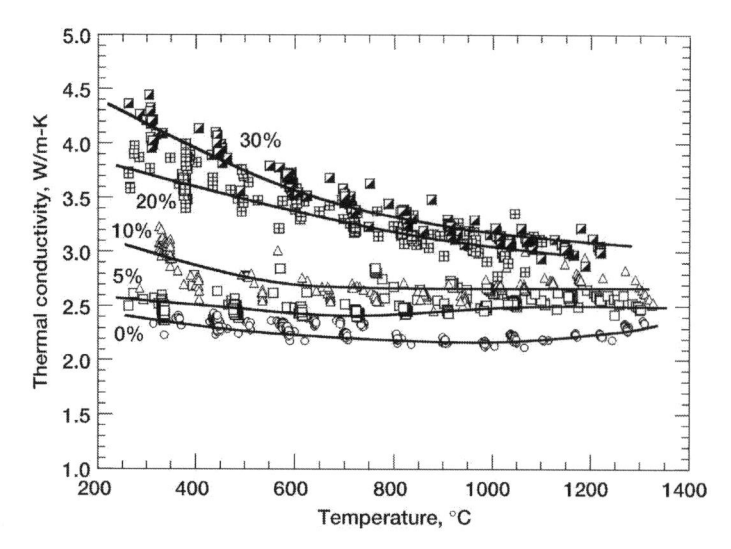

FIGURE 18. Thermal conductivity as a function of temperature for 10-YSZ/alumina particulate composites containing 0 (10-YSZ), 5, 10, 20 and 30 mol% alumina, determined by a steady-state laser heat flux technique [30].

20 and 30-mol% alumina showed a sharper decrease in thermal conductivity with increasing temperature. Typical values of thermal conductivity as a function of alumina content determined at 1000°C are shown in Table 1.

7. CHOICE OF MATERIAL FOR STRUCTURAL RELIABILITY/LIFE

As seen from the aforementioned properties, the maximum flexure strength at 1000°C was achieved for the 30 mol% particulate composite, while the maximum fracture toughness at 1000°C was attained for the 30 mol% platelet composite. The resistance to SCG susceptibility was greater in the 30 mol% platelet composite with a higher SCG parameter of $n = 33$ than in the 30 mol% particulate composite with a lower SCG parameter of $n = 6$.

From the structural reliability/life point of view, a composite which gives long life and is strongest (in strength), toughest (in fracture toughness), stiffest (in elastic modulus) and lightest (in weight) is certainly the best choice for a fuel-cell component material. Elastic modulus increased and density decreased with increasing alumina content. It is obvious from Figures 5, 9, 15, and 4 that the 30 mol% particulate composite is the best material choice based on strength, fracture toughness, elastic modulus and density that were evaluated at *ambient temperature*. Since operating temperatures of typical SOFCs are within or close to 1000°C, the choice of a candidate material should not solely be based on ambient temperature properties but based on elevated temperature properties, particularly slow crack growth which controls life of fuel cell components.

With respect to elevated-temperature strength, the 30 mol% particulate composite is better than the platelet counterpart. In contrast, with regard to fracture toughness and SCG resistance, the platelet composite is better than the particulate counterpart. Hence, a unified choice of *a material* to satisfy all the important requirements – strong, tough, long life – can hardly be made. A case-by-case selection, depending on the types of operation/service conditions (temperature, loading (continuous or intermittent), environment, and components configurations, etc.), is needed. For example, if the components are subjected to a continuous, isothermal type of operations without frequent interruption (thus encountering little thermal shock loading, etc), then the 30 mol% platelet composite would be a good candidate since the material would give longer service life of components.

8. SUMMARY

At ambient temperature, both flexure strength and fracture toughness increased with increasing alumina content. For a given alumina content, strength of particulate composites was greater than that of platelet composites. Fracture toughness increased with alumina content with little difference in fracture toughness of the two types of composites.

At 1000°C, the 30 mol% particulate composites yielded the maximum strength, whereas the 30 mol% platelet composite exhibited the maximum fracture toughness. Fracture toughness was approximately 16% greater in the platelet composites than in the particulate composites.

The susceptibility to slow crack growth was high for both 0 mol% and 30 mol% particulate composites with lower SCG parameters of $n = 6-8$, whereas the susceptibility to SCG was low for the 30 mol% platelet composite with its higher SCG parameter of $n = 33$.

No significant difference in elastic modulus and density was observed for a given alumina content between the particulate and platelet composites. In both cases, the prediction from the rule of mixture was in good agreement with the experimental data. Elevated-temperature elastic modulus of both composite systems was characterized with a well-defined transition around 400°C, below which elastic modulus decreased monotonically and above which it remained almost unchanged up to 1000°C. Vickers microhardness of the particulate composites increased with increasing alumina content; while Vickers microhardness of the platelet composites followed an opposite trend, in which a significant decrease in hardness resulted at higher alumina contents. Thermal conductivity increased with increasing alumina content. Thermal conductivity showed slight change with temperature for 0, 5, and 10 mol% alumina compositions, whereas it decreased with temperature for composites containing 20 and 30 mol% alumina.

Thermal cycling/fatigue up to 10 cycles between 200 to 1000°C did not show any strength degradation of the 30 mol% platelet composites, indicative of negligible influence of CTE mismatch on residual stresses and/or microcracking between YSZ and alumina grains.

ACKNOWLEDGEMENTS

The authors are grateful to Ralph Pawlik for mechanical testing, John Setlock for materials processing, and Ralph Garlick for X-ray diffraction analysis. This work was funded by Zero CO_2 Emission Technology (ZCET) Project of the Aerospace Propulsion and Power Program, NASA Glenn Research Center, Cleveland, OH.

REFERENCES

1. N. Q. Minh, Ceramic fuel cells, *J. Am. Ceram. Soc.*, **76** [3] 563–588 (1993).
2. S. R. Choi and N. P. Bansal, Strength and fracture toughness of zirconia/alumina composites for solid oxide fuel cells," *Ceram. Eng. Sci. Proc.*, **23** [3] 741–750 (2002).
3. N. P. Bansal and S. R. Choi, Processing of alumina-toughened zirconia composites, NASA/TM—2003-212451, National Aeronautics and Space Administration, Glenn Research Center, Cleveland, OH (2003).
4. S. R. Choi and N. P. Bansal, Processing and mechanical properties of various zirconia/alumina composites for fuel cells applications, NASA/TM—2002-211580, National Aeronautics and Space Administration, Glenn Research Center, Cleveland, OH (2002); also CIMTEC 2002 Conference, paper no. G1:P03, June 14–18, 2002, Florence, Italy.
5. S. R. Choi and N. P. Bansal, Strength, fracture toughness, and slow crack growth of zirconia/alumina composites at elevated temperature, NASA/TM—2003-212108, National Aeronautics and Space Administration, Glenn Research Center, Cleveland, OH (2002).
6. S. R. Choi and N. P. Bansal, High-temperature flexure strength, fracture toughness, and fatigue of zirconia/alumina composites, *Ceram. Eng. Sci. Proc.*, **24** [3] 273–279 (2003).
7. ASTM C 1161, Test method for flexural strength of advanced ceramics at ambient temperature, *Annual Book of ASTM Standards*, Vol. 15.01, American Society for Testing & Materials, West Conshohocken, PA (2001).
8. (a) S. R. Choi and J. P. Gyekenyesi, 'Ultra'-fast fracture strength of advanced structural ceramics at elevated temperatures: an approach to high-temperature 'inert' strength," pp. 27–46 in *Fracture Mechanics of Ceramics*, Vol. 13, Edited by R. C. Bradt, D. Munz, M. Sakai, V. Ya. Shevchenko, and K. W. White, Kluwer Academic/Plenum Publishers, New York (2002); (b) S. R. Choi and J. P. Gyekenyesi, Elevated-temperature 'ultra'-fast fracture strength of advanced ceramics: an approach to elevated-temperature 'inert' strength, *ASME J. Eng. Gas Turbines & Powers*, **121** 18–24 (1999).
9. ASTM C 1211, Test method for flexural strength of advanced ceramics at elevated temperatures, *Annual Book of ASTM Standards*, Vol. 15.01, American Society for Testing and Materials, West Conshohocken, PA (2001).
10. F. F. Lange, Transformation toughening; Part 4. Fabrication, fracture toughness and strength of Al_2O_3-ZrO_2 composites,"*J. Mater. Sci.*, **17** 247–254 (1982).
11. I. K. Cherian and W. M. Kriven, Alumina-platelet-reinforced 3Y-TZP, *Am. Ceram. Soc. Bull.*, **80** [12] 57–63 (2001).
12. J. Kübler, (a) Fracture toughness of ceramics using the SEVNB method: preliminary results, *Ceram. Eng. Sci. Proc.*, **18** [4] 155–162 (1997); (b) Fracture toughness of ceramics using the SEVNB method; round robin, VAMAS Report No. 37, EMPA, Swiss Federal Laboratories for Materials Testing & Research, Dübendorf, Switzerland (1999).
13. ASTM C 1421, Test methods for determination of fracture toughness of advanced ceramics at ambient temperature, *Annual Book of ASTM Standards*, Vol. 15.01, American Society for Testing and Materials, West Conshohocken, PA (2001).
14. (a) S. R. Choi and V. Tikare, Crack healing of alumina with a residual glassy phase: strength, fracture toughness and fatigue, *Matl. Sci. Eng.* **A171** 77–83 (1993); (b) S. R. Choi and V. Tikare, Crack healing in silicon nitride due to oxidation, *Ceram. Eng. Sci. Proc.*, **12** [9–10] 2190–2202 (1991).
15. J. E. Srawley and B. Gross, Side-cracked plates subjected to combined direct and bending forces, pp. 559–579 in *Cracks and Fracture*, ASTM STP 601, American Society for Testing and Materials, Philadelphia (1976).
16. ASTM C 1465, Test method for determination of slow crack growth parameters of advanced ceramics by constant stress-rate flexural testing at elevated temperatures, *Annual Book of ASTM Standards*, Vol. 15.01, American Society for Testing and Materials, West Conshohocken, PA (2001).
17. S. M. Wiederhorn, Subcritical crack growth in ceramics, pp. 613–646 in *Fracture Mechanics of Ceramics*, Vol. 2, Edited by R. C. Bradt, D. P. H. Hasselman, and F. F. Lange, Plenum Press, New York (1974).
18. F. F. Lange, High-temperature strength behavior of hot-pressed Si_3N_4: evidence of subcritical crack growth, *J. Am. Ceram. Soc.*, **57** 84–87 (1974).

19. J. E. Weston and P. L. Pratt, Crystallization of grain boundary phases in hot-pressed silicon nitride materials, *J. Mater. Sci.*, **13** 2147–2156 (1978).
20. R. L. Tsai and R. Raj, The role of grain boundary sliding in fracture of hot-pressed Si_3N_4 at high temperatures, *J. Am. Ceram. Soc.*, **63** [1–2] 513–517 (1980).
21. K. Jakus T. Service, and J. E. Ritter, High-temperature fatigue behavior of polycrystalline alumina, *J. Am. Ceram. Soc.*, **64** 4–7 (1980).
22. A. G. Evans, Slow crack growth in brittle materials under dynamic loading condition, *Int. J. Fracture*, **10** [2] 251–259 (1974).
23. ASTM C 1368, Test method for determination of slow crack growth parameters of advanced ceramics by constant stress-rate flexural testing at ambient temperature, *Annual Book of ASTM Standards*, Vol. 15.01, American Society for Testing and Materials, West Conshohocken, PA (2001).
24. S. R. Choi and J. P. Gyekenyesi, Specimen geometry effect on the determination of slow crack growth parameters of advanced ceramics in constant flexural stress-rate testing at elevated temperatures, *Ceram. Eng. Sci. Proc.*, **20** [3] 525–534 (1999).
25. N. N. Nemeth, L. M. Powers, L. A. Janosik, and J. P. Gyekenyesi, Time dependent reliability analysis of monolithic ceramic components using the CARES/LIFE integrated design program, pp. 390–408 in *Life Prediction Methodologies and Data for Ceramic Materials*, ASTM STP 1201, Edited by C. R. Brinkman and S. F. Duffy, American Society for Testing and Materials, Philadelphia (1994).
26. ASTM C 1259, Test method for dynamic Young's modulus, shear modulus, and Poisson's ratio for advanced ceramics by impulse excitation of vibration, *Annual Book of ASTM Standards*, Vol. 15.01, American Society for Testing and Materials, West Conshohocken, PA (2001).
27. J. W. Adams, R. Ruh, and K. S. Mazdiyasni, Young's modulus, flexural strength, and fracture of yttria-stabilized zirconia versus temperature, *J. Am. Ceram. Soc.*, **80** [4] 903–908 (1997).
28. ASTM C 1327, Test method for Vickers indentation hardness of advanced ceramics, *Annual Book of ASTM Standards*, Vol. 15.01, American Society for Testing and Materials, West Conshohocken, PA (2001).
29. D. Zhu, N. P. Bansal, K. N. Lee, and R. A. Miller, Thermal conductivity of ceramic thermal barrier and environmental barrier coating materials, NASA/TM—2001-211122, National Aeronautics and Space Administration, Glenn Research Center, Cleveland, OH (2001).
30. N. P. Bansal and D. Zhu, Thermal conductivity of alumina-toughened zirconia composites, NASA/TM-2003-212896, National Aeronautics and Space Administration, Glenn Research Center, Cleveland, OH (2003).

Part V

Glass and Glass-Ceramic Composites

19

Continuous Fibre Reinforced Glass and Glass-Ceramic Matrix Composites

Aldo R. Boccaccini
Department of Materials, Imperial College London, London SW7 2AZ, UK
Tel.: 44 207 5946731, Email: a.boccaccini@imperial.ac.uk

ABSTRACT

Glass and glass-ceramic matrix composites with continuous fibre reinforcement are considered in this Chapter, covering aspects of their fabrication, microstructural characterisation, properties and applications. The great variety of composite systems developed during the last 30 years is discussed and their outstanding thermomechanical properties and high technological potential are highlighted. These composites constitute a new family of high-temperature capability, lightweight structural materials exhibiting quasi-ductile fracture behaviour, characteristics which allow their application as critical components, e.g. in the aerospace sector, energy-conversion systems, metallurgy processes, precision and microengineering, special machinery as well as in automotive components and chemical plants. Remaining challenges for the future development of these composites which currently limit their wider commercial exploitation are discussed.

1. INTRODUCTION

The use of fibre reinforcement in silicate glass and glass-ceramic matrices started at the end of the decade after 1960 [1, 2]. A great variety of composite systems has been developed since then employing numerous matrix compositions and many types of ceramic and metallic fibre reinforcements. Interest in these inorganic composites, which constitute a particular class of ceramic matrix composites (CMCs), arises from their outstanding thermomechanical properties and potential use at temperatures up to about 1200°C. These characteristics allow their application in critical components for use in severe conditions involving relatively high temperatures, high stress levels and aggressive environments [3, 4]. The purpose of this Chapter is to give an up-to-date appreciation of the underlying science and engineering of glass and glass-ceramic composite materials with fibre reinforcement, focussing on applications, processing and thermomechanical properties. The most recent literature reviews on these composite systems were published in 1995 [4, 5] and 2001 [6].

2. APPLICATIONS

High-temperature, aerospace and impact resistant applications

The original objective for research on fibre reinforced glass and glass-ceramic composites was to develop light materials with suitable mechanical and thermal properties for applications in gas turbines and other specialised areas demanding high oxidation and corrosion resistance and high temperature capability. Since these materials have approximately half the density of most Ni-based alloys, significant weight savings in components are possible. Thus, the high strength-to-weight-ratio, coupled with the high oxidation resistance make these composite systems candidates for general high-temperature aerospace applications, including thermal protection shrouds, leading-edges and rocket nozzle inserts [7, 8]. The application of fibre-reinforced glass and glass-ceramic composites in gas turbines for military and commercial propulsion has also been considered [9–11]. However, cost-effective manufacturing technologies required to fabricate real component shapes (e.g. turbine blades) and complex structures have not been developed yet. For example, only limited work has been devoted to the development of machining techniques [12, 13] and the joining of composite parts to ceramics and metals has also received little attention [14].

Another application area of fibre-reinforced glasses and glass-ceramics is in ballistic-protection materials and similar impact resistant structures at room and high temperatures [15, 16]. In this application area, fibre reinforced-composites having higher toughness and spalling resistance than monolithic ceramics can offer a significant advantage.

Automotive and special machinery applications

A wide range of application possibilities exists for glass and glass-ceramic matrix composites in conventional technologies, i.e. at low to moderate temperatures and under low to moderate stresses. These applications include: components for pump manufacture (e.g. bearings and seals), automotive applications (e.g. brake and gear systems) and construction of special machinery (e.g. tools and components for use in hot environments). Fibre-reinforced glass matrix composite products are being used with commercial success in the handling of hot materials, e.g. low-melting point metals and glasses [17], providing adequate performance in the extreme conditions of high working temperature, thermal shock loading and aggressive tribological interactions.

Successful applications in corrosive and dusty environments, both at room and elevated temperatures, may require the development of protective coating systems. Several studies have shown that the hot corrosion of glass-ceramic matrix composites, for example under liquid sodium sulfate at 900°C in air and argon environments, is poor: the composites show more corrosion damage than monolithic glass-ceramics [18, 19]. This behaviour imposes a serious limitation for uses in naval gas turbine engines where the aggresive environment of sea water and fuel will be active. Oxide coatings have been proposed to tackle this problem [20]. Coated glass and glass-ceramic matrix composites could find application in the chemical process industry.

Glass matrix composites, in particular with carbon fibre reinforcement, have been also proposed for a variety of applications which require thermal dimensional stability, i.e. materials with multidimensional near zero thermal expansion coefficients such as support structures for laser mirrors [21, 22].

Electronic, biomedical and other functional applications

Nicalon® fibre-reinforced glass matrix composites have been proposed for producing substrates for electronic packages [23]. The optimum thermal conductivity and dielectric constant at 1 MHz, coupled with high fracture strength and toughness, are attractive properties for this application. Low dielectric loss fibre-reinforced glass matrix composites have been also proposed for Rf transmission window applications [24]. Analysis of the literature reveals, however, that no further research has been conducted regarding functional applications of fibre-reinforced glass and glass-ceramic composites. An exploration of this research area thus remains an interesting challenge.

The biomedical field offers another potential area for broader application of composites with biocompatible glass and glass-ceramic matrices. Fibre reinforcement could be used to enhance the mechanical properties of components made of bioctive glass with the aim of fabricating load-bearing implants [25, 26]. Although fibre reinforcement may provide the necessary structural integrity, the biocompatibility of the composite must equally be achieved [27].

A futuristic application for fibre-reinforced glass matrix composites is related to the use of lunar materials for future space construction activities. Glass/glass composites in which both the fibre and the matrix are made of fused lunar soil have been proposed [28]. These materials, obtained so far on a laboratory scale, show great promise for providing large quantities of basic structural materials for cost-effective outer-space construction.

3. OVERVIEW OF CONTINUOUS FIBRE-REINFORCED GLASS AND GLASS-CERAMIC MATRIX COMPOSITES

A great variety of silicate matrices has been considered for the fabrication of fibre-reinforced glass and glass-ceramic matrix composites [4–6, 29–31]. Typical matrices investigated are listed in Table 1. Table 2 gives an overview of different composite systems developed and some of the most remarkable properties achieved.

TABLE 1. Some glass and glass-ceramic matrices commonly used to fabricate fibre-reinforced composites [6]

Material	Elastic Modulus (GPa)	Thermal Expansion Coefficient ($10^{-6}°C^{-1}$)	Fracture Strength (MPa)	Density (g/cm^3)	Softening point (°C)
GLASS MATRICES:					
Silica glass	84	1.8	70–105	2.5	1300
Borosilicate	63	3.3	70–100	2.2	815
Aluminium silicate	90	4.1	80	2.6	950
GLASS-CERAMIC MATRICES:					
Lithium aluminosilicate	100	1.5	100–150	2.0	n.a.
Cordierite (magnesium aluminosilicate)	119	1.5–5.5	110–170	2.6–2.8	1450
Barium magnesium aluminosilicate	125	1.2	140	2.8	1450
Calcium aluminosilicate	110	n.a.	100–130	3.0	1550

Carbon fibre reinforced glass matrix composites

The first paper on the fabrication and characterisation of carbon fibre-reinforced glass matrix composites was published in 1969 [1]. Major developments in these composite systems were carried out during the 70s and 80s, especially in USA, England and Germany [2, 29, 30, 32, 46]. There are still considerable research efforts worldwide in this area [6, 43, 47–54]. A recent development is the use of nitride glass matrices, for example Y-Si-Al-O-N glass [42].

The interfacial properties in carbon fibre composites depend primarily on the physical structure and chemical bonding at the interface and on the type of carbon fibre used [48]. Moreover, the interfacial strength in these composites can be influenced by changing the chemistry of the matrix [55]. The major disadvantage of these composites is the limited temperature capability in oxidising atmospheres at high temperatures [49]. The oxidation behaviour under different conditions has been investigated [50, 56, 57].

3.1. Nicalon® and Tyranno® fibre-reinforced glass and glass-ceramic matrix composites

Glass and glass-ceramic matrix composites reinforced with SiC-based fibre of the type Nicalon® or Tyranno® combine strength and toughness with the potential for high temperature oxidation resistance. A great variety of silicate matrices has been reinforced with these fibres [4–6, 29, 32, 34, 11, 13, 58–73] (see also Table 2). Several products have reached commercial exploitation, such as the material FORTADUR® (Schott Glas, Germany) [34] and the composite Tyrannohex® (Ube Industries, Japan) [72, 73]. Glass matrices such as silica, borosilicate and aluminosilicate have been used, as well as glass-ceramic matrices, such as lithium aluminosilicate (LAS), magnesium aluminosilicate (MAS), calcium aluminosilicate (CAS), barium aluminosilicate (BAS), barium magnesium aluminosilicate (BMAS), calcium magnesium aluminosilicate (CMAS), yttrium aluminosilicate (YAS), lithium magnesium aluminosilicate (LMAS) and yttrium magnesium aluminosilicate (YMAS). Oxynitride glass matrices (e.g. of $Y_{44}Si_{81}Al_{48}O_{240}N_{40}$ and $Li_{40}Si_{80}Al_{40}O_{216}N_{16}$

TABLE 2. Overview of fibre-reinforced glass and glass-ceramic matrix composites (σ : ultimate fracture strength, K_{Ic} : fracture toughness*, 2-D: 2 dimensional reinforcement) [6]

Matrix / Fibre	Properties investigated	Reference
Borosilicate / SiC (Nicalon®)	$\sigma = 1200$ MPa, $K_{Ic} = 18$ MPm$^{1/2}$ (up to 600°C)	[32]
Magnesium aluminosilicate / SiC-monofilament	$\sigma = 600$ MPa, (up to 1000°C) $\sigma = 700$ MPa (up to 1100°C in air)	
Lithium aluminosilicate / SiC- (Nicalon)	$\sigma = 1100$ MPa (up to 1100°C in argon)	
Silica / carbon	$\sigma = 800$ MPa, (at room temperature)	[31]
Borosilicate / Metal (Ni-Si-B alloy)	$\sigma = 225$ MPa (up to 500°C)	[33]
Borosilicate / SiC (Nicalon®)	$\sigma = 840$ MPa, $K_{Ic} = 25$ MPm$^{1/2}$ (up to 530°C)	[29, 34]
Aluminosilicate / SiC (Nicalon)	$\sigma = 1200$ MPa, $K_{Ic} = 36$ MPm$^{1/2}$ (up to 700°C)	
Borosilicate / carbon	$\sigma = 800$–1000 MPa, $K_{Ic} = 35$ MPm$^{1/2}$ (up to 400°C)	
Borosilicate / mullite	$\sigma = 150$ MPa, $K_{Ic} = 2.5$ MPm$^{1/2}$ (up to 600°C)	[35]
Ca-aluminosilicate / SiC (Nicalon®)	$\sigma = 290$ MPa, (up to 1000°C)	[36]
Silica / SiC (Nicalon®), 2-D	$\sigma = 205$ MPa, (up to 1000°C)	[37]
Magnesium aluminosilicate / SiC (Nicalon®)	$\sigma = 1057$ MPa, (up to 500°C) $\sigma = 414$ MPa, (up to 700°C)	[38]
Barium-magnesium aluminosilicate / SiC (Nicalon®)	$\sigma = 900$ MPa, (up to 1100°C)	[39]
Barium-magnesium aluminosilicate / SiC (Tyranno®)	$\sigma = 700$ MPa (up to 1100°C)	[40]
YSiAlON, LiSiAlON / carbon	$\sigma = 323$ MPa, $K_{Ic} = 12$ MPm$^{1/2}$	[41, 42]
YSiAlON, LiSiAlON / SiC (Nicalon®)	$\sigma = 323$ MPa, $K_{Ic} = 12$ MPm$^{1/2}$	
Y-magnesium aluminosilicate / carbon	$\sigma = 1100$ MPa	[43]
Ba-Sr-aluminosilicate / SiC (Hi-Nicalon®)	$\sigma = 900$ MPa, E = 165 GPa	[44]
Mica glass-ceramic / SiC (Nicalon®)	High machinability	[45]

* K_{Ic} values are often employed to quantify composite toughness. These values must be treated with caution, however, because, strictly speaking, linear elastic fracture mechanics cannot be applied to these complex composite systems. The values serve for comparison purposes

compositions) have also been considered [41]. The use of refractory glass-ceramic matrices has the objective to develop materials with high temperature capability (>1000°C).

A distinct member of the family of oxycarbide fibre materials is Tyrannohex®, a composite developed on the basis of bonded Tyranno® fibres [72, 73]. The microstructure of this composite is shown in Figure 1. The fibre content reaches values up to 90 vol%. Due to the near absence of a matrix phase, the material retains high strength up to very high temperatures (195 MPa at 1500°C) and it has a very high creep resistance. The new generation Tyrannohex® material, fabricated with novel sintered high-performance SA-Tyranno® fibre has shown improved thermomechanical properties up to 1600°C [74].

Detailed high-resolution electron microscopy and microanalytical investigations have been conducted to elucidate the phase structure and microchemical composition of the fibre/matrix interfacial region in a variety of composite systems [36, 71, 75]. These investigations showed that the interfacial zone is occupied by a carbon-rich thin layer of thickness 10–50 nm, depending on the matrix/fibre combination. Figure 2 is a transmission

FIGURE 1. SEM micrograph showing the microstructure of a Tyrannohex® composite [72]. The fibre volume fraction of these composites is very high in comparison to conventional composites (e.g. in Figure 1). (Micrograph reproduced with permission from Ref. [6]).

electron microscopy (TEM) image showing the interfacial region in a Nicalon® fibre reinforced borosilicate glass matrix composite [76]. The carbon-rich interfacial layer is clearly observed. High-resolution electron microscopy of these interfaces has revealed that the layers contain graphitic carbon textured to varying degrees [75]. This carbonaceous layer is weaker than the matrix so that the fibres are effective in deflecting matrix cracks and promote fibre pull-out during composite failure. As explained below (Section 5), this is the main mechanism leading to the high fracture toughness and flaw-tolerant fracture behaviour of this class of composites. At the same time, the interface layer allows load transfer from matrix to fibre so that strengthening takes place.

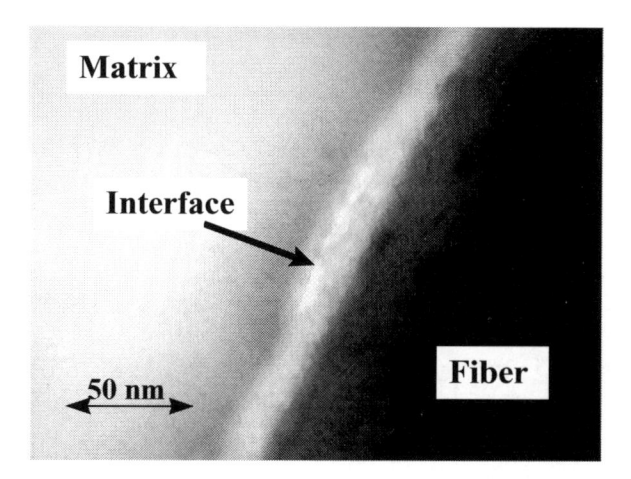

FIGURE 2. TEM micrograph showing the interfacial zone in a SiC-Nicalon® fibre-reinforced borosilicate glass matrix composite. The carbonaceous interface is clearly observed. (Micrograph reproduced with permission from Ref. [6]).

The mechanisms of formation of this carbon-rich layer at the fibre-matrix interfaces have been studied [36, 71, 75, 77]. It has been proposed that the carbon layer is the result of a fortunate combination of silicate matrix chemistry and non-stoichiometric/non-crystalline fibre structure [71]. The solid-state reaction between the SiC in the fibre and oxygen from the glass and the fibre surface can be written as:

$$SiC(s) + O_2 \rightarrow SiO_2(s) + C(s)$$

When using SiC-based fibres of the type Nicalon® and Tyranno®, the temperature capability of the composites depends strongly on the environment [4–6]. In oxidising atmospheres (i.e. hot air), the stability of the carbon-rich fibre-matrix interface represents the limiting factor. One way to alleviate the problem of degradation of the carbon layer is to use protective coatings on the fibres. These coatings should act as a diffusion barrier layer and a low decohesion layer. A variety of single-layer and two-layer coatings produced by chemical vapour deposition (CVD) have been tried in attempts to replicate the mechanical response of the carbon-rich interfaces, including C, BN, BN(+C), BN/SiC and C/SiC coatings [39, 71, 78, 79].

SiC and boron monofilament-reinforced glass matrix composites

Monofilament SiC fibres, produced by chemical vapour deposition (CVD), are normally fabricated in diameters ranging from 100 to 140 μm. Since these monofilaments are less flexible than the fibres derived from organo-silicon polymers, only simple shape components can be produced. Composites were fabricated using up to 65 vol.% SiC monofilaments (type SCS-6) in a borosilicate glass matrix [32]. The higher elastic modulus of SiC monofilament results in composites significantly stiffer than those fabricated using Nicalon® or Tyranno® fibres. A major disadvantage of using these filaments is their large diameter, which may lead to extensive microcracking in the matrix and, therefore, to unacceptable low off-axis strength.

SiC monofilaments have been used in borosilicate and aluminoslicate glass matrices for the fabrication of model composites [80], including composites with a transparent matrix [81]. These were conveniently used to investigate the development of matrix microcracking under flexure stresses since the formation of the microcracking pattern in the transparent glass matrix could be observed in-situ.

Researchers at NASA Lewis Research Centre have used SiC monofilaments for the reinforcement of glass-ceramic matrices of refractory compositions, including strontium aluminosilicate (SAS) [82] and barium aluminosilicate (BAS) [83], with temperature capability of up to 1600°C.

Boron monofilaments (about 20 vol.%) were used by Tredway and Prewo to reinforce borosilicate glass [84]. They also introduced carbon fibre yarn to fill in the glass-rich regions between monofilaments in order to toughen the matrix and provide additional structural integrity of the composite.

Glass matrix composites with oxide fibres

Alumina fibres have been incorporated into LAS glass-ceramics [69] while aluminosilicate and mullite (Nextel®-type) fibres have been used in borosilicate and other glass matrices [35, 85, 86]. Nextel® fibre-reinforced alumina/silica composites have been suggested for applications as divergent flaps and seals in advanced aerospace turbines [87]. A variety of glass matrix composites containing oxide fibres have also been prepared as model composite materials to investigate interfacial phenomena or development of residual stresses [88–91]. In recent research a mesh-like Al_2O_3-ZrO_2 minicomposite fibre structure was used to reinforce borosilicate glass matrices in order to obtain optically transparent composites (optomechanical composites) [92].

This group of composites has been developed primarily for applications above 1000°C. However, due to the strong bond which usually occurs as a result of chemical reactions at the fibre/matrix interface, the composites showed a less fibrous fracture surface and hence, a poorer flaw-tolerant behaviour in comparison to Nicalon® or Tyranno® fibre reinforced materials. To overcome this problem several coating approaches for the fibres were tried, with the aim of providing relatively weak fibre-matrix interfaces. Such a coating (or interphase) must act as a diffusion barrier layer at the intended application temperature, i.e. impeding the chemical reaction at the oxide fibre/glass interface. Moreover, the coating should also serve to deflect or arrest cracks in order to enable the fibres to act as toughening elements. A further requirement for the coating material is that it must have thermal expansion compatibility with the fibre and matrix. SnO_2 is a refractory material proposed for coating alumina and aluminosilicate fibres [85, 93]. Carbon has also been considered as a coating material for oxide fibres used to reinforce glasses [94]. However, the use of carbon coatings causes the principal advantage of oxide fibre reinforced composites, which is their intrinsic oxidation resistance, to be lost.

Glass/glass-ceramic fibre – glass matrix composites

The first reports on the preparation of this class of composites were published by Japanese researchers using *chopped* SiCaON glass fibres and matrices [95]. The fabrication of model composites consisting of a single optical fibre embedded in borosilicate glass [96] and of continuous oxynitride glass fibre reinforced glass matrix composites with a SiO_2-B_2O_3-La_2O_3 glass matrix has been also reported [97]. Further work on silicate glass matrix composites with continuous silicate glass fibre reinforcement has been conducted in Germany [86, 98] and the UK [99]. In a similar way as when using crystalline fibres, there is need for engineering the interface in glass/glass composites in order to avoid strong bonding which results from chemical reactions during composite fabrication. A declared goal of further research in the area of glass/glass composites is the development of transparent or translucent materials showing high fracture toughness and adequate flaw tolerant behaviour. Figure 3 shows the optical transparency of soda-lime glass matrix composites containing basalt or AR-glass fibres [99]. A major challenge in the development of such composites is to be able to incorporate interfaces which are optically, chemically, thermally and mechanically compatible with the matrix and fibres. One suggested way to tailor the interfaces is to use dense, transparent (or translucent) nanosized oxide coatings. In this regard, translucent tin dioxide may be a good candidate [85, 93]. Other suggested oxide coating on silicate fibres used in transparent glass matrix composites is titanium dioxide produced by the sol-gel method [99]. Glass fibre-reinforced glass matrix composite materials may lead to interesting products for replacement of laminate glass in applications requiring relatively high fracture

FIGURE 3. Images of silicate fibre-reinforced soda-lime glass matrix composites demonstrating their light transmittance and transparency: (a) basalt fibre, (b) AR-glass fibre reinforcement [99].

strength and toughness, impact and thermal shock resistance as well as optical transparency at high temperatures.

Metal fibre-reinforced glass matrix composites

An advantage of these composites is the increased resistance to fibre damage during composite processing which results from the intrinsic ductility of metallic fibres and the possibility of exploiting their plastic deformation for composite toughness enhancement [100]. A penalty is paid due to the relatively low thermal capability and poor chemical resistance of the metallic fibre reinforcement, limiting the application temperature and environment.

In an earlier study, Ducheyne and Hench fabricated composite materials with a bioactive glass (Bioglass®) matrix and stainless steel fibre reinforcement by an immersion technique [25]. Donald et al. have reported on the fabrication of glass and glass-ceramic matrix composites reinforced by stainless steel and Ni-based alloy filaments with diameters in the range 4 to 22 μm [100, 101]. Glass-encapsulated metal filaments prepared by the Taylor-wire process were used for the fabrication of the composites [100].

FIGURE 4. SEM micrograph of the fracture surface of a soda-lime glass matrix composite reinforced by stainless steel fibres. The fracture surface exhibits fibre pull-out and partial plastic deformation of the fibres. (Micrograph reproduced with permission from Ref. [6]).

Russian researchers have demonstrated the use of 2-dimensional metal fibre structures to reinforce glass and glass-ceramics [102]. However, only a limited number of glass matrices and metallic fibre reinforcements were tried. More recently, Boccaccini and co-workers have used electrophoretic deposition to fabricate a number of glass matrix composites containing 2-dimensional metal fibre reinforcement [103, 104]. In particular, soda-lime, borosilicate, cathode-ray tube recycled glass and bioactive glass were used as matrices and a variety of commercially available stainless steel fibre mats were used as reinforcement. Figure 4 shows a scanning electron microscopy image of the fracture surface of a soda-lime glass matrix composite reinforced by 2-D stainless steel fibre fabric [103]. The fracture surface exhibits fibre pull-out and partial plastic deformation of the fibres, which indicates flaw-tolerant behaviour of the composite.

4. PROCESSING

The ease of processing in comparison to polycrystalline ceramic matrix composites is one of the outstanding attributes of glass and glass-ceramic matrix composites. This is due to the ability of glass to flow at high temperatures in a similar way to resins, which is exploited in different fabrication strategies as discussed below.

Slurry infiltration and hot-pressing

Hot-pressing of infiltrated fibre tapes or fabric lay-ups is the most extensively used technique to fabricate dense fibre-reinforced glass matrix composites. Figure 5 shows a

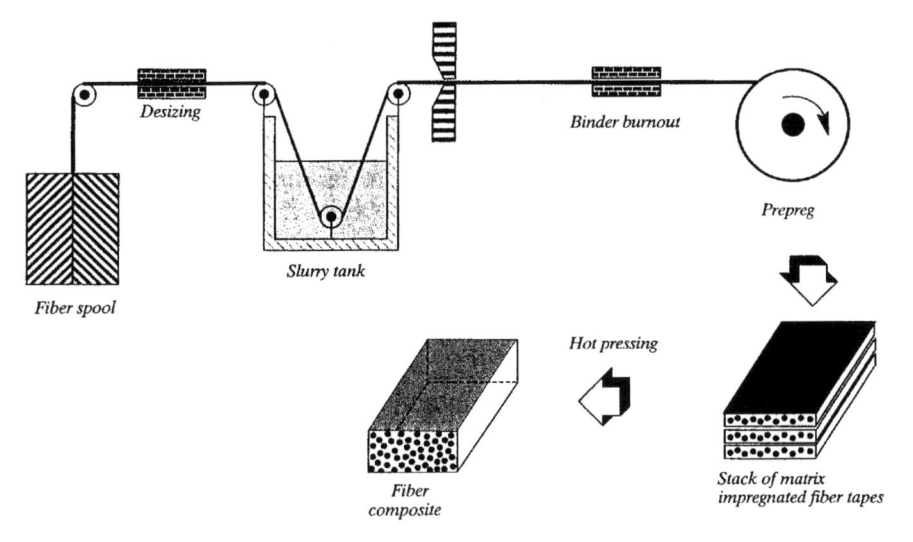

FIGURE 5. The standard slurry impregnation and hot-pressing route to fabricate fibre-reinforced glass matrix composites [105].

schematic diagram of the standard fabrication process [105]. For proper densification, the time-temperature-pressure schedule during hot-pressing must be optimised so that the glass viscosity is low enough to permit the glass to flow into the spaces between individual fibres within the tows. The processing temperature must be chosen after taking into account the possible occurrence of crystallisation of the glass matrix and the "in-situ" formation of the carbonaceous interface, as explained in Section 3, when oxycarbide fibres (e.g. Nicalon® or Tyranno®) are used. The pressure (of about 10–20 MPa) is usually applied after the temperature reaches the softening point of the matrix glass. In this manner almost fully dense composites can be fabricated. A typical microstructure of a borosilicate glass matrix composite reinforced with 40 vol% unidirectional SiC-Nicalon® fibres fabricated by the described technique is shown in Figure 6. The carbonaceous interface in the composite, shown in Figure 3, is formed during hot-pressing at about 1100°C [71].

When a glass-ceramic matrix is required, the densified composites are subjected to a "ceraming" heat-treatment after densification. In this case, an optimised time-temperature "window" must be found where densification takes place by viscous flow of the glass matrix before the onset of crystallisation. Ideally, the temperature range at which densification occurs at maximum rate lies between the softening temperature of the glass and the onset of crystallisation temperature [106]. A post-densification heat-treatment leads to the desired crystalline, refractory microstructure of the matrix. In some systems, the crystallisation of the glass-ceramic matrix can be achieved during hot-pressing [68].

The possibility for net shape fabrication of composite components using hot-pressing has been shown in the literature [34, 60, 69, 107]. Figure 7 shows, as examples, a variety of fibre reinforced glass matrix composite parts with different shapes, fabricated by hot-pressing (Schott Glas, Germany).

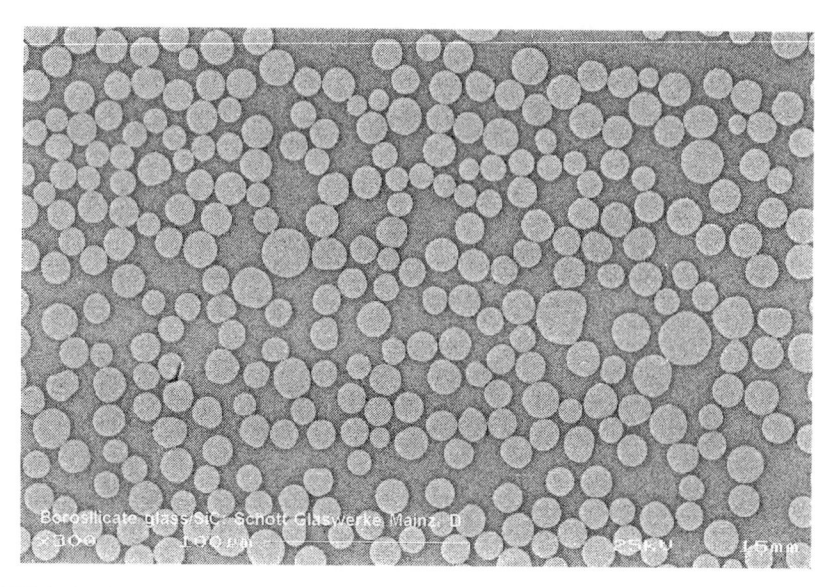

FIGURE 6. A typical microstructure of a borosilicate glass matrix composite reinforced with unidirectional SiC-Nicalon® fibres (40 vol%) fabricated by hot-pressing (Schott Glas, Mainz, Germany). The microstructure is characterised by the absence of porosity and a fairly homogenous distribution of the fibres [6].

FIGURE 7. Fibre-reinforced glass matrix composite parts with different shapes all fabricated by hot pressing. (Photograph courtesy of Schott Glas, Germany, reproduced with permission from Ref. [6]).

Tape casting

When using large-diameter monofilaments as reinforcement (e.g. SCS SiC), the slurry infiltration route shown in Figure 5 is not appropriate due to the lack of flexibility of the filaments. In these composites, a tape casting route is usually employed to fabricate the "green" bodies [62, 98]. The densification takes place by uniaxial hot-pressing or by hot-isostatic-pressing [108, 109]. A very uniform array of the fibres in the matrix can be achieved by this processing technique.

Sol-gel, colloidal routes and electrophoretic deposition

Using the sol-gel approach, matrices are produced from metal alkoxide solutions or colloidal sols as precursors [110–113]. A great variety of glass and glass-ceramic matrices has been prepared by sol-gel processing, including borosilicate, and lithium- (LAS), magnesium- (MAS), barium- (BAS), calcium- (CAS) and sodium- (NAS) aluminosilicates. The composites are fabricated by drawing the fibres through a sol in order to deposit a gel layer on the fibre surfaces. After gelation and drying, the prepregs are densified by hot-pressing, but pressureless sintering densification is also possible. Potential advantages of the sol-gel method over slurry processing include: more effective fibre infiltration enabling a reduction of the hot-pressing temperature, which in turn leads to limitation of damage to the fibres, and the ability to tailor matrix composition in order to control thermal expansion mismatch and fibre-matrix interfacial chemistry [114]. Another advantage of using sol-gel processing is the possibility of infiltrating complex fibre architectures and to develop near-net size and shape manufacturing technologies. Disadvantages of the sol-gel method are associated with large matrix shrinkage during drying and densification due to low solid content of the sols, leading to matrix cracking and residual porosity. Moreover, the process is time-consuming, usually requiring a large number of infiltration/drying steps.

Another way to fabricate glass matrix composites is to combine the sol-gel and slurry approaches [115]. This can overcome the large shrinkages inherent in the sol-gel process while maintaining the advantages of ease of infiltration and lower fabrication temperatures.

An alternative process to fabricate glass and glass-ceramic matrix composites involves electrophoretic deposition (EPD) to infiltrate the fibre preform with matrix material followed by conventional pressureless sintering or hot-pressing for densification of the composite [103]. A schematic diagram of the electrophoretic deposition cell is shown in Figure 8 [6, 103]. If the deposition electrode is replaced by a conducting fibre preform, the suspended charged (nano) particles will be attracted into and deposited within it. Using EPD, a range of glass and glass-ceramic matrix composites containing 2-dimensional fibre reinforcement, including composites of tubular shape, has been fabricated [37, 103, 104, 116–118].

Other fabrication methods

A technique suggested to fabricate complex shaped articles is the matrix transfer molding technique [69]. Woven structures used as reinforcement are arranged inside a mold cavity. Fluid matrix is transferred at high temperature into the mold cavity to fill the void space around the reinforcement structure. In this way, for example, thin-walled cylinders can be fabricated. Another proposed method for obtaining structures of complex shapes uses a

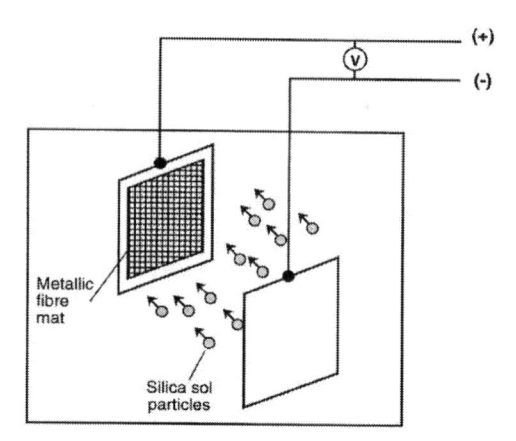

FIGURE 8. Schematic diagram of the electrophoretic deposition cell for obtaining silica deposits on metallic fibre mats. The metallic fibre mat acts as the positive electrode when the particles in the sol are negatively charged. (Figure reproduced with permission from Ref. [6]).

superplastically deformed foil [119]. The foil is used to partially encapsulate the composite sample and a part holding die, which are both contained in a rigid box held in a press. This method allows near-net-shape manufacturing but it involves complex and time consuming operations [119].

Efforts have also been made to evaluate the utility of the polymer precursor method for processing fibre reinforced glass-ceramics [120]. In general, the precursor approach provides access to glass-ceramics via low temperature processing methods with good control of chemical and phase homogeneity [120]. Because the polymer precursor method has the potential to yield near-net-shape products at relatively low temperatures, this processing route represents an interesting alternative to the established slurry and hot-pressing technique.

Another simple method of fabricating unidirectional fibre reinforced glass matrix composite rods is based on pulltrusion [121]. The technique is simple and can potentially be used for a wide range of fibres and matrices, it enables unidirectional composites in the form of rods with different cross-sections to be fabricated. A related processing method yielding composite shapes in the form of longitudinal rods or cylinders is the extrusion technique developed by Klein and Röeder [122] (see also Chapter 20).

The densification of Nicalon® fibre-reinforced borosilicate glass matrix composites by microwave heating has also been investigated [123]. The results demonstrated that microwave processing could be a highly efficient method, saving time and reducing costs, in comparison with traditional hot-pressing densification.

5. DISCUSSION OF PROPERTIES

The main objective of research in fibre-reinforced glass and glass-ceramics is the development of strong, tough, reliable and corrosion-resistant materials for high-temperature applications [4–6]. Figure 9 offers a comparison of fibre-reinforced glass and glass-ceramic

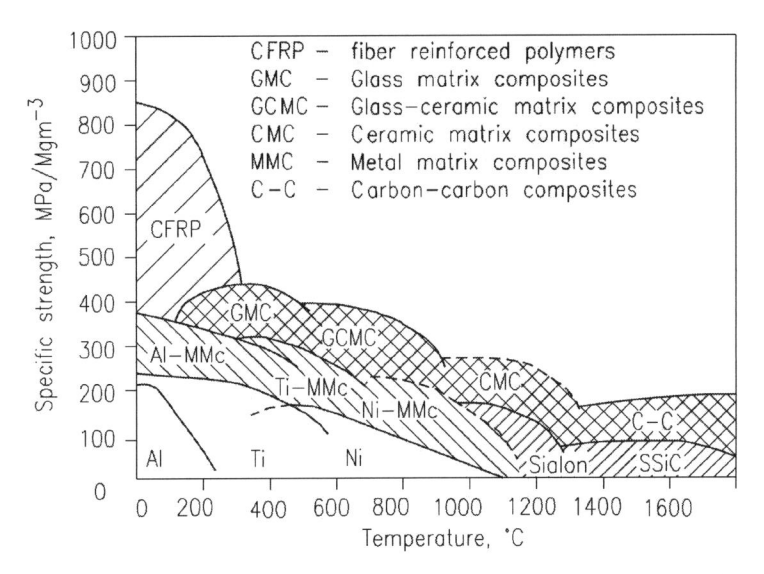

FIGURE 9. Specific strength and temperature capability of engineering materials showing the relative position of fibre-reinforced glass and glass-ceramic matrix composites. (After de Voorde [124]).

matrix composites and other engineering materials in terms of specific strength versus temperature capability [124]. The focus in the last 20 years has been placed on understanding the composites' thermomechanical properties and fracture behaviour under different conditions and environments, which is reflected by the large number of publications on the subject. Table 2 shows a summary of composite systems and associated mechanical properties. In this Section, the significant thermomechanical properties are discussed briefly and relevant references are provided. The discussion is focused on systems with oxycarbide fibre reinforcement.

Toughnening

The primary aim of fibre reinforcement of glass and glass-ceramics is to achieve high toughness while retaining high fracture strength. The basic approach to increasing the resistance of brittle materials to catastrophic failure is by increasing the crack propagation area introducing weak interfaces. The quantitative principles of interface design in fibre-reinforced glass matrix composites have been discussed in the literature and the toughening mechanisms that are active in these composites have been studied extensively [71, 125, 126]. The major processes contributing to the fracture toughness (work of fracture) of fibre-reinforced glass matrix composites are crack bridging, fibre debonding and fibre pull-out [126]. For the composites reinforced by oxycarbide fibres, a carbon-rich phase at the fibre/matrix interface usually forms *in-situ* during processing (see Figure 2). The carbonaceous layer is weaker than the matrix and the fibres and is therefore effective in deflecting matrix cracks and promoting fibre pull-out during composite failure. As an example, a typical fracture surface of a SiC (Nicalon®) fibre-reinforced borosilicate glass matrix composite is shown in Figure 10.

FIGURE 10. A typical fracture surface of a borosilicate glass matrix composite reinforced by Nicalon® fibres exhibiting extensive fibre pull-out.

High Temperature Properties

The high-temperature mechanical properties of fibre-reinforced glass and glass-ceramic matrix composites have been frequently reported in the literature for materials tested both under tensile and bending loadings [3, 6, 39, 40, 71, 74, 128–130]. In glass matrix composites, the temperature capability of the materials is determined primarily by the softening and viscous flow deformation of the glass matrix. In glass-ceramic matrix composites, in contrast, oxidation resistance of the fibres and interfaces determine the high-temperature behaviour of the composites.

Embrittlement and weakening, i.e. loss of fracture toughness and strength, in composites tested in air at temperatures between about 450 and 1000°C are the result of interfacial oxidation, as observed by several authors [3, 5, 6, 42, 65, 128–130]. The generic problem for composites reinforced by oxycarbide fibres is the oxidation of the carbon-rich interlayer at the interface, a process which occurs either due to channelled reaction from exposed fibre ends or because of microcracks in the matrix [3, 40, 126].

The difference in thermal expansion between matrix and fibre reinforcement is an important parameter influencing the mechanical behaviour of the composites because it determines the residual stress distribution after fabrication [105]. In glass matrix composites, these residual stresses can be measured by photoelastic techniques [127].

Oxygen reacts with carbon according to the reaction:

$$C + O_2 \rightarrow CO_2$$

Following interface oxidation, the gap is filled by SiO_2, which is produced by oxidation of the SiC, according to:

$$SiC + 2O_2 \rightarrow SiO_2 + CO_2$$

As a consequence of this reaction, a strongly bonded SiO_2 oxidation layer is formed at the interface impeding crack deflection and fibre pull-out, thus rendering the composite brittle [3, 126].

Analysis of the literature reveals that sufficient understanding of the thermomechanical behaviour of fibre-reinforced glass and glass-ceramic matrix composites at high-temperatures exists, including adequate knowledge of their thermal shock and thermal ageing behaviour [58, 65, 131–136], as well as high-temperature fatigue and creep resistance [64, 65, 137–139]. Interestingly, glass and glass-ceramic matrix composites exhibit an ability to heal microcracks developed in the matrix (as a result for example of thermal shock damage) by means of a suitable annealing heat-treatment by exploiting viscous flow of the glass [58, 132, 140]. It appears as if these composites are smart materials, able to heal their own flaws. Thus, the early non-destructive detection of damage in the form of matrix microcracks (induced for example by severe thermal shock) acquires a major practical significance. Upon detecting a significant change in Young's modulus or damping capacity, for example, a given component could be subjected to a heat-treatment to restore its structural integrity by healing matrix microcracks, thus prolonging its service life [58].

Other properties: impact resistance, wear and erosion behaviour

The resistance of fibre-reinforced glass and glass-ceramic matrix composites to damage under impact loads at room and elevated temperature has been reported in the literature [64, 141, 142]. In particular, carbon and Nicalon® fibre-reinforced borosilicate glass and LAS glass-ceramic matrix composites were tested. When compared to monolithic ceramics, such as silicon nitride, it was shown that the impact resistance of the glass-ceramic composite was over 50 times greater. Work has been also carried out on ballistic impact resistance of carbon and Nicalon® fibre-reinforced composites using gas guns and stainless steel projectiles [15, 142].

The tribological properties (dry sliding and abrasive wear behaviour) of carbon and SiC fibre-reinforced glass matrix composites against different ceramics, metals and abrasive counterparts have been investigated [6, 47, 52–54, 143, 144]. Carbon fibre-reinforced glass and glass-ceramic matrix composites present a low coefficient of friction and exhibit relatively low wear rates which can be attributed to the solid lubricant properties of carbon fibres. In SiC (Nicalon®-type) fibre-reinforced composites, the presence of a carbon-rich interface layer provides adequate lubrication in dry sliding conditions.

The erosion resistance of carbon and SiC (Nicalon®) fibre-reinforced composites has been studied using an erosive jet of sand and other solid particles at both room and elevated temperatures (up to 750°C) [21, 145]. At temperatures above about 500°C oxidation of the carbonaceous interface and subsequent development of silica bridges should occur, which has a significant detrimental effect on erosion behaviour. The development of oxide protective coatings seems to be a necessary step to allow glass and glass-ceramic matrix composites to be employed in erosive environments (e.g. dusty environments) at high temperatures [20].

6. FINAL REMARKS

Analysis of the relevant literature reveals that a number of challenges remain for future developments in the area of fibre-reinforced glass and glass-ceramic matrix composites. While considerable effort has been expended to improve the materials, much less work has been done on the development of component design methods and data bases of engineering properties. Additionally, there is a need for the development of cost-effective, strictly repeatable and consequently largely automated processing methods. The highly innovative work required in the area of composite manufacturing technology also includes the development of suitable machining and joining techniques.

As previously highlighted in the literature [32], it is fundamental for materials scientists and engineers to change the usual approach towards this type of composites. According to traditional belief, the most interesting and lucrative applications exist at very high temperatures and under high stresses, however, there is a wide scope for other application areas, including low-temperature and low-stress structural, non-structural and functional applications. It is evident that there is a need within the industrial sector to recognise the full potential of these materials. Overall, as in many other recent materials technologies, it is of great importance to seek a closer collaboration and interrelationship between materials scientists, engineers, designers, technologists and end-users.

REFERENCES

1. I. Crivelli-Visconti and G. A. Cooper, Mechanical Properties of a New Carbon Fibre Material, *Nature* **221**, 754–755 (1969).
2. R. A. J. Sambell, D. H. Bowen and D. C. Phillips, Carbon Fibre Composites with Ceramic and Glass Matrices, Part II: Continuous Fibres, *J. Mat. Sci.* **7**, 676–681 (1972).
3. M. H. Lewis, Ceramic Matrix Composites, in *Mechanical Behaviour of Materials at High Temperatures*, C. Moura Branco, R. Ritchie, V. Sklenicka eds., Kluwer Academic Publishers, Dordrecht (1996) 599–624.
4. R. L. Lehman, Glass and Glass-Ceramic Matrix Fibre Composites, in *Handbook on Continuous Fiber-Reinforced Ceramic Matrix Composites*, R. L. Lehman, S. K. El-Rahaiby and J. B. Wachtman eds., Purdue University Press, West Lafayette, USA (1995) 527–545.
5. I. W. Donald, Preparation, Properties and Applications of Glass and Glass-Ceramic Matrix Composites, *Key Eng. Mat.* **108–110**, 123–144 (1995).
6. A. R. Boccaccini, Glass and Glass-Ceramic Matrix Composite Materials. A Review, *J. Ceram. Soc. Japan* **109** [7] S99-S109 (2001).
7. C. P. Beesley, The Application of CMCs in High Integrity Gas Turbine Engines, *Key Eng. Mat.* **127–131**, 165–174 (1997).
8. J. Vicens, G. Farizy, J.-L. Chermant, Microstructure of Ceramic Composites with Glass-ceramic Matrices Reinforced by SiC-Based Fibres, Aerospace Sci. Technol. **7**, 135–146 (2003).
9. D. A. Clarke, Fabrication Aspects of Glass Matrix Composites for Gas Turbine Applications, in *Inst. Phys. Conf. Ser.* No. 111 (1990) 173–183.
10. D. Carruthers and L. Lindburg, Critical Issues for Ceramics for Gas Turbines, in *Proc. 3rd Int. Symp. on Ceramic Materials and Components for Engines*, V. J. Tennery ed., The American Ceramic Society, Westerville (1989) 1258–1272.

11. K. Igashira, Y. Matsuda, G. Matsubara, A. Imamura, Development of the Advanced Combustor Liner Composed of CMC/GMC Hybrid Composite Material, in *High Temperature Ceramic Matrix Composites*, W. Krenkel, R. Naslain, H. Schneider, eds., Wiley-VCH, Weinheim, New York, (2001) 789–796.
12. I. P. Tuersley, A. P. Hoult and I. R. Pashby, The Processing of a Magnesium Alumino-Silicate Matrix, SiC Fibre Glass-Ceramic Matrix Composite Using a Pulsed Nd-YAG Laser, *J. Mat. Sci.* **31**, 4111–4119 (1996).
13. J. A. Ridealgh, R. D. Rawlings and R. D. West, Laser Cutting of Glass Ceramic Matrix Composite, *Mat. Sci. Technol.* **6**, 395–398 (1990).
14. M. Nakamura, M. Mabuchi, N. Saito, Y. Yamada, M. Nakanishi, K. Shimojima and I. Shigematsu, Joining of a Si-Ti-C-O Fiber-Bonded Ceramic and an Fe-Cr-Ni Stainless Steel with a Ag-Cu-Ti Brazing Alloy, *J. Ceram. Soc. Japan* **106**, 927–930 (1998).
15. R. D. Rawlings, Glass-Ceramic Matrix Composites, *Composites* **25**, 372–379 (1994).
16. I. W. Donald and P. W. McMillan, Review: Ceramic Matrix Composites, *J. Mat. Sci.* 11 (1976) 949–972.
17. Beier, W., Hart in nehmen, *Schott Information* **73**, 3–6 (1995).
18. A. Kumar and A. G. Fox, Microstructural Study of the Hot Corrosion of a Calcium Aluminosilicate Glass-Ceramic and a Si-C-O-Fiber-Reinforced Calcium Aluminosilicate Matrix Composite via Sodium Sulfate in Air and Argon at 900°C, *J. Am. Ceram. Soc.* **81**, 613–623 (1998).
19. S.-W. Wang, R. W. Kowalik and R. Sands, High Temperature Behaviour of Salt Coated Nicalon Fiber Reinforced Calcium Aluminosilicate Composite; *Ceram. Eng. Sci. Proc.* **15**, 465–474 (1994).
20. L. Bonhomme-Coury, M. Najman, F. Babonneau and P. Boch, Multicomponent Oxide Coatings via Sol-Gel Process, in *Ceramic Microstructures: Control at the Atomic Level*. A. P. Tomsia and A. Glaeser eds., Plenum Press, New York (1998) 513–525.
21. K. M. Prewo, J. F. Bacon and D. L. Dicus, Graphite Fiber Reinforced Glass Matrix Composites, *SAMPE Quarterly* **10** [4], 42–47 (1979).
22. T. Klug, V. Fleischer and R. Brückner, Thermal Expansion Behaviour of Fibre-Reinforced DURAN Glass, *Glastech. Ber.* **66**, 201–206 (1983).
23. P. N. Kumta, Processing Aspects of Glass-Nicalon Fibre and Interconnected Porous Aluminium Nitride Ceramic and Glass Composites, *J. Mat. Sci.* **31**, 6229–6240 (1996).
24. J. A. Rice, C. S. Hazelton and M. J. Haun, Optimisation of the Electrical and Mechanical Properties of a Low Dielectric Loss, Continuous Fibre Ceramic Composite, *Ceram. Trans.* **74**, 497–508 (1996).
25. P. Ducheyne and L. L. Hench, The Processing and Static Mechanical Properties of Metal Fibre Reinforced Bioglass, *J. Mat. Sci.* **17**, (1982) 595–606.
26. P. Ducheyne, Bioglass Coatings and Bioglass Composites as Implant Materials, *J. Biomed. Mat. Res.* **19**, 273–291 (1985).
27. Schepers, E., Ducheyne, P., De Clercq, M., Interfacial Analysis of Fibre–Reinforced Bioactive Glass Dental Root Implants, *J. Biomed. Mat. Res.* 23 (1989) 735–752.
28. W. B. Goldsworthy, Composites Fibres and Matrices from Lunar Regolith, *Space Studies Institute Report*, The Space Studies Institute, Princeton, New Jersey (1985).
29. R. Brückner, Glass Composites, in: *Proc. 16th Int. Congress on Glass, Bol Soc. Esp. Ceram. Vid.* **31** C [1] 97–118 (1992).
30. K. M. Prewo, The Development of Fibre Reinforced Glasses and Glass-Ceramics, in *Mat. Sci. Research Vol. 20: Tailoring Multiphase and Composite Ceramics*, R. E. Tressler, G. Messing, C. Pantano and R. Newnham eds., Plenum Press, New York (1986) 529–547.
31. A. R. Hyde, Ceramic Composites, a New Generation of Materials for Mechanical and Electrical Applications, *GECJ. Res.* **7**, 65–71 (1989).
32. K. M. Prewo, J. J. Brennan and G. K. Layden, Fiber Reinforced Glasses and Glass-Ceramics for High Performance Applications, *Ceram. Bull.* **65**, 305–322 (1986).
33. B. L. Metcalfe and I. W. Donald, The Production of Some Glass and Glass-Ceramic Matrix Composites, *Silicates Industriels* **5–6**, 99–102 (1991).
34. W. Beier, J. Heinz and W. Pannhorst, Langfaserverstärkte Gläser und Glaskeramiken: eine neue Klasse von Konstruktionswerkstoffen, *VDI Berichte* **1021**, 255–267 (1993).
35. A. Wolfenden, J. Gill, V. Thomas, A. J. Giacomin, L. S. Cook, K. K. Chawla, R. Venkatesh and R. U. Vaidya, The Relation of Dynamic Elastic Moduli, Mechanical Damping and Mass Density to the Microstructure of some Glass-Matrix Composites, *J. Mat. Sci.* **29**, 1670–1675 (1994).
36. R. F. Cooper and K. Chyung, Structure and Chemistry of Fibre-Matrix Interfaces in Silicon Carbide Fibre-Reinforced Glass-Ceramic Composites: an Electron Microscopy Study, *J. Mat. Sci.* **22**, 3148–3160 (1987).

37. T. J. Illston et al., The Manufacture of Woven Fibre Ceramic Matrix Composites Using Electrophoretic Deposition, in *Third Euroceramics* **1**, 419–424 (1993).
38. B. L. Metcalfe, I. W. Donald, D. J. Bradley, Preparation and Properties of a SiC Fibre-Reinforced Glass-Ceramic Matrix Composite, *Br. Ceram. Trans.* **92**, 13–20 (1993).
39. E. Y. Sun, S. R. Nutt and J. J. Brennan, Interfacial Microstructure and Chemistry of SiC/BN Dual Coated Nicalon Fibre Reinforced Glass-Ceramic Matrix Composites, *J. Am. Ceram. Soc.* **77**, 1329–1339 (1994).
40. K. P. Plucknett, S. Sutherland, A. M. Daniel et al., Environmental Aging Effects in a Silicon Carbide Fibre-Reinforced Glass-Ceramic Matrix Composite, *J. Microsc.* **177**, 251–263 (1995).
41. E. Zhang and D. P. Thompson, Fracture Behaviour of SiC Fibre-Reinforced Nitrogen Glass Matrix Composites, *J. Mat. Sci.* **31**, 6423–6429 (1996).
42. E. Zhang and D. P. Thompson, Carbon Fibre Reinforcement of Nitrogen Glass, *Composites Part A* **28A**, 581–586 (1997).
43. V. Bianchi, P. Goursat, W. Sinkler, M. Monthioux and E. Ménessier, Carbon-Fibre-Reinforced (YMAS) Glass Ceramic Matrix Composites. I. Preparation, Structure and Fracture Strength, *J. Europ. Ceram. Soc.* **17**, 1485–1500 (1997).
44. N. P. Bansal and J. A. Setlock, Fabrication of Fiber-Reinforced Celsian Matrix Composites, *Composites Part A* **32**, 1021–1029 (2001).
45. Frühauf, S., Oestreich, C., Müller, E., Untersuchungen im glaskeramischen Verbundsystem Nicalon/Na_2O-K_2O-MgO-Al_2O_3-SiO_2-F, in: *Proc. Conference Verbundwerkstoffe und Werkstoffverbunde*, K. Friedrich, Editor, DGM-Informationsgesellschaft, Frankfurt, Germany (1997) 277–282.
46. D. C. Phillips, R. A. J. Sambell and D. H. Bowen, The Mechanical Properties of Carbon Fibre Reinforced Pyrex Glass, *J. Mat. Sci.* **7**, 1454–1463 (1972).
47. V. Bianchi, P. Fournier, F. Platon and P. Reynaud, Carbon Fibre-Reinforced (YMAS) Glass-ceramic Matrix Composites: Dry Friction Behaviour, *J. Europ. Ceram. Soc.* **19**, 581–589 (1999).
48. V. Bianchi, P. Goursat, W. Sinkler, M. Monthioux and E. Menessier, Carbon Fibre-Reinforced (YMAS) Glass-Ceramic Matrix Composites. III. Interfacial Aspects, *J. Europ. Ceram. Soc.* **19**, 317–327 (1999).
49. R. G. Iacocca and D. J. Duquette, The Effect of Oxygen Partial Presure on the Oxidation Behaviour of Carbon Fibres and Carbon Fibre/Glass Matrix Composites, *J. Mat. Sci.* **29**, 4294–4299 (1994).
50. R. G. Iacocca and D. J. Duquette, The Effects of Matrix Microcracking on the Oxidation Behaviour of Carbon Fibre/Glass Matrix Composites, *J. Mat. Sci.* **28**, 4749–4761 (1993).
51. Y.-M. Sung, S. Park, Thermal and Mechanical Properties of Graphite Fibre-Reinforced Off-Stoichiometric $BaO.Al_2O_3.2SiO_2$ Glass-Ceramic Matrix Composites, *J. Mat. Sci. Lett.* **19**, 315–317 (2000).
52. J. M. McKittrick, N. S. Sridharan and M. F. Amateau, Wear Behaviour of Graphite-Fibre-Reinforced Glass, *Wear* **96**, 285–299 (1984).
53. E. Minford and K. M. Prewo, Friction and Wear of Graphite-Fiber-Reinforced Glass Matrix Composites, *Wear* **102**, 253–264 (1985).
54. A. R. Boccaccini and G. Gevorkian, Carbon Fibre Reinforced Glass Matrix Composites: Self-lubricating Materials for Wear Applications in Vacuum, *Glastech. Berichte. Glass Sci. Technol.* **74**, 17–21 (2001).
55. P. M. Benson, K. E. Spear and C. G. Pantano, Thermochemical Analysis of Interface Reactions in Carbon-Fiber Reinforced Glass Matrix Composites, in *Ceramic Microstructures '86. Role of Interfaces*, ed. by J. A. Pask and A. G. Evans, Plenum Press, New York (1987) 415–425.
56. K. M. Prewo and J. A. Batt, The Oxidative Stability of Carbon Fibre Reinforced Glass-Matrix Composites, *J. Mat. Sci.* **23**, 523–527 (1988).
57. K. L. Page and D. J. Duquette, The Oxidation Behaviour of Carbon Reinforced Glass Matrix Composites, *Ceram. Int.* **23**, 209–213 (1997).
58. A. R. Boccaccini, C. B. Ponton and K. K. Chawla, Development and Healing of Matrix Microcracks in Fibre Reinforced Glass Matrix Composites: Assessment by Internal Friction, *Mat. Sci. Eng.* A **241**, 142–150 (1998).
59. B. Fankhänel, E. Müller, U. Mosler, W. Siegel and W. Beier, Electrical Properties and Damage Monitoring of SiC-Fibre-Reinforced Glasses, *Comp. Sci. Technol.* **61**, 825–830 (2001).
60. A. Briggs and R. W. Davidge, Borosilicate Glass Reinforced with Continuous SiC Fibres: A New Engineering Ceramic, *Mat. Sci. Eng.* A **109**, 363–372 (1989).
61. K. M. Prewo and J. J. Brennan, Silicon Carbide Yarn Reinforced Glass Matrix Composites, *J. Mat. Sci.* **17**, 1201–1206 (1982).

62. K. M. Prewo, B. Johnson and S. Starrett, Silicon Carbide Fibre-Reinforced Glass-Ceramic Composite Tensile Behaviour at Elevated Temperature, *J. Mat. Sci.* **24**, 1373–1379 (1989).
63. A. L. Ham, J. A. Yeomans and J. F. Watts, Effect of Temperature and Particle Velocity on the Erosion of a Silicon Carbide Continuous Fibre Reinforced Calcium Aluminosilicate Glass-Ceramic Matrix Composite, *Wear* **233–235**, 237–245 (1999).
64. J. J. Brennan and K. M. Prewo, Silicon Carbide Fibre Reinforced Glass-Ceramic Matrix Composites Exhibiting High Strength and Toughness, *J. Mat. Sci.* **17**, 2371–2383 (1982).
65. G. West, D. M. R. Taplin, A. R. Boccaccini, K. Plucknett and M. H. Lewis, Mechanical Behaviour and Environmental Stability of Continuous Fibre Reinforced Glass-Ceramic Matrix Composites, *Int. J. Glass Sci. Tech. Glastech. Ber.* **69**, 34–43 (1996).
66. S. Widjaja, K. Jakus, J. E. Ritter, E. Lara-Curzio, T. R. Watkins, E. Y. Sun and J. J. Brennan, Creep-Induced Residual Stress Strengthening in a Nicalon-Fiber-Reinforced BMAS-Glass-Ceramic Matrix Composite, *J. Am. Ceram. Soc.* **82** 657–664 (1999).
67. S. M. Bleay, V. D. Scott, B. Harris, R. G. Cooke and F. A. Habib, Interface Characterization and Fracture of Calcium Aluminosilicate Glass-Ceramic Reinforced with Nicalon Fibres, *J. Mat. Sci.* **27**, 2811–2822 (1992).
68. N. P. Bansal and J. I. Eldrige, Hi-Nicalon Fibre-Reinforced Celsian Matrix Composites: Influence of Interface Modification, *J. Mat. Res.* **13**, 1530–1537 (1998).
69. J. J. Brennan, Glass and Glass-Ceramic Matrix Composites, in *Fibre Reinforced Ceramic Composites*, Materials, Processing and Technology, K. S. Mazdiyasni ed., Noyes Publications, New Jersey (1990) 222–259.
70. K. Chyung and S. B. Dawes, Fluoromica Coated Nicalon Fibre Reinforced Glass-Ceramic Composites, *Mat. Sci. Eng.* **A162**, 27–33 (1993).
71. M. H. Lewis and V. S. R. Murthy, Microstructural Characterisation of Interfaces in Fibre-Reinforced Ceramics, *Comp. Sci. Technol.* **42**, 221–249 (1991).
72. T. Ishikawa, S. Kajii, K. Matsunaga, T. Hogami and Y. Kohtoku, Structure and Properties of Si-Ti-C-O Fibre-Bonded Ceramic Material, *J. Mat. Sci.* **30**, 6218–6222 (1995).
73. M. Drissi-Habti, J. F. Després, K. Nakano and K. Suzuki, Preliminary Assessment of the Thermomechanical Behaviour of Tyrannohex Composites in Relation with the Evolution of the Microstructure, *Key Eng. Mat.* Vols. **164–165**, 341–348 (1999).
74. M. Drissi-Habti, T. Ishikawa, L. P. Zawada, Tyrannohex Composites: High Potential Materials for Structural Applications, in *High Temperature Ceramic Matrix Composites*, W. Krenkel, R. Naslain, H. Schneider eds., Wiley-VCH, Weinheim, New York, (2001) 866–872.
75. A. Hähnel, E. Pippel and J. Woltersdorf, Nanostructure of Interlayers in Different Nicalon Fibre/Glass Matrix Composites and their Effect on Mechanical Properties, *J. Microscopy* **177**, 264–271 (1995).
76. J. Janczak-Rusch and A. R. Boccaccini, Characterisation of Fibre/Matrix Interfaces in Glass Matrix Composites Using a SEM-Push-Out Indentation Apparatus, *Microscopy and Analysis* **60**, 13–14 (1999).
77. G. Qi, K. E. Spear and C. G. Pantano, Carbon-Layer Formation at Silicon-Carbide-Glass Interfaces, *Mat. Sci. Eng.* **A162** 45–52 (1993).
78. N. Ricca, A. Guette, G. Camus and J. M. Jouin, SiC (ex-PCS)/MAS Composites with a BN Interphase: Microstructure, Mechanical Properties and Oxidation Resistance, in *High Temperature Ceramic Matrix Composites*, Vol. 1, R. Naslain, J. Lamon and D. Doumeingts eds., Woodhead Publ. Ltd. (1993) 455–462.
79. J. J. Brennan, Interfacial Design and Properties of Layered BN(+C) Coated Nicalon Fiber-Reinforced Glass-Ceramic Matrix Composites, in *Ceramic Microstructures: Control at the Atomic Level*, A. P. Tomsia and A. Glaeser eds., Plenum Press, New York (1998) 705–712.
80. G. Rausch, B. Meier and G. Grathwohl, A Push-out Technique for the Evaluation of Interfacial Properties of Fibre-Reinforced Materials, *J. Europ. Ceram. Soc.* **10**, 229–235 (1992).
81. Y. Sun and R. N. Singh, The Generation of Multiple Matrix Cracking and Fibre-Matrix Interfacial Debonding in a Glass Composite, *Acta Mater.* **46**, 1657–1667 (1998).
82. N. P. Bansal, Mechanical behaviour of Silicon Carbide Fiber-Reinforced Strontium Aluminosilicate Glass-Ceramic Composites, *Mat. Sci. Eng.* **A231** 117–127 (1997).
83. N. P. Bansal, CVD SiC Fibre-Reinforced Barium Aluminosilicate Glass-Ceramic Matrix Composite, *Mat. Sci. Eng.* **A220**, 129–139 (1996).
84. W. K. Tredway and K. M. Prewo, Improved Performance in Monofilament Fibre Reinforced Glass Matrix Composites Through the Use of Fibre Coatings, *Mat. Res. Symp. Proc.* **170** 215–221 (1990).

85. R. U. Vaidya, J. Fernando, K. K. Chawla and M. K. Ferber, Effect of Fibre Coating on the Mechanical Properties of a Nextel-480-Fibre-Reinforced Glass Matrix Composite, *Mat. Sci. Eng.* **A150**, 161–169 (1992).
86. T. Leutbecher and D. Hülsenberg, Oxide Fibre Reinforced Glass: a Challenge to New Composites, *Adv. Eng. Mat.* **2**, 93–99 (2000).
87. R. John, L. P. Zawada and J. L. Kroupa, Stresses Due to Temperature Gradients in Ceramic-Matrix Composite Aerospace Components, *J. Am. Ceram. Soc.* **82**, 161–168 (1999).
88. R. J. Young and X. Yang, Interfacial Failure in Ceramic Fibre/Glass Composites, *Composites Part A* **27A**, 737–741 (1996).
89. K. K. Chawla, M. K. Ferber, Z. R. Xu and R. Venkatesh, Interface Engineering in Alumina/Glass Composites, *Mat. Sci. Eng.* **A162**, 35–44 (1993).
90. D. Banerjee, H. Rho, H. E. Jackson and R. N. Singh, Characterisation of Residual Stresses in a Sapphire-Fiber-Reinforced Glass-Matrix Composite by Micro-Fluorescence Spectroscopy, *Comp. Sci. Technol.* **61**, 1639–1647 (2001).
91. R. W. Goettler and K. T. Faber, Interfacial Shear Stresses in SiC and Al_2O_3 Fibre Reinforced Glasses, *Ceram. Eng. Sci. Proc.* **9** [7–8], 861–870 (1988).
92. A. F. Dericioglu, S. Zhu and Y. Kagawa, Improvement of Fracture Resistance in a Glass Matrix Optomechanical Composite Reinforced by Al_2O_3-ZrO_2 Minicomposite, *Ceram. Eng. Sci. Proc.* **23** [3], 485–492 (2002).
93. A. Maheshwari, K. K. Chawla and T. A. Michalske, Behavior of Interface in Alumina/Glass Composite, *Mat. Sci. Eng.* **A107**, 269–276 (1989).
94. R. L. Lehman and C. A. Doughan, Carbon Coated Alumina Fibre/Glass Matrix Composites, *Comp. Sci. Tech.* **37**, 149–164 (1990).
95. H. Iba, T. Chang, Y. Kagawa, H. Minakuchi and K. Kanamaru, Fabrication of Optically Transparent Short Fibre-Reinforced Glass Matrix Composites, *J. Am. Ceram. Soc.* **79**, 881–884 (1996).
96. Y. Kagawa and T. Yamada, Tensile Properties of Optical Fibre-Glass Matrix Sensor-Based Composite, *J. Mat. Sci. Lett.* **13**, 1403–1405 (1994).
97. H. Iba, T. Naganuma, K. Matsumura and Y. Kagawa, Fabrication of Transparent Continuous Oxynitride Glass Fibre-Reinforced Glass Matrix Composite, *J. Mat. Sci.* **34**, 5701–5705 (1999).
98. B. Fankänel, E. Müller, K. Weise and G. Marx, Translucent Fibre Reinforced Glass, *Key Eng. Mat.* **206–213**, 1109–1112 (2002).
99. A. R. Boccaccini, S. Atiq and G. Helsch, Optomechanical Glass Matrix Composites, *Comp. Sci. and Technol.* **63**, 779–783 (2003).
100. I. W. Donald and B. L. Metcalfe, The Preparation, Properties and Applications of Some Glass-Coated Metal Filaments Prepared by the Taylor-Wire Process, *J. Mat. Sci.* **31**, 1139–1149 (1996).
101. I. W. Donald, B. L. Metcalfe and A. D. Bye, Preparation of a Novel Uni-directionally Aligned Microwire-Reinforced Glass Matrix Composite, *J. Mat. Sci. Lett.* **7**, 964–966 (1988).
102. A. E. Rutkovskij, P. D. Sarkisov, A. A. Ivashin and V. V. Budov, Glass-Ceramic-Based Composites, in *Ceramic- and Carbon-Matrix Composites*, V. I. Trefilov ed., Chapman and Hall, London (1995) 255–285.
103. A. R. Boccaccini, J. Ovenstone and P. A. Trusty, Fabrication of woven metal fibre reinforced glass matrix composites, *Applied Composite Materials* **4**, 145–155 (1997).
104. C. Kaya, A. R. Boccaccini and P. A. Trusty, Processing and Characterisation of 2-D Woven Metal Fibre-Reinforced Multilayer Silica Matrix Composites Using Electrophoretic Deposition and Pressure Filtration, *J. Europ. Ceram. Soc.* **19**, 2859–2666 (1999).
105. K. K. Chawla, *Ceramic Matrix Composites*, 2nd Edition, Kluwer Academic Publishers, Boston, Dordrecht, London (2003).
106. C. Zhao, K. Lambrinou and O. Van der Biest, Hot Pressing "Window" for (P_2O_5, B_2O_3) – Containing Magnesium Aluminosilicate Reinforced with SiC Fibres, *J. Mat. Sci.* **34**, 1865–1871(1999).
107. G. Larnac, P. Lespade, P. Peres and J. M. Donzac, High Temperature Glass-Ceramic Matrix Composites: from the Thermal and Mechanical Behaviour to the Realisation of Structures of Complex Shapes, in *High Temperature Ceramic Matrix Composites*, Vol. 1, R. Naslain, J. Lamon and D. Doumeingts eds., Woodhead Publ. Ltd. (1993) 777–784.
108. K. Cho, R. J. Kerans, K. A. Jepsen, Selection, Fabrication and Failure Behaviour of SiC Monofilament-Reinforced Glass Composites, *Ceram Eng. Sci. Proc.* 815–823 (1995).
109. C. M. Gustafson, R. E. Dutton and R. J. Kerans, Fabrication of Glass Matrix Composites by Tape Casting, *J. Am. Ceram. Soc.* **78**, 1423–1424 (1995).

110. G. H. Cullum, Ceramic Matrix Composite Fabrication and Processing: Sol-Gel Infiltration, *Handbook on Continuous Fiber-Reinforced Ceramic Matrix Composites*, R. L. Lehman, S. K. El-Rahaiby, J. B. Wachtman eds., Purdue University Press, West Lafayette, USA (1995) 185–204.
111. K. S. Mazdiyasni, Ceramic Composite Matrices from Metal-Organic Precursors, *Mat. Sci. Eng.* **A144**, 83–90 (1991).
112. Ph. Colomban and N. Lapous, New Sol-Gel Matrices of Chemically Stable Composites of BAS, NAS, and CAS, *Comp. Sci. Technol.* **56**, 739–746 (1996).
113. V. Gunay, P. F. James, F. R. Jones and J. E. Bailey, Continuous Fibre Reinforced Glass and Glass Ceramic Matrix Composites by Sol-Gel Processing, in: *Inst. Phys. Conf. Ser.* Nr. 111, IOP Publishing Ltd (1990) 217–226.
114. D. Qi and C. Pantano, Sol-Gel Processing of Carbon Fibre Reinforced Glass Matrix Composites. In *Ultrastructure Processing of Advanced Ceramics*. J. D. Mackenzie and D. R. Ulrich ed., J. Wiley and Sons, New York (1987) 635–649.
115. W. Pannhorst, M. Spallek, R. Brückner, H. Hegeler, C. Reich, G. Grathwohl, B. Meier and D. Spelmann, Fibre-Reinforced Glasses and Glass-Ceramics Fabricated by a Novel Process, *Ceram. Eng. Sci. Proc.* **11** [7–8] 947–963 (1990).
116. A. R. Boccaccini, C. Kaya, K. K. Chawla, Use of Electrophoretic Deposition in the Processing of Fibre Reinforced Ceramic and Glass Matrix Composites. A Review, *Composites Part A* **32**, 997–1006 (2001).
117. A. R. Boccaccini, C. Kaya, Glass and Glass-Ceramic Matrix Composites: From Model Systems to Useful Materials, In: *MRS Proceedings, Vol. 702, Advanced Fibers, Plastics, Laminates and Composites*, F. T. Wallenberger, N. Weston, K. K. Chawla, R. Ford, R. P. Wool, eds., Materials Research Society, Warrendale (2002), p. 277–288.
118. C. Kaya, F. Kaya, A. R. Boccaccini, Fabrication of Stainless-Steel Fibre-Reinforced Cordierite Matrix Composites of Tubular Shape Using Electrophoretic Deposition, *J. Am. Ceram. Soc.* **85**, 2575–2577 (2002).
119. R. W. Goettler, *A Method of Densifying a Glass or Glass Composite Structure*, US Patent Nr. 5,122,176, Jun. 16, 1992.
120. T. R. Hinklin, S. S. Neo, K. W. Chew and R. M. Laine, Precursor Impregnation and Pyrolysis (PIP) Processing of Barium Aluminosilicate-Nicalon Composites, *Ceram. Trans.* **74**, 117–128 (1996).
121. S. M. Johnson, D. J. Rowcliffe and M. K. Cinibulk, Continuous SiC Fibre/Glass Composites, in: *Ceramic Microstructures '86: Role of Interfaces*, J. A. Pask and A. G. Evans eds., Plenum Press, New York (1987) 633–641.
122. E. Roeder, N. Klein and K. Langhans, Manufacture of glass composites reinforced with long and short fibres by extrusion, *Glastech. Ber.* **61**, 143–148 (1988).
123. Y. Zhou and O. Van der Biest, Microwave Processed Glass Composite Materials, *Sil. Indust.* **7/8** 163–169 (1996).
124. de Voorde, V. M. H., Developments of Continuous Fibre Reinforced Ceramic Matrix Composites (CFCC) in Europe, *Sil. Ind.* **63**, 59–67 (1999).
125. A. G. Evans and D. B. Marshall, Overview Nr. 85. The Mechanical Behaviour of Ceramic Matrix Composites, *Acta Metall.* **37**, 2567–2583 (1989).
126. A. G. Evans and F. W. Zok, Review. The Physics and Mechanics of Fibre-Reinforced Brittle Matrix Composites, *J. Mat. Sci.* **29**, 3857–3896 (1994).
127. J. W. Krynicki, D. C. Nagle and R. E. Green Jr., Photoelastic Measurement of Residual Thermomechanical Stress in SiC–Reinforced Glass Composites, *J. Am. Ceram. Soc.* **75**, 2225–2231 (1992).
128. S. Sutherland, K. P. Plucknett and M. H. Lewis, High Mechanical and Thermal Stability of Silicate Matrix Composites, *Comp. Eng.* **5**, 1367–1378 (1995).
129. T. Mah, M. G. Mendiratta, A. P. Katz, R. Ruh and K. S. Mazdiyasni, High-Temperature Mechanical Behaviour of Fiber-Reinforced Glass-Ceramic-Matrix Composites, *J. Am. Ceram. Soc.* **68**, C-248, C-251 (1985).
130. A. G. Evans, F. W. Zok, R. M. McMeeking and Z. Z. Du, Models of High-Temperature, Environmentally Assisted Embrittlement in Ceramic-Matrix Composites, *J. Am. Ceram. Soc.* **79**, 2345–2352 (1996).
131. A. R. Boccaccini, J. Janczak-Rusch, D. H. Pearce and H. Kern, Assessment of thermal shock induced damage in fibre reinforced glass matrix composites, *Composites Science and Technology* **59**, 105–112 (1999).
132. T. Klug, J. Reichert and R. Brückner, Thermal Shock Behaviour of SiC-Fibre Reinforeced Glasses, *Glastech. Ber.*, **65**, 41–49 (1992).
133. Y. Kagawa, N. Kurosawa, T. Kishi, Y. Tanaka, Y. Iamai and H. Ichikawa, Thermal Shock Behavior of SiC Fiber-(Nicalon) Reinforced Glass, *Ceram. Eng. Sci. Proc.*, **10** [9–10], 1327–1336 (1989).

134. M. J. Blisset, P. A. Smith and J. A. Yeomans, Thermal Shock Behaviour of Unidirectional Silicon Carbide Fibre Reinforced Calcium Aluminosilicate, *J. Mat. Sci.* **32**, 317–325 (1997).
135. C. Labrugere, L. Guillaumat, A. Guette and R. Naslain, Mechanical and Microstructural Change in 0-90° SiC/MAS-L Composites after Thermal Ageing: Vacuum, Ar and CO Atmosphere, *J. Europ. Ceram. Soc.* **19**, 17–31 (1999).
136. A. Kumar and K. M. Knowles, Effect of Oxidation Heat Treatments on the Mechanical Behaviour of a Si-C-O-Fiber-Reinforced Magnesium Aluminosilicate, *J. Am. Ceram. Soc.* **79**, 2375–2378 (1996).
137. C. H. Weber, J. P. Lofvander, A. G. Evans, Creep Anisotropy of a Continuous Fibre-Reinforced Silicon Carbide/Calcium Aluminosilicate Composite, *J. Am.Ceram. Soc.* **77**, 1745–1752 (1994).
138. J. Vicens, F. Doreau and J. L. Chermant, Microstructural Characterisation of SiCf/YMAS Composites in the As-Received State and After Thermomechanical Tests, *Composites Part A* **27A**, 723–727 (1996).
139. X. Wu and J. W. Holmes, Tensile Creep and Creep-Strain Recovery Behaviour of Silicon Carbide Fiber/Calcium Aluminosilicate Matrix Ceramic Composites, *J. Am. Ceram. Soc.* **76**, 2695–2700 (1993).
140. R. C. Wetherhold and L. P. Zawada, Heat Treatments as a Method of Protection for a Ceramic Fiber-Glass Matrix Composite, *J. Am. Ceram. Soc.* **74**, 1997–2000 (1991).
141. D. F. Hasson, S. G. Fishman, Impact Behavior of Fibre Reinforced Glass Matrix Composites, *Mat. Res. Soc. Symp. Proc.* **120**, 285–290 (1988).
142. D. C. Phillips, N. Park, R. J. Lee, The Impact Behaviour of High Performance, Ceramic Matrix Fibre Composites, *Comp. Sci. Technol.* **37**, 249–265 (1990).
143. R. Reinicke, K. Friedrich, W. Beier and R. Liebald, Tribological Properties of SiC and C-Fibre Reinforced Glass Matrix Composites, *Wear* **225–229**, 1315–1321 (1999).
144. A. Skopp, M. Woydt, K.-H. Habig, T. Klug and R. Brückner, Friction and Wear Behaviour of C- and SiC-Fibre Reinforced Glass Composites Against Ceramic Materials, *Wear* **169**, 243–250 (1993).
145. P. Saewong and R. D. Rawlings, Erosion of Carbon Fibre Reinforced Glass Matrix Composites, *J. Mat. Sci. Lett.* **18**, 1915–1919 (1999).

20

Dispersion-Reinforced Glass and Glass-Ceramic Matrix Composites

Judith A. Roether* and Aldo R. Boccaccini**

*Department of Dental Biomaterials Science
GKT Dental Institute, London SE1 9RT, UK
Tel: 44 20 7955 4281
E-mail: judith.roether@kcl.ac.uk
**Department of Materials
Imperial College London, London SW7 2AZ, UK
Tel.: 44 207 1881 789
Email: a.boccaccini@imperial.ac.uk

ABSTRACT

Glass and glass-ceramic matrix composites with dispersion reinforcement, namely those containing whisker, platelet, chopped fibres or particles, are considered in this Chapter, reviewing their fabrication, microstructural characterisation, properties and applications. The Chapter is mainly devoted to composite systems developed to improve the thermo-mechanical properties of the glass (or glass-ceramic) matrix, in order to fabricate novel materials suitable for structural and load-bearing applications. The use of the composites in mechanical engineering and machinery, electronics, aerospace, chemical, high-temperature and wear resistance applications as well as in the biomedical field and in industrial recycling technologies is discussed.

1. INTRODUCTION

The development of dispersion reinforced glass matrix composites with improved mechanical properties for structural applications was first reported in 1962 [1]. Since then, hundreds of composite systems have been developed by incorporating a great variety of

TABLE 1. Examples of dispersion-reinforced glass and glass-ceramic matrix composite systems

Matrix/reinforcement	Mechanical properties tested	Investigator (year) [Ref.]
Sodium borosilicate Al_2O_3 particulates (40 vol%)	$\sigma = 165$ MPa $E = 180$ GPa	Binns (1962) [1]
Sodium borosilicate Al_2O_3 particulates (50 vol%)	$\sigma = 175$ MPa $E = 160$ GPa	Hasselman, Fulrath (1965) [2]
Borosilicate/Tungsten spheres (50 vol%)	$\sigma = 76$ MPa $E = 150$ GPa	Hasselman, Fulrath (1965) [3]
Sodium borosilicate/Nickel chopped fibre	$\sigma = 112$ MPa	Einmahl (1966) [4]
Li-sodium aluminosilicate/Al spheres (20 vol%)	$K_{IC} = 6.2$ MPa$\sqrt{m}$	Krstic (1983) [5]
Sodium-zinc glass (Code 6810)/ zirconia (40 vol%)	$\sigma = 60$ MPa $E = 90$ GPa	Frey, Mackenzie (1967) [6]
Borosilicate/Fe-Ni-Co alloy, spherical particles (65 vol%)	$E = 95$ GPa $K_{IC} = 5.3$ MPa$\sqrt{m}$	Jessen, et al. (1986) [7]
Aluminiumsilicate (Code 1723)/ SiC whiskers	$\sigma = 337$ MPa $K_{IC} = 3.4$ MPa$\sqrt{m}$ $E = 141$ GPa	Gadkaree, Chyung (1986) [8]
Borosilicate/thoria sphere (10 vol%)	$\sigma = 88$ MPa	Davidge, Green (1968) [9]
Sodium lime borosilicate/Al_2O_3 particulates (70 vol%)	$\sigma = 400$ MPa $K_{IC} = 4.4$ MPa$\sqrt{m}$ $E = 200$ GPa	Haber et al. (1989) [10]
Sodium lime silica/Al alloy particles (40 vol%)	$\sigma = 150$ MPa	Trocziynsky et al. (1989) [11]
Borosilicate/Nb chopped fibres (30 vol%)	$K_{IC} = 4.7$ MPa$\sqrt{m}$	Lucas et al. (1980) [12]
Sodium lime borosilcate/Au-Pt Alloy spheres (35 vol%) Au-In alloy spheres (35 vol%)	$K_{IC} = 1.84$ MPa$\sqrt{m}$ $K_{IC} = 2.13$ MPa$\sqrt{m}$	Baran et al. (1990) [13]
Fused silica/W chopped fibre (30 vol%)	$\sigma = 275$ MPa	Dungan (1973) [14]
Sodium borosilicate/Al_2O_3 particulates (35 vol%)	$K_{IC} = 2.5$ MPa$\sqrt{m}$ $E = 140$ GPa	Swearengen et al. (1978) [15]
Borosilicate/Hastelloy chopped fibre (15 vol%)	$\sigma = 97$ MPa $E = 90$ GPa $K_{IC} = 2.5$ MPa$\sqrt{m}$	Boccaccini et al. (1994) [16]
Borosilicate/Al_2O_3 platelets (30 vol%)	$E = 102$ GPa $\sigma = 150$ MPa $K_{IC} = 3.6$ MPa$\sqrt{m}$ $Hv = 7.1$ GPa	Boccaccini, Trusty (1996) [17]
SiO_2/AlN particulates (30 vol%)	$K_{IC} = 2.96$ MPa$\sqrt{m}$ $\sigma = 200$ MPa	Wu et al. (1999) [18]
Lithium-ziconium silicate (LZS) glass-ceramic/Al_2O_3 particulates (20 vol%)	$E = 120$ GPa $\sigma = 160$ MPa	N. de Oliveira et al. (1999) [19]
Bioactive glass-ceramic/Ag particles (10 vol%)	$K_{IC} = 1.6$ MPa$\sqrt{m}$ $\sigma = 160$ MPa	Claxton et al. (2002) [20]
Barium aluminosilicate (BAS) glass-ceramic/SiC platelets (30 vol%)	$K_{IC} = 3.2$ MPa$\sqrt{m}$ $\sigma = 180$ MPa	Ye et al. (2003) [21]

ceramic or metallic inclusions into different glass or glass-ceramic matrices. Table 1 gives an overview of some dispersion-reinforced glass and glass-ceramic matrix composite systems developed in the last forty years, indicating the great variety of matrices considered [2–21]. Besides the typical silicate matrices listed in Table 1, phosphate and oxynitride systems have been employed [22–24]. The objective in most cases has been to increase the mechanical properties, i.e. fracture strength (σ), fracture toughness (K_{IC}) and Young's modulus (E) of the brittle matrices. The improvement of other engineering properties such as hardness, thermal shock and wear resistance has also been considered, as has been the attainment of given functional properties, such as electric, magnetic, dielectric or optical properties.

Composites with enhanced strength and toughness

Ceramic reinforcement

Ceramic particles, chopped fibres, whiskers and platelets have been used as discontinuous reinforcement in glass and glass-ceramic matrices.

Mainly alumina and SiC particles have been considered, however other ceramic particles including zirconia, mullite, silicon nitride, silica, cordierite, chromia, titania, boron nitride, aluminium nitride and thoria particles have also been used [2,6,8–10,15,17–19,25].

At the beginning of the seventies the first reports on chopped carbon fibre-reinforced glasses were published [26]. The development of chopped carbon fibre-reinforced glass matrix composites has resulted in compliant, high failure strain materials with low thermal expansion coefficient and thus high dimensional thermal stability and high thermal shock resistance [27–29]. SiC-based chopped fibres have also been used to reinforce a variety of glass and glass-ceramic matrices [30]. Other types of short fibres, such as silica, alumina, zirconia and mullite have found only limited use as reinforcement [31,32]. SiC-whiskers have found extensive application as reinforcement in different glass and glass-ceramic matrices [8,33–37], while Si_3N_4-whiskers have been considered to a lesser extent [38]. Because of the problems associated with the health hazards of whiskers, ceramic platelets, mainly alumina and SiC platelets, are increasingly being used as reinforcing elements [17,21,39–42]. The phenomenon of transformation toughening using the rapid stress-induced structural transformation of zirconia has been considered by incorporating different volume concentrations of zirconia particles in glass or glass-ceramic matrices [43,44]. In recent research, it has been shown that the incorporation of a piezoelectric particulate phase, e.g. lead-zirconate-titanate particles, in a glass matrix may lead to toughening by exploiting the piezoelectric effect (domain switching mechanism) in the dispersed phase [45]. Nanocomposites consisting of SiC nanoparticles (5–10 vol%) incorporated into cordierite glass-ceramic matrices [46] and oxynitride glass matrices [47] have also been investigated.

In recent developments, carbon nanotubes have been explored as possible reinforcing elements in silicate matrices, including fused silica [48,49] and borosilicate glass [50]. The SEM micrograph (Figure 1) shows the fracture surface of a multi-wall carbon nanotube reinforced borosilicate glass matrix composite fabricated by hot-pressing. Research in these systems has been very limited so far, the main difficulty being the attainment of a homogeneous distribution of the nanotubes in a dense glass matrix.

Metallic reinforcements

The incorporation of a metallic phase, such as nickel particles or stainless steel chopped fibres, was the first approach intended specifically to develop tough glass matrix composites

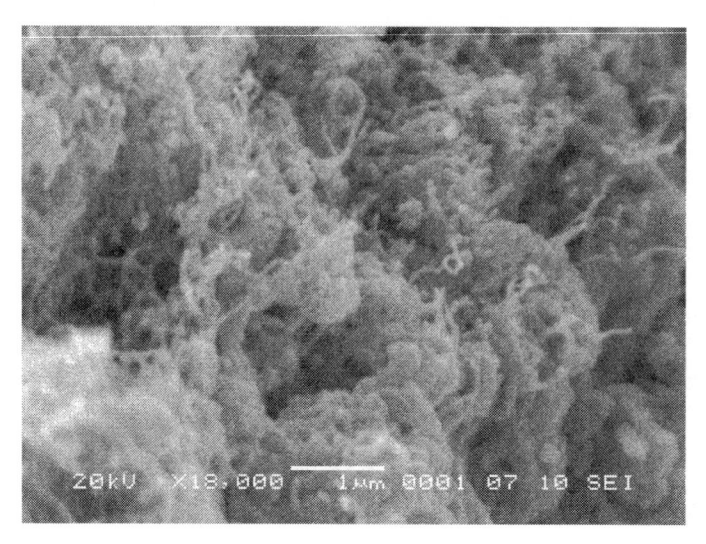

FIGURE 1. SEM image showing the fracture surface of carbon nanotube-reinforced borosilicate glass matrix composite [50].

by exploiting the ductility and plastic deformation properties of the metallic phase [3,4]. This concept has been further developed during the last thirty years using different metals such as Al, Ni, Ag, Au, Fe, Ti, V, Mo, Sb, Cu, Pb, Nb, stainless steel, Fe-Ni-Co and other alloys and intermetallics (e.g. Ni_3Al) in the form of particles, chopped fibres, flakes and ribbons or combinations of these in a variety of glass and glass-ceramic matrices [5,7,11–14,20,24,51–55].

Composites developed to improve other engineering properties and functional composites

Further research on dispersion-reinforced glass matrix composites has been conducted specifically to improve other engineering properties, such as the thermal shock resistance, impact strength, thermal expansion and thermal conductivity, hardness, machinability and wear resistance. An important group of research activities has dealt with high-temperature properties. For example, the effect of inclusions on the rheological behaviour, creep resistance, compression strength and viscoelastic behaviour of glass and glass-ceramic matrix composites has been investigated [56–58].

Numerous glass matrix composites have been developed in which the inclusion phase has a functional purpose not related to mechanical properties. Some examples are: glass matrix/ceramic particle mixtures for multilayer ceramic circuit boards in high-performance microelectronic packages [59,60], thick film resistors [61,62], glasses and glass-ceramics containing ferroelectric phases (e.g. lead zirconate titanate, barium titanate) [63,64], semiconductor nanoparticles (e.g. CdS, ZnSe, etc.) in silica [65] and in borosilicate matrices [66], ZnO-glass composite varistors [67], sol-gel derived carbon powder containing silica matrix composites [68], and metallic nanoparticles/nanoclusters in silica and other glassy host matrices for optical [69], decoration [70] and high-dielectric strength applications [71]. A complete review on the formation and properties of glass matrices containing ultrafine particles for optical and other specialised functional applications has been published by Mackenzie and Bescher [72]. The incorporation of organic materials in glass matrices has also been explored for achieving specific functional properties [73,74].

2. APPLICATIONS

Table 2 gives an overview of the application areas for dispersion-reinforced glass and glass-ceramic matrix composites. These composite materials offer a combination of structural, electrical and thermal properties not usually available in monolithic glasses or glass-ceramics. The materials are less dense than most structural metals and are generally hard and of high stiffness. Through dispersion reinforcement glass composites gain mechanical reliability and attain attractive strength and toughness values for structural applications requiring moderate loading, while maintaining the resistance to chemical attack and oxidation of the glass matrix. Moreover, the incorporation of a second phase allows tailoring of electrical, magnetic and thermal properties for selected applications [75]. In general, dispersion-reinforced glass matrix composites may be used in several of the application areas suggested for glass-ceramics.

One important field of application for dispersion-reinforced glass matrix composites is as biomaterials for implants and in orthopaedics. Bioactive silicate and phosphate glasses and glass-ceramics are toughened by incorporating biocompatible second phases, for example titanium, silver or stainless steel particles or chopped fibres [20,24,54,76,77].

TABLE 2. Applications of dispersion-reinforced glass and glass-ceramic matrix composites

Electrical Engineering Applications:
Induction heating equipment, microwave components, connectors, high-voltage electron tube bases, neon lamp holders and insulators, insulation plates, capacitor end plates, power station panel components, feed-through insulators, relay spacers, computer components, high frequency insulation components, arc barriers, microwave components, substrates, seals.

High-Temperature and Environmental Resistance Applications:
Thermal insulators, infrared components, supports in furnaces, precision spacers, motor armatures and end bells, high-T instrumentation, thermal switches, heat barrier requirements, brush holders, soldering fixtures, hermetic seal feed-through, thermocouple and thermostat housings, high-T terminal blocks, electrical heater terminals, heater coil supports, fracture-tough sealing materials, vacuum envelopes, sound insulation, chemical plant devices.

Dimensional-Stability Applications:
Laser components/housing, radar insulators, mechanical supports, precision spacers in plugs and jacks, jigs and fixtures for heating, welding, brazing, telemetering devices, potentiometer insulators, electromagnetic shields, mirror substrates.

Other Applications:
Thermal insulation pads, glass handling jaws, vacuum pump insulators, fixtures and components in X-ray equipment, asbestos replacement, antenna insulators, precision crystal holders, abrasion resistance piece tools, body armour, other low load-bearing structural components.

The composites are used in their bulk form or as coatings on metallic implants. It has been shown that besides the improvement of strength and toughness, the incorporation of the ductile phase led to a significant reduction in slow crack gowth. The inclusions also decrease the sensitivity of the glass to stress corrosion cracking [78]. In related research on biomaterials, composites consisting of a dispersed alumina phase in glassy matrices are being increasingly considered as dental materials [79] as are ZrO_2 containing glass-ceramics with dispersion-type microstructure [80] and machinable mica-containing glass-ceramics for dental restorations [81].

High-temperature composite coatings with high corrosion and ablation resistance is another important area of application of these materials. The potential application of tungsten particle reinforced silica in thermal protection systems, for example for the ablating surfaces of rocket nozzles and re-entry bodies, was reported 30 years ago [82]. High-performance coatings using ceramic particle reinforced glass-ceramic matrices have been developed [83]. Moreover glass composite coatings consisting of SiC, carbon or intermetallic fillers in glass matrices have been developed for thermal protection of carbon/carbon components [84]. Novel thermal barrier coatings for high-temperature components have been proposed on the basis of glass-metal (NiCoCrAlY) composites [85].

In enamel technology, ceramic particles can be added to improve one or several of the critical properties of enamel, such as chemical and wear resistance, impact strength and optical properties [86]. The quality of heat resistant enamels for protection of chromium-nickel steels can thus be improved by incorporating different oxide fillers, e.g. NiO, Al_2O_3, ZrO_2, Cr_2O_3 and Fe_2O_3, into the basic silicate enamel composition [87].

Glass matrix composites reinforced by crystalline particles are also considered for the production of glazes with high performance. The incorporation of up to 30 vol% of zircon ($ZrSiO_4$) particles in a transparent glassy matrix to fabricate glazes of high abrasion resistance for floor tiles has been reported [88].

Other application areas for partially crystallised glass-ceramics and particle reinforced glass matrix composites are those of sealing and joining of components and structures [89,90]. Glass-ceramic composites containing metallic fillers are good candidates for applications as anti-friction materials. The incorporation of metallic particles increases the thermal conductivity of the composites, thus facilitating the heat removal from the contact zone within the bulk of the material [83]. A related application area is in the handling of molten non-ferrous metals. Composite materials with high corrosion resistance to molten aluminium have been fabricated by the incorporation of Si_3N_4 particles into cordierite glass-ceramics [91].

Glass-ceramic materials are candidates for armour applications, including transparent armoured windscreens in vehicles and windows in buildings, as well as armour with antiballistic properties for personnel protection. Particle-reinforced glass and glass-ceramic matrix composites have potential use in this application area. The condition for transparency is either that the inclusions within the glass matrix be much smaller than the wave length of the incident light, or that the refractive indices of the dispersed crystalline inclusions and the glass matrix be nearly equal [92]. Transparent glass-ceramics and glass matrix composites are also suitable for use in electromagnetic windows, both at microwave wavelengths, and at visible and near infrared wavelengths. Other applications of transparent glass-ceramics with a matrix type of microstructure, i.e. a glass matrix containing dispersed crystals of nanometre dimensions, have been reviewed [92]. In addition to transmission, in several electromagnetic window applications, the materials must have high resistance to elevated temperatures, high strength and resistance to abrasion and thermal shock, properties that cannot be readily met by unreinforced glasses. Interesting examples of transparent glass matrix composites are chopped glass fibre reinforced glass matrix composites [93,94]. These composites showed a combination of adequate optical properties in the UV-visible wavelength region and moderate fracture toughness. Transparency was achieved by choosing fibres and glass matrices with matching refractive indices.

Lightweight and isotropic near-zero thermal expansion materials can be obtained by adding metallic particles to a negative thermal expansion glass-ceramic matrix [95]. These materials find applications in mirrors and general optics, sensors, microwave components and antennae because of their impressive thermal and dimensional stability.

Metal particle-reinforced glass matrix composites have also been proposed as heat-resistant, high-strength electromagnetic wave-absorbing components [96,97]. Random stainless steel fibre-reinforced magnesium aluminosilicate glass matrix composites have been developed for applications in electromagnetic interference shielding [97]. These products offer a similar electromagnetic shielding capability to polymer matrix composites while exhibiting a significantly higher strength and temperature resistance.

An innovative potential application area for glass matrix composites is in the recycling of silicate residues both with and without hazardous elements (e.g. heavy metals). Here both economical and ecological issues are combined. Industrial wastes, such as fly ash from coal power stations, slag and ashes from waste incinerator plants, slag from iron and steel production, waste water treatment sludge and cullet glass, contain a significant amount of SiO_2 and they can be melted (vitrified) to yield glass or glass-ceramic products [98]. A multibarrier system using glass matrix composites has been proposed to recycle such residues [99]. The idea is to create a biostable glass matrix using residues free from hazardous elements and to incorporate the hazardous residue (e.g. containing heavy-metals) into the glass matrix in particulate form. Thus the stable matrix provides a true physical barrier to the leaching of the heavy metals and improves the chemical durability of the products. The dispersed phase should also contribute to improving the mechanical properties and reliability of the material, thus broadening the application potential of the products.

Several sintered glass and glass-ceramic matrix composites obtained from recycled silicate waste have been reported in the literature [100–106]. These are dense or porous products with potential application as building, decoration or architectural materials, such as wall partition blocks, pavements, wall and floor tiles, thermal insulation, fire protection elements, roofing granules and acoustic tiles. Other possible uses include abrasive media for blasting and polishing applications.

Powder glass sintering has been also proposed as an alternative technology to the traditional glass melting method for immobilisation of high-level nuclear waste [107]. The composite approach can be used in this application to enhance the thermomechanical properties of the sintered components: this was shown with reference to improving the thermal shock resistance of radioactive waste borosilicate glass by incorporating aluminum titanate dispersions [108]. The concept of using glass matrix composites to immobilise nuclear waste has recently been further developed by considering the incorporation of pyrochlore crystalline particles (e.g. lanthanum zirconate) into a lead silicate glass matrix of matched thermal expansion coefficient [109].

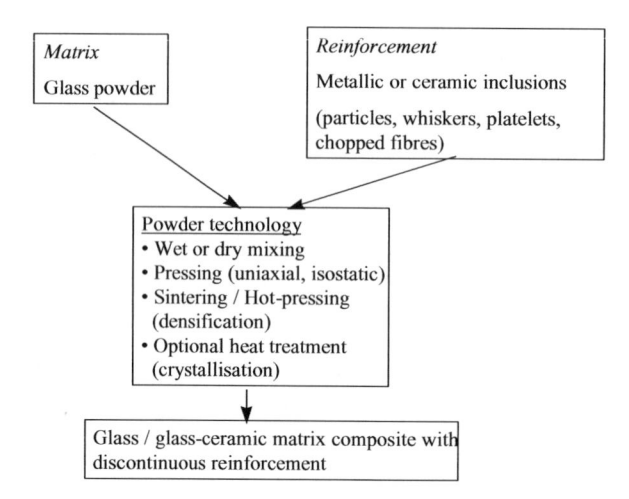

```
┌─────────────────────┐      ┌──────────────────────────────┐
│ *Matrix*            │      │ *Reinforcement*              │
│ Glass powder        │      │ Metallic or ceramic inclusions│
└─────────────────────┘      │ (particles, whiskers, platelets,│
            \                │ chopped fibres)              │
             \               └──────────────────────────────┘
              \                    /
               ↓                  ↓
         ┌──────────────────────────────────┐
         │ Powder technology                │
         │ • Wet or dry mixing              │
         │ • Pressing (uniaxial, isostatic) │
         │ • Sintering / Hot-pressing       │
         │   (densification)                │
         │ • Optional heat treatment        │
         │   (crystallisation)              │
         └──────────────────────────────────┘
                      ↓
         ┌──────────────────────────────────────────┐
         │ Glass / glass-ceramic matrix composite with│
         │ discontinuous reinforcement              │
         └──────────────────────────────────────────┘
```

FIGURE 2. The powder technology route to fabricate dispersion-reinforced glass and glass-ceramic matrix composites.

3. PROCESSING

Powder technology, sintering, hot-pressing

Dispersion-reinforced glass matrix composite materials are usually fabricated by powder technology. This is a relatively simple and cost-effective technology, allowing the use of standard ceramic processing techniques such as powder mixing, compacting and sintering, as schematically shown in Figure 2.

A critical step in this fabrication approach is the mixing of the glass matrix powders and the reinforcing elements. The homogeneous distribution of these in the glass matrix is a fundamental requirement for obtaining high-quality composites with optimised properties. Inhomogeneous distribution of inclusions, forming of particle clumps or agglomerates or inclusion-inclusion interactions, may lead to microstructural defects such as pores and cracks, which will have a negative effect on the mechanical properties of the products. The improvement of mixing techniques includes optimising the particle sizes and the size distributions of the powders and the use of wet-mixing routes, i.e. mixing in water or isopropanol with the addition of binders such as PVA coupled with ultrasonic or magnetic stirring. Adequate mixing of the powder matrix and the reinforcing elements is particularly problematic when chopped fibres are used, and careful control of the processing variables, including slurry viscosity, fibre content and stirring velocity is required. A typical microstructure of a mixture of a glass powder matrix and chopped SiC fibres obtained by a wet-mixing route is shown in Figure 3. A fairly homogeneous glass powder-chopped fibre mixture has been achieved, which could not have been reached by dry-mixing techniques. Another approach to improve homogeneity of the mixture is the coating of the reinforcing elements by a thin layer of the matrix material [110]. This approach presupposes the development of a technique to synthesise sub-micrometric glass particles.

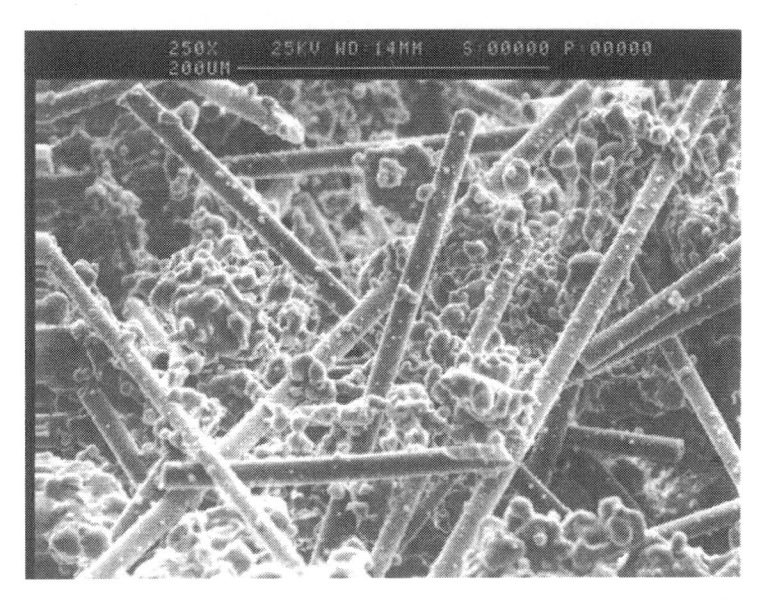

FIGURE 3. Microstructure of a SiC chopped fibre reinforced glass matrix composite showing the fairly homogeneous mixture of the composite constituents attained by wet-mixing.

The most economical process involves densification of the green bodies by simple pressureless sintering at temperatures between the glass transition and the melting temperatures of the glass matrix. Sintering in glass matrices occurs by a viscous flow mechanism [111]. During sintering, reactions at the inclusion/matrix interfaces or degradation of the inclusions may take place, especially when the inclusions are non-oxide ceramics (e.g. SiC) or metallic. If the glass matrix used is prone to crystallisation or the aim is to produce a glass-ceramic matrix, the sintering procedure must be optimised in order to avoid the onset of crystallisation before the densification by viscous flow has been fully completed. One way to achieve this is by increasing the heating rate during sintering in order to delay the nucleation and growth of crystalline phases. Thus an ideal heating schedule to produce glass-ceramic matrix composites should include three independent stages: densification, nucleation and crystallisation.

The presence of inclusions can affect the formation of crystalline phases in a glass matrix. For example, it has been shown that aluminium-containing ceramic inclusions, e.g. alumina, mullite or aluminium nitride, may suppress cristobalite formation in borosilicate glass during sintering [112]. In general, the presence of rigid inclusions will jeopardise the densification process driven by viscous flow sintering, a problem that has been well studied both experimentally [113,114] and theoretically [115,116]. In practice, the maximum volume fraction of inclusions suitable to yield high-quality, dense composites by pressureless sintering is about 15 vol%. For higher contents, the consolidation of the composite powder mixtures is usually conducted by hot-pressing. This technique involves the sintering of the composite glass powder mixture under uniaxial pressure, usually in the range 5–20 MPa, in a die. While allowing the fabrication of composites with a high volume fraction of inclusions (of up to 90 vol%) and without porosity, this technique is cost-intensive and has limitations regarding the shape and complexity of the parts that can be produced. Figures 4 (a–c) show typical microstructures of dispersion-reinforced borosilicate glass matrix composites obtained by hot-pressing, including alumina-platelet, chopped metal fibre and molybdenum particle reinforced systems. The fairly homogeneous distribution of the reinforcing elements and the absence of porosity, cracks or other microstructural defects is evident, which indicates improved mechanical properties in these composites.

Alternative processing methods

Alternative fabrication procedures for dispersion-reinforced glass matrix composites via wet routes have been developed such as slip and tape casting [117,118], as well as colloidal processing and sol-gel techniques [74].

The extrusion of viscous glass/reinforcement mixtures has also been considered to produce glass matrix composites. This process is shown schematically in Figure 5. It has been applied to the fabrication of composites with aligned chopped SiC-fibre and C-fibre [30] as well as SiC-platelet [42] and alumina platelet [119]. The technique enables the fabrication of components with large length-to-diameter ratios.

The fabrication of dispersion-reinforced glass-ceramic matrix composites by in-situ growing of whiskers in the glassy matrix has also been explored. In particular, in barium-aluminosilicate (BAS) glass-ceramic matrices containing additions of Si_3N_4, the formation of β-Si_3N_4 whiskers during sintering to form whisker-reinforced BAS glass-ceramic matrix composites has been reported [38]. The in-situ growth of hollandite whiskers in a BAS matrix [120] and TiO_2 whiskers in lithia-alumina-silica [121] matrices has also been investigated.

A novel method to fabricate dispersion-reinforced glass matrix composites includes the use of microwave heating. This approach has been used to fabricate SiC whisker-reinforced calcium magnesium aluminosilicate glass-ceramic matrix composites [122]. Almost fully dense composites could be obtained at 850 °C after heating for 15 minutes, which represents a significant reduction of heating time in comparison with conventional heating and reduced whisker oxidation. The inverse heating gradient in a microwave furnace has also been exploited to produce porous Mo particle-reinforced borosilicate glass matrix composites [123] and a variety of other metal particle reinforced borosilicate glass composites [124]. The application of microwave heating seems to be a favourable technique for rapid production of dispersion-reinforced glass matrix composites with tailored microstructures.

It is likely that the fabrication of dispersion-reinforced glass matrix composites will benefit from advances in the so called "ultrastructure" processing of glasses and ceramics in the future [125]. In particular techniques based on chemical principles such as those of organic chemistry and surface chemistry, are playing an increasingly significant role in glass and ceramic manufacture to overcome the limitations of traditional powder technology. This is manifested in the continuing development of methods to manipulate sols, gels and organometallic precursors at low temperatures, and to synthesise and manipulate ultrafine, submicrometre particles and colloids with the aim to create more homogeneous microstructures, and thus components of high reliability. Adaptation of these novel techniques for the processing of high-quality dispersion-reinforced glass matrix composites will be one of the most promising areas expanding traditional glass composite research in the future.

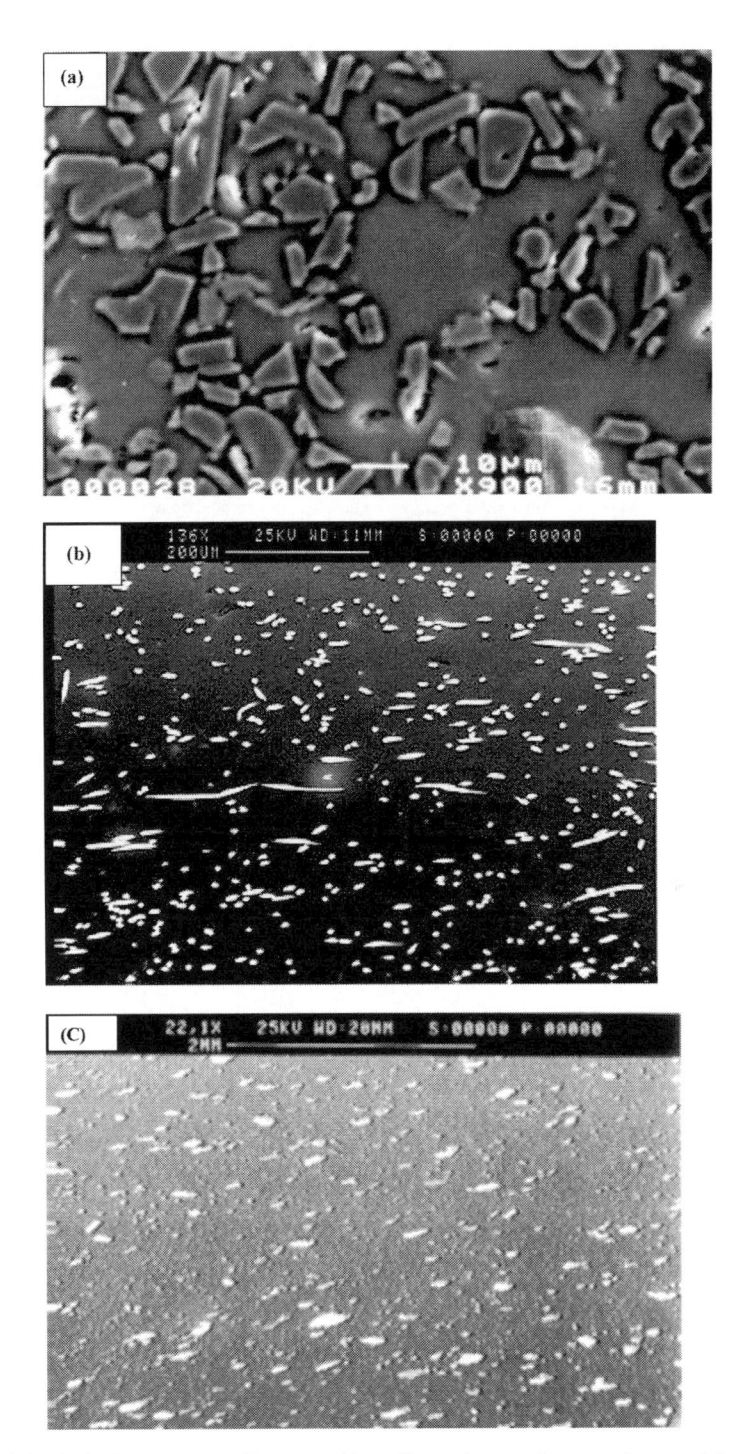

FIGURE 4. Typical microstructures of hot-pressed borosilicate glass matrix composites containing a) alumina platelet, b) chopped stainless steel fibres and c) molybdenum particles as reinforcement [16,17,55].

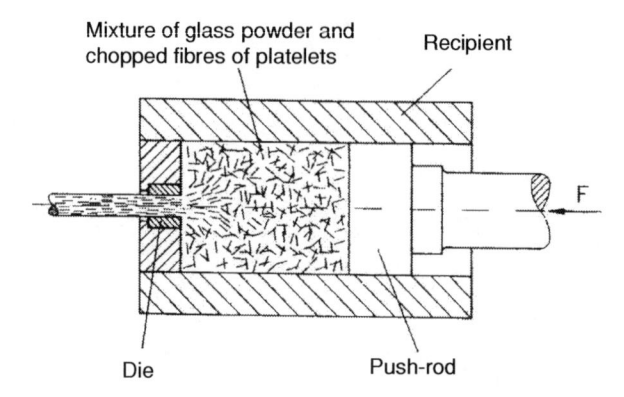

FIGURE 5. The extrusion process for the fabrication of glass matrix composites with chopped fibre or platelet reinforcement ([42]).

4. PROPERTIES

Mechanical properties

Table 1 presented a summary of mechanical properties achieved in typical glass and glass-ceramic composites with dispersion reinforcement. The values for fracture strength (σ) and fracture toughness (K_{IC}) quoted in Table 1 are consistently higher than typical values for unreinforced silicate glasses, which are usually: $\sigma < 50$–60 MPa and $K_{IC} = 0.7$–0.8 MPa$\sqrt{m}$ [126].

In general, the relationship between fracture strength, properties of constituents and microstructure cannot be reliably predicted in brittle matrix composites. Knowledge of a number of additional parameters is needed to allow the description of the composite mechanical behaviour.

The following factors have been identified as important in determining the fracture strength and fracture toughness values of dispersion-reinforced glass matrix composites:

- Volume fraction of the dispersed phase
- Microstructural parameters, such as distribution, shape, orientation and size of the inclusions
- Residual porosity in the matrix
- Different elastic properties, mechanical strength and toughness of the matrix and the inclusions
- Thermal expansion mismatch between the matrix and the inclusions
- Properties of the matrix/inclusion interfaces, i.e. bond strength.

The combination of these factors affects a number of strengthening and toughening mechanisms which can be exploited in glass and glass-ceramic matrix composite materials.

Strengthening mechanims

Load sharing occurs in a dispersion of a higher modulus second phase in a lower modulus glass matrix, provided that strong particle-matrix bonding exists [127]. If the components of

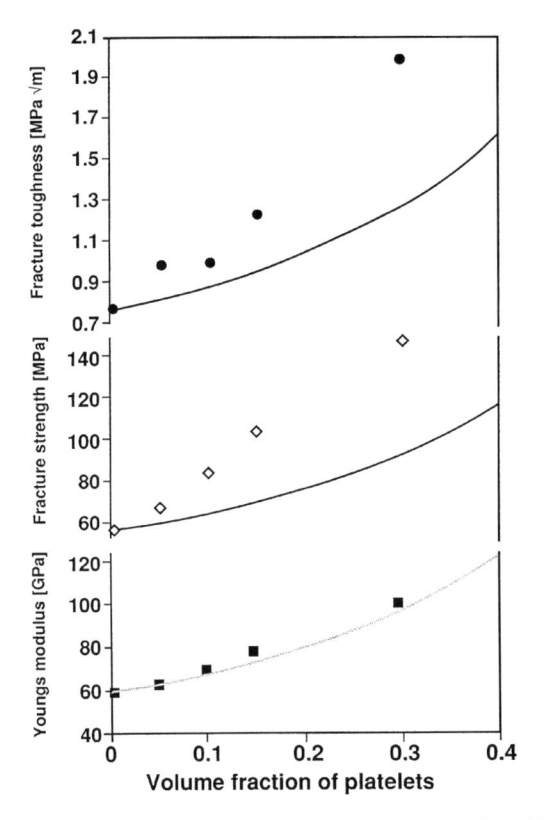

FIGURE 6. The dependence of the mechanical properties of alumina platelet reinforced borosilicate glass matrix composites on the platelet volume fraction: (a) Young's modulus, (b) fracture strength and (c) fracture toughness [17,128].

the system share the applied load in proportion to their elastic modulus, both the inclusions and the glass matrix will deform equally, i.e. the strains in both components in unidirectional tension will be the same and one can write [127]:

$$\sigma_c = \sigma_m \frac{E_c}{E_m} \qquad (1)$$

where σ_c and σ_m are the fracture stress of the composite and the matrix respectively, and E_c and E_m are the Young's moduli of the composite and the matrix respectively. It follows from Eq. 1 that when E_c increases with increasing volume fraction of the dispersed phase, the load to failure and consequently the fracture strength of the composite must increase proportionally.

The variation of the elastic modulus, fracture strength and fracture toughness of a model alumina platelet-reinforced borosilicate glass matrix composite with the volume fraction of platelets is shown in Figure 6 [17,128]. The material exhibited a pore-free matrix, uniform distribution of the platelets and strong matrix/platelet interfacial bonding. In Figure 6 both the experimentally measured data and the theoretical curves calculated by means of the simple load-sharing model are represented.

To explain the enhanced strength of particle-reinforced glass matrix composites, Hasselman and Fulrath proposed a theory based on the limitation of flaw size by interparticle spacing [129] and invoking the Griffith's flaw theory. They postulated that fine dispersions may limit the size of Griffith flaws, thereby raising the stress required to initiate or propagate a crack. They showed that strengthening occurred whenever the interparticle spacing λ became small enough to limit the size of surface flaws from where fracture began, to less than that present in the glass containing no inclusions. For very low fraction of inclusions, implying a large value of λ, no strengthening occurred. In general, however, the prediction of the mechanical strength of particulate reinforced glass matrix composites, from a knowledge of the component properties and microstructure features, is still imprecise. There is thus a clear need for further theoretical and experimental research in this area.

The role of internal stresses

A relevant aspect which must be taken into account when evaluating the mechanical behaviour of composite materials is the presence of internal stresses. These arise as the result of thermal expansion mismatch between the matrix and the reinforcing phase, and usually develop upon cooling from the fabrication temperature [130,29]. In addition, differences in the elastic properties of the constituents of a composite can lead to localised stress concentrations around the inclusions under an externally applied stress [131]. Both effects will control the path of a running crack and will determine the interaction of the crack with the reinforcing inclusions.

Residual thermal stresses may lead to spontaneous microcracking of glass matrix composites provided the size of the embedded particles is greater than the so-called critical particle size D_c [9]. Several approaches have been developed to determine D_c. In the model of Davidge and Green, for example, D_c is given by [9]:

$$D_c = \frac{4\gamma_m [E_i (1 + \nu_m) + 2E_m (1 - 2\nu_i)]}{(\Delta\alpha\Delta T)^2 E_m E_i} \quad (2)$$

where γ_m is the fracture energy of the glass matrix, E_m and E_i are the Young's moduli, ν_m and ν_i are the Poisson's ratios of matrix and inclusions respectively, ΔT is the temperature range over which stress development is not relaxed by inelastic processes during cooling (usually the difference between fabrication and room temperature) and $\Delta\alpha$ is the differential thermal expansion of the two phases.

If the thermal expansion mismatch is such that compressive stresses are developed in the matrix, these will contribute to the strength of the composite since they will oppose any applied tensile stresses.

Toughening mechanisms

A variety of mechanisms related to the crack/inclusion interaction can be exploited to increase the toughness of dispersion-reinforced glass matrix composites. In fact, toughening rather than strengthening has frequently been the goal when developing this kind of composite systems. Toughening mechanisms are also affected by the presence of residual stresses in the material.

In general, a great number of toughening mechanisms have been indentified in particulate-reinforced brittle matrix composites, such as crack bowing, crack deflection, crack bridging, crack branching, increase of Young's modulus, inclusion fracture, inclusion debonding and pull-out, formation of matrix microcracks near the crack tip, ductile deformation of the inclusions (for metallic inclusions) and the direct effect of the residual stresses, as described by Matthews and Rawlings [130]. Only those mechanisms most frequently reported to be active in the composites of interest in this Chapter will be briefly described here. If interactions between different toughening mechanisms are ignored, the total fracture toughness can be given by simple summation of the individual contributions [132].

Toughening by residual stresses

This effect is achieved when the hydrostatic residual stress state in the glass matrix is compressive: in this case, the effective crack opening stress at the crack tip is reduced, resulting in an increase in fracture toughness ΔK given by [133]:

$$\Delta K = 2\sigma_h \sqrt{\frac{2(\lambda - d)}{\pi}} \quad (3)$$

where σ_h is the hydrostatic compressive stress in the glass matrix, d is the average inclusion size and λ is the average interparticle spacing, which depends on the inclusion volume fraction [133].

Crack deflection

The crack deflection mechanism [134] has been invoked several times to explain the fracture toughness increment of dispersion-reinforced glass matrix composites [10,15,17 30,128]. For the alumina platelet reinforced borosilicate glass matrix composite considered above [33] (see Figure 6), observation of the fracture surfaces led to the conclusion that the crack deflection process was responsible for the increment of toughness achieved [128]. Numerical computations based on the OOF finite element based method have confirmed the experimental observations [135]. As Figure 7 shows, the roughness of the fracture surfaces increases with the volume fraction of particles, thus indicating increasing crack path tortuosity as the crack is deflected by the platelets. It has been shown that the level of increment of fracture surface roughness, as measured by profilometer techniques, parallels the increasing values of fracture toughness in these composites [136], thus confirming the toughening effect of crack deflection.

In the platelet reinforced glass composite referred to above, the thermal expansion coefficient of the inclusions (alumina, $\alpha_i = 8 \; 10^{-6}/°C$) is higher than that of the borosilicate matrix ($\alpha_m = 3.3 \; 10^{-6}/°C$), resulting in a hoop compression stress in the matrix around the platelets [128]. Such a stress field will deflect the crack and cause it to travel around the platelet. These residual compressive stresses in the matrix will also contribute to the strengthening of the material (see Figure 6) since they will reduce the effect of any externally applied tensile stresses.

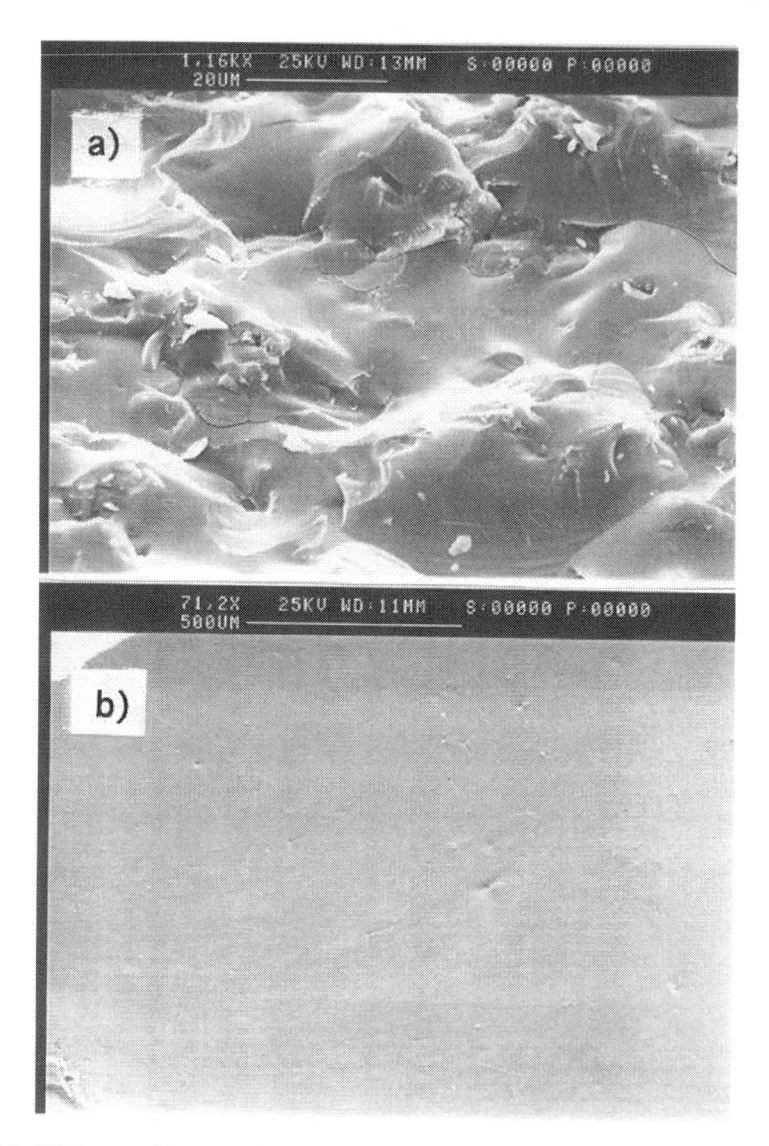

FIGURE 7. SEM images of fracture surfaces showing the different topography (roughness) for (a) glass matrix composite containing 15 vol% platelets and (b) unreinforced glass [17].

Other toughening mechanisms: microcracking, inclusion debonding and pull-out

The creation of a stress field in a two-phase brittle matrix material can be exploited for toughening since it will induce microcracks on application of an external load. Microcracks are created at the vicinity of the tip of a macrocrack forming a process zone [130]. Using this mechanism, the elastic modulus within the process zone is reduced, the stress intensity at the crack tip is lowered and the toughness is increased.

Other toughening mechanisms are related to the debonding and pull-out of the inclusions during fracture. These mechanisms are especially effective in whisker and short-fibre reinforced glass composite materials [35].

The phenomenon of transformation toughening by means of the rapid stress-induced structural transformation of zirconia particles dispersed in glass matrices has also been considered [44]. Moreover, toughening glass matrices by the piezoelectric effect has been recently proposed [45]. Lead-zirconate-titanate particles were introduced in a lead silicate glass of matching thermal expansion coefficient. An increase of fracture toughness of $>50\%$ was achieved [45]. However further work is required to determine the real potential of piezoelectric toughening in glass matrices.

Crack bridging by ductile inclusions

When using metallic inclusions the ductility of the metallic phase may be exploited to increase the overall toughness of the composite. In this case, an optimal interfacial bonding and a low level of internal stresses are required in order to ensure that a running crack will propagate through the metallic inclusion rather than be deflected or propagate along the interface. The toughness increment of the composite results from the mechanism of crack bridging, in which the ductile phase is stretched to failure between the crack surfaces [5,16,20,31,53–55].

Figure 8 shows a typical fracture surface of a glass matrix composite reinforced by short stainless steel fibres [16]. It is apparent that the fibre was intercepted and sectioned by the propagating crack and there is some evidence of plastic deformation.

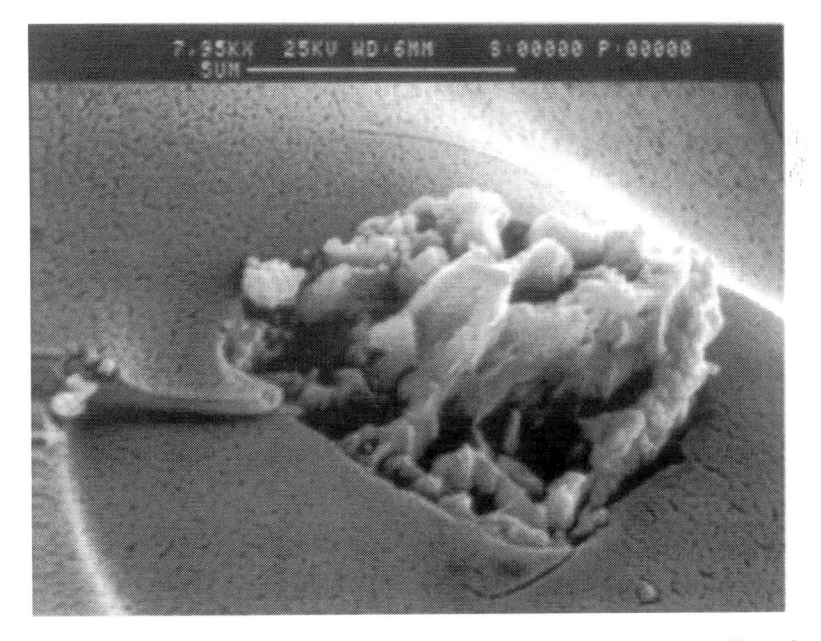

FIGURE 8. Fracture surface of a glass matrix composite reinforced by chopped stainless steel fibres, showing a fibre that has been fractured with limited plastic deformation [16].

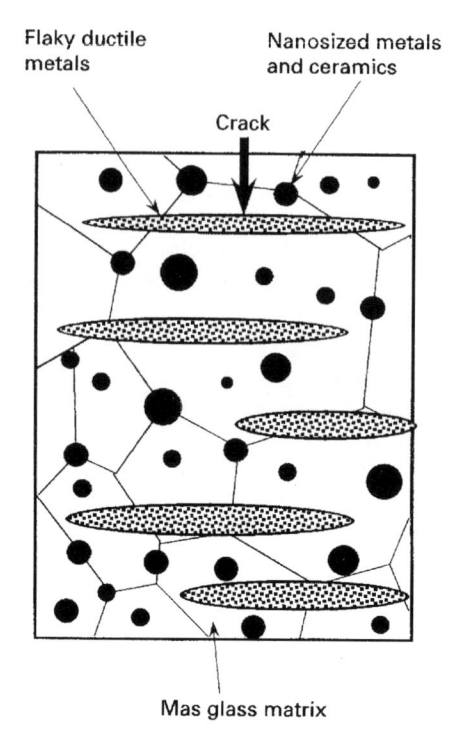

FIGURE 9. A composite approach for strengthening and toughening of glass by metallic inclusions [137].

Using molybdenum inclusions in MAS glass, it has been demonstrated that the incorporation of both nanosized particles and bigger, microscopic flaky particles of high aspect ratio (>17), as shown schematically in Figure 9, can lead to considerable toughness and strength increments [137]. This favourable mechanical behaviour was linked to the plastic deformation of the flaky metallic particles at the crack tip during fracture.

Other properties

Hardness and Brittleness

The incorporation of a ceramic dispersion phase in a glass matrix leads to an increase in the effective hardness in comparison with that of the unreinforced matrix. On the other hand, metallic inclusions will usually have the opposite effect. The simultaneous modification of hardness (H) and fracture toughness (K_{IC}) in composite materials results in variations in the brittleness of the material, which can be quantified by the brittleness index (B), with $B = H/K_{IC}$ [138].

Machinability, wear and erosion resistance

Borosilicate glass matrix composites with chopped graphite fibres are highly machineable materials [29]. Machinable glass-ceramics containing interlocking platelet-like and easily cleavable crystals of fluorophlogopite mica dispersed in a glassy matrix are well-known [139].

The wear resistance of glasses and glass-ceramics can be increased by incorporating hard and tough particles [140–142] or SiC whiskers [36]. Erosion resistant materials can also be fabricated on the basis of glass and glass-ceramic matrix composites. For example, fused silica matrix composites reinforced with high purity silica short fibres have been developed for hypersonic radome applications [142]. Erosion and wear resistant particle reinforced glass-ceramic matrix composites have been developed by incorporating TiC particle (6–22 vol%) into CaO-MgO-Al_2O_3-SiO_2 glass-ceramic [143]. Glass-ceramic matrix composites in the system LiO_2-ZrO_2-SiO_2 containing alumina particles have been developed as tiles with high wear resistance [19].

Thermal shock resistance

The thermal shock resistance of brittle materials can be improved by incorporating a dispersion phase of lower elastic modulus and lower thermal expansion coefficient than that of the matrix. Antimony and aluminium titanate (Al_2TiO_5) particles [108] have been used in borosilicate and aluminosilicate matrices to obtain glass matrix composites with improved thermal shock behaviour. Short carbon fibres [28] and SiC particles [100] have been incorporated in borosilicate and glass-ceramic matrices leading to composite materials of high thermal shock resistance. A critical temperature difference $\Delta T_c = 600$ K for microcracking development in the composite has been reported [28], whereas ΔT_c for the unreinforced glass matrix was about 350 K.

The improvement of the thermal shock behaviour of the composites may be attributed to a combination of a lower coefficient of thermal expansion and the higher thermal conductivity expected on the incorporation of metallic, SiC or aluminium titanate particles or of carbon fibres into the silicate matrices [108].

5. SUMMARY

It was shown that by reinforcing the low-modulus, low-strength glass matrix with a high-modulus, high-strength and/or high-ductility second constituent in the form of whiskers, platelets, particles or short fibres, dispersion-reinforced glass and glass-ceramic matrix composites with improved mechanical properties can be fabricated. These materials can be used under moderate loads and at intermediate and elevated temperatures (up to about 1200 °C in case of refractory glass-ceramic matrices). Clearly, the improvement of mechanical properties in these composite systems is not large when compared with that achieved using continuous fibre reinforcement (see Chapter 19); however, the use of relatively simple and cost-effective fabrication routes and the possibility of tailoring other (functional) properties by addition of the dispersed phase, justifies research and development efforts into these systems.

The cost of glass and glass-ceramic matrix composites is strongly dependent on the fabrication procedures employed and on the type of reinforcement used [75], while the matrix has a lower cost impact. Since thermal treatments to densify the components are normally conducted at relatively low temperatures (in any case lower than those necessary to densify technical ceramics such as alumina, zirconia, silicon carbide or mullite), glass and glass-ceramic matrix composite production will usually incur lower costs.

The current uses and future potential of these composites were reviewed with emphasis on mechanical engineering and machinery, electronics, aerospace, chemical, high-temperature and wear resistance applications. Current and potential applications in medicine and in industrial waste recycling technologies were also considered. Interesting application possibilities arise for these composite materials when the inclusion phase imparts a combination of both the desired functional properties and a higher mechanical reliability. As for all new material developments, a close co-operation of materials scientists, design engineers and potential end-users is required if a greater number of these composite systems are to find widespread applications.

6. REFERENCES

1. D. B. Binns, Some Physical Properties of Two-Phase Crystal-Glass Solids, in *Science of Ceramics*, Vol. 1. G. H. Stewart, ed. Academic Press, New York (1962). pp. 315–334.
2. D. P. H. Hasselman and R. M. Fulrath, Effect of Dispersions on Young's Modulus of a Glass, *J. Am. Ceram. Soc.* **48**, 218–219 (1965).
3. D. P. H. Hasselman and R. M. Fulrath, Effect of Spherical Tungsten Dispersions on Young's Modulus of a Glass, *J. Am. Ceram. Soc.* **48**, 548–549 (1965).
4. G. Einmahl, *Strength in a Two-Phase Model System with Fiber Reinforcement*, M. S. Thesis, University of California, Berkeley (1966).
5. V. D. Krstic, On the Fracture of Brittle-Matrix/Ductile-Particle Composites, *Phil. Mag.* **A48**, 695–708 (1983).
6. W. J. Frey and J. D. Mackenzie, Mechanical Properties of Selected Glass-Crystal Composites, *J. Mater. Sci.* **2**, 124–130 (1967).
7. T. L. Jessen, J. J. Mecholsky and R. H. Moore, Fast and Slow Fracture in Glass Composites Reinforced with Fe-Ni-Co Alloy, *Ceram. Bull.* **65**, 377–381 (1986).
8. K. P. Gadkaree and K. Chyung, Silicon Carbide Whisker-Reinforced Glass and Glass-ceramic Matrix Composites, *Am. Ceram. Soc. Bull.* **65**, 370–376 (1986).
9. R. W. Davidge and T. J. Green, The Strength of Two-Phase Ceramic/Glass Materials, *J. Mater. Sci.* **3**, 629–634 (1968).
10. R. A. Haber, G. E. Hannon and J. B. Wachtman, Fracture Strength and Toughness in Particulate Reinforced Glass Composites: Effects of Microstructural Variations, in: *Ceramic Materials and Components for Engines*, V. S. Tennery ed., Am. Ceram. Soc. (1989), pp. 927–936.
11. T. B. Troczynski, P. S. Nicholson and C. E. Rucker, Inclusion-Size-Independent Strength of Glass/Particulate-Metal Composites, *J. Am. Ceram. Soc.* **71**, C-276–C-279 (1988).
12. J. P. Lucas, L. E. Toth and W. W. Geberich, A Novel Technique for Producing Fine Metal Fibres for Enhancing Mechanical Properties of Glass Matrix Composites, *J. Am. Ceram. Soc.* **63**, 280–285 (1980).
13. G. Baran, M. Degrange, C. Roques-Carmes and D. Wehbi, Fracture Toughness of Metal Reinforced Glass Composites, *J. Mat. Sci.* **25**, 4211–4215 (1990).
14. R. H. Dunga, J. A. Gilbert and J. C. Smith, Preparation and Mechanical Properties of Composites of Fused SiO_2 and W Fibres, *J. Am. Ceram. Soc.* **56**, 345 (1973).
15. J. C. Swearengen, E. K. Beauchamp and R. J. Eagan Fracture Toughness of Reinforced Glasses, in *Fracture Mechanics of Ceramics*, Vol. IV. R. C. Bradt D. P. H. Hasselman and F. F. Lange ed., Plenum Press, New York, London (1978) pp. 973–987.
16. A. R. Boccaccini, G. Ondracek and C. Syhre, Borosilicate Glass Matrix Composites Reinforced with Short Metal Fibres, *Glastech. Ber. Glass Sci. Technol.* **67**, 16–20 (1994).
17. A. R. Boccaccini and P. A. Trusty, Toughening and Strengthening of Glass by Al_2O_3 Platelets, *J. Mat. Sci. Lett.* **15**, 60–63 (1996).
18. J. Wu, B. Li, J. Guo, The Influence of Addition of AlN Particles on Mechanical Properties of SiO_2 Matrix Composites Doped with AlN Particles, *Mat. Lett.* **41**, 145–148 (1999).

19. A. P. N. de Oliveira, T. Manfredini, G. C. Pellacani, A. Bonamartini Corradi and L. Di Landro, Al_2O_3 Particulate-Reinforced LZS Glass-Ceramic Matrix Composites, in: *9th CIMTEC World Ceramics Congress, Part C*, P. Vincenzini ed., Techna Srl (1999) pp. 707–714.
20. E. Claxton, B. A. Taylor, R. D. Rawlings, Processing and Properties of a Bioactive Glass-ceramic Reinforced with Ductile Silver Particles, *J. Mat. Sci.* **37**, 3725–3732 (2002).
21. F. Ye, J. C. Gu, Y. Zhou, M. Iwasa, Synthesis of $BaAl_2Si_2O_8$ Glass-ceramic by a Sol-gel Method and the Fabrication of $SiC_{pl}/BaAl_2Si_2O_8$ Composites, *J. Europ. Ceram. Soc.* **23**, 2203–2209 (2003).
22. T. Rouxel, B. Baron, P. Verdier and T. Sakuma, SiC Particle Reinforced Oxynitride Glass: Stress Relaxation, Creep and Strain-Rate Imposed Experiments, *Acta Mater.* **46**, 6115–6130 (1998).
23. E. Zhang and D. P. Thompson, Fracture Behaviour of SiC Fibre-Reinforced Nitrogen Glass Matrix Composites, *J. Mat. Sci.* **31**, 6423–6429 (1996).
24. F. Pernot and R. Rogier, Mechanical Properties of Phosphate Glass-Ceramic-316L Stainless Steel Composites, *J. Mat. Sci.* **28**, 6676–6682 (1993).
25. G. Wen, G. L. Wu, T. Q. Lei, Y. Zhou, Z. X. Guo, Co-enhanced SiO_2-BN Ceramics for High-Temperature Dielectric Applications, *J. Europ. Ceram. Soc.* **20**, 1923–1928 (2000).
26. R. A. J. Sambell, D. H. Bowen and D. C. Phillips, Carbon Fibre Composites with Ceramic and Glass Matrices, Part I: Discontinuous Fibres, *J. Mat. Sci.* **7**, 663–675 (1972).
27. D. Qi and C. G. Pantano, Effects of Composite Processing on the Performance of Carbon Fibre/Glass Matrix Composites, *Ceram. Eng. Sci. Proc.* **13**, 863–872 (1992).
28. K. Ogi, N. Takeda and K. M. Prewo, Fracture Process of Thermally Shocked Discontinuous Fibre-Reinforced Glass Matrix Composites Under Tensile Loading, *J. Mat. Sci.* **32**, 6153–6162 (1997).
29. K. M. Prewo, A Compliant, High Failure Strain, Fibre-Reinforced Glass-Matrix Composite, *J. Mat. Sci.* **17**, 3549–3563 (1982).
30. K. Langhans and E. Roeder, Manufacture of Short-Fibre Reinforced Glasses by Extrusion and Examinations Regarding their Structure and their Mechanical Properties, *Glastech. Ber.* **65**, 103–111 (1992).
31. M. J. Pascual, A. Duran, L. Pascual, Sintering Behaviour of Borosilicate glass-ZrO_2 Fibre Composite Materials, *J. Europ. Ceram. Soc.* **22**, 1513–1524 (2002).
32. J. S. Lyons and T. L. Starr, Strength and Toughness of Slip-Cast Fused-Silica Composites, *J. Am. Ceram. Soc.* **77**, 1673–1675 (1994).
33. K. P. Gadkaree, Whisker Reinforcement of Glass-Ceramics, *J. Mat. Sci.* **26**, 4845–4854 (1991).
34. H. M. Jang, K. S. Kim and Ch. J. Jung, Development of SiC-whisker-reinforced lithium aluminosilicate matrix composites by a mixed colloidal processing route, *J. Am. Ceram. Soc.* **75**, 2883–2886 (1992).
35. F. D. Gac, J. J. Petrovic, J. V. Milewski and P. D. Shalek, Performance of Commercial and Research Grade SiC Whiskers in a Borosilicate Glass Matrix, *Ceram. Eng. Sci. Proc.* **7**, 978–982 (1986).
36. V. S. R. Murthy, K. Srikanth and C. B. Raju, Abrasive Wear Behaviour of SiC Whisker-Reinforced Silicate Matrix Composites, *Wear* **223**, 79–92 (1998).
37. F. Ye, J. M. Yang, L. T. Zhang, W. C. Zhou, Y. Zhou, T. C. Lei, Fracture Behaviour of SiC-Whisker-Reinforced Barium Aluminosilicate Glass-ceramic Matrix Composites, *J. Am. Ceram. Soc.* **84**, 881–883 (2001).
38. F. Yu, C. R. Ortiz-Long and K. W. White, The Microstructural Characterization of an in situ Grown Si_3N_4 Whisker-Reinforced BAS Glass-Ceramic Matrix Composite, *Ceram. Trans.* **74**, 203–214 (1996).
39. I. Wadsworth and R. Stevens, Strengthening and Toughening of Cordierite by the Addition of Silicon Carbide Whiskers, Platelets and Particles, *J. Mat. Sci.* **26**, 6800–6808 (1991).
40. A. R. Hyde and G. Partridge, Fabrication of Particulate, Platelet, Whisker and Continuous Fibre Reinforced Glass, Glass-ceramic and Ceramic Materials, *Br. Ceram. Proc.* **45**, 221–227 (1990).
41. R. Chaim and V. Talanker, Microstructure and Mechanical Properties of SiC Platelet/Cordierite Glass-Ceramic Composites, *J. Am. Ceram. Soc.* **78**, 166–172 (1995).
42. E. Roeder H.-J. Mayer and M. Huber, Theoretische und experimentelle Analyse des Ausrichtungsverhaltens plättchenförmiger Verstärkungskomponenten beim Strangpressen von Verbundkörpern mit Glasmatrix, *Mat.-wiss. und Werkstofftech.* **27**, 37–44 (1996).
43. Y. Cheng and D. P. Thompson, ZrO_2 Toughened Glass-Ceramic Composites Prepared by Hot-Pressing Route, *Silicate Industriels* **12**, 5–14 (1991).
44. M. Nogami and M. Tomozawa, ZrO_2-Transformation Toughened Glass-Ceramics Prepared by the Sol-gel Process from Metal Alkoxides, *J. Am. Ceram. Soc.* **69**, 99–102 (1986).
45. A. R. Boccaccini D. H. Pearce, Toughening of Glass by a Piezoelectric Secondary Phase, *J. Am. Ceram. Soc.* **86**, 180–182 (2003).

46. D. O'Sullivan, C. Courtois, A. Leriche and B. Thierry, Properties of Cordierite-Silicon Carbide Nanocomposites, *Key Eng. Mat.* Vols. **132–136**, 1997–2000 (1997).
47. N. K. Schneider, C. Mooney, B. Baron and S. Hampshire, Oxynitride Glass Composites Containing Nanosize SiC, *Br. Ceram. Proc.* **60** [1], 401–402 (1999).
48. J. DiMaio, S. Rhyne, Z. Yang, K. Fu, R. Czerw, J. Xu, S. Webster, Y.-P. Sun, D. L. Carroll and J. Ballato, Transparent Silica Glasses Containing Single Walled Carbon Nanotubes, *Information Sciences* **149**, 69–73 (2003).
49. T. Seeger, T. Köhler, T. Frauenheim, N. Grober, M. Rühle, M. Terrones, G. Seifert, Nanotube Composites: Novel SiO_2 Coated Carbon Nanotubes, *Chem. Commun.* **1**, 34–35 (2002).
50. A. R. Boccaccini, D. B. Acevedo, G. Brusatin, and P. Colombo, Borosilicate glass matrix composites containing multi-wall carbon nanotubes, *J. Europ. Ceram. Soc.* (2004) in press.
51. B. R. Zhang, M. Ferraris and F. Marino, Borosilicate Glass-Ceramic Composites Reinforced by Ni_3Al Ribbons and Particles, *J. Europ. Ceram. Soc.* **17**, 1381–1386 (1997).
52. D. R. Biswas, Strength and Fracture Toughness of Indented Glass-Nickel Compacts, *J. Mat. Sci.* **15**, 1696–1700 (1980).
53. R. U. Vaidya, C. Norris and K. N. Subramanian, Interfacial Effects in a Metallic-Glass Ribbon Reinforced Glass-Ceramic Matrix Composite, *J. Mat. Sci.* **27**, 4957–4960 (1992).
54. E. Verné, M. Ferraris, A. Ventrella, L. Paracchini, A. Krajewski and A. Ravaglioli, Sintering and Plasma Spray Deposition of Bioactive Glass-Matrix Composites for Medical Applications, *J. Europ. Ceram. Soc.* **18**, 363–372 (1998).
55. I. Dlouhy, M. Rheinish, A. R. Boccaccini and J. F. Knott, Fracture Characteristics of Borosilicate Glasses Reinforced by Ductile Metallic Particles, *Fatigue and Fracture of Engineering Materials and Structures* **20**, 1235–1253 (1997).
56. A. Tewari, V. S. R. Murthy and G. S. Murty, Rheological Behaviour of SiC (Particulate)-Borosilicate Composites at Elevated Temperatures, *J. Mat. Sci. Lett.* **15**, 227–229 (1996).
57. C. H. Drummond III, Deformation of Microstructures Containing a Glassy Phase, *J. Noncryst. Sol.* **196**, 326–333 (1996).
58. A. B. R. Verma, V. S. R. Murthy and G. S. Murty, Microstructure and Compressive Strength of SiC Platelet Reinforced Borosilicate Composites, *J. Am. Ceram. Soc.* **78**, 2732–2736 (1995).
59. R. R. Tummala, Ceramic and Glass-Ceramic Packaging in the 1990s, *J. Am. Ceram. Soc.* **74**, 895–908 (1991).
60. J.-J. Shyu and C.-T. Wang, Sintering and Properties of Li_2O-Al_2O_3-$4SiO_2$-Borosilicate Glass Composites, *J. Mater. Res.* **11**, 2518–2527 (1996).
61. T. Yamaguchi and K. Lizuka, Microstructure Development in RuO_2-Glass Thick-Film Resistors and Ist Effect on the Electrical Resistivity, *J. Am. Ceram. Soc.* **73**, 1953–1957 (1990).
62. H. Shiomi and H. Furukawa, Effect of Addition of Ag on the Microstructures and Electrical Properties of Sol-Gel Derived SnO_2 Glass Composites, *J. Mat. Sci. Mat. Elect.* **11**, 31–37 (2000).
63. K. Watanabe and K. Hoshi, Crystallisation Kinetics of Fine Barium Hexaferrite, $BaFe_{12}O_{19}$, Particles in a Glass Matrix, *Phys. Chem. Glasses* **40**, 75–78 (1999).
64. F. Duan, C. Fang, Z. Ding and H. Zhu, Properties and Applications of Piezoelectric Glass-Crystalline Phase Composite in the BaO-SrO-TiO_2-SiO_2 Sytem, *Mat. Lett.* **34**, 184–187 (1998).
65. S.-Y. Chang, L. Liu and S. A. Asher, Preparation and Processing of Monodisperse Colloidal Silica-Cadmium Sulfide Nanocomposites, *Mat. Res. Symp. Proc.* **346**, 875–880 (1994).
66. S. H. Risbud and V. J. Leppert, Nanometer Level Characterisation of Rapidly Densified Ceramics and Glass-Semiconductor Composites, in: *Ceramic Microstructures: Control at the Atomic Level*, A. P. Tomsia and A. Glaeser eds., Plenum Press, New York (1998) pp. 199–207.
67. Y.-S. Lee and T.-Y. Tseng, Correlation of Grain Boundary Characteristics with Electrical Properties in ZnO-Glass Varistors, *J. Mat. Sci.: Mat. in Electr.* **9**, 65–76 (1998).
68. K. Fujiki T. Ogasawara and N. Tsubokawa, Preparation of a Silica Gel-Carbon Black Composite by the Sol-Gel Process in the Presence of Polymer-Grafted Carbon Black, *J. Mat. Sci.* **33**, 1871–1879 (1998).
69. P. Chakraborty, Review. Metal Nanoclusters in Glasses as Non-linear Photonic Materials, *J. Mat. Sci.* **33**, 2235–2249 (1998).
70. T. Burkhart, M. Mennig, H. Schmidt and A. Licciulli, Nano-sized Pd Particles in a SiO_2 Matrix by Sol-Gel Processing, *Mat. Res. Soc. Symp. Proc.* **346** 779–784 (1994).
71. G. C. Vezzoli, M. F. Chen and J. Caslavsky, New High Dielectric Strength Materials: Micro/Nanocomposites of Suspended Au Clusters in SiO_2/SiO_2-Al_2O_3-Li_2O Gels, *Ceram. Int.* **23**, 105–108 (1997).

72. J. D. Mackenzie and E. Bescher, Formation and Properties of Ultrafine Particles in Glass, in *Proc. XVIII International Congress on Glass*, M. K. Choudhary N. T. Huff and Ch. H. Drummond III eds., San Francisco (USA), July 5–10, 1998, The American Ceramic Society, Ohio.
73. A. Das, T. T. Srinivasan and R. E. Newnham, Ceramic/Polymer Nanocomposite Properties for Microelectronic Packages, *Mat. Res. Soc. Symp. Proc.* **167**, 165–175 (1990).
74. A. Nazeri, E. Bescher and J. D. Mackenzie, Ceramic Composites by the Sol-Gel Method: A Review, *Ceram. Eng. Sci. Proc.* **14** [11–12], 1–19 (1993).
75. A. R. Hyde, Ceramic Composites, a New Generation of Materials for Mechanical and Electrical Applications, *GEC J. Res.* **7**, 65–71 (1989).
76. P. Van Landuyt, D. Michel, L. Nicks, J.-M. Streydio, E. Munting and F. Delannay, Processing and Characterisation of a Biocompatible Glass Matrix Composite Reinforced with Titanium Fibres, *Silicates Industriels* **9–10**, 257–259 (1995).
77. T. B. Troczynski and P. S. Nicholson, Fracture Mechanics of Titanium/Bioactive Glass-Ceramic Particulate Composites, *J. Am. Ceram. Soc.* **74**, 1803–1806 (1991).
78. T. B. Troczynski and P. S. Nicholson, Stress Corrosion Cracking of Bioactive Glass Composites, *J. Am. Ceram. Soc.* **73**, 164–166 (1990).
79. S.-J. Lee, W. M. Kriven and H.-M. Kim, Shrinkage-Free, Alumina-Glass Dental Composites via Aluminium Oxidation, *J. Am. Ceram. Soc.* **80**, 2141–2147 (1997).
80. W. Hölland, Biocompatible and Bioactive Glass-Ceramics: State of the Art and New Directions, *J. Non-Cryst. Solids* **219**, 192–197 (1997).
81. B. R. Lawn, N. P. Padture, H. Cai and F. Guiberteau, Making Ceramics "Ductile", *Science* **263**, 1114–1116 (1994).
82. S. A. Dunn, Viscous Behaviour of Silica with Tungsten Inclusions, *Ceram. Bull.* **47**, 554–559 (1968).
83. A. E. Rudovskij, P. D. Sarkisov, A. A. Ivashin and V. V. Budov, Glass Ceramic-Based Composites, in *Ceramic- and Carbon-Matrix Composites*. V. I. Trefilov ed., Chapman and Hall, London (1995) 255–285.
84. T. Kachi, T. Furuhata, N. Arai, M. Iwata and S. Kato, Innovative Glass Coating with High Emittance at High Temperatures, *Ceram. Trans.* Vol. **99**, 137–146 (1998).
85. M. Dietrich, V. Verlotski, R. Vassen and D. Stoever, Metal-Glass Based Composites for Novel TBC-Systems, *Mat.-wiss. u. Werkstofftech.* **32**, 669–672 (2001).
86. B. Rödicker, R. Weber, P. Hellmold, P. Seidel and H. A. Maaouf, Korrelationen zwischen Gefüge und Eigenschaften von Emails, in: *Proc. Conference Verbundwerkstoffe und Werkstoffverbunde*, K. Friedrich ed., DGM-Informationsgesellschaft, Frankfurt, Germany (1997) 715–720.
87. G. V. Berdova, V. E. Gorbatenko, A. P. Zubekhin and T. A. Ionina, Synthesis and Dilatometric Investigation of High-Endurance Composite Glass-Enamel Heat Resistant Coatings, *Glass and Ceramics* **53**, 48–50 (1996).
88. A. P. Novaes de Oliveira and O. E. Alarcon, Microstructural Design Concepts Applied to Ceramic Glazes, *Tile & Brick Int.* **15**, 90–94 (1999).
89. C. Swearengen, R. J. Eagan, Mechanical Properties of Molybdenum-Sealing Glass-Ceramics, *J. Mat. Sci.* **11**, 1857–1866 (1976).
90. G. Partridge, Joining Glass-Ceramics to Metals, in *Joining of Ceramics*, M. G. Nicholas ed., Chapman and Hall, London, New York (1990) 31–55.
91. I. Penkov, R. Pascova and I. Drangajova, A New Glass Ceramic Material with High Resistance to Molten Aluminium, *J. Mat. Sci. Lett.* **16**, 1544–1546 (1997).
92. G. H. Beall and L. R. Pinckney, Nanophase Glass-Ceramics, *J. Am. Ceram. Soc.* **82**, 5–16 (1999).
93. H. Iba, T. Chang, Y. Kagawa, H. Minakuchi and K. Kanamaru, Fabrication of Optically Transparent Short Fibre-Reinforced Glass Matrix Composites, *J. Am. Ceram. Soc.* **79**, 881–884 (1996).
94. A. R. Boccaccini, S. Atiq, G. Helsch, Optomechanical Glass Matrix Composites, *Comp. Sci. and Technol.* **63**, 779–783 (2003).
95. E. G. Wolff, Thermal Expansion in Metal/Lithia-Alumina-Silica (LAS) Composites, *Int. Journal of Thermophysics* **9**, 221–232 (1988).
96. Y. Waku, M. Suzuki, Y. Oda and Y. Kamitoku, Electromagnetic Shielding Material, *Japanese Patent* 09074298 A (1997).
97. R. L. McGee and S. Yalvac, Random Stainless Steel Fibre Reinforced Magnesia-Alumina-Silicate Glass Matrix Composites, in *Int. SAMPE Symp. Exhib.* (1990) 520–532.
98. Ch. H. Drummond, R. D. Blume, P. Nevatia, Z. Gao and A. B. Sarko, Vitrified Glass-Ceramic Product Development From Industrial Wastes, *Key Eng. Materials* **132–136**, 2220–2223 (1997).

99. A. R. Boccaccini, J. Janczak, D. M. R. Taplin and M. Köpf, The Multibarriers-System as a Materials Science Approach for Industrial Waste Disposal and Recycling: Application of Gradient and Multilayered Microstructures, *Environmental Technology* **17**, 1193–1203 (1996).
100. C. B. Von Schweitzer, R. D. Rawlings and P. S. Rogers, Processing and Properties of Silceram Glass-ceramic Matrix Composites Prepared by the Powder Route, in *Third Euroceamics* Vol. 2, P. Durán and J. F. Fernández eds., Faenza Editrice Ibérica, 1139–1144 (1993).
101. A. R. Boccaccini, M. Bücker, J. Bossert and K. Marszalek, Glass Matrix Composites from Coal Flyash and Waste Glass *Waste Management* **17**, 39–45 (1997).
102. N. M. P. Low, Fabrication of Cellular Structure Composite Material from Recycled Soda-Lime Glass and Phlogopite Mica Powders, *J. Mat. Sci.* **15**, 1509–1517 (1980).
103. A. R. Boccaccini, M. Köpf and W. Stumpfe, Glass-Ceramics from Filter Dusts From Waste Incinerators, *Ceramics International* **21**, 231–235 (1995).
104. I. Rozenstrauha, R. Cimdins, L. Berzina, D. Bajare, A. R. Boccaccini, Sintered Glass-Ceramic Matrix Composites Made from Latvian Silicate Wastes, *Glass Sci. and Technol. Glastech. Ber.* **75**, 132–139 (2002).
105. M. Ferraris, M. Salvo, F. Smeacetto, L. Augier, L. Barbieri, A. Corradi, I. Lancellotti, Glass Matrix Composites from Solid Waste Materials, *J. Europ. Ceram. Soc.* **21**, 453–460 (2001).
106. E. Bernardo, G. Scarinci, S. Hreglich, Mechanical Properties of Metal-Particulate Lead-Silicate Glass Matrix Composites Obtained by Means of Powder Technology, *J. Europ. Ceram. Soc.* **23**, 1819–1827 (2003).
107. M. A. Audero, A. M. Bevilacqua, N. B. de Bernasconi, D. O. Russo and M. E. Sterba, Immobilization of Simulated High-Level Waste in Sintered Glasses, *J. Nucl. Mat.* **223**, 151–156 (1995).
108. A. R. Boccaccini, and K. Pfeiffer, Preparation and Characterisation of a Glass Matrix Composite Containing Aluminium Titanate Particles with Improved Thermal Shock Resistance, *Glass Sci. Technol. Glastech. Ber.* **72**, 352–357 (1999).
109. A. R. Boccaccini, S. Atiq and R. W. Grimes, Hot-Pressed Glass Matrix Composites Containing Pyrochlore Phase Particles for Nuclear Waste Encapsulation, *Adv. Eng. Mat.* **5**, 501–508 (2003).
110. M. A. Harmer, H. Bergna, M. Saltzberg and Y. H. Hu, Preparation and Properties of Borosilicate-Coated Alumina Particles from Alkoxides, *J. Am. Ceram. Soc.* **79**, 1546–1552 (1996).
111. G. W. Scherer, Sintering of Low-Density Glasses. Part I. Theory, *J. Am. Ceram. Soc.* **60**, 239–243 (1977).
112. Y. Imanaka, S. Aoki, N. Kamehara and K. Niwa, Cristobalite Phase Formation in Glass/Ceramic Composites, *J. Am. Ceram. Soc.* **78**, 1265–1271 (1995).
113. M. N. Rahaman and L. C. De Jonghe, Effect of Rigid Inclusions on the Sintering of Glass Powder Compacts, *J. Am. Ceram. Soc.* **70** C-348–C-351 (1987).
114. A. R. Boccaccini and E. A. Olevsky, Effect of Rigid Inclusions on Sintering Anisotropy of Composite Glass Powder Compacts, *J. Mat. Proc. Technol.* **96**, 92–101 (1999).
115. G. W. Scherer, Sintering with Rigid Inclusions, *J. Am. Ceram. Soc.* **70**, 719–725 (1987).
116. R. K. Bordia, G.W. Scherer, On Constrained Sintering, Parts I, II, III, *Acta Metall.* **36**, 2393–2416 (1988).
117. I. W. Donald and P. W. McMillan, Review: Ceramic Matrix Composites, *J. Mat. Sci.* **11**, 949–972 (1976).
118. M. Kinoshita, M. Satou and K. Uematsu, Dispersant Affects Glass-Based Multicomponent Slurries, *Ceram. Bull.* **76** [10], 55–58 (1997).
119. E. J. Minay, V. Desbois, A. R. Boccaccini, Innovative Manufacturing Technique for Glass Matrix Composites: Extrusion of Recycled TV Set Screen Glass Reinforced with Al_2O_3 Platelets, *J. Mat. Proc. Technol.* **142**, 471–478 (2003).
120. V. Lansmann, M. Jansen, Application of the Glass-ceramic Process for the Fabrication of Whisker Reinforced Celsian-Composites, *J. Mat. Sci.* **36**, 1531–1538 (2001).
121. K. H. Lee, D. A. Hirschfeld and J. J. Brown, In Situ Reinforced Glass-Ceramic in the Lithia-Alumina-Silica System, *Ceram. Trans.* **30**, 293–301 (1992).
122. L. Chen, C. Leonelli, T. Manfredini, C. Siligardi, Processing of Silicon Carbide Whisker-Reinforced Glass-Ceramic Composite by Microwave Heating, *J. Am. Ceram. Soc.* **80**, 3245–3249 (1997).
123. A. R. Boccaccini, P. Veronesi, C. Leonelli, Microwave processing of glass matrix composites containing controlled isolated porosity, *J. Europ. Ceram. Soc.* **21**, 1073–1080 (2001).
124. E. J. Minay, A. R. Boccaccini, P. Veronesi, V. Cannillo, C. Leonelli, Processing of Novel Glass Matrix Composites by Microwave Heating, in: *Proc. Int. Conf. On Advances in Materials and Processing Technologies, AMPT 2003*, A. G. Olabi, S. J. Hashmi, eds., Dublin City University (2003) pp. 1277–1280.
125. L. L. Hench, Concepts of Ultrastructure Processing, in *Ultrastructure Processing of Ceramics, Glasses and Composites*, L. L. Hench D. R. Ulrich eds., J. Wiley & Sons, New York (1984), 3–5.

126. A. R. Boccaccini, R. D. Rawlings and I. Dlouhy, Reliability of Chevron-Notch Technique for Fracture Toughness Determination in Glass, *Mat. Sci. Eng. A* **347**, 102–108 (2003).
127. M. P. Borom, Dispersion-Strengthened Glass Matrices – Glass-Ceramics, A Case in Point, *J. Am. Ceram. Soc.* **60**, 17–21 (1977).
128. R. I. Todd, A. R. Boccaccini, R. Sinclair, R. B. Yallee and R. J. Young, Thermal Residual Stresses and Their Toughening Effect in Al_2O_3 Platelet Reinforced Glass, *Acta Materialia* **47** 3233–3240 (1999).
129. D. P. H. Hasselman and R. M. Fulrath, Proposed Fracture Theory of a Dispersion-Strengthened Glass Matrix, *J. Am. Ceram. Soc.* **49**, 68–72 (1966).
130. F. L. Matthews and R. D. Rawlings, *Composite Materials: Engineering and Science*, Chapman and Hall, London (1994).
131. A. G. Evans, The Role of Inclusions in the Fracture of Ceramic Materials, *J. Mat. Sci.* **9**, 1145–1152 (1974).
132. V. Laws, The Efficiency of Fibrous Reinforcement of Brittle Matrices, *J. Phys. D: Appl. Phys.* **4**, 1737–1746 (1971).
133. M. Taya, S. Hayashi, A. S. Kobayashi and H. S. Yoon, Toughening of a Particulate-Reinforced Ceramic Matrix Composite by Thermal Residual Stress, *J. Am. Ceram. Soc.* **73**, 1382–1391 (1990).
134. Faber, K. T. and Evans, A. G., Crack Deflection Processes – I. and II., *Acta Metall.* **31**, 565–584 (1983).
135. V. Cannillo, G. C. Pellacani, C. Leonelli, A. R. Boccaccini, Numerical Modelling of the Fracture Behaviour of a Glass Matrix Composite Reinforced with Alumina Platelets, *Composites Part A* **34**, 43–51 (2003).
136. A. R. Boccaccini, V. Winkler, Fracture Surface Roughness and Toughness of Al_2O_3-Platelet Reinforced Glass Matrix Composites, *Composites Part A* **33**, 125–131 (2002).
137. Y. Waku, M. Suzuki, Y. Oda and Y. Kohtoku, Improving the Fracture Toughness of $MgO-Al_2O_3-SiO_2$ Glass/Molybdenum Composites by the Microdispersion of Flaky Molybdenum Particles, *J. Mat. Sci.* **32**, 4549–4557 (1997).
138. A. R. Boccaccini, Machinability and Brittleness of Glass-Ceramics, *J. Materials Processing Technology* **65**, 302–304 (1997).
139. D. S. Baik, K. S. No, J. S. Chun, Y. J. Yoon and H. Y. Cho, A Comparative Evaluation Method of Machinability for Mica-Based Glass-Ceramics, *J. Mat. Sci.* **30**, 1801–1806 (1995).
140. N. Bobkova, S. Barentceva and S. Gailevich, Composites on Glass-Ceramic Basis, in *Proc. 17. Int. Glass Congress* (China) (1995), 338–342.
141. N. D. Nazarenko, A. I. Yuga, L. F. Kolesnichenko, M. S. Shevchuk and N. I. Vlasko, Frictional Properties of Sitall-Metallic-Filler Composites, *Poroshk. Metall.* **7**, 498–501 (1980).
142. F. P. Meyer, G. D. Quinn and J. C. Walck, Reinforcing Fused Silica with High Purity Fibres, *Ceram. Eng. Sci. Proc.* **6**, 646–656 (1985).
143. P. Saewong and R. D. Rawlings, Erosion of TiC Reinforced Silceram Glass-Ceramic Composites, in *Proc. Materials Solutions 97 on Wear of Engineering Materials*, Indianapolis, USA, J. A. Hawk ed., ASM International (1998) 133–136.

21

Glass-Containing Composite Materials. Alternative Reinforcement Concepts

Aldo R. Boccaccini
Department of Materials
Imperial College London
London SW7 2AZ, UK
Tel.: 44 207 5946731
Email:a.boccaccini@imperial.ac.uk

ABSTRACT

Glass-containing composites with interpenetrating, graded or layered microstructures as well as hybrid glass and glass-ceramic matrix composites are discussed. Aspects of their fabrication, microstructural characterisation, properties and applications are reviewed. These materials have advantages regarding ease of processing and/or special properties which can be achieved, in comparison with conventional dispersion-reinforced and fibre-reinforced glasses and glass-ceramics. The use of these materials in specific areas is expected to increase in the future, including structural, functional and biomedical applications.

1. INTRODUCTION

Glass and glass-ceramic composite materials with alternative designed microstructures have received much less attention in the past than their "classic" counterparts, namely dispersion-reinforced and continuous fibre-reinforced glass and glass-ceramic matrix composites (discussed in the previous Chapters 20 and 19, respectively). Due to their particular microstructure, these composites may display a range of properties not achievable with standard reinforcing concepts and thus they become attractive for a number of specific applications. The composites to be discussed in this Chapter are:

i) Composites with interpenetrating microstructures,
ii) Composites with layered and graded microstructures
iii) Hybrid composites, defined as those containing both fibre and dispersion reinforcement.

In all cases, the composite systems to be considered contain a silicate glass or glass-ceramic phase. Aspects of their processing, microstructural characterisation, properties and applications are discussed and relevant references are provided.

2. GLASS-CONTAINING COMPOSITES WITH INTERPENETRATING MICROSTRUCTURES

The concept of interpenetrating microstructures

Composites with interpenetrating microstructures are defined as multiphase composites in which each phase is continuous, i.e. topologically interconnected throughout the microstructure [1]. In terms of percolation theory, each phase spans or percolates throughout the microstructure. Such three-dimensional (3-D) microstructures have potential for designing glass composites with multifunctional characteristics, i.e. each phase contributing its own properties to the macroscopic composite behaviour. With regard to mechanical properties and fracture behaviour, the greatest benefit of interpenetrating materials is their resistance to crack propagation as a result of interconnected reinforcement, irrespective of the direction in which the crack is driven. Composites with interpenetrating microstructures have therefore no "weak" direction, as for example is the case in unidirectionally reinforced composites, i.e. they behave like isotropic materials. Once the glass phase has cracked in these materials, the unbroken second phase will enhance the fracture resistance by a crack bridging mechanism added to which the interconnected phases can constrain one another mechanically. It is possible to place one of the phases (usually the weaker glass phase) in compression, further enhancing fracture strength.

Although materials with an interpenetrating phase microstructure are relatively common in nature (for example mammals' bones and trunks and branches in plants), they are still a novelty as synthetic materials. A complete review of composite materials with interpenetrating microstructures was published by Clarke, emphasising on materials characterisation and models for property prediction [1].

3-D fibre reinforcement

Three-dimensional fibre-reinforced materials constitute a special class of composites with interpenetrating microstructure.

In fibre-reinforced glass matrix composites, 3-D parts offer greater potential than unidirectionally reinforced composites (in 1-D or 2-D fibre lay-up), discussed in Chapter 19, because of their potentially better interlaminar shear strength and isotropic properties. A

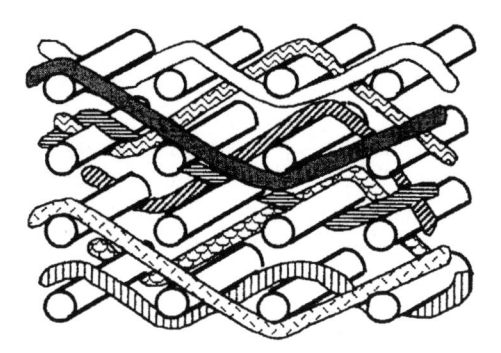

FIGURE 1. Schematic diagram showing the layer to layer 3-D angle interlock fibre preform used to fabricate 3-D silica matrix composites [7].

variety of 3-D fibre architectures is available including 3-D woven, braided and knitted structures [2].

Very limited research has been reported however on the fabrication of this kind of glass matrix composite. The challenge is in process chemistry: how to incorporate the glass matrix into a tightly bound fibre structure, given the poorly connected porous structure characteristic of 3-D preforms, without damaging the fibres [3]. Special consideration should also be given to tailoring the interfaces: if the reinforcement by its very 3-D geometry already ensures damage tolerance and mechanical integrity (see below), then the appropriate interface should be strong rather than weak as it is in unidirectional or 2-D brittle matrix composites [4]. The application of the electrophoretic deposition technique, as described in Chapter 19 for 2-D reinforced glass and glass-ceramic matrix composites [5], could be the best processing route for the fabrication of composites with three-dimensional fibre architecture but no significant work has been carried out in this area so far.

Three-dimensional braid silica-silica composites have been investigated by Brazel [6]. It was found that the 3-D fibre architecture significantly improved the fracture toughness of the composite at the expense of tensile strength, which was as low as 27 MPa. Liu has fabricated 3-D carbon fibre reinforced silica matrix composites by a sol-gel route [7]. Infiltration was conducted using a pressure infiltration technique. The 3-D angle interlock preform used is shown schematically in Figure 1. The relationship of the permeability of the fibre preform to sol infiltration, the processing time and pressure, and the silica particle size, were investigated [7]. The results were satisfactory in terms of silica infiltration into the fibre preform, but no high-temperature densification of the porous matrix was reported and consequently the mechanical properties of such a composite remain unknown. Some results on the mechanical properties of 3-D carbon fibre-reinforced borosilicate glass matrix composites have been presented by Chou et al. [8], but no details were reported on the fabrication procedure for these composites.

In general, manufacture of 3-D structures becomes difficult when using oxycarbide fibres. For example, Nicalon® fibre is inherently brittle and available yarns cannot be bent to curvatures with radii of less than 1–2 mm without fracturing single fibres and causing yarn damage. Woven and knitted structures made of these fibres therefore usually contain large quantities of broken fibre debris, which poses handling problems for manufacturers [1].

In spite of these problems, three-dimensional composites consisting of LAS glass-ceramic matrix containing braided Nicalon® fibre architecture as reinforcement have been fabricated [2, 9] by the slurry infiltration and hot-pressing method. With an intensive three-dimensional fibre network, the three-dimensional braid showed an order of magnitude improvement in strength across the thickness of the 0/90 cross-ply composites, without incurring inferior in-plane tensile strength. The flexural strength of the three-dimensional braided composite was found to be comparable with that of the unidirectional composites, however, the fracture strength of the three-dimensional composites was always lower than the tensile strength of unidirectional composites in fibre direction. Inter-fibre contact points in braided composites are detrimental to composite performance since they result in surface abrasions, which act as fracture initiation sites [9].

The most significant difference between 3-D braided and other composites (unidirectionally reinforced and cross-ply) is the absence of the shear failure in the former, which is the major advantage of three-dimensional reinforcement. Another advantage of 3-D fibre-reinforced composites is their resistance to impact loads and their outstanding damage tolerance [2]. The explanation for this behaviour lies in the way the spatial distribution of damage is controlled by the geometry of the reinforcement, in particular its topology and irregularity.

Finally, it should be pointed out that the tensile strength of a 3-D fibre-reinforced composite can be enhanced by additionally incorporating unidirectionally orientated fibres. This was carried out for example in LAS glass-ceramic matrix composites reinforced by 3-D braid SiC yarns [2]. Both Nicalon® and SiC monofilaments in the axial (0°) direction were used and a very high tensile strength (656 MPa) was achieved by placing SCS filaments in the interior of the composite. The Nicalon® braiding yarns, which were wrapped around the SCS filaments, were placed on the outer surface of the composite so that under tensile loading, the composite strength was dominated by the SCS filaments.

Another way to fabricate glass composites with 3-dimensional fibre reinforcement has been developed recently [10]. These composites consist of a quasi-continuous network of chopped Tyranno® fibres in an aluminosilicate glass matrix, and were fabricated by the standard slurry and hot-pressing route described in Chapter 19. Instead of preparing unidirectional prepregs, the glass-coated fibres were cut at lengths of 20–40 mm, randomly placed in a graphite die and hot-pressed. The high fibre volume fraction and homogeneous fibre distribution were due to the presence of the glass layer on the individual fibres.

The microstructure and mechanical properties of these composites have been investigated [10]. Figure 2 shows the fracture surface of a sample broken under compressive loads, demonstrating the high fibre volume fraction and random orientation of the fibre bundles. The 3-D fibre arrangement depicted in Figure 2 has never been reported before for glass matrix composites and is the result of the particular slurry impregnation technique used. The preparation of 3-D fibre reinforced materials poses two major difficulties, namely to achieve sufficient separation of the fibres in the bundle and to get adequate coating of the single fibres by glass powder particles. As can be seen in Figure 2, the impregnation process used in this work has produced sufficient glass particle packing both around the fibres and in the inter- and intra-tow region, to allow materials free from porosity to be fabricated by hot-pressing. Figure 2 also shows that the fibres are homogeneously distributed in the glass matrix. Very few matrix-rich regions, which could represent weak areas in the material, were observed. The composite behaves isotropically: for example the Young's modulus,

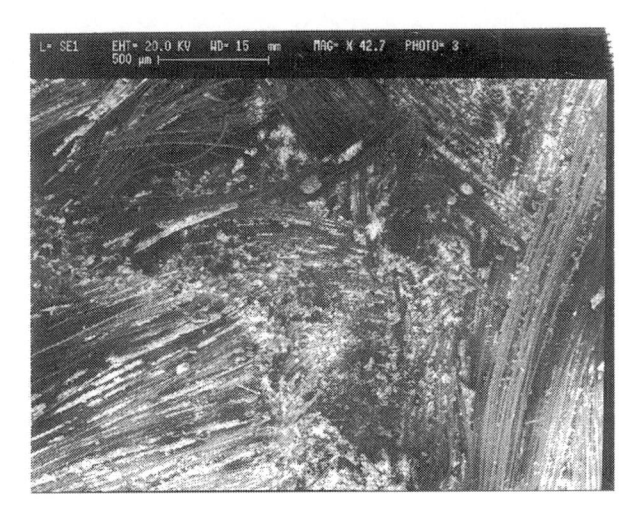

FIGURE 2. SEM image of the fracture surface of 3-D Tyranno® fibre-reinforced aluminosilicate glass matrix composite tested in compression [10].

tensile strength and compressive strengths, 90 GPa, 170 MPa and 320 MPa, respectively, were very similar in all testing directions.

Other 3-D reinforced composites

An interesting approach for fabricating composites with a 3-dimensional interpenetrating microstructure is by infiltration of a porous preform, e.g. metallic or ceramic, with a liquid precursor, such as a glass slurry, or with molten glass. This requires fabrication of materials with interconnected porosity to be used as substrates. Numerous processing techniques have been developed to fabricate highly porous ceramics, including ceramic foams, ceramic filters, catalyst supports, sound insulators and gas sensors [11]. A common way of producing porous ceramic bodies is by partial sintering of particles, platelets or fibres. The simplest way is to sinter the particles to merely form necks between them, hence creating a ceramic skeleton. Another more versatile processing approach used to fabricate foam-like structures is by casting a ceramic slip onto a sacrificial open-cell polymeric material, which is then burned away by firing in an oxidising atmosphere [12]. The use of polysilanes and polycarbosilanes to create SiC-based ceramic foams has also been reported [13], as well as the application of foaming or blowing agents as means to fabricate highly porous ceramic bodies [11]. Preparation of ceramic substrates with graded pore structure by centrifugal deposition from mixed, diluted sol-powder suspensions has been also reported [14]. These porous structures may be used to fabricate composites with gradient microstructures (see Section 3). The preparation of porous metallic preforms by powder metallurgy or by sintering metallic fibres and felts is well known. Metal foams are also suitable porous structures to use as the reinforcing element in 3-D glass matrix composites.

To fabricate glass-containing composites with interpenetrating microstructures either a glass slurry or molten glass is infiltrated into the porous preform or a metallic or polymer

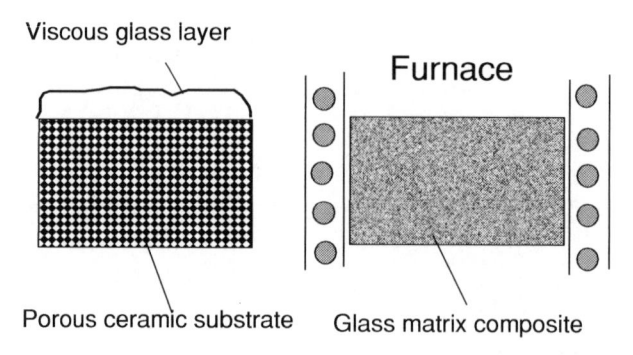

FIGURE 3. Schematic diagram of the fabrication of glass composites with 3-D reinforcement by infiltration of molten glass into a porous ceramic preform.

liquid is infiltrated onto a porous glass preform. The infiltration of a ceramic preform by molten glass is the technique most frequently used. The process is shown schematically in Figure 3, and it is similar to a process reported for the fabrication of glass coatings on porous alumina substrates [15].

The well known In-Ceram process [16, 17] and its modifications [18–22] constitute advanced processing methods for the near-net shape fabrication of composites by infiltration of molten glass onto an alumina skeleton. Typically, the volume fraction of the alumina phase is greater than 75%. Interpenetrating alumina/glass composites exhibiting impressive mechanical strength (as high as 600 MPa) and improved fracture toughness (about 4 MPam$^{1/2}$) were successfully introduced commercially in 1989. A very important advantage of these materials when used in all-ceramic dental restorations is that the porous alumina skeleton can be shaped easily prior to the glass infiltration and so complex shapes such as crowns are possible creating a good fit between the restoration and the tooth [21]. The glasses used are composed of lanthanum aluminosilicate and infiltration takes place at about 1100°C for up to 8 hours.

Figure 4 is a TEM micrograph of the microstructure of an In-Ceram composite sample [16] which shows that the interface between the glass and alumina phases is free from pores and microcracks. Even extremely low-angle triple points and narrow channels between adjacent alumina particles (marked by arrows in the Figure) are filled with the glass phase, indicating the excellent wetting behaviour.

Work has been carried out to understand the mechanical behaviour of these composites [16, 21, 22]. It was found that dispersion strengthening is not the only mechanism contributing to the high fracture strength; but also the absence of porosity and microcracking in this material are important factors (see Figure 4). It has been shown that the internal stresses developed on cooling from the fabrication temperature are compensated for by dislocation formation [16]. This behaviour is not usually exhibited by particulate composites fabricated by powder processing, in which internal stresses usually lead to microcracking at interfaces between the inclusions and the glass matrix (see Chapter 20).

Satisfactory clinical performance of In-Ceram dental restorations has been reported [23], and further improvement of the chemical durability and wear behaviour of these glass/alumina composites has been achieved by using diamond-like carbon coatings [17].

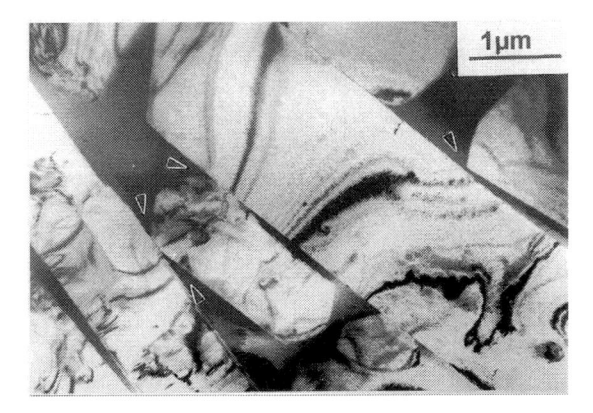

FIGURE 4. TEM micrograph of the microstructure of a glass/alumina In-Ceram composite showing no microcracking development at the interface between the glass (dark) and alumina (light) phases [16]. (Photo courtesy of Dr. H. Hornberger).

In addition, the incorporation of zirconia has been proposed to enhance the mechanical properties of these alumina/glass composites [24].

Results have also been published on borosilicate glass/AlN interconnected composites intended for substrates in electronic packaging applications [25]. This material combination was chosen because of the low dielectric constant of borosilicate glass together with the electrical insulating capability and relatively high thermal conductivity of AlN ceramic, an optimal property combination for the fabrication of advanced substrates for electronic packaging. The porous AlN body containing about 28 vol% of interconnected porosity was fabricated by sintering. Hot infiltration with a borosilicate glass yielded about 100 μm penetration of the glass in the pores. Incorporation of the glass into the ceramic provides a good hermetic seal, which is important if the composite is to be useful as a substrate material [25, 26].

Infiltration of molten glass into a porous ceramic substrate has also been used to fabricate borosilicate glass/zirconia composites [27]. The infiltration was performed at 1100°C over 3 hours but no further details on the microstructure, properties and applications of these composites were given. Glass-spinel composites with interpenetrating microstructure have been also fabricated by infiltration of molten La_2O_3-Al_2O_3-SiO_2 glass onto $MgAl_2O_4$ pre-sintered porous preforms at 1080 °C [28]. A similar technique was used to fabricate 3-D composites consisting of aluminosilicate glass and SiC reinforcement, however the infiltration of the SiC porous preforms with the glass melt was carried out at higher temperatures (1400°C) under gas pressures of up to 3 MPa [29].

The use of whisker preforms for the fabrication of 3-D reinforced glass-ceramic matrix composites has been reported by Travitzky [30]. Mullite whisker preforms were infiltrated with molten magnesium aluminosilicate glass at a temperature of 1650°C for 30 minutes, following the processing approach shown in Figure 3. Attractive mechanical properties were achieved namely fracture strength values of 330 MPa and fracture toughness values of 3.3 MPam$^{1/2}$ [30]. The use of whisker preforms has the advantage of eliminating the problems which can arise from the inhomogeneous distribution of individual whiskers in the matrix as well as the health hazard associated with whiskers. More complex 3-D preforms, comprising mixtures of SiC particulates and whiskers, have also been used to produce glass-ceramic (strontium feldspar) matrix composites with an interpenetrating microstructure by melt infiltration [31, 32].

When conducting infiltration using slurries consisting of glass particles and liquid, additional issues related to the drying and densification of the glass phase must be considered. Both mechanisms pose practical problems: the creation of cracks and disruption of the material during drying as well as incomplete densification due to constrained sintering. Optimisation of this processing approach can be achieved by employing colloids, which allow control of the rheology of powder slips through manipulation of the interparticle forces so that higher particle densities can be achieved [33].

Interpenetrating metal/ceramic composites can also be fabricated by chemical means [34]. In this approach based on sol-gel processing, a metal and a ceramic precursor are mixed intimately during synthesis. Despite successful results for fabrication of alumina/metal composites [34], the technique has not as yet been widely exploited for fabricating silicate glass-containing composites.

Another method suitable to obtain glass and glass-ceramic composites with interpenetrating microstructures is by in-situ growth of a 3-D network of a crystalline phase, ideally with a high aspect ratio. This has been demonstrated in mixtures of Si_3N_4 and barium aluminosilicate (BAS) glass [35, 36]. After suitable heat-treatment, the final microstructure of the composite, which shows enhanced fracture toughness, is composed of a BAS matrix crystallised into hexacelsian with a 3-D reinforcing network of β-Si_3N_4 whiskers.

3. GLASS-CONTAINING COMPOSITES WITH GRADED AND LAYERED MICROSTRUCTURES

Functionally Graded Materials

Continuously graded microstructures can be designed to enhance a wide variety of mechanical properties in brittle materials [37]. The particular benefits of microstructure grading include the following: improvement in interfacial bonding between dissimilar materials, minimisation and optimisation of thermal stresses in structural components, suppression of the onset of microcracking and reduction of the effective driving force for crack extension and fracture. Most work on functionally graded materials has been devoted to ceramic/ceramic and ceramic/metal composite systems with only limited research dealing with glass-containing composites.

In order to study the effect of graded elastic modulus on contact damage under indentation loads, graded alumina/glass composites have been fabricated [38]. An in-situ processing method involving impregnation of a dense, fine-grained alumina by an aluminosilicate glass was employed. The process is shown schematically in Figure 5 and was carried out at 1690°C for 2 hours. The glass chosen (Code 0317, Corning, N. Y.) had a thermal expansion coefficient matched to alumina in order to eliminate the development of thermal stresses upon cooling. When dense, sintered alumina comes in contact with a silicate glass, alumina crystals come apart and float in the glass [39]. The technique led to a graded composite having a monotonic, unidirectional variation of the alumina content and consequently a similar variation of the Young's modulus below the surface. The indentation study showed

GLASS-CONTAINING COMPOSITE MATERIALS

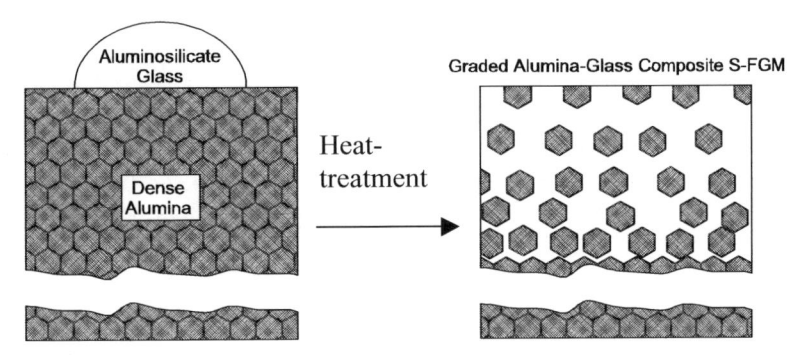

FIGURE 5. Schematic diagram showing the in-situ processing method for the fabrication of graded alumina-glass composite. (Adapted from Ref. [38]).

that unidirectional variation of Young's modulus under the indenter can fully suppress the formation of Hertzian cone cracks [38]. Thus, the development of graded surfaces is an attractive approach for enhancing contact damage resistance in glass composites.

Functionally graded bioactive coatings on pure titanium or Ti-6Al-V alloys consisting of hydroxyapatite-glass composites have also been fabricated for bioactive surgical implants [40–43]. A schematic illustration showing a cross-section of such a graded coating system is shown in Figure 6. Typically, the coating consists of a continuous bioactive glass phase containing dispersed hydroxyapatite particles and pores. The volume fractions of hydroxyapatite particles and porosity vary gradually throughout the thickness of the coating. The glass composition used was 67.7% SiO_2, 10.4% B_2O_3, 5.2% Al_2O_3, 8.3% Na_2O, 4.2% K_2O, 2.1% Li_2O, 1.05% ZrO_2 and 1.05% TiO_2 by weight, and it was chosen not only for its adhesion and compatibility between glass and titanium, or titanium alloy, substrate but also for its chemical compatibility with hydroxyapatite during the sintering process which was conducted at 950°C for 5 min [40, 41]. The total thickness of the layer was about 100 μm. It is claimed that these functionally graded composite coatings provide a stronger interfacial adhesion at the metal-coating interface than conventional monolithic

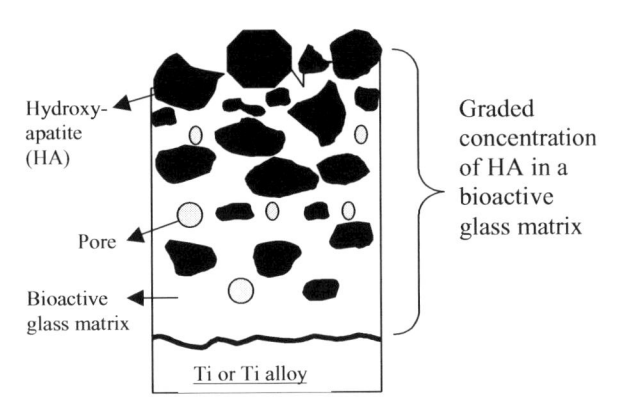

FIGURE 6. Schematic illustration showing the cross-section of hydroxyapatite-glass-titanium functionally graded composite [40, 41].

hydroxyapatite coatings [40]. The application of the coated substrates as implants was also demonstrated. A bioactive surface layer was prepared by etching the composite coating with a mixed solution of 3% HF and 5% HNO_3 in order to remove the glass phase. Implantation in a dog's femur was carried out and it was found that the implants were well integrated into the surrounding cortical and cancellous bone [40]. The electrochemical corrosion behaviour of the hydroxyapatite-glass-titanium composites was also tested in order to assess stability under *in vivo* conditions [41]. Similar graded coatings for biomedical applications have been prepared on alumina substrates using a bioactive glass matrix and a graded concentration of zirconia particles [44].

Alternatively to the most commonly used powder technology route for fabricating graded materials, i.e. by compaction and sintering, high-temperature colloidal processes are also being considered for fabricating glass-containing composites [39]. Using this approach, glass/metal functionally graded materials can be obtained from high melting-point metal particles dispersed in viscous glass matrices by controlled settling, under gravity [39].

Composites with layered structures

Composites with layered or sandwich-type structures exhibit discrete changes in their composition throughout the thickness of the component. They consist of multiple layers of different compositions, or different materials, stacked in a predetermined sequence to achieve the desired properties. Layered or laminated composites can have improved mechanical performance compared to monolithic materials [45]. An interesting example of this is provided in nature: the hierarchical structure of shells such as abalones is formed by aragonite layers of about 1 μm thickness joined by a mortar of proteins. Because of this particular microarchitecture, abalone shells show about 10 times higher bending strength and toughness than aragonite single crystals.

The strength of brittle materials can also be increased and surface flaws rendered harmless by the introduction of a tougher layer below the surface. This layer would ablate surface damage by blunting the cracks when they reach the underlying toughened region, which can contain whiskers, metallic particles or chopped fibres. In addition, in composite laminates the presence of residual stresses on the surface which are due to thermal expansion mismatch between the layers, can give rise to higher values of strength compared to their monolithic counterparts [46]. In this context, the deliberate introduction of residual stresses into the layered or laminated composites enhances the resistance to crack extension, and results in increased fracture toughness.

High-temperature corrosion resistance can also be achieved by the layered composite concept. For example, a corrosion-resistant layer could be used on the exterior surface with layers tailored to high-temperature strength in the interior. The increase of thermal shock resistance of brittle materials by designing laminated structures, including ceramic/metal multilayered composites [47, 48], has also been demonstrated. The dominant behaviour is the absence of interaction between the biaxial cracking mechanisms in adjacent ceramic layers, and the confining of damage to those most exposed layers only.

Laminated composites containing glass or glass-ceramic layers have received relatively limited attention, apart from standard laminate polymer/glass security panels for building, automotive and other glazing applications [49, 50].

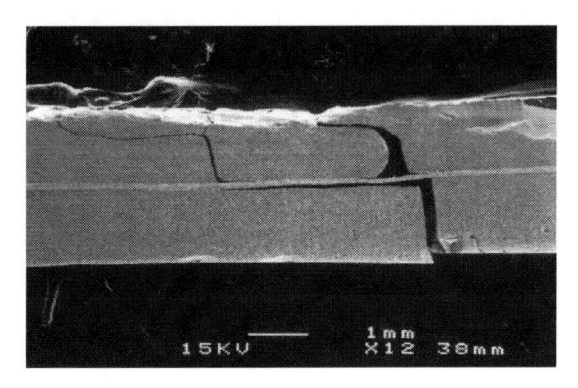

FIGURE 7. SEM micrograph of a bioactive laminate composite fractured in flexure. The tensile side was located at the bottom of the micrograph. The outer layers (tape-cast sintered Bioglass®) are bioactive, whereas the middle metallic layer provides ductility [53, 54]. (Micrograph courtesy of Prof. J. J. Mecholsky Jr, University of Florida, USA).

Novel systems investigated include multilayer soda-lime glass layers of different thicknesses and either epoxy or polyvinyl butyral (PVB) layers, which exhibit high failure loads and considerable displacement at failure [51].

Model glass/glass-ceramic bilayer laminates with a well-defined and strongly bonded interface have been fabricated by hot-pressing soda-lime glass and Macor® glass-ceramic bars [52]. The objective was to investigate the microstructural design of laminated materials with alternating hard layers (for wear resistance) and tough layers (for fracture resistance). Damage development was studied using Hertzian and Vickers indentation tests. In Hertzian contact, the glass produces cone fractures typical of ideal brittle materials, whereas the mica containing glass-ceramic (e. g. Macor® type material) produces the subsurface quasi-plastic damage more typical of plastic solids. A major distinguishing feature of the model materials is the incorporation of a strong interface, so that cracks are inhibited by the underlayer rather than deflected between the layers. The hard component is placed on the outer surface, to provide wear resistance. The tough component forms the underlayer to provide fracture resistance if crack penetration occurs, or to suppress penetration by energy absorption within a sublayer damage zone [52].

Bioactive glass (Bioglass®)/metal multilayer composites for structural applications in the biomedical field have been developed [53, 54]. The materials were fabricated by tape-casting using metallic layers of stainless steel 316 L or titanium and Bioglass® tapes. The laminates were hot-pressed under vacuum at temperatures between 700 and 1250°C and pressures up to 50 MPa resulting in a composite structure consisting of a thin metallic layer between two Bioglass® layers. Strengths over 100 MPa and fracture toughness values one order of magnitude greater that that of monolithic Bioglass® were achieved. It was shown that the middle metallic layer provides flaw-tolerant behaviour [53, 54]. Figure 7 shows a SEM micrograph of a bioactive laminate composite fractured in flexure.

Laminate composites have also been developed [55–57] in which some of the layers themselves are fibre-reinforced glass or glass-ceramic matrix composites. For example composite laminates alternating dense ceramic sheets (either SiC or Si_3N_4) and Nicalon®

fibre-reinforced glass matrix composite layers have been fabricated. Several matrices were considered, including aluminosilicate, magnesium aluminosilicate and calcium aluminosilicate. The multilayered composites were densified by hot-pressing and the constituent layers were selected on the basis of thermal expansion compatibility. The mechanical behaviour, thermal shock resistance and evolution of damage were studied for different material combinations and composite architectures [55–57]. The strategy behind this concept is to exploit the high strength and stiffness of the ceramic layers to delay the onset of matrix cracking within the composite layers [56] which is the limiting factor for high-temperature applications of fibre-reinforced glass and glass-ceramic matrix composites (as discussed in Chapter 19). Placing the ceramic layers on the outer surfaces of the laminates will also provide protection from mechanical abrasion and inhibit oxygen ingress into the composite. To achieve a high strain-to-failure rate and good resistance to microcracking, a strong bond must exist between the ceramic and glass composite layers; otherwise delamination is likely to occur between the layers. Although the composites fabricated by Cutler et al. exhibited attractive mechanical properties under both mechanical and thermal loading conditions [56, 57], no extensive research on this type of composite laminate has been performed, and the concept provides an interesting opportunity for further work.

A similar concept involving fibre-reinforced glass matrix composite layers was reported by Jessen et al. [55]. They proposed the incorporation of a fibre-reinforced composite "skin" over a bulk, low cost monolithic ceramic core, focusing on the influence of the skin/core interfacial bonding on the mechanical properties of the laminate system. Incompatibility of thermal expansion and elastic moduli incompatibility between the layer materials were minimised. The composite had a glass-ceramic (Pyroceram®) core bonded to a composite skin by means of a room temperature curing epoxy. The skin was a zirconium titanate matrix composite reinforced by 50 vol% unidirectionally aligned Nicalon® fibres. The most important result was that co-fired laminate composites, in which the thermomechanical properties of the skin and core are matched, can produce equivalent mechanical performance to systems consisting of 100% "skin" material [55]. Since the skin is made of the more expensive fibre-reinforced composite, this approach may lead to low-cost, high-performance materials. A similar approach involving layered glass-matrix composites and the concept of surface-reinforcement was proposed by Jessen and Lewis [58, 59]. The systems studied consisted of Ni particles dispersed in S-glass matrix and Fe-Ni-Co alloy particles in a borosilicate glass matrix. The glass/metal combinations showed no thermal expansion mismatch. For the preparation of the laminate composites, blended composite powders of different metal loadings were layered into a hot-press die and consolidated into discs. Composites of various relative layer thicknesses were produced. It was shown that tough surface layers obtained by the incorporation of a higher metallic particle content could be used effectively to increase the apparent toughness of lower metal volume fraction bulk materials [58, 59]. This result is important in designing surface-reinforced laminate systems, where an optimal surface layer thickness can be employed to provide maximum mechanical improvement while minimising both component weight and material costs.

Other layered and graded glass composites

Some other material concepts involving layered and gradient glass and glass-ceramic composites have been proposed, as reviewed in this section.

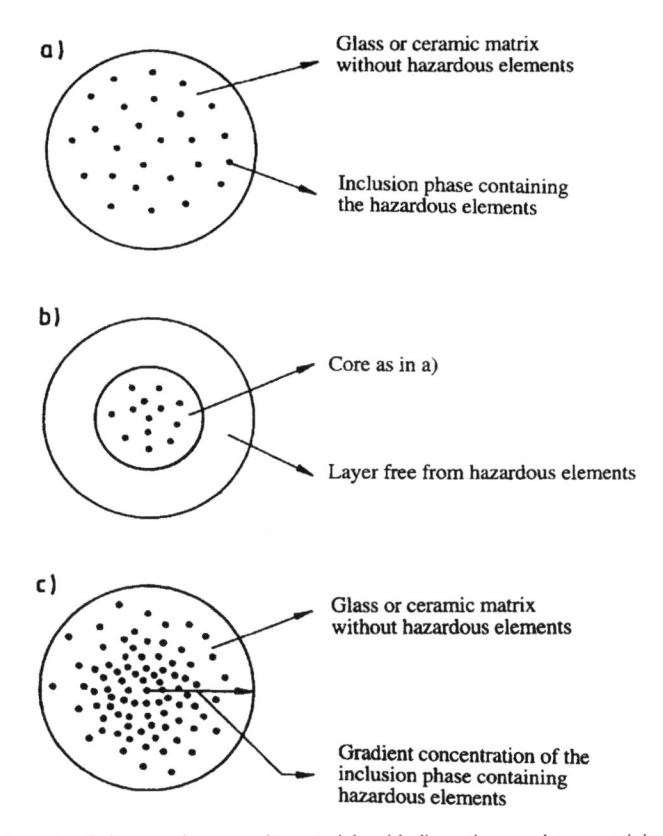

FIGURE 8. Schematic of glass matrix composite materials with discontinuous phase containing hazardous elements, i.e. heavy metals: a) microstructure after conventional processing, b) layered microstructure comprising a core with a high content of hazardous elements surrounded by shale free from dangerous components, c) optimised gradient microstructure in which the fraction of the hazardous phase progressively decreases from the highly loaded centre of the component to its surface, allowing maximum efficiency in terms of hazardous element content and safe leaching behaviour [60].

A multilayered glass composite concept has been considered for the safe recycling of hazardous waste containing silicate phases [60]. In this concept, the use of layered and graded microstructures in multiphase glass-containing composites has two major objectives [60]:

1) to increase the leaching resistance and chemical durability of products containing hazardous waste by maintaining the highest possible waste content in the component;
2) to improve the effective engineering and technological properties of the products in relation to potential industrial applications.

As shown in Figure 8, the optimised concept for recycling hazardous components in a glass matrix involves layered and graded microstructures. The basic concept (Fig 8-a) consists of a well dispersed discontinuous inclusion phase containing the hazardous elements

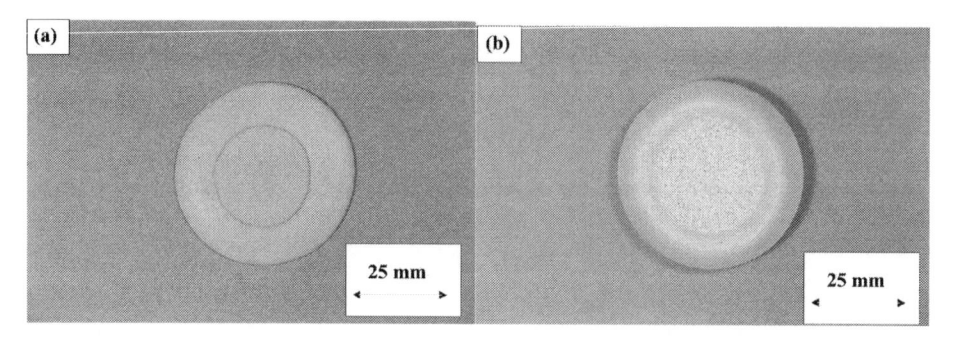

FIGURE 9. Example of layered glass composite samples fabricated by powder technology [60]. The core of the samples is made of waste incinerator filter ash containing hazardous elements and the outer layer comprises cullet glass (a) and vitrified incinerator filter ash (b).

(e.g. heavy metals) in a glass matrix free from hazardous components. As an alternative to the use of coatings to prevent chemical reactions of the external surface with the environment, layered and graded microstructures can be designed. These composite microstructures may exhibit progressive change of composition of the hazardous component (e.g. a heavy metal-containing phase) in a discrete (layered) or continuous (gradient) form, as shown in Figures 8-b and 8-c, respectively. The application of these microstructures is the most efficient approach to increasing the leaching resistance of a composite material by keeping the hazardous element content as high as possible.

The feasibility of the powder technology approach for obtaining layered cylindrical compacts from industrial silicate wastes, e. g. cullet glass and vitrified incinerator filter ash, was demonstrated [60], as shown in Figure 9. The materials underwent pressureless sintering after being compacted uniaxially in a die. They did not exhibit processing microcracks or other microstructural defects at the core/outer layer interfaces.

Another possibility for fabricating glass containing gradient and layered structures is by differential crystallisation, i.e. by the glass-ceramic approach. Glass-ceramics with discrete or continuous spatial variation of the crystalline content and/or type of crystal phase can be fabricated by a combination of melting and powder technologies. The last route allows the production of layered structures by "sandwiching" layers of different composition. By optimisation of the heat-treatment parameters (i.e. heating time and temperature), and/or by selective addition of nucleating agents, it is possible to design graded glass-ceramics with a varying degree of crystallisation or a spatially variable concentration of different crystals [60]. Such graded and layered microstructures, which have the advantage of being created "in-situ", can improve the mechanical and fracture properties in different ways because of the beneficial elastic and thermal mismatches developed. The fracture strength is increased due to the higher Young's modulus of the crystalline phases present in comparison to the original amorphous glass, such as in dispersion-reinforced glass matrix composites (see Chapter 20). The internal stresses created between the different layers upon cooling from the fabrication temperature can cause crack deflection, and thus increase fracture toughness, as shown for multilayer ceramic materials [45]. The concept of layered structures can also be used to increase chemical and wear resistance of components; protective layers of a pre-determined crystalline composition, i.e. comprising hard crystals, could be produced for instance by adding the appropriate nucleation agents to the outermost layer in a multilayer structure obtained by powder technology [60].

In related developments, the strengthening of glass-ceramic laminates by differential densification has been proposed [61] in order to take advantage of the volume change that occurs on crystallisation. For this mechanism to occur, there must be i) an adequate densification differential between the interior and the surface of the body during crystallisation, and ii) at the same time the viscosity must be sufficiently high to resist complete stress relaxation. Chyung has demonstrated this mechanism in glass-ceramic laminates in the system $Li_2O-Al_2O_3-SiO_2$ [61].

Another example of layered composites obtained by the glass-ceramic route has been presented by Hing et al. [62], who developed cordierite glass-ceramic/$TiSi_2$ materials for applications in thick-film heater plates. The composites were fabricated by powder technology and sintering. An insulating external layer was formed in-situ, depending on the sintering temperatures used and the composite comprised an integral surface layer of electrically non-conducting glass-ceramic, with cristobalite, rutile, anatase and indialite as major phases, surrounding an electrically conducting core. It was also found that the external layer provided very good oxidation resistance. Ceramic/glass layered composites made of alumina/silicate glass [63] and SiC/rare-earth silicate glass [64] have been also reported.

The use of layered glass coatings for corrosion and oxidation protection has also been proposed for a variety of applications. For example, Chen et al. developed a two-layer glass coating as corrosion protection for lead telluride (PbTe) elements used in thermoelectric power generation plants [65]. Smeacetto et al. have reported on multilayer barium alumino borosilicate glass-ceramic coatings with different concentrations of boron carbide, molybdenum disilicide and yttrium oxide particles for oxidation protection of carbon-carbon composites [66]. Other multilayered glass and glass-ceramic coatings for high-temperature electrical insulation, fabricated by thermal spray coating, have been reported [67].

Graded porous glasses as limiting case of graded glass composites

The presence of porosity in brittle materials has a pronounced effect on their physical and mechanical properties [11]. While the majority of the mechanical properties of glass, including fracture strength, elastic constants and fracture toughness, decrease with volume fraction of pores, it has been shown that porosity can be beneficial for thermal shock behaviour [68]. However, an adequate control of the total porosity as well as its distribution and structure, i.e. pore shape and orientation, is required. A further improvement in thermal shock resistance can be achieved by incorporating a spatially varying pore concentration in the material. The thermal stresses developed under conditions of thermal shock, i.e. after an abrupt temperature change, are related to the material properties such as fracture strength, thermal expansion coefficient, thermal conductivity and Young's modulus [68]. It has been demonstrated theoretically that a variable porosity leads to a continuous variation of the thermal conductivity and Young's modulus thus having a major effect in decreasing the magnitude of the maximum thermal stress induced as a consequence of a sudden thermal shock [69]. Similar theoretical studies on laminated brittle composites [70] have shown that the presence of porosity in some of the layers led to lower Young's modulus values in the laminate, resulting in an effective decrease in induced thermal stresses.

The presence of porosity will usually have a negative effect on the fracture strength of the material. In a component under thermal stress however, the negative effect of porosity on strength can be largely reduced with controlled variable porosity throughout the section of the component. This can be achieved if the material is fully dense in the region of maximum tensile thermal stress and has the highest possible porosity in the region where compressive thermal stresses develop [69]. If the component is cylindrical the region of high thermal stress will be along the centre, while the near-surface region will be under compressive stress.

4. GLASS-CONTAINING HYBRID COMPOSITES

Hybrid composites are created by simulateous use of aligned continuous fibres and dispersion reinforcement [71]. The primary goal is to improve the transverse mechanical properties and to increase the resistance to microcracking of unidirectionally reinforced composites (see Chapter 19) without compromising the original damage tolerance capability. The few hybrid glass and glass-ceramic matrix composites proposed in the literature include: barium magnesium aluminosilicate glass-ceramic (BMAS) composites containing Nicalon® fibres and SiC whiskers [72, 73], cordierite glass-ceramic matrix composites containing SiC monofilament and SiC whiskers [71], aluminosilicate glass composites containing Nicalon® fibres and SiC particles [74], fused silica containing carbon fibres and

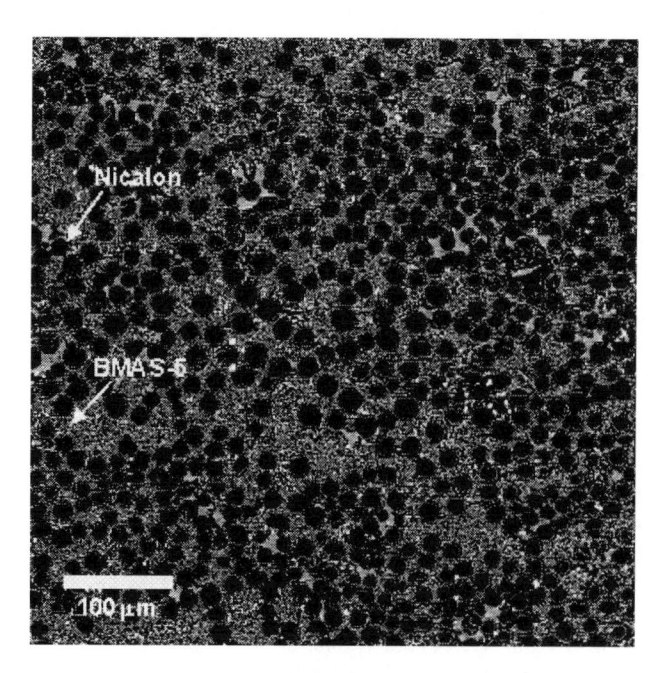

FIGURE 10. Hybrid barium magnesium aluminosilicate glass-ceramic matrix composite with Nicalon® fibre and SiC-whisker reinforcement. The white dots are SiC whiskers distributed in the glass-ceramic matrix. (Micrograph courtesy of Prof. K. Chawla, University of Alabama at Birmingham, USA).

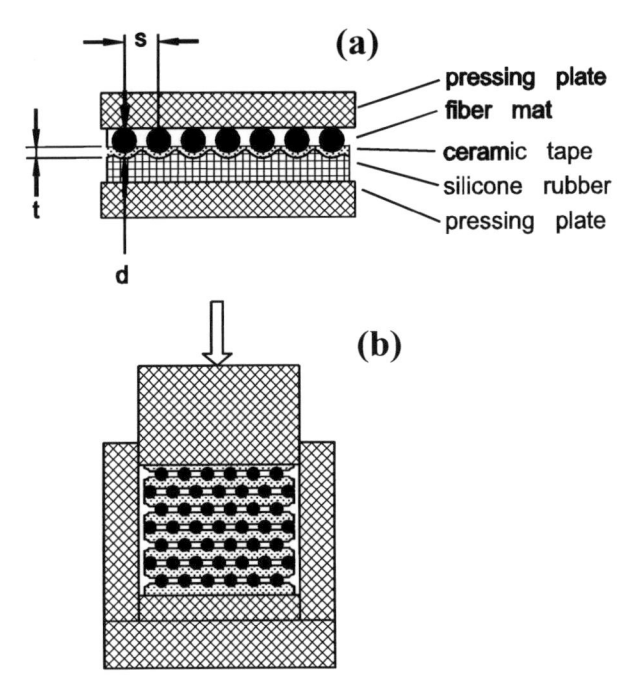

FIGURE 11. Fabrication of hybrid glass matrix composites: (a) setup for pressing the fibres into the tape and (b) laminating the layers in a graphite hot press [71].

silicon nitride particles [75] and borosilicate glass containing SiC-Nicalon® fibres, carbon fibres and various incorporated fillers in the form of ceramic particles, such as alumina, zirconia, carbon [76–78]. Figure 10 shows the microstructure of a composite consisting of Nicalon® fibre and SiC whisker in a BMAS glass-ceramic matrix. Hybrid composites such as the example shown in Figure 10 are fabricated by conventional uniaxial hot-pressing while a tape casting and lamination approach has been used to prepare composites containing large diameter SiC monofilament (SCS fibre) and whisker reinforcement [71]. The setup for pressing the fibres into the tape is shown schematically in Figure 11-a. Subsequently, the composite green body can be fabricated by stacking the appropriate number of layers required to produce the desired specimen thickness. Densification is carried out in a hot-press, as Figure 11-b shows [71]. Besides increasing the first microcracking stress resistance, the addition of dispersion reinforcement may be beneficial for other reasons: for example, the thermal shock behaviour of fibre-reinforced glass-ceramic matrix composites can be increased by incorporating SiC-whiskers [72], while the wear resistance of borosilicate glass matrix composites was shown to increase by the incorporation of ceramic particles in the matrix, including alumina, zirconia and carbon [76, 77]. Since very little research has been conducted in the area of hybrid glass matrix composites, there is scope for further investigations. It is important to assess the full potential of a dispersion-reinforcing phase for optimisation of the overall combined properties of fibre-reinforced glass and glass-ceramic composites, considering both mechanical and functional behaviour. In these composites,

TABLE 1. Selected mechanical properties and area of application of some glass-containing composite systems with interpenetrating, graded or layered microstructure and hybrid composites. (The fracture strength values quoted are flexural strength data unless otherwise stated).

Composite system	Selected properties	Area of application	Reference
LAS glass-ceramic matrix composite reinforced by 3-D braid Nicalon® yarns and SiC monofilaments in the axial direction.	Tensile strength: 656 MPa	Aerospace	[2]
Aluminosilicate glass matrix composite reinforced by 3-D chopped Tyranno® fibre network	Isotropic properties Young's modulus: 90 GPa, Fracture strength: 170 MPa, Compressive strength: 320 MPa	Aerospace Special machinery Hot glass handling	[10]
Interpenetrating alumina/glass composites (In-Ceram process)	Fracture strength: 600 MPa, Fracture toughness: 4 MPam$^{1/2}$	All-ceramic dental restorations	[16, 17]
Magnesium aluminosilicate glass composites with 3-D mullite whisker reinforcement	Fracture strength: 330 MPa, Fracture toughness: 3.3 MPam$^{1/2}$	Structural applications requiring isotropic mechanical properties	[30]
Si_3N_4/barium aluminosilicate (BAS) glass composites with 3-D β-Si_3N_4 whisker reinforcement	Fracture strength: 916 MPa, Fracture toughness: 6.6 MPam$^{1/2}$	Structural applications	[35, 36]
La_2O_3-Al_2O_3-SiO_2 glass-spinel ($MgAl_2O_4$) composites with interpenetrating microstructure	Biaxial fracture strength: 317 MPa, Fracture toughness: 3.5 MPam$^{1/2}$	Structural applications	[28]
Bioactive glass (Bioglass®)/metal multilayer composites	Fracture strength: 100 MPa Work of fracture: 5.7 10^{-3} N.m	Structural applications in the biomedical field	[53, 54]
Hybrid aluminosilicate glass matrix composite with Nicalon® fibre and SiC particulate reinforcement	Fracture strength: 778 MPa – 827 MPa (depending of SiC particulate content)	High-temperature aerospace applications	[74]

the fibres contribute to improved mechanical properties and enhanced structural integrity while the dispersed phase may be used to achieve desired electric, dielectric, magnetic or thermal properties.

5. PROPERTIES

Table 1 shows a compilation of typical mechanical properties of different glass-containing composite systems discussed above. The properties should be compared with those of traditional glass and glass-ceramic matrix composites with dispersion or fibre reinforcement, discussed in Chapters 20 and 19, respectively. Although the principal effort has been in developing composites with improved mechanical properties, in particular fracture strength and fracture toughness, other properties have also been improved by the concepts of interpenetrating, layered and graded microstructures as well with hybrid composites. Other properties investigated in specific systems have been hardness, thermal shock, wear and impact resistance as well as tailored functional properties for selected applications in a wide range of fields including electronics, biomedical and recycling technologies.

6. SUMMARY

The design of particular microstructures in different glass-containing composite systems, as discussed in this Chapter, leads to composites with specific properties targeted for selected applications' in a variety of fields. The variety of processing techniques developed to produce composites with interpenetrating, graded and layered microstructures as well as hybrid composites was reviewed. A number of specific composite systems was described and their typical properties were presented.

The main interest in developing composite systems with interpenetrating, graded and layered microstructures as well as hybrid composites is the improvement of thermomechanical properties. However the composites are usually designed for specific applications where functional properties are also required, i.e. electric, magnetic, dielectric, etc., or a combination of them.

The use of glass-containing composites exhibiting particular microstructures was discussed, which involves applications in special machinery, electronics, aerospace, chemical plants, including corrosion, high-temperature and wear resistant components. Specific applications in medicine and in industrial waste recycling were also considered. Many of the composite systems reviewed are not widely known by the industry, therefore a closer co-operation of materials scientists, technologists and potential end-users is needed for broadening the commercial exploitation of these materials.

REFERENCES

1. D. R. Clarke, Interpenetrating Phase Composites, *J. Am. Ceram. Soc.* **75**, 739–759 (1992).
2. F. K. Ko, Preform Fibre Architecture for Ceramic-Matrix Composites, *Am. Ceram. Soc. Bull.* **68**, 401–414 (1989).
3. Ph. Colomban and M. Wey, Sol-Gel of Matrix Net-Shape Sintering in 3D Fibre Reinforced Ceramic Matrix Composites, *J. Europ. Ceram. Soc.* **17**, 1475–1483 (1997).
4. B. N. Cox, A View of 3D Composites, in *The Processing, Properties and Applications of Metallic and Ceramic Materials*, M. H. Loretto and C. J. Beevers eds., University of Birmingham (1992). pp. 1087–1098.
5. A. R. Boccaccini, C. Kaya, K. K. Chawla, Use of Electrophoretic Deposition in the Processing of Fibre Reinforced Ceramic and Glass Matrix Composites. A Review, *Composites Part A* **32**, 997–1006 (2001).
6. J. P. Brazel, Multidirectionally Reinforced Ceramics, in *Engineers Material Handbook, Vol. 1: Composites*, C. A. Dostal ed., American Society for Metals International, Metals Park, OH (1988), pp. 933–940.
7. H.-K. Liu, Investigation on the Pressure Infiltration of Sol-Gel Processed Textile Ceramic Matrix Composites, *J. Mat. Sci.* **31**, 5093–5099 (1996).
8. T.-W. Chou and J.-M. Yang, Structure-Performance Maps of Polymeric, Metal and Ceramic Matrix Composites, *Metall. Trans. A* **17A**, 1547–1559 (1986).
9. A. S. Fareed, M. J. Koczak, F. Ko and G. Layden, Fracture of SiC/LAS Ceramic Composites, in *Advances in Ceramics, Vol. 22: Fractography of Glasses and Ceramics*, ed., by V. D. Frechette and J. R. Varner, American Ceramic Society (1988), pp. 261–278.

10. A. R. Boccaccini, R. Liebald, W. Beier and K. K. Chawla, Fabrication, Mechanical Properties and Thermal Stability of a Novel Glass Matrix Composite Material Reinforced by Short Oxycarbide Fibres, *J. Mat. Sci.* **37**, 4379–4384 (2002).
11. R. W. Rice, Fabrication of Ceramics with Designed Porosity, *Ceram. Eng. Sci. Proc.* **23** (4), 149–158 (2002).
12. F. F. Lange, B. V. Velamakanni and A. G. Evans, Method for Processing Metal-Reinforced Ceramic Composites, *J. Am. Ceram.Soc.* **73**, 388–393 (1990).
13. P. Colombo and J. R. Hellmann, Ceramic Foams from Preceramic Polymers, *Mater. Res. Innovations* **6**, 260–272 (2002).
14. T. Moritz, G. Werner, G. Tomandl, M. Mangler and H. Eichler, Sintering of Thin Ceramic Layers with a Graded Pore Structure, *Br. Ceram. Proc.* No. 60, Vol. 2 (1999) 245–246.
15. K. Morinaga, H. Takebe and Y. Kuromitsu, Interactions between Al_2O_3 Substrate and Glass Melts, in *Ceramic Microstructures: Control at the Atomic Level*, A. P. Tomsia and A. Glaeser eds., Plenum Press, New York (1998) pp. 535–542.
16. H. Hornberger, P. M. Marquis, S. Christiansen and H. P. Strunk, Microstructure of a High Strength Alumina Glass Composite, *J. Mat. Res.* **11**, 855–858 (1996).
17. S. Christiansen, M. Albrecht, H. P. Strunk, H. Hornberger, P. M. Marquis and J. Franks, Mechanical Properties and Microstructural Analysis of a Diamond-Like Carbon Coating on an Alumina/Glass Composite, *J. Mat. Res.* **11**, 1934–1942 (1996).
18. S.-J. Lee, W. M. Kriven and H.-M. Kim, Shrinkage-Free, Alumina-Glass Dental Composites via Aluminium Oxidation, *J. Am. Ceram. Soc.* **80**, 2141–2147 (1997).
19. D.-J. Kim, M.-H. Lee and C.-E. Kim, Mechanical Properties of Tape-Cast Alumina-Glass Dental Composites, *J. Am. Ceram. Soc.* **82**, 3167–3172 (1999).
20. J.-M. Tian, Y.-L. Zhang, S.-X. Zhang X.-P. Luo, Mechanical Properties and Microstructure of Alumina-Glass Composites, *J. Am. Ceram. Soc.* **82**, 1592–1594 (1999).
21. Q. Zhu, G. de With, L. J. M. G. Dortmans and F. Feenstra, Subcritical Crack Growth Behaviour of Al_2O_3-Glass Dental Composites, *J. Biomed. Mater. Res. Part B: Appl. Biomater.* **65B**, 233–238 (2003).
22. D. Y. Lee, D.-J. Kim, B.-Y. Kim and Y.-S. Song, Effect of Alumina Particle Size and Distribution on Infiltration Rate and Fracture Toughness of Alumina-Glass Composites Prepared by Melt Infiltration, *Mat. Sci. Eng.* **A341**, 98–105 (2003).
23. L. Pröbster, Four Year Clinical Study of Glass-Infiltrated, Sintered Alumina Crowns, *J. Oral Rehabil.* **23**, 147–151 (1996).
24. M. Stephan, Nickel, K. G., Haftung und Festigkeit von ZrO_2-verstärkter Dentalkeramik als Verbundwerkstoff, in *Werkstoffwoche 98, Band IV*, H. Planck and H. Stallforth eds., Wiley-VCH, Weinheim (1999), 281–286.
25. P. N. Kumta, Processing Aspects of Glass-Nicalon Fibre and Interconnected Porous Aluminium Nitride Ceramic and Glass Composites, *J. Mat. Sci.* **31**, 6229–6240 (1996).
26. P. N. Kumta, T. Mah, P. D. Jero and R. J. Kerans, Processing of 3D Interconnected Porous Aluminium Nitride Composites for Electronic Packaging, *Mat. Lett.* **21**, 329–333 (1994).
27. C. Ries, Herstellung und Eigenschaften von glasinfiltrierter ZrO_2-Keramik, in *DKG Jahrestagung, Kurzreferate* (1995) pp. 21–22.
28. D. Y. Lee, D.-J- Kim and Y.-S. Song, Properties of Glass-Spinel Composites Prepared by Melt Infiltration, *J. Mat. Sci. Lett.* **21**, 1223–1226 (2002).
29. K. Langguth, S. Böckle, E. Müller and G. Roewer, Polysilane-Derived Porous SiC Preforms for the Preparation of SiC-Glass Composites, *J. Mat. Sci.* **30**, 5973–5978 (1995).
30. Travitzky, N. A., Mechanical Properties and Microstructure of Mullite Whisker-Reinforced Magnesium Aluminosilicate Glass with Cordierite Composition, *J. Mat. Sci. Lett.* **17** (1998) 1609–1611.
31. W. B. Hillig, Melt Infiltration Approach to Ceramic Matrix Composites, *J. Am. Ceram. Soc.* **71**, C-96–C-99 (1988).
32. M. K. Brun, W. B. Hillig, H. C. McGuigan, High Temperature Mechanical Properties of a Continuous Fibre Reinforced Composite Made by Melt Infiltration, *Ceram. Eng. Sci. Proc.* **10** [7–8], 611–621 (1989).
33. Lange, F. F., Powder Processing Science and Technology for Increased Reliability, *J. Am. Ceram. Soc.* **72**, 3–15 (1989).
34. E. D. Rodeghiero, Tse, O. K., Gianelis, E. P., Interconnected Metal-Ceramic Composites by Chemical Means, *JOM*, **34** [3], 26–28 (1995).
35. Y. Fang, F. Yu and K. W. White, Microstructural Modification to Improve Mechanical Properties of a 70%Si_3N_4-30%BAS Self-Reinforced Ceramic Composite, *J. Mat. Sci.* **37**, 4411–4417 (2002).

36. F. Ye, S. Chen, M. Iwasa, Synthesis and Properties of Barium Aluminosilicate Glass-Ceramic Composites Reinforced with in situ grown Si_3N_4 whiskers, *Scipta Mat.* **48**, 1433–1438 (2003).
37. S. Suresh and A. Mortensen, Functionally Graded Metal and Metal-Ceramic Composites. Part 2, Thermomechanical Behaviour, *Int. Mater. Rev.* **42**, 85–116 (1997).
38. J. Jitcharoen, N. P. Padture, A. E. Giannakopoulos and S. Suresh, Hertzian-Crack Suppression in Ceramics with Elastic-Modulus-Graded Surfaces, *J. Am. Ceram. Soc.* **81**, 2301–2308 (1998).
39. L. Esposito, E. Saiz, A. P. Tomsia and R. M. Cannon, High Temperature Colloidal Processing for Glass/Metal and Glass/Ceramic FGM's, in *Ceramic Microstructures: Control at the Atomic Level*, A. P. Tomsia, A. Glaeser eds., Plenum Press, New York (1998), pp. 503–512.
40. S. Maruno, S. Ban, Y.-F. Wang, H. Iwata and H. Itoh, Properties of Functionally Gradient Composite Consisting of Hydroxyapatite Containing Glass Coated Titanium and Characters for Bioactive Implant, *J. Ceram. Soc. Japan* **100**, 362–367 (1992).
41. S. Ban, J. Hasegawa and S. Maruno, Electrochemical Corrosion Behaviour of Hydroxyapatite-Glass-Titanium Composite, *Biomaterials* **12**, 205–209 (1991).
42. K. Yamada, K. Imamura, H. Itoh, H. Iwata and S. Maruno, Bone Bonding Behaviour of the Hydroxyapatite Containing Glass-Titanium Composite Prepared by the Cullet Method, *Biomaterials* **22**, 2207–2214 (2001).
43. J. M Gomez-Vega, E. Saiz, A. P. Tomsia, T. Oku, K. Suganuma, G. W. Marshall, and S. J. Marshall, Novel Bioactive Functionally Graded Coatings on Ti6Al4V, *Adv. Mat.* **12**, 894–898 (2000).
44. C. Vitale Brovarone, E. Verne, A. Krajewski, and A. Ravaglioli, Graded Coatings on Ceramic Substrates for Biomedical Applications, *J. Europ. Ceram. Soc.* **21**, 2855–2862 (2001).
45. J. S. Moya, Layered Ceramics, *Adv. Mat.* **7**, 185–189 (1995).
46. P. Z. Cai, D. J. Green and G. L. Messing, Mechanical Characterization of Al_2O_3/ZrO_2 Hybrid Composites, *J. Europ. Ceram. Soc.* **18**, 2025–2034 (1998).
47. D. Sherman and D. Schlumm, The Mechanical Behaviour of Ceramic-Metal Laminate Under Thermal Shock, *J. Mater. Res.* **14**, 3544–3551 (1999).
48. Z. Chen and J. J. Mecholsky Jr., Damage-Tolerant Laminated Composites in Thermal Shock, *J. Mat. Sci.* **28**, 6365–6370 (1993).
49. S. J. Bennison, A. Jagota and C. A. Smith, Fracture of Glass/Poly(vinyl butyral) (Butacite®) Laminates in Biaxial Flexure, *J. Am. Ceram. Soc.* **82**, 1761–1770 (1999).
50. P. Weigt and G. Helmich, Security Glazing. Choosing the Most Cost Effective Glazing to Meet Every Security Need, in *Proc. 29th Int. Symposium in Automotive Technology and Automation*, D. Röller ed., Automotive Automation Ltd., Croydon (1997), pp. 1369–1381.
51. A. Seal, S. K. Dalui, A. K. Mukhopadhyay, K. K. Phani, H. S. Maiti, Mechanical Behaviour of Glass Polymer Multilayer Composites, *J. Mat. Sci.* **38**, 1063–1071 (2003).
52. S. Wuttiphan, B. R. Lawn and N. P. Padture, Crack Suppression in Strongly Bonded Homogeneous/Heterogeneous Laminates: A Study on Glass/Glass-ceramic Bilayers, *J. Am. Ceram. Soc.* **79**, 634–640 (1996).
53. D. C. Clupper and J. J. Mecholsky, Toughening of Tape Cast Bioglass® by Lamination with Stainless Steel 316 L, *J. Mat. Sci. Lett.* **20**, 1885–1888 (2001).
54. D. C. Clupper, J. J. Mecholsky, G. P. LaTorre and D. C. Greenspan, Bioactivity of Bioglass®-Steel and Bioglass®-Titanium Laminate Composites, *J. Mat. Sci. Lett.* **20**, 959–960 (2001).
55. T. L. Jessen, B. A. Bender and D. Lewis III, Mechanical Performance of Model and Surface-Reinforced SiC Fiber /CMC Laminate Systems, *Key Eng. Mat.* **108–110**, 195–202 (1995).
56. W. A. Cutler, F. W. Zok and F. F. Lange, Mechanical Behaviour of Several Hybrid Ceramic-Matrix-Composite Laminates, *J. Am. Ceram. Soc.* **79**, 1825–1833 (1996).
57. W. A. Cutler, F. W. Zok, F. F. Lange and P. G. Charalambides, Delamination Resistance of Two Hybrid Ceramic-Composite Laminates, *J. Am. Ceram. Soc.* **80**, 3029–3037 (1997).
58. T. L. Jessen and D. Lewis III, Effect of Composite Layering on the Fracture Toughness of Brittle Matrix/Particulate Composites, *Composites* **26**, 67–71 (1995).
59. T. L. Jessen and D. Lewis III, Fracture Toughness of Graded Metal-Particulate/Brittle-Matrix Composites, *J. Am. Ceram. Soc.* **73**, 1405–1408 (1990).
60. A. R. Boccaccini, J. Janczak, D. M. R. Taplin and M. Köpf, The Multibarriers-System as a Materials Science Approach for Industrial Waste Disposal and Recycling: Application of Gradient and Multilayered Microstructures, *Environmental Technology* **17**, 1193–1203 (1996).

61. K. Chyung, Strengthening of Glass-Ceramic Laminates by Differential Densification, in *Advances in Ceramics, Vol. 4: Nucleation and Crystallisation in Glasses*, J. H. Simmons, D. R. Uhlmann and G. H. Beall eds., The American Ceramic Society, Ohio (1982), pp. 341–352.
62. P. Hing and A. Adotey, The Sinterability and Electrical Conductivity of Some $TiSi_2$/Glass Systems, *J. Mat. Proc. Technol.* **38**, 465–482 (1993).
63. M. Sroda, and L. Stoch, Glasses for Alumina Laminates, in *Proc. Int. Congress on Glass, Volume 2. Extended Abstracts*, Society of Glass Technology, Sheffield, UK (2001), p. 978–979.
64. Y. Goto, M. Kato, T. Fukusawa, T. Kameda, Microstructure of Rare-Earth Silicate/Silicon Carbide Layered Composites, *J. Mat. Sci. Lett.* **21**, 121–124 (2002).
65. Chen, L., Goto, T., Tu, R., Guo, Ch., Hirai, T., Preparation and Corrosion Resistance of Graded Glass Coating on PbTe, *Mat. Sci. Forum*, 308–311, 256–261 (1999).
66. F. Smeacetto, M. Ferraris, M. Salvo, Oxidation Barrier Multilayer Coatings for Carbon-Carbon Composites, *Ceram. Eng. Sci. Proc.* **23** (4), 477–481 (2002).
67. R. Gadow, A. Killinger, C. Li, Ceramic on Glass and Glass-Ceramic Layer Composites for Industrial Applications, *Ceram. Eng. Sci. Proc.* **23** (4), 125–131 (2002).
68. M. Arnold, A. R. Boccaccini and G. Ondracek, Theoretical and experimental considerations on the thermal shock resistance of sintered glasses and ceramics using modeled microstructure-property correlations, *J. Mat. Sci.* **31**, 463–469 (1996).
69. K. Satyamurthy and D. P. H. Hasselman, Effect of Spatially Varying Porosity on Magnitude of Thermal Stresses During Steady State Heat Flow, *J. Am. Ceram. Soc.* **62**, 431–432 (1979).
70. A. R. Boccaccini, Incorporation of Porosity to Control the Residual Thermal Stresses in Ceramic Composites and Laminates, *The European Physical Journal. Applied Physics* **2**, 197–202 (1998).
71. S. B. Haug, L. R. Dharani and D. R. Carroll, Fabrication of Hybrid Ceramic Matrix Composites, *Appl. Comp. Mater.* **1**, 177–181 (1994).
72. N. Chawla, K. K. Chawla, M. Koopman, B. Patel, C. Coffin and J. I. Eldridge, Thermal-shock behavior of a Nicalon-fiber-reinforced hybrid glass-ceramic composite, *Comp. Sci. Tecnol.* **61**, 1923–1930 (2001).
73. L. A. Lewinshon, Hybrid Whisker-Fiber-Reinforced Glass-Matrix Composites with Improved Transverse Toughness, *J. Mat. Sci, Lett.* **12**, 1478–1480 (1993).
74. K. P. Gadkaree, Particulate-Fibre-Reinforced Glass Matrix Hybrid Composites, *J. Mat. Sci.* **27**, 3827–3834 (1992).
75. D. C. Jia, Y. Zhou, T. C. Lei, Ambient and Elevated Temperature Mechanical Properties of Hot-pressed Fused Silica Matrix Composite, *J. Europ. Ceram. Soc.* **23**, 801–808 (2003).
76. R. Reinicke, K. Friedrich, W. Beier and R. Liebald, Tribological Properties of SiC and C-Fibre Reinforced Glass Matrix Composites, *Wear* 225–229, 1315–1321 (1999).
77. R. Reinicke, K. Friedrich, W. Beier and R. Liebald, SiC- and C-Fiber Reinforced Glass Matrix Composites for Tribological Applications, *Adv. Eng. Mat.* **3**, 423–427 (2001).
78. I. Dlouhy, Z. Chlup, D. N. Boccaccini, S. Atiq, A. R. Boccaccini, Fracture Behaviour of Hybrid Glass Matrix Composites: Thermal Ageing Effects, *Composites Part A* **34**, 1177–1185 (2003).

Index

aerospace applications, glass composites, 462
agglomeration, SiC whiskers, 315
aircraft engines, $MoSi_2$-base composites applications in, 195
alkoxide process for mullite composites, 327
Almax fiber, 14
Almax/Al_2O_3 composites. *See also* Hi-Nicalon/Al_2O_3 composites; Nextel/Al_2O_3 composites; Nicalon/Al_2O_3 composites
   fracture surfaces, 301-303
   mechanical properties, 297
   microstructure, 295
alpha-alumina fibers
   Almax fiber, 14
   Fiber FP, 14
   Nextel 610, 15
   properties
      creep behavior, 25
      creep rate, 24
      creep resistance, 14
      strength, 23
      tensile, 28
   saphikon fiber, 28
Altex fiber, 13, 27. *See also* Nextel fibers
alumina crystals for whiskers, 17
alumina fibers. *See also* alumina-zirconia fibers; aluminosilicate fibers; silica fibers
   alpha. *See* alpha-alumina fibers
   based continuous fiber, 5
   continuous fine oxide fibers, 11
   continuous monocrystalline filaments, 16
   discontinuous oxide-based
      melt-spun aluminosilicate fibers, 9
      Saffil fiber, 10
   property and microstructural changes, 21
   silica content and strength of, 12
   YAG, 7, 16-, 28–29
alumina grains, growth of, 11
alumina matrix
   Al_2O_3-mullite matrix, 387–388
   for porous-matrix oxide-oxide composites, 387
   SiC whiskers in, 311
   WHIPOX CMCs, 430
alumina matrix composites, fiber reinforced
   directed metal oxidation fabrication, 279
   oxide fiber reinforced, 278, 295
   perform fabrication, 279
   SiC reinforced
      2-D Hi-Nicalon™/Al_2O_3 composites, 287
      2-D Nicalon™/ Al_2O_3 composites, 281, 282
      BN/SiC fiber-matrix interface role, 280
      turbine engine test, 294
alumina-mullite matrix, 387–388
alumina reinforced YSZ composites, 438. *See also* YSZ composites
   hot pressing, 438
   mechanical properties
      CTE, 444, 452, 453
      density, 440
      elastic modulus, 450
      flextural strength, 440
      flexure toughness, 440, 447
      fracture toughness, 444–447
      microcracks, 440
      microhardness, 451
      slow crack growth, 447
      thermal conductivity, 453
      thermal fatigue, 452
      Vickers indentation, 447
      Weibull modulus, 444
   microstructural analysis
      SEM, 439
      TEM, 439
      XRD, 439
alumina reinforced zirconia composites. *See also* 8 alumina reinforced YSZ composites
   material choice for structural reliability/life, 454
   processing, 438
alumina, SiC whisker reinforced
   microstructure, 315
   properties, 318
   toughening behavior of, 309

alumina whiskers, 17
alumina zirconia fibers. *See also* alumina fibers; silica fibers
   Nextel 650 fiber, 15
   PRD-166 fiber, 15
aluminosilicate based glass-ceramic matrices
   BAS, 464. *See also under* BAS
   BMAS, 464
   CAS, 464
   CMAS, 464
   LAS, 464
   LMAS, 464
   MAS, 464
   YAS, 464
   YMAS, 464
aluminosilicate fibers. *See also* alumina fibers; silica fibers
   continous, 13
   continuous single crystal oxide fibers, 27
   creep behavior, 24
   melt-spun alumino-silicate, 20
   melt-spun, 9, 20
   produced from precursors, 21
   tensile properties, 23
aluminosilicate matrix
   glass matrices, 467
   porous-matrix oxide-oxide composites, 387
   WHIPOX all oxide CMCs, 430
anisotropic thermal expansion, SiC/RBSN composites, 156
anisotropy, continuous single crystal oxide fibers properties, 28
annealing in air, high-temperature
   Hi-Nicalon/celsian composites, effect of, 243
   Nicalon/celsian composites, effect of, 233
arc jet testing, 218. *See also* furnace oxidation testing
automotive applications, glass composites, 462

BAS (barium aluminosilicate). *See also* aluminosilicate based glass-ceramic matrices; silicon nitride-BAS composites
   celsian formation, 230
   composition, 254
   crystal structure, 254
   glass ceramics, 464, 467
   glass composites, 473
   glass matrix composites, 494
   matrix crystallization, 263
   phase transitions, 257
   polymorphs
      hexacelsian, 254, 255, 256, 257
      monoclinic, 255
      orthorhombic, 255, 256, 257
biomedical applications of glass composites, 463
Blackglas composites

Blackglas 312 for SiOC composites, 348
Blackglas 493-type preceramic resin, 352
Blackglas/Nextel 312 applications
   high performance brake, rotors, and pads, 372
   jet engine tailcone, 371
curing, 351, 352
densification, 352
dielectric properties, 369, 370
infiltration, 351, 352
mechanical properties
   after 4000-hrs oxidation, 362–363
   after 600°C oxidation, 355–360
   compression, 353, 360
   creep rupture tests, 364–365
   flextural strength, 362–363
   low cycle fatigue, 366
   shear, 353, 359
   tensile, 353–356
oxidation exposure testing, 354
PDC technology
   Blackglas polymer and SiOC, 350
   fabrication, 351
   Nextel 312 fiber with BN coating, 350, 351
pyrolysis, 351–352
SiOC/Nextel 312 BN composite system, 352
specimen characterization
   after 600°C oxidation, 355
   as-prepared, 354
thermal properties
   conductivity, 368
   diffusivity, 367
   expansion, 368
   heat capacity, 367
blade outer airseal component (BOAS), 187
bonding, carbon/carbon composites, 118. *See also* debonding
boride ceramics, 200
borides. *See also* diborides
   hafnium, 199
   structural stability in UHTC composites, 199
   thermal conductivities, 208
   zirconium, 199
boron monofilament fiber reinforced glass composites, 467
boron nitride (BN).*See also* silicon nitride
   BN/SiC fiber-matrix interface, 280
   for SiC fiber-reinforced silicon nitride composites, 150
boron nitride coatings. *See also* SiOC-Nextel 312 BN composites
   in-situ, 84
   interfacial coatings
      N24-A CMC system, 83, 84
      N24-B CMC system, 85
      N24-C CMC system, 86

INDEX

MI-CMCs, 101, 103
  Nextel 312 fiber with, 350, 351
borosilicate glass, 468. *See also* aluminosilicate based glass-ceramic matrices; 3-D carbon fiber-reinforced composites
  matrix, 467
  matrix composites, 495
brittle composite behavior, oxide-oxide composite, 382
brittleness, dispersion-reinforced glass-matrix composites, 502

C/C-SiC composites, 117, 119. *See also* carbon fiber reinforced SiC composites
  gradient, 144
  homogeneous, 144
  properties
    crack deflection, 137
    frictional, 143
    high temperature, 138
    mechanical and thermophysical, 140
C/SiC composites, 117, 119. *See also* carbon fiber reinforced SiC composites
  CVI, 135
  LPI, 135
  properties
    high processing temperature, 137
    mechanical and thermal, 138–139
    shrinkage, 137
calcium aluminosilicate. *See* CAS
calcium oxide containing fibers, 5
carbides structural stability in UHTC composites, 199
carbon fiber reinforced composites. *See also* carbon/carbon composites
  3-D, 512–515
  glass matrix composites, 464
  SiC. *See* carbon fiber reinforced SiC composites
carbon fiber reinforced plastics (CFRP), 130
carbon fiber reinforced SiC composites. *See also* C/C-SiC composites; C/SiC composites
  applications
    advanced friction systems, 123
    calibrating bodies, 125
    energy and power station engineering, 126
    furnace engineering, 127
    low-expansion structures, 124
    mechanical, thermal, and corrosion resistance, 127
    satellite communication system, 126
    space vehicle's TPS and hot structures, 119
    vanes, nozzles and flaps of rocket motors and jet engines, 121
  microcracks, 144
  processing
    combined processing methods, 133
    CVI, 128, 129
    LPI, 128, 130
    LSI, 131, 132
    PIP, 128
  tribological tests, 145
carbon fibers. *See also* alumina fibers; silicon fiber
  high modulus (HM), 132
  high tenacity (HT), 132
carbon/carbon composites. *See also* carbon fiber reinforced SiC composites
  advantages, 118
  C/C-SiC, 117–119, 134–138, 140–144
  C/SiC, 117–119, 134–139
  microcracks, 132
  oxidation resistance, 118
  processing, 131, 132
  stiffness, 118
CAS (calcium aluminosilicate). *See also* BAS; LAS; MAS; SAS
  glass-ceramic matrices, 464
  glass composites, 473
celsian
  BAS usage, 230
  polymorphs and crystal structure
    features, 230
    hexacelsian, 228
    monoclinic, 230
  properties, 231
  SAS usage, 230
celsian composites
  Hi-Nicalon/celsian composites, 237
  Nicalon/celsian composites, 232
  SiC fiber-reinforced
    CMC processing, 231
    crystal structure, 228
    Hi-Nicalon/celsian composites, 237
    Nicalon/celsian composites, 232
ceramic borides
  hafnium, 199
  structural stability in UHTC composites, 199
  zirconium, 199
ceramic carbides, 199
ceramic fibers, silicon carbide-based, 33. *See also* alumina fibers; silica fiber
ceramic matrix composites. *See* CMC
ceramic nitrides, 199. *See also* ceramic borides; ceramic carbides
ceramic reinforcement, dispersion-reinforced glass matrices, 487
CFRP. *See* carbon fiber reinforced plastics
CG/HW tests. *See* furnace oxidation testing
Charpy V-notch (CVN) impact tests, 183
chemical vapor deposition. *See* CVD
chemical vapor infiltration. *See* CVI

CMC, 33
  applications, 49, 57
  failure, 80
  fiber reinforced (FRCMC), 77, 150. *See also*
      Blackglas composites
  high temperature applications, 78
  low cost ceramic matrix composites (LC^3)
      program, 349
  melt-infiltrated. *See* MI-CMCs
  PDC technology for, 349
  polymer-derived technology for, 349
  properties
    creep resistance, 79
    matrix cracking strength, 79
    proportional limit stress (PLS), 79
    rupture life, 79
    tensile stress-strain behavior, 79
    thermal conductivity, 79
    ultimate tensile strength, 79
  reinforcement, 34
  reinforcements based
    C/C-SiC, 117
    C/SiC, 117
  silicon melt-infiltrated. *See* MI-CMCs
  WHIPOX, 423
CMC system, SiC/SiC based, 78
CMC systems, NASA-developed
  creep rupture testing, 95
  creep rupture, 88
  debonding behavior, 92
  interphase issues, 95
  N22, 81, 82
  N24-A, 81, 83
  N24-B, 81, 85
  N24-C, 81, 86
  N26-A, 81
  stress-rupture testing, 95
  stress-strain, 88
  tensile testing, 95
  upper use temperature, 95
CMC systems, SiC reinforced
  for high temperature applications, 80
  NASA developed, 80–86
  properties of, 88
coefficient of thermal expansion (CTE)
  alumina-reinforced zirconia composites, 444, 452, 453
  C/C-SiC composites, 144
  carbon fiber reinforced SiC composites, 141
  carbon/carbon composites, 119
  Hi-Nicalon/celsian composites, 244
  MI-CMCs, 101
  $MoSi_2$-base composites, 175
  NASA-developed CMC systems, 90
  Nicalon™/Al_2O_3 composites, 282
  silicon nitride-BAS composite, 271
  UHTC composites, 207
coefficients of friction (CoF), carbon fiber reinforced
      SiC composites, 143
COI fiber composites. *See* high temperature
      oxide-oxide composites
cold gas/hot wall tests. *See* furnace oxidation testing
compression properties
  Blackglas composites, 353
  creep properties of $MoSi_2$-Si_3N_4 matrix composite, 178
  $MoSi_2 - \beta Si_3N_4$ composite, 190
  SiOC-Nextel 312 BN, 360
consolidation, oxide-oxide composites processing
  hot isotatic pressing, 409
  hot pressing, 409
  pressureless sintering, 409
continuous alumino-silicate fibers. *See also*
      aluminosilicate fibers
  Altex fiber, 13
  Nextel fiber, 13
continuous fiber reinforced ceramic composites,
      nonaerospace applications of, 57
continuous fiber reinforced glass composites
  applications
    aerospace, 462
    automotive, 462
    biomedical, 463
    electronic, 463
    functional, 463
    high temperature, 462
    impact resistant, 462
  boron monofilaments, 467
  carbon fiber reinforced, 464
  fracture toughness, 465
  glass/glass-ceramic fiber reinforced, 468
  high temperature properties, 476
  impact resistance, 477
  metal fiber reinforced, 469
  Nicalon® fiber reinforced, 464
  processing
    electrophoretic deposition, 473
    hot pressing, 470
    matrix transfer molding, 473
    polymer precursor method, 474
    pulltrusion, 474
    slurry infiltration, 470
    sol-gel, colloidal routes, 473
    tape casting, 473
  SiC monofilaments, 467
  toughnening, 475
  tribology, 477
  Tyranno® fiber reinforced, 464
  wear and erosion behavior, 477
  with oxide fibers, 467

INDEX

continuous fine oxide fibers, 11
continuous glass fibers, 9
continuous monocrystalline filaments
  crystal growth and orientation, 16
  grain boundaries in, 16
  manufacturing, 16
  molybdenum die material for, 17
continuous single crystal oxide fiber properties
  anisotropy, 28
  creep rate, 28
  tensile failure stress, 27
crack behavior. *See also* creep behavior; microcracking
  alumina reinforced zirconia composites, 446
  crack opening displacement (COD), 446
  CVI SiC/SiC composites, 65–72
  dispersion-reinforced glass-matrix composites, 501
  mullite-SiC whisker composites, 338, 339
  porous-matrix composites, 382
  SiC whisker-reinforced alumina, 319
crack deflection
  dispersion-reinforced glass-matrix composites, 499
  oxide-oxide composites, 380, 382
creep behavior. *See also* creep rupture behavior
  alpha-alumina fibers, 24, 25
  aluminosilicate fibers, 24
  continuous single crystal oxide fibers, 28
  CVI SiC/SiC composites, 73
  high temperature oxide-oxide composites, 401
  interface controlled diffusional (ICD), 42
  MI-CMCs, 110
  $MoSi_2$-Si_3N_4 matrix composites, 178
  mullite-SiC whisker, 339
  mullite-zirconia-SiC whisker composites, 340
  SiC/Si_3N_4 composite, 168
  silicon carbide fibers, 40, 43
creep resistance
  alpha-alumina fibers, 14
  CMC, 79
  N24-C CMC system, 86
  N26-A CMC system, 87
  SiC/RBSN, 162
  silicon carbide fiber, 41
creep rupture behavior
  MI-CMCs, 110
  Prepreg HiPerComp™ composites, 111
creep rupture tests
  NASA-developed CMC system, 88, 95
  SiOC-Nextel 312 BN, 364–365
crystal growth and orientation, in continuous monocrystalline filaments manufacturing, 16
crystallization, BAS matrix, 263
CTE. *See* coefficient of thermal expansion
curing, Blackglas composites, 351, 352

CVD (chemical vapor deposition). *See also* CVI
  alumina whiskers production, 17
  derived silicon carbide fibers, 35, 39
  interface coating processing, 405
  or prepeg MI-CMCs, 102
  oxide-oxide composites processing, 410
CVI (chemical vapor infiltration). *See also* liquid polymer infiltration; liquid silicon infiltration
  advantages and disadvantages, 55–56
  BN fiber coating for N24-C CMC System, 86
  BN reactor, 83
  carbon fiber reinforced SiC composites processing, 128, 129
  history, 59
  infiltrated silicon carbide, 55
  oxide-oxide composites processing, 410
  SiC fiber-reinforced silicon nitride composites, 150
  slurry-cast process and, 103
CVI processes
  forced (F-CVI), 60
  isothermal-isobaric (I-CVI), 60
  pulsed (P-CVI), 61
CVI SiC composite system, 83, 86
CVI SiC matrix
  N22 CMC system, 83
  N24-B CMC system, 85
  N24-C CMC system, 86
  N26-A CMC system, 87
  penetration properties of NASA developed CMC systems, 89
  reactor, 83
CVI SiC/SiC composites
  applications, 57
  cracking behavior, 65, 66, 67, 68, 70, 71, 72
  fiber/matrix interactions controlling aspects, 56
  high temperature properties, 59
  mechanical properties
    creep, 73
    damage mechanisms, 65
    fatigue behavior, 72
    flexural strength, 69
    fracture toughness, 72
    interface properties, 70
    reliability, 68
    tensile stress-strain behavior, 63
    ultimate failure, 67
  microcracks, 65
  microstructure, 65
  oxidation resistance, 61
  processing
    fibers coating, 60
    fibrous perform preparation, 59
    SiC matrix infiltration, 60
  superiority over metals, 62
  thermophysical

high temperature behavior, 73
    thermal shock, 72
    Young's modulus, 67
CVN. *See* Charpy V-notch impact tests
cyclic fatigue
    SiOC-NEXTEL 312 BN, 366
    WHIPOX CMCs, 429
cyclic oxidation, $MoSi_2$-βSi_3N_4 composite, 194

damage mechanisms in composites
    matrix cracking, 66
    microcracks, 65
damage tolerance
    carbon/carbon composites, 118
    MI-CMCs, 113
    WHIPOX CMCs, 431
deagglomeration, SiC whiskers, 314
debonding behavior
    2-D HI-Nicalon/Al2O3 composites, 293
    dispersion-reinforced glass-matrix composites, 500
    MI-CMCs, 108
    NASA-developed CMC systems, 92
    NicalonTM/Al2O3 composites, 283
    WHIPOX CMCs,, 428
    whisker reinforced alumina, 309
dense-matrix composites, 382. *See also* porous-matrix composites
dense SiC/silicon nitride composites. *See also* SiC/RBSN
    application, 169
    creep properties, 168
    impact resistance, 169
    physical and mechanical properties, 167, 168
densification
    Blackglas composites, 352
    dispersion-reinforced glass-matrix composites, 493
    fiber-reinforced glass matrix composites, 471
    oxide-oxide composites, 409
    prepeg MI-CMC, 102
    SiC whisker composites
        pressure-assisted sintering, 317
        pressure-less sintering, 317
    silicon nitride-BAS composite, 258
    slurry-casted MI-CMC, 103
density, alumina reinforced YSZ composites, 440
diboride powders
    hafnium, 204
    zirconium, 204
diborides, 200–204
    hafnium
        properties, 207–211, 217
        processing, 219
    zirconium
        properties, 203, 210–211, 219

high temperature testing, 212
    processing, 204
dielectric properties
    silicon nitride-BAS composite, 273
    SiOC-Nextel 312 BN, 369, 370
diesel engine glow plug, $MoSi_2$-base composites applications in, 194
directed metal oxidation (DIMOX). *See also* reaction-bonding process
    $Nicalon^{TM}/Al_2O_3$, 280, 294
    oxide fiber reinforced alumina composites, 279, 296
discontinuous oxide fibers. *See also* fine continuous oxide fibers; glass fibers
    melt-spun aluminosilicate fibers, 9
    Saffil fiber, 10
dispersion-reinforced glass matrix composites. *See also* fiber reinforced glass matrix composites
    applications, 489, 490, 491
    engineering properties improvement, 488
    erosion resistance and machinability, 502
    hardness and brittleness, 502
    mechanical properties, 486
        crack bridging by ductile inclusions, 501
        crack deflection, 499
        internal stresses, 498
        toughening by residual stresses, 499
        toughening, 498
    processing
        alternative methods, 494
        hot pressing, 492
        powder technology, 492
        sintering, 492
    thermal shock resistance, 503
    with enhanced strength and toughness
        ceramic reinforcement, 487
        metallic reinforcement, 487
dispersion, SiC whiskers, 314, 315
Double Notch Shear (DNS) test
    SiOC-Nextel 312 BN, 359, 360
dry pressing, SiC whisker composites, 316
dynamic fatigue, alumina reinforced YSZ composites, 447

EBC. *See* environmental barrier coatings
Edge-defined Film-Fed Growth (EFG), 385
EDM. *See* electron discharge machining
elastic modulus
    alumina reinforced zirconia composites, 450
    SiC/RBSN, 154
    silicon nitride-BAS composite, 265, 266
elastic properties, MI-CMCs, 106
electrical properties. *See also* mechanical properties; thermal properties
    high temperature oxide-oxide composites, 396

INDEX

MoSi$_2$-βSi$_3$N$_4$ composite, 189
mullite composites, 343
SiC whisker-reinforced alumina, 322
UHTC composites, 210
electromechanical testing, alumina reinforced zirconia composites, 444
electron discharge machining, UHTC composites, 207
electronic applications of glass composites, 463
electrophoretic deposition, 408
   fiber-reinforced glass matrix, 473
   interface coatings processing, 405
environmental barrier coatings, 80
environmental resistance, MoSi$_2$-βSi$_3$N$_4$ composite, 193
environmental stability, SiC/RBSN composites, 162
EPD. *See* electrophoretic deposition
erosion behavior. *See also* wear behavior
   dispersion-reinforced glass-matrix composites, 502
   glass and glass-ceramic matrix composites, 477
   MoSi$_2$-Si$_3$N$_4$ matrix composites, 179
eutectic fiber, 16

failure behavior
   CVI SiC/SiC composites, 67
   MI-CMCs, 110
   porous-matrix composites, 382
   WHIPOX all oxide CMCs,, 428
fatigue behavior
   alumina reinforced YSZ composites, 447
   CVI SiC/SiC composites, 72
   dynamic, 447
   high temperature oxide-oxide composites, 402
   MI-CMCs, 110
   WHIPOX all oxide CMCs, 429
fatigue tests
   alumina reinforced zirconia composites, 452
   Nicalon/Al$_2$O$_3$ composites, 287
F-CVI. *See* forced-CVI process
fiber coating, for oxide-oxide composites design, 380. *See also* boron nitride coatings
fiber/matrix bonding
   carbon fiber reinforced SiC composites, 137
   carbon/carbon composites, 118
   polymer composites, 132
fiber/matrix interface, 56
   fiber-reinforced aluminum oxide matrix composites, 280
   MI-CMCs, 101
   oxide-oxide composites, 380
fiber/matrix ratio, WHIPOX all oxide CMCs, 428
fiber reinforced composites. *See also* CMC
   alumina matrix composites
      oxide fiber reinforced, 295
      SiC-fiber reinforced, 280
   ceramic matrix composites, 150

celsian, 227
glass for
   C- and D-glasses, 8
   E-glass, 8
   R- and S-glasses, 8
   Z-glass, 8
glass matrix composites. *See* fiber reinforced glass matrix composites
fiber reinforced glass composites
   aluminosilicate, 464
   applications
      aerospace, 462
      automotive, 462
      biomedical, 463
      electronic, 463
      functional, 463
      high temperature, 462
      impact resistant, 462
   boron monofilament fiber reinforced, 467
   carbon fiber reinforced, 464
   Nicalon® fiber reinforced, 464
   SiC fiber reinforced, 467
   Tyranno® fiber reinforced, 464
fiber reinforced glass matrix composites. *See also* dispersion-reinforced glass matrix composites
   3-D, 512
   boron monofilament fiber, 467
   glass/glass-ceramic fiber, 468
   high temperature properties, 476
   impact resistance, 477
   metal fiber-reinforced, 469
   oxycarbide fiber for, 465, 471, 476
   processing
      densification, 471
      electrophoretic deposition, 473
      hot pressing, 470
      matrix transfer molding, 473
      polymer precursor method, 474
      pulltrusion, 474
      slurry infiltration, 470
      sol-gel, colloidal routes, 473
      tape casting, 473
   SiC-fiber, 467
   toughnening, 475
   tribology, 477
   with oxide fibers, 467
fiber reinforcement in silicate glass, 461
fibrous perform, CVI SiC/SiC composites, 59
filaments
   continuous monocrystalline, 16
   glass, 4
   monocrystals, 17
   single crystal oxide, 7
   spinning glass, 4
fine continuous oxide fibers

alpha alumina fibers
   Almax Fiber, 14
   FP-fiber, 14
   Nextel 610, 15
alumina zirconia fibers
   Nextel 650 fiber, 15
   PRD-166 Fiber, 15
continuous alumino-silicate fibers
   Altex Fiber, 13
   Nextel Fibers, 13
continuous monocrystalline filaments, 16
manufacture, 11
nano-oxide fibers, 18
whiskers, 17
flextural strength
   2-D Nicalon^{TM}/Al$_2$O$_3$ composites, 283
   alumina reinforced zirconia composites, 440-441, 447
   CVI SiC/SiC composites, 69
   high temperature oxide-oxide composites, 398
   Hi-Nicalon/celsian composites, 237
   MoSi$_2$-βSi$_3$N$_4$ composite, 191
   mullite-SiC whisker
      after thermal exposure, 334-335
      high temperature strength, 338
   mullite-zirconia composites, 334
   mullite-zirconia-SiC whisker composites
      after thermal exposure, 334–336
      high temperature strength, 338
   Nicalon/celsian composites, 232
   oxide/oxide composites, 298
   SiC whisker reinforced alumina, 320
   silicon nitride-BAS composite, 268
   SiOC-Nextel 312 BN, 362, 363
flexure test, YSZ composites, 439
FMB. *See* fiber/matrix bonding
forced-CVI process, 60
fracture behavior
   MI-CMCs, 107
   WHIPOX all oxide CMCs, 429–430
fracture surface analysis
   oxide/oxide composites
      Alman/Al$_2$O$_3$, 300
      Nextel/Al$_2$O$_3$, 300
   SiC whisker reinforced alumina, 319
fracture toughness
   2-D Nicalon^{TM}/Al$_2$O$_3$ composites properties, 283
   alumina reinforced zirconia composites
      at 1000º C, 446
      crack opening displacement, 446
      SEPB method, 445
      SEVNB method, 444, 445
   CVI SiC/SiC composites, 72
   dispersion-reinforced glass-matrix composites, 499

glass and glass-ceramic composites, 475
glass matrix composites, 465
glass-ceramic matrix composites, 465
MoSi$_2$-Si$_3$N$_4$ matrix composites, 179
MoSi$_2$-βSi$_3$N$_4$ composite, 191
mullite composites, 340
   mullite-SiC whisker, 342
   mullite-zirconia-SiC whisker, 342
SCS-6/MoSi$_2$-Si$_3$N$_4$ matrix composites, 181
SiC/RBSN, 154
SiC whisker reinforced alumina, 318
silicon nitride-BAS composite, 267, 268
whisker reinforced alumina, 309
FRCMC, 150. *See also* CMC
frictional properties, carbon fiber reinforced SiC composites
   C/C-SiC, 143
   C/SiC, 143
fugitive coatings. *See also* porous coatings; weak oxide coatings
   carbon coatings, 394
   molybdenum, 394
functionally-graded bioactive coatings, 519
functionally-graded materials
   glass-containing composites with, 518
   mechanical properties, 518
furnace oxidation testing, 216. *See also* arc jet testing

GE GEN-IV fiber composites. *See* high temperature oxide-oxide composites
glass-ceramic matrices, 461
   barium magnesium aluminosilicate, 464
   calcium aluminosilicate, 464
   lithium aluminosilicate, 464
   microcracking, 467
glass-ceramic matrix composites. *See also* glass composites; glass matrix composites
   applications
      aerospace, 462
      automotive, 462
      biomedical, 463
      electronic, 463
      functional, 463
      high temperature, 462
      impact resistant applications, 462
   with oxide fibers, 467
glass-ceramic matrix composites, dispersion-reinforced. *See also* glass-ceramic matrix composites, fiber-reinforced; glass matrix composites, dispersion-reinforced
   applications, 489–491
   engineering properties improvement, 488
   with enhanced strength and toughness
      ceramic reinforcement, 487
      metallic reinforcement, 487

INDEX

glass-ceramic matrix composites, fiber-reinforced. *See also* glass-ceramic matrix composites, dispersion-reinforced; glass matrix composites, fiber-reinforced
  carbon fiber reinforced, 464
  fracture toughness, 465
  high temperature properties, 476
  impact resistance, 477
  Nicalon® fiber reinforced, 464
  processing
    electrophoretic deposition, 473
    hot pressing, 470
    matrix transfer molding, 473
    polymer precursor method, 474
    pulltrusion, 474
    slurry infiltration, 470
    sol-gel, colloidal routes, 473
    tape casting, 473
  toughening, 475
  tribology, 477
  Tyranno® fiber reinforced, 464
  wear and erosion behavior, 477
glass composites (fiber reinforced)
  aluminosilicate fiber reinforced
    BAS, 464
    BMAS, 464
    CAS, 464
    CMAS, 464
    LAS, 464
    LMAS, 464
    MAS, 464
    YAS, 464
    YMAS, 464
  applications, 462–463
glass-containing composites
  with graded and layered microstructures
    functionally graded materials, 518
    layered materials, 520
    mechanical properties, 518
    multilayers structures, 523
  hybrid composites, 526
  with interpenetrating microstructures
    3-D fiber reinforcement, 512, 515
    In-Ceram process, 516
    infiltration, 518
    interpenetrating microstructures, 512
    mechanical properties, 513
    whisker preforming, 517
glass fibers, 3, 4. *See also* discontinuous oxide fibers; fine continuous oxide fibers
  continuous, 9
  processing
    crystallization, 9
    drawing conditions, 9

fiber strength aspects, 9
melting point aspects, 7
properties
  chemical, 19
  mechanical, 19
  physical, 19
glass filaments, 4
glass matrices
  aluminium silicate, 464
  borosilicate, 464
  silica glass, 464
glass matrix composites. *See also* glass-ceramic matrix composites
  carbon fiber reinforced, 464
  fracture toughness, 465
  Nicalon® fiber reinforced, 464
  Tyranno® fiber reinforced, 464
glass matrix composites, dispersion-reinforced. *See also* glass-ceramic matrix composites, dispersion-reinforced; glass matrix composites, fiber-reinforced
  erosion resistance, 502
  hardness and brittleness, 502
  mechanical properties
    crack bridging by ductile inclusions, 501
    crack deflection, 499
    internal stresses, 498
    toughening, 498–499
  processing
    alternative methods, 494
    hot pressing, 492
    powder technology, 492
    sintering, 492
  thermal shock resistance, 503
  wear behavior, 502
glass matrix composites, fiber-reinforced. *See also* glass-ceramic matrix composites, fiber-reinforced
  3-D, 512
  boron monofilament fiber, 467
  glass/glass-ceramic fiber, 468
  high temperature properties, 476
  impact resistance, 477
  metal fiber-reinforced, 469
  oxycarbide fiber for, 465, 471, 476
  processing
    densification, 471
    electrophoretic deposition, 473
    hot pressing, 470
    matrix transfer molding, 473
    polymer precursor method, 474
    pulltrusion, 474
    slurry infiltration, 470
    sol-gel, colloidal routes, 473
    tape casting, 473

SiC-fiber, 467
toughnening, 475
tribology, 477
with oxide fibers, 467
glass matrix powders, 492
global load sharing (GLS), 398
graded glass composites, 525
graded microstructures, 518. *See also* layered microstructures
graded porous glasses, 525
gradient C/C-SiC composites. *See also* homogeneous C/C-SiC composites, 144
grain boundaries
　continuous monocrystalline filaments, 16
　diffusion property in silicon carbide fibers, 42
　sliding property, silicon carbide fiber, 41
grain size properties
　silicon carbide fibers, 39
　silicon nitride-BAS composite, 269
Griffith's flaw theory, 498

hafnium borides, 199
hafnium diborides. *See also* zirconium diborides
　high temperature testing
　　arc jet, 219
　　furnace oxidation, 216
　oxidation studies, 212
　powder for UHTC composites, 204
　processing, 204
　properties
　　electrical, 210
　　mechanical, 210
　　optical, 211
　　thermal, 207, 219
　　thermodynamic, 203
hardness
　dispersion-reinforced glass-matrix composites, 502
　SiC whisker-reinforced alumina, 321
　silicon nitride-BAS composite, 265, 266
heat capacity, SiOC-Nextel 312 BN, 367
heat engines, 379
hexacelsian, 228. *See also* monoclinic celsian
　barium aluminum silicate, 255–257
　crystal structure, 229
HG/CW tests. *See* arc jet testing
high modulus (HM) carbon fibers, 132
high temperature annealing in air, effect of
　Hi-Nicalon/celsian composites, 243
　Nicalon/celsian composites, 233
high temperature applications
　CMC, 78
　glass composites, 462
high temperature behavior. *See also* high temperature testing
　CVI SiC/SiC composites, 73

glass and glass-ceramic composites, 476
rupture, NASA-developed CMC system, 95
high temperature oxide-oxide composites. *See also* oxide-oxide composites; porous matrix oxide-oxide composites
　creep behavior, 401
　electrical properties, 396
　fatigue, 402
　interface coatings
　　fugitive coatings, 394
　　porous coatings, 393
　　weak oxides, 391
　interface control, 390
　long term thermal exposure, 400
　mechanical properties, 396
　next generation enabling of, 411, 412
　notch sensitivity and toughness, 401
　off axis properties, 403
　physical characteristics, 395
　thermal properties, 396
　weak oxide coatings
　　monazites, 391
　　xenotimes, 391
high temperature testing
　UHTC composites, 212
　arc jet testing, 218
　furnace oxidation testing, 216
high tenacity (HT) carbon fibers, 132
Hi-Nicalon fiber, for NASA developed CMC systems, 82, 90
Hi-Nicalon/Al_2O_3 composites, 287, *See also* Almax/Al_2O_3 composites; Nextel/Al_2O_3 composites; Nicalon/Al_2O_3 composites
　debonding, 293
　residual tensile strength, 292
　tensile stress-strain, 292
　thermal conductivity, 292
Hi-Nicalon/celsian composites. *See also* Nicalon/celsian composites
　flexure strength, 237
　high-temperature annealing in air, effect of, 243
　shear strength, 240
　tensile strength, 239
　thermo-oxidative stability
　　degradation mechanism at 1200º C, 245
　　high temperature annealing in air, 243
Hi-Nicalon/$MoSi_2$-Si_3N_4 composites
　mechanical properties, 185
　by melt infiltration, 186
HIP. *See* hot-isostatic pressing
HiPerComp™ composites. *See also* MI-CMCs
　applications, 100
　damage tolerance, 114
　elastic properties, 106
　fatigue and creep behavior, 110

fracture strength, 107
MI-CMC constituents, 101
prepeg fabrication, 102–105
slurry cast fabrication, 103–105
thermal properties, 105
thermal stability, 108
homogeneous C/C-SiC composites, 144.
    See also gradient C/C-SiC composites
hot gas/cold wall tests. *See* arc jet testing
hot isotatic pressing (HIP). *See also* hot pressing
    $MoSi_2$-base composites, 174
    oxide-oxide composites, 409
    SiC fiber-reinforced silicon nitride composites, 150
    SiC whisker composites, 317
hot pressing
    dispersion-reinforced glass-matrix composites, 492
    fiber-reinforced glass matrix composites, 470
    oxide-oxide composites, 409
    SiC fiber-reinforced silicon nitride composites, 150
    SiC whisker composites, 317
    UHTC composites, 205
    YSZ/alumina composites, 438

I-CVI. *See* isothermal-isobaric CVI process
impact behavior, SCS-6/$MoSi_2$-Si_3N_4 matrix composites, 183. *See also* Charpy V-notch (CVN) impact tests
impact resistance
    glass composites applications, 462
    glass and glass-ceramic matrix composites, 477
    SiC/Si_3N_4 composite, 169
impedance, mullite composites, 343
In-Ceram process, glass-containing composites with interpenetrating microstructures, 516
infiltration procedure, Blackglas composites, 351, 352. *See also* CVI
inorganic fibers, 3. *See also* organic fibers
in-situ BN coatings
    N24-A CMC system, 84
    N24-B CMC system, 85
    N24-C CMC System, 86
in-situ reinforced silicon nitride, 251, 253. *See also* silicon nitride-BAS composites
    barium aluminum silicate, 254
    XRD analysis, 258
interface coatings
    for NASA developed CMC System, 83–86
    high temperature oxide-oxide composites
        fugitive coatings, 394
        porous coatings, 393
        weak oxides, 391
    oxidation resistant, 384, 389
    processing
        CVD, 405
        electrophoretic deposition, 405

interface control. *See also* weak oxide coatings
    mechanisms, 390
    oxide-oxide composites design
        crack deflection aspects, 380
        weak fiber-matrix interface, 380
interface controlled diffusional (ICD) creep, 42
interface properties, CVI SiC/SiC composites, 70
interfacial shear strength, NASA developed CMC systems, 93
interfacial shear stresses, CVI SiC/SiC composites, 70
interpenetrating microstructures, glass-containing composites with, 512
interphase issues, NASA-developed CMC system, 95
isothermal-isobaric CVI process, 60. *See also* CVI processes

LAS (lithium aluminosilicate). *See also* BAS; CAS; LAS; MAS
    glass ceramics, 467
    glass composites, 473
layered microstructures. *See also* graded microstructures
    glass-containing composites with, 520
    mechanical properties, 520
LC^3. *See* low cost ceramic matrix composites
liquid polymer infiltration, 128–130. *See also* CVI
liquid silicon infiltration, 123, 131–132
low cost ceramic matrix composites, 349, 352.
    See also SiOC/Nextel 312 BN composites
low cycle fatigue (LCF), carbon fiber reinforced SiC fibers, 140

machinability, dispersion-reinforced glass-matrix composites, 502
machining, UHTC composites, 207
magnesia containing fibers, 5
MAS, 473. *See also* SAS
matrix agglomerations, WHIPOX CMCs, 427–432
matrix cracking strength
    CMC, 79
    MI-CMCs, 106, 108
    SiC/RBSN, 154
matrix cracking, 66–72
    MI-CMCs, 111
    microcracking, Nicalon/Al_2O_3 composites, 285
    $MoSi_2$-base composites, 175
matrix infiltration, oxide-oxide composites
    EPD, 408
    pre-peg processing, 407
    pressure infiltration, 407
matrix transfer molding, fiber-reinforced glass matrix, 473
mechanical properties. *See also* electrical properties; thermal properties
    2-D Nicalon/Al_2O_3 composites

flextural strength, 283
fracture toughness, 283
3-D fiber-reinforced glass containing composites, 514
alumina reinforced zirconia composites
  elastic modulus, 450
  flextural strength, 440, 441
  fracture toughness, 444
  microhardness, 451
  slow crack growth, 447
  thermal fatigue, 452
Blackglas composites
  compression, 360
  creep-rupture tests, 364, 365
  flextural strength, 362, 363
  low cycle fatigue tests, 366
  shear, 359
  tensile, 355, 356
carbon fiber reinforced SiC composites, 138–140
dispersion-reinforced glass-matrix composites, 486, 496–498
glass and glass-ceramic composites, 475, 476
high temperature oxide-oxide composites, 396
Hi-Nicalon/$MoSi_2$-Si_3N_4 composites, 185
layered structures, 520
$MoSi_2$-Si_3N_4 matrix composites
  compressive creep, 178
  erosion behavior, 179
  fracture toughness, 179
$MoSi_2$-βSi_3N_4 composites
  compression strength, 190
  flextural strength, 191
  fracture toughness, 191
  weibull behavior, 191
mullite composites, 333
oxide/oxide composites, 297
SCS-6/$MoSi_2$-Si_3N_4 matrix composites
  fracture toughness, 181
  impact behavior, 183
  tensile behavior, 181
SiC/RBSN composite, 153
SiC/Si_3N_4 composite, 167, 168
UHTC composites
  hafnium diborides, 210
  zirconium diborides, 210
WHIPOX all oxide CMCs, 427
melt filtration, Hi-Nicalon/$MoSi_2$-Si_3N_4 composites, 186
melt infiltration (MI)
  carbon fiber reinforced SiC composites processing, 132
  CMC. See MI-CMCs
  NASA developed CMC systems, 86
melt-spun aluminosilicate fibers, 9, 20. See also Saffil fiber

melting point, for making glass, 7
mesostructure, WHIPOX CMCs, 427
metal oxidation processing
  fiber reinforced alumina matrix composites, 279, 296
  oxide-oxide composites
    directed metal oxidation, 410
    reaction bonding, 411
metallic reinforcement, dispersion-reinforced glass matrices, 487
methylchlorosilane, 60
methyltrichlorosilane. See MTS
MI-CMCs
  applications, 100
  creep behavior, 110
  damage tolerance, 113
  elastic properties, 106
  fabrication process
    prepeg, 102
    slurry cast, 102, 103
  fatigue behavior, 110
  fracture strength, 107
  HiPerComp™, microstructure study of, 100
    Prepreg, 103
    Slurry Cast, 104
  oxidation resistance, 102
  processing
    BN coating, 101
    fiber matrix composite constituents, 101
    fiber-matrix interphase, 101
  stress-strain behavior, 107
  Sylramic fibers for, 101
  tensile creep, 110
  thermal and physical properties
    prepeg, 105
    slurry cast, 105
  thermal stability, 108
microcracking, 72
  dispersion-reinforced glass-matrix composites, 500
  fiber reinforced glass matrices, 467
  $MoSi_2$-base composites, 175
  SiC/RBSN composites, 175
microcracks, 67. See also crack behavior
  alumina reinforced YSZ composites, 440
  C/C composite, 132
  carbon fiber reinforced SiC composites, 144
  CVI SiC/SiC composites, 65
  glass and glass-ceramic composites, 477
microhardness. See also mechanical properties
  alumina reinforced zirconia composites, 451
  mullite-SiC whisker, 342
microstructural analysis
  mullite composites, 329
  YSZ composites
    SEM, 439

INDEX

TEM, 439
XRD, 439
WHIPOX all oxide CMCs, 430
microstructure
  interpenetrating, 512
  NicalonTM/Al_2O_3 composites, 281
  prepreg HiPerComp^{TM}, 103
  SiC whisker reinforced alumina, 315
  silicon nitride-BAS composite, 260
  slurry cast HiPerComp^{TM}, 104
molybdenum die material, for Continuous Monocrystalline Filaments, 17
monazite coatings. *See also* scheelite coatings; xenotimes coatings
  for porous matrix Nextel 610-based composites, 393
  in-situ, 393
  Nextel 312 composites, 391
  Nextel 610/(porous) alumina composites, 391
Monkman-Grant diagrams. *See under* silicon carbide fiber properties
monoclinic celsian. *See also* hexacelsian
  BAS, 230
  formation, 230
  SAS, 230
monoclinic structure, BAS, 255
monofilament fiber reinforced glass composites
  boron, 467
  SiC, 467
monolithic ceramics
  non-oxide, 424
  oxide ceramics, 423
$MoSi_2$-base composites
  as-consolidated $MoSi_2$-Si_3N_4 matrix, 175
  Hi-Nicalon/$MoSi_2$-Si_3N_4 composite
    by melt-filtration, 186
    mechanical properties, 185
  mechanical properties ($MoSi_2$-Si_3N_4 matrix composite)
    compressive creep, 178
    erosion behavior, 179
    fracture toughness, 179
  mechanical properties ($MoSi_2$-βSi_3N_4 composite)
    compression strength, 190
    flexural strength, 191
    fracture toughness, 191
    Weibull Behavior, 191
  mechanical properties (SCS-6/$MoSi_2$-Si_3N_4 matrix composites)
    fracture toughness, 181
    impact behavior, 183
    tensile behavior, 181
  microstructural properties, 175
  $MoSi_2$-Si_3N_4 matrix composites, 178

$MoSi_2$-βSi_3N_4 composites
  electrical conductivity, 189
  environmental resistance, 193
  microstructure, 188
  physical properties, 188
  processing, 188
$MoSi_2$-βSi_3N_4 composites, applications of
  aircraft engines, 195
  diesel engine glow plug, 194
oxidation behavior
  cyclic oxidation tests, 178
  high temperature, 178
  low temperature, 175
processing
  hot isotatic pressing, 174
  vacuum hot pressing, 174
SCS-6 reinforced, 174
SCS-6/$MoSi_2$-Si_3N_4 matrix composites, 181
SiC/$MoSi_2$-Si_3N_4 composite, 187
MTS (methylchlorosilane), 60, 129
mullite composites
  Al_2O_3 phase in, 329
  alkoxide process for, 327
  electrical properties, 343
  fracture toughness, 340
  materials characterization, 327, 328
  mechanical properties, 333
  phase and microstructural analysis, 329
  physical properties, 331
  SiC phase in, 329
  SiC whiskers, 328
  SiO_2 phase in, 329
  suspensions preparation for matrix, 328
  without zirconia phase, 328
  zirconia addition, 327
  zirconia phase in, 329
mullite matrix
  porous-matrix oxide-oxide composites, 387, 388
  WHIPOX all oxide CMCs, 431
mullite, properties, 327
mullite-SiC whisker composites, 326–328
  flextural strength, 333
    after thermal exposure, 334, 335
    high temperature strength, 338
  fracture toughness, 342
  high temperature crack growth, 338, 339
  microhardness, 342
  permitivity property
    high frequency, 343
    low frequency, 343, 344
  specific heat, 332
  thermal diffusivity, 332
  thermal shock resistance, 341, 343
mullite-zirconia composites
  flexural strength after thermal exposure, 334

high temperature deformation, 339
permittivity properties, 343
SEM analysis, 330
specific heat, 332
thermal diffusivity of, 332
thermal expansion, 331
mullite-zirconia-SiC whisker composites
   flexural strength
      after thermal exposure, 334–336
      high temperature strength, 338
   fracture toughness, 342
   high temperature creep, 340
   preparation, 328
   specific heat, 334
   thermal diffusivity of, 332
   thermal expansion, 331
   thermal shock resistance, 341, 343
multilayered glass composite concept, 523

N22 CMC system. *See also* NASA developed CMC systems
   CMC thermostructural properties, 82
   CVI BN reactor for, 83
   CVI SiC matrix in, 83
   Hi-Nicalon SiC fiber for, 82
N24-A CMC system
   BN interfacial coating for, 83, 84
   ceramic fiber types for, 83
   interphase coating, 83
   microstructure analysis, 83
   Sylramic fibers for, 83
N24-B CMC system
   BN interphase coating, 85
   CMC reliability, 85
   CVI SiC matrix, 85
   damage tolerance, 86
   Sylramic fiber, 85
   thermal resistance, 86
N24-C CMC system
   BN interphase coating, 86
   creep resistance, 86
   CVI BN fiber coating, 86
   CVI SiC fiber coating, 86
   CVI SiC matrix in, 86
   interphase coating, 86
   melt-infiltration, 86
   Sylramic fiber for, 86
   thermal conductivity, 86
   ultimate strength, 87
N26-A CMC system
   creep resistance, 87
   CVI SiC matrix in, 87
   melt-infiltration, 87
   polymer infiltration and pyrolysis, 87
   thermal conductivity, 87

nano-oxide fibers, 18
NAS glass composites, 473. *See also* BAS; glass composites
NASA developed CMC systems, 78
   creep rupture testing, 95
   creep rupture, 88
   debonding behavior, 92
   interphase issues, 95
   N22, 81, 82
   N24-A, 81, 83
   N24-B, 81, 85
   N24-C, 81, 86
   N26-A, 81, 87
   properties
      CVI SiC matrix penetration, 89
      mechanical, 91
      Nicalon and Sylramic fiber types, 90
      physical, 90
      tensile stress strain curves, 92
      thermal expansion, 89, 90
      ultimate tensile strength, 93
   stress-rupture testing, 95
   stress-strain, 88
   tensile testing, 95
   upper use temperature, 91, 95
NDE. *See* non-destructive evaluation
$NdPO_4$ coatings, 393
Nextel fibers, 13. *See also* Altex fibers
   Nextel 312 boron nitride/silicon oxycarbide composite system, 348, 352
   Nextel 312 fiber with boron nitride coating, 350, 351
   Nextel 480, 27
   Nextel 610 fiber, 15
   Nextel 650 fiber, 15, 26
   Nextel 720, 27
   for porous-matrix oxide-oxide composites, 385
   for WHIPOX all oxide CMCs, 425
Nextel/Al_2O_3 composites. *See also* Almax/Al_2O_3 composites; Hi-Nicalon/Al_2O_3 composites; Nicalon/Al_2O_3 composites
   fracture surfaces, 303
   mechanical properties, 297
   microstructure, 295
Nicalon fiber, 38. *See also* silicon carbide fiber
Nicalon® fiber, 464
Nicalon/Al_2O_3 composites. *See also* Almax/Al_2O_3 composites; Hi-Nicalon/Al_2O_3composites; Nextel/Al_2O_3 composites
   coefficient of thermal expansion, 282
   debonding, 283
   directed metal oxidation fabrication, 280
   fatigue tests, 287
   flexural strength, 283
   fracture toughness, 283

INDEX

matrix microcracking, 285
microstructure, 281
tensile strength, 286
thermal and physical, 282, 283
thermal cycles and, 287
Nicalon/celsian composites. *See also*
 Hi-Nicalon/celsian composites
 high-temperature annealing in air, effect on, 233
 flexure strength, 232
 rupture strength, 235
 shear strength, 235
 tensile strength, 234
 tensile stress, 235
 thermal-oxidative stability, 233
nitrided composites
 SiC/RBSN, 151
 SiC/Si_3N_4, 151, 166
nitrides structural stability in UHTC composites, 199
non-destructive evaluation, MI-CMCs, 114
non-oxide ceramic fibers
 silicon carbide-based, 33–34
 silicon nitride-based, 34
non-oxide fibers. *See also* oxide fibers
 applications, 34
 ceramic matrix composites for, 34
 processing, 34
notch sensitivity, 401

off axis properties, 403
optical properties, UHTC composites, 211
organic fibers, 3. *See also* inorganic fibers
orthorhombic structure, BAS, 255–257
oxidation behavior
 $MoSi_2$-base composites
  high temperature oxidation, 178
  low temperature oxidation, 175
 $MoSi_2$-βSi_3N_4 composites
  high temperature, 194
  low temperature, 193
oxidation exposure testing, Blackglas composites
 dielectric properties, 369
 mechanical properties
  compression, 360
  creep rupture, 364
  flextural strength, 362
  low cycle fatigue, 366
  shear, 359
  tensile, 355
oxidation protection methods for SiC/RBSN composites, 165
oxidation resistance
 carbon/carbon composites, 118
 CVI SiC/SiC composites, 61
 MI-CMCs, 102

$MoSi_2$-base composites, 178
UHTC composites
 hafnium diboride, 213
 zirconium diboride, 213
oxidation resistant coatings
 interface coatings, 384, 389
 oxide-oxide composites, 381
oxide ceramics, monolith, 423
oxide fiber reinforced alumina matrix composites.
  See also SiC fiber reinforced alumina matrix composites
 directed metal oxidation, 296
 oxide/oxide composites
  fracture surface analysis, 300
  mechanical properties, 297
  microstructure, 295, 296
oxide fibers. *See also* non-oxide fibers
 development, 4
 discontinuous, 9
 fine continuous oxide fibers
  alpha alumina fibers, 14
  alumina zirconia fibers, 15
  continuous alumino-silicate fibers, 13
  continuous monocrystalline filaments, 16
  nano-oxide fibers, 18
  whiskers, 17
 glass fibers, 4
 polycrystalline, 6
 processing
  discontinuous oxide fibers, 9
  fine continuous oxide fibers, 11
  glass fibers, 7
 produced by sol-gel processes, 5
 properties, 18
oxide-oxide composites, 377. *See also* high temperature oxide-oxide composites; porous-matrix oxide-oxide composites
 Almax/Al_2O_3 composites, 295
 applications
  heat engines, 379
  thermal proptection system (TPS), 379
  turbine engines, 379
 brittle composite behavior, 382
 fracture surface analysis, 300
 mechanical properties, 297
 microstructure, 295, 296
 next generation enabling of
  higher temperature capability, 411, 412
  oxidation resistant interface control, 411, 412
 Nextel/Al_2O_3 composites, 295
 oxidation resistant coatings, 381
 tough behavior, design for
  interface control, 380
  weak interface (fiber coating), 380
  weak matrix (porous matrix), 380, 382

oxide-oxide composites, processing methods, 404
  consolidation
    CVD, 410
    CVI, 410
    hot isotatic pressing, 409
    hot pressing, 409
    pressureless sintering, 409
  interface coatings processing, 405
  matrix infiltration
    EPD, 408
    pre-peg processing, 407
    pressure infiltration, 407
  metal oxidation processing
    directed metal oxidation, 410
    reaction bonding, 411
oxycarbide fibers. *See also* SiOC
  for glass matrices
    Nicalon® fiber, 471
    Tyranno® fiber, 465, 471
  reinforced glass matrices, 475–476

P-CVI. *See* pulsed-CVI process
PDC. *See* polymer-derived ceramic technology
perform, CVI SiC/SiC composites fibrous
  1D, 59
  2D, 59
  3D, 59
perform fabrication, fiber-reinforced alumina matrix composites, 279
permeability, mullite composites, 343
permittivity, mullite composites
  high frequencies, 343
  low frequencies, 343, 344
pesting, 193
phase transformation
  mullite composites, 329
  silicon nitride-BAS composite, 258, 259
PIP. *See* polymer infiltration and pyrolysis
PMC. *See* polymer matrix composites
PMMA for SiC/RBSN, 152
polycrystalline fibers
  oxide, 6
  SiC, 35
polymer-derived ceramic technology, 349–350. *See also* Blackglas composites
polymer infiltration and pyrolysis (PIP), 87
  carbon fiber reinforced SiC composites, 128
  SiC fiber-reinforced silicon nitride composites, 150
polymer matrix composites, 128
polymer precursor method, fiber-reinforced glass matrix, 474
polymethylmethacralate for SiC/RBSN, 152
porosity
  high temperature oxide-oxide composites, 396

SiC/RBSN, 154
porous coatings. *See also* fugitive coatings; weak oxide coatings
  rare-earth aluminates, 393
  zirconia, 393
porous glasses, graded, 525
porous matrix composites. *See also* dense-matrix composites
  concept, 424
  crack behavior, 382
  failure behavior, 382
  matrix sintering limitation of, 383
  mechanical behavior, 382
  oxide-oxide composites design, 380, 382
porous matrix oxide-oxide composites, 384. *See also* high temperature oxide-oxide composites
  fibers
    eutectic, 385
    Nextel, 385
    sapphire, 385
  matrix systems
    alumina, 387
    alumina-mullite, 387, 388
    aluminosilicate, 387
    WHIPOX, 388
porous oxide fiber coatings, 382
porous preform, MI-CMCs for, 102
powder glass sintering, 491
powder technology, dispersion-reinforced glass-matrix composites, 492
PRD-166 fiber, 15, 26
preform fabrication
  during chemical vapor infiltration, 128
  liquid silicon infiltration, 132
Prepreg HiPerComp™, 103–105
  creep rupture behavior, 111
  fatigue and creep behavior, 110
  thermal stability, 108
prepreg MI-CMCs
  elastic properties, 105
  thermal properties, 106
prepreg processing
  comparison with slurry cast process, 104
  MI-CMCs
    BN-based interface coating for, 102
    CVD for, 102
  oxide-oxide composites, 407
pressure-assisted sintering. *See also* pressureless sintering
  HIP, 317
  hot pressing, 317
pressure infiltration, oxide-oxide composite processing, 407
pressureless sintering. *See also* pressure-assisted sintering

oxide-oxide composites, 409
SiC whisker composites, 317
proportional limit stress (PLS), 79
pulltrusion, fiber-reinforced glass matrix, 474
pulsed-CVI process, 61. See also forced-CVI process; isothermal-isobaric CVI process
pyrocarbon, 56, 60
pyrolysis
    Blackglas composites, 351, 352
    during chemical vapor infiltration, 130
    during liquid polymer infiltration, 130
    during liquid silicon infiltration, 132
pyrolytic carbon, for SiC fiber-reinforced silicon nitride composites, 150

RBSN (reaction-bonded silicon nitride). See also SiC/RBSN; silicon nitride composites
  matrix, 151
  processing temperatures, 151
  SiC-based, 151, 152
reaction-bonding process, 411. See also directed metal oxidation
refractory fibers, 5
reliability property, CVI SiC/SiC composites, 68
residual stresses, 498–499
residual tensile strength
  2-D Nicalon^{TM}/Al$_2$O$_3$ composites, 282
  Hi-Nicalon/Al$_2$O$_3$ composites, 292
resistivity, mullite composites, 343
rupture behavior
  CMC, 79
  Prepreg HiPerComp^{TM} composites, 111
  silicon carbide fibers, 42
rupture strength
  Nicalon/celsian composites, 235
  SiC fibers, 44
rupture tests, Nicalon/Al$_2$O$_3$ composites, 286

Saffil fiber, 10. See also melt-spun aluminosilicate fibers
Saphikon fiber, alpha-alumina, 28
SAS (strontium aluminosilicate). See also BAS; glass composites
  for celsian formation, 230
  glass ceramics, 467
scheelite (CaWO$_4$) coatings, 393. See also monazite coatings
SCS-6 reinforced MoSi$_2$ composites, 174
SCS-6/MoSi$_2$-Si$_3$N$_4$ matrix composites, mechanical properties of
  fracture toughness, 181
  impact behavior, 183
  tensile creep tests, 182
  tensile stress-strain behavior, 181
SENB, 429

SEVNB method, 444
SHARP-B1 vehicle. See UHTC composites, NASA Ames working on
shear strength
  Hi-Nicalon/celsian composites, 240
  Nicalon/celsian composites, 235
  SiOC-Nextel 312 BN, 359
  WHIPOX all oxide CMCs, 429
shear stresses, interfacial, 70
shear tests, Blackglas composites, 354
SiC based ceramic matrix composites, melt-infiltrated. See MI-CMCs
SiC composites, carbon fiber reinforced. See also SiC/SiC composites
  applications
    advanced friction systems, 123
    calibrating bodies, 125
    energy and power station engineering, 126
    furnace engineering, 127
    low-expansion structures, 124
    mechanical, thermal, and corrosion resistance, 127
    satellite communication system, 126
    space vehicle's TPS and hot structures, 119
    vanes, nozzles and flaps of rocket motors and jet engines, 121
  C/C-SiC, 117
  C/SiC, 117
  microcracks, 144
  processing
    combined processing methods, 133
    CVI, 128, 129
    LPI, 128, 130
    LSI, 131, 132
    PIP, 128
  tribological tests, 145
SiC fiber. See also CMC
  applications, 34
  chemical vapor infiltrated, 55
  continuous-length polycrystalline, 33
  creep properties, 40
  CVD-derived, 35, 39
  grain-boundary diffusion mechanism, 42
  Nicalon and Tyranno Lox M fibers, 38
  non-oxide, 33–34
  polycrystalline, 35
  polymer-derived, 34, 39
  processing, 34
  properties
    creep resistance, 41
    creep strength, 43
    environmental effect on, 48
    for CMC applications, 49
    grain boundary sliding, 41
    grain size, 39

interface controlled diffusional (ICD) creep, 42
mechanical and physical, 38
Monkman-Grant diagrams, 45, 46
rupture bahavior, 42, 44, 47
strain limit, 43
stress rupture, 43
temperature-dependent, 40
reinforced silicon nitride composites, 149–151, 169
SCS-6 and Sylramic fibers, 41
tensile strength, 40
SiC fiber reinforced alumina matrix composites. *See also* oxide-fiber reinforced alumina matrix composites
2-D Hi-Nicalon™/Al_2O_3 composites, 287
2-D Nicalon™/Al_2O_3 composites, 281, 282
BN/SiC Fiber-Matrix Interface role, 280
SiC fiber reinforced celsian composites
CMC processing, 231
crystal structure, 228
Hi-Nicalon/celsian composites, 237
Nicalon/celsian composites, 232
SiC fiber reinforced glass composites, 467
SiC fiber reinforced silicon nitride composites
reaction-bonded silicon nitride. *See* SiC/RBSN
SiC/$MoSi_2$-Si_3N_4composite, 187
SiC/Si_3N_4, 149, 151, 166
SiC reinforced CMC systems, NASA developed for high temperature applications, 80
N22, 81, 82
N24-A, 81, 83
N24-B, 81, 85
N24-C, 81, 86
N26-A, 81
properties
CVI SiC matrix penetration, 89
mechanical, 91
Nicalon and Sylramic fiber types, 90
physical, 90
tensile stress strain curves, 92
thermal expansion, 89, 90
ultimate tensile strength, 93
SiC whisker
characteristics
bonding, 314
bulk chemistry, 312, 313
defects, 314
elastic moduli, 314
growth processes, 311
surface chemistry, 313
health effect of, 318
mullite composites preparation, 328
vapor-liquid-solid mechanism, 309
SiC whisker composites
mullite-based, 332

mullite-SiC, 326
mullite-ZrO_2-SiC, 326
SiC whisker composites fabrication
agglomeration, 315
deagglomeration, 314
densification
pressure-assisted sintering, 317
pressure-less sintering, 317
dispersion, 314, 315
dry pressing, 316
sintering, 316
slip casting, 316
SiC whisker reinforced alumina
applications, 320
microstructure, 315
properties
crack propagation, 319
electrical resistivity, 322
flextural strength, 320
thermal conductivity, 321
Thermal expansion coefficient, 322
toughness, 309, 319
SiC/$MoSi_2$-Si_3N_4composite
applications, 187
testing for blade outer airseal component (BOAS), 187
SiC/RBSN
applications, 169
oxidation protection methods, 165
SiC monofilament-reinforced RBSN composite
creep properties, 162, 163
environmental stability, 162
physical and mechanical properties, 153, 154, 155
processing, 152
thermal properties, 156
sub-element testing, 166
with tow fibers
processing, 163, 164
properties, 164
SiC/Si_3N_4. *See* dense SiC/silicon nitride composites
SiC/SiC composites
based CMC system, 77–78
chemical vapor infiltrated. *See* CVI SiC/SiC composites
silica based glasses, 8
silica containing fibers, 5
silica fibers. *See also* alumina fibers
continuous fine oxide fibers, 11
discontinuous oxide-based
melt-spun aluminosilicate fibers, 9
Saffil fiber, 10
silica ratio for melt-spun aluminosilicate fibers, 9
silica rich matrices, WHIPOX CMCs, 430
silicate glass matrix composites, 468

INDEX

silicon based polymers, 349
silicon melt-infiltrated ceramic composites. *See* MI-CMCs
silicon nitride*See also* SiC/RBSN
  based non-oxide fiber, 34
  in-situ reinforced, 251, 253
    barium aluminum silicate, 254
    XRD analysis, 258
  polymorphs
    amorphous, 253
    cubic, 253
silicon nitride composites
  dense SiC/silicon nitride, 166
  SiC fiber-reinforced
    applications, 169
    boron nitride for, 150
    CVI for, 150
    hot pressing, 150
    hot-isostatic pressing, 150
    PIP, 150
    pyrolytic carbon for, 150
    reaction-bonded. *See* SiC/RBSN
  SiC/Si_3N_4, 166
silicon nitride-BAS composites
  crystallization, 263
    HRTEM analysis, 265
    TEM analysis, 263-265
  densification, 258
  EDS analysis, 263
  microstructure evolution, 260
  phase transformation, 258, 259
  processing, 258
  properties
    dielectric, 273
    flextural strength, 268
    fracture toughness, 267, 268
    grain size distribution, 269
    hardness and elastic modulus, 265, 266
    strength and toughness, 266
    thermal shock resistance, 270
    tribology, 271
    Weibull modulus, 269
  sintering mechanism, 260
  whisker morphology, 260, 262
silicon oxycarbide ceramic. *See* SiOC
siliconizing, during liquid silicon infiltration, 133
single crystal
  fibers, 16
  oxide filaments, 7
  Saphikon, 28
single edge notched beam (SENB) tests, WHIPOX CMCs, 429
single edge precracked beam method, for alumina reinforced zirconia composites testing, 445
single edge v-notched beam test, 444

sintered glass, 491
sintering. *See also* pressureless sintering
  based porous-matrix limitation, 383
  dispersion-reinforced glass-matrix composites, 492
  powder glass, 491
  SiC whisker composites, 316
  silicon nitride-BAS composite, 260
  UHTC composites, 204
SiOC, 350. *See also* Blackglas polymer
SiOC-Nextel 312 BN composites
  dielectric properties, 369, 370
  properties after 4000-hrs oxidation, 362, 363
  properties after 600°C oxidation
    compression properties, 360
    Double Notch Shear Strength, 359, 360
    shear properties, 359
    tensile failure strain, 357
    tensile modulus, 358
    tensile strength, 355, 357
  properties at 566°C
    creep rupture tests, 364, 365
    low cycle fatigue, 366
  specimen preparation, 352
  stress-strain curves, 356
  tensile properties, 356, 357
  thermal properties
    conductivity, 368
    diffusivity, 367
    expansion, 368
    heat capacity, 367
slip casting, SiC whisker composites, 316
slow crack growth, alumina reinforced YSZ composites, 447
slurry cast HiPerComp™, 104
slurry cast MI-CMCs
  BN-based interphase coating for, 103
  elastic property, 106
  thermal properties, 105
slurry cast process. *See also* prepeg processing
  chemical vapor infiltration and, 103
  comparison with prepreg system, 104
  MI-CMCs, 103
slurry infiltration, fiber-reinforced glass matrix composites, 470
SOFC (solid oxide fuel cells), 437–438
sol-gel processing, fiber-reinforced glass matrix, 473
space vehicle's TPS and hot structures, 119
specific heat
  mullite-SiC whisker composites, 332
  mullite-zirconia composites, 332
  mullite-zirconia-SiC whisker, 334
spinning glass filaments, 4
stiffness, carbon/carbon composites, 118
strain rates, MI-CMCs, 108
stress rupture tests

NASA-developed CMC system, 95
Nicalon/Al$_2$O$_3$ composites, 286
stress strain curves
  HI-Nicalon/Al$_2$O$_3$ composites, 292
  MI-CMCs, 107
  NASA developed CMC systems, 88, 92
  Nicalon/Al$_2$O$_3$ composites, 286
  oxide/oxide composites, 300
  SCS-6/MoSi$_2$-Si$_3$N$_4$ matrix composites, 181
  SiC/RBSN, 154
  SiOC-Nextel 312 BN, 356
stress-strain behavior, CVI SiC/SiC composites
  damage insensitive, 64
  damage sensitive, 64
strontium aluminosilicate. *See* SAS
Sylramic fibers.*See also* silicon carbide fibers
  MI-CMCs, 101
  N24-A CMC system, 83
  N24-B CMC system, 85
  N24-C CMC system, 86
  NASA developed CMC systems, 90
synthetic fibers
  inorganic, 3
  organic, 3

tailcone, 371
tape casting, fiber-reinforced glass matrix, 473
tensile properties. *See also* mechanical properties
  alpha-alumina fibers, 28
  aluminosilicate fibers, 23
  continuous single crystal oxide fibers, 27
  MI-CMCs, 107, 110
  oxide/oxide composites, 299
  SCS-6/MoSi$_2$- Si$_3$N$_4$ matrix composites, 182
  SiC/Si$_3$N$_4$ composite, 168
  SiOC-Nextel 312 BN, 355, 366
  YAG-alumina fibers, 29
tensile strength. *See also* flexture strength; shear strength
  Hi-Nicalon/Al$_2$O$_3$ composites, 292
  Hi-Nicalon/celsian composites, 239
  Nicalon/celsian composites, 234
  NicalonTM/Al$_2$O$_3$ composites, 286
  Saphikon single crystal, 28
  silicon carbide fibers, 40
  SiOC-Nextel 312 BN, 357
tensile stress
  2-D Nicalon^{TM}/Al$_2$O$_3$ composites properties, 282
  mullite-SiC whisker, 338
  Nicalon/Al$_2$O$_3$ composites, 287
  Nicalon/celsian composites, 235
tensile stress-strain behavior
  CMC behavior, 79
  CVI SiC/SiC composites, 63
    damage insensitive, 64
    damage sensitive, 64
  high temperature oxide-oxide composites, 397
  HI-Nicalon/Al$_2$O$_3$ composites, 292
  MI-CMCs, 107
  NASA developed CMC systems, 92
  Nicalon/Al$_2$O$_3$ composites, 286
  oxide/oxide composites, 300
  SCS-6/MoSi$_2$-Si$_3$N$_4$ matrix composites, 181
  SiC/RBSN, 154
  SiOC-Nextel 312 BN, 356
tensile test
  Blackglas composites, 353
  NASA-developed CMC system, 95
  Nicalon/Al$_2$O$_3$ composites, 283
thermal aging, effect on WHIPOX CMCs, 430
thermal and physical properties, MI-CMCs, 105
thermal conductivity. *See also* thermal properties
  alumina-reinforced zirconia composites, 453
  carbon fiber reinforced SiC composites, 142, 143
  HI-Nicalon/Al$_2$O$_3$ composites, 292
  N24-B CMC system, 86
  N24-C CMC System, 86
  N26-A CMC System, 87
  prepeg, 106
  SiC whisker-reinforced alumina, 321
  SiC/RBSN composites, 158
  SiOC-Nextel 312 BN, 368
  slurry cast, 106
  UHTC composites, 207
thermal cycles
  alumina reinforced zirconia composites, 452
  Nicalon/Al$_2$O$_3$ composites, 287
  SiC/RBSN, 158
thermal diffusivity
  mullite-SiC whisker composites, 332
  mullite-zirconia composites, 332
  SiOC-Nextel 312 BN, 367
thermal expansion. *See also* coefficient of thermal expansion
  celsian, 228
  glass and glass-ceramic composites, 476
  mullite-zirconia compositions, 331
  NASA developed CMC systems, 89, 90
  SiC whisker-reinforced alumina, 322
  SiOC-Nextel 312 BN, 368
thermal fatigue
  alumina reinforced zirconia composites, 452
  WHIPOX CMCs, 429
thermal heat capacity, SiOC-Nextel 312 BN, 367
thermal properties. *See also* electrical properties; mechanical properties
  alumina-reinforced zirconia, 452
  Blackglas composites

conductivity, 368
diffusivity, 367
expansion, 368
heat capacity, 367
carbon fiber reinforced SiC composites
C/C-SiC composite, 140
C/SiC composite, 139
high temperature oxide-oxide composites, 396
MI-CMCs, 105–106
mullite-based composites
diffusivity, 332
thermal shock, 341–343
NASA developed CMC systems, 86–87
NicalonTM/Al$_2$O$_3$ composites, 283
SiC/RBSN
conductivity, 158
expansion, 156–157
shock resistance, 160
SiOC/Nextel 312 BN, 367
UHTC composites
CTE, 207
thermal conductivity, 207
thermal protection systems, 119, 379, 380
thermal shock
2-D Nicalon/Al$_2$O$_3$ composites properties, 283
CVI SiC/SiC composites, 72
mullite-SiC whisker composites, 341, 343
mullite-zirconia-SiC whisker composites, 341, 343
WHIPOX all oxide CMCs, 429
thermal shock resistance
dispersion-reinforced glass-matrix composites, 503
SiC/RBSN composites, 160
silicon nitride-BAS composite, 270
thermal stress
dispersion-reinforced glass-matrix composites, 498
resistance, SiC/RBSN, 160
UHTC composites
hafnium diborides, 219
zirconium diborides, 219
thermomechanical behavior, glass and glass-ceramic composites, 477
thermo-oxidative stability
Hi-Nicalon/celsian composites
degradation mechanism at 1200° C, 245
high temperature annealing in air, 243
Nicalon/celsian composites, 233
3-D carbon fiber-reinforced glass containing composites, 512–515
toughening behavior
dispersion-reinforced glass-matrix composites, 498–499
glass and glass-ceramic composites, 475
whisker reinforced composites, 309
toughness property
high temperature oxide-oxide composites, 401

SiC whisker reinforced alumina, 319
silicon nitride-BAS composite, 266
TPS. *See* thermal protection systems
tribology
glass and glass-ceramic matrix composites, 477
silicon nitride-BAS composite, 271
tests for carbon fiber reinforced SiC composites, 145
turbine engine, 379
turbine engine test, 294
Tyranno® fiber, 464. *See also* Nicalon® fiber

UCSB fiber composites. *See* high temperature oxide-oxide composites
UHTC composites
application, 198
atomic structure, 203
bonding and structure
anti-bonding states, 202
bonding levels, 202
ceramic borides stability, 199
ceramic carbides stability, 199
ceramic nitrides stability, 199
hybridization aspects, 201
high temperature testing, 212
arc jet Testing, 218
Furnace Oxidation Testing, 216
NASA Ames working on, 220–221
processing
hot pressing technology, 205
machining, 207
raw materials, 204
SiC additions and borides densification aspects, 205
sintering, 204
properties
electrical, 210
mechanical, 210
optical, 211
thermal, 207
thermal stress, 219
thermodynamic properties, 203
ultimate failure
of matrix infiltrated tow, 67
of tow, 67
ultimate tensile strain
NASA developed CMC systems, 93
SiC/RBSN, 155
ultimate tensile strength (UTS)
CMC, 79
high temperature oxide-oxide composites, 398
NASA developed CMC systems, 93
oxide/oxide composites, 298
SiC/RBSN, 155

ultra high temperature ceramics composites. *See* UHTC composites
UUT (upper use temperature), NASA-developed CMC systems, 91, 95

Vicker's Hardness, property of silicon nitride-BAS composite, 266
Vickers indentation, alumina reinforced YSZ composites, 447
Vickers microhardness, alumina reinforced zirconia composites, 451
vitreous fibers, 5
VLS mechanism, 309
volume fraction, SiC/RBSN, 154
weak oxide coatings. *See also* fugitive coatings; porous coatings
   monazites, 391
   scheelite, 393
   xenotimes, 391
wear behavior. *See also* erosion behavior; mechanical properties
   dispersion-reinforced glass-matrix composites, 502
   glass and glass-ceramic matrix composites, 477
Weibull behavior, $MoSi_2$-βSi_3N_4 composites, 191
Weibull model, 69
Weibull modulus, 69
   alumina reinforced zirconia composites, 444
   silicon nitride-BAS composite, 269
WHIPOX, 423
WHIPOX CMCs
   cyclic fatigue behavior, 429
   damage-tolerant fracture behavior, 431
   debonding, 428
   fabrication, 424
   failure mechanisms, 428
   fatigue behavior, 429
   fiber orientation, 428
   fracture behavior, 429
   matrix agglomerations, 427, 428
   matrix compositions
      alumino-rich matrix, 430
      aluminosilicate matrix, 430
      fiber/matrix ratio, 428
      Interlaminate matrix-rich areas, 427
      mullite matrix, 431
      silica-rich matrices, 430
   mechanical properties, 427
   mesostructure, 427
   procesing, 426
   SENB tests, 429
   shear strength, 429
   thermal aging effect on, 430
   thermal fatigue, 429
   thermal shock, 429
whisker characteristics, SiC
   bonding, 314
   bulk chemistry, 312, 313
   defects, 314
   elastic moduli, 314
   growth processes, 311
   surface chemistry, 313
whisker composites fabrication, SiC
   agglomeration, 315
   deagglomeration, 314
   densification, 317
   dispersion, 314, 315
   dry pressing, 316
   mullite-SiC, 326
   mullite-zirconia-SiC, 326
   sintering, 316
   slip casting, 316
whisker morphology, silicon nitride-BAS composite, 260, 262
whisker reinforced composites
   alumina, 308
   SiC whisker reinforced, 318
   toughening behavior, 309
whiskers
   alumina, 17
   properties, 29
   SiC whiskers effect on health, 318
wound highly porous oxide ceramic composite. *See* WHIPOX
xenotimes coatings, 391. *See also* monazites coatings
YAG-alumina fibers, 7, 16, 28–29
Young's modulus
   CVI SiC-SiC composites, 67
   SiC/RBSN, 154
YSZ (yttria-stabilized zirconia) composites. *See also* alumina reinforced YSZ composites
   cubic, 439
   flexure test, 439
   monoclinic, 439
   tetragonal, 439
zirconia composites, alumina-reinforced, 437, 438. *See also* YSZ composites
zirconia phase in mullite whiskers, 327
zirconium borides. *See also* hafnium diborides
   bonding and structure, 199
   electrical properties, 210
   high temperature testing
      arc jet testing, 219
      furnace oxidation testing, 216
   mechanical properties, 210
   optical properties, 211
   oxidation studies, 212
   powder for UHTC composites, 204
   processing, 204
   thermal properties, 207
      stress, 219
      thermodynamic, 203